金国藩　院士

金国藩院士与夫人段淑贞教授合影

金国藩院士在做实验

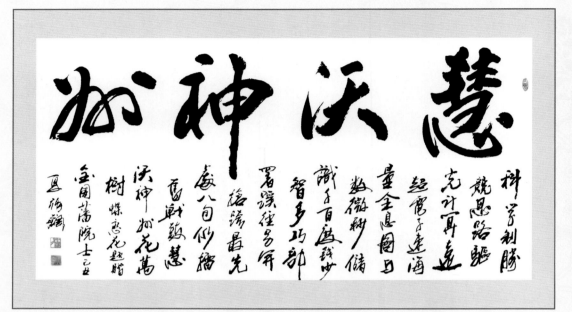

书法家夏鹤龄赠金国藩院士的自作词并书写的匾

蝶恋花
慧沃神州

科学制胜竟思路,
驱光计算远超电子速。
海量全息图与数,
微秒储识千百度。

钱少智多巧部署,
蹊径另开抢跨最先处。
八旬仍擂奋战鼓,
慧沃神州花万树。

赠金国藩院士
乙丑*
夏鹤龄

注:*乙丑是 2009 年

中国工程院 院士文集

Collections from Members of the Chinese Academy of Engineering

金国藩文集

A Collection from Jin Guofan

北京
冶金工业出版社
2016

内 容 提 要

本书为《中国工程院院士文集》之一。书中精选了金国藩院士 20 世纪 50 年代至 2013 年发表的学术论文 60 篇,重点论述了光学信息处理、二元光学、体全息存储以及光学仪器。其中,光学信息处理包括光学子波变换、光学形态学数字图像处理、基于体全息存储的相关器及其在高分辨率遥感图像、人脸、指纹识别中的应用等,为在我国开展光学信息处理研究作出了重要贡献;二元光学包括各种功能器件的设计理论、设计方法、器件研制与系统应用等,在我国独树一帜;体全息存储包括各种复用技术、曝光时序以及超高容量体全息存储系统与体全息光盘等,引领我国的"数字体全息光学存储和识别技术"研究;光学仪器包括光学面形检测、金属纳米颗粒几何特征测量与多种光谱仪器研制等,努力促进国产仪器的发展。此外,文集还收录了金国藩院士关于我国仪器仪表及相关领域的科技发展方面的报告 10 篇。

本书可供相关专业科研人员、工程技术人员和大专院校师生学习、参考。

图书在版编目(CIP)数据

金国藩文集/金国藩著. —北京:冶金工业出版社,2016.6
(中国工程院院士文集)
ISBN 978-7-5024-7164-4

Ⅰ.①金… Ⅱ.①金… Ⅲ.①全息存储—文集 Ⅳ.①TP333.4

中国版本图书馆 CIP 数据核字(2016)第 090198 号

出 版 人 谭学余
地 址 北京市东城区嵩祝院北巷 39 号 邮编 100009 电话 (010)64027926
网 址 www.cnmip.com.cn 电子信箱 yjcbs@cnmip.com.cn
策 划 任静波 责任编辑 卢 敏 任静波 美术编辑 彭子赫
版式设计 孙跃红 责任校对 李 娜 责任印制 牛晓波
ISBN 978-7-5024-7164-4
冶金工业出版社出版发行;各地新华书店经销;三河市双峰印刷装订有限公司印刷
2016 年 6 月第 1 版,2016 年 6 月第 1 次印刷
787mm×1092mm 1/16;40 印张;2 彩页;925 千字;628 页
280.00 元

冶金工业出版社 投稿电话 (010)64027932 投稿信箱 tougao@cnmip.com.cn
冶金工业出版社营销中心 电话 (010)64044283 传真 (010)64027893
冶金书店 地址 北京市东四西大街 46 号(100010) 电话 (010)65289081(兼传真)
冶金工业出版社天猫旗舰店 yjgycbs.tmall.com

(本书如有印装质量问题,本社营销中心负责退换)

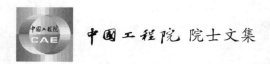

《中国工程院院士文集》总序

2012年暮秋，中国工程院开始组织并陆续出版《中国工程院院士文集》系列丛书。《中国工程院院士文集》收录了院士的传略、学术论著、中外论文及其目录、讲话文稿与科普作品等。其中，既有院士们早年初涉工程科技领域的学术论文，亦有其成为学科领军人物后，学术观点日趋成熟的思想硕果。卷卷文集在手，众多院士数十载辛勤耕耘的学术人生跃然纸上，透过严谨的工程科技论文，院士笑谈宏论的生动形象历历在目。

中国工程院是中国工程科学技术界的最高荣誉性、咨询性学术机构，由院士组成，致力于促进工程科学技术事业的发展。作为工程科学技术方面的领军人物，院士们在各自的研究领域具有极高的学术造诣，为我国工程科技事业发展做出了重大的、创造性的成就和贡献。《中国工程院院士文集》既是院士们一生事业成果的凝炼，也是他们高尚人格情操的写照。工程院出版史上能够留下这样丰富深刻的一笔，余有荣焉。

我向来认为，为中国工程院院士们组织出版院士文集之意义，贵在"真、善、美"三字。他们脚踏实地，放眼未来，自朴实的工程技术升华至引领学术前沿的至高境界，此谓其"真"；他们热爱祖国，提携后进，具有坚定的理想信念和高尚的人格魅力，此谓其"善"；他们治学严谨，著作等身，求真务实，科学创新，此谓其"美"。《中国工程院院士文集》集真、善、美于一体，辩而不华，质而不俚，既有"居高声自远"之澹泊意蕴，又有"大济于苍生"之战略胸怀，斯人斯事，斯情斯志，令人阅后难忘。

读一本文集，犹如阅读一段院士的"攀登"高峰的人生。让我们翻开《中国工程院院士文集》，进入院士们的学术世界。愿后之览者，亦有感于斯文，体味院士们的学术历程。

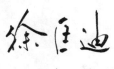

2012年7月

前 言

本文集中的作品是作者从20世纪50年代至2013年发表的300余篇文章中精选出来的，共70篇。

现代科学技术日新月异，各种技术交错融合，仅靠个人的聪明才智或个人奋斗，根本不可能取得像样的成果。我能取得成绩，完全和我有一支富有创新和牺牲精神、团结奋斗、脚踏实地苦干的团队分不开。邬敏贤教授、严瑛白教授和何庆声教授多年来一直和我在一起，完成了一项又一项国家科研项目，培养了一名又一名博士、硕士，取得了一个又一个研究成果，发表了一篇又一篇研究论文，我非常珍惜我们之间的合作与友谊，同时由衷地感谢他们对我的支持。

在我的学术生涯中，无论是光学并行处理器件、光学神经网络、光学子波变换、光学形态学数字图像处理与高分辨率遥感图像处理等研究，还是各种二元光学器件的研制、功能拓展与系统应用；无论是动态散斑体全息存储新概念的提出与超高容量体全息光盘系统的构架，还是光学自由曲面、金属纳米颗粒几何特征的高精度测量与多种光谱仪器的研制；无论是脚踏实地的研究，还是统揽全局的管理；一路走来，既不断延续和发展，又有一定的阶段性。因此，我将文章归纳为光学信息处理、二元光学、体全息存储、光学仪器以及关于我国仪器仪表的科技发展等五部分。每部分不分中文和英文，都以发表的时间为顺序。

本文集收录的各篇论文的原发表刊物、发表年代以及署名的合作者，读者可从每篇文章头一页的下注中查到。

余兴龙研究员负责文集的策划，并为我的自述执笔。谭峭峰副研究员、曹良才副研究员和顾华荣助理研究员与余兴龙研究员一起完成文集的整理工作。其中，谭峭峰副研究员负责前四部分论文的整理、精选和录入，顾华荣助理研究员负责最后一部分文章的整理和录入，谭峭峰副研究员和曹良才副研究员对文集论文进行了校订和格式编排。李岩教授大力支持文集的出版，光电工程研究所的众多教师对文集的出版提供了多种形式的支持与帮助。在此，我对他们表示十分感谢。

中国工程院和清华大学精密测试技术及仪器国家重点实验室提供出版资助，我深表感谢。

作　者
2016 年 1 月

目 录

院士传略

» 金国藩简介 ··· 3
» 金国藩自述 ··· 4

光学信息处理

» 使用全息光学元件的合成孔径雷达信号处理器 ·· 15
» 利用改进的符号数算法和光学符号代换实现矩阵计算 ································· 25
» Optical Binary Image Algebra Processor (OBIAP) ··· 33
» 光学子波变换（1）基本理论 ··· 41
» 光学子波变换（2）实验结构及技术 ··· 50
» 光学子波变换（3）应用 ·· 55
» Multiobject Recognition in a Multichannel Joint-transform Correlator ············ 64
» Joint Wavelet-transform Correlator for Image Feature Extraction ··················· 69
» 实时光学模糊关联记忆神经网络 ··· 80
» One-step Implementation of the Optical Hit-miss Transform ························· 87
» Developed, Binary, Image Processing in a Dual-channel, Optical, Real-time Morphological Processor ··· 98
» 基于相似性度量的灰度图像光学匹配运算 ··· 111
» Volume Holographic Wavelet Correlation Processor ····································· 117
» Sidelobe Suppression in Volume Holographic Optical Correlators by Use of Speckle Modulation ·· 128
» Experiment on Parallel Correlated Recognition of 2030 Human Faces Based on Speckle Modulation ·· 132

- Improving Accuracy of Multichannel Volume Holographic Correlators by Using a Two-dimensional Interleaving Method ········· 140
- Phase-modulated Multigroup Volume Holographic Correlator ········· 146
- Fast Associative Filtering Based on Two-dimensional Discrete Walsh Transform by a Volume Holographic Correlator ········· 152
- Multi-sample Aarallel Estimation in Volume Holographic Correlator for Remote Sensing Image Recognition ········· 159
- Optical Fingerprint Recognition Based on Local Minutiae Structure Coding ········· 171

二元光学

- 计算机源生的全息光学元件（COHOE）的合成及优化设计 ········· 189
- 电子束计算全息图的制作 ········· 195
- Optimization of Grating Multi-beamsplitters ········· 199
- The Fabrication of a 25×25 Multiple Beam Splitter ········· 206
- 平行传输阵列光斑器件的研制 ········· 214
- 实现ICF均匀照明的二元光学器件的混合优化设计 ········· 223
- Aberration Theory of Arrayed Waveguide Grating ········· 229
- Theories for Design of Diffractive Superresolution Elements and Limits of Optical Superresolution ········· 238
- Theories for the Design of a Hybrid Refractive-diffractive Superresolution Lens with High Numerical Aperture ········· 255
- Broadband Polarizing Beam Splitter Based on the Form Birefringence of a Subwavelength Grating in the Quasi-static Domain ········· 277
- Statistic Analysis of Influence of Phase Distortion on Diffractive Optical Element for Beam Smoothing ········· 282
- High Quality Light Guide Plates that can Control the Illumination Angle Based on Microprism Structures ········· 292
- Polarized Light-guide Plate for Liquid Crystal Display ········· 298
- 基于PWC方法的折衍混合红外物镜设计 ········· 306
- Enhancement of the Light Output of Light-emitting Diode with Double Photonic Crystals ········· 313
- Achromatic Generation of Radially Polarized Beams in Visible Range Using Segmented Subwavelength Metal Wire Gratings ········· 320
- Security Enhanced Optical Encryption System by Random Phase Key and

- Permutation Key ·········· 325
- » Achromatic Phase Retarder Applied to MWIR & LWIR Dual-band ·········· 339
- » Experimental Demonstration of Tunable Directional Excitation of Surface Plasmon Polaritons with a Subwavelength Metallic Double Slit ·········· 349
- » Design Method of Surface Contour for a Freeform Lens with Wide Linear Field-of-view ·········· 355

体全息存储

- » 1000幅数字图像的晶体体全息存储与恢复 ·········· 373
- » Dynamic Speckle Multiplexing Scheme in Volume Holographic Data Storage and Its Realization ·········· 376
- » Exposure-schedule Study of Uniform Diffraction Efficiency for DSSM Holographic Storage ·········· 382
- » $10Gb/cm^3$ 小型化体全息数据存储及相关识别系统 ·········· 389
- » Orthogonal Polarization Dual-channel Holographic Memory in Cationic Ring-opening Photopolymer ·········· 397
- » Improving Signal-to-noise Ratio by Use of a Cross-shaped Aperture in the Holographic Data Storage System ·········· 406
- » Orthogonal-reference-pattern-modulated Shift Multiplexing for Collinear Holographic Data Storage ·········· 417
- » Improvement of Volume Holographic Performance by Plasmon-induced Holographic Absorption Grating ·········· 424

光学仪器

- » ВП-4-ЭИ型三向电感式车削测力仪的介绍 ·········· 433
- » 两种利用计算全息检测非球面的方法 ·········· 438
- » 二维光学传递函数测量 ·········· 445
- » 中医舌诊自动识别方法的研究 ·········· 451
- » Resolution Enhancement by Combination of Subpixel and Deconvolution in Miniature Spectrometers ·········· 461
- » Design of Freeform Mirrors in Czerny-Turner Spectrometers to Suppress Astigmatism ·········· 468
- » Generalized Method for Calculating Astigmatism of the Unit-magnification

- Multipass System ·········· 484
- Approximate Analytic Astigmatism of Unit-magnification Multipass System ·········· 498
- CGH Null Test of a Freeform Optical Surface with Rectangular Aperture ·········· 517
- Fast Statistical Measurement of Aspect Ratio Distribution of Gold Nanorod Ensembles by Optical Extinction Spectroscopy ·········· 527
- Phase Extraction from Interferograms with Unknown Tilt Phase Shifts Based on a Regularized Optical Flow Method ·········· 544
- Accurate Geometric Characterization of Gold Nanorod Ensemble by an Inverse Extinction/Scattering Spectroscopic Method ·········· 560

论仪器仪表科技发展

- 国外光计算的进展 ·········· 577
- 信息时代的光学器件——透镜仅针尖那么大 ·········· 591
- 仪器仪表的微小化、集成化和智能化 ·········· 592
- 21世纪是信息与生命科学的时代 ·········· 595
- 二元光学 ·········· 596
- 超高密度光存储技术的现状和今后的发展 ·········· 602
- 我国仪器仪表产业发展之路 ·········· 611
- 测量技术是信息技术的源头——谈王大珩院士的仪器科学思想 ·········· 618
- 我国当代仪器仪表的发展 ·········· 620
- 我国平板显示产业面临的迫切科学技术问题 ·········· 626

院士传略

金国藩简介

金国藩（1929—），浙江绍兴人。光学仪器与光学信息处理专家。他是我国光学信息处理的奠基人之一、"计算全息"与"二元光学"研究的开拓者和"数字体全息光学存储和识别技术"研究的引领者，也是将"光学信息处理"课程全面引入到国内的第一人。1994年，金国藩当选中国工程院首批院士。20世纪60年代，负责完成了我国第一台"三坐标光栅劈锥测量机"的研制，打破封锁并为国防建设作出贡献；70年代末，率先应用计算机制全息图技术检测非球面和制作凹面光栅，促进企业技术进步；80年代后期，跟踪前沿，开拓光计算技术研究新领域；90年代，抓住新方向，开展"二元光学"研究独树一帜；90年代后期至今，立足创新，引领数字体全息光学存储和识别技术研究健康发展。科研成果突出，荣获国家技术发明二等奖2项和国家科技进步三等奖1项等。曾任清华大学原精密仪器与机械学系系主任和机械工程学院院长、国家自然科学基金委员会副主任、教育部科学技术委员会常务副主任、亚太地区仪器与控制学会主席、世界光学委员会（ICO）副主席等。现担任中国仪器仪表学会名誉理事长、中国光学工程学会名誉理事长，且被国际光学工程学会（SPIE）、美国光学学会（OSA）、中美光电子学会等选为资深会员（Fellow）。

金国藩自述

成长过程

1929年1月8日，我降生在一个书香门第之家。父亲是"庚子赔款"的第一期赴美留学生，著名的铁路工程师与桥梁专家。我从小就受到良好的家庭教育。天下父母总是想把自己的子女培养成"超人"，我的父亲更是如此，对我倍加教育。我和兄弟姐妹们是学校放了暑假，紧接着家教"开学"，学古文，学英文，学算术……若在校学习较差，父亲则"恨铁不成钢"，有时免不了让我们遭受皮肉之苦。严格的家教使我在少年时期就养成了爱学习、勤思考和求上进的习惯。我父亲还常以"己所不欲，勿施于人""宁人负我，我勿负人""立大志，不怕受胯下之辱"等哲理教育我们，更重要的是父亲的严格认真和一丝不苟的学风，潜移默化，使我逐步塑造完美人格，终生自觉践行受益。

中学时代，我几乎是在日本帝国主义的铁蹄下度过的。日军的入侵使得我的家境每况愈下，我便从私立学校转到了市立学校。私立学校很少学日语，转至市立学校后，课时增多，教日语又是日本教官，很严厉，我本来就对日语一点兴趣都没有，这样对日语就更反感了。一次，因日语很差，我遭到教官毒打，至今刻骨铭心。青年时代的亡国奴生活激发了我要为民族争气和为国家争光的抱负。

我的大学时代是在"民主堡垒"的北京大学度过的。亲眼目睹国民党政府的腐败，中共地下党组织的教育，使我认识到"争民主，争自由"的重要性，逐步懂得了只有共产党才能救中国的革命道理，便从只知钻研学问转变到投身民主革命运动中。我曾担任北京大学机械系学生自治会主席，组织同学满腔热情迎接解放。同时在党团组织的教育下不断进步，我成为新中国的第一批新民主主义青年团团员。

新中国成立后的第一个夏天，我从北京大学毕业。我至今仍清楚地记得，入学时全班有40人，毕业时只剩下12人，我是其中的幸运者之一。我格外珍惜这段经历，总是想着要为新生的中国多作贡献。当时，国家百废待兴，教育尤为突出，我毅然服从分配，留校任教。祖国的新生给我们每个青年知识分子带来了无限的希望，1952年院系调整，我来到了清华大学机械系，讲授画法几何、机械制图、切削原理、航空仪表等课，带专业实习，搞工程设计，满腔热情，矢志不渝，学术生涯由此愉快地开始了。

三坐标光栅劈锥测量机的研制

20世纪60年代初，我国面临严峻的国际形势，国防建设急需劈锥这种零件，国内没有制造能力。苏联有意卡我们的脖子，只卖火炮，不卖解算装置用的劈锥。我们自己按照国外的样子生产了劈锥，当时没有测量仪器，不知好坏。国家打算通过港商从英国购买测量机，可价格惊人，1台200万英镑。因此，国防工办公室决定自力更生、自行研制，并把任务下到清华大学。学校组织精仪系和自控系联合攻关，研制数控劈锥铣床与测量机，由光学仪器教研组承担测量机的研制。国防工办的任务下达后，1965年上半年，系领导便将我从陀螺仪器教研组调到光学仪器教研组，任课题负责人，主持"劈锥测量机"的研制。

我学的是机械制造，对光学技术一窍不通，接受任务后面临巨大的压力。当时，国外又对有关技术严密封闭，如何选择方案就是首先遇到的难题。要实现高精度自动测量，就必须要有精密的传感器。我与研制组的教师们一起，仔细分析了已有的码盘和磁尺等技术，发现其精度都不能满足设计要求。怎么办？我们硬是憋着一股气上，带领着青年教师到工厂调研，向工人和技术人员请教，从实践中得真知；同时，仔细翻阅资料，反复联系实际，认真琢磨思索。我们从数控机床上得到了启发：既然它是用光栅作为传感器，能达到微米的测量精度，那不也可同样选用光栅传感器，满足"劈锥测量机"的微米精度要求吗？随之，我们将课题命名为"三坐标光栅劈锥测量机"。

从机械专业转到光学专业，对光学的掌握自然要有个过程。我们一方面认真阅读有关技术书籍和文献，另一方面虚心向实践和有经验的人学习。在这过程中，我带领全组找来参考书和资料，细细琢磨，弄清原理，边设计边修改；没有试验条件，就自己动手改造和研制；没有加工条件，就到工厂，利用工厂的设备，与工人同吃同干。就这样，一个个困难被一一克服，一种种技术问题被逐一解决。

这个项目是1965年上半年启动的，第二年就遇上了"文化大革命"。不管外面多乱，研制组始终想着的是早点完成研制任务，冲破国外封锁，为加强国防建设作贡献。1968年，项目进入关键时刻，校内闹派性，搞武斗，实验室所在的楼成了"武斗"据点，这也未能干扰我们的研究。为了争取时间，大家带馒头作夜餐，一干就是通宵。一次次试验，一步步前进；一次次失败，一点点提高。经过四年多的努力，终于在1969年国庆节前夕，试制成功，测试数据自动输出，X、Y两个坐标的分辨率为0.025mm，角分辨率为1/4分。精度和自动化程度都达到了当时的国际先进水平。不仅如此，"程控劈锥铣床"和"劈锥测量机"两种设备，研制成本只用了60万人民币，远远低于从国外购买1台测量机的价格。当年，前联邦德国总理参观后，惊叹道："我真没有想到，你们也能研制出精度和自动化程度这么高的光机电结合的仪器。"

计算机制全息图的应用研究

十多年的光学仪器教学与科研实践，使我深刻地领悟到：传统光学已经面临许多障碍，只有掌握新理论和新技术，才能跨越；光不仅仅是几何量测的"标尺"，还可以是信息的载体，信息光学将会蓬勃兴起。其中，计算机全息图是制作全息图的一种方法，它利用数字计算机去生成全息图，不需要物体的实际存在，只要把物光波的数学描述输入计算机处理，就能生成出世界上不存在的物体的全息图，非常适用于信号处理中的波面补偿或空间滤波器的生成，生成特殊的光学样板，用于光学元件的面形检测。

1978年，关闭多年的国门打开了，我有幸出国进修，也是怀着这种悟性出去的。曾任国际光学学会（ICO）主席的前联邦德国爱尔兰根大学的罗曼（A. W. Lohmann）教授把通信理论中的抽样定理应用于空间滤波器的制作，奠定了计算全息图制作的理论基础。因此，我选择他的实验室作为进修点，也非常珍惜这个机会，不仅如饥似渴地阅读最新文献，而且动手操作各种先进仪器。五个月后，既通晓其原理，又掌握了技术，并亲手制作出计算全息图。

在德国进修的日子里，有件事我一直无法忘记。一次，我到德国蔡司工厂参观，一台高大的实验装置引起了我的兴趣。这是利用计算全息技术检验直径2.1m大透镜的装置，计算全息检测技术的核心是计算全息图，当我提出要看计算全息图时，厂方婉言拒绝，这使我的心灵受到很大伤害。我暗下决心，回国自己搞，非争这口气不可！

回国后，我得知218厂有一项非球面检测任务，恰好可以运用计算全息技术予以解决。我喜出望外，多次下厂落实，并安排研究生开展研究。这是一种实际需要的新全息干涉图，与我在德国所制作的全然不同，还需一切从零开始。从数学建模到程序设计，从建实验装置到实际检测，每个环节都慎之又慎。经过一年努力，解决了厂里生产中的难题，圆满完成了任务。这个项目与计算全息用于综合孔径雷达研究等成果一起，获得了1992年国家教委科技进步二等奖。

上海光学仪器厂生产光谱仪，凹面光栅是其中的核心元件。可是，制造凹面光栅一直沿用国外的传统方法，精度不理想。要提高光栅制作精度，只有对有关加工设备提出严格甚至苛刻的要求，可是国内无法满足。我了解此事后，进行了认真分析，认为凹面光栅加工的难点正是计算全息能发挥作用的独到之处，可以利用计算全息消除像散等像差，得到满意的精度。于是，我主动与厂家联系，为厂家制作全息图，并共同制作出经像差补偿的全息凹面光栅，使光栅的分辨率得到很大提高，这是一个独创。

需要指出的是，我还利用全息滤波器实现综合孔径雷达信号处理，并培养了我国光学仪器的第一位博士；同时，运用通信中的脉冲调制理论解释计算全息的物理意义，提出了"空域滤波"等崭新概念。我又在总结科研的基础上，与学生虞祖良合作编写了专著《计算机制全息图》。

光计算的部署与器件研制

现行的计算机都是基于冯·诺依曼（John von Neumann）原理，要提高运算速度，就需突破。然而，这是一个"瓶颈"，要跨越，又存在很多问题。尽管各国都想方设法提高运算速度，但始终越不过极限。峰回路转，美国从基于电子转向基于光子，开展对光计算机的研究。理论分析表明：电子传输信息速度不如光子快，相差几个数量级；电子载频能力不如光子大，相差1000倍以上；电子带负电，互相排斥；光子为中性，具有高度的相容性和天然的并行处理与互联能力。我敏感地注意到这个科技发展动向，认为这是光学与计算机科学结合的一个新契机，能使光学在计算机中发挥重要作用，促进光学发展。

于是，我在1987年特意赴英国赫里奥特-瓦特大学（Heriot-Watt University）工作了半年，学习国外的先进经验，准备涉足该领域。当时，国内有人说，我国电子计算机还是跟在别人后面，搞光计算机谈何容易。我详细地分析了有关资料，认识到光计算是光学信息处理最高水平的综合体现，具有很强的带动性，必须尽早开展研究；同时，又清楚地看到国内已在光学信息处理方面有了一定的基础，具备初步条件，应该抓紧部署。我就及时组织申请了第一批"863"光计算的课题，在国内率先开展了光计算的研究。

如同电子计算机一样，光计算机也包含理论、系统、器件和算法等四个方面，研究如何下手呢？我领导课题组细致分析、认真论证，既认识到研究基础和条件与发达国家的差距，要在理论和系统方面跟踪并超越有困难，又认识到四个方面是有机地联系，其中器件不仅是构成系统的基础，而且可独立验证有关理论和实现相关算法，还有可能独立应用。从器件研制入手，完全可以发挥工科的优势，形成特色。接着，我巧妙部署，相继开展了光学并行处理器件、光学神经网络、光学子波变换、联合变换相关器以及光学形态学数字图像处理等研究，开拓了光计算技术研究的新领域，并取得了多项成果，例如：提出了改进型二值图像代数MBIA的基本理论，构建了硬件系统：双通道光学形态学实时处理器；提出了采用空间互补编码方法和相干光相关器来一步实现二值形态变换，构建了非相干光人脸自动识别系统等。由我负责并和严瑛白教授一起指导的博士生王文陆，研究出色，成果显著，学术水平高，其博士学位论文《光学子波变换及其在图像处理中的应用》被评为1999年全国首届百篇优秀博士学位论文。这些成果还得到了国内外一些单位的重视和关注，为我国光计算研究赶上世界先进水平做出了贡献。

二元光学的研究

20世纪80年代中期，美国MIT林肯实验室威尔得坎普（W. B. Veldkamp）领导的研究组在设计新型光学传感系统时，率先提出了"二元光学"的新概念。光波衍射理论是其理论基础，然而器件的制作方法却完全不同于依靠铣、磨和抛光等传统机械加工方式。器件的特点是表面带有浮雕结构，可利用计算机辅助设计，借助超大规模集成电路制作工艺，在片基上（或传统光学器件表面）刻蚀产生两个或多个台阶，如同浮雕，是一种纯相位器件，同轴再现，具有极高衍射效率。它不仅容易小型化、阵列化和集成化，而且光学功能特殊，可实现如非球面、环状面、锥面和镯面等奇特的光波面，从而有利于促进光学系统实现微型化、阵列化和集成化，将引起传统光学设计理论及其加工工艺的一次革命。

面对崭露头角的这一光学领域前沿科学，我深刻理解其意义，毅然决定立即开展"二元光学"的研究，力求在国际上有一席之地。我很清楚，我们已经有了设计与制作计算全息图的丰富经验，开展二元光学的研究可以说是顺理成章的事，并非一切从零开始。

器件设计和加工是研究的首要任务，我们选择多相位菲涅耳透镜阵列作为研究开端。在研究中发现，按常规，在它的焦点处应得到与透镜数目相同的聚焦光斑，其实不然，在某些特定距离处，可以得到更多的聚焦光斑。常规和实际的差异就是科学研究的突破点。我和同事们格外注意这个现象，决定首先从理论分析入手。透镜阵列可以看成是周期性物体，除了本身的聚焦特性之外，还具有泰伯（H. F. Talbot）效应及分数泰伯效应，这可能是产生光斑倍增现象的基础。接着，倒过来从像面到物面，反复进行理论研究，解释这种现象。理论研究使我们深刻理解了这种现象，同时开阔了思路，产生了新的设计思想：可以利用较少的透镜阵列得到更多的聚焦光斑数。

二元光学器件刻蚀深度的精密检测是一项关键技术。没有精密检测，就谈不上精密加工，更谈不上制作出精密器件。当时的台阶仪精度不高，纵向分辨率只有 0.01mm，且为接触式测量，会产生划痕，根本不能用。我深知，只在理论上理解是很不够的，还必须有可靠的技术和工艺来保证。测量问题不解决，将会直接影响二元光学器件的研制。因此，立即组织大家认真研究和分析各种测量技术，并从电视图像处理技术中得到了启发：将其与干涉测量显微镜相结合。白光干涉仪产生的是彩色干涉条纹，可利用其零级黑条纹随二元光学器件的台阶而产生的错位，由此不就可以推算出两个台阶之间的相位差即台阶深度吗？这样，只需用摄像机摄入干涉条纹，再经图像处理，测量分辨率可达到 $\lambda/120$，重复精度为 2nm。方法简单，技术现成，既是非接触，又达到纳米级，实在是二元光学研究的极好检测仪器。

就是这样，经过四年艰苦奋斗，我们获得了丰硕的成果：深入研究了基于菲涅耳衍射理论及周期物体的泰伯自成像效应及分数泰伯效应；提出了利用达曼（H. Dammann）光栅实现多通道光学形态学中并行邻域操作原理；提出了多相位泰伯光栅型阵列发生器以及光波分束与聚焦功能合一的二元光学器件的设计原理与方法；研制了 8×8 伽伯（D. Gabor）透镜阵列、50m 准直用长焦深器件、光盘用多功能器件和 9 相位 50×50 的菲涅耳透镜阵列等器件。达曼光栅等器件研制出来后，就在国内外 10 多个单位得到应用，其中有美国宾夕法尼亚州立大学、斯坦福大学、贝尔实验室和 NEC 实验室等。

"二元光学技术与器件研究"获 1994 年国家教委科技进步二等奖，"高衍射效率二元光学器件的设计与制备技术"获 1996 年国家科技进步三等奖。同时，结合任务培养了 8 名博士生，还与严瑛白教授、邬敏贤教授等合作编写了专著《二元光学》。

数字体全息光学存储和识别技术的研究

光盘以存储量大、保存时间长、抗干扰能力强和快捷方便等著称，发展迅速，从科技信息的存储到娱乐音像的记录播放，广为流行。虽然光盘号称是海量存储，但是实际上远未达到。存储大容量数据需用许多张光盘，遇上图像还需要对图像进行压缩和解码，传输速度也不理想，因而在气象、地质、公安和军事等领域中的应用受到诸多限制。如何革新图像存储技术，真正实现海量存储，这是下一代光学存储技术所面临的难题。

面对挑战，我们细致分析了各种光存储机理和介质，清楚地看到，体全息存储是一种三维数据存储技术，存储密度高，一块 $10mm^3$ 的铌酸锂晶体可存储 1 万幅以上高分辨率图像。同时，体全息存储基于"页面"方式对数据进行读取，访问速度极快，图像可即选即现，不必经过计算机就可直接与有关图像传输系统对接，进行存储，如卫星侦察图像系统。不仅如此，体全息存储还能作为光学相关器，实现图像的相关识别，对存储的数据进行快速内容寻址。我们认真总结了20多年来从事光学信息处理研究经验，结合非线性光学晶体材料的研究进展，综合国内外文献，清醒地认识到体全息存储是光信息存储和处理最重要的研究领域之一，其意义重大，应该紧紧抓住不放。同时，我也十分清楚，一个新的研究方向出现之时，既面临激烈的竞争，同时也提供了创新的良机。重要的是选好创新的突破点。我们认真分析，充分了解其内容，决定从构建一套晶体体全息数据存储及相关识别系统入手。经过不懈的努力，完成了系统的构建，在国内首先实现了晶体中公共体积内1000幅数据页的存储容量。

1999年8月，我赴美出席"第十八届世界光学委员会"国际学术会议，在会上宣读了题为"应用光学子波变换改善体全息相关器的质量"论文，引起了很大反响。我负责指导的博士生冯文毅研究出色，创新性强，其博士学位论文《光学子波并行处理技术及应用》被评为2001年全国百篇优秀博士学位论文。

我一直认为，只有原创性的技术发明才具有引领性和扩展性，才有可能形成一个产业，带动国民经济发展。并且，清醒地看到：研究与国外同步，这意味着我们也有获得原创性技术的机会，可是还有差距，绝不能满足于已取得的成果。接着，我又联合国内相关优势单位，争取到了国家重点基础研究计划"973"项目"新型超高密度、超快速光信息存储与处理的基础研究"，并负责承担其中的"体全息存储机理研究"课题，带领课题组继续前进，使研究更上一层楼。

就是这样，我们朝着追求技术原创性的研究目标，齐心协力，脚踏实地，不急不躁，坚持不懈，攻克难题，登上一个又一个高地，取得了一项又一项令人振奋的成果。其中突出的有：

提出动态散斑体全息存储的新概念，采用动态散斑-角度混合复用技术方案，使复用灵敏度比单纯角度复用灵敏度提高了8倍，为进一步实现超高密度体全息存储提供了有效途径。

采用约束递减曝光时序、散斑调制和子波变换等技术，研制了1台小型化晶体体全息存储与相关识别系统，有效地抑制了旁瓣噪声，实现了体全息相关识别，识别准确率大于95%，识别速率大于4500幅/s，存储及相关识别的综合性能指标都达到了国际先进水平，其中并行相关识别通道数属国际领先水平。

提出基于正交偏振双通道记录方案，构建了聚合物体全息光盘存储系统，独创了在有机聚合物全息光盘系统中实现正交偏振双通道体全息存储，使存储介质直径和厚度分别为120mm和300μm的商用化全息光盘的信息容量达到1.6Tb，性能指标达到了国际先进水平。

这些研究成果有效地解决了超高密度超快速体全息存储中的关键问题，推动了体全息存储的实用化进程，荣获2005年北京市科学技术奖一等奖。

光学体全息存储既是理论基础研究的热点，更是开发应用的竞争高地。要使其能得到应用，在有了一定的理论研究基础后，就要十分重视应用基础研究。为此，我们在保证光学体全息存储机理研究前提下，将研究重点转移到存储技术、系统集成以及全息存储器应用等方面。

首先，研究多通道并行相关图像识别技术，实现了单一存储位置6000并行相关通道，库图像识别准确率为100%，存储密度达到了60Gb/cm³，相关速率为150000幅/s，运算速率达到1.18×10^{11}MAC/s，全面提升了相关器的性能。

接着，面向TB量级存储容量和Gbps量级数据传输率，研发出并行光学存储系统，实现数据加密、内容寻址和特殊运算等功能，可用于海量数据存储与处理。

开展了高分辨率遥感图像快速处理的研究，提出了二维随机交织和多样本并行估计等一系列方法，使图像处理精度可以达到亚像素量级；还将体全息光学图像处理技术应用在指纹图像识别，能够在20万枚指纹中进行快速光学并行识别。上述工作将体全息光学图像处理技术推向了实用化，研究成果获得了2013年北京市科学技术奖一等奖。

多年来，我们先后获得国家自然科学基金、"863"、"973"、武器装备预研重点基金等10余项研究课题。目前，正在研究和开发面向实用的高性能体全息数据存储器与光学图像处理器，争取有更大的作为，对国防建设做出实实在在的贡献。

感　悟

时至今日，我已在清华工作了64年，大大超过了马约翰教授提出"为祖国健康工作五十年"的标准，感到很自豪。生命的意义在于不断地学习，不断地奋斗，不断地创新，不断地为祖国作贡献。我是个幸运者，我的学术生涯完全与新中国共命运。1950年夏天，中华人民共和国建立还不到一年我就大学毕业了，立即投身发展祖国高教事业；1952年院系调整，我来到清华大学，从此开始了工科的学术生涯；国民经济恢复时期，我边讲课边带领学生实习，精心培养急需人才；国家面临外国封锁急需大力加强国防建设时期，我被从陀螺仪器教研组调到光学仪器教研组，负责研制"劈锥测量机"。改革开放时期，焕发了我的科学研究青春，坚韧不拔地攀登科学高峰。同时，还逐步走上教学和科研管理岗位，先后任精密仪器与机械学系主任和机械工程学院院长、国家自然科学基金委员会副主任和教育部科学技术委员会常务副主任等。回首我的学术生涯，有苦闷，有喜悦；有生气，有振奋；有崎岖，有坦途；有挫折，有成功；感慨万千，深切感悟：

祖国的迫切需要就是第一志愿，这是我学术生涯永葆青春的根本；忠实地为祖国作贡献，那是我信心和力量的源泉。大学毕业就遇上国家急需大学老师来培养建设人才，我毫不犹豫留校；国家急需加强国防建设，我从机械制造转向光学仪器；"文革"后百业待兴，国门刚打开我就有幸出国进修开阔眼界；国家重视发展高科技，我一直不断地申请到多项"973"和"863"项目。国家建设持续发展，新的需求不断提出；我义无反顾地响应，学术生涯便常新常青。我受过日本帝国主义的欺凌，目睹国民党的腐败，喜迎新中国的诞生，切身感受使我深深地爱着新中国，并立志献身祖国的教育和科技发展的事业。在这65年的学术生涯中，无论是在科研第一线，还是授课和指导博士生；无论是站在国际学术讲坛上，还是身处教学科研管理岗位，一心想着的是多为祖国作贡献，多为祖国争光，因而总能不气馁，不畏缩，精力充沛，勇往直前。

科学研究应该跟踪前沿，站在前沿，引领前沿。一旦前沿在眼前出现时，如果畏难放弃，就会使遇上的机会失之交臂，这不是一位科学工作者应有的态度。"跟踪前沿"，不应着眼在"跟"字上，而应瞄准前沿，先"跟"后"追"，独辟蹊径，扬我所长，有所作为，抢占科学高地。如盲目跟从，他做我学，简单模仿，将永远落后。

作为一名清醒的科技工作者，应该具有敏锐的目光，洞察并及时发现新的研究方向和科学发展中出现的新事物，开拓科学研究新天地；同时，还要能抓住并驾驭新的研究，这样才能在激烈的国际竞争中取得主动权。一个新的研究方向出现之时，既面临激烈的竞争，同时也提供了创新的良机，重要的是选好创新的突破点。

大学教授应该既是知识的传授者，更是知识的创造者，尤其是在研究生的培养中要重在能力培养，包括创新能力、阅读能力、写作能力和表达能力。学生做什么课题并不重要，一个课题做了几年，并不可能就成为该领域的专家，也不可能将来一定做这方面的工作。但是，鼓励学生在国际著名杂志上发表论文十分必要，这可表明他的研究具有创新性；还有，工科学生最重要的是一定要做出实际的成果，实实在在地解决实际科学和技术问题。

科学发现或技术发明都是世界性的，只有第一，没有第二。尤其是技术发明，只有原创性的才具有引领性和扩展性，才有可能形成一个产业，带动国民经济发展。

现代科学技术高度进步，迅速发展，相关技术交错，仅靠个人的聪明才智或个人奋斗，根本不可能取得像样的成果。我能取得点成绩，完全和我有一支富有创新和牺牲精神、团结奋斗、脚踏实地苦干的团队分不开，没有他们，就没有我的今天。

（金国藩口述　余兴龙执笔）

光学信息处理

使用全息光学元件的合成孔径雷达信号处理器[*]

摘 要 本文提出了两种新的以计算机源生的全息光学元件作为空间变元件的 SAR 处理器模型,并讨论了系统设计、像质评价等问题,文后给出了一个实际的实验结果。

1 引言

合成孔径雷达(Synthetic Aperture Radar,简称 SAR)是一种主动式的微波遥感手段。在航天技术、资源探测以及军事侦察等方面获得了广泛的应用。

现有的机载 SAR,一般都将雷达信号经飞点扫描管变为光信号存储在感光胶片上,这个胶片称为 SAR 数据片。但 SAR 数据片并不是直观的地面几何像,而是某种特殊的编码。光学处理器的目的,就是将 SAR 数据片还原为地面几何像。

2 SAR 数据片的光学特征

SAR 数据片是由许多单元图形叠加而成。单元图形的数学表达式如下:

$$T_H(x_H, y_H) = b_0 + t_\alpha + t_\beta \tag{1}$$

式中,$t_\alpha = a\exp\{-jk(\phi_{0H} + \phi_{xH} + \phi_{yH})\}$;$t_\beta = t_\alpha^*$;* 是共轭符号;$\phi_{0H} = x_H\sin\theta$;$\theta$ 是 SAR 数据片的偏频角;$\phi_{xH} = (x_H - x_{H0})^2/[2f_x(y_{H0})]$;$\phi_{yH} = (y_H - y_{H0})^2/(2f_r)$;$x_{H0}$、$y_{H0}$ 是单元图形的中心坐标;f_x、f_r 分别是数据片的航向及舷向焦距。

在 SAR 数据片 x_H 方向记录的信息我们称为航向信息,与其正交的方向记录的是舷向信息。当用一束相干的平面波照射 $T(x_H, y_H)$ 时,波面将发生衍射(图 1),舷向的信息会聚在一条与光轴成 θ 角的焦线上,焦线到 (x_{H0}, y_{H0}) 的距离称为航向焦距,记作 f_x。舷向信息则会聚在另一条正交的焦线上,该段距离称为舷向焦距,记作 f_r,共轭的衍射分别用 $-f_x$,$-f_r$ 表示。

理论推导声明 f_x 基本上是 y_H 的线性函数。因此,就整个 SAR 数据片而论,航向的焦面是倾斜的,且与舷内焦面分离了一段距离。由标量波的衍射原理可知必然存在共轭级组。

SAR 光学处理器的任务之一就是要把分离的焦面合并到一起。由航向焦面的倾斜可知,这个系统必然是一个空间变的光学系统。

在 SAR 的成像过程中,x_H 向及 y_H 向的缩图比是不一样的,分别记作 P 和 q。定义

---

[*] 本文合作者:陆达。原发表于《仪器仪表学报》,1995,6(4):382~391。

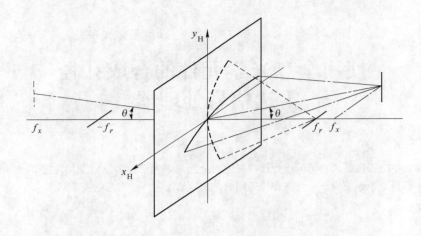

图 1 SAR 数据片的衍射

$K \triangleq q/p$,K 通常被称为纵横比。SAR 处理器也应具备纵横比校正的功能。

3 使用全息元件的 SAR 处理器

为了处理 SAR 图像,曾经出现过多种 SAR 光学处理器,主要有锥透镜处理器[1]、斜平面处理器[2]、匹配滤波处理器[3,4]等。这几种处理器的一个共同缺点就是体积过于庞大。为此,我们提出了两种使用全息光学元件的 SAR 光学处理器,主要的目的在于缩小体积以便于机载。与目前已报道过的 SAR 光学处理器相比,我们系统的主要特点在于使用了计算机源生的全息光学元件(Computer Originated Holographic Optical Elements,简称 COHOE)作为空间变元件。COHOE 是一种高效能的光学元件[5],可以集多种光学功能于一身,产生各种复杂的波面。SAR 光学处理器引入 COHOE 后,可以使系统趋于简化。关于 COHOE 制备问题将另文发表。

3.1 处理器光路

我们设计的第一种系统如图 2 所示。由上节的讨论可知,SAR 数据片航向、舷向具有不同的光学性质。为简化设计,可以认为它们是相互独立的[10],因此有

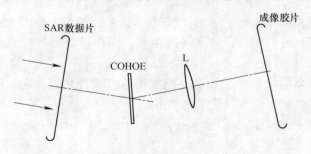

图 2 使用全息元件的 SAR 处理器

必要按两个不同方向将光学系统分别展开，以利进一步的讨论。展开图见图 3。其中图 3（a）称为航向光学系统，其透镜焦距称为航向焦距，放大率记作 M_x；图 3（b）称为舷向光学系统，其透镜焦距称为舷向焦距，放大率记作 M_y。舷向放大率与航向放大率之比记作 K_0，$K_0 \triangleq M_y/M_x$，展开图左侧为输入的相干平行光，$p_1(x_1, y_1)$ 代表 SAR 数据片，$p_2(x_2, y_2)$ 代表 COHOE，$p_3(x_3, y_3)$ 代表球面镜，$p_4(x_4, y_4)$ 是像面。

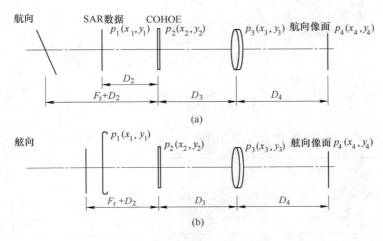

图 3　SAR 光学处理系统展开图

3.2　处理器系统方程

航向光学系统中作为物的是倾斜的航向焦面，其焦距记为 $f_x(y_1)$，COHOE 的航向焦距记为 $f_{xx}(y_2)$。在舷向光学系统中，作为输入的是舷向焦面，数据片的舷向焦距用 f_r 表示，COHOE 的舷向焦距记为 f_y。系统设计首先应满足：

$$D_{4x} = D_{4y} \tag{2}$$

$$KM_x = M_y \tag{3}$$

分析表明，图 2 所示的系统中不可能使各个点均满足式（3），因此在设计中式（3）仅作为核算的条件。

设计的结果表明，COHOE 在 x 方向为一个曲母线的锥透镜，而在 y 方向则为一个柱透镜。这个波面的立体示意图见图 4。图 5 给出了航向锥透镜焦距 $f_{xx}(y_2)$ 与 y_2 的函数关系，在图中用"○"符号标出。图中标有"△"的曲线是 $y_2 - \dfrac{\mathrm{d}f_{xx}(y_2)}{\mathrm{d}y_2}$ 曲线，由曲线可以看出锥透镜各点导数不等，即锥透镜的母线是弯曲的。y_2 与 $f_{xx}(y_2)$ 的典型值见表 1。

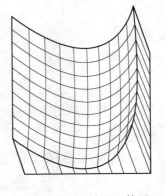

图 4　COHOE 波面立体图

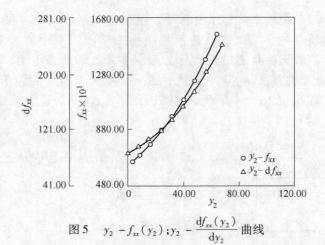

图 5 $y_2 - f_{xx}(y_2)$；$y_2 - \dfrac{df_{xx}(y_2)}{dy_2}$ 曲线

表 1 $f_x(y_1)$，$f_{xx}(y_2)$，$\dfrac{df_{xx}(y_2)}{dy_2}$ 的典型值

$f_x(y_1)$	$f_{xx}(y_2)$	$\dfrac{df_{xx}(y_2)}{dy_2}$
-4937.98	10036.0	143.75
-5632.15	13258.5	195.70
-6333.19	17766.2	281.64

如前所述，这个系统并不满足式（3），即成像有畸变。在图 2 的系统中，M 是一个实常数，而 M_x 则是 y_1 的函数。M_x 随 y_1 变化的趋势见图 6 中的"△"曲线，"○"曲线给出的是 $y_1 - K_0$ 的关系。

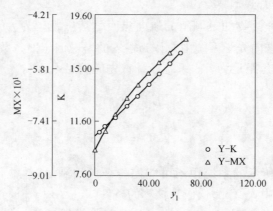

图 6 $M_x - y_1$，$K_0 - y_1$ 曲线

参照锥透镜处理器的方法在成像胶片前设置一狭缝，可使畸变减小到可以接受的程度。设计得到的结构参数为：$D_2 = 80\text{mm}$，$D_3 = 120\text{mm}$，$D_4 = 317.623\text{mm}$。结构是比较紧凑的。

4 具有全面纵横比校正的 SAR 光学处理器

为克服前述系统不能全面校正纵横比的缺点，我们又设计了使用两块 COHOE 的 SAR 光学处理器。为区别起见，上节的系统称为系统 I，本节的系统称为系统 II。从光学功能上看，系统Ⅱ与斜平面处理器类似，但光路和设计思想完全不同。

由上节的分析可知，使用一块 COHOE 难以同时满足式（2），式（3），必须引入新的变量。为此，我们提出了图 7 所示的光路。为便于区别，第一块全息元件称 COHOE1，其航向焦距记作 $f_{xx1}(y_1)$，舷向焦距 f_{y1}；第二块全息元件称 COHOE2，其航向焦距为 $f_{xx2}:(y_3)$，无舷向光焦度。系统中各参数名称的约定与系统Ⅰ一致。

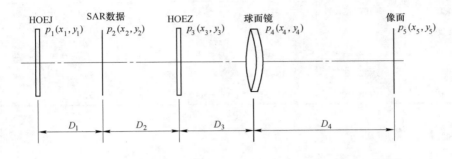

图 7　系统Ⅱ光路展开图

系统Ⅱ的基本设计方程与系统Ⅰ一样，但此时式（3）作为强制的约束条件引入，因此设计方法与系统Ⅰ不尽相等。

设计的结果表明，COHOE1 仍为一个柱面镜与锥透镜的组合，COHOE2 是一个锥透镜，这两个锥透镜的焦距均呈复杂的分布，焦距存在间断点。y_1-f_{xx1} 及 y_2-f_{xx2} 的曲线分别见图 8 和图 9。COHOE1 所产生的波面示意图见图 10。

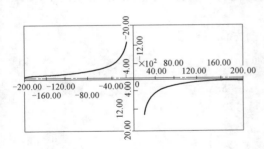
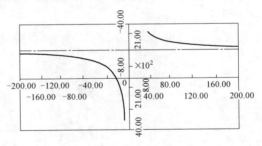

图 8　y_1-f_{xx1} 曲线　　　　　　　　　图 9　y_2-f_{xx2} 曲线

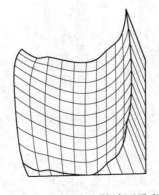

图 10　COHOE1 产生的波面示意图

设计的结构参数为：$D_2 = 150\text{mm}$，$D_3 = 350\text{mm}$，$D_4 = 100\text{mm}$，$D_5 = 321.790\text{mm}$。表 2 列出的是系统 Ⅱ 的典型值。这样复杂的波面只能靠 COHOE 产生，一般的光学元件是无能为力的。

表 2 系统 Ⅱ 典型值

y_1	$f_{xx1}(y_1)$	y_2	$f_x(y_2)$	y_3	$f_{xx2}(y_3)$
−14.00	−1078.81	−31.02	3805.87	−33.52	3473.44
−8.26	−2231.54	−18.60	4263.35	−19.76	∞
−0.48	∞	−1.083	4889.90	−1.18	−2536.23
8.00	2967.13	18.03	5583.59	19.17	−1203.52
14.00	1936.61	31.62	6081.21	33.52	−874.08

5 系统 Ⅰ 的像质评价问题

为了达到结构紧凑的目的，我们在系统 Ⅰ 及系统 Ⅱ 中使用了全息光学元件，并以离散点的形式给出了它们所应产生的波面。在一般情况下，全息元件的制作条件与使用条件总是有差异的，因此有必要在系统设计时将全息元件置于实际使用的光路中进行理论上的像质评价，以充分估计该系统的性能和潜力。这也是使用全息元件的 SAR 光学系统设计与普通光学系统设计的区别之一。

评价全息光学元件像质的方法有许多种[6~9]。由于我们的系统是空间变系统，波面也已用离散点的形式给出，因此采用了光栅方程追迹的方法，将结果用点列图标出，作为评价的依据。

光波通过全息光学元件的行为可用如下光栅方程组来描述[5]：

$$\phi_1 = \phi_c \pm (\phi_0 - \phi_R) \tag{4a}$$

$$\frac{\partial \phi_1}{\partial x} = \frac{\partial \phi_c}{\partial x} \pm \mu \left(\frac{\partial \phi_0}{\partial x} - \frac{\partial \phi_H}{\partial x} \right) \tag{4b}$$

$$\frac{\partial \phi_1}{\partial y} = \frac{\partial \phi_c}{\partial y} \pm \mu \left(\frac{\partial \phi_0}{\partial y} - \frac{\partial \phi_H}{\partial y} \right) \tag{4c}$$

$$n_1 = \pm \sqrt{1 - \left(\frac{\partial \phi_1}{\partial x}\right)^2 - \left(\frac{\partial \phi_1}{\partial y}\right)^2} \tag{4d}$$

上式中的脚标 1、c、0、R 分别代表成像波、再现波、物波及参考波，$\mu = \lambda_c / \lambda_0$ 是再现光波与构造光波的波长之比；式 (4a)、式 (4b)、式 (4c) 中的 " + " 号表示选 +1 级衍射，" − " 号表示 −1 级，式 (4d) 中的 " + " 号代表光波沿光轴正向传播，反之则用 " − " 号表示。

系统 Ⅰ 的追迹坐标系见图 11，图中 $x_1 - y_1$，$x_2 - y_2$，$x_3 - y_3$，$x_4 - y_4$，平面分别代表 SAR 数据片，COHOE 球面镜及像面。

一般的全息光学元件都是离轴工作的，所以在含全息元件的平面必须引入旋转变换，又由于 SAR 数据片存在一定的偏频角（见图 1），因此在其后续的平面上要做平移变换，我们设计的是一个空间变系统，若像一般的光学系统那样仅追迹少数几个点不

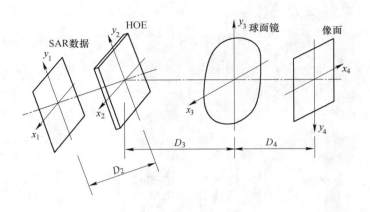

图 11　系统 I 追迹坐标系

足以评价整个系统的性质，因此必须采用较多的点数。系统 I 追迹了 405 个点，系统 II 189 个点。凡此种种，构成了我们的光路追迹方法及公式与通常光学追迹的差别，以上论述也同样适用于系统 II。

下面讨论点列选取的问题。设对 $x_4 - y_4$ 中 K 个物点进行追迹，每个物点记作 $P_4^{(k)}(x_{40}^{(k)}, y_{40}^{(k)})$，$k = 1, 2, \cdots, K$。

与之相应的 SAR 数据片的有效信息则在以 $\vec{X}_1^{(k)} - \vec{Y}_1^{(k)}$ 为中心的矩形中，矩形尺寸为 $b_x^{(k)} \times b_y$，见图 12。其中 $x_{10}^{(k)} = x_{40}^{(k)}/M_x$，$y_{10}^{(k)} = y_{40}^{(k)}/M_y$。$M_x$、$M_y$ 分别是系统的航向及舷向放大率。

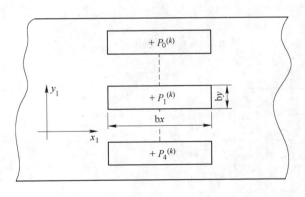

图 12　以 $P_1^{(k)}(x_{10}^{(k)}, y_{10}^{(k)})$ 为中心的 SAR 数据有效区域

将图 12 中的每个矩形域分割成 $p \times q$ 个结点，对每个结点实施方程组（4），则由节点发出的光线经 $x_2 - y_2$，$x_3 - y_3$ 后，应当汇聚在 $P_4^{(k)}(x_{40}^{(k)}, y_{40}^{(k)})$ 附近。其他平面上矢量的定义内容与书写形式均与此相同。

在矢量表达式中，实施方程组（4）将表现为一系列的矩阵变换。若理想成像，则 $\vec{X}_4^{(k)}$、$\vec{Y}_4^{(k)}$ 是一个常数与单位向量的乘积，在一般情况下，$\vec{X}_4^{(k)}$、$\vec{Y}_4^{(k)}$ 描述了像点弥散情况，现将 $\vec{X}_4^{(k)}$、$\vec{Y}_4^{(k)}$ 各分量至中心的平均距离定义为像点弥散的数量标度，记作 D_{iss}。

$$D_{iss}^{(k)} \triangleq \{[(\vec{X}_4^{(k)} - x_c\vec{I})(\vec{X}_4^{(k)} - x_c\vec{I})^T + (\vec{Y}_4^{(k)} - y_c\vec{I})(\vec{Y}_4^{(k)} - y_c\vec{I})^T]\vec{I}\cdot\vec{I}/M\}^{\frac{1}{2}} \quad (5)$$

式中，$x_c = \vec{X}_4^{(k)} \cdot \vec{I}/M; y_c = \vec{Y}_4^{(k)} \cdot \vec{I}/M; I \triangleq (1,1,\cdots,1)^T$；"·"表示点积。

图13给出了$\vec{X}_1^{(k)}$、$\vec{Y}_1^{(k)}(k=1,\cdots,9)$点列分布，DISS诸值见表3。本例中$k=9$，但本文只给出其中的三个结果。

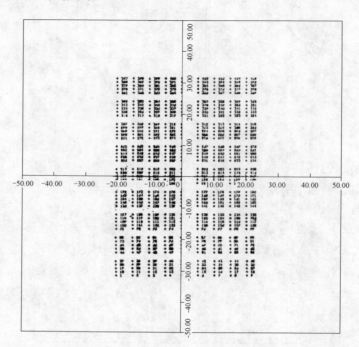

图13　$\vec{X}_1^{(k)} - \vec{Y}_1^{(k)}$点列分布

表3　系统Ⅰ追迹结果

编　号	理论值		实际值		DISS
	X_0	Y_0	X_G	Y_G	
1	-0.015	-30.131	-0.015	-30.189	0.00810
5	0	0	0	-0.037	0.00508
9	-0.015	30.131	-0.015	30.105	0.00379

6　系统Ⅱ的像质评价问题

系统Ⅱ像质评价的方法及原则与系统Ⅰ基本一致，以下仍沿用上节的符号及定约。系统Ⅱ追迹与系统Ⅰ的主要差别在于间断点的处理。由于这个问题的存在，系统Ⅱ追迹的计算机程序文本比系统Ⅰ长一倍以上。

系统Ⅱ光路追迹用坐标系见图7及图11。图14是$x_1 - y_1$平面的出发点列，图15是$x_s - y_s$平面的点列分布。表4无校正一栏给出了追迹结果。

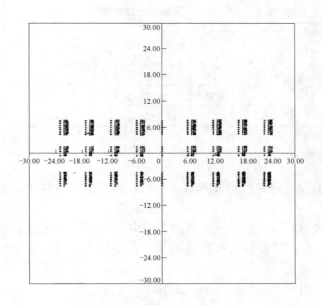

图 14　$x_1 - y_1$ 平面点列分布

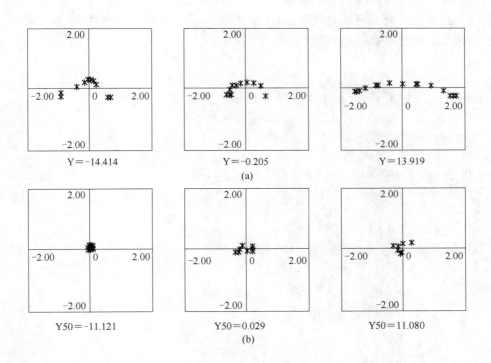

图 15　$x_6 - y_6$ 平面点列分布

由追迹的结果可知，系统Ⅱ的像质远劣于系统Ⅰ，因此必须进行优化校正，使像质得到改善。结果见表 4 有校正一栏及图 15（b）。优化设计的方法将另文发表，不再赘述。

表4 系统Ⅱ追迹结果

编号	理论值		无校正			有校正		
	X_0	Y_0	X_c	Y_c	DISS	X_c	Y_c	DISS
1	-0.259	-14.118	-0.345	-14.365	0.063	-0.263	-14.100	0.0190
2	0	0	-0.009	-0.172	0.058	-0.003	0.011	0.0370
3	-0.261	14.118	-0.237	13.982	0.133	-0.236	14.129	0.0303

7 后滤波技术

SAR 处理器的输入端是数据胶片 A，存在着较大的散射噪声，对成像质量有很大的影响。

为降低散射噪声，可在图 2 或图 7 的系统中引入一个 4f 的光学系统，在 4f 系统的频谱面上安放一个空间滤波器就可以达到降低散射噪声的目的。显然，这个装置是比较庞大的。

理论分析和实验都表明在系统Ⅰ及Ⅱ中的球面镜与成像胶片之间有这样一个位置，在这个位置上舷向信息被压缩在一条水平的窄带上，而散射光则弥散在它的四周。这个位置恰是舷向光学系统的焦面位置。在此置一个水平窄缝，即可放过有用信息而最大限度地拦掉散射光。目视的观察表明，此时的噪声水平仍稍高于用 4f 光学系统的预滤波的情况，但比不做任何滤波要好得多。

参 考 文 献

[1] L. J. Cutrona, et al. Pro. of the IEEE 54, 8, (1966), 1626.
[2] A. Kozma, et al. Appl. Opt., 11, 8, (1972), 1768.
[3] W. H. Lee, AD-771792, (1974).
[4] J. R. Fienup, SPLE, 118.
[5] J. R. Fienup, AD-A085219, (1980).
[6] W. N. Mohon, AD-782509, (1974).
[7] W. C. Sweatt, Appl. Opt., 1978, 17, 8, 1220.
[8] E. A. Magerum, AD-A027043, (1975).
[9] I. Weingartner, et al. Optik, Band57, 1, (1980), 103.
[10] E. N. Leith, J. O. S. A. 1973, 63, 2, 119.

利用改进的符号数算法和光学符号代换实现矩阵计算[*]

摘　要　本文提出了一种利用改进的符号数算法和多窗口解码光学符号代换法则实现多值矩阵计算的光学方法，并给出两个多比特改进的符号数矩阵外积计算的实验结果。这一方法具有精度高、速度快等特点。

关键词　改进的符号数；多窗口解码；光学符号代换；光矩阵计算

1　引言

由于光矩阵计算在许多光学变换和信号处理中所起的重要作用，在国内外得到广泛的研究[1,2]。光的并行处理和空间互不干扰特性非常适合于矩阵计算二维信息处理。迄今为止，人们已经提出并实现了许多光学矩阵计算方法，但总的说来，精度不高，速度慢，大多只能实现二值（0，1）矩阵的计算。

本文提出一种利用改进的符号数（MSD）算法和多窗口解码光学符号代换法则（MW—OSSR）实现矩阵计算的方法，改进符号数编码增加了操作数数值范围且隐含一种并行处理方式，而与操作数长度无关，多窗口解码光学符号代换法则大大减少了识别过程所需的通道数。这一方法的特点是精度高、速度快以及可完成多值矩阵的计算。

2　基于改进符号数算法的多值矩阵计算

对于多值矩阵的计算，为增加多值数数值范围，必须对其进行编码，常见的二进制码由于其算法上的时序性而不适合用光学实现，为此，我们采用可并行处理的改进符号数编码方式。

任意一个十进制数 D 均可由下式表示成改进符号数[3]

$$D = \sum_{k=0}^{N-1} a_k 2^k = a_{N-1} 2^{N-1} + a_{N-2} + \cdots + a_1 2^1 + a_0 2^0 = [a_{N-1} a_{N-2} \cdots a_k \cdots a_1 a_0]_{MSD} \quad (1)$$

式中，$a_k \in \{1, 0, \bar{1}\}^*$，$N$ 为改进符号数数字的表达精度。例如，十进制数 13 在五位精度的改进符号数数字系统可表示成

$$(13)_D = (1)2^4 + (0)2^3 + (-1)2^2 + (0)2^1 + (1)2^0 = [10\bar{1}01]_{MSD}$$

或

$$(13)_D = (0)2^4 + (1)2^3 + (1)2^2 + (0)2^1 + (1)2^0 = [01101]_{MSD} \quad (2)$$

上述两种表示方式在数值上完全等价，仅仅是编码形式不同而已，不影响运算结

---

[*]　本文合作者：周少敏，邹敏贤。原发表于《光学学报》，1991，11(1)：43～50。

果。改进符号数码同时还可以携带符号信息，一个改进符号负数是其对应的改进符号数正数的逻辑补（定义：$1' = \bar{1}, 0' = 0, \bar{1}' = 1$，"¯"表示取补），所以

$$(-13)_D = [\overline{1}010\overline{1}]_{MSD} = [0\overline{1}\ \overline{1}01]_{MSD} \tag{3}$$

这样便可以方便地将改进符号数减法转化为改进符号数加法。

下面，以两个 2×2 阶多值矩阵为例进行讨论，设

$$A = \begin{pmatrix} a_{11} & a_{12} \\ a_{21} & a_{22} \end{pmatrix}_D = \begin{pmatrix} a_{11}^{N-1} a_{11}^{N-2} \cdots a_{11}^1 a_{11}^0 & a_{12}^{N-1} a_{12}^{N-2} \cdots a_{12}^1 a_{12}^0 \\ a_{21}^{N-1} a_{21}^{N-2} \cdots a_{21}^1 a_{21}^0 & a_{22}^{N-1} a_{22}^{N-2} \cdots a_{22}^1 a_{22}^0 \end{pmatrix}_{MSD} \tag{4}$$

$$B = \begin{pmatrix} b_{11} & b_{12} \\ b_{21} & b_{22} \end{pmatrix}_D = \begin{pmatrix} b_{11}^{N-1} b_{11}^{N-2} \cdots b_{11}^1 b_{11}^0 & b_{12}^{N-1} b_{12}^{N-2} \cdots b_{12}^1 b_{12}^0 \\ b_{21}^{N-1} b_{21}^{N-2} \cdots b_{21}^1 b_{21}^0 & b_{22}^{N-1} b_{22}^{N-2} \cdots b_{22}^1 b_{22}^0 \end{pmatrix}_{MSD} \tag{5}$$

则

$$A + B = \begin{pmatrix} a_{11} + b_{11} & a_{12} + b_{12} \\ a_{21} + b_{21} & a_{22} + b_{22} \end{pmatrix} = \begin{pmatrix} a_{11}^{N-1} a_{11}^{N-2} \cdots a_{11}^1 a_{11}^0 & a_{12}^{N-1} a_{12}^{N-2} \cdots a_{12}^1 a_{12}^0 \\ + b_{11}^{N-1} b_{11}^{N-2} \cdots b_{11}^1 b_{11}^0 & + b_{12}^{N-1} b_{12}^{N-2} \cdots b_{12}^1 b_{12}^0 \\ a_{21}^{N-1} a_{21}^{N-2} \cdots a_{21}^1 a_{21}^0 & a_{22}^{N-1} a_{22}^{N-2} \cdots a_{22}^1 a_{22}^0 \\ + b_{21}^{N-1} b_{21}^{N-2} \cdots b_{21}^1 b_{21}^0 & + b_{22}^{N-1} b_{22}^{N-2} \cdots b_{22}^1 b_{22}^0 \end{pmatrix}_{MSD} \tag{6}$$

$$A \cdot B = \sum_{l=1}^{2} A_l \cdot B_l = (a_{11}^{N-1} a_{11}^{N-2} \cdots a_{11}^1 a_{11}^0 \quad a_{21}^{N-1} a_{21}^{N-2} \cdots a_{21}^1 a_{21}^0)_{MSD}^T \cdot$$
$$(b_{11}^{N-1} b_{11}^{N-2} \cdots b_{11}^1 b_{11}^0 \quad b_{21}^{N-1} b_{21}^{N-2} \cdots b_{21}^1 b_{21}^0)_{MSD} +$$
$$(a_{12}^{N-1} a_{12}^{N-2} \cdots a_{12}^1 a_{12}^0 \quad a_{22}^{N-1} a_{22}^{N-2} \cdots a_{22}^1 a_{22}^0)_{MSD}^T \cdot$$
$$(b_{21}^{N-1} b_{21}^{N-2} \cdots b_{21}^1 b_{21}^0 \quad b_{22}^{N-1} b_{22}^{N-2} \cdots b_{22}^1 b_{22}^0)_{MSD} \tag{7}$$

式中，$A_l \cdot B_l$ 称为 $A \cdot B$ 的第 l 个部分积（$l = 1, 2$），T 表示转置。

在实现 $A + B$ 运算时，仅仅需要进行三次变换而与操作数长度（N）无关，其中隐含了一个进位方式，使得两个 N bit 改进符号数的加法可并行地完成。Fig. 1（a）示出了变换规则以及 Fig. 1（b）示出其计算步骤，Table 1 为一个计算实例。

在实现 $A \cdot B$ 运算时，先用外积算法进行部分积阵列的计算，其中所遵循的变换规则（M）如 Fig. 2 所示，然后对部分积阵列实行移位相加，依次每两行为一组（当表达矩阵元素的改进符号数精度 N 为奇数时，部分积阵列由奇数行组成，因此，每两行相加后，剩下最后一行为单独一组），经过 $N/2$ 次（N 为偶数）或 $(N/2)+1$ 次（N 为奇数）循环后，将得到最后结果。

下面，再以矩阵求逆 A^{-1} 为例来讨论改进符号数矩阵除法。在线性方程组及线性微分方程的求解中，矩阵求逆是必需的，而矩阵求逆又可转化成矩阵除法 A^*/d（A^* 称为 A 的伴随矩阵，d 是 A 的行列式值）。本文利用 Hwang 等人提出的一种改进符号数并行收敛算法[4]来实现上述过程。设 X、Y 分别为被除数和除数，且满足 $0.5 \leq |X| \leq 1$，通过寻找一系列的乘法因子 $m_i (i = 0, 1, \cdots, N)$，使得 $Y \cdot \prod_{i=1}^{N} m_{i \to 1}$，置 $X_0 = X, Y_0 = Y$，利用递推式

$$X_{i+1} = X_i \cdot m_i \tag{8}$$

$$Y_{i+1} = Y_i \cdot m_i \tag{9}$$

则有

$$\left. \begin{aligned} Q &= \frac{X}{Y} = X \cdot \prod_{i=0}^{N} m_i \\ m_i &= 2 - Y_{io} \end{aligned} \right\} \tag{10}$$

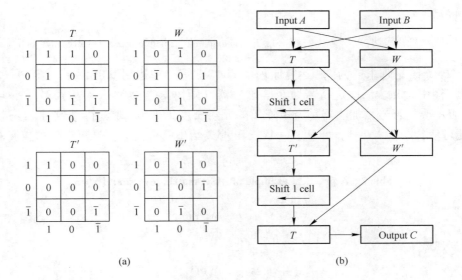

Fig. 1 (a) Transformation rule; (b) operation stop

Table 1 A practical computation example

Input Matrix: $A = \begin{pmatrix} 17 & 24 \\ 9 & 20 \end{pmatrix} = \begin{pmatrix} 10001 & 11000 \\ 01001 & 11100 \end{pmatrix}_{MSD}$ $B = \begin{pmatrix} 7 & 13 \\ 28 & 10 \end{pmatrix} = \begin{pmatrix} 01001 & 01101 \\ 11100 & 10110 \end{pmatrix}_{MSD}$				
Step 1:		T		W
	11000	11101	11000	10101
	11101	11110	10101	01010
Step 2:		T'		W'
	000000	000000	101000	101111
	000000	001000	101111	110110
Step 3:		T		Output Matrix:
	0101000	0101111		$C = A + B = \begin{pmatrix} 24 & 37 \\ 37 & 30 \end{pmatrix}$
	0101111	0110110		

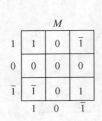
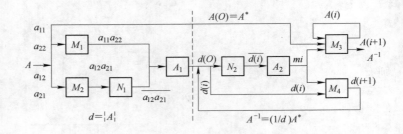

Fig. 2 Transformation rule M 　　Fig. 3 Computation principle of 2×2 matrix inversion

以 2×2 阶矩阵求逆为例，运算原理如 Fig. 3 所示，左边用以求 $|A|$ 之值，右边用来求 A^{-1}。图中，M 表示改进符号数矩阵乘法器（矩阵×常数）；N 表示 2'；补码运算器；A 表示改进符号数加法器。当矩阵 A 的元素长度为 N bit 改进符号数时，经过 N 次迭代便可使$||A|^{(N)} \to 1$，而 $A^{(N)} \to A^{-1}$。Table 2 给出一个改进符号数矩阵求逆实例（$N=3$）。

Table 2 A practical example of MSD matrix inversion ($N=3$)

Original matrix：
$$A = \begin{pmatrix} 0.25 & 0.5 \\ -0.25 & 0.5 \end{pmatrix}_D = \begin{pmatrix} 0.01 & 0.10 \\ 0.01 & 0.10 \end{pmatrix}_{MSD}$$

Accompany matrix：
$$A^* = \begin{pmatrix} 0.10 & 0.10 \\ 0.01 & 0.01 \end{pmatrix}_{MSD} = A^{(0)}$$

$$d = |{}_1^1 A_1^1| = (0.25)_D = (0.01)_{MSD} = d^{(0)}$$

Iteration step (1) M_1	$d^{(1)}$	$A^{(1)}$
$m_0 = 2 - d^{(0)}$ $= 10.00 - 0.01$ $= (1.11)_{MSD}$ $= (1.75)_D$ $m_1 = 2 - d^{(0)}xm_0$ $= 10.00 - 0.01 \times 1.11$ $= (1.1001)_{MSD}$ $= (1.5625)_D$ $m_2 = 2 - d^{(0)}xm_0xm1$ $= 10.0000 - 0.0100 \times$ 1.1100×1.1001 $= (1.01010001)_{MSD}$ $= (1.31640625)_D$ $m_3 = 2 - d^{(0)}xm_0xm_1xm_2$ $= 10.00000000 - 0.01000000 \times$ $1.11000000 \times 1.10010000$ $= (1.0001100110)_{MSD}$ $= (1.099609375)_D$ ——————————1	$d^{(1)} = d^{(0)}xm_0$ $= 0.01 \times 1.11$ $= (0.1001)_{MSD}$ $= (0.4375)_D$ $d^{(2)} = d^{(1)}xm_1$ $= 0.1001 \frac{1}{x} 1.1001$ $= (0.10101111)_{MSD}$ $= (0.68359375)_D$ $d^{(3)} = d^{(2)}xm_2$ $= 0.1010111 \times 1.01010001$ $= (0.1110011001)_{MSD}$ $= (0.899414063)$ $d^{(4)} = d^{(3)}xm_3$ $= 0.1110011001 \times 1.0001100110$ $= (0.1111110101)_{MSD}$ $= (0.989257813)_D$	$A^{(1)} = A^{(0)}xm_0$ $\begin{pmatrix} 0.1110 & 0.1110 \\ 0.1001 & 0.1001 \end{pmatrix}_{MSD}$ $\begin{pmatrix} 0.875 & -0.875 \\ 0.4375 & 0.4375 \end{pmatrix}_D$ $A^{(2)} = A^{(1)}xm_1$ $\begin{pmatrix} 1.0101110 & 1.0101110 \\ 0.10101111 & 0.1010111 \end{pmatrix}_{MSD}$ $\begin{pmatrix} 1.3671875 & -1.798828125 \\ 0.899414063 & 0.899414063 \end{pmatrix}_D$ $A^{(4)} = A^{(3)}xm_3$ $\begin{pmatrix} 1.111111010 & 1.11111101 \\ 0.111111010 & 0.11111101 \end{pmatrix}_{MSD}$ $\begin{pmatrix} 1.98046875 & -1.98046875 \\ 0.98828125 & 0.98828125 \end{pmatrix}_D$ ——————————A^{-1}

3 多窗口解码光学符号代换法则用于改进符号数矩阵外积计算

符号代换法则以二维模式的相关识别和传递为基础，是一种并行算法。与布尔逻辑相比它具有并行处理与扇入扇出的优点，它不仅能识别位的状态，而且能识别位的空间分布。对于这种二维算法，显然，用光学方式大大优于用电子处理方式。近年来，光学符号代换法则作为光计算研究中的一种有效算法引起人们极大的重视[5,6]。为了解决存在的多通道处理问题，作者曾提出过一种所谓的多窗口解码光学符号代换法则[7]，其实质是通过在识别输出面上设置多窗口的解码掩模，使得那些具有相同移位规则的模式组合能被同时识别出来。设每一输入图形中含有 M 种不同模式，则两个输入图形间共有 M^2 种不同的模式组合，一般需要 M^2 个独立通道加以识别和代换，采用多窗口解码光学符号代换法则后，可将通道数减少

$$P = 2M - 1 \tag{11}$$

个。Fig. 4 和 Fig. 5 分别为该法则原理及其相应的光学系统。原理图中示出了五进制乘法运算时的输入、本位和及进位。

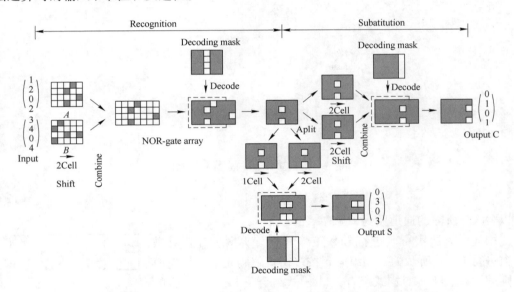

Fig. 4　Principle of MW-OSSR（Input: 2*4, Sum: 3, Carry: 1）

显然，A、B 两个输入中的 13 和 24 组合由于具有相同的移位规则（间隔相同），所以被同时得到识别（图中的 NOR 门阵列后）。类似地，A、B 间具有相同数值间隔 d 具有 Δx（$d=0$，$\pm =$，$\pm \pm$，…）的模式组合均可分别被同时识别出来，Δx 表示相邻两数的间隔。Fig. 5 中，A、B 分别从四边形光路的两臂输入，经透镜 L_1 成像及反射镜 M_1、半透反射镜 BS_1、BS_2 移位，合成到实时器件 PROM 上（使用波长为 48800nm 的 Ar^+ 激光写入）。He-Ne 激光（63280nm）用来对 PROM 进行读出，由多窗口解码掩模丑解码后可得到多个识别结果，刀一般由分别位于 P 个代换系统中的 P 个子解码掩模 d_1，d_2，…，d_p 来代替。各识别结果在相应的代换通道进行符号代换，由反射镜和半透反射镜或由衍射光栅 $G_1 \sim G_p$ 来完成，它们的输出再次合成后便得到所需的结果。

改进符号数算法中的几种变换规则可以认为是几种符号代换法则，输入图形中含有 9 种待识别模式组合：11、10、$1\bar{1}$、01、00、$0\bar{1}$ $\bar{1}1$、$\bar{1}0$ 及 $\bar{1}\bar{1}$，代换结果为 1.0 或 1。实现改进符号数矩阵加减运算时，$M=3$，$P=5$；实现改进符号数矩阵外积计算时，考虑到含"0"的模式组合其识别结果在光强上对代换输出不作贡献，则式（9）可改写成

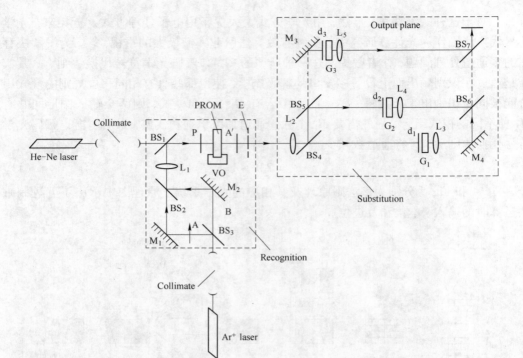

Fig. 5 Optical system for MW-OSSR

$$P = 2(M-2) + 1_0 \quad (M \geq 2) \tag{12}$$

所以,这时所需的通道数为 $P=3$。

用光学方法实行改进符号数矩阵外积计算时,首先对操作数进行 Fig. 6 所示的二则编码,按多窗口解码光学符号代换法则的原理,11,$\bar{1}\bar{1}$ 模式组合,$1\bar{1}$ 模式组合及 $\bar{1}1$ 模式组合的识别各需一个通道。对同一实例(见 Table 1),则有

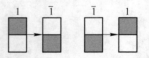

Fig. 6 Dual rail coding

$$A_1 \cdot B_1 = (1000101011\bar{1})^T (0100\bar{1}01101)_{MSD}$$

$$= \begin{pmatrix} 0 & 1 & 0 & 0 & \bar{1} & 0 & 1 & 1 & 0 & 1 \\ 0 & 0 & 0 & 0 & 0 & 0 & 0 & 0 & 0 & 0 \\ 0 & 0 & 0 & 0 & 0 & 0 & 0 & 0 & 0 & 0 \\ 0 & 0 & 0 & 0 & 0 & 0 & 0 & 0 & 0 & 0 \\ 0 & 0 & 0 & 0 & \bar{1} & 0 & 1 & 1 & 0 & 1 \\ 0 & 0 & 0 & 0 & 0 & 0 & 0 & 0 & 0 & 0 \\ 0 & 0 & 0 & 0 & \bar{1} & 0 & 1 & 1 & 0 & 1 \\ 0 & 0 & 0 & 0 & 0 & 0 & 0 & 0 & 0 & 0 \\ 0 & 0 & 0 & 0 & \bar{1} & 0 & 1 & 1 & 0 & 1 \\ 0 & \bar{1} & 0 & 0 & 1 & 0 & \bar{1} & \bar{1} & 0 & \bar{1} \end{pmatrix}_{MSD} \tag{13}$$

$$A_2 \cdot B_2 = (1100011\bar{1}00)^T (11100101\bar{1}\bar{1}0)_{MSD}$$

$$= \begin{pmatrix} 1 & 1 & 1 & 0 & 0 & 1 & 0 & \bar{1} & \bar{1} & 0 \\ 1 & 1 & 1 & 0 & 0 & 1 & 0 & \bar{1} & \bar{1} & 0 \\ 0 & 0 & 0 & 0 & 0 & 0 & 0 & 0 & 0 & 0 \\ 0 & 0 & 0 & 0 & 0 & 0 & 0 & 0 & 0 & 0 \\ 0 & 0 & 0 & 0 & 0 & 0 & 0 & 0 & 0 & 0 \\ 1 & 1 & 1 & 0 & 0 & 1 & 0 & \bar{1} & \bar{1} & 0 \\ 1 & 1 & 1 & 0 & 0 & 1 & 0 & \bar{1} & \bar{1} & 0 \\ \bar{1} & \bar{1} & \bar{1} & 0 & 0 & \bar{1} & 0 & 1 & 1 & 0 \\ 0 & 0 & 0 & 0 & 0 & 0 & 0 & 0 & 0 & 0 \\ 0 & 0 & 0 & 0 & 0 & 0 & 0 & 0 & 0 & 0 \end{pmatrix}_{MSD} \quad (14)$$

A_1B_1 及 A_2B_2 的二则编码结果如 Fig. 7 所示, Fig. 8 给出了外积计算时三个识别通道的实验结果。三个识别通道的输出经代换后重叠在一起便可得到部分积阵列。

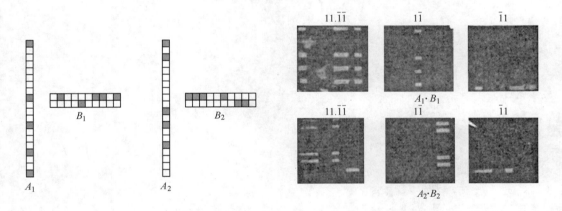

Fig. 7　The dual rial coding results of $A_1 B_1$ and $A_2 B_2$

Fig. 8　The experimental results of three recognition channel

4　结论

利用改进符号数算法和多窗口解码光学符号代换法则可以实现快速、高精度的多元多值矩阵列计算,当使用响应时间为 5ms 的 BSO-PROM 作为 NOR 门阵列(有效口径和实用分辨率分别为 25mm×25mm 和 10l/mm)时,系统可实现的等效运算能力是每秒进行 26 次 10×10 阶 32bit 的改进符号数矩阵加(减);或 3 次 5 改进阶 8bit 的改进符号数矩阵相乘;或 1 次 6 改进阶 8bit 的改进符号数矩阵求逆。改进符号数数字表达精度 N 决定了最大操作数大小和计算精度,设整数部分需 N_1bit 表示,则 $N-N_1$bit 表示小数部分,当 $N=5$, $N_1=2$ 时,计算精度为 0.125;当 $N=32$, $N_1=16$ 时,计算精度可达 10^{-5}。此外,为了使实际系统小型化,还可以用体全息片(重铬酸明胶等)和计算机制全息图来代替反射镜、透镜及半透反射镜。

参 考 文 献

[1] 陈岩松,等. 物理, 1988, 17, No. 10(Oct), 626~630.
[2] R. P. Bocker. Opt. Eng., 1984, 23, No. 1(Jan), 26~33.
[3] B. L. Drake, et al. Opt. Eng., 1986, 25, No. 1(Jan), 38~43.
[4] K. Hwag, et al. Opt. Eng., 1989, 28, No. 4(Apr), 364~372.
[5] K. H. Brenner, et al. Appl. Opt., 1986, 25, No. 18(Sep), 3054~3060.
[6] 周少敏,等. 应用激光联刊, 1989, 8, No. 4(Aug), 181~184.
[7] 周少敏,等. 仪器仪表学报, 1990, 11, No. 1(Feb), 34~39.

Optical Binary Image Algebra Processor(OBIAP) *

Abstract A novel optical implementation of several fundamental operations of morphological processing including complement, union dilation and erosion has been developed. Also an optical binary image algebra processor (OBIAP) with a multi-beam splitter (MBS), a liquid crystal television (LCTV), 2-D detector array and a microprocessor is established. It is characterized by its simple structure, capability of feedback and real-time imaging.

1 Introduction

To develop highly-parallel optical computing, much attention has been paid to computing approaches based on binary image transformation. Optical symbolic substitution[1], digital morphological image processing[2] and truth-table look-up processing[3], etc, are some of the examples. A 2-D image can be considered as a 2-D set. Binary image processing is equivalent to the transformation of this set. Huang et al have proposed the synthesis of the binary image algebra (BIA)[4] in which studies of boolean logic, theory of collection and digital image processing have been combined. In BIA the parallel binary digital image processing algorithms (including parallel numerical computation) can be written as compact algebraic expressions where an algebraic symbol represents an image (not a pixel) or an image operator (not a pixel-wise operation). Any image transformation and 2-D digital computation can be easily performed by using BIA.

2 Binary image algebra(BIA)

Contemporary electronic computers are based on boolean logic while BIA is formed by five elementary images (I, A, A^{-1}, B, B^{-1}) and three fundamental operations: complement($\ominus$), union ($\cup$) and dilation($\oplus$).

$$BIA = [P(W); \oplus, \cup, -, I, A, A^{-1}, B, B^{-1}] \tag{1}$$

where $P(W)$ is the image space and $F = (\oplus, \cup, -, I, A, A^{-1}, B, B^{-1})$ is the family of operations.

The three fundamental operations are defined as follows:

(1) Complement of an image X

$$\overline{X} = \{(x,y) \mid (x,y) \in W \wedge (x,y) \notin X\} \tag{2}$$

(2) Union of two images X and R

---

* Copartner: Wu Minxian, Zhou Shaomin, Caijianhong. Reprinted from *Optics Communications*, 1991, 86(6), 454-460.

$$X \cup R = \{(x,y) \mid (x,y) \in X \lor (x,y) \in R\} \tag{3}$$

(3) Dilation of two images X and R

$$X \oplus R = \begin{cases} \{(x_1 + x_2, y_1 + y_2) \in W \mid (x_1, y_1) \in X, (x_2, y_2) \in R\} \\ (X \neq \emptyset)(R \neq \emptyset) \\ \{\emptyset, \text{otherwise}\} \end{cases} \tag{4}$$

where $-$ means complement, $\lor$ means OR, $\land$ means AND. W is the universal image set

$$\begin{aligned} W &= \{(x,y) \mid x \in Z_n, y \in Z_n\} \\ Z_n &= \{0, \pm 1, \pm 2, \cdots, \pm n\}, n \text{ positive integer} \end{aligned} \tag{5}$$

3 Optical implementation of three fundamental operations of BIA

The optical implementation of digital morphology and the fundamental operations of BIA has interested many scientists[4-6].

In BIA, the transformation operations are sequential processes, so that a 2-D detector array for image processing and devices for providing real-time feedback are required. A multiple beam splitter, LCTV, 2-D detector array and a microcomputer are used in our optical system.

3.1 The optical implementation of dilation

Dilation is one of the operations in BIA that is most important but difficult to realize. Optical implementation of it has been discussed by several authors[5,6]. Here we developed a novel method by using a multi-beam splitter (Dammann. grating) in our optical scheme (see Fig. 1). $LCTV_1$ is used for input reference image R, MBS is located at plane P_2, $LCTV_2$ which is located at plane P_3 is used for input image X, P_4 is the image plane of R. At the spectral plane the Fourier transform of reference R and MBS can be obtained. Each spectrum point of the MBS contains complete information of R. At the image plane P_4 a set of image R can be formed. If the image X is located at specrum plane and some imaging conditions are satisfied, then the dilation of image X by reference image R is formed according to the multiple imaging principle.

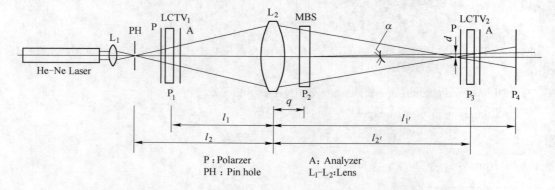

Fig. 1 Optical scheme based on MBS and LCTV

The multiple beam splitter has been designed such that the center $N \times N$ spectrum points have the same intensity[7]. We can simulate the spectrum of MBS by the product of the functions $\mathrm{Comb}(x/a, y/a)$ and $\mathrm{Rect}[x/a(N-1), y/a(N-1)]$, where a is the sample spacing at plane P_3. The parameter a is determined by grating spacing d, incident wavelength λ and parameters ($l_i, l'_i, \cdots$) of that optical system. Fig. 1 shows that

$$\Delta\theta = \lambda/d, \quad a = \Delta\theta(l'_2 - q)$$
$$1/l_i + 1/l'_i = 1/f, \quad i = 1, 2 \tag{6}$$

Because of the short distance between P_3 and P_4, the propagation of light from P_3 to P_4 can be approximately considered as a projection process instead of Fresnel diffraction. It only affects the distribution of the intensity but not the output geometric relations of image X. The point spread function $h(x, y)$ can be expressed as follows

$$O(x, y) = R(-x/M, -y/M) \times h(x, y)$$
$$= \left[b^2 \sum_{n=-\infty}^{\infty} \sum_{m=-\infty}^{\infty} X(na, ma) \right] \cdot R\left(-\frac{x - nb}{M}, -\frac{y - mb}{M} \right) \cdot$$
$$\mathrm{Rect}[x/b(N-1), y/b(N-1)] \tag{7}$$

where $R(-x/M, -y/M)$ is the output image given by the geometric optical theory $M = l'_i/l_i$ is amplification. From Eq. (7) we can see that the $N \times N$ multiple images are located at the distance of $b(\Delta x = \Delta y = b)$ from each other. When the parameters of the optical system (such as d, l_1, l_2, q, f) are such arranged that the multiple images of R overlap with each other, each being shifted by one cell in a different dlirection, the dilation of image X can be obtained. It is convenient when monochromatic light is used. The experimental result is given in Fig. 2a. R is a pattern of 3×3 cells, X is the English letter H.

In the experiment a 9×9 MBS is utilized. The paramelers of this Dammann grating are

$$x_0 = 0d, \quad x_1 = 0.0666d, \quad x_2 = 0.4745d, \quad x_3 = 0.6072d$$
$$x_4 = 0.7789d, \quad x_5 = 0.9377d, \quad x_6 = 1d$$

$d = 150\mu\mathrm{m}$ (grating constant), $D = 20\mathrm{mm} \times 20\mathrm{mm}$ (effective aperture). The other parameters are

$$f = 308\mathrm{mm}, \quad l'_2 = 620\mathrm{mm}, \quad l_2 = 612\mathrm{mm}, \quad q = 120\mathrm{mm}, \quad l'_2 = 500\mathrm{mm},$$
$$\Delta\theta = \lambda/d = 4.22 \times 10^{-3}, \quad a = \Delta\theta(l'_2 - q) = 2.1\mathrm{mm}.$$

Let $l'_1 = l'_2 + 200 = 820\mathrm{mm}$, then $l_1 = 493\mathrm{mm}$ and $b = \Delta\theta(l'_1 - q) = 2.95\mathrm{mm}$.

The distance between the neighbor cells of R should be $b/M = 1.8\mathrm{mm}$. The cell size of R is $1.2\mathrm{mm}$.

The LCTV's (Seiko Model LVD-501) used for the experiment have a resolution of 320×220 pixels and effective aperture of $68\mathrm{mm} \times 42\mathrm{mm}$. A maximum number of 56×35 cells can actually be realized in the system.

Notice that in this dilation operation the intensity of every sample point of the output image is not the same. That is because the number of superimposed image cells is different. The ratio of the maximum and minimum of intensity is possibly $s^2:1$. s means the number of structure element(cells) of reference image R. Thus, the output image must be thresholded.

3.2 The optical implementation of complement(-) and union(∪)

It is easier to implement these two fundamental operations in our system. The input image X can be presented on LCTV that is illuminated by collimated light. When the analyzer is adjusted perpendicular to the axis of polarizer, then we can get the complement operation of X. The experimental result is shown in Fig. 2(b).

Two symmetric spectrum points are used for realizing union operation. In this case, reference R and image X are inputs at different areas of the same LCTV. When we adjust the optical system such that the R of the first image superposes with the X of the second image, the $U \cup R$ is obtained in the central area of the multiple image plane. The experimental result is shown in Fig. 2(c).

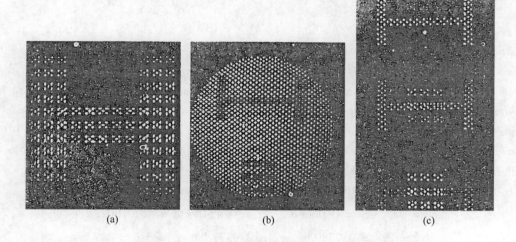

Fig. 2 The experimental results of three kinds of BIA fundamental operations
(a) dilation of image X(image X is a letter H); (b) complement of X(the upper one) and R (the lower one); (c) union of X and R

4 Optical binary image algebra processor(OBIAP)

We have discussed the optical implementation of three fundamental operations and based on it we have developed an optical binary image processor(OBIAP). The scheme of the OBIAP is shown in Fig. 3. There are two optical arms for implementing the dilation and union operations. The electronic part consists of detectors, instruction control, input, output and display. In Fig. 3 the single line is the instruction set and the double line is the data flow. Any binary image transformation can be completed by using this system. The experimental arrangement is shown in Fig. 4. For dilation operation $LCTV_1$ and $LCTV_2$ are used for input R and X respectively. For union operation

$LCTV_1$ is used for both R and X. The complement operation is implemented by inverting the input signal on $LCTV_2$ by computer(or by adjusting the analyzer perpendicular to the axis of polarizer). Two electronic shutters(SH_1 and SH_2) are controlled by microprocessor for selecting different operations. A 2-D detector array is used for receiving the intermediate image which is necessary for feedback. The final processed result can be displayed on CRT.

Some experiments for image processing have been made by using the OBIAP.

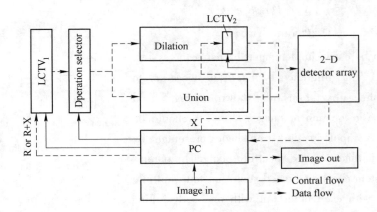

Fig. 3 Block diagram of the optical binary image algebra processor (OBIAP)

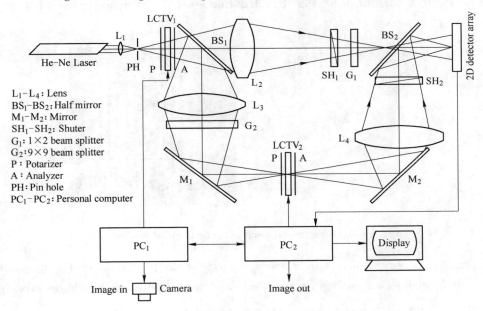

Fig. 4 Experimental arrangement of OBIAP

4.1 Erosion($\ominus$) of X by R

The erosion is an important operation in morphology although it can be expressed by dilation and complement.

$$X \ominus R = \overline{\overline{X} \oplus \overset{\vee}{R}} \tag{8}$$

where $\vee$ means reflection.

In fact the erosion of X by R can be expressed also by the following equation similar to the dilation expression

$$X \ominus R = \begin{cases} \{(x_1 - x_2, y_1 - y_2) \in W \mid (x_1, y_1) \in X, (x_2, y_2) \in R\} \\ (X \neq \emptyset)(R \neq \emptyset) \\ \{\emptyset, \text{otherwise}\} \end{cases} \tag{9}$$

Eq. (9) can be easily implemented by using the system shown in Fig. 1. The difference is that it is necessary to use the symbolic substitution rule shown in Fig. 5 and dark-truelogic. The location of images R and X should be exchanged X is input in dark-true logic presented by LCTV$_1$, and R is placed at the spectrum plane. The parameters of the system can be adjusted such that four multiple images overlap with each other but with one cell shifting respectively in four directions. The erosion operation in dark-true logic can be realized on the P_4 plane. The experimental result is shown in Fig. 6. X is an English letter H in dark-true logic. The reference image R is shown in Fig. 6(b).

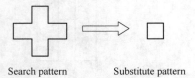

Search pattern Substitute pattern

Fig. 5 The symbolic substitution rule used for erosion

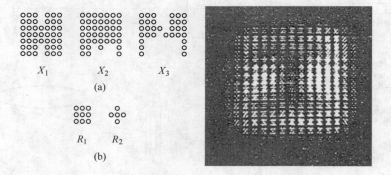

Fig. 6 Experimental result of the first kind of erosion

As Eq. (8), the erosion of a X by a reference image R can be implemented as the complement of the dilation of background by the reflection of the reference image R. In the experiment a 3×3 array is used as reference image which is symmetric, so that $\overset{\vee}{R}$ just is the original image R.

4.2 Difference(/) of X by R

$$X/R = \{(x, y) \in X \mid (x, y) \notin R\} = X \cap \overline{R} = \overline{\overline{X} \cup R} \tag{10}$$

The difference operation can be implemented by applying complement twice and union once.

4.3 Opening (○) and closing (●) of X by R

$$X \circ R = (X \ominus R) \oplus R = \overline{\overline{X \oplus \overset{\vee}{R}} \oplus R} \qquad (11)$$

$$X \bullet R = (X \oplus R) \ominus R = \overline{\overline{X \oplus R} \oplus \overset{\vee}{R}} \qquad (12)$$

The above two kinds of operations can be implemented by applying complement twice, dilation twice and reflection once.

4.4 Edge detection

There are two kinds of methods for implementing the edge detection:

$$X \oplus R / X = (X \oplus R) \cap \overline{X} = \overline{\overline{X \oplus R} \cup X} \qquad (13)$$

or
$$X / (X \ominus R) = X \cap \overline{(X \oplus \overset{\vee}{R})} = \overline{\overline{X} \cup (X \oplus \overset{\vee}{R})} \qquad (14)$$

The difference of the above two kinds of equation is that the first one will enlarge the original image by one cell while the second one will get an edge which is just the one of the original image. We have chosen the first one in the experiment.

Fig. 7 give the experimental results of erosion according to the Eq. (8) ~ Eq. (13). They are difference, opening, closing and edge detection respectively.

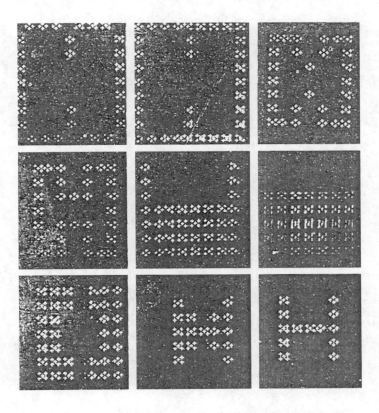

Fig. 7 Experimental results of several image processing
(a) original images; (b) reference images (structure element); (c) experimental results

5 Conclusion

We have presented an optical binary image algebra processor (OBIAP) based on multi-beam splitter(MBS), LCTV and microprocessor. The experimental results of three kinds of BIA fundamental operations and some basic image processing are given. This OBIAP has the feature of spatial processing fan-in and fan-out. It is convenient not only for edge enhancement, smoothing, associative recognition but also for symbolic substitution, Chinese character recognition, matrix calculation and neural computing.

References

[1] A. Huang. IEEE Tenth International Optical Computing Conference. 1983:13.
[2] S. D. Goodman, W. T. Rhodes. Appl. Optics 27(1988).
[3] M. M. Mirsalehi, T. K. Gaylord. Appl. Optics 25(1986).
[4] K. S. Huang, B. K. Jenkins, Sawchuk. Computer vision, graphics, and image processing 45(1989) 295.
[5] Y. Li, A. Kostrzewski, D. Kim, G. Eichmann. Optics Lett. 14(1989)981.
[6] D. Casasent, E. Botha. Appl. Optics 27(1988) 3806.
[7] J. Turunen, X. Bian, G. Jin, Acta Optica Sinica 8(1988) 946.

光学子波变换（1）基本理论*

摘 要 本文介绍了从傅里叶变换到盖伯变换，以及子波变换的发展。分析了它们的区别和联系。描述了连续和离散的子波变换，讨论了子波函数成立的容许性条件和子波变换的一些基本性质。

1 引言

近几十年来现代光学的最新进展之一是光学信息处理和数字光计算的飞速发展[1]。自 20 世纪 70 年代起数字光计算研究便逐渐成为举世瞩目的研究领域。到目前为止主要从事了以下几方面的探索性研究：

（1）光学神经网络[2~4]：它是以光学方法实现模拟人脑的神经逻辑处理过程。网络的信息和运算结果分布式存储在神经元的互连中，并以一种并行方式工作。神经网络由于考虑了大量神经元的互连集合效应，使得它不仅可以完成联想识别，而且具有自学习、自适应和自组织方面的功能。

（2）真值表查询[5,6]：这也是实现以人脑的标的一种形式，其结果和运算过程都以一种列表的方式给出，根据输入信息可以很快地从表格中查出结果。

（3）细胞元阵列[7]：其标法以细胞元的逻辑结构来实现。每一个细胞元是一个小的处理单元，各个单元的功能几乎完全相同。

（4）二进制图像代数[8,9]：以图像作为代数变量，把复杂的图像简化为简单的图像指令来运算。可由三种基本操作：扩"$\oplus$"、并"$\cup$"、补"$-$"，以及 5 种基本图像来实现。

（5）数学形态学[10]：该理论是在 20 世纪 70 年代分析地质结构，石油勘探等领域得出的一种解决物理与空间关系的非常有效的理论。要解决怎样从物质的空间结构去知道其物理性质，或者相反的问题。由蚀、扩、并三个基本的及关、闭等操作来实现复杂的形态变换。这些基本操作都易于以光学实现，所以目前光学形态学的研究是一个十分活跃的方向。该理论已经在数字图像处理、计算机视觉、遥感图像处理等许多领域获得了应用。

（6）光学铸影系统[11]：通过编码的输入，以不同角度投影的重叠来实现各种逻辑运算。

（7）符号代换[12]：是一种空间光学逻辑；逻辑值不仅取决于信息的编码，还取决于信息符号的位置。处理过程中要完成符号的识别和移位，并根据一定替换规则实现并行处理。

以上这些方面的研究都体现了光学处理的优点，并取得了丰硕的研究成果，且

---

* 本文合作者：王文陆，严瑛白，邬敏贤。原发表于《光电子·激光》，1994，5(4)：193~212。

有些已投入实际应用；有些还处于实验室阶段，需要更深入的探索。近期内，由于器件、材料与算法等诸多因素的限制，全光计算机的梦想还难以实现。目前科技人员除了在原有领域更深入地开展研究，还在如何发挥光的并行性的优势方面进行深入的探讨。

光学子波变换也是在学科的交叉、数字计算的子波运算的基础上发展起来的，近年来已得到迅猛发展。子波变换，它是一种很有效的数学处理工具，即由扩展和平移参数标定的基元函数把任意一个函数展开成光滑局域的分布。与此种展开相联系的设想和技术，其实在相当长的一段时间内已存在，并已被广泛应用于数学分析、理论物理及工程中[13]。由于子波变换的许多优越性，它已经引起了众多学科的研究人员的注目。但是在子波变换处理中，每一输入坐标（空间、时间等）都有两个输出坐标，这样一个一维的信号就产生了一个二维的子波变换表达；二维的信号会产生四维的子波变换表达。这样用电子数字处理来实现子波变换都是较慢的。光学的高度并行处理能力将能在子波变换处理中发挥其优越性。所谓光学子波变换就是利用光学技术与方法来实现对信号图像的子波变换处理，得到所预期的结果。另外，子波变换是从傅里叶变换发展和演变而来的，光学傅里叶变换能力和其在光学信息处理中所发挥的作用是众所周知的。由此可知以光学技术来进行子波变换有其理论基础和实践应用的可能。

本文简要地介绍了从傅里叶变换到盖伯变换，以及子波变换的发展，探讨了它们的区别和联系。描述了子波变换中的一些基本内容，讨论了子波函数的容许性条件和子波变换的一些基本性质。

2　从傅里叶变换到盖伯变换（Gabor transform）

2.1　傅里叶变换（Fourier transform）

19世纪初，法国科学家傅里叶（Fourier）在向巴黎科学院呈交的关于热传导的著名论文中提出了傅里叶级数，其意义是无法估量的。今天，傅里叶分析方法已经成为了各种信号数据处理中最基本的数学工具。对于连续信号 $s(t)$ 的傅里叶变换为 $S(\omega) = <c^{i\omega t}, s(t)t>$。$S(\omega)$ 为信号 $s(t)$ 各频率成分的权重因子。由于傅里叶变换的基函数在时间轴上是无限扩展的，任何时间局域的信息被扩展到整个频率轴上。因而信号的频谱 $S(\omega)$ 只能刻画 $s(t)$ 在整个时间轴上的频谱特征，而不能反映出信号在时间轴上的局部区域的频率特征。适用于分析稳定的慢变信号。而在不少实际问题中，我们所关心的恰恰是信号的局部特征，信号中的实变。例如，对地震波信号的记录，人们关心的是什么位置出现什么样的反射波；图像处理中的边缘检测关心的是灰度突变部分的位置。对这些信号的傅里叶分析，会有很多的高频模式。当进行信号重构时，仅仅截取有限项数来逼近时，就会引入高频噪声。从信号传递的角度来考虑，当然希望输出的信号尽可能逼近原信号。光学系统中的点物成点像仅是几何光学的理想，没有考虑由于孔径有限造成的衍射效应，限制了系统的分辨力，故不可能传递物体的全体细节。

2.2　盖伯变换（Gabor transform）

为了研究信号在局部范围内的频率特性，D. Gabor 于1946年引进了"窗口"傅里叶变换，也叫短时傅里叶变换[14]。用固定窗口的余弦函数作为基函数。对信号$s(t)$的盖伯变换定义为：

$$G_s(\omega,\tau) = \int_{-\infty}^{\infty} e^{-j\omega t} g(t-\tau) S(t) \mathrm{d}t \qquad (1)$$

式中的$g(t)$是一适当的窗口函数如高斯函数。由上式可见，对信号的盖伯变换就是用平移了的窗函数去乘信号再作傅里叶变换。可以等价地看做盖伯变换的基函数是窗口函数的调制形式，如图1（a）所示。盖伯函数一般采用高斯函数乘以一个调制函数的形式，因为高斯函数在时间和频率域中的形式是一致的，有比较好的局域性。盖伯变换的主要优点是当信号的大部分能量处于$t=\tau$时间间隔$[-T,T]$和频率间隔$[-\Omega,\Omega]$，那么它的盖伯变换将局限在区域$[-T,T]\times[-\Omega,\Omega]$内，在此以外的空间，盖伯因子趋近于零。但是盖伯变换中所用的窗口对所有的频率都是固定的，这样在时间—频率平面中的所有位置分析的分辨率都是一样的，如图1（b）所示。这样会造成高频时取样不足，因而达不到所要求的分辨率。

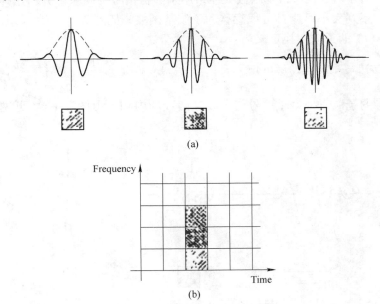

图1　盖伯变换的基函数和时间频率分辨率
(a) 基函数；(b) 时间频率平面中的收敛性

3　子波变换（wavelet transform）

在对信号的分析中，我们有时既不想单独分析其在时间域中的行为，也不想单独分析其在频率域中的行为。因为这样的分析会完全丢掉信号在另一空间中的信息。例如傅里叶变换把信号在频率域中展开后，只得到信号的频率信息，而不知道对应频率的时间信息。在很多实际问题中，对信号的局域特征比较感兴趣，需要了解对应一定时间间隔的信号的频率特征。盖伯变换（也叫窗口傅里叶变换）提供了很好的局域分析方法。但是海森堡的不确定性原理限制了任意提高时间域和频率域中分辨率的可能性，因为基元函数的时间带宽积有一个最小的下限。两个共变域中的分辨率是相互牵制的。显然若所采用的基函数的窗口大小可变，就可以以牺牲一个域中的分辨率来提高另一个域中的分辨率。子波变换正是提供了这样的一种很有效的数学工具，要分析信号中的不连续性时采用短的高频基函数，而分析缓变部分时采用较长的低频基函数。它使得时间域和频率域中的分辨率服从测不准原理的要求，获得了良好的综合。

3.1 子波变换的定义

子波变换的基函数是通过一个原型函数的平移（translation）和扩（dilation）得到：

$$h_{a,b}(t) = \frac{1}{\sqrt{a}} h\left(\frac{t-b}{a}\right) \tag{2}$$

参数 $a>0$ 称为扩因子，b 为实数，称为平移因子。对于较大的 a，基函数扩，用来分析低频部分；对于较小的 a，基函数收缩，用来分析信号的高频部分。这样经过扩和平移就产生了一系列大小可变且可任意移动子波基函数，如图2（a）所示。

对信号 $S(t)$ 的子波变换定义为：

$$W_s(a,b) = <h_{a,b}(t), s(t)> = \int_{-\infty}^{\infty} h_{a,b}^*(t)s(t)\,dt = \frac{1}{\sqrt{a}}\int_{-\infty}^{\infty} h^*\left(\frac{t-b}{a}\right)s(t)\,dt \tag{3}$$

即对信号的子波变换就是信号与子波基函数的内积。由图2（a）可见子波基函数

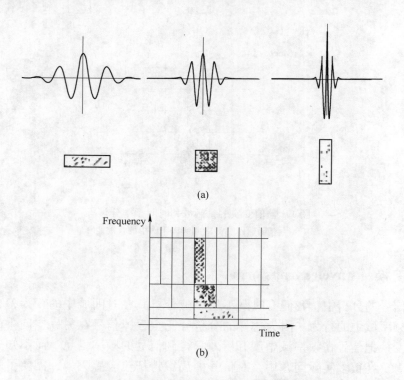

图2 子波变换的基函数和时间-频率分辨率
（a）子波基函数；（b）时间-频率平面的收敛性

的形状是一致的，每一个子波基函数都有相同个数的振荡数。由测不准原理，使我们在时间域和频率域中的分辨率有一个最优的折中方案：在一个域中的分辨率变好，在另一个域中就要变差，如图 2（b）所示。不同的扩因子可以得到不同分辨率的分解。这要看我们要从被分析的信号中提取何种特征而定。

下面我们给出一个子波变换与盖伯变换区别的简单例子[15]，如图 3 所示。由图可见盖伯变换在所有的频率范围其局域性都是一样的；而子波变换在不连续、边和畸性的地方具有收敛的能力，在趋于高频时域性变窄，信号的突变的位置可以很明显地从时间-频率平面中的局域性表现出来。

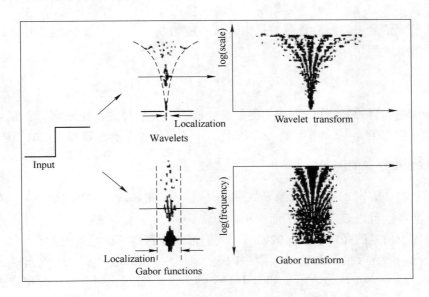

图 3　子波变换与盖伯变换对直边信号输入的比较，
子波变换在边的位置有很好的收敛能力

3.2　子波函数的限制

3.2.1　子波函数的容许性条件

为了根据信号的子波系数精确地恢复信号，以及为了直接根据信号的子波系数能够计算能量或其他一些不变量，至少应该有一个重构公式，即逆子波变换，定义为[16]：

$$s(t) = \frac{1}{C_h} \iint W_s(a,b) h_{a,b}(t) \frac{da}{a^2} db$$

$$= \frac{1}{C_h} \iint <h_{a,b}(t), S(t)> h_{a,b}(t) \frac{da}{a^2} db \quad (4)$$

其中的归一化常数 C_h 为：

$$C_h = 2\pi \int_{-\infty}^{\infty} |H(\omega)|^2 |\omega|^{-1} d\omega < \infty \quad (5)$$

$H(\omega)$ 为函数 $h(t)$ 的频谱。式（5）称为子波函数的容许性条件。它意味着当 $\omega = 0$ 时

$|H(\omega)|^2 = 0$，于是就有 $\int h(t)\mathrm{d}t = 0$，这就是说子波母函数一定是振荡型的，在时间域中的积分面积为零，它的傅里叶频谱中的直流成分为零。

3.2.2 子波函数的正则性条件

任何满足容许性条件的平方可积函数都可以成为一个子波函数。但在实际应用中对子波函数还有正则性方面的要求。子波变换是一种时间和频率域中的局域算子。子波函数和它的傅里叶变换在两个域中都应有较好的光滑性和收敛性。当扩因子 a 减小时，子波函数在时间域上是收敛的。在频率域中我们已有 $H(o) = 0$，并且我们希望当超过一定频率的 $H(\omega)$ 衰减为零。所以子波变换正好是一个带通滤波器。

为了研究信号中的高阶导数的行为，子波应当具有直至 M 阶的消去性，以便使它不与信号的低阶变化起作用，即：

$$M_P = \int_{-b}^{x} t^P h(t)\mathrm{d}t = 0 \quad (P = 0,1,2,\cdots,M) \tag{6}$$

M_P 称为子波的 P 阶矩。这个要求使得子波变换消去了信号的最正则部分，仅对函数的高阶变化（信号的 m 阶导数以上）起作用，并导致子波变换在频率域中的局域性。

3.3 子波变换系数的图示

从式（3）可以看到子波变换系数 $W_s(a,b)$ 与尺度参数 a 和时间维参数 b 有关。这样子波变换系数可以表示在半平面（$b \in R, a > 0$）上，使 a 轴朝下，b 轴向右。因为小尺度相应于信号的高频，使 a 轴朝下，我们可以看到高频在低频的上面。如图 4（a）所示。当我们要在尺度参数的较大范围内表示时，一般采用对 b 为线性而对 a 为对数的全平面 $\{b, -\log a\}$ 来表示。如图 4（b）所示。一般来说对 a 的线性尺度表示是不常采用的，因为这时小尺度行为（这是对信号的子波变换最有兴趣进行研究的）将完全被拉平。通常情况下对尺度 a 的对数表示是很自然的，因为它对应于伸缩参量 a 的倍增性质。例如，对正交子波的情形，将取以 2 为底的对数，因为 a 一般是 2 的幂，因此应当一个倍频道一个倍频道地（octave by octave）表示子波变换系数。子波变换系数一般是复值函数：

$$W_s(a,b) = |W_s(a,b)|e^{i\varphi(a,b)} \quad (0 \leq \varphi(a,b) < 2n)$$

它的幅值归一化后进行彩色编码来表示，而其相位值用黑点的密度来表示[17]。

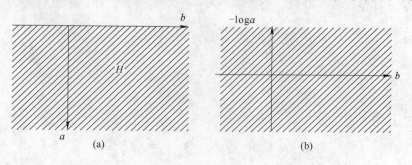

图 4 （a）$\{b \in R, a > 0\}$ 的半平面；（b）$\{b, -\log a\}$ 的平面

3.4 离散的子波变换

当参数 (ω,τ) 和 (a,b) 是连续的,盖伯变换和子波变换都是高度冗余的,即它们各自的基函数之间是相互重叠的。问题是是否存在一个离散化的格点使得基函数集合构成一个标准正交基,这样变换就是无冗余的。对于盖伯变换仅当窗函数 $g(t)$ 在时间域和频率域的局域性很差时才行,这也就是盖伯变换通常是"过密采样"(over sampled)的原因。对于子波变换,可通过设计"较好"的函数 $h(t)$ 使得 $h(t)$ 的扩展和平移的函数集构成一个标准正交基。"较好"意味着该函数至少是连续的,可能还有连续的高阶导数,也即满足子波函数的容许性和正则性条件。

把式(2)的子波函数的参数 a,b 离散化有

$$h_{mm}(t) = a_0^{-m/2} \cdot h(a_0^{-m}t - nb_0) \quad (m,n \in Z, a_0 > 1, b_0 \neq 0)$$

相应于 $a = a_0^m, b = na_0^m b_0$。注意到平移步长依赖于扩参数,因为伸展的子波以较大的步距前进,而较小的子波步距也就较小。在这一离散的格点上,子波变换成为:

$$W_s(m,n) = a_0^{-m/2} \int_{-\infty}^{\infty} h^*(a_0^{-m}t - nb_0) s(t) dt \tag{7}$$

当 $a_0 = 2$, $b_0 = 1$ 时,离散化在一个二进的网络上,$h_{mm}(t)$ 是标准正交的,即:

$$<h_{mm}(t), h_{kl}(t)> = \delta_{mk}\delta_{nl} \tag{8}$$

此时的基函数覆盖整个平方可积空间 $L^2(R)$,而没有相互重叠。

3.5 子波变换的一些基本性质

(1) 线性:子波变换是线性的。

$$W_{s_1+s_2}(a,b) = W_{s_1}(a,b) + W_{s_2}(a,b) \tag{9}$$

(2) 平移和伸缩的共变性:

$$W_{s(t-t_0)}(a,b) = W_{s(t)}(a, b - t_0)$$

$$W_{s(\alpha t)}(a,b) = \alpha^{-1} W_{s(t)}(\alpha a, \alpha b) \tag{10}$$

平移共变性表明一个谐振信号的频率可以从子波变换系数的相位读出,在 a 为常数的线上相位为零的数目给出信号的频率。这个性质与所选的子波无关。子波变换的伸缩共变性表明一个幂指数函数的子波变换可以通过任何 a 为常数的线完全确定。常数相位的那些线指出函数的可能奇点[13]。

(3) 能量守恒:

$$\int |S(t)|^2 dt = C_h \iint |W_s(a,b)|^2 \frac{da}{a^2} db \tag{11}$$

它意味着信号作子波变换后能量没有损失,保持着信号的全部信息。

(4) 微分:

$$W \frac{\partial^m S}{\partial t^m}(a,b) = (-1)^m \int_{-\infty}^{\infty} S(t) \frac{\partial^m}{\partial t^m} [h_{a,b}^*(t)] dt \tag{12}$$

(5) 局域性:

子波函数 $h(t)$ 在时间域中的局域性导致了子波变换系数空间中局域性的守恒。也即若 $h(t)$ 在间隔 $[t_{\min}, t_{\max}]$ 以外的空间都为零，那么对于位置 t_0 的子波系数将完全包含在

$$t \in [t_0 - a \cdot t_{\min}, t_0 + a \cdot t_{\max}] \tag{13}$$

所确定的"影响锥"内。这个影响锥对应于点 t_0 处所有伸缩子波的空间支撑。如图 5（a）所示。如果用对数坐标则如图 5（b）所示。

如果子波函数的频谱 $H(\omega)$ 在频率间隔 $[\omega_{\min}, \omega_{\max}]$ 以外的空间都为零，则信号 $s(t)$ 的频谱 $S(\omega_0)$ 对 $\{b, a\}$ 平面所能影响的区域为：$\dfrac{\omega_{\min}}{\omega_0} < a < \dfrac{\omega_{\max}}{\omega_0}$ 所确定的水平带。如图 5（c）所示。

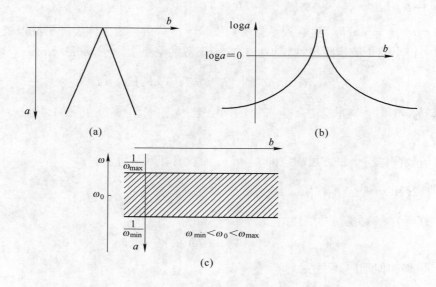

图 5　局域性分析
（a）时间局域性的线性坐标；（b）时间局域性的对数坐标表示；（c）频率局域性

（6）再生核（Reproducing Kernel）：

在连续子波变换的情形下，子波变换系数形成一个过度完全基（over complete basis）。这意味着子波系数之间有一个相关关系。这个相关关系又对应于与子波 h 有关的一个再生 K_h 的存在。

$$K_h(a_1, a_2, b_1, b_2) = \frac{1}{C_h} \int_{-\infty}^{\infty} h_{a_1 b_1}(t) h_{a_2 b_2}(t) \, \mathrm{d}t \tag{14}$$

它表征了连续半平面 $\{b, a > 0\}$ 的两个不同点之间的相关关系。它的结构取决于子波的选取；实际上这个再生核度量了每个子波 h_{ab} 的空间和尺度的选择性，因此将非常有助于选择最适合于给定问题的子波函数。

4　结束语

本文设想读者对这个课题尚不熟悉，从而从傅里叶变换出发，介绍了盖伯变换到

子波变换的发展，并阐明它们之间的区别和联系。尽可能全面地介绍了连续子波变换和离散的子波变换，包括子波函数的限制条件，子波变换系数的表达和子波变换的基本性质等，以便使读者能对子波变换理论有一个大致的了解，从而可使用这个有效的信息处理工具。

参 考 文 献

[1] D. G. Feirelson. Optical computinga survey for computer scientists, the MIT press, 1988.

[2] N. H. Farhat, D. Psaltis, A. Prata, et al. AIPP. OPI., 24, 1469(1985).

[3] J. M. Kinser, H. L. Cauifeld, J. Shamir. Appl. Optl., 27, 3992(1988).

[4] A. D. Fisher, W. L. Lippincott, N. N. Lee. Appl. Opt., 26, 5039(1987).

[5] T. K. Gaylord, M. M. Mirsalehi, C. C. Guest. Opr. nEg., 24, 48(1985).

[6] F. J. Taylor. ComPUler, 17, 50(1984).

[7] J. Taboury, J. M. Wang, P. Chavel, et al. AIPP. Opt., 27, 1643(1988).

[8] K. S. Huang, B. K. Jenkins, A. A. Sawehuk. Computer VisionGraPhics& image Processing, 45, 295(1989).

[9] K. S. Huang, B. K. Jenkins, A. A. Sawchuk. Appl. Opt., 28, 1263(1989).

[10] L. Liu. Opt. Lett., 14, 482(1989).

[11] Y. Ichioka, J. Tanida. Proc. IEEE, 72, 787(1984).

[12] K. H. Brenner, A. Huang, N. Streibl. Appl. Opt., 25, 3054(1986).

[13] Marie Farge. 力学进展, 22(3,4),(1992).

[14] D. Gabor. J. Inst. Elect. Eng. 93(3),(1946).

[15] M. O. Freeman. Optics & Photonics News, August, 8-14(1993).

[16] H. Szu. Yunlong Sheng, J. Chen. Appl. Opt., 31, 3267(1992).

[17] J. M. Combes, A. Grossmann and Ph. Tchamitchian, eds, wavelelets. 2nded. (springer-Verlag), Berlin, (1990).

光学子波变换（2）实验结构及技术*

摘　要　本文描述了子波变换在频谱域中的表达，讨论了光学子波变换的特性。着重介绍了一维的和二维的光学子波变换实验系统和实验技术。并报道了我们用光学子波变换方法对一维信号的分割和对二值图像特征提取的实验结果。

1　引言

子波变换在频率域中可以写为：

$$W_s(a,b) = a^{1/2}\int H^*(af)\exp(j2\pi fb)S(f)df = <Ha,b(f),S(f)> \quad Ha \qquad (1)$$

$b(f)$ 和 $S(f)$ 分别为子波函数 a，$b(t)$ 和 $S(t)$ 的傅里叶变换，其中

$$Ha,b(f) = \int Ha,b(t)\exp(-j2\pi ft)dt = a^{1/2}H(af)\exp(-j2\pi fb)$$

由式（1）可知，子波变换在光学系统的频谱域中实现与带通滤波器的操作相类似，只要在傅里叶频谱面放置一系列的带通滤波器就可实现，可以用典型的光学 $4f$ 匹配滤波器来实现，如图 1 所示。

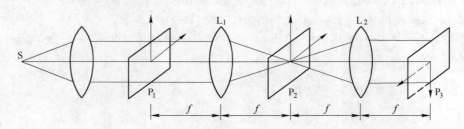

图 1　光学 $4f$ 匹配相关系统实现子波变换

本文就一维信号和二维信号的光学子波变换的实验系统结构及实验技术进行了分析说明，并给出我们对一维模拟信号的光学子波变换结果，以及采用角度复合技术利用 Harr 子波对二值化图像进行光学子波变换得到边、角特征等的实验结果。最后讨论了光学子波变换中应该注意的问题和一些基本特性。

2　光学子波变换的实验系统

2.1　一维信号的光学子波变换系统

在图 1 的基本结构上，用声光盒作为输入装置把一维信号转化为空间分布的信号，

---

* 本文合作者：王文陆，严瑛白，邬敏贤。原发表于《光电子·激光》，1994，5(5)：268～272。

用空间调制器显示一系列的带通滤波器 $H(af)$，采用两个柱面傅氏透镜作正、逆一维傅氏变换，就可实现对一维信号的光学子波变换。如图 2 的实验结构为 H. Szu 等[1] 和 Yunlong Sheng 等[2] 所采用的典型的一维信号光学子波变换系统。在这一实验结构上，我们采用 Ronchi 光栅和狭缝来近似高斯余弦函数子波：

$$h(t) = \cos(Wt)\frac{1}{\sqrt{2\pi}\sigma}\exp\left(-\frac{t^2}{2\sigma^2}\right)$$

用不同的光栅周期和狭缝宽度来近似代表不同的扩参数 a 所相对应的子波。实验中 a 是以 2 的整数次幂变化的。光栅周期分别为 100 线/mm，50 线/mm，12.5 线/mm，相应的狭缝宽度分别为 1mm，2mm，8mm，对应于 a 分别为 1，2 和 8。并把它们的傅氏谱记录在透明胶片上，记录时挡掉 0 级，±2，±3 等高级次，如图 3（a）所示。不同周期的模拟合成一维信号记录在照相底片上，代替声光盒作为输入，如图 3（b）所示。实验结果如图 3（c）所示。可见不同频率的信号被很好地分割开，得到多重分辨率分析的结果。

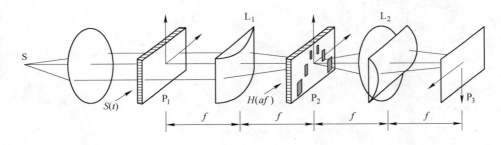

图 2　一维信号的光学子波变换系统

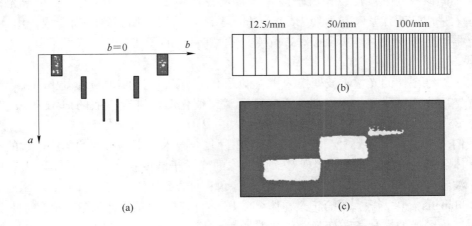

图 3　子波滤波器系列（a），输入合成信号（b），输出结果（c）

YaoLi 和 Yang Zhang[3] 建议了带有逆子波变换的实验系统，在输出面加全反镜，经图 2 所示系统反变换后在声光盒的前方加半反半透镜把逆变换的结果引出来。他们提出了对长时间信号和短时间信号的光学子波变换系统应有所区别，在对长时间信号的子波变换展开时，先把 1-D 信号用栅状扫描或基于余数理论等方法转换成 2-D 分布，在处理结果中再作 2-D 到 1-D 的转换。并认为，虽然这种并行方法减少了信号的带宽，但是它能处理带宽有限的极长序列的信号。

2.2 二维信号的光学子波变换系统

光学技术的并行处理能力和光学系统内在的特性，使得用光学技术来处理图像等二维信号极富意义，也更能显示出光学技术的优越性。下面介绍对二维图像的光学子波变换实验系统和实验手段。

如图 4 为 Xiang Yong 等[4]报道的在空间域中基于铸影法原理对二维二值化图像的 Harr 子波变换实验原理结构。不同的扩参数的子波函数可以由子波在光轴上平移（改变 d_1）来产生。

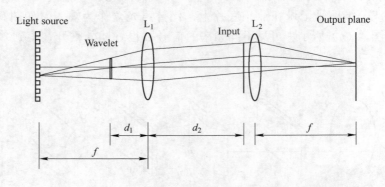

图 4 基于铸影法的 2-D 子波变换

用阵列点光源时可以进行离散的子波变换，若用分布光源时则可获得连续的子波变换。不同参数的子波可以做成子波库直接放在透镜 L_1 后面，类似于非相干光相关器。输入采用透明片紧贴在透镜 L_2 上，Harr 子波函数用微偏振器阵列合成，用 Ar 离子激光器产生阵列光源，实现了对二值英文母"T-"的子波变换，获得了角、边等特征信息。

对于二维图像的光学子波变换的实验报道还有很多。比如：Thomas J. Burns 等[5]采用磁光空间光调制器和孔径遮挡技术实现了对二值化图像进行离散的 Harr 子波变换；Yun long sheng 等[6]采用 N^4 多通道相关器结构的 2-D 全息子波变换系统。

我们采用如图 5 所示的实验光路，利用角度复合技术把不同参数的子波函数的频谱记录在体全息上。在频谱域上实现对二值图像的光学子波变换[7]。图 6（a）为利用角 Harr 子波和直边 Harr 子波同时提取出二值图像角和直边的特征；图 6（b）为利用角 Harr 子波和 Roberts 滤波函数[8]复合记录在体全息上同时提取出二值图像的角和轮廓特征。

对二维图像的光学子波变换，结果是以四维形式的子波变换因子，在系统的输出面是以二维阵列式公布的，它们事实上就是对输入图像的四维空间/频率联合表示。由于光学系统带宽的限制，使得二维光学子波变换系统每次能处理的通道数目（也即子波滤波器的数目）是有限的，这样只有采用分时处理或高分辨率的空间光调制器来实现。若考虑到子波函数本身的一些对称性，可以减少子波变换因子的维数，来克服带宽限制的困难。

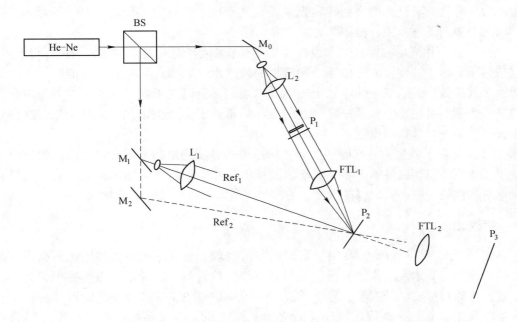

图 5 体全息角度复合技术实现 2-D 子波变换

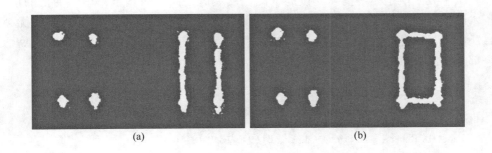

图 6 光学实验结果
（a）利用角 Harr 子波和直边 Harr 子波；（b）利用角 Harr 子波和 Roberts 滤波器函数

3 光学子波变换的特性讨论

（1）一般来说，子波函数及其频谱都是复值函数，这样通常采用光学全息记录法或计算机制全息图来制作光学子波变换中的子波滤波器系列 $H(af)$，以便表示复值。

（2）光学 $4f$ 匹配相关器的平移不变性，可以使子波变换中滤波器的平移因子 $b = 0$。也就是说，光学子波变换不依赖于起始时间的选择。并且，许多子波函数是实函数，还具有对称或反对称的性质，可以用来消除子波滤波器中的位相部分。更易于光学方法实现，例如 Harr 子波函数是台阶型的实值函数，很易于光学或光电混合方法实现。

（3）光学子波变换的处理精度不高，也不可能达到用数字电子技术来实现子波变换的处理精度。若不考虑所用的时间，数字电子技术可以生成任何精度的窗函数或子波函数,并且具有极大的灵活性。光学模拟处理技术潜在的不精确性,使得对实时光学输入装置的动态范围,分辨率等的要求很高。

(4) 在利用光学技术或光电混合技术实现子波变换时,对子波函数的选择很重要。这依赖于光学子波变换的应用领域和器件的技术水平。因为有些数学性能好的子波函数由于其函数形式的复杂性不易于光学技术表达,所以在光学子波变换中不一定是理想的。而那些可用光学技术精确表达的函数却不一定能产生最优的处理结果。这样对子波函数的选择就显得很重要。

(5) 光学子波变换的性能还受到光学系统带宽积的限制。若要对二维图像进行光学子波变换并作四维显示,需要对输入图像的带宽有一定的限制。并且通常采用具有全局对称性的子波函数,以减少显示的维数,例如采用高斯余弦函数。

4 结束语

光学子波变换可以实现实时、快速和并行处理。可以用来对大量原始数据进行预处理,比如:对输入图像的特征提取,目标位置确定等。在光学子波变换中,对子波函数的选择很重要,既依赖于光学子波变换的应用对象,又依赖于光学器件的技术水平。并且光学子波变换的性能还受到光学系统带宽积以及系统的调整精度等因素的限制。

光学子波变换是在傅里叶变换的基础上发展起来的,而基于傅里叶变换的信号处理技术已经得到了极成功的应用。可以想象,光学子波变换不仅具有同样的优越性,而且还因为其自身的特性,其应用领域是极广泛的。这个连载的第三部分,也就是最后一部分将主要讨论光学子波变换的应用。

参 考 文 献

[1] H. Szu, Y. Sheng, Jin Chen. Appl. Opt., 31(7), 3267-3277(1992).
[2] Y. Sheng, D. Roberge, H. Szu. Opt. Eng., 31(9), 1840-1845(1992).
[3] Yao Li, Yang Zhang. Opt. Eng., 31(9), 1865-1885(1992).
[4] X. Yang, H. Szu, Y. Sheng, et al. Opt. Eng., 31(9), 1846-1851(1992).
[5] T. J. Burns, K. H. Fielding, S. K. Rogers, et al. Opt. Eng., 31(9), 1852-1858(1992).
[6] Y. Sheng, T. Lu, D. Roberge, et al. Opt. Eng. 31(9), 1859-(1992).
[7] G. Jin, Y. Yan, W. Wang, et al. SPIE, 2034, 371-380, San Diego 1993.
[8] 王文陆, 严瑛白, 金国藩, 等. 光电子器件与集成年会, 南京, 1994.

光学子波变换（3）应用*

摘 要 本文综述了子波变换的发展概况。总结了光学子波变换的一些应用领域，重点论述了在纹理分割中的应用。提出了联合子波变换相关器及其应用于二值图像的特征提取。给出了多通道盖伯变换在纹理分割中的应用设想以及初步的研究结果。

1 子波变换的发展概况

从1822年Joseph Fourier在解决热传导问题时提出傅里叶分析以来，随着科研和生产实践要求的提高，信号处理的工具得到了进一步的发展和完善，逐步发展了盖伯变换（gabor transform）和子波变换（wavelet transform）。子波变换最早是于1984年由法国科学家J. Morlet等在进行地震数据分析工作时开创起来的[1]。围绕子波概念的数学框架是由"french school"构造起来的[2]。在Meyer最近出版的一本书中证明了这一点，并表明了子波概念与早期算子理论的关系[3]。多分辨率分析方法已经普遍应用于计算机视觉问题中，从范围探测到运动估计[4]。多分辨率分析方法在图像编码中的一个重要应用是由P. J. Burt和E. H. Adelson开发的谱系结物的金字塔（pyramid）方法来表示和分析图像[5]，这种方法与子带编码（subband coding）以及子波具有紧密的联系，是向子波基元概念迈出的重要一步。

S. Mallat用多分辨率分析的概念来定义子波[6]。Ingrid Daubechies于1988年构造了紧支撑正交子波（compactly supported orthonormal wavelet）。它们是建立在离散滤波器的迭代上，并发明了"Multiplier2"紧支撑离散子波变换理论，获得了由"乘子2"子波序列的平滑化和多项式表达的条件[7]。H. L. Resnikoff[8]认为Daubeches的论文是20世纪后半叶最主要的数学贡献，具有巨大的理论和应用意义。Daubeches的工作建立了数字信号处理和函数分析之间的紧密联系。S. Mallat于1989年建立了离散子波变换与Danbeches的标准正交基之间的联系，以及当子波理论应用于数字取样信号时必须有限子波变换的计标，称为Malat变换[9]。他构造了二维可分离的标准正交子波基，并且用来描述图像，然后为数字图像建立了一个层次递减的正交性，无冗余，可逆的表达方法。这恰好是金字塔滤波方法所追求的目标。美国防御高技术研究计划局支持了一个国家子波研究计划[8]，它大大激发了子波变换在美国其他单位开展这方面的研究[10]。1990年Y. Meyer：出版了《Ondelettes》一书[3]，这是子波理论的第一本书，集中讨论了无约束子波的数字性质。1992年M. Vetterli和C. Herler研究了多分辨率信号处理、滤波器库和子波之间的关系[11]。在假定滤波器库服从光滑性的限制下，可以用来计算离散的子波变换和推导连续的子波基。M. J. Shensa研究了Malat算法和"A, Torus"的结合，建立了离散子波变换的系统框架，并导出了可精确计算连续子波变

---

* 本文合作者：王文陆，严瑛白，邬敏贤。原发表于《光电子·激光》，1994，5(6)：331~339。

换的条件[12]。此后（确切地说从1990年以后），有关子波变换的文章在许多著名的杂志上不断地有报道，《Optical Engneering》于1992年9月出了一本专辑[13]，报道了光学子波变换的研究成果，并掀起了光学子波变换的研究热潮；《IEEE TRANSACTIONS ON SIGNAL PROCESSING》于1993年12月出了一本专辑[14]，分六大类报道了子波变换的理论以及在信号处理中的应用的杰出成果；《Applied Optics》也于1994年出版一本包括盖伯变换、子波变换，以及相关的信号与图像变换在内的专辑[15]。从1992年开始，SPIE及OSA年会都有子波变换的专题，并有大量的文章报道，每年都有相应的论文集出版[16]。可以说，子波变换的研究在世界范围内受到各个领域的研究者的重视，并形成了子波变换研究的高潮，相应的科研成果将会不断地出现。

我们在这方面的研究并不落后，在诸如：纯数学、模式识别和信号处理、自动控制、语音合成和识别、地震信号分析等许多领域的科研人员都开展了这方面的研究工作[17~19]。我单位于1992年就已开始了光学子波变换的研究工作[20]。

综上所述，子波变换技术是最近发展起来的非常有效的信息处理工具，已引起了众多科研人员的重视。光学子波变换与成熟的光学信息处理技术的结合将在光计算和光学图像处理领域中发挥其独特的作用。

2 光学子波变换的应用

（1）子波函数的局域性（在时间域与频率域）以及子波变换在相空间（时间-频率域，空间-空间频率域）中的表达，使得子波变换在对非稳定传输信号的分析时既可以得到信号的时间域中的信息，又可以得到其频率域中的信息，可以很好地确定信号的位置和频率特性[20]。C. Hirlimann和J. F. Morhange把子波分析用于短光脉冲的特性研究[21]。他们用维纳-维列变换分析超短光脉冲在时间域中和频率域中的特性来研究非线性光学效应。当超短光脉冲通过晶体后的线性色散自相位调制及时间展宽、功率谱展宽，脉冲的再成形、瞬时频率时间变化等特性都可获得。Y. Zhang和Y. Li把盖伯变换应用于传递信号分析[22]，用声光调制器作为信号输入器件，液晶空间光调制器来产生盖伯窗和子波窗，很好地确定了不同传递信号的起始位置。

（2）图像特征提取是光学信息处理中所研究的主要问题之一。图像的特征是图像中变化比较激烈的所在，这可由子波变换的局域分析特性来研究。X. Yang等采用铸影法利用Haar子波变换来提取二值图像的特征，缩放变化是通过改变Haar子波与光源的相对位置来实现的[23]。Y. Sheng等人把光学子波变换应用于平移不变的模式识别中[24]，通过适当选取子波滤波器的参数来获得模式的边沿增强，并把这些变换因子做匹配相关操作来达到平移不变的模式识别。与通常的方法相比，这种方法改善了分辨能力和信噪比。最早有关光学子波变换的应用报道是1990年在物理评论快报上，E. Freysz等[25]把光学子波变换用于分维聚类体（Fractal Aggregates）的分析。所用的分析子波为中心对称的墨西哥帽子波，对雪花状分维体和生长模型（Dla分维体）进行了不同尺度的分析，获得的实验结果与数值计算结果相一致。他们的这种分析方法不仅适用于静态的分析，而且在动态分析中也是很有效的。这种快速分析方法奠定了动态现象中时间分辨率的研究的道路。

（3）因为所有的子波基元函数（Daughter Wavelet）之间是正交的，当某一子波基元函数捕获到一个期望的信号时，不会由于该信号与子波变换的基元函数作内积而引入交扰串言。这一特性使得子波变换在信号的去噪声中很有效。人耳是一个具有子波变换特性的装置，所以当人们置身于一个闹哄哄的大型酒会中时，对酒杯掉到地上等这样的事情很敏感，这也就是所谓的"鸡尾酒会效应"。若在音乐录音中要把其中的"咳嗽"声去掉，就可采用适当的"咳嗽"子波去处理，这样就可避免重新录音的麻烦。S. J. Schiff 采用光学子波变换来去除时间序列信号中的噪声[26]。把带高斯白噪声的信号经子波变换后，在相空间中信号的局域性很明显，和噪声很好地分离开，设计一定的噪声模型后就可以得到很好的信号。

（4）由于子波变换的局域分析特性，并存在严格的逆变换，可以用较少的子波因子重构原始信号，使得子波变换在图像数据压缩方面很有发展前途。据美国的科学前沿报道[27]。把一幅头像照片经子波变换技术进行数据压缩，只用5%的信息就可以很好地恢复原图像。

（5）光学子波变换在图像纹理分割中的应用。纹理是重要的表面特征，在图像分析和计算机视觉中是非常重要的信息。因为用它既可以分辨表面又可以依据二维纹理图像的纹理特征来测量高维信息，如形状和运动。在纹理背景的目标识别方面，纹理信息也是极其重要的。纹理分析的首要任务是把纹理特征从原始图像提取出来，用来描述和分类图像，进而把原始的纹理图像按纹理特征分割成性质相同的区域。通常纹理分割的方法中包含纹理特征提取和分割是非常关键的。纹理特征的描述对完成纹理分割是非常关键的。纹理特征一般用统计的方法和空间-空间频率域联合表达的方法来描述。也就是说纹理特征既要反映局域的信息又要反映全局的信息，图像中任何点的特性是由该点与其周围像点的相互关系来确定的。这样就要用一个移动的窗口来测量纹理特征，所用的窗口越大测得的特征越精确并且噪声越低。然而，在区域的边缘部分所用的窗口越大意味着不同纹理性质之间有更多的平均效应，结果就造成分割算法很难精确地确定不同纹理区域之间的边界。可见，纹理特征的提取和分割的精度是相互矛盾的，并服从测不准原理。子波变换技术提供多尺度和多分辨率分析的手段，可以很好地解决纹理分割分辨中，并表明了盖伯变换与人类视觉系统的响应相似[28]。P. Veronin 等人把子波滤波技术与神经网络相结合用于光学图像分割[29]。通过适当选取子波滤波器的参数，比如滤波孔径的大小，分离的距离，以及滤波孔对的取向等，可以很好地把目标从纹理背景中分割出来，并可得到目标的边沿增强像。图1是M. O. Freeman 等把光学子波变换应用于纹理分割的比较典型的实验光路[30]。他们把滤波器做成楔环状，径向只用了一个拉普拉斯高斯函数子波滤波器，角度方向运用四个通带的滤波器平行提取输入图像的角度信息。缩放变化是对通过输入图像经反馈光路实现的。图2（a）是一幅具有4个取向的纹理图像，图2（b）是利用图1进行多分辨率分割的结果。

当然，光学子波变换作为一种有效的信息处理手段，其应用领域不光是局限在上述几个方面，是极其广泛的，在许多方面的研究还刚刚开始，有待进一步的探索和完善。比如光学子波变换在图像分割、图像信息压缩，在各种不变性的模式识别，以及在光通讯中的信息、交换等领域中的应用是很值得研究的。

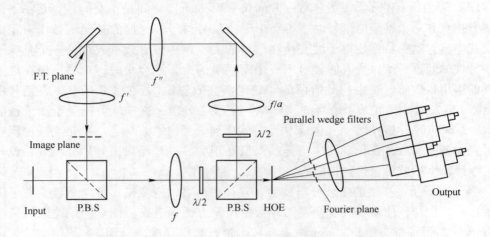

图 1　多通道子波处理器

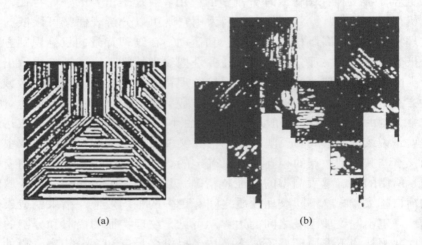

图 2　纹理图像（a）和光学子波变换处理器分割结果（b）

3　联合子波变换相关器

联合变换相关器（Joint Transform Correlator，JTC）由 Weaver 和 Goodman 在 20 世纪 60 年代提出来，用来实现两个函数的光学相关[31]。通常 JTC 是用来实现匹配模式识别，或在多物体的背景中寻找某一物体，或跟踪某一运动物体等[32]。此处我们把它用来实现子波变换[33]，把子波函数和物体放在联合变换的输入面，在输出面就可以观察到物体的子波变换结果。我们所建议的联合子波变换相关器如图 3 所示。利用联合变换相关器来实现子波变换具有许多优越性[34]，比如可以实现平移不变的子波变换，不必再制作一系列的复匹配滤波器，并避免了在 4f 系统中所遇到的对滤波器精确复位的要求等。图 4（a）是一个"+"字与 Haar 子波作为联合输入图像；图 4（b）是用 2×2 个像素的角 Haar 子波得到"+"字的角特征；图 4（c）是用 4×4 个像素的角 Haar 子波得到"+"字的角特征。比较图 4（b）与图 4（c）可以看出，子波变换的多分辨率分析特性，图 4（b）的角点比图 4（c）的角点所占的像素要少。在作计算机模拟中我们对相关面输出的零级都作阈值处理，以便 ±1 级能够突出出来。很显然，不同参数的子波函数可以顺序显示在联合变换输入面上，或者利用系统的空间带宽积，在输入面的不同位置同时显示几个子波函数，那么在输出面可以同时得到几个相应的子波变换结果。如果图 5（a）是把 Roberts 滤波器[33]的两个互补函数与一个圆放在一起作为联合输入图像，图 5（b）是相应的结果。

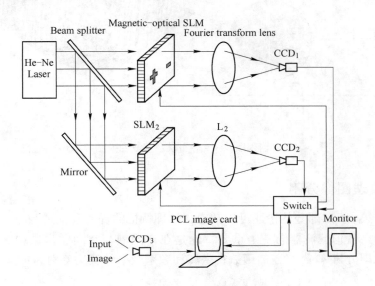

图 3　联合子波变换相关器

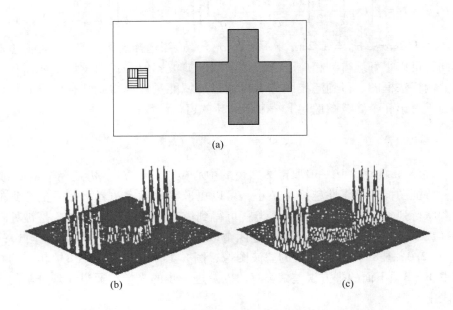

图 4　联合输入图像（a），利用 2×2 个像素的角 Haar 子波（b），
利用 4×4 像素的角 Haar 子波（c）

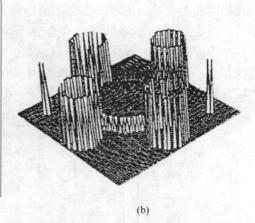

图 5　联合输入图像（a）和联合子波变换结果（b）

4　多通道纹理图像分割

由第 2 节第（5）部分的论述可以知道，纹理特征的提取和描述与分割算法的精度是相互矛盾的。盖伯滤波器的一个重要特性是在空间域与空间频率域中具有最优的联合局域性，也即分辨率[34]。盖伯函数在空域和频率域中的表达分别为：

$$h(x,y) = \exp\left[-\frac{1}{2}\left(\frac{x^2}{\sigma_x^2} + \frac{y^2}{\sigma_x^2}\right)\right]\cos(2\pi U_o x) \tag{1}$$

$$H(u,V) = A\left\{\exp\left\{-\frac{1}{2}\left[\frac{(u-u_o)^2}{\sigma_u^2} + \frac{v^2}{\sigma_x^2}\right]\right\}\right\} \times \exp\left\{-\frac{1}{2}\left[\frac{(u+u_o)^2}{\sigma_u^2} + \frac{v^2}{\sigma_v^2}\right]\right\} \tag{2}$$

其中，$\sigma_u = 1/2\pi\sigma_x$，$\sigma_v = 1/2\pi\sigma_y$，$A = 2\pi\sigma_x\sigma_y$。适当选择这些参数、中心频率和滤波器的取向，可以很好地得到相应于各纹理区域的滤波图像。我们设计了如图 6 的多通道纹理分割算法框图。得到滤波图像后再通过局域能量计算得到特征图像。局域能量计算相当于对第几个滤波图像卷上一个 $M \times M$ 的窗口函数：

$$e_n(i,j) = \frac{1}{M^2}\sum_{kj \in W_{ij}} |J_n(k.l)| \tag{3}$$

W_{ij} 是一个 $M \times M$ 的窗，M 一般取奇数，窗的中心位于 (i,j)。M 是一个重要的参数，M 越大，特征图像的可靠性越好，M 小，可以更准确地确定边界。然后把得到的特征图像并考虑行列坐标进行平方误差聚类就可得到纹理分割的结果。图 7 是被处理的纹理图像（128×128 像素）；图 8（a）是四个取向的盖伯滤波器，其中 $u_0 = 10$；$\sigma_u = 2.5$；$\sigma_v = 2.0$；图 8（b）是相应的滤波图像；图 8（c）是与图 8（b）相应的特征图像，其中 $M = 3$。下面的平方误差聚类算法以及进一步的光学多通道纹理分割的工作还在进行之中。

图 8（c）是对图 8（b）进行能计算所得到的特征图像，每一通道较好地代表了图 7 中的一种纹理特征。

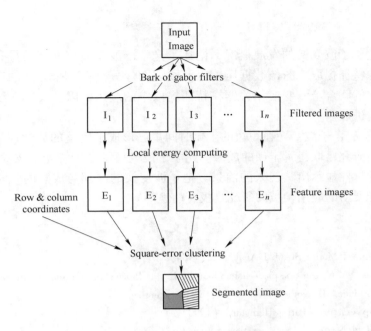

图 6 多通道盖伯变换纹理分割算法框图

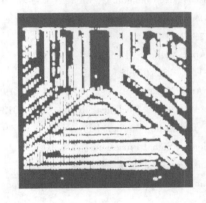

图 7 被处理的纹理图像

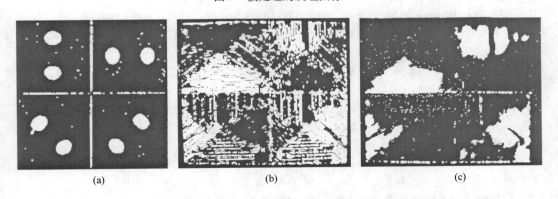

图 8 （a）四个取向的盖伯滤波器，（b）相应的滤波图像，（c）相应的特征图像

5 结束语

本文概述了子波变换的发展概况，论述了光学子波变换在动态信号分析、信号的去噪声、图像特征提取、图像数据压缩以及纹理图像分割等几个方面的应用。并报道了我们所提出的联合子波变换相关器在特征提取中的应用，以及我们在纹理图像分割研究中的最新进展。

从1984年起，子波变换在短短的十年时间里已形成了完整的数学理论，并在许多领域中的应用研究已取得了丰硕的成果。光学子波变换结合传统光学技术的平行处理，高速度等的优越性显示出其迷人的前景。把子波变换的原理与成熟的光学信息处理技术相结合，开展新的应用领域的研究是极有前途的。

参 考 文 献

[1] A. Grossmann，J. Morlet. SIAM. J. Anal.，15(1984).

[2] J. M. Combes，A. Grossmann. ph. Tchamitchian (Eds.)，Proceedings of the International Conference，Marseille，France，December14-18，1987，Springer-Verlag.

[3] Y. Meyer. Op'eratcurs，Paris：Hermann，(1990).

[4] E. A. Rosenfeld. NewYork，Springer-Verlag. (1984).

[5] P. J. Burt，E. H. Adelson. IEEE Trans. Commun. COM-31，532-540 (1983).

[6] S. Mallat. Techniquereport，Department of Computer Information Science，University of Pennsylvania，PiladePhia，PA，September 1987.

[7] I. Daubechies. Commun. PureAppl. Math.，XLI，909-996(1988).

[8] H. L. Resnikoff. Opt. Eng.，31.6，1229(1992).

[9] S. Mallat. IEEE Trans. PatternAnal. Mach. Intell.，11，7，674-693(1989).

[10] W. Lawton，H. L. Resnikoff. AwareInc.，Technical Report，AD910130，Cambridge. Mass，(1991).

[11] M. Vetterli，C. Herler. IEEE Transaction on Signal Processing，40，9，2207-2232 (1992).

[12] M. J. Shensa. IEEE Transaction on Signal Processing，40，10，2462-2482 (1991).

[13] Special Issue of Wavelet Transform，Opt. Eng.，31，9，(1992).

[14] Special Issue of Wavelers and Signal Processing IEEE Transaction on Signal Processing，40，12，(1993).

[15] Special Issue on Applications of Limited-Extent Waves，Appl. Opt.，(1994).

[16] OSA AnnualMeeting，1992，1993，1994. SPIE Annual Meeting，1992，1993，1994.

[17] 邓东皋，彭立中. 小波分析，数学进展，第20卷，第3期，1991.

[18] 李世雄. 小波理论及其应用(1997).

[19] 吴波，何岭松，蔡志强，等. 基于小波变换的波形特征抽取与识别，华中理工大学学报，第21卷，第1期，1993.

[20] Guofan Jin，Yingbai Yan，Wenlu Wang，et al. SPIE. 2304 371-380 (1993).

[21] C. Hirlimann，J. F. Morhange. Appl. Opt.，31，7，(1992).

[22] Y. Zhang，Y. Li. Opt. Lett，16，13，1031-1033 (1991).

[23] Xiangyang Yang，H. Szu，YunlongSheng，et al. Opt. Eng.，31，5，1946 (1992).

[24] Yunlong Sheng，D. Roberge，H. Szu，et al. Opt. Lett.，18，4，299-231 (1993).

[25] E. Freysz，B. Pouligny，F. Argoul et al. Phys. Rev. Lett. 64，7，(1990).

[26] S. J. Schiff. Opt. Eng.，31，11，2492 (1992).

[27] A. Greenwood, et al. National Academy Press, Washington DC USA, 164-167 (1992).
[28] M. R. Turner. Biol. Cybern., 55, 71-82(1986).
[29] P. Veronin, L. Priddy, S K. Rogers, et al. Opt. Eng., 31, 2, 287(1992).
[30] M. O. Freeman, A. Fedor, B. Bock, et al. SPIE. 1772, 241-250(1992).
[31] C. S. Weaver, J. W. Goodman. Appl. Opt., 5, 1248-1249(1966).
[32] F. T. S. Yu, S. Jutamulia, Kai Chang (Editor). Wiley-Interscience. NewYork. 203-248 (1992).
[33] 王文陆,金国藩,严瑛白,等. 联合变换实现子波变换的理论分析,光学学报,第 14 卷,第 10 期,1994.
[34] J. G. Daugman. J. Opt. Soc Amer., 2, 1160-1169,(1985).
[35] A. K. Jain, F. Farroknia. Patt. Rec. Rec., 24, 12, 1167-1186 (1991).

Multiobject Recognition in a Multichannel Joint-transform Correlator*

Abstract A multichannel joint-transform correlator that incorporates a Dammann grating as a beam splitter is described. The Dammann grating splits a single incident beam into a two-dimensional array of equal-intensity beams, which form several channels for correlation. Each channel contains both a single target and a single reference. Optical experimental results are presented.

Real-time pattern recognition is one of the most important applications of optical information processing. The VanderLugt matched spatial filter[1] and the joint-transform correlator[2] (JTC) are the two major types of coherent optical correlator used in optical pattern recognition. The JTC can perform optical image correlation without the need for matched-filter fabrication and thus is more suitable for operation in real-time mode. In the time since the joint-transform correlator was proposed by Weaver and Goodman,[2] it has been studied extensively, and several modifications have been made to improve its performance. These include the binary (or nonlinear) JTC, the amplitude-compensation JTC, the chirp-encoded JTC, and the preprocessed JTC[3-6]. Among these, the binary or nonlinear JTC,[3] the chirp-encoded JTC,[4] the fringe-adjusted JTC,[5] and the preprocessed JTC[6] are used to study multi-object recognition.

In general, a JTC can perform multiobject recognition in two ways. One technique is to put one reference image and one target image in the input plane and subsequently update the target image. The other is to put a reference image together with multiple target images in the input plane. Multi-object detection that uses the first method is relatively slow. The time delay is especially obvious in the JTC that uses a liquid-crystal light valve (LCLV) as a square-law detector for power spectrum conversion, because the response time of a LCLV is long. Further, a real-time single-channel JTC based on a LCLV has another drawback, which is that the effective area of a LCLV is not used sufficiently. Because the dynamic range of the joint-frequency spectrum is much greater than that which a LCLV can detect, only a small center portion of the joint-frequency spectrum, written on the LCLV, can be read out for further use. Multiobject detection that uses the second method can be implemented in parallel, but it may yield unwanted cross correlations between the target images to be detected. These cross correlations will yield false alarms and misses that make detection difficult. In this Letter we describe a multichannel JTC that uses a Dammann grating as a beam splitter to avoid the unwanted cross correlation between the target images for multiobject detection. Each channel contains a single target and a

---

* Copartner: J. H. Feng, M. X. Wu, S. H. Yan, and Y. B. Yan. Reprinted from *Optics Letters*, 1995, 20(1):82-84.

single reference. The joint-transform correlation goes on independently in each channel. The proposed system can perform correlations at high speed and in parallel.

A binary-phase grating, which can split an incident laser beam into a one-or two-dimensional array of equal-intensity light spots, first was presented by Dammann and Gortler[7] in 1971. It is well known that the design and analysis of a so-called Dammann grating have so far been based on scalar diffraction theory. Several techniques can be used to fabricate a Dammann grating[8].

In this Letter we discuss a multichannel JTC in which a Dammann grating is used as a beam splitter. The implementation of the multichannel JTC with an optically addressed spatial light modulator (SLM) is shown in Fig. 1. An expanded and collimated laser beam impinges upon Dammann grating 1. Suppose that the Dammann grating is so designed that it forms multiple channels without overlapping parts and that each channel covers just a single scene. The amplitude transmittance behind Dammann grating 1 can be expressed by

$$g(x, y) = \sum_{j=-N}^{N} \sum_{k=-N}^{N} c_{jk} \exp\left[i2\pi\left(\frac{jx}{d} + \frac{ky}{d}\right)\right] \quad (1)$$

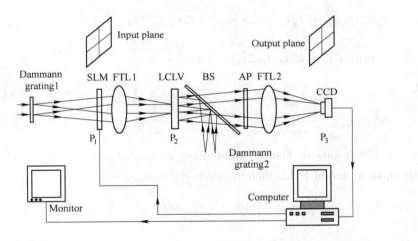

Fig. 1 Multichannel JTC that uses two Dammann gratings. BS, beam splitter; AP, analyzer-polarizer

where c_{jk} is set to be equal to a constant. Eq. (1) shows that the transmitted light can be considered the superposition of $2N \times 2N$ quasi-plane-wave light with different diffraction angles. P_1 is the input plane, which contains $2N \times 2N$ input scenes. Each scene consists of a single reference image and a single target image. The amplitude transmittance of P_1 is described by

$$t(x, y) = \sum_{j=-N}^{N} \sum_{k=-N}^{N} [r_{jk}(x - a_j + x_0, y - b_k + y_0)] + t_{jk}(y - a_j - x_0, y - b_k - y_0) \quad (2)$$

where $r_{jk}(x_j - a_j + x_0, y_k - b_k + y_0,)$ and $t_{jk}(x_j - a_j - x_0, y_k - b_k - y_0)$ are the reference image and the target image in the (j, k)th scene, respectively. These are separated by $2x_0$ and $2y_0$ from the center point (a_j, b_k) of the (j, k)th beam channel along the x and y directions, respectively. Multiple input scenes are Fourier transformed, and their frequency spectrums are written on the input plane of a LCLV. The joint-frequency spectrum in each channel can be expressed by

$$G_{jk}(u,v) = R_{jk}(u,v)\exp[i2\pi(u2x_0 + v2y_0)] \times \exp\left[+i2\pi\left(u\frac{j\lambda f}{d} + v\frac{k\lambda f}{d}\right)\right] +$$

$$T_{jk}(u,v)\exp[i2\pi(u2x_0 + v2y_0)] \times \exp\left[-i2\pi\left(u\frac{j\lambda f}{d} + v\frac{k\lambda f}{d}\right)\right]$$

$$j, k = -N, -N-1, \cdots, -1, 1, \cdots, N-1, N \tag{3}$$

where $R(u, v)$ and $T(u, v)$ are the Fourier transforms of $r_{jk}(x, y)$ and $t_{jk}(x, y)$, respectively; u and v are mutually independent frequency-domain variables scaled by a factor $1/\lambda f$; λ is the wavelength of the collimating light; and f is the focal length of Fourier-transforming lens FTL1. The joint-frequency spectrum is written on the LCLV and is converted into the joint-power spectrum (JPS) in each channel. These JPS's in the multiple channels are read out from the LCLV in parallel. The JPS in each channel is expressed by

$$E_{jk}(u,v) = |G_{jk}(u,v)|^2 \exp\left[i2\pi\left(u\frac{j\lambda f}{d} + v\frac{k\lambda f}{d}\right)\right]$$

$$= R_{jk}^2(u,v) + T_{jk}^2(u,v) \times \exp\left[+i2\pi\left(u\frac{j\lambda f}{d} + v\frac{k\lambda f}{d}\right)\right] + R_{jk}(u,v) + T_{jk}^*(u,v) \times$$

$$\exp[+i2\pi(u2x_0 + v2y_0)] \times \exp\left[+i2\pi\left(u\frac{j\lambda f}{d} + v\frac{k\lambda f}{d}\right)\right] + R_{jk}^*(u,v) +$$

$$T_{jk}(u,v) \times \exp[-i2\pi(u2x_0 + v2y_0)] \times \exp\left[i2\pi\left(u\frac{j\lambda f}{d} + v\frac{k\lambda f}{d}\right)\right]$$

$$j, k = -N, -N-1, \cdots, -1, 1, \cdots, N-1, N \tag{4}$$

Then the JPS is inverse Fourier transformed by Fourier-transform lens FTL2, whose focal length is equal to that of FTL1, thereby yielding the correlation output $f_c(x,y)$, given by

$$f_{jk}(x_3, y_3) = r_{jk}\left(x_3 - \frac{j\lambda f}{d}, y_3 - \frac{k\lambda f}{d}\right) \otimes r_{jk}\left(x_3 - \frac{j\lambda f}{d}, y_3 - \frac{k\lambda f}{d}\right) + t_{jk}\left(x_3 - \frac{j\lambda f}{d}, y_3 - \frac{k\lambda f}{d}\right)$$

$$\otimes t_{jk}\left(x_3 + \frac{j\lambda f}{d}, y_3 + \frac{k\lambda f}{d}\right) + r_{jk}\left(x_3 - 2x_0 - \frac{j\lambda f}{d}, y_3 - 2y_0 - \frac{k\lambda f}{d}\right)$$

$$\otimes {}^* t_{jk}\left(x_3 - 2x_0 - \frac{j\lambda f}{d}, y_3 - 2y_0 - \frac{k\lambda f}{d}\right) + r_{jk}^*\left(x_3 + 2x_0 - \frac{j\lambda f}{d}, y_3 + 2y_0 - \frac{k\lambda f}{d}\right)$$

$$\otimes t_{jk}\left(x_3 + 2x_0 - \frac{j\lambda f}{d}, y_3 + 2y_0 - \frac{k\lambda f}{d}\right)$$

$$j, k = -N, -N+1, \cdots, -1, 1, \cdots, N-1, N \tag{5}$$

The correlation output is described by Eq. (5), where the first dc term represents the autocorrelation of the reference image and the second dc term represents the autocorrelation of the target image located at the center $(j\lambda f/d, k\lambda f/d)$ of the output plane in each channel. The third and fourth terms correspond to the cross-correlation between the reference and the target at distances $(2x_0, 2y_0)$ and $(-2x_0, -2y_0)$ from the center of the output plane $(j\lambda f/d, k\lambda f/d)$ in each channel. The experimental setup is shown in Fig. 1. An expanded and collimated He-Ne laser beam impinges upon Dammann grating 1 and is split into an array of 2 ×2 beams to form four channels. Dammann grating 1 and Dammann grating 2 were designed in our laboratory. These four light beams pass through a photographic transparency containing four input scenes. In each channel the scene consists of a single reference image and a single target image. Fig. 2 shows the results of the correlation test of the multiobject detection system. The four input scenes are shown in Fig. 2(a). There are three different targets in the four scenes: two are an airplane, one is a rocket, and one is a tank. The reference object is an airplane. FTL1 behind the transparency transforms the four input scenes, and their frequency spectra are simultaneously written on the input plane of a LCLV. Another He-Ne laser beam impinges upon Dammann grating 2 and is split into an array of 2 ×2 beams to read out the JPS's in corresponding channels at the same time. The reflected beams pass through a beam splitter and an analyzer-polarizer. FTL2 is placed behind the analyzer-polarizer to form correlations. These correlation outputs are detected with a CCD camera interfaced with a computer through an image board and are displayed on a monitor. Figs. 2(b) and 2(c) present photographs of the same correlation signals displayed on the monitor at low threshold and at high threshold, respectively. In Fig. 2(b) there are two groups of obvious autocorrelations and two groups of diffusible cross correlations. In Fig. 2(c) there are only two groups of obvious autocorrelations, corresponding to the input scenes in Fig. 2(a).

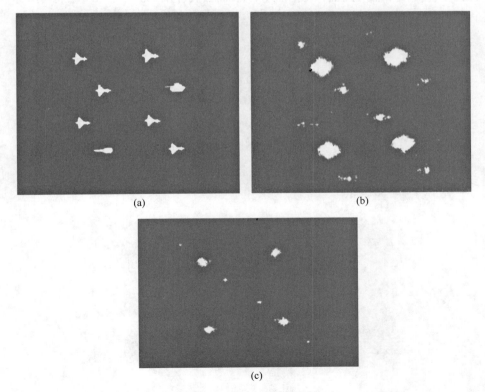

Fig. 2 (a) Four scenes for multiobject detection; (b) correlation output (low threshold); (c) correlation output (high threshold)

In conclusion, we have presented a multichannel joint-transform correlation using a Dammann grating as a beam splitter to form multiple channels. The multiple channels formed by a Dammann grating are independent and have no overlapping parts. Each scene consists of one single reference image and one single target image. In each channel the reference and the target are Fourier transformed. Another Dammann grating is used as a beam splitter to form multiple channels to read out the corresponding joint power spectra. The channels are independent, so the images between different channels will not yield cross correlation. This technique can avoid false alarms caused by cross correlation between multiple targets in multitarget detection, and it can take full advantage of the parallelism of optics to handle high space-bandwidth signals.

References

[1] A. VanderLugt. IEEE Trans. Inf. Theory IT-10,139(1964).
[2] C. J. Weaver,J. W. Goodman. Appl. Opt. 5,1248(1966).
[3] B. Javidi,S. F. Odeh. Opt. Eng. 27,295(1988).
[4] Q. Tang,B. Javidi. Appl. Opt. 32,5079(1993).
[5] F. Cheng,P. Andres,F. T. S. Yu,et al. Appl. Opt. 32,4357(1993).
[6] M. S. Alam,O. Perez,M. A. Karim,Appl. Opt. 32,3102(1993).
[7] H. Dammann,K. Gortler. Opt. Commun. 3,312(1971).
[8] Xueneng Lu,Ying Wang,Minxian Wu,et al. Opt. Commun. 72,157(1989).

Joint Wavelet-transform Correlator for Image Feature Extraction[*]

Abstract We describe a joint wavelet-transform correlator in which the wavelet function is combined with the input image as the input joint image to realize the wavelet transform of the objective image. The Haar wavelet and the Roberts filter are chosen as the wavelet functions to extract the features of the objective image. The relationship of the Haar wavelet and the Roberts filter is analyzed mathematically based on admissible condition of the wavelet. Computer simulations are provided to verify the theory and to illustrate the performance of this correlator.

Key words   wavelet transform; joint transform correlator; feature extraction

1 Introduction

The wavelet transform (WT) is an efficient tool for signal processing such as multiresolution image analysis, data compression and pattern recognition, fractal aggregation analysis, and turbulence phase transience[1]. The wavelet analysis is performed by use of a single prototype function called a wavelet, which can be thought of as a bandpass filter. Fine temporal analysis is done with contracted (high-frequency) versions of the wavelet, while frequency analysis uses dilated (low-frequency) versions. Simply put, the WT of a signal is the inner product between the signal and the dilated wavelets, which is strikingly similar to Vander Lugt matched filters in optics. The wavelet expansion is linear and is a coordinate-doubling-type operation (see Eqs. (3) and (4)) that requires a large number of addition and multiplication combinations. Thus a conventional digital electrical processor may have difficulties implementing this wavelet expansion at high speed.[2] On the other hand the optical implementation takes advantage of massive parallel processing and three-dimensional interconnecting capabilities. Furthermore, Fourier optics can map shift continuously into the lightwave complex phase information that becomes invariant under the square-law intensity detector, while a slight error in the digital computing of shift variables can produce a large error in wavelet coefficients[2]. Recently the importance of applying various optical methods to implement this tool has been recognized. Many types of optical WT processors have been proposed based on the classical optical 4-f matched-filtering system[2-4]. However, it is generally necessary to fabricate a bank of matched filters and to realign the filters in the following filtering operations. Furthermore, it is difficult to realize the wavelet transform in real time and to display the four-dimensional (4-D) WT coefficients of a two-dimensional (2-D) image. All these drawbacks may be overcome by a joint transform correlator

---

[*] Copartner: Wenlu Wang, Yingbai Yan, and Minxian Wu. Reprinted from *Applied Optics*, 1995, 34(2):370-376.

(JTC), a technique originally proposed by Weaver and Goodman[5] for optically correlating two functions. Conventionally a JTC is used to recognize the matched pattern, [6] to find the location of an object in a multiobject scene, [7] or to trace the movement of the object. [8] Lu et al. have reported a JTC with a WT, where they used a set of ring-type bandpass filters on the readout path of the JTC to increase the ability for discriminating similar patterns[9].

In this paper we proposed a joint wavelet-transform correlator (JWTC) in which the object is combined with the wavelet functions to form the input joint image. The wavelet transform of the object is observed at the output plane of the JWTC. The chief advantages of the JWTC are that it provides a shift-invariant wavelet transform, it overcomes the high-resolution requirement of the spatial light modulators (SLM's), it is not necessary to produce a bank of complex filters, it is easy to implement with optics, and it is practical to realize an optical wavelet transform in real time. In addition, the accurate optical alignment of the filters required in spatially matched filtering is not necessary in a JWTC. The Haar wavelet[10] and the Roberts numerical gradient filters[7] are chosen as the wavelet functions for extracting different features of input images.

In Subsection 2. A we describe the basic concept of the WT. In Subsection 2. B the details of the proposed JWTC are discussed. The relationship between the Haar wavelet and the Roberts filter is analyzed in Subsection 2. C. In Section 3, computer-simulation results are provided to verify the theory and to illustrate the performance of this system. Finally, conclusions are presented in Section 4.

2 Theory

2.1 Wavelet Transform

An appropriate square integratable function $h(t)$ can be named as a wavelet if it satisfies the admissible condition

$$\int |H(f)|^2 |f|^{-1} df < \infty \tag{1}$$

where $H(f)$ is the Fourier spectrum of $h(t)$. Condition (1) implies that $\int h(t) dt = 0$; a mother wavelet must oscillate to have a zero-integrated area. A complete orthogonal set of daughter wavelets $h_{a,b}(t)$ may be generated from the mother wavelet $h(t)$ by dilation and translation operations:

$$h_{a,b}(t) = \frac{1}{(a)^{1/2}} h\left(\frac{t-b}{a}\right), \quad a,b \in R, a > 0 \tag{2}$$

$h_{a,b}(t)$ is then the basis of the WT. From Eq. (2) it is clear that each daughter wavelet may have a different length (parameter a) and location (parameter b), but all daughter wavelets have shapes identical to that of the mother wavelet $h(t)$. Stretched versions can be used to analyze the low-frequency part of a signal, while contracted versions, whose spectrum concentrates of the high-frequency part, can be used to interpret the high-frequency part of the signal. This satisfies the demand in signal processing in which the higher the signal frequency, the narrower the analysis window, in order to obtain finer resolution in the time domain (or in the spatial domain).

The WT of a signal is defined as the inner product in the Hilbert space of the L^2 norm:

$$W_s(a,b) = <h_{a,b}(t), s(t)> = \frac{1}{(a)^{1/2}} \int h^*\left(\frac{t-b}{a}\right) S(t) \mathrm{d}t \tag{3}$$

as the dilation factor a approaches to zero, the wavelet $h_{a,b}(t)$ becomes more concentrated about $t = b$ by compression, but it still has the same energy as the original function because the factor $1/(a)^{1/2}$ becomes high. $W_s(a, b)$ then displays the small-scale high-frequency features of the signal. This performance is useful in extracting different features of a 2-D image.

The WT of 2-D image $i(x, y)$ is a 4-D function, which can be written in the frequency domain:

$$W_i(a_x, b_x; a_y, b_y) = (a_x a_y)^{1/2} \iint H^*(a_x f_x, a_y f_y) I(f_x, f_y) \times \exp[-j2\pi(b_x f_x + b_y f_y)] \mathrm{d}f_x \mathrm{d}f_y \tag{4}$$

where $I(f_x, f_y)$ and $H(f_x, f_y)$ are Fourier transforms of $i(x, y)$ and $h(x, y)$, respectively. Eq. (4) shows that the WT of an image is the inner product of its Fourier spectrum and the spectrum of the wavelets. This can be implemented in an optical processing system through spatial filtering in the spectrum domain.

2.2 Joint Wavelet-Transform Correlator

The proposed real-time JWTC is shown in Fig. 1, where the dilated wavelets are generated by a computer and displayed on a magnetic-optical SLM, and the input image is detected by a CCD and introduced to the magnetic-optical SLM through a computer. This typical architecture has been reported in Refs. [7] and [8]. However, in the present case we use it to realize the wavelet transform of the input image and to extract the features of the image. Assume that $i(x, y)$ re-

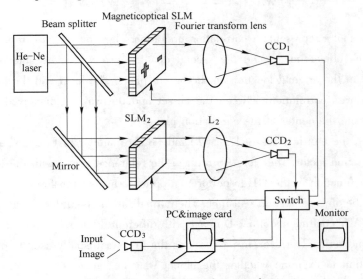

Fig. 1 Joint wavelet-transform correlator

presents the input image and $h(x, y)$ represents the wavelet function. The input joint image $g(x, y)$ can be expressed by

$$g(x,y) = i(x,y) + h\left(\frac{x - b_x}{a_x}, \frac{y - b_y}{a_y}\right) \qquad (5)$$

The corresponding joint power spectrum (JPS) in the Fourier plane of the Fourier lens is

$$G(f_x, f_y) = I(f_x, f_y) + a_x a_y H(a_x f_x, a_y f_y) \times \exp[-j2\pi(b_x f_x + b_y f_y)] \qquad (6)$$

where $G(f_x, f_y)$, $I(f_x, f_y)$, and $H(f_x, f_y)$ are the Fourier transforms of $g(x, y)$, $i(x, y)$, and $h(x, y)$, respectively. The f_x and f_y are mutually independent frequency-domain variables scaled by a factor $2\pi/\lambda f_l$. λ is the wavelength of the collimating light, and f_l is the focal length of the Fourier-transform lens.

The JPS can be detected at the Fourier plane by a square-law device such as a CCD detector array (CCD$_1$ in Fig. 1). The corresponding intensity distribution is given by

$$|G(f_x, f_y)|^2 = |I(f_x, f_y)|^2 + |a_x a_y H(a_x f_x, a_y f_y)|^2 +$$
$$a_x a_y I(f_x, f_y) H^*(a_x f_x, a_y f_y) \exp[j2\pi(b_x f_x + b_y f_y)] +$$
$$a_x a_y I^*(f_x, f_y) H(a_x f_x, a_y f_y) \exp[-j2\pi(b_x f_x + b_y f_y)] \qquad (7)$$

where the asterisk indicates a complex conjugate operation. The JPS is delivered to a second SLM to perform the final inverse Fourier transform, thereby yielding the correlation output $g_c(x, y)$:

$$g_c(x,y) = i(x,y) \star i^*(x,y) + h\left(\frac{x}{a_x}, \frac{y}{a_y}\right) \star h^*\left(\frac{x}{a_x}, \frac{y}{a_y}\right) +$$
$$i(x,y) \star h^*\left(\frac{x - b_x}{a_x}, \frac{y - b_y}{a_y}\right) + i(x,y) \star h^*\left(\frac{x + b_x}{a_x}, \frac{y + b_y}{a_y}\right) \qquad (8)$$

where the symbol $\star$ represents a correlation operation. The first dc term represents the autocorrelation of the object image, and the second dc term represents the autocorrelation of the wavelet located at the original location on the output plane. The third and the fourth terms correspond to the cross correlation between the object and the wavelet displayed at the distance ($\pm b_x, \pm b_y$) from the center of the correlation plane.

The third and the fourth terms in Eqs. (7) and (8) are similar to Eqs. (3) and (4) except for the normalization factor. Therefore we replace the reference signal with the wavelet function or another filter function in the JTC to perform an optical wavelet transform for extracting different features of the object image in real time. The normalization operation is performed by a computer. The 4-D WT coefficients of a 2-D image with different dilated wavelets are displayed on a monitor in a time series. However, in past reports the JTC was usually used to perform matched pattern recognition to examine whether the object was the same as the reference, to perform multiobject recognition to find the location of the object, or to trace a moving object.

2.3 Haar Wavelet and Roberts Filter

Unlike the Fourier transform, the kernel of the WT is not unique. The choice of a transform kernel is highly dependent on the application for which it is intended. The demand on the kernel is that it should satisfy the admissible condition (1). However, in practical application the wavelet should have a good localization property not only in the physical space but also in the Fourier space in order to represents a signal with few WT coefficients and to extract the features effectively. The Haar wavelet was commonly used in early wavelet applications. It is a bipolar step function in one dimension[2]:

$$h(t) = \begin{cases} 1 & 0 \leq t < 1/2 \\ -1 & 1/2 \leq t < 1 \\ 0 & \text{other} \end{cases} \quad (9)$$

which is real and antisymmetric about $t = 1/2$. The combination of one-dimensional Haar wavelets can produce several types of 2-D Haar wavelets for extracting different features, such as corners, vertical edges, and horizontal edges from a 2-D image.[10]

A 2-D corner Haar wavelet and its Fourier spectrum are shown in Fig. 2(a). It is clear that the spectrum amplitude converges to zero very slowly for the irregularity of $h(x, y)$. Therefore from the point of view of a mathematician, the Haar wavelet is of no use as a kernel for analyzing highly continuous images or signals. However, it has been proven that Haar's set is periodical, orthogonal, complete, and satisfies the wavelet admissible condition[11]. A complete and orthogonal Haar-wavelet basis can be constructed by the combination of one-dimensional Haar wavelets with different dilation and translation, which is a 4-D matrix in 2-D display, shown in

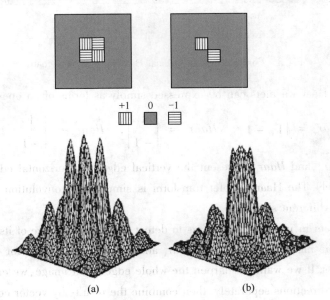

Fig. 2 (a) 2-D corner Haar wavelet and its Fourier spectrum;
(b) 2-D Roberts filter R_+ and its Fourier spectrum

Fig. 3. From Fig. 3 the Haar transform of a 2-D image may induce the sampling from coarse to fine and result in multiresolution analysis of the image. The submatrices in the first vertical column are the horizontal-edge wavelets, the submatrices in the last line are the vertical-edge wavelets, and the others are the corner wavelets with different dilation and translation factors. Furthermore, the Haar wavelet is easy to represent with binary electro-optic devices because of its bipolar character, and it is useful to detect edges and corners. Thus it is used frequently in optical processing systems.

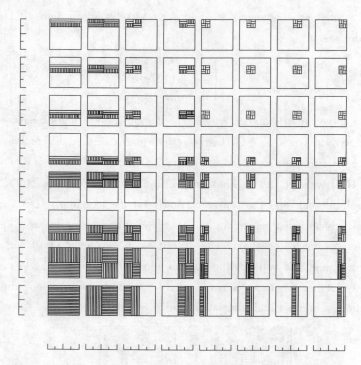

Fig. 3 Complete and orthogonal Haar-wavelet basis matrix

In comparison, Haar wavelets can be expressed simply as forms of an operator:

$$Haar_v = |1, -1|, \quad Haar_h = \begin{vmatrix} 1 \\ -1 \end{vmatrix}, \quad Haar_c = \begin{vmatrix} 1 & -1 \\ -1 & 1 \end{vmatrix} \quad (10)$$

where $Haar_v$, $Haar_h$, and $Haar_c$ represent the vertical edge, the horizontal edge, and the corner wavelet, respectively. The Haar wavelet transform is simply the convolution of the image with Haar operators of different size.

A typical problem in image processing is to detect the abrupt change of its gray scale, revealing the edge of the image. The operators $Haar_v$ and $Haar_h$ are designed for detecting only two special orientations. If we want to sharpen the whole edge of an image, we can filter the image in two orthogonal directions separately, then combine the results by vector computing. The Roberts gradient is efficient in detecting edges. It is performed by convolving an image with two 2×2 kernels to approximate the horizontal and the vertical strength of the edge at each pixel location:

$$R_+ = \begin{vmatrix} 1 & 0 \\ 0 & -1 \end{vmatrix}, \quad R_- = \begin{vmatrix} 0 & 1 \\ -1 & 0 \end{vmatrix} \quad (11)$$

Each pixel in the input image is replaced with the larger of the absolute values of these two operators[7]:

$$i_{\text{edge}}(x,u) = \max\{|R_+ \star i(x,y)|, |R_- \star i(x,y)|\} \quad (12)$$

The Roberts edge kernels are convolved with the input image to create an edge-enhanced or a sharpened image. A 2-D display of R_+ and its Fourier spectrum are shown in Fig. 2(b). Compared with the Haar wavelet (see Fig. 2(a)), the Roberts filter is inversely antisymmetric about its center, it has as good a localization in the space domain as that of the Haar wavelet, its area integration is zero, and it fulfills the admissible condition of a wavelet in condition (1). Its spectrum amplitude also converges to zero slowly. From the above description we can see that the Roberts filter is similar to the Haar wavelet both in the space domain and in the frequency domain. Therefore the Roberts filter can be used as a wavelet function, and we adopt it in the JWTC for extracting the edges of an image.

3 Simulation Results

To verify the theory and to investigate the performance of the proposed JWTC, we use the Haar wavelet and the Roberts filter in our simulation for extracting image features. The simulation procedure is the same as that of the optical JTC experiment. The wavelet function is combined with an objective image to form a 64×64 pixel input joint image $g(x,y)$, where the objective image $i(x,y)$ occupies 16×16 pixels and is located at the center or at a distance of $bx/2$ from the center. The wavelet is larger than 1×2 pixels, is determined by the dilation parameters a_x and a_y, and is located at a distance of b_x (or $b_x/2$) from the center. The other pixels are zero padded.

The simulation tests are performed by use of a 2-D fast Fourier transform. The results are normalized according to Eq. (3) and plotted by use of a three-dimensional subroutine or displayed on a monitor by an image card. Three different images are processed in our JWTC: a cross, a rectangle, and a circle. Fig. 4(a) shows a cross and a corner Haar wavelet as an input joint image, Fig. 4(b) is the simulation result of Fig. 4(a), where the corner Haar wavelet is of 2×2 pixels, and the corners of the cross are extracted clearly at one pixel level. Fig. 4(c) is the simulation result of Fig. 4(a), where the corner Haar wavelet is of 4×4 pixels and the corner points extracted from the cross are larger than that in Fig. 4(b). A cross and a Roberts filter as an input joint image is shown in Fig. 5(a); Fig. 5(b) and Fig. 5(c) are the simulation results of Fig. 5(a) with scaling of the Roberts filter of 2×2 and 434 pixels, respectively. Apparently the edge of the cross in Fig. 5(b) is finer than that in Fig. 5(c), where the edge is 4 pixels wide. These multiscale analysis results are exactly the fundamental property of the wavelet transform.

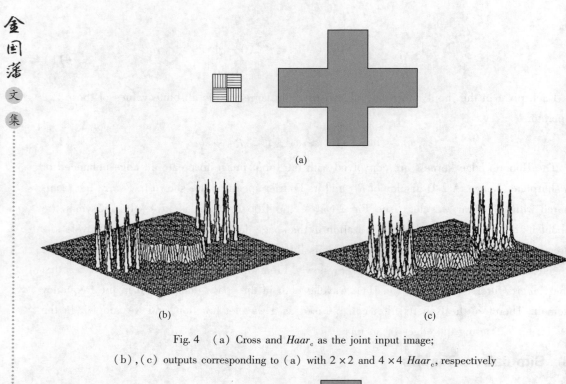

Fig. 4 (a) Cross and $Haar_c$ as the joint input image;
(b),(c) outputs corresponding to (a) with 2×2 and 4×4 $Haar_c$, respectively

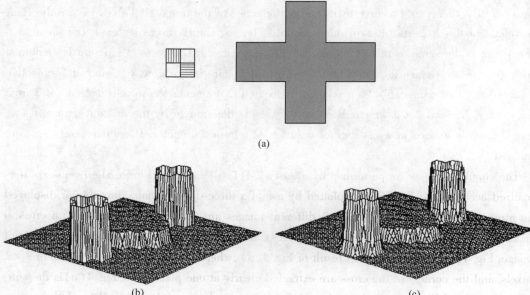

Fig. 5 (a) Cross and R_+ as the joint input image;
(b),(c) outputs corresponding to (a) with 2×2 and 4×4 R_+, respectively

The wavelets $Haar_h$ and $Haar_v$ can extract the horizontal edge and the vertical edge of an image, respectively. We may combine the two directional results to get the whole edge of the image by vector computing, as we stated before. There are shift-invariant properties in the JWTC;

therefore we put $Haar_h$ and $Haar_v$ exactly beside a rectangle as the input joint image, as shown in Fig. 6(a), in order to obtain the superimposed result of these two operators in one operation. Fig. 6(b) is the sharpened image of the rectangle, as we expected. It shows that the combination of the horizontal and vertical Haar wavelets can be used to detect the edge of an image, which is the same as the effect of the Roberts gradient, as shown in Fig. 5(b), and the proposed scheme can achieve the shift-invariant recognition. The cross-correlation peak of $Haar_h$ and $Haar_v$ can also be seen in Fig. 6(b). The Roberts filter R_1 and R_2 are put in the orthogonal directions beside a circle as the input joint image, as shown in Fig. 7(a). The corresponding correlation output is shown in Fig. 7(b), which shows that the Roberts filter R_1 and R_2 are complementing operators and can be used as wavelets for extracting the edge of an image in our JWTC. The combination of the complementing operators according to Eq. (12) can obtain the continuous edge of the input image, which is also verified in our simulation test. Fig. 8(a) shows two 18×18 pixel circles put together with R_1 and R_2 in the orthogonal directions as the input joint image with 128×128 pixels. Fig. 8(b) is the corresponding result to Fig. 8(a) with 2×2

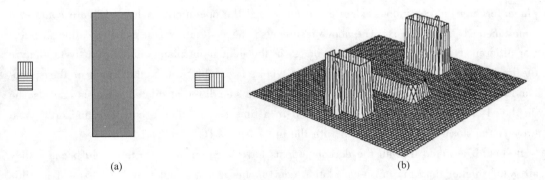

Fig. 6 (a) Rectangle with 1×2 $Haar_v$ and 2×1 $Haar_h$ as the joint input image;
(b) output corresponding to (a)

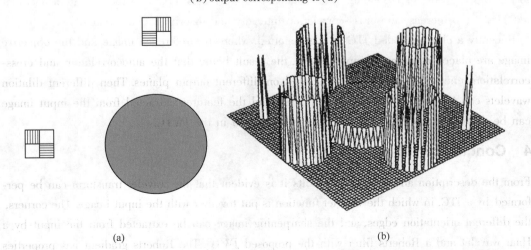

Fig. 7 (a) 2×2 R_+ and R_- combined with a circle as the joint input image;
(b) output corresponding to (a)

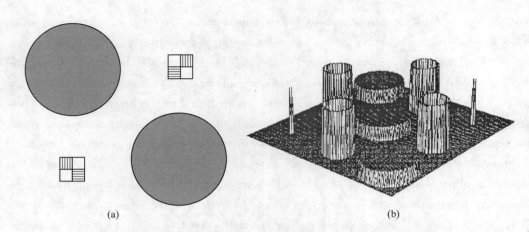

Fig. 8 (a) 2×2 R_+ and R_- and two circles as the joint input image with 128×128 pixels;
(b) output corresponding to (a)

pixels R_1 and R_2, where the results of R_1 and R_2 to the circle are superimposed in the output plane, forming the continuous edge of the circle. All the operations in Eq. (12) are completed simultaneously in the JWTC. The above results also suggest that we can put the scaling wavelets (or different types of wavelet) with an image in the joint input plane simultaneously to analyze the image with multiscale (or with different types of wavelet) within the limits of the space-bandwidth product of the optical system. Needless to say, different dilated wavelets can also be displayed on the input magneto-optical SLM in a time series; thus a complete and orthogonal wavelet transform can be performed with the proposed JWTC.

It should be noted that all the dc components have been cut off by a threshold because they are much higher than the cross-correlation components in the simulations. Therefore it is difficult to perform an optical wavelet transform by a general JTC. Some measures must be adopted to the JPS or to the experimental structure, such as the binary JTC[12] or the amplitude-modulated JTC[13], to increase the cross-correlation intensity and discrimination sensitivity.

Recently a chirp-encoded JTC[14] was reported, where the reference image and the objective image are placed at different input planes, the result being that the autocorrelation and cross-correlation functions of the JTC are focused on different output planes. Then different dilation wavelets can be put at different input planes, and the features extracted from the input image can be detected separately if this structure is adopted in the JWTC.

4 Conclusion

From the description and simulation results it is evident that the wavelet transform can be performed by a JTC, in which the wavelet function is put together with the input image. The corners, the different orientation edges, and the sharpening image can be extracted from the input by a Haar wavelet and a Roberts filter with the proposed JWTC. The Roberts gradient has properties similar to those of the Haar wavelet and can be used as a wavelet function. The whole edge of an image can be extracted by appropriate placement of the $Haar_h$ and $Haar_v$ or the R_+ and R_- with the image in the joint input plane in our proposed JWTC with only one operation. In order to make full use of the space-bandwidth product of the optical system, we can combine the scaling wavelets and/or different types of wavelet functions with an image in the joint input plane of the JWTC to analyze the image simultaneously with multiresolution and/or with multiwavelets.

Different wavelet functions or series of dilated wavelets can be generated by computer, and the orthogonal and complete WT of an image can be realized with the JWTC in real time. Although the dc component is too large to perform optical WT with a classical JTC, the chirp-encoded JTC can separate the cross correlation from the autocorrelation and can focus on separate output planes. Therefore it is practical to extract different features from an input image with the proposed hybrid JWTC in real time.

This research was supported by the 863 High Technology Program and the National Natural Science Foundation of China.

References

[1] J. M. Combes, A. Grossman, and Ph. Tchamitchian, eds., Wavelets, Time-Frequency Methods, and Phase Space (SpringerVerlag, New York, 1989), 68-158.
[2] Special issue on wavelet transform, Opt. Eng. 31, 1821-1916(1992).
[3] E. Freysz, B. Pouligny, F. Argoul, and A. Arneodo, "Optical wavelet transform of fractal aggregates," Phys. Rev. Lett. 64, 745-748(1990).
[4] M. O. Freeman, "Wavelets: signal representations with important advantages," Opt. Photon. News 4(8), 8-14(1993).
[5] C. S. Weaver and J. W. Goodman, "Technique for optically convolving two functions," Appl. Opt. 5, 1248-1249(1966).
[6] O. Perez and M. A. Karim, "An efficient implementation of joint Fourier transform correlation using a modified LCTV," Microwave Opt. Technol. Lett. 2, 193-196(1989).
[7] M. S. Alam, O. Perez, and M. A. Karim, "Preprocessed multiobject joint transform correlator," Appl. Opt. 32, 3102-3107(1993).
[8] F. T. S. Yu and S. Jutamulia, "Hybrid-optical signal processing," in OpticlSignal Processing, Computing, and NeuralNetworks, K. Chang, ed. (Wiley, New York, 1992), 203-248.
[9] X. J. Lu, A. Katz, E. G. Kanterakis, and N. P. Caviris, "Joint transform correlator that uses wavelet transforms," Opt. Lett. 17, 1700-1702(1992).
[10] X. Yang, H. Szu, Y. Sheng, and H. J. Caufield, "Optical Haar wavelet transforms of binary images," Opt. Eng. 31, 1846-1851(1992).
[11] M. Vetterli and C. Herley, "Wavelets and filter banks: theory and design," IEEE Trans. Acoust. Speech Signal Process. 40, 2207-2232(1992).
[12] B. Javidi and C. Kuo, "Joint image transform correlation using a binary spatial light modulator at the Fourier plane," Appl. Opt. 27, 663-665(1988).
[13] D. Feng, H. Zhao, and S. Xia, "Amplitude-modulated JTC for improving correlation discrimination," Opt. Commun. 86, 260-264(1991).
[14] Q. Tang and B. Javidi, "Technique for reducing the redundant and self-correlation terms in joint transform correlators," Appl. Opt. 32, 1911-1918(1993).

实时光学模糊关联记忆神经网络[*]

摘　要　构建了实时光学模糊关联记忆神经网络系统，采用空间面积编码实现了光学模糊逻辑，提出了分时处理技术，为实现更多神经元的网络提供了一条有效途径；并进一步论证了网络作为灵活实时互连功能模块的可行性；给出了模拟运算为实验结果。

关键词　神经网络；模糊逻辑；关联记忆；光互联

1　引言

模糊逻辑和神经网络都基于对人脑思维的模拟，有其固有的相似性[1,2]。理论分析表明[3]：任一连续的多层前馈网络都可以用一个（离散的）模糊输出系统来无限趋近；任一连续的（离散的）模糊输出系统都可以用一个3层前馈网络来无限趋近。模糊理论的引入为光学人工智能和神经网络的研究开辟了一条蹊径，然而利用光学技术实现模糊逻辑和运算尚无成熟的器件和方法。Mada等人用液晶光阀实现了连续值逻辑[4]，刘立人基于多重成像系统，用空间编码后取闭的方法实现了光学模糊逻辑[5]。最近，刘树田、张树群等人分别利用光电子混合和阴影投射法实现了模糊逻辑7种基本运算[6,7]。这些研究为模糊理论在光学领域的应用提供了有益的借鉴。在此基础上，本文以液晶电视为主体构造了一个实时光学模糊关联记忆神经网络系统，着重探讨了网络的联想功能、互连功能和分时处理技术。

2　模糊关联记忆模型

模糊关联记忆（FAM）是由 Kosko 提出的一种能存储任意 $[0,1]$ 间模糊值模式对 (A,B) $A=(a_1,a_2,\cdots,a_n)$，$B=(b_1,b_2,\cdots,b_m)$ 的神经网络[8]，其拓扑结构见图1。图1中 A，B 分别构成输入和输出模式的模糊特征集，W 为互连权重矩阵。

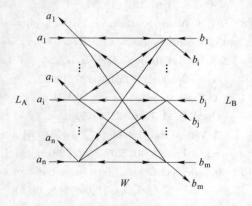

图1　FAM 的拓扑结构

网络的学习和联想是由模糊 Hebb 规则完成的。互连权重矩阵 $W=[\omega_{ij}]_{n\times m}$ 由下式确定：

$$\omega_{ij}=a_i\wedge b_j=\min(a_i,b_j)$$
$$(i=1,2,\cdots,n;j=1,2,\cdots,m) \quad (1)$$

该网络具有双向联想特性，既可由 A 联想出 B，也可由 B 联想出 A。联想方程为：

---

[*] 本文合作者：冯文毅，温志庆，严瑛白。原发表于《红外与毫米波学报》，1995，14(1)：39～46。

(1) 当以 A 联想 B 时，
$$B = AoW \quad (2)$$
即 $$b_j = \bigvee_{i=1}^{n}(a_i \wedge \omega_{ij}) = \max_{i=1}^{n}[\min(a_j,\omega_{ij})] \quad (i=1,2,\cdots,m) \quad (3)$$

(2) 当以 B 联想 A 时，
$$A = BoW^T \quad (4)$$
即 $$a_i = \bigvee_{j=1}^{m}(b_j \wedge \omega_{ij}) = \max_{j=1}^{m}[\min(b_j,\omega_{ij})] \quad (i=1,2,\cdots,n) \quad (5)$$

W^T 为 W 的转置矩阵。

网络还具有一定的容错能力。联想方程(2)与模糊关系方程 $B = XoW$ 具有相同的形式，其中 X 与 A 等价。模糊理论表明[9]：求解所得的 X 值通常在一定范围内均满足关系方程。解的范围越大，意味着吸引域越大，容错和联想能力越强。网络既可实现异联想 ($A \neq B$)，也可实现自联想 ($A = B$)。由于 FAM 网络具有双向联想特性和一定的容错能力，可以采取反馈方式用异联想实现自联想。

若输入模式 $\tilde{A}$ 是 A 的畸变或残缺模式，$W_{(AoB)}$ 是由 (A,B) 模式对决定的互连权重矩阵，则联想过程为 $\tilde{A} \to W_{(AoB)} \to B \to W_{(AoB)} \to A$，如图2所示，该过程等效于自联想 $\tilde{A} \to W_{(AoA)} \to A$。

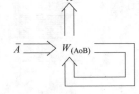

图2 异联想实现自联想示意图

以上讨论的仅存贮 1 对一维模糊矢量 (A,B) 的情形可以推广到二维模糊模式对，一种方法是先将二维模式转化为一维矢量表示，经一维网络处理后，将输出结果转换回二维模式；另一种方法是对二维模式直接以矩阵形式输入，相应的互连权重则由二维矩阵变为四阶张量，输入与权重经三维互连后，获得矩阵形式的输出。

当模糊关联记忆网络中存入多对模糊模式 $\{(A^k,B^k)\}|A^k=(a_1^k,a_2^k,\cdots,a_n^k), B^k=(b_1^k,\cdots,b_2^k,\cdots,b_m^k), k=1,2,\cdots,p\}$ 时，每一对 (A^k,B^k) 的关联记忆矩阵 w^k 按式（1）规则确定，将各个 w^k 按模糊"并"运算合成后，即可获得网络最终互连权重矩阵 $W = [\omega_{ij}]_{n \times m}$，即
$$\omega_i = \bigvee_{k=1}^{p}\omega_{ij}^k = \bigvee_{k=1}^{p}(a_i^k,b_j^k) \quad (i=1,2,\cdots,n, j=,2,\cdots,m) \quad (6)$$

其联想方程与式（2）和式（4）相同。

3 模糊关联记忆的光学实现

模糊关联记忆中的取大-取小运算比普通神经网络中的积和运算更难以光学方法实现。图3是本文提出的基于空间面积编码并以液晶电视为主体构成的实时光学模糊关联记忆神经网络系统。

由于模糊特征值均在 [0，1] 之间，因此可采用图4(a)所示的空间面积编码来表示，即令编码单元的总长度为1，则透光部分的长度便为相应的模糊值。若将图4（a）所示的模糊值 X 和 Y 的编码单元重叠，并以平行光照射，则透光部分长度即为 X、Y 的

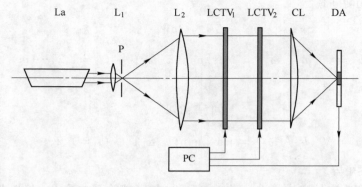

图 3 实时光学模糊关联记忆神经网络光学系统
La—激光器；L—透镜；P—针孔；LCTV—液晶电视；CL—柱面透镜；DA—探测器阵列

取小运算 $X \wedge Y$（见图4(b)）。若以平行光分别照射 X 和 Y 的编码单元，并使两者的透射部分在输出像面重叠，则非全暗区长度即为 X、Y 取大运算 $X \vee Y$（见图4(c)）。

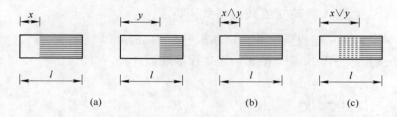

图 4 （a）空间面积编码；（b）取小运算；（c）取大运算

在图 3 系统中，激光经扩束、滤波、准直后，投射到两个平行放置的液晶电视 $LCTV_1$ 和 $LCTV_2$ 上，两者分别作为输入矢量和互连权重矩阵的空间面积编码显示器件。为了充分利用液晶电视的二维可编程特性，将输入的一维矢量矛直接展宽并输入 $LCTV_1$。平行光依次穿过 $LCTV_1$ 与 $LCTV_2$，即实现取小运算，$a_i \wedge \omega_{ij}(i = 1,2,\cdots,n; j = 1,2,\cdots,m)$。后经柱面镜 CL 单向会聚，则可在输出面 DA 上实现取大运算，$\bigvee_{i=1}^{n}(a_i \wedge \omega_{ij}), (j = 1,2,\cdots,m)$。探测器阵列 DA 的各输出编码经单位编码元比较后获得输出矢量，从而完成了联想过程。两个液晶电视的编码输入和探测阵列的比较输出均由一台 PC 机控制。

在我们的实验中，网络系统存贮了一对模糊模式 (A, B)，即 $A = (0.8, 0.4, 0.3, 0.6)$，$B = (0.8, 0.4, 0.6, 0.4)$，则网络的互连权重矩阵 W 为

$$\begin{bmatrix} 0.8 & 0.4 & 0.6 & 0.4 \\ 0.4 & 0.4 & 0.4 & 0.4 \\ 0.3 & 0.3 & 0.3 & 0.3 \\ 0.6 & 0.4 & 0.6 & 0.4 \end{bmatrix}$$

当以模式 A 输入时，可联想出模式 B。当以 A 的相似模式 $\tilde{A} = (1.0\ 0.4\ 0.3\ 1.0)$

输入系统时，仍能联想出 B。图 5 给出了具体实验结果（输出图像在稍离焦面处拍摄），验证了 FAM 算法与本实验系统的可行性。

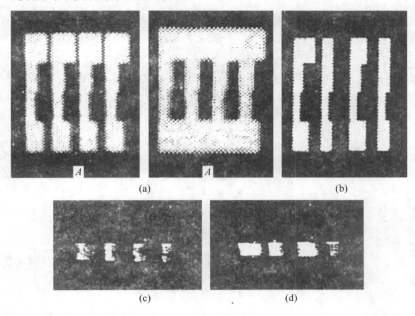

图 5 实验结果

（a）输入模式 A 和 A；（b）权重矩阵 W；（c）A 模式的相应输出模式；（d）$\tilde{A}$ 模式的相应输出模式

该系统充分利用了光的高速并行性，但模糊值的面积编码及送入液晶电视过程仍耗费较长时间，影响了系统的整体处理速度。

4 模糊关联记忆的分时处理技术

要构建一个一维（N）模糊关联记忆网络，需要二维 $N \times N$ 的互连权重矩阵，而构建一个二维（$N \times N$）网络，将需要更大规模的光互连。现有的空间光调制器（SLM），如液晶电视（LCTV）等，还不能满足光互连的过高要求。为了解决网络存贮容量与器件之间的矛盾，本文提出并采用了一种分时处理技术。

如图 3 所示，在 $LCTV_1$ 和 $LCTV_2$ 上分别输入一维矢量 A 的列向展宽矩阵和互连权重矩阵 $W_{\min}$，令

$$L_1 = [A_{n\times 1}, A_{n\times 1}, \cdots, A_{n\times 1}] = \begin{bmatrix} a_1 & \cdots & a_1 \\ \vdots & & \vdots \\ a_n & \cdots & a_n \end{bmatrix}_{n\times m}, L_2 = W_{n\times m}$$

由矩阵代数，对 L_2 进行如下分割，每一个元素都是 $n_i \times m_j (i = 1, \cdots, k)$ 维矩阵，即

$$\begin{bmatrix} W_{n1\times m1} & \cdots & W_{n1\times mk} \\ \vdots & & \vdots \\ W_{nL\times m1} & \cdots & W_{nL\times mk} \end{bmatrix}_{n\times m} \quad n_1 + n_2 + \cdots + n_L = n, m_1 + m_2 + \cdots + m_k = m$$

对应于 L_2，对 L_1 进行同样形式的分割，

$$\begin{bmatrix} A_{n1\times m1} & \cdots & A_{n1\times mk} \\ \vdots & & \vdots \\ A_{nL\times m1} & \cdots & A_{nL\times mk} \end{bmatrix}_{n\times m} \quad n_1+n_2+\cdots+n_L=n, m_1+m_2+\cdots+m_k=m$$

运算过程中的“取小”即对应子矩阵之间的“取小”，各列间“取大”亦即子矩阵按列“取大”后的诸结果再“取大”。数学表述为

$$AoW=\bigvee_{列}(L_1\wedge L_2)=\bigvee_{列}\begin{bmatrix} \bigvee\limits_{列}(A_{n1\times m1}\wedge W_{n1\times m1}) & \cdots & \bigvee\limits_{列}(A_{n1\times mk}\wedge W_{n1\times mk}) \\ \vdots & & \vdots \\ \bigvee\limits_{列}(A_{nL\times m_1}\wedge W_{nL\times m1}) & \cdots & \bigvee\limits_{列}(A_{nL\times mk}\wedge W_{nL\times mk}) \end{bmatrix}$$

$$=\bigvee_{列}\begin{bmatrix} A_{n1\times m1}oW_{n1\times m1} & \cdots & A_{n1\times mk}oW_{n1\times mk} \\ \vdots & & \vdots \\ A_{nL\times m1}oW_{nL\times m1} & \cdots & A_{nL\times mk}\wedge W_{nL\times mk} \end{bmatrix}=\bigvee_{列}\begin{bmatrix} B^1_{1\times m1} & \cdots & B^1_{1\times mk} \\ \vdots & & \vdots \\ B^L_{1\times m1} & \cdots & B^L_{1\times mk} \end{bmatrix}_{L\times m}$$

$$\overset{*}{=}(b_1,b_2,\cdots,b_m)=B$$

由此可知对应子矩阵之间运算与矩阵相同，对原 5 矩阵进行适当分割后，仍用原光路完成各子矩阵运算，并将其结果依次存于计算机中，待所有矩阵处理完毕，将结果再次编码送入 $LCTV_2$ 中，同时使 $LCTV_1$ 全通，即可获得联想输出。当输入特征元素在一次分时处理后，其子矩阵的存贮结果仍庞大时，可在“＊”等式前再次进行分割和分时处理，但无需进行取小运算，而只按列取大。模拟运算如下：设网络存贮一对模糊模式 (A,B)，即 $A=(0.1, 0.2, 0.3, 0.4, 0.5, 0.6, 0.7, 0.8)$，$B=(0.1, 0.5, 0.2, 0.6, 0.3, 0.7, 0.4, 0.8)$，分时处理过程如图 6 所示。

$$\begin{array}{cc|cc|cc|cc} 0.1 & 0.1 & 0.1 & 0.1 & 0.1 & 0.1 & 0.1 & 0.1 \\ 0.2 & 0.2 & 0.2 & 0.2 & 0.2 & 0.2 & 0.2 & 0.2 \\ \hline 0.3 & 0.3 & 0.3 & 0.3 & 0.3 & 0.3 & 0.3 & 0.3 \\ 0.4 & 0.4 & 0.4 & 0.4 & 0.4 & 0.4 & 0.4 & 0.4 \\ \hline 0.5 & 0.5 & 0.5 & 0.5 & 0.5 & 0.5 & 0.5 & 0.5 \\ 0.6 & 0.6 & 0.6 & 0.6 & 0.6 & 0.6 & 0.6 & 0.6 \\ \hline 0.7 & 0.7 & 0.7 & 0.7 & 0.7 & 0.7 & 0.7 & 0.7 \\ 0.8 & 0.8 & 0.7 & 0.7 & 0.7 & 0.7 & 0.7 & 0.7 \end{array}$$

(a)

$$\begin{array}{cc|cc|cc|cc} 0.1 & 0.1 & 0.1 & 0.1 & 0.1 & 0.1 & 0.1 & 0.1 \\ 0.2 & 0.2 & 0.2 & 0.2 & 0.2 & 0.2 & 0.2 & 0.2 \\ \hline 0.1 & 0.3 & 0.2 & 0.3 & 0.3 & 0.3 & 0.3 & 0.3 \\ 0.1 & 0.4 & 0.2 & 0.4 & 0.3 & 0.4 & 0.4 & 0.4 \\ \hline 0.1 & 0.5 & 0.2 & 0.5 & 0.3 & 0.5 & 0.4 & 0.5 \\ 0.1 & 0.6 & 0.2 & 0.6 & 0.3 & 0.6 & 0.4 & 0.6 \\ \hline 0.1 & 0.6 & 0.2 & 0.6 & 0.3 & 0.7 & 0.4 & 0.7 \\ 0.1 & 0.6 & 0.2 & 0.6 & 0.3 & 0.7 & 0.4 & 0.8 \end{array}$$

(b)

$$\begin{array}{cc|cc|cc|cc} 0.1 & 0.1 & 0.1 & 0.1 & 0.1 & 0.1 & 0.1 & 0.1 \\ 0.1 & 0.2 & 0.2 & 0.2 & 0.2 & 0.2 & 0.2 & 0.2 \\ \hline 0.1 & 0.3 & 0.3 & 0.3 & 0.3 & 0.3 & 0.3 & 0.3 \\ 0.1 & 0.4 & 0.2 & 0.4 & 0.3 & 0.4 & 0.4 & 0.4 \\ \hline 0.1 & 0.5 & 0.2 & 0.5 & 0.3 & 0.5 & 0.4 & 0.5 \\ 0.1 & 0.5 & 0.2 & 0.6 & 0.3 & 0.6 & 0.4 & 0.6 \\ \hline 0.1 & 0.6 & 0.2 & 0.6 & 0.3 & 0.7 & 0.4 & 0.7 \\ 0.1 & 0.6 & 0.2 & 0.6 & 0.3 & 0.7 & 0.4 & 0.8 \end{array}$$

(c)

$$\begin{array}{cc|cc|cc|cc} 0.1 & 0.2 & 0.2 & 0.2 & 0.2 & 0.2 & 0.2 & 0.2 \\ 0.1 & 0.4 & 0.2 & 0.4 & 0.3 & 0.4 & 0.4 & 0.4 \\ \hline 0.1 & 0.5 & 0.2 & 0.6 & 0.3 & 0.6 & 0.4 & 0.6 \\ 0.1 & 0.5 & 0.2 & 0.6 & 0.3 & 0.7 & 0.4 & 0.8 \end{array}$$

(d)

$$\begin{array}{cc|cc|cc|cc} 0.1 & 0.4 & 0.2 & 0.4 & 0.3 & 0.4 & 0.4 & 0.4 \\ 0.1 & 0.5 & 0.2 & 0.6 & 0.3 & 0.7 & 0.4 & 0.8 \end{array}$$

(e)

$$\begin{array}{cc|cc|cc|cc} 0.1 & 0.5 & 0.2 & 0.6 & 0.3 & 0.7 & 0.4 & 0.6 \end{array}$$

(f)

图 6　分时处理模拟运算结果

(a) L_1；(b) L_2；(c) $L_1\wedge L_2$；(d) 一次分割结果；(e) 二次分割结果；(f) 三次分割结果（输出模式）

设图元个数为 2^N，则进行 N 次分割（子矩阵为 2×2）分时处理，即可得到联想结果，而光路中只需 2×2 个编码单元实现互连，这种分时处理算法在一定程度上类似于快速傅里叶变换（FFT）。当然这种方法是以牺牲运算速度为代价的，但在目前缺乏大容量实时空间光调制器的情形下，仍不失为一种较好的途径。

5 模糊关联记忆的可编程互连功能

图 3 所示实时模糊关联记忆神经网络系统不仅具有较强的联想功能，而且具有灵活的互连特性，尤其是对混洗操作方式，通常的混洗光路只能按一定规则进行固定互连，而基于 FAM 的实时系统则可按多种规则进行任意互连。其理论依据为：如果存贮的模糊模式对 (A,B) 中的 A 是 B 经任意次序重排后的矢量，则异联想 $A \Leftrightarrow B$ 总可以正确进行。对这一命题可作如下证明：令 $A = (a_1, a_2, \cdots, a_i, \cdots, a_n)$，经任意次序重排后，得 $B = (b_1, b_2, \cdots, b_j, \cdots, b_n)$，设 A 中任一元素 a_i 重排后放置在 B 中任一位置 b_j 上，即 $a_i = b_j$（i,j 取 $1 \sim n$ 中任意值）由 Hebb 规则，权重矩阵为

$$W = \begin{bmatrix} a_1 \wedge b_1 & a_1 \wedge b_2 & \cdots & a_1 \wedge b_j & \cdots & a_1 \wedge b_n \\ \vdots & & & & & \vdots \\ a_i \wedge b_1 & a_i \wedge b_2 & \cdots & a_i \wedge b_j & \cdots & a_i \wedge b_n \\ \vdots & & & & & \vdots \\ a_n \wedge b_1 & a_n \wedge b_2 & \cdots & a_n \wedge b_j & \cdots & a_n \wedge b_n \end{bmatrix}$$

由 $A \to B$ 的联想过程为

$$A \circ W = (a_1, a_2, \cdots, a_i, \cdots, a_n) \circ \begin{bmatrix} a_1 \wedge b_1 & a_1 \wedge b_2 & \cdots & a_1 \wedge b_j & \cdots & a_1 \wedge b_n \\ \vdots & & & & & \vdots \\ a_i \wedge b_1 & a_i \wedge b_2 & \cdots & a_i \wedge b_j & \cdots & a_i \wedge b_n \\ \vdots & & & & & \vdots \\ a_n \wedge b_1 & a_n \wedge b_2 & \cdots & a_n \wedge b_j & \cdots & a_n \wedge b_n \end{bmatrix}$$

考察第 j 个输出元素，$\vee (a_1 \wedge (a_1 \wedge b_j) \cdots a_i \wedge (a_i \wedge b_j) \cdots a_n \wedge (a_n \wedge b_j))$，即 $\vee (a_1 \wedge (a_1 \wedge b_j) \cdots b_i \cdots a_n \wedge (a_n \wedge b_j))$，显然有 $a_k \vee (a_k \wedge b_j) \leqslant b_j (k=1,2,\cdots,n)$，必有 $\vee (a_1 \wedge (a_1 \wedge b_j) \cdots b_i \cdots a_n \wedge (a_n \wedge b_j)) = b_j = a_i$。由于 a_i 为 A 中任一元素，b_j 为 B 中任一位置，故原命题成立。

实时 FAM 互连网络较固定互连方式有较大优越性。首先，它具有灵活的可编程互连特性。对一输入序列，若想获得任意次序重排后的互连输出，只需将输入和输出模式离线学习获得的互连权重矩阵送入液晶电视。该过程不改变光路，只编程改变互连权重矩阵，因此具有较大的灵活性。其次，该网络可采用分时处理技术，提高了互连总数，以适应大规模互连的要求。其三，网络具有较强的联想功能和处理模糊信息的能力，从而使互连系统具有一定的容错性和鲁棒性，在外界干扰和器件制造误差等不确定因素引起的模糊性影响下，仍能保证互连的正确性。

6 结语

模糊理论和神经网络的结合是当前最引人注目的动向之一,本文研究的特性初步展示了这种结合的优越性。FAM 网络是最具代表性的一种模糊神经网络,其光学实现具有一定的开拓性。本文在其关联记忆的基础上开发的互连功能,FAM 互连网络灵活可靠,经小型化、集成化后,可以获得一个可编程模糊互连功能模块,缓解了模糊光学硬件缺乏的矛盾,具有较为广泛的应用前景。

参 考 文 献

[1] Witold Pedrycz. Fuzzy Sets and Systems,1993,56(1):1.
[2] 李晓钟. 模糊系统与数学,1990,4(2):53.
[3] Buckley Janes, et al. Fuzzy Sets and Systems,1993,53(2):129.
[4] Mada H. In:Technical Digest of the 1990 International Topical Meeting on Optical Computing,Japan,1990.
[5] Liu L. Opt. Commun.,1989,73:183.
[6] 刘树田,等. 中国激光,1992,19(4):310.
[7] 张树群,等. 中国激光,1993,20(7):520.
[8] Kosko B,Kande A eds. Fuzzy Expert Systems Reading,MA:Addison-wesley,1987:299.
[9] 汪培庄. 模糊集合论及其应用,上海:上海科学技术出版社,1983.

One-step Implementation of the
Optical Hit-miss Transform*

Abstract A spatial-encoding scheme for the one-step implementation of the optical hit-miss transform (HMT) is presented. The rank-order HMT, which yields better performance in the presence of noise and clutter, can also be carried out by the use of this scheme. Numerical simulations and experimental results have proved the scheme to be feasible.

1 Introduction

Morphological transformations are useful in many problems of image processing and computer vision[1-4]. The morphological hit-miss transform (HMT) is a basic morphological transform that can be used to perform pattern recognition efficiently[2-4]. Casasent et al[5,6] and we[7] have used a coherent and an incoherent optical correlator, respectively, to realize the optical HMT operation based on the definition of the HMT. The research can be regarded as an improvement of the pattern-recognition capability of a coherent or an incoherent optical correlator. The two HMT implementations both need to be executed for at least three operations (two erosions and an intersection) sequentially.

To realize a HMT in one step, a two-opticalcorrelator scheme may be used. We have used two photorefractive correlators to perform a HMT for the implementation of holographic associative memory with accurate addressing[8,9]. Along this direction, Yao et al[10] have presented a HMT implementation that uses two coherent optical convolvers to provide two parallel-processing channels. The two-opticalcorrelator scheme needs double optical elements, but then the optical systems are very cumbersome. Recently, Liu[11] has proposed a method that uses only one optical correlator to perform a two-channel correlation.

For implementing the morphological HMT in one step with only one optical correlator in a much simpler way, we present in this paper a scheme based on a spatial-encoding method[12-22] that has been widely used for digital optical computing. Spatial encoding was first proposed by Tanida and Ichioka[12] for performing optical logic-array processing. Since then, spatial encoding has become a useful method for optical logic-array operations.[13,14] To perform symbolic substitution, which is a promising architecture of optical digital computing, Brenner et al[15] proposed a dual-rail encoding technique (called complementary encoding by us) to perform pattern searching. The technique has been improved to prevent cross talk in optical symbolic substitution[16]. Dual-rail complementary encoding has been mentioned and used in various optical

---

* Copartner: Shifu Yuan, Minxian Wu, Yingbai Yan. Reprinted from *Applied Optics*, 1996, 35(35):6881-6887.

symbolic-substitution and cellular-logic systems[17-22]. However, all these systems confined the dual-rail complementary-encoding technique to the search of simple patterns in symbolic substitution for performing digital optical computing. In fact, dual-rail complementary encoding can be widely used in intensity correlation recognition. In this paper, we use the dual-rail complementary-encoding technique to perform the morphological HMT operation for multi object recognition. We show that the dual-rail complementary-encoding correlation is a HMT operation and propose that the rank-order HMT can also be implemented on the basis of dual-rail complementary encoding for fault-tolerant recognition. We also give numerical simulations and experimental results to demonstrate the suggested optical HMT scheme.

2 Morphological HMT Based on Dual-Rail Complementary Encoding

We denote a binary image by X and its complementary image by $\overline{X}$. The standard HMT of a binary image X by a binary image P that is defined by a binary image pair (F, B) is denoted as

$$X \circledast P = (X \ominus F) \cap (\overline{X} \ominus B) \tag{1}$$

where the symbol ($\circledast$) denotes the HMT operation, the symbol ($\ominus$) denotes erosion, and the symbol ($\cap$) denotes intersection. The binary image pair (F, B) is generally defined as

$$F = P \tag{2}$$

$$B = M \cap \overline{P} \tag{3}$$

where M is a binary image mask and P is the binary image to be recognized.[3] Images P and M define the binary image pair (F, B), in which F and B are the foreground and background of the pattern P, respectively. Fig. 1 shows some binary images according to the preceding definitions. Fig. 1(a) is an original image X that contains four patterns (the characters T, L, O, and

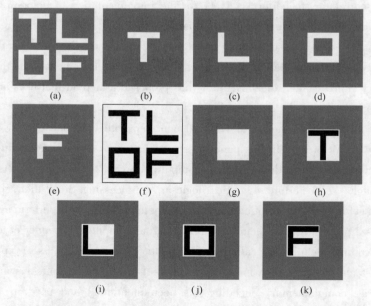

Fig. 1 Some binary images for the definition of a standard HMT
(a)X; (b)P_1; (c)P_2; (d)P_3; (e)P_4; (f)$\overline{X}$; (g)M; (h)B_1; (i)B_2; (j)B_3; (k)B_4

F). Figs. 1(b)-1(e) indicate four foreground structuring-element images P_1, P_2, P_3, and P_4, respectively. Fig. 1(f) represents the complementary image $\overline{X}$ of image X. Fig. 1(g) shows the binary image mask M. Figs. 1(h)-1(k) illustrate the corresponding background structuring-element images B_1, B_2, B_3, and B_4, respectively.

The HMT of an image X by a pattern $P = (F, B)$ recognizes the pattern P in three steps: (1) recognizing the foreground of pattern P by the use of the erosion operation $X \ominus F$, (2) recognizing the background of pattern P by the use of the erosion operation $\overline{X} \ominus B$, and (3) obtaining the final recognition result by means of the intersection operation. The reason that the HMT needs to perform two erosions and an intersection is that the background information is not included in the original image. Dual-rail complementary encoding is a useful spatial-encoding method for digital optical computing[12-22]. By the adoption of the method to code the input images and the structuring elements, both the background image and the foreground image are included in the coded image, so that the HMT can be performed with only one erosion operation.

In dual-rail complementary encoding each pixel is represented by two cells. Fig. 2 shows the

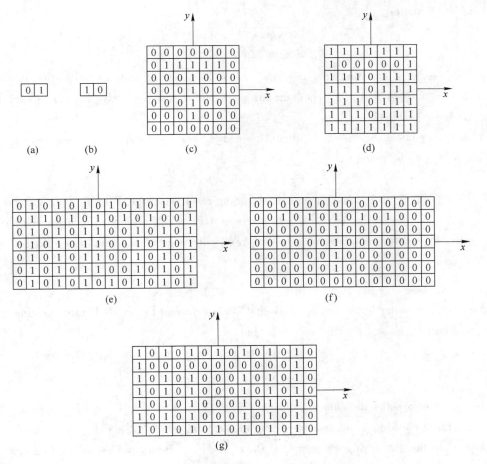

Fig. 2 Dual-rail complementary-encoding technique
(a)0;(b)1;(c)I;(d)$\bar{I}$;(e)I_e;(f)I_{fe};(g)I_{be}

dual-rail complementary codes and an encoding example for a binary image. Pixels with a value of 0 are represented with the code shown in Fig. 2(a), in which the left-hand cell is dark (0) and the right-hand cell is bright (1). In contrast to the code shown in Fig. 2(a), pixels with a value of 1 are represented with the code shown in Fig. 2(b), in which the left-hand cell is bright (1) and the right-hand cell is dark (0). The complementary-encoding process for each pixel of an original image I can be denoted as $CE(I)$. Fig. 2(c) illustrates a 7×7 pixel binary image I to be encoded, and Fig. 2(d) shows the complementary image $\bar{I}$. Fig. 2(e) exhibits the dual-rail complementary-encoding resultant image $CE(I)$ of the image I shown in Fig. 2(c). By the use of the binary image algebra (BIA) image-representation method proposed by Huang et al.,[3] the encoded image I_e can be represented as

$$CE(I) = I_{fe} \cup (I_{be} \oplus A) \tag{4}$$

where the symbol $\cup$ denotes the union operation, A is an elementary image that consists of only one image point (pixels with a value of 1) to the right of the origin, i.e., $A = \{1, 0\}$, and I_{fe} and I_{be} are the coded foreground and background of image I, respectively. The images I_{fe} and I_{be} have the forms of

$$I_{fe} = \{(m, 2n) \mid (m, n) \in I\} \tag{5}$$

$$I_{be} = \{(m, 2n) \mid (m, n) \notin I\} \tag{6}$$

$I_{be} \oplus A$ is an image-shifting operation that makes the image I_{be} shift a pixel to the right along the x axis. Figs. 2(f) and 2(g) show the images I_{fe} and I_{be}, respectively, that correspond to the image I shown in Fig. 2(c).

For an input image X, each pixel should be coded, then the complementary-encoding resultant image X_e can be written as

$$X_e = CE(X) = X_{fe} \cup (X_{be} \oplus A) \tag{7}$$

Fig. 3(a) illustrates the complementary-encoding resultant image X_e of the image X shown in Fig. 1(a). For a structuring element, only pixels in the mask M are coded with the complementary codes. Thus the complementary-encoding resultant image P_e of structuring element P can be denoted as

$$P_e = CE(P) \cap M_e \tag{8}$$

where M_e is another binary image mask that is twice size of the mask M in the x dimension. According to Eq. (4), P_e can be represented as

$$P_e = [P_{fe} \oplus A \cup (P_{be})] \cap M_e = (P_{fe} \cap M_e) \cup [(P_{be} \oplus A) \cap M_e]$$
$$= P_{fe} \cup [(P_{be} \cap M_e) \oplus A] \tag{9}$$

Fig. 3(b) represents the binary image mask M_e. Figs. 3(c)-3(f) are illustrations of the complementary-encoding resultant structuring elements P_{e1}, P_{e2}, P_{e3}, and P_{e4}, respectively. They correspond to the structuring elements P_1, P_2, P_3, and P_4 shown in Figs. 1(b)-1(e), respectively. We then denote the complementary-encoding HMT as

$$(X_e \ominus P_e) \cap M_d \tag{10}$$

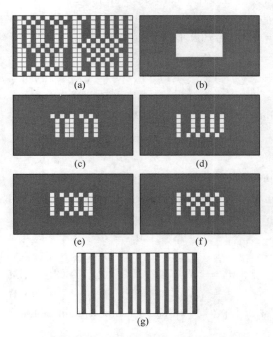

Fig. 3 Complementary-encoding resultant images
(a)X_e; (b)M_e; (c)P_{e1}; (d)P_{e2}; (e)P_{e3}; (f)P_{e4}; (g)M_d

Where M_d denotes a decoding mask, as shown in Figs. 3(g). Substituting Eqs. (7) and (9) for X_e and P_e, respectively, into Eq. (10), we have

$$(X_e \ominus P_e) \cap M_d = ([X_{fe} \cup (X_{be} \oplus A)] \ominus \{P_{fe} \cup [(P_{be} \cap M_e) \oplus A]\}) \cap M_d$$
$$= (\{[X_{fe} \cup (X_{be} \oplus A)] \ominus P_{fe}\} \cap \{[X_{fe} \cup (X_{be} \oplus A)] \ominus$$
$$[(P_{be} \cap M_e) \oplus A]\}) \cap M_d \qquad (11)$$

Generally, $(X \cup Y) \ominus R \neq (X \ominus R) \cup (Y \ominus R)$. However, because of the special encoding technique, here we have

$$[X_{fe} \cup (X_{be} \oplus A)] \ominus P_{fe} = (X_{fe} \ominus P_{fe}) \cup [(X_{be} \oplus A) \ominus P_{fe}]$$
$$= (X_{fe} \ominus P_{fe}) \cup (X_{be} \ominus P_{fe} \oplus A] \qquad (12)$$

$$[X_{fe} \cup (X_{be} \oplus A)] \ominus [(P_{be} \cap M_e) \oplus A]$$
$$= \{X_{fe} \ominus [(P_{be} \cap M_e) \oplus A]\} \cup \{(X_{be} \oplus A) \ominus [(P_{be} \cap M_e) \oplus A]\}$$
$$= \{X_{fe} \ominus [(P_{be} \cap M_e) \oplus A]\} \cup [X_{be} \ominus (P_{be} \cap M_e) \ominus A \oplus A]$$
$$= [X_{fe} \ominus (P_{be} \cap M_e) \ominus A] \cup [X_{be} \ominus (P_{be} \cap M_e)] \qquad (13)$$

Substituting Eqs. (12) and (13) into Eq. (11), we can obtain

$$(X_e \ominus P_e) \cap M_d = (\{(X_{fe} \ominus P_{fe}) \cup (X_{be} \ominus P_{fe} \oplus A)\} \cap \{[X_{fe} \ominus (P_{be} \cap M_e) \ominus A]$$
$$\cup [X_{be} \ominus (P_{be} \cap M_e)]\}) \cap M_d$$
$$= (S_1 \cup S_2 \cup S_3 \cup S_4) \cap M_d \qquad (14)$$

where
$$S_1 = (X_{fe} \ominus P_{fe}) \cap [X_{be} \ominus (P_{be} \cap M_e)] \quad (15)$$
$$S_2 = (X_{fe} \ominus P_{fe}) \cap [X_{fe} \ominus (P_{be} \cap M_e) \ominus A] \quad (16)$$
$$S_3 = (X_{be} \ominus P_{fe} \oplus A) \cap [X_{fe} \ominus (P_{be} \cap M_e) \ominus A] \quad (17)$$
$$S_4 = (X_{be} \ominus P_{fe} \oplus A) \cap [X_{be} \ominus (P_{be} \cap M_e)] \quad (18)$$

Because of the encoding of images X_{fe}, X_{be}, P_{fe}, and P_{be}, we can see that
$$S_2 = \Phi \quad (19)$$
$$S_4 = \Phi \quad (20)$$
where the symbol Φ represents a null that has no image points. S_3 is the cross-talk term that should be removed by the use of the decoding mark M_d. Because S_3 has image points at only points $(2m+1, n)$, then
$$S_3 \cap M_d = \Phi \quad (21)$$
Accordingly,
$$(X_e \ominus P_{fe}) \cap M_d = (X_{fe} \ominus P_{fe}) \cap [X_{be} \ominus (P_{be} \cap M_e)] \cap M_d$$
$$= (X_{fe} \ominus P_{fe}) \cap [X_{be} \ominus (P_{be} \cap M_e)] \quad (22)$$

Comparing Eq. (22) with Eq. (1), we can see that it is, in fact, a morphological HMT operation, that is,
$$X \circledast P = (X_e \ominus P_e) \cap M_d \quad (23)$$

It should be pointed out that on many occasions the decoding mask is not necessary. Eq. (17) can be rewritten as
$$S_3 = \{X_{fe} \ominus [(P_{be} \cap M_e) \oplus A]\} \cap [X_{be} \ominus (P_{fe} \ominus A)] \quad (24)$$
When
$$[(P_{be} \cap M_e) \oplus A] \cap (P_{fe} \ominus A) \neq \Phi \quad (25)$$
then $S_3 \neq \Phi$, and the decoding mask M_d is not necessary. When
$$[(P_{be} \cap M_e) \oplus A] \cap (P_{fe} \ominus A) \neq \Phi \quad (26)$$
for example, $P_{be} \neq \Phi$ or $P_{fe} \neq \Phi$, the decoding mask M_d may be required for decoding so that correct recognition can be achieved.

The numerical simulation of the complementary-encoding HMT has been performed, and the simulation results are given in Fig. 4. Figs. 4(a)-4(d) show the simulated results of $X_e \ominus P_{e1}$, $X_e \ominus P_{e2}$, $X_e \ominus P_{e3}$, and $X_e \ominus P_{e4}$, respectively. From the simulation results, we can see that com-

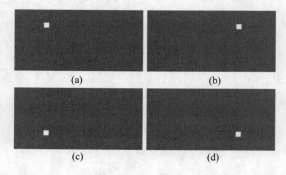

Fig. 4 Simulation results of the complementary-encoding HMT
(a) $X_e \ominus P_{e1}$; (b) $X_e \ominus P_{e2}$; (c) $X_e \ominus P_{e3}$; (d) $X_e \ominus P_{e4}$

plementary-encoding HMT's yield the same results as the standard HMT's.

3 Complementary-Encoding Rank-Order HMT

Casasent and Sturgill[5] have demonstrated a rank-order HMT that gives a better performance in the presence of noise and clutter. The rank-order HMT can also be performed by the use of the dual-rail complementary-encoding scheme. A set-theoretic definition of a binary rank-order filter is

$$X\square_r B = \{a: |X \cap B_a| \geq r\} \quad (27)$$

where the symbol ($\square$) denotes rank-order filtering, $|B|$ is the cardinality of B (the number of image points in the binary image B), and the r is the threshold of the rank order of the filter. We assume that the structuring element B has N image points, i.e., $N = |B|$. Note that if $r = N$, then $X\square_r B$ is the binary erosion $X \ominus B$; for $r = 1$, $X\square_r \check{B}$ is the binary dilation $X \ominus B$, i.e.,

$$X\square_{r=N} B = X \ominus B \quad (28)$$

$$X\square_{r=1} \check{B} = X \oplus B \quad (29)$$

where $\check{B}$ is a reflective image of B. The reflective image $\check{B}$ is defined as

$$\check{B} = \{(-m, -n) | (m,n) \in B\} \quad (30)$$

In complementary-encoding rank-order (CERO) HMT the erosion represented in Eq. (23) is substituted with the rank-order filtering operation. With the threshold q, the complementary-encoding q thrank HMT of X by P results in

$$X \otimes P = (X_e \square_q P_e) \cap M_d \quad (31)$$

where q is the rank order of the complementary-encoding image filter.

If the binary image mask M is assumed to have $n \times n$ image points, the binary image mask Me in complementary encoding has $2n \times n$ image points. No matter how many image points the structuring element P has, the structuring element Pe after complementary encoding always has $N = |Pe| = n \times n$ image points. When $q = N$, the CERO HMT results in the standard HMT.

Because the rank-order HMT can recognize patterns with partial information if there are proper ranks,[6,10] a CERO HMT can also recognize patterns with partial information by the selection of a different threshold q. Thus, the CERO HMT has fault tolerance that can yield a better performance than does the general HMT operation in cases with noise, distortion, and clutter. With the threshold level q the CERO HMT can recognize patterns that have $e\%$ errors, where

$$e = [(n \times n - q)/(n \times n)] \times 100 \quad (32)$$

4 Optical Implementation of the Complementary-encoding Rank-order Hit-miss Transform

Erosion is the basic operation for the implementation of the HMT. Optical implementations of erosion have been widely discussed.[3,5-7,23-25] On the basis of correlation, there are two methods for erosion that can be expressed as

$$X \ominus R = (X * \overset{\vee}{R})|_{T=N} = (X \star R)|_{T=N} \qquad (33)$$

$$X \ominus R = \overline{\overline{X} \oplus \overset{\vee}{R}} = \overline{\overline{X} * \overset{\vee}{R}}|_{T=1} = \overline{\overline{X} \star \overset{\vee}{R}}|_{T=1} \qquad (34)$$

where the asterisk denotes convolution, the star denotes correlation, T is a threshold value, and N is the cardinality of the structuring element R. Accordingly, the rank-order filter can also be expressed with the methods in Eqs. (33) and (34):

$$X_e \square_q P_e = (X_e \star P_e)|_{T=q} \qquad (35)$$

$$X_e \square_q P_e = \overline{(\overline{X}_e \star P_e)}|_{T=N-q+1} \qquad (36)$$

Then the CERO HMT has the following two forms:

$$X \circledast P = (X_e \star P_e)_{T=q} \cap M_d \qquad (37)$$

$$\overline{X \circledast P} = (\overline{X}_e \star P_e)_{T=N-q+1} \cup M_d \qquad (38)$$

respectively. Using Eq. (37) to perform pattern recognition yields the recognition result determined by pixels with a value of 1, whereas with Eq. (38) the recognition result is represented with pixels with a value of 0.

The optical setup for the complementary-encoding HMT is shown in Fig. 5, which has been used to perform the HMT by us[7] and by Liu[11]. Here we prefer to use the method described by Eq. (37). Because the decoding mask M_d is not necessary in our experiments, no decoding mask was used for the data shown in Fig. 5. A diffuse screen (G) is placed just before the input plane (IP). An incoherent, uniform light source (S) is used to illuminate the diffuse screen (G). The structuring-element mask (M) is put close in front of the imaging lens (L). At the back focal plane of lens L there is a correlation of the input image and the structuring element.[23-25]

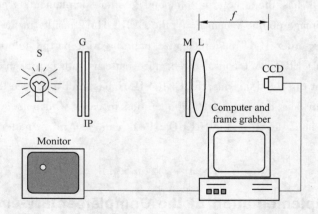

Fig. 5 Optical setup for the complementary-encoding HMT

In our experiments, the complementary-encoding image X_e, shown in Fig. 3(a), is used as the input image for recognition; the coded structuring elements P_{e1}, P_{e2}, P_{e3} and P_{e4} are the corresponding patterns to be recognized. The correlation image of input image X_e and structuring element P_e is formed at the back focal plane of lens L and acquired by a CCD camera attached to the frame grabber. Performing electrical thresholding with a proper threshold level permits the CERO HMT result to be obtained. The correlation image acquired by the frame grabber is a digital gray-scale image with 256 gray levels. To implement the HMT, we choose here a threshold value of $q = 145$. Fig. 6(a) shows the experimental result of $(X_e \star P_{e1})\mid_{T=145}$, which is the final HMT result of $X \circledast P_1$. Figs. 6(b)-6(d) demonstrate the results of $(X_e \star P_{e2})\mid_{T=145}$, $(X_e \star P_{e3})\mid_{T=145}$, and $(X_e \star P_{e4})\mid_{T=145}$, respectively, and these are the results of the corresponding HMT operations. From these results we can see that the four characters are recognized.

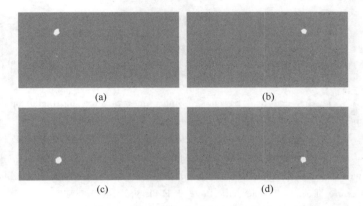

Fig. 6　Experimental results of the complementary-encoding HMT
(a) $X_e \star P_{e1}\mid_{T=145}$; (b) $X_e \star P_{e2}\mid_{T=145}$; (c) $X_e \star P_{e3}\mid_{T=145}$; (d) $X_e \star P_{e4}\mid_{T=145}$

5　Discussion and Conclusion

In comparison with the existing schemes for the HMT implementation, the complementary-encoding HMT has some obvious advantages over the others. The complementary-encoding HMT can perform the HMT easily, with only one optical correlator in one optical iteration step. The complementary-encoding HMT needs to code input images and structuring elements before performing optical correlation, which then seems to be an additional task. However, complementary encoding is a very easy task for electrical hardware and can be realized rapidly at just the same time as an input scene is acquired. In other HMT schemes, although complementary encoding is not necessary, there is another complement operation that should be performed to obtain the complementary images. In addition, decoding, which possibly is needed on some occasions, is also an easy task for optical implementation. Decoding is an intersection operation that can be performed with a decoding mask in the correlation-detection plane. Thus encoding and decoding have no influence on the processing rate or the processing complexity. As a result of complementary encoding, the processing area of an augmented image is doubled, so that the optical-system volume is doubled. Because of the parallelism of optics, an increase in the processing-image area may make full use of the inherent parallelism of optical processing.

In conclusion, we have presented a novel scheme for the one-step implementation of the optical HMT that uses a dual-rail complementary-encoding technique. The rank-order HMT, which provides better performance in cases with noise and clutter, can also be performed with this scheme. By the use of an incoherent optical correlator, pattern recognitions based on the morphological HMT are demonstrated and the experimental results are given. Although an incoherent optical correlator has been used, the complementary-encoding HMT can also be performed by the use of a coherent optical correlator. The complementary-encoding HMT can be widely used in correlation-pattern recognition, such as holographic associative memory and joint-transform correlation.

References

[1] J. Serra, *Image Analysis and Mathematical Morphology* (Academic, New York, 1982).

[2] P. Maragos, "Tutorial on advances in morphological image processing and analysis," Opt. Eng. 26, 623-632(1987).

[3] K. S. Huang, B. K. Jenkins, and A. A. Sawchuk, "Binary image algebra and digital optical cellular image processor design," Comput. Vision Graphics Image Process. 45, 295-345(1989).

[4] E. R. Dougherty, *An Introduction to Morphological Image Processing*, Vol. TT09 of SPIE Tutorial Text Series (SPIE Optical Engineering Press, Bellingham, Wash., 1992).

[5] D. Casasent and R. Sturgill, "Optical hit-miss morphological transforms for ATR," in *Applications of Digital Image Processing* XIII, A. G. Tescher, ed., Proc. SPIE 1153, 500-510(1990).

[6] D. Casasent, R. Schaefer, and R. Sturgill, "Optical hit-miss morphological transform," Appl. Opt. 31, 6255-6263(1992).

[7] S. Yuan, L. Chen, and J. Hong, "Pattern recognition based on morphological transformations and its optical implementation," in *Optics, Illumination, and Image Sensing for Machine Vision* VII, D. J. Svetkoff, ed., Proc. SPIE 1822, 50-58(1992).

[8] S. Yuan, J. Zhang, X. Zhang, K. Xu, and L. Chen, "Addressing accuracy improvement of holographic associative memory," in *Practical Holography* IX, S. A. Benton, ed., Proc. SPIE 2176, 289-294(1994).

[9] S. Yuan, G. Jin, M. Wu, Y. Yan, J. Zhang, K. Xu, and L. Chen, "Holographic associative memory with accurate addressing," Opt. Eng. 34, 2115-2119(1995).

[10] Z. Yao, M. Wu, G. Jin, G. Huang, and Y. Yan, "New optoelectronic morphological scheme for multi-object recognition," Opt. Eng. 33, 3727-3732(1994).

[11] L. Liu, "Morphological hit-or-miss transform for binary and gray-tone image processing and its optical implementation," Opt. Eng. 33, 3447-3455(1994).

[12] J. Tanida and Y. Ichioka, "Optical logic array processor using shadowgrams," J. Opt. Soc. Am. 73, 800-809(1983).

[13] T. Yatagai, "Cellular logic architectures for optical computers," Appl. Opt. 25, 1571-1577(1986).

[14] J. Tanida, M. Iwata, and Y. Ichioka, "Extended coding for optical array logic," Appl. Opt. 33, 3663-3669 (1994).

[15] K. H. Brenner, A. Huang, and N. Streibl, "Digital optical computing with symbolic substitution," Appl. Opt. 25, 3054-3060(1986).

[16] H.-I. Jeon, M. A. G. Abushagur, A. A. Sawchuk, and B. K. Jenkins, "Digital optical processor based on symbolic substitution using holographic matched filtering," Appl. Opt. 29, (1990).

[17] A. K. Cherri and M. A. Karim, "Symbolic substitution-based operations using holograms: multiplication and histogram equalization," Opt. Eng. 28,638-642(1989).
[18] R. Thalmann, G. Pedrini, and K. J. Weible, "Optical symbolic substitution using diffraction gratings", Appl. Opt. 29,2126-2134(1990).
[19] G. Eichmann, A. Kostrzewski, D. H. Kim, and Y. Li, "Optical higher-order symbolic recognition", Appl. Opt. 29,2135-2147(1990).
[20] S. D. Goodman and W. T. Rhodes, "Symbolic substitution applications to image processing", Appl. Opt. 27,1708-1714(1988).
[21] L. Chen, S. Yuan, S. Qian, and S. Duan, "Complementaryencodingtruth table look-up processing for calculation of complex multinomials", (in Chinese) Chin. J. Lasers 19,63-67(1991).
[22] S. Yuan, S. Zhao, X. Zhang, L. Chen, and J. Hong, "Optical parallel logic gates with a liquid crystal light valve and their applications", Optik(Stuttgart)97,149-159(1994).
[23] S. Yuan, S. Zhao, X. Zhang, and L. Chen, "Optical implementation of binary image morphological transformations using a liquid crystal light valve (LCLV)", in *Computer Vision for Industry*, D. W. Braggins, ed., Proc. SPIE 1989,402-409(1993).
[24] K. S. O'Neill and W. T. Rhodes, "Morphological transformations by hybrid optical-electronic methods", in *Hybrid Image Processing*, D. P. Casasentand A. G. Tescher, eds., Proc. SPIE 638,41-44(1986).
[25] L. Liu, "Optoelectronic implementation of mathematical morphology", Opt. Lett. 14,482-484(1989).

Developed, Binary, Image Processing in a Dual-channel, Optical, Real-time Morphological Processor*

Abstract A developed, binary, image-processing technique is proposed, and a dual-channel, optical, real-time morphological processor is developed. Nine binary image processings can be realized fully in parallel. The measures for compensating scale and rotation distortion for pattern recognition are provided. Some applications of optical, morphological binary image processing are studied and experimental results are listed.

1 Introduction

Mathematical morphology[1] is an efficient algorithm for image parallel processing. With optoelectronic architecture a massive number of parallel operations in morphological image processing can be performed simply. A morphological definition of image algebra is a complete and unified algebraic approach to image processing and image analysis that can be used to represent a broad class of nonlinear and linear operations with a minimal combination of fundamental operations[1-9]. Generally speaking, the smaller the number of fundamental operations and the simpler the hardware architectures, the better the parallelism and the universalism. Binary image algebra (BIA) suggested by Huang et al[2]. is most closely related to morphological image algebra, and any binary image processing (BIP) can be implemented with the appropriate structure elements based on three fundamental operations: complement, union, and dilation. In its optical implementations there must be three hardware units to perform three such independent operations. In one-operation image algebra proposed by Liu[8] there is only one dilation followed by a logic operation. One can perform the logic operation by using an optoelectronic cellular two-layer logic array.

Here we present our efforts to search for a new parallel architecture more suitable to optical implementation and simpler in algebraic representation. A developed, binary, image-processing (DBIP) technique is proposed, which can be performed by two convolutions of parallel superposition followed by thresholding. A dual-channel, optical, real-time morphological processor was constructed, with which nine common binary image processings (CBIP's) can be realized fully in parallel. Using DBIP and the dual-channel, optical, real-time morphological processor, we studied some applications of BIP, such as noise removal, character extraction, and pattern recognition. We present the computer simulations and the experimental results. A novel method of detecting the convolution interlaced minimum is proposed. Based on this method pattern rec-

---

* Copartner: Cuoliang Huang, Minxian Wu, and Yingbai Yan. Reprinted from *Applied Optics*, 1997, 36(23):5675-5681.

ognition can be realized in parallel. The scale distortion of the pattern recognition can be compensated efficiently by adjusting the processor amplification. The scale rate change allowed in our experiment ranges from 0. 5 to 2. Because we changed the thresholding value of the convolution superposition, the rate of rotation dispersion is meaningless. The change of rotation angle allowed ranges from $-25°$ to $+25°$.

2 Principle of Developed Binary Image Processing

BIA is a simple, unified, complete parallel-image-processing theory that comprises three fundamental operations ($\oplus, \cup, -$) and five elementary images (I, A, A^{-1}, B, B^{-1}).[2] Any image can be generated by five elementary images, and any BIP can be formed by three fundamental operations with the appropriate structure elements. The three fundamental operations are expressed as follows:

(1) The complement of image X,

$$\overline{X} = (x,y) \mid (x,y) \in W \wedge (x,y) \notin X$$

where $\wedge$ means and, $\in$ means belongs to, $\notin$ means does not belong to, and W is a universal image that defines the domain of our images. The complement of X is an image whose foreground points are the background points of X and whose background points are the foreground points of X. The complement of X is presented in Fig. 1.

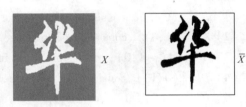

Fig. 1 Complement of image X

(2) The union of two images X and R,

$$X \cup R = (x,y) \mid (x,y) \in X \vee (x,y) \in R$$

where $\vee$ means or. The union of two images X and R is a set with all foreground points of X and R. The union of two images X and R is illustrated in Fig. 2.

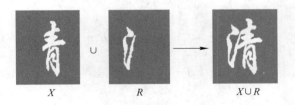

Fig. 2 Union of two images X and R

(3) The dilation of image X by structure element S,

$$X \oplus R = \begin{cases} (x_1+x_2, y_1+y_2) \in W \mid (x_1,y_1) \in X, (x_2,y_2) \in S, (X \neq \varnothing) \wedge (S \neq \varnothing) \\ \varnothing \qquad \qquad \qquad \qquad \text{otherwise} \end{cases}$$

where $\varnothing$ means null, X usually represents an input or data image, and S is a structure element that is a simple binary image. The dilation of nonnull image X by nonnull structure element S is the union of all translation of X by all image points in S, which can increase the size of regions, decrease or fill in holes and cavities, and bridge gaps in X. The dilation of image X by structure element S is shown in Fig. 3.

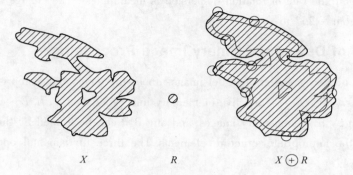

Fig. 3 Dilation of image X by structure element S

The five elementary images are defined as follows:
(1) $I = \{(0,0)\}$, consisting of an image point at the origin;
(2) $A = \{(1,0)\}$, consisting of an image point right of the origin;
(3) $A^{-1} = \{(-1,0)\}$, consisting of an image point left of the origin;
(4) $B = \{(0,1)\}$, consisting of an image point above the origin;
(5) $B^{-1} = \{(0,-1)\}$, consisting of an image point below the origin.

One advantage of morphological image processing is the convenience of optical implementation. In terms of set notation binary image $X = \{(x,j) \mid x(i,j) = 1\}$ corresponds to function $x(i,j)$. If we assume that $s(i,j) = 1$, at and only at n points that correspond to structure element S with n image points, the convolution of $x(i,j)$ and $s(i,j)$ with a thresholding value of $T = 0$ is

$$\begin{aligned} F(X,S) &= X * S \big|_{T=0} \\ &= \{(i,j) \mid \sum_{k,m} x(k,m) \cdot s(i-k,j-m) > 0\} \\ &= \{(i+k,j+m) \mid \sum_{k,m} x(k,m) \cdot s(i,j) > 0\} \\ &= \{(i+k,j+m) \mid (i,j) \in X, (k,m) \in S\} \\ &= X \oplus S \end{aligned} \qquad (1)$$

where $*$ means convolution, the output of the threshold is defined as 1 if $x(i,j) \cdot s(i,j) > 0$ and is 0 otherwise. Eq. (1) indicates that dilation $X \oplus S$ is the same as the addition of a threshold value of $T = 0$ to the convolution sum.

Instead of three fundamental operations in BIA, a DBIP technique with two convolution parallel superpositions followed by thresholding is suggested, which is expressed as

$$F(X_1, X_2, S_1, S_2, T) = [(X_1 * S_1) + (X_2 * S_2)] S \big|_T \qquad (2)$$

where $+$ means an incoherent superposition, X_1 and X_2 are the images to be processed, and S_1

and S_2 are the structure elements. T is the thresholding that is illustrated in Fig. 4, in which $T < 1$ indicates a low-pass filter that retains all pixels whose intensities are less than 1 and removes all others, $T = 0$ is a high-pass filter that retains all pixels whose intensities are greater than (or equal to) 1 and removes all others, $T = K - 1$ is a high-pass filter that retains all pixels whose intensities are greater than (or equal to) K and removes all others, and $K_1 < T < K_2$ is a bandpass filter that retains only the pixels whose intensities are between K_1 and K_2. When appropriate structure elements and thresholding are chosen, Eq. (2) can be used to derive three fundamental operations of BIA:

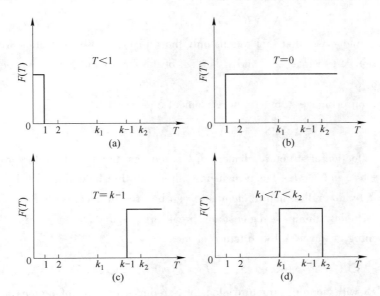

Fig. 4 Diagrams of the threshold formats

(1) When $S_2 = \emptyset$, $S_1 = I$, $X_2 = \emptyset$, and $T < 1$, then

$$F(X_1, X_2, S_1, S_2, T) = [(X_1 * S_1) + (X_2 * S_2)]|_T$$
$$= [(X_1 * I) + (\emptyset * \emptyset)]|_{T<1} \quad (3)$$
$$= (X_1)|_{T<1} = \overline{X_1}$$

where $T < 1$ indicates low thresholding, which is the same as a low-pass filter.

(2) When $S_1 = I$, $S_2 = I$, and $T = 0$, then

$$F(X_1, X_2, S_1, S_2, T) = [(X_1 * S_1) + (X_2 * S_2)]|_T$$
$$= [(X_1 * I) + (X_2 * I)]|_{T=0}$$
$$= (X_1 + X_2)|_{T=0} \quad (4)$$
$$= X_1 \cup X_2$$

(3) When $S_1 = S$, $S_2 = \emptyset$, $X_2 = \emptyset$, and $T = 0$, then

$$F(X_1, X_2, S_1, S_2, T) = [(X_1 * S_1) + (X_2 * S_2)]|_T$$
$$= [(X_1 * S) + (\varnothing * \varnothing)]|_{T=0}$$
$$= (X_1 * S)|_{T=0} \tag{5}$$
$$= X_1 \oplus S$$

Most BIP can be presented by the DBIP technique shown in Eq. (2). In addition to the complement, union, and dilation, the other eight common operations are the following:

(1) The intersection of two images X and R,
$$X \cap R = F(X, R, I, I, T = 1) \tag{6}$$
where $T = 1$ is a high-pass filter that retains only those pixels whose intensities are greater than 1. The intersection of two images X and R is a set of the foreground points that belong not only to X but also to R.

(2) The erosion of image X by structure element S,
$$X \ominus S = F(X, \varnothing, \overset{\vee}{S}, \varnothing, T = k - 1) \tag{7}$$
where $\overset{\vee}{S}$ is the reflection of structure element S, k is the total number of the foreground points of structure element S, and $T = k - 1$ is a high-pass filter with threshold value $k - 1$. The erosion of nonnull image X by nonnull structure element S can be used to decrease the size of regions, increase the size of holes, eliminate regions, and break bridges in X.

(3) The opening of image X by structure element S,
$$X \bigcirc S = F[F(X, \varnothing, \overset{\vee}{S}, \varnothing, T = k - 1), \times \varnothing, S, \varnothing, T = 0] \tag{8}$$
Opening $X \bigcirc S$ with nonnull structure element S reduces the size of regions and eliminates some image points by removing all the features in X that cannot contain structure element S.

(4) The closing of image X by structure element S,
$$X \bullet S = F[F(X, \varnothing, S, \varnothing, T = 0), \varnothing, \overset{\vee}{S}, \varnothing, T = k - 1] \tag{9}$$
Closing $X \bullet S$ with nonnull structure element S increases the size of regions and eliminates some background points by filling in all the background areas that cannot contain structure element S.

(5) The edge detection of image X,
$$\mathrm{MED}(X) = F[X, F(X, \varnothing, I, \varnothing, T < 1), \times S_1, S_2, T = m] \tag{10}$$
where S_1 and S_2 are two structure elements and m is a variable threshold value corresponding to the needs of different types of edge detection, such as inner edge, outer edge, and especially direction edge and multiple-layer edge. The edge detection of image X is used to extract all or part of the profile character of X.

(6) The difference of image X by structure element S,
$$X/S = F[X, F(S, \varnothing, I, \varnothing, T < 1), I, I, T = 1] \tag{11}$$
The difference of image X by structure element S is used to detect defects in the foreground

of tested image X.

(7) The symmetric difference of two images X and R,

$$X \Delta R = F(X, R, I, I, 0 < T < 2) \tag{12}$$

where $0 < T < 2$ is a bandpass filter with a range of threshold value of 0 to 2. The symmetric difference is an obvious approach to detecting defects in the foreground and the background of tested image X.

(8) The pattern recognition of image X by reference object S (as a structure element),

$$\text{MHOM}(X) = F[X, F(X, \varnothing, I, \varnothing, T < 1), \times F(\overset{\vee}{S}, \varnothing, I, \varnothing, T < 1, \overset{\vee}{S}, T < 1)] \tag{13}$$

Eq. (13) is derived from the hit-or-miss operation of BIA but is simpler. The pattern recognition of image X by structure element S is used to search for an interesting object from multiple-object image X, which is the same as reference object S.

With Eq. (2) presenting CBIP, the combination number is greatly reduced, as shown in Table 1.

Table 1 Combination Operation Number of the CBIP in DBIP

Type of Image Processing	Combination Number of Image Processing DBIP	Type of Image Processing	Combination Number of Image Processing DBIP
Union	1	Difference	2 (Including one complement)
Complement	1	Opening	2
Intersection	1	Closing	2
Dilation	1	Symmetric difference	1
Erosion	1	Pattern recognition	3 (Including two complements)
Edge detection	2 (Including one complement)		

The convolution and the incoherent superposition are performed in parallel; their time costs are much smaller, but computer thresholding is time-consuming, so the combination number of every CBIP is considered approximately the same as the number of the thresholding. From Table 1 it is obvious that DBIP is a more suitable parallel architecture for BIP: the combination number is reduced and the speed of image processing is increased significantly. Only one fundamental operation function and a single hardware unit are required.

As a comprehensive comparison, BIA has three operators, and the implementation of a given morphological processing function needs more cascade operation, which makes it more difficult for the optical system.

3 Dual-Channel, Optical, Real-Time Morphological Processor

3.1 Dammann Grating

A Dammann grating[7] is an important device in our dual-channel, optical, real-time morphological processor because it can split an incident light into $n \times n$ arrays uniformly. Each beam in these arrays contains all the information on incident light, but only its intensity is $1/(n \times n)$ of the incident intensity. One application of a Dammann grating is the production of multiple images for the convolution operation in a coherent optical system, which is shown in Fig. 5. Binary image X[shown in Fig. 6(a)] is placed at plane P_1, and its pixel size is Δx. A Dammann grating is placed behind plane P_1. A filter, which is used as image R with 1×3 pixels [shown in Fig. 6(b)], is placed at the spectrum plane P_2 of the system. The pixel size of R is Δr, which equals the distance between two spectrum points of the Dammann grating. The focal lengths L_1 and L_2 are f. Plane P_3 is the convolution output of two images X and R, in which $\Delta = 2f \times \Delta x / (\Delta r - \Delta x)$ (see Appendix A). When a coherent beam illuminates lens L_1, we can obtain the multiple-image projections of X in free space between lens L_2 and plane P_3, shown in Fig. 6(c). The overlapping image part at plane P_3 is the convolution output of two images X and R. Fig. 6(d) shows the morphological dilation operation.

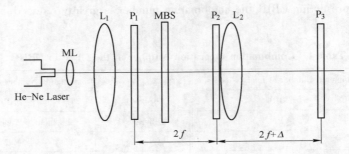

Fig. 5 Implementation of the convolution operation with a Dammann grating: ML, an objective lens; L_1 and L_2, lenses; P_1, P_2, and P_3, planes; MBS, Dammann grating

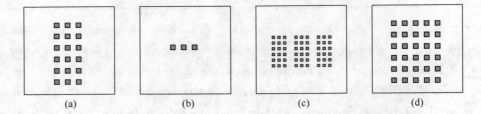

Fig. 6 Experiment of the multiple images with a Dammann grating

3.2 Dual-Channel, Optical, Real-Time Morphological Processor

To perform DBIP efficiently, we constructed a dual-channel, optical, real-time morphological processor, as shown in Fig. 7, which is the hardware unit of the developed fundamental operation in DBIP. Two 33×33 Dammann gratings are used. Liquid crystal TV_1, divided into two parts, is used to input real-time processed images X_1 and X_2. Liquid crystal TV_2 is on the spectrum plane of the processor and is also divided into two parts to input the structure elements S_1 and S_2. The CCD collects the convolution superposition of $X_1 * S_1$ and $X_2 * S_2$ from two channels and puts them into the computer for thresholding. CCD_1 is used to capture an image to be processed in real time. The computer is used to execute the memories, thresholding, feedback, and some other control instructions. The monitor is used to display the processing result.

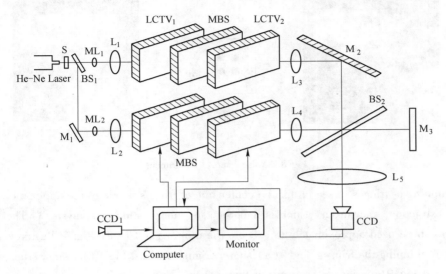

Fig. 7 Dual-channel, optical, real-time morphological processor: S, an attenuator; ML_1 and ML_2, objective lenses; BS_1 and BS_2, beam splitters; L_1, L_2, L_3, L_4, and L_5, lenses; $LCTV_1$ and $LCTV_2$, liquid crystal TV's; MBS, Dammann grating

With the processor shown in Fig. 7, one can easily achieve the parallel architecture of DBIP. When appropriate structure elements S_1 and S_2 and the thresholding are chosen, six DBIP's, such as the union, complement, intersection, dilation, erosion, and symmetric difference, can be performed in a parallel manner by only one step of operation. If one uses computer preprocessing to obtain the complement of X, the edge detection and the difference can also be realized in parallel by only one step of operation. Finally, if the complement of X and the complement of R can be obtained first by computer preprocessing, the pattern recognition to search for an interesting object of image R from a multiple-object scene of image X can also be performed in parallel by only one step of operation.

If the result of dilation is fed back to liquid crystal TV_1 and a subsequent erosion is performed, one can obtain closure. Similarly, opening can be realized by an erosion followed by a dilation.

4 Applications

Using the principle of DBIP and its operation hardware unit—the dual-channel, optical, real-time morphological processor—we can study some applications of BIP, such as noise removal, edge detection, skeleton extraction, and pattern recognition.

The noise removal of a binary image can be realized by using the closing or the opening operation of DBIP. Fig. 8 illustrates an experiment of noise removal. Fig. 8(a) shows a binary image with some concavity noise, Fig. 8(b) shows the chosen structure element, and Fig. 8(c) shows the result of the noise removal from Fig. 8(a).

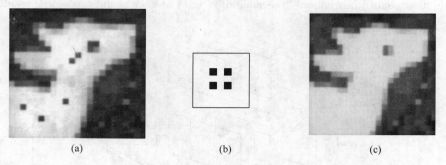

Fig. 8 Noise removal experiment

The thinning of fringes to search for the center line of fringes is one of the applications of the skeleton extraction of an image, which is important for interpolation or analysis.[3] The erosion operation can be used approximately to execute the thinning of fringes. Fig. 9 illustrates the experiment of thinning the fringes. Fig. 9(a) shows original fringes, Fig. 9(b) shows the structure element, and Fig. 9(c) shows the result of thinning the fringes.

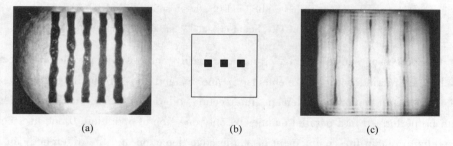

Fig. 9 Thinning the fringes experiment

The edge detection of an image is important to stress the character and make the structure of the image simple and clear. With the dual-channel, optical, real-time morphological processor, the edge detection of an image can be realized in parallel by only one step of operation. In Fig. 10 we present the experimental result of edge detection. Fig. 10(a) shows the original image, and Figs. 10(b) and (c) show the results of edge detection with two different structure el-

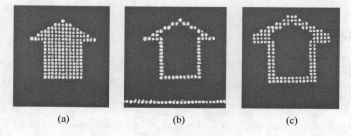

Fig. 10 Edge detection experiment

ements.

Pattern recognition is of importance to machine vision, industrial inspection, radar navigation, and object tracking. By using DBIP and our processor, pattern recognition can be realized by one step of operation. Fig. 11 illustrates the experiment of pattern recognition. Fig. 11(a) shows a three-object scene with noise, Fig. 11(b) shows a reference object, Fig. 11(c) shows another reference object, Fig. 11(d) shows the result of an interesting object recognition that corresponds to the reference object in Fig. 11(b), and Fig. 11(e) shows the result of another interesting object recognition that corresponds to the reference object in Fig. 11(c).

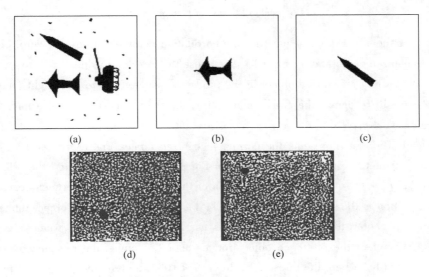

Fig. 11 Pattern recognition experiment

Three distortion invariances of pattern recognition, shift invariance, scale invariance, and rotation invariance, must be considered. Because the convolution of two images has the advantage of shift invariance, in our experiment an object shifting along any direction is recognized without adjusting any position of the processor components. However, if a scale-variant object is to be located, we need to change the amplification of the processor by adjusting the relative position of the Dammann grating, lens L_3, and CCD. Then the effective pixel-sampling period of the reference object in the spectrum plane is changed to match the interesting object of the multiple-object scene. So scale invariance can be achieved by adjusting the processor amplification. Fig. 12 illustrates the experiment for the study of scale distortion. Fig. 12(a) shows a reference object (missile) that is 1.3 times the interesting object (missile) of the scene in Fig. 11(a). Without adjustment it is difficult to discover the position of the interesting object (missile) [see Fig. 12(b)]. After changing the amplification of the processor, we find that the recognized result is the same as that in Fig. 12(c), in which we can easily discover the position of the interesting object (missile). In our experiment the scale range is from 0.5 to 2.

For a rotation-distortion object within a limited angle, we can obtain satisfactory recognition by changing threshold value T. For an object without any variation, the convolution superposi-

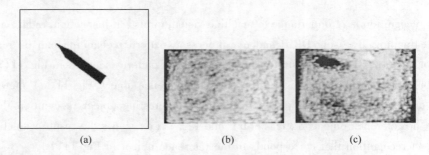

Fig. 12 Scale distortion of pattern recognition

tion of two channels in Eq. (13) must have a correlation point whose level is zero, which indicates the recognized object. So for Eq. (13) we chose the threshold value $T < 1$, and a dark spot represents the correct recognized object in the processed scene. If there is a slight rotation of an object, we can still recognize the correct interesting object by choosing the threshold T slightly greater than 1, such as $T < n$, where n is a positive integer. However, in this case a small dark region instead of a spot represents the correct recognized object. Fig. 13 shows the experiment for the study of rotation distortion. The scene in Fig. 11(a) is still used, and the reference object in Fig. 11(c) is also used. Without rotation of the reference, the experimental result of the pattern recognition is the same as that of Fig. 11(d). When the reference object rotates at a 5° angle around its center, the recognized result is as shown in Fig. 13(a), where the threshold value is $T < 30$. When the reference object rotates at a 15° angle, the recognized result is as shown in Fig. 13(b), where the threshold value is $T < 100$. The reference object rotates again at a 30° angle; the object cannot be recognized yet, and the result of pattern recognition is shown in Fig. 13(c), where the three dark regions correspond to the position of the tank, the plane, and the missile in the scene if threshold value T is greater than 150. Fig. 14 shows the simulated computing results that correspond to the above experimental results, where the x and y coordinates represent the space position of the scene and the z coordinate indicates the relative intensity distribution in the scene. From Fig. 14 it is obvious that, when the reference object rotates at a 30° angle, the relative intensity in the position of three objects of the scene is approximately equal, so it is difficult to recognize the interesting object by changing threshold value T.

The above results indicate that we can efficiently compensate for the slight rotation distortion in the pattern recognition by changing threshold value T.

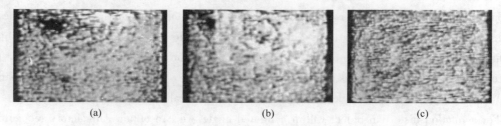

Fig. 13 Rotation distortion of pattern recognition

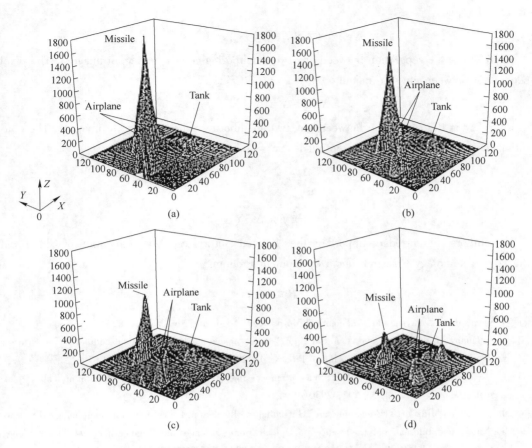

Fig. 14 Simulated computation of the rotation distortion of pattern recognition

5 Conclusion

DBIP is an ideal parallel architecture more suitable to optical implementation and simpler in algebraic representation of BIP. The combination number of the CBIP's is reduced, and the speed of image processing is increased significantly. A dual-channel, optical, real-time morphological processor can perform this DBIP efficiently; any nine CBIP's can be realized in parallel. We can compensate for the scale distortion of pattern recognition by adjusting the amplification of the processor. In our experiment the allowed range of relative rate between the interesting object and the reference object is from 0.5 to 2. The rotation distortion of pattern recognition is meaningless because we changed the threshold value. The allowed range of rotation angle is from $-25°$ to $+25°$.

Appendix A

In Fig. 5 any cell x_1 of image X at plane P_1 produces an imaging cell x_1' at plane P_3, where $x_1'/x_1 = (2 \times f + \Delta)/(2 \times f)$. Cell r_1 of image R at plane P_2 also produces a sample reflected cell r_1', where $r_1'/r_1 = \Delta/(2 \times f)$. The pixel separation between any two neighboring cells, such as x'_1 and x'_2, is

$$\Delta x' = (2 \times f + \Delta) \times \Delta x/(2 \times f)$$

where Δx is the separation between any two neighboring cells x_1 and x_2 in image X. The pixel separation between any two neighboring cells r'_1 and r'_2 is

$$\Delta r' = \Delta \times \Delta r/(2 \times f)$$

where Δr is the separation between any two neighboring cells r_1 and r_2 in image R. The condition for the correct convolution output is $\Delta x' = \Delta r'$. It follows that

$$(2 \times f + \Delta) \times \Delta x/(2 \times f) = \Delta \times \Delta r/(2 \times f)$$

So

$$\Delta = 2 \times f \times \Delta x/(\Delta r - \Delta x)$$

The authors acknowledge support by the National Natural Science Foundation of China and the High Technology Research and Development Program.

References

[1] J. Serra, Image Analysis and Mathematical Morphology (Academic, New York, 1982).

[2] K. S. Huang, B. K. Jenkins, and A. A. Sawchuk, "Binary image algebra and optical cellular logic processor design", Comput. Vision, Graphics, Image Process. 45, 295-345 (1989).

[3] G. Huang, M. Wu, G. Jin, and Y. Yan, "Fast preprocessing interference fringes based on an optical morphological scheme", Opt. Eng. 33, 2927-2930 (1994).

[4] J. Guofan, W. Minxian, Y. Zhongbing, and H. Guoliang, "Incoherent hybrid real time morphological processor", in Second International Conference on Optoelectronic Science and Engineering '94, D.-H. Wang, A. Consortini, and J. B. Breckinridge, eds., Proc. SPIE 2321, 565-567 (1994).

[5] E. R. Dougherty and R. P. Loce, "Optimal mean-absolute-error hit-or-miss filters: morphological representation and estimation of the binary conditional expectation", Opt. Eng. 32, 815-827 (1994).

[6] D. P. Casasent, A. Ye, J.-S. Smokelin, and R. H. Schaefer, "Optical correlation filter fusion for object detection", Opt. Eng. 33, 1757-1766 (1994).

[7] J. Jahns, M. M. Downs, M. E. Prise, N. Streibi, and S. J. Walker, "Dammann gratings for laser beam shaping", Opt. Eng. 28, 1267-1275 (1989).

[8] L. Liu, Z. Zhang, and X. Zhang, "One-operation image algebra and optoelectronic cellular two-layer logic array", J. Opt. Soc. Am. A 11, 1789-1797 (1994).

[9] L. Liu, "Morphological hit-or-miss transform for binary and gray-tone image processing and its optical implementation", Opt. Eng. 33, 3447-3455 (1994).

基于相似性度量的灰度图像光学匹配运算*

摘　要　利用非相干光相关操作处理和空间移位编码方法,实现了基于相似性度量的灰阶图像匹配处理。针对不同程度的噪声干扰情况,分析了空间移位编码方法中字长的选择问题,提高了识别的抗畸变能力,给出了基于非相干相关系统的实验示例。

关键词　相似性度量;空间移位编码方法;光学相关器;光学模式识别;灰阶图像

1 引言

图像识别大体上可以分为特征匹配和图像匹配两种方法。特征匹配的匹配效率高,稳定性好,对图像的各种非本质变化(旋转、缩放和光照强度变化等等)不敏感,是计算机采用的主要方法。图像匹配能利用图像中的所有信息,区分不同对象的能力强,精度高,特别是在十分复杂的图像环境下有效地工作,是光学识别的主要方法。

光学图像识别进行图像匹配处理具有一定的优越性,如并行性、速度快和容量大等,是当前的研究热点[1]。Vanderlut滤波器、联合变换相关器和非相干相关器是3种典型的用于执行图像识别的光学结构,最近几年人们又提出了许多用于模式识别的实时光学相关器[2~10]。相干光的相关器基于傅氏变换和频域处理,结构比较复杂庞大,且有很大的相干噪声,为了实现紧凑型的识别系统,已大量研究非相干光相关系统[8~10]。

在非相干光相关系统中,两幅图像之间的"相关运算"往往无法得到正确的结果。在某些特定情况下,输入图像与参考图像进行相关可以找出相匹配的模式。但是在大多数情况下,直接进行相关会产生误判现象,例如:一个灰阶图像与一幅全白图像的互相关峰一定大于灰阶图像的自相关峰。在二值图像处理中,人们引入了归一化处理[1~5]和形态学击中击不中变换[4~9]有效地解决了这个问题,使得只有目标图像与参考图像达到最近匹配相关峰才达到最大值,但是这两种改进方法在应用到灰阶图像匹配识别时却很麻烦。

本文利用空间移位编码方法在非相关结构上,实现了基于相似性度量的灰阶图像的匹配处理,克服了直接相关出现的误判现象;针对不同程度的噪声干扰情况,分析了空间移位编码方法中字长的选择问题,最后给出计算机模拟和实验结果。

2 基于相似性度量的图像匹配技术和空间移位编码方法

本文用 $X5R$ 来表示输入图像 X 和模板 R 之间的图像匹配运算。通常 X 尺寸大于 R 尺寸。将 R 在 X 中进行全方位的平移,当 R 处于某一位置时,即和相应 X 的窗口图像

---

* 本文合作者:成罡,邹敏贤,何庆声,刘海松,严瑛白。原发表于《红外与毫米波学报》,1998,17(5):369~374。

进行匹配度量操作 Match ( )，输出值就反映了当前位置上的匹配关系，如

$$[X5R](s,t) = \text{Match}(W(X,s,t,p,q),R) \tag{1}$$

式中，$[X5R](s,t)$ 为匹配结果中坐标为 (s,t) 的值；p 和 q 分别为 R 的宽度和高度；$W(X,s,t,p,q)$ 表示 X 上中心点位置为 (s,t) 的窗口图像，其宽度和高度等于 R 的宽度和高度。当模板 R 在整幅图像 X 上搜索一遍后，输出图像就标明 X 的不同位置与 R 匹配的情况。

如何定义灰阶图像之间的匹配度量操作 Match ( ) 是实现图像匹配的关键。本文引入相似性度量 $SM(W,R)$ 以衡量灰阶窗口图像 W 和模板 R 之间的相同性。$SM(W,R)$ 值越大表示 W 与 R 越相似，有

$$SM(W,R) = \sum_{i=1}^{p}\sum_{j=1}^{q} W(i,j) \# R(i,j) \tag{2}$$

式（2）中符号 # 表示两个灰阶像素点之间"模糊相似度关系"，定义为

$$x \# y = \begin{cases} m-n, & |x-y| = n \leq m \\ 0, & |x-y| = n > m \end{cases} \tag{3}$$

式中 m 为给定变量。按照式（3），不同取值的灰阶像素之间的相互关系构成了一个对角线上元素为每行每列中最大值的模糊矩阵 C，并且每行每列均满足三角形隶属函数关系，即

$$C_{iy} = \begin{cases} i/m + (m-y)/m, & y-m \leq i \leq y \\ -i/m + (m+y)/m, & y \leq i \leq y+m \\ 0, & \text{其他} \end{cases}$$

$$C_{y} = \begin{cases} j/m + (m-x)/m, & x-m \leq i \leq x \\ -j/m + (m+x)/m, & x \leq j \leq x+m \\ 0, & \text{其他} \end{cases} \tag{4}$$

式（4）中 C_{iy}，C_{xj} 分别表示矩阵 C 中第 y 和第 x 行的系数。m 的选择对于匹配处理中的抗畸变能力有很大的关系。首先，当需要进行图像 X 和模板 R 之间的精确匹配时，X 和 R 受到噪声等畸变干扰的影响很小，相应像素点的灰度值保持不变，采用 $m=1$ 来实现输入图像和模板之间的点匹配处理，从而模糊关系矩阵 C 转变为一个布尔关系矩阵，这时匹配的精确度最高，但是容错能力最低。其次，当图像和模板之间图像亮度的动态范围有很小程度的变化时，图像和模板之间相应像素点的灰度值可能会有小的差异，可以采用 $m=1$ 或 2 等较小值来改善系统的抗畸变的能力，这时的匹配精确度就降低了。最后，当图像亮度的动态范围有相当程度的变化时，采用 $m=L$ 来改善系统的抗畸变能力，这时的匹配精确度最低，但是容错能力最高。

将相似性度量 $SM(W,R)$ 代入式（1），得到基于相似性度量的灰阶图像匹配表示式

$$[X5R](s,t) = SM(W(X,s,t,p,q),R) \tag{5}$$

如何利用光学相关结构来实现相似性度量，是本文需要解决的关键问题。我们首先提出一种空间移位编码方法，设数字图像 I 的灰阶级次为 L，其中灰阶值为 $i(0 \leq i \leq$

$L-1$)的像素点可以用字长为 $L+m-1$ 的空间移位编码 $[b_{L+m-2} \quad b_{L+m-3} \quad \cdots \quad b_1 b_0]$ 来表示,其中从第 i 位到第 $i+m-1$ 位为1,长度为 m,其余的码位上都是0。例如:灰阶级次为16的图像中,灰度值为10的像素点的空间移位编码方法(字长为19,$m=4$)的表示为 [00 0001 1110 0000 00000]。编码方法中的"0"和"1"表示光线通过调制模板时,分别为"断"和"通"状态。当灰阶像素点 x 和 y 分别采用字长为 $L+m-1$ 的空间移位编码来表示时,得到编码向量 X_e 和 Y_e,则它们之间按位相乘和叠加运算满足如下关系:

$$[X_e] \times [Y_e]^T = \begin{cases} m-n, |x-y| = n \leq m \\ 0, \quad |x-y| = n \geq m \end{cases} \quad (6)$$

从式(3)和式(6)可以看到,采用字长为 $L+m-1$ 的空间移位编码方法,不同取值的灰阶像素点之间按位相乘和叠加运算满足模糊相似度关系。实际上,按位相乘和叠加运算就是光学相关运算。因此,将待处理的灰阶图像采用字长为 $L+m-1$ 的空间移位编码方法来表示,则可以利用光学相关系统实现基于相似性度量的灰阶图像匹配识别。

3 灰阶图像匹配处理

设灰阶图像 X 和模板 R 分别采用字长为 $L+m-1$ 的空间移位编码方法表示,得到编码图像 X_e 和 R_e,则 X 与 R 之间基于相似性度量的灰阶图像匹配识别操作可表示为

$$X5^{SM}R = (X_e \star R_e) \cap M_d \quad (7)$$

式中,5^{SM} 表示基于相似性度量的灰阶图像匹配识别操作;M_d 为取决于编码格式的解码模板。图1为一个基于相似性度量的灰阶图像匹配识别操作的例子,图1(a)和(b)分别给出输入的灰阶图像 X 和灰阶参考图像 R,其中的灰阶层次为6(从0~5),图1(c)给出了字长为8的空间移位编码方法,图1(d)和(e)分别表示编码后得到的图像 X_e 和 R_e,图1(f)为解码模板 M_d,图1(g)给出了匹配识别结果 $(X_e \star R_e) \cap$

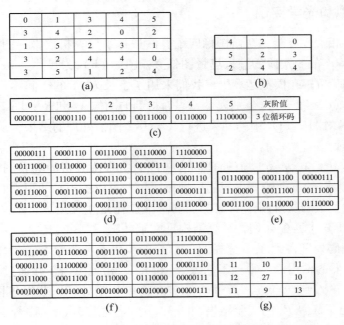

图1 基于相似性度量的图像匹配识别实例

M_d。对输出结果进行阈值分割，就可以反映图像中具有不同程度相关性特征点的位置分布情况，表示为

$$[X5^{SM}R]|_{T=\lambda} = \begin{cases} X5^{SM}R(m,n) = 1, X5^{SM}R(m,n) \geq \lambda \\ X5^{SM}R(m,n) = 0. X5^{SM}R(m,n) < \lambda \end{cases} \quad (8)$$

式中，$[X5^{SM}R]|_{T=\lambda}$ 表示对 $X5^{SM}R$ 的结果进行阈值为 λ 的分割处理，而 $X5^{SM}R(m,n)$ 表示 $X5^{SM}R$ 中位置为 (m,n) 的点的数值，显然，$[X5^{SM}R]|_{T=mpq}$ 得到的是图像 X 和模板 R 的完全匹配结果。匹配结果中只在完全匹配的地方出现最高相关峰，标志着匹配的位置。一般情况下，最高相关峰位于解码板的有效读出位内，为了简化处理步骤，不需要解码处理过程 M_d，可将式（7）和式（8）简化为

$$[X5^{SM}R]|_{T=\lambda} = (X_e \star R_e)|_{T=\lambda} \quad (9)$$

对于灰阶图像匹配，输出结果受到匹配像素数目和像素匹配灰度值两个变量的影响，当图像和模板之间图像亮度的动态范围有很小程度的变化时，即受到噪声等畸变干扰时，图像和模板之间相应像素点的灰度值可能会有小的差异，我们通过降低分割阈值的方法来改善系统的抗畸变能力。当 $\lambda < mpq$ 时，$[X5^{SM}R]|_{T=\lambda}$ 可以匹配具有差异为 e 的图像，其中

$$e = \frac{mpq - \lambda}{mpq} \times 100\% \quad (10)$$

按照待处理图像的具体情况，适当选择式（9）中阈值 λ，可以获得较好的抗畸变干扰性能。虽然通过降低阈值 λ 可以获得较好的抗畸变能力，但是我们必须认识到基于模板匹配方法的抗畸变能力是很有限的，无法和综合判别函数方法相媲美。

4 计算机模拟和光学实现

本文对基于相似性度量的灰阶图像匹配算法进行了计算机模拟。图 2（a）显示了一个尺寸为 64×64 的 8 灰阶的输入场景图像 X；图 2（b）为尺寸为 32×32 的 8 灰阶的待匹配桥梁图像 R；在模拟运算中，我们采用了字长为 11 的空间移位编码方法，见图 2（c）；图 2（d）和（e）分别表示编码后得到的图像 X_e 和 R_e；图 2（f）为模拟结果；图 2（g）为对图 2（f）中的结果进行 $\lambda = 3072$ 阈值处理的结果，从中可以清楚地看到匹配点的位置。

我们同样模拟了 $\lambda < mpq$ 阈值处理的容错能力，将待分析的场景图像 X 加入 30% Gauss 噪声干扰和 8°的旋转畸变，我们发现桥梁仍然可以被正确的识别。

基于相似性度量的灰阶图像匹配处理的基本操作是相关运算，本文选用非相干光相关器实现式（9）中匹配处理，系统结构见文献［8，9］。实验中，带有 8°旋转畸变的图像 X 进行编码处理得到 X_e。选择图 2（f）的编码参考图像 R_e 作为待识别的目标，在非相干光相关器的输出面上得到了输入 X_e 和 R_e 的相关图像，并通过一个 CCD 摄像机和图像采集卡输入到计算机，通过软件实现阈值分割处理，获得了图像匹配结果。在图 3（a）给出了 $X_e \star R_e$ 的实验结果，图 3（b）是对图 3（a）进行阈值分割后得到的结果，从中可以清楚地看到正确的识别结果。

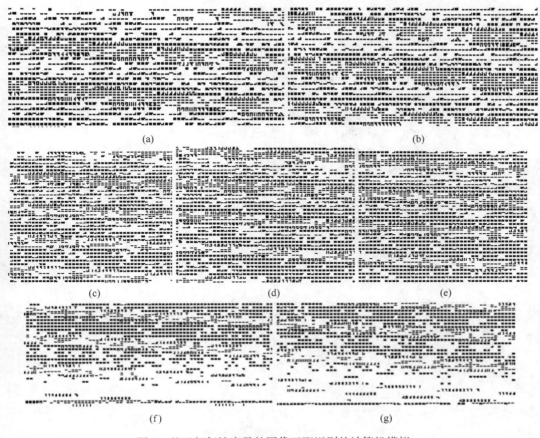

图 2　基于相似性度量的图像匹配识别的计算机模拟

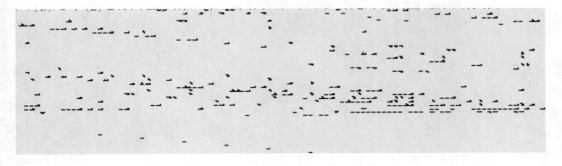

图 3　基于相似性度量的图像匹配识别的光学实验结果

5　结语

我们利用字长为 $L+m-1$ 的空间移位编码方法实现了基于相似性度量的灰阶图像匹配处理，针对不同程度的噪声干扰情况，分析了字长的选择问题，从而提高了灰阶目标识别抵抗各种畸变干扰的能力。本方法适用相干光和非相干光相关系统，本文给出非相干光相关器的实验示例。由于运算中采用的是二值编码形式的循环码，模板的实时写入很容易，可以很方便地构建紧凑型实时灰阶图像识别系统。

参 考 文 献

[1] Goodman J W. Introduction to Fourier Optics. New York McGraw-Hill, 1968.
[2] YU F T, SIU X L. Opt. Commun, 1984, 52(1), 10-16.
[3] Javide B, Horner J L, Walkup J F. Opt Eng., 1994, 33(6), 1752-1756.
[4] CHENG Gang, et al. Acta Optica Sinica（成罡，等. 光学学报），1997, 17(3), 318-324.
[5] Dickey F M, Romero L A. Opt. Lett., 1991, 16(15), 1186-1189.
[6] Casasent D, Sturgill R. Proc. SPIE, 1989, 1153, 500-510.
[7] Casasent D, Schaefer R, Sturgill R. Appl. Opt, 1992, 31(29), 6225-6263.
[8] LIU L. Opt. Eng., 1994, 33(10), 3447-3455.
[9] CHENG Gang, et al. High Technology（成罡，等. 高技术通讯），1997, 3(2), 18-21.
[10] WU Min-Xian, YUAN Shi-Fu, YAN Ying-Bai, et al. Proc. SPIE, 1996, 2751, 264-270.

Volume Holographic Wavelet Correlation Processor*

Abstract A volume holographic wavelet correlation processor is proposed and constructed for correlation identification. It is based on the theory of wavelet transforms and the mechanism of angle-multiplexing volume holographic associative storage in a photorefractive crystal. High parallelism and discrimination are achieved with the system. Our research shows that cross-talk noise is significantly reduced with wavelet filtering preprocessing. Correlation outputs can be expanded from one dimension in a conventional system to two dimensions in our system. As a result, the parallelism is greatly enhanced. Furthermore, several advantages of wavelet transforms in improving the discrimination capability of the system are described. The conventional correlation between two images is replaced by wavelet correlation between main local features extracted by an appropriate wavelet filter, which provides a sharp peak with low side lobes. Theoretical analysis and experimental results are both given to support our conclusions. Its preliminary application to human-face recognition is studied.

1 Introduction

Image-processing systems based on volume holographic storage are becoming increasingly important because of their large storage capacity, high transfer speed, and fast location seeking[1,2]. Such systems fully show the advantages of optical information processing in speed and parallelism. The development of volume holographic image-processingtechniques is driven by two applications. One is data storage and recovery[3,4], and the other is correlation identification[5-9]. The two applications are different in function but similar in many basic problems. With the maturing of some key elements such as storage media, spatial light modulators (SLMs), and detector arrays, it is reported that 10,000 images have been recorded and restored in a single crystal[10]. Development of volume holographic storage systems boosts applications to volume holographic correlation identification in practice. Some practical systems, such as areal-time vehicle navigation system[11], have been proposed and constructed. Once information is stored in a volume-holographic medium, it can be used as a database for pattern recognition or other processing. The essential principle of volume holographic storage is that a certain referencebeam satisfying the Bragg condition can only read out one corresponding image. According to the associative characteristic of volume holographic storage, a set of beams with different transmitting angles as reference beams will be re stored when an object beam modulated by an input image is used to read out the multiple holograms. The intensities of the beams stand for correlation results between the input image and all the recorded patterns. A correlation identification system

---

* Copartner: Wenyi Feng, Yingbai Yan, Minxian Wu, Qingsheng He. Reprinted from *Society of Photo-Optical Instrumentation Engineers*, 2000, 39(9):2444-2450.

based on this framework is capable of instant multichannel parallel processing.

Parallelism and discrimination capabilities are important for a volume holographic correlation identification system. The higher the parallelism, the more correlation peaks will be obtained from a single system output. The higher the discrimination capability, the more accurate the identification will be. Recently, wavelet transforms have been introduced into a volume holographic correlation identification system. A preliminary result shows that a correlation peak becomes sharper with wavelet filtering[12]. Our results show that the introduction of wavelet transforms not only improves the parallelism of the system but also enhances its discrimination at the same time. The volume holographic wavelet correlation processor is described first in Sec. 2 as a basis for analysis. Cross-talk noise of the system is simulated to demonstrate enhancement of parallelism in Sec. 3. In Sec. 4, the utility of wavelet transformation for discrimination improvement is studied by experiments. Preliminary applications of the system to human face recognition are studied in Sec. 5. Conclusions are given in the last section.

2 Volume Holographic Wavelet Correlation System

If two images are described by functions $f(x, y)$ and $s(x, y)$, their wavelet correlation can be defined as

$$W_f(x,y) \otimes W_s(x,y) = [f(x,y) \otimes h_a(x,y)] \otimes [s(x,y) \otimes h_a(x,y)]$$

$$= \int_{-\infty}^{+\infty} \int_{-\infty}^{+\infty} F(u,v) H^*(a_x u, a_y v) \times S^*(u,v) H(a_x u, a_y v) \times$$

$$\exp[i2\pi(xu + yv)] du dv \quad (1)$$

where $\otimes$ is the correlation operator, $h_a(x, y) = [1/(a_x a_y)^{1/2}] h(x/a_x, x/a_y)$ is the function describing a wavelet filter, $a = (a_x, a_y)$ is the dilation factor of the wavelet function, and $F(u, v)$, $S(u, v)$, and $H(a_x u, a_y v)$ are the Fourier transforms of $f(x, y)$, $s(x, y)$, and $h_a(x, y)$. According to the Eq. (1), the wavelet correlation is the correlation between features of two images extracted by the same wavelet filter. The Mexican-hat wavelet function is adopted in the system to fabricate the filter. It is the second derivative of a Gaussian function and also one of the most useful edge detection functions[13]:

$$h(x,y) = [1 - (x^2 + y^2)] \exp\left(-\frac{x^2 + y^2}{2}\right) \quad (2)$$

In the frequency domain, it is

$$H(u,v) = 4\pi^2 (u^2 + v^2) \exp[-2\pi(u^2 + v^2)] \quad (3)$$

It is a real positive function, which is beneficial for fabrication of the filter. Fig. 1 shows 3-D views of a Mexican-hat wavelet in the space and frequency domains.

In Fig. 2, conventional autocorrelation is compared with wavelet autocorrelation by simulation. Apparently, the performance of wavelet correlation is better than that of conventional correlation. A sharp peak with low side lobes is obtained with wavelet correlation.

Fig. 3 shows the volume holographic wavelet correlation processor, which is proposed and

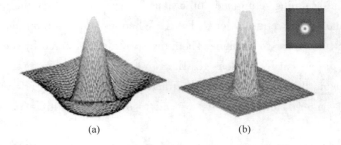

Fig. 1 3-D views of a Mexican-hat wavelet
(a) in the space domain; (b) in the frequency domain

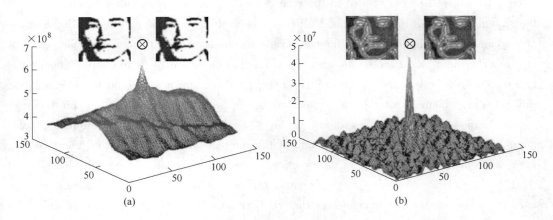

Fig. 2 Comparison of conventional correlation and wavelet correlation
(a) conventional autocorrelation; (b) wavelet autocorrelation

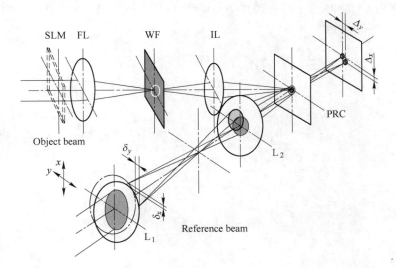

Fig. 3 Schematic drawing of the volume holographic wavelet correlation processor

constructed to realize multichannel wavelet correlation. Linearly polarized light from a laser is divided into two parts, a reference beam and an object beam, after being expanded, filtered, and collimated. A beam-scanning setup made up of lenses L_1 and L_2 controls the reference beam before it projects onto a photorefractive crystal PRC. When a pattern is input on the SLM, the lens FL forms its spectrum on the focal plane. Then, the spectrum is filtered by the wavelet filter WF and imaged onto the crystal by the lens IL. The two beams interfere to form a volume hologram in the crystal. Moving the lens L_1 to alter the transmitting angle of the reference beam and replacing the input pattern simultaneously, we can record angle-multiplexing holograms in the crystal.

When all the patterns are stored in the crystal, it can be used as a ROM and the system is ready for correlation identification.

Only the object beam is needed in the recognition step. An image for identification is fed to the system to read out the holograms. Multichannel wavelet correlation outputs will be detected at the convergent plane O of the reference beams and transferred to the computer for post processing. The input image can be recognized according to the intensity distribution of the correlation peaks. Because the time-consuming recording process is made prior and the identification process is implemented instantly, the system can be used for real-time recognition and tracking.

The beam-scanning setup, as shown in Fig. 4, is important for angle-multiplexing holograms. To ensure that the size and position of the reference beam are stable at the recording plane and only permit alteration of its direction, we should make the plane of the lens L_1 and the recording plane satisfy the imaging relation of the lens L_2. A spherical reference beam is used in our system for a more compact structure. Correlation outputs are directly obtained at the plane O without adding another lens to create the inverse Fourier transform of the beams recovered from the crystal. The scanning intervals δ_x and δ_y of the lens L_1 along the x and y directions will cause corresponding movements Δ_x and Δ_y of the correlation peak at the output plane; their relationship can be expressed as

$$\Delta_x = \frac{f_2}{l_1 - f_1 - f_2}\delta_x, \quad \Delta_y = \frac{f_2}{l_1 - f_1 - f_2}\delta_y \qquad (4)$$

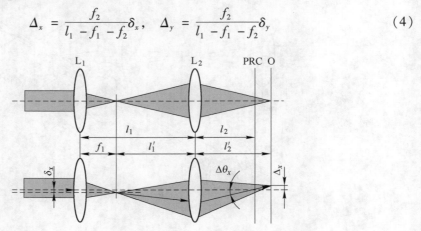

Fig. 4 The beam-scanning setup for angle-multiplexing holograms

Correspondingly, the deflection angles of the reference beam projecting on the crystal will change by

$$\Delta\theta_x = \tan^{-1}\left(\frac{l_1 - f_2}{f_1 f_2}\delta_x\right) \approx \frac{l_1 - f_2}{f_1 f_2}\delta_x$$

$$\Delta\theta_y = \tan^{-1}\left(\frac{l_1 - f_2}{f_1 f_2}\delta_y\right) \approx \frac{l_1 - f_2}{f_1 f_2}\delta_y$$

(5)

where f_1 and f_2 are the focal lengths of L_1 and L_2, and l_1 is the distance between L_1 and L_2.

3 Parallelism Enhancement with Wavelet Filtering

In the volume holographic wavelet correlation system, correlation peaks between an input image and all the stored patterns are obtained at the output plane simultaneously. The parallelism of the system is defined as the number of correlation peaks at the output plane, and is mainly determined by cross-talk noise of the system. The model shown in Fig. 5 is used for simulation[14]. The input image is placed in the plane (x_0, y_0), and illuminated with a plane wave. To simplify calculation, plane waves generated at different points in the plane (x_m, y_m) are adopted as reference beams. The interference pattern formed between the Fourier transform of the input image and the reference beams is recorded in a thick holographic medium centered at the plane (x, y). The correlation of a particular input image $f(x_0, y_0)$ and all the stored patterns $f_m(x_0, y_0)$ is obtained at the plane (x_c, y_c).

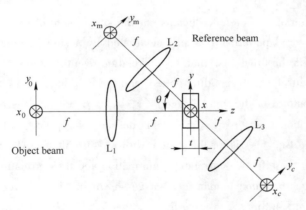

Fig. 5 The model for simulation

In a system without wavelet filtering, Ref. [14] gives us the diffractive field at the output plane. It is

$$E(x_c, y_c) = t\sum_{m=-M}^{M}\iint dx_0 dy_0 f(x_0, y_0) \times f_m^*(x_0 + \xi, y_0 + \eta) \times \mathrm{sinc}\left\{\frac{t}{\lambda f}\left[\frac{\xi(2x_0 + \xi) + \eta(2y_0 + \eta)}{2f} - \psi\right]\right\}$$

(6)

where

$$\xi = x_c + x_m, \eta(y_c + y_m)\cos\theta + [(x_c^2 + y_c^2 - x_m^2 - y_m^2)/2f]\sin\theta$$

$$\psi = (y_c + y_m)\sin\theta + [(x_m^2 + y_m^2 - x_c^2 - y_c^2)/2f]\cos\theta$$

The meanings of the parameters in Eq. (6) are shown in Fig. 5, except that $2M+1$ is the number of patterns recorded in the crystal. Replacing the input image at the plane (x_0, y_0) with its wavelet-filtered image, we can get the diffractive field of the system with wavelet filtering. It is

$$E(x_c, y_c) = t \sum_{m=-M}^{M} \iint dx_0 dy_0 f(x_0, y_0) \times W_{f_m}^*(x_0 + \xi, y_0 + \eta) \times \mathrm{sinc}\left\{\frac{t}{\lambda f}\left[\frac{\xi(2x_0 + \xi) + \eta(2y_0 + \eta)}{2f} - \psi\right]\right\} \quad (7)$$

According to Eqs. (6) and (7), the output of the system is

$$NSR = \frac{t \sum_{m \ne i} \iint dx_0 dy_0 f(x_0, y_0) f_m^*(x_0 + \xi, y_0 + \eta) \mathrm{sinc}\left\{\frac{t}{\lambda f}\left[\frac{\xi(2x_0 + \xi) + \eta(2y_0 + \eta)}{2f} - \psi\right]\right\}}{\iint dx_0 dy_0 f(x_0, y_0) f_i^*(x_0, y_0)} \quad (8)$$

$$NSR = \frac{t \sum_{m \ne i} \iint dx_0 dy_0 W_f(x_0, y_0) W_{f_m}^*(x_0 + \xi, y_0 + \eta) \mathrm{sinc}\left\{\frac{t}{\lambda f}\left[\frac{\xi(2x_0 + \xi) + \eta(2y_0 + \eta)}{2f} - \psi\right]\right\}}{\iint dx_0 dy_0 W_f(x_0, y_0) W_{f_i}^*(x_0, y_0)} \quad (9)$$

neither the restoration of the reference beams nor the simple sum of the correlation of the input image and all the stored patterns. It is still modulated by a sinc function. Another conclusion can be extracted from the functions: that any correlation result contains both the effective correlation information with the corresponding pattern and the correlation noise with other patterns. If the corresponding pattern of the input image $f(x_0, y_0)$ is $f_i(x_0, y_0)$, the cross-talk noise at the center of the correlation $(x_c = -x_i, y_c = -y_i)$ is defined as Eq. (8) is for the system without wavelet filtering, and Eq. (9) is for the system with wavelet filtering. Their numerators and denominators stand for the effective correlation information and the correlation noise, respectively.

It is supposed that the same human-face image shown in Fig. 2 is recorded in the crystal, whose size is 4.8mm × 4.8mm. The recording number is 21 ($M = 10$). Other parameters of the system in simulation are $\lambda = 0.6328\mu m$, $f = 250$mm, $\theta = 90°$, $t = 3$mm. Two different instances are analyzed below.

When the reference beam is scanning only along the y direction, the cross-talk noise curves of the system with and without wavelet filtering at all the centers of the correlation outputs $(x_c = 0, y_c = -y_m)$ are shown in Fig. 6. It can be concluded that the cross-talk noise along the y direction is very small, and the introduction of wavelet filtering decreases it further.

When the reference beam is scanning only along the x direction, the cross-talk noise curves of the system with and without wavelet filtering at all the centers of the correlation outputs $(x_c = -x_m, y_c = 0)$ are shown in Fig. 7. It can be concluded that the cross-talk noise of the system without wavelet filtering is very large, and the introduction of wavelet filtering significantly reduces the cross-talk noise along the x direction.

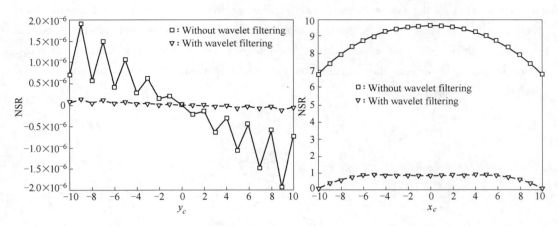

Fig. 6　Cross-talk noise along y direction　　Fig. 7　Cross-talk noise along x direction

The cross-talk noise along the x direction, also called degeneration noise, is caused by degeneration of volume gratings while recording along the x direction[15]. Hence, one-dimensional scanning is adopted in the system without wavelet filtering. Introduction of scanning on another dimension will lead to large cross-talk noise. Fig. 8 shows the experimental results of large cross-talk noise, where the thickness of the crystal is 2mm and the number of patterns recorded in the crystal is 4×40 (4 rows along the x direction and 40 columns along the y direction).

Two-dimensional scanning can be used (with caution) in the system with wavelet filtering

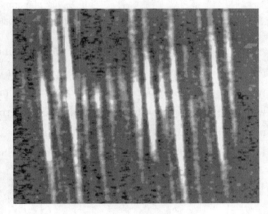

Fig. 8　Experimental result of large cross-talk noise

because the degeneration noise is significantly reduced. The parallelism of the volume holographic wavelet correlation system is then enhanced greatly.

4　Discrimination Enhancement with Wavelet Filtering

The reason wavelet filtering is effective in reducing the cross-talk noise can be explained as follows. The Mexican-hat wavelet filter here acts as an amplitude-modulated band pass filter, whose scale can be optimized by a certain neural network[16]. The wavelet filter filters the public zero-grade spectrum, and the most diverse spectra among different patterns are passed in the recording step. The passing spectra and reference beams with different directions form volume gratings in the crystal. The process is shown in Figs. 9(a) and 9(b). Two volume gratings are drawn for simplicity. In Fig. 9(c), a filtered spectrum is used to read out the gratings. Only the gratings formed by similar spectra satisfy the Bragg condition, while other gratings are not effective. In contrast with the system without wavelet filtering, the grating formed by the zero-order spectrum will be always read out. The degeneration gratings will also be read out. This leads to large cross-talk noise. On the contrary, the introduction of wavelet filtering eliminates the zero-order spectrum and reduces the degeneration gratings. As a result, the cross-talk noise is decreased.

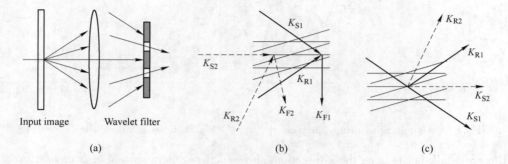

Fig. 9 Explanation of wavelet filtering
(a) function of the wavelet filter; (b) recorded volume grating; (c) bragg matching and mismatching

The decrease of cross-talk noise can also improve the discrimination capability of the system. It is concluded that autocorrelation output of the main features extracted by a wavelet filter will generate a sharper peak with little side-lobes. On the contrary, hetero correlation outputs will diminish. We store four human-face patterns in the crystal with the reference beam scanning along the x direction and the y direction. The first pattern is chosen as the input image. Fig. 10 shows the system outputs in our experiments.

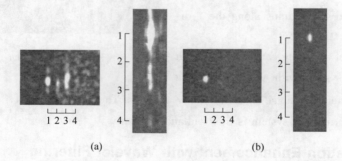

Fig. 10 Wavelet filtering for discrimination enhancement
(a) results along x and y directions without wavelet filtering; (b) corresponding results with wavelet filtering

Fig. 10(a) shows the experimental results without wavelet filtering. Correlation peaks with high intensity appear not only at the position corresponding to the first human-face pattern but also in other places. Wrong identification will probably happen if a simple postprocessing method, such as seeking the peak with maximum intensity or using a threshold, is adopted. Fig. 10(b) shows the results with wavelet filtering. There is only one distinct correlation peak at the corresponding position. Thus, the discrimination capability is improved with wavelet filtering.

Furthermore, the introduction of wavelet filtering will solve the problem when there is no input image; i. e., the crystal is illuminated with a plane wave limited by a constant frame. Fig. 11 shows the experimental results. It is difficult to choose a threshold to exclude the correlation peaks in Fig. 11 (a). However, it is easy in Fig. 11 (b), where no distinct correlation peak appears.

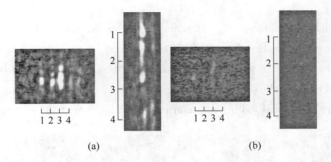

Fig. 11 Outputs of the system while there is no input image
(a) results along x and y directions without wavelet filtering; (b) corresponding results with wavelet filtering

A general volume holographic wavelet correlation processor has only one wavelet-filtering channel, which can process the input image with one wavelet scale at a time. Wavelet filters with different scales are needed in many situations of pattern recognition[17]. Features of some patterns are very similar to each other on some wavelet scales, where wavelet correlation peaks are close and hard to distinguish. However, they are different and can be distinguished on other scales. Hence, it is effective to synthesize the correlation outputs on different wavelet scales for high recognition accuracy. Based on that idea, we have proposed and constructed an improved system with multiple wavelet-filtering channels[18]. In the system, wavelet transforms of all the patterns with different wavelet scales are stored in the crystal. That of an input image can be used to obtain wavelet correlation outputs between the image and all the patterns on different wavelet scales simultaneously. Higher recognition accuracy is obtained by synthesizing the outputs. On the other hand, the effect on performance invariance of using wavelet domain correlation has also been studied[19]. Our results show that shift invariance and rotation invariance are key problems for practical applications of the processor to pattern recognition. The focal length of the lens "FL" is the main factor affecting shift invariance of the processor. Shift invariance would be improved if the focal length were increased. With regard to rotation invariance, a novel mechanism to recognize input images at any rotation angle has been proposed and testified by experiments. Such a processor is more practical with the modifications mentioned in this paper. Research based on these points is currently being pursued.

5 Application to Human-Face Recognition

The multichannel instant parallel processing of the volume holographic wavelet correlation system provides a suitable mechanism for real-time human-face recognition. The power of the He-

Ne laser used in the experiment is 30 mW. The photorefractive crystal is Fe: $LiNbO_3$, and its size is 8mm × 8mm × 3mm. The choice of the wavelet parameters is a key problem for human-face recognition. In our system, the parameters are chosen by calculating the matched effects of all the face patterns with different wavelet parameters. Because the Mexican-hat wavelet filter in the frequency domain is real and positive, we use a 3600-dot/in. laser printer to fabricate the filter on a high quality film.

An experiment on the recognition of 120 human faces has been done to test the validity of our system. The faces are recorded as 4 rows along the x direction and 30 columns along the y direction. All the patterns are stored in the crystal by scanning from top to bottom and left to right. Fig. 12 shows the system outputs with faces 1, 40, 80, and 120 as the input images, respectively. The correlation peak with maximum intensity appears at the correct position.

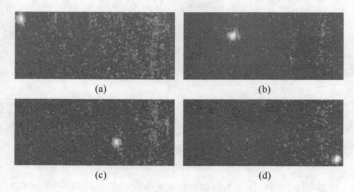

Fig. 12　Experimental results on the recognition of 120 human faces
(a)-(d) show the system outputs with faces
1, 40, 80, and 120 as the input image, respectively

6　Conclusion

A multichannel correlation system based on wavelet transforms and volume holographic associative memory has been constructed and studied. Multichannel wavelet correlation identification can be implemented by the system instantly. Simulation and experiments verify the utility of wavelet filtering in the enhancement of parallelism and discrimination. The calculation of crosstalk noise shows that the introduction of wavelet transforms reduces the crosstalk noise not only along the y direction but also, significantly, along the x direction. Two-dimensional scanning can be used in recording. As a result, the correlation peaks on the output plane expand from one dimension to two dimensions. Our experimental results also show advantages of wavelet filtering in enhancing the discrimination of the system. Its performance in human-face recognition is promising, though the invariance of the system needs to be investigated along with discrimination improvement. Studies in these directions are in progress.

References

[1] J. P. van Heerden. "Theory of optical information storage in solids", Appl. Opt. 2, 393-400(1963).

[2] J. Heanue, M. Bashaw, and L. Hesselink, "Volume holographic storage and retrieval of digital data", Science 265, 749-752(1994).

[3] B. J. Goertzen and P. A. Mitkas, "Volume holographic storage for large relational databases", Opt. Eng. 35, 1847-1853(1996).

[4] G. Barbastathis, M. Levene, and D. Psaltis, "Shift multiplexing with spherical reference waves", Appl. Opt. 35, 2403-2417(1996).

[5] Z. Wen and X. Yang, "Multichannel photorefractive correlator for rotation-invariant optical pattern recognition", Opt. Commun. 135, 212-216(1997).

[6] W. Y. Feng, G. G. Huang, Y. B. Yan, and G. F. Jin, "Multichannel wavelet correlators by the use of associative storage in a photorefractive material", Proc. SPIE 3554, 149-154(1998).

[7] F. T. S. Yu, S. Wu, A. W. Mayers, and S. Rajan, "Wavelength multiplexed reflection matched spatial filters using $LiNbO_3$", Opt. Commun. 81, 343-347(1991).

[8] C. Gu, J. Hong, and S. Campbell, "2-D shift-invariant volume holographic correlator", Opt. Commun. 88, 309-314(1992).

[9] F. T. S. Yu and S. Yin, "Bragg diffraction-limited photorefractive crystal-based correlation", Opt. Eng. 34, 2224-2231(1995).

[10] C. W. Burr, F. H. Mok, and D. Psaltis, "Large scale volume holographic storage in the long interaction length architecture", Proc. SPIE 2297, 402-414(1994).

[11] A. Pu, R. Denkewalter, and D. Psaltis, "Real time vehicle navigation using a holographic memory", Opt. Eng. 36, 2737-2746(1997).

[12] M. Wen, S. Yin, P. Purwardi, and F. T. S. Yu, "Wavelet matched filtering using a photorefractive crystal", Opt. Commun. 99, 325-330(1993).

[13] I. Ouzieli and D. Mendlovic, "Two-dimensional wavelet processor", Appl. Opt. 35, 5839-5846(1996).

[14] C. Gu, H. Fu, and J. R. Lien, "Correlation patterns and cross-talk noise in volume holographic optical correlators", J. Opt. Soc. Am. A 12, 861-868(1995).

[15] H. Lee, X. Gu, and D. Psaltis, "Volume holographic interconnections with maximal capacity and minimal cross talk", J. Appl. Phys. 65, 2191-2193(1989).

[16] W. Y. Feng, Y. B. Yan, G. G. Huang, G. F. Jin, and M. X. Wu, "Optimization of wavelet filters to improve recognition accuracy of a volume holographic correlator", Proc. SPIE 3813, 949-956(1999).

[17] R. A. Maestre, J. Garcia, and C. Ferreira, "Pattern recognition using sequential matched filtering of wavelet coefficients", Opt. Commun. 133, 401-414(1997).

[18] W. Y. Feng, Y. B. Yan, G. F. Jin, M. X. Wu, and Q. S. He, "Dual multichannel optical wavelet transform processor", Proc. SPIE 3804, 249-256(1999).

[19] W. Y. Feng, Q. S. He, Y. B. Yan, G. F. Jin, and M. X. Wu, "Realtime human face recognition system with high parallelism", Proc. SPIE 3817, 108-115(1999).

Sidelobe Suppression in Volume Holographic Optical Correlators by Use of Speckle Modulation[*]

Abstract By use of speckle modulation on the object beam of a volume holographic optical correlation system, we show that the sidelobes of the correlation patterns along the vertical direction, as well as those along the horizontal direction, are well suppressed. A theoretical explanation and experimental results are presented and discussed.

Because of their fast parallel process, high storage density, and content addressability, volume holographic correlators are becoming increasingly important and generating more and more applications[1-4]. It has been reported that 10,000 data pages can be stored in each of 16 locations in a single crystal, for a total of 160,000 holograms[5]. These holograms can serve as a library for pattern recognition, content addressing, and associative recall. In previous research sidelobes of correlation peaks and the cross-talk noise caused by them were often mentioned[4,6,7]. Because of the anisotropic Bragg condition, the sidelobes along the vertical direction (the Bragg-degenerate direction) are much more remarkable than those along the horizontal direction (in the plane defined by the object beam and the reference beam). Vertical sidelobes form an important source of cross-talk noise, because they often overlap adjacent correlation spots.

In previous studies of joint transform correlators, random-phase masks were used to enhance the performance of correlation[8] and were applied as input patterns for validation and security verification[9]. They were also used in volume holographic correlators to solve the problem of material oversaturation[10]. In a similar way we introduce a diffuser to speckle modulate the object beam, and we find that the sidelobes along the vertical direction, as well as the horizontal direction, are well suppressed.

With the scalar diffraction theory and Born's approximation the theoretical formulation of correlation patterns and cross-talk noise in volume holographic optical correlators without speckle modulation has been investigated in detail by Gu et al.[6] In this Letter we present an analysis of the special correlators for which speckle modulation is introduced. According to the conclusion in Ref. 6, the diffraction field on the output plane of conventional volume holographic correlators can be expressed as

$$g(x_c, y_c) \propto \sum_{m=-M}^{M} \iint dx_0 dy_0 f'(x_0, y_0) f_m^*(x_0 + \xi, y_0 + \eta) \times$$

---

[*] Copartner: Chuan Ouyang, Liangcai Cao, Qingsheng He, Yi Liao, Minxian Wu. Reprinted from *Optics Letters*, 2003, 28 (20): 1972-1973.

$$t\mathrm{sinc}\left\{\frac{t}{2\pi}\left[k_{mz} - k_{dz} + \frac{\pi}{\lambda}\frac{\xi(2x_0 + \xi) + \eta(2y_0 + \eta)}{f^2}\right]\right\} \tag{1}$$

where $g, x_0, y_0, x_c, y_c, f'(\ ), f_m^*(\ ), t, \xi, \eta, f, k_{mz}$, and k_{dz} are defined in Ref. 6. If we place a diffuser close to the object plane, the object beam is speckle modulated, and we can use $s(x_0, y_0) = f(x_0, y_0)a(x_0, y_0)$ to replace the object function, where $a(\ )$ denotes the random modulation function of the diffuser. Additionally, we use ensemble averaging to replace the random value in the integration, denoted by $<\ >$, so that the term $s'(x_0, y_0)s_m^*(x_0 + \xi, y_0 + \eta)$ in relation (1) can be described as $f'(x_0, y_0)f_m^*(x_0 + \xi, y_0 + \eta)<a(x_0, y_0)a^*(x_0 + \xi, y_0 + \eta)>$. According to the second statistical properties of speckle patterns the autocorrelation function of the speckle field $<a(x_0, y_0)a^*(x_0 + \xi, y_0 + \eta)>$ can be regarded as a δ function.[11] Thus we get

$$g(x_c, y_c) \propto \sum_{m=-M}^{M}\iint dx_0 dy_0 f'(x_0, y_0) \times f_m^*(x_0 + \xi, y_0 + \eta)\delta(\xi, \eta)$$
$$t\mathrm{sinc}\left\{\frac{t}{2\pi}\left[k_{mz} - k_{dz}\frac{\pi}{\lambda}\frac{\xi(2x_0 + \xi) + \eta(2y_0 + \eta)}{f^2}\right]\right\} \tag{2}$$

As shown in relation (2), volume holographic correlation is modified by two functions, namely, the δ function and the sinc function. The former is because of speckle modulation, and the latter is because of the thickness of the volume holographic medium. Both contribute to the suppression of side-lobes. However, the modification of the sinc function is anisotropic: The suppression effect along the x_c direction (the vertical direction) is much weaker than along the y_c direction (the horizontal direction), which is the reason for the much more remarkable vertical sidelobes.[6] However, the modification of the δ function is isotropic, and the degree of suppression by the δ function is much larger than that by the sinc function. Only when $t \to \infty$ can the sinc function be regarded as a δ function.

Furthermore, by use of the sifting property of the δ function, the specific expressions of the symbols defined in Ref. 6, and the fact that $\theta = \pi/2$ in our system, we find that $g(x_c, y_c)$ is nonzero only when $(x_c = -x_m, y_c = -y_m)$ and $(x_c = -x_m, y_c = y_m)$. According to the specific meanings, the former are so-called correlation spots, whose amplitudes are

$$g(-x_m, y_m) \propto t\int dx_0 dy_0 f'(x_0, y_0)f_m^*(x_0, y_0) \tag{3}$$

They are proportional to the inner product of two object functions, $f'(x_0, y_0)$ and $f_m(x_0, y_0)$.

The experimental setup for the holographic storage and correlation system is shown in Fig. 1. A diode-pumped solid-state laser ($\lambda = 532$nm) is employed as the light source, and holograms are angle-fractal multiplexed in a 2mm thick Fe: LiNbO$_3$ crystal. CCD$_1$ receives the reconstructions of recorded holograms, while CCD$_2$ reads out correlation patterns.

As shown in Fig. 2(a), we use a simple pattern of four squares to serve as the object, and we get its autocorrelation pattern (Fig. 2(b)) while speckle modulation is not used. In Fig. 2(b), which shows a similar result to that presented in Ref. 6, we find obvious sidelobes along the vertical direction. However, if a diffuser is placed close to the spatial light modulator (SLM), we

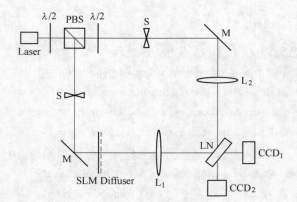

Fig. 1 Experimental setup for the volume holographic storage and
correlation system on which speckel modulation is used

PBS—polarzing beam splitter; SLM—spatial light modulator; S—shutter; LN—lithium
niobate crystal; L_1, L_2—lenses; M—mirror; $\lambda/2$—half wave plate

get the autocorrelation pattern shown in Fig. 2(c). It is shown that the experimental results are in good qualitative agreement with theoretical predictions.

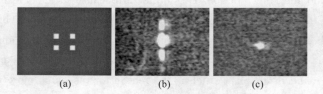

Fig. 2 (a) Object pattern; (b) Autocorrelation pattern without speckle
modulation; (c) Corresponding results with speckle modulation

In further experiments under different conditions we angle-fractal multiplex 20 random data pages (the ratio of white and black pixels is 3 : 7) in a 10mm × 10mm × 15mm photorefractive crystal.

In Fig. 3 we use a white image (all the pixels of the SLM are set on) to serve as the input. In Fig. 3(a), while speckle modulation is not used, correlation spots and sidelobes (especially along the vertical direction) overlap, and the signal and noise are indistinguishable. The traditional solution to this problem is to separate the correlation spots widely, which reduces the parallel-processing capacity of the correlator. But in Fig. 3(b), by introducing speckle modulation on the object pattern instead, we find a much clearer spot array in comparison.

To verify the correlation performance of the system with speckle modulation, the tenth object data page is used as the input. In Fig. 4(a) and in the digital processing in Fig. 5(a), which show the correlation results from the system on which speckle modulation is not used, several peaks with their sidelobes appear, and we cannot discriminate the autocorrelation peak from cross-correlation peaks. However, as shown in Figs. 4(b) and 5(b), when we use speckle-modulation, only one sharp peak corresponding to the tenth page is focused on the output plane.

Thereby cross-talk noise and the cross-correlation signal are remarkably suppressed in our experimental demonstration.

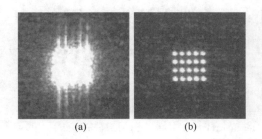 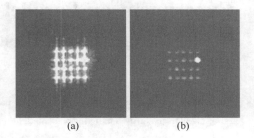

Fig. 3 Correlation spot array read out by a white image
(all the pixels of the SLM are set on)
(a) without speckle modulation; (b) with speckle modulation

Fig. 4 Experimental results for the correlation
of the tenth page as the input pattern
(a) without speckle modulation;
(b) with speckle modulation

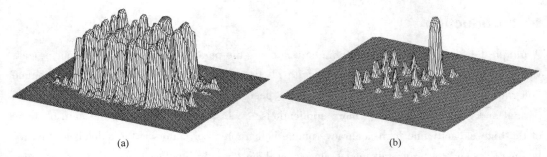

Fig. 5 Corresponding output profile of Fig. 4 after digital processing

In summary, the method of speckle modulation could remarkably enhance the performance of volume holographic optical correlators. Theoretical formulation and experimental results demonstrate the effect of sidelobe suppression.

References

[1] D. Psaltis and F. Mok, Sci. Am. 273, 70(1995).

[2] B. J. Goertzen and P. A. Mitkas, Opt. Eng. 35, 1847(1995).

[3] P. A. Mitkas, G. A. Betzos, S. Mailis, and A. Vainos, Proc. SPIE 3388, 198(1998).

[4] G. W. Burr, S. Kobras, H. Hanssen, and H. Coufal, Appl. Opt. 38, 6779(1999).

[5] G. W. Burr, F. H. Mok, and D. Psaltis, Proc. SPIE 2297, 402(1994).

[6] C. Gu, H. Fu, and J. Lien, J. Opt. Soc. Am. A 12, 861(1995).

[7] W. Feng, Y. Yan, G. Jin, M. Wu, and Q. He, Opt. Eng. 39, 2444(2000).

[8] V. M. Fitio, L. I. Muravsky, and A. I. Stefansky, Proc. SPIE 2647, 224(1995).

[9] B. Javidi and J. L. Horner, Opt. Eng. 33, 1752(1994).

[10] F. Grawert, S. Kobras, G. W. Burr, H. Coufal, H. Hanssen, M. Riedel, C. M. Jefferson, and M. Jurich, Proc. SPIE 4109, 177(2000).

[11] J. Goodman, in Laser Speckle and Related Phenomena, J. C. Dainty, ed. (Springer-Verlag, Berlin, 1975), pp. 9-76.

Experiment on Parallel Correlated Recognition of 2030 Human Faces Based on Speckle Modulation*

Abstract In this paper, the experiment on parallel correlated recognition of 2030 human faces in Fe: LiNbO$_3$ crystal is detailedly presented, a very clear correlation spots array was achieved and the recognition accuracy is better than 95%. According to the experiment, it is proved that speckle modulation on the object beam of volume holographic correlators can well suppress the crosstalk, so that the multiplexing spacing is markedly reduced and the channel density is increased 10 times compared with the traditional holographic correlators without speckle modulation.

1 Introduction

A unique benefit of volume holographic correlators is the parallel nature of the read-out process where an input object can be compared with all the stored images simultaneously, so volume holographic correlators as next-generation high speed correlators are becoming increasingly important and generated more and more applications[1-3]. In previous research, the main obstacle of the high-capacity and high-accuracy volume holographic correlators is the sidelobes of correlation peaks and the cross-talk noise caused by them[2,4,5]. In our last paper, O. Chuanets proposed a simple yet effective method by using a holographic diffuser to modulate the object pattern with a high-frequency speckle pattern, and we find that the sidelobes along the vertical direction, as well as the horizontal direction, are well suppressed[6]. In subsequent researches, we find that speckle modulation can reduce the multiplexing spacing which is one-half along the horizontal direction and one-fifth along the vertical direction, so that the channel density can be increased 10 times compared with the traditional holographic correlators without speckle modulation, however, the correlated recognition accuracy is greatly improved. In our experiment, we angular-fractal multiplexed 2030 human faces, using 70 angles in the horizontal direction and 29 lines in the vertical direction. The experimental result shows that the sidelobes are well suppressed and a very homogeneous correlation spots array can be acquired and the recognition accuracy is better than 95%.

2 Principle

The theoretical formulation of traditional correlation patterns has been investigated by C. Gu et al.[4], from which we can see that the modulation of the sinc function along the vertical direc-

---

* Copartner: Yi Liao, Yunbo Guo, Liangcai Cao, Xiaosu Ma, Qingsheng He. Reprinted from *Optics Express*, 2004, 12(7): 4047-4052.

tion is much weaker than that along the horizontal direction, hence the sidelobes along the vertical direction is much more remarkable. A simple yet effective method to suppress the sidelobes is using a high-frequency speckle pattern to modulate the object beam[6]:

$$g(x_c, y_c) \propto \sum_{m=-M}^{M} \iint dx_0 dy_0 f'(x_0, y_0) f_m^*(x_0 + \xi, y_0 + \eta) \times <$$
$$a(x_0, y_0) a^*(x_0 + \xi, y_0 + \eta) \times \qquad (1)$$
$$t \mathrm{sinc}\left\{\frac{t}{2\pi}\left[k_{mz} - k_{dz} + \frac{\pi}{\lambda}\frac{\xi(2x_0 + \xi) + \eta(2y_0 + \eta)}{f^2}\right]\right\}$$

If the frequency of the speckle pattern is high enough, then we can approximately describe the autocorrelation function of the speckle field as a δ function according to the second statistical properties of speckle[7], that is $<a(x_0, y_0) a^*(x_0 + \xi, y_0 + \eta)> = \delta(\xi, \eta)$. So the correlation pattern is modulated by the δ function and the sinc function, however, the degree of modulation by the δ function is much larger. And, another advantage is that the modulation of the δ function is isotropic, so the crosstalk along the vertical direction as well as the horizontal direction can be both well suppressed. And the δ function is normalized, so the integral of the δ function over the correlation plane is a unity, it will not change the absolute value of the correlation peak intensity.

Fig. 1 is the numerical simulation of the correlation intensity varied with the size of the speckle grain, it can be obtained that the smaller the grain of the speckle, the higher the frequency of the speckle field, so the sharper the correlation peak. When the aperture of the speckle field is large enough and speckle grain is small enough, the autocorrelation of the random speckle function would approach the ideal δ function, the correlation peak will become very sharp.

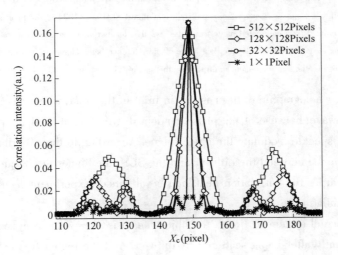

Fig. 1 The correlation intensity along the vertical direction with different sizes of the speckle grain
(from 1 × 1 to 512 × 512 pixels)

So the crosstalk caused by the sidelobes will be well suppressed with speckle modulation. And the multiplexing spacing can be reduced, that means the parallelism of the correlation system is increasing; at the same time, the accuracy of the correlated recognition does not decrease, it notably increases.

3 Experiment

As shown in Fig. 2, our correlation system use a diode-pumped solid-state laser(λ = 532nm) as the light source, and all of the holograms are angular-fractal multiplexed at a coherent volume of 74mm^3 in a 17mm × 17mm × 25mm Fe: LiNbO$_3$ crystal which is immersed in the NaCl solution to suppress the influence of a photovoltaic dc field[8]. And the correlation patterns can be read out by CCD$_2$. The shutters, SLM, CCD, and translate stage are controlled by the computer.

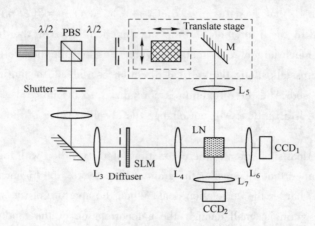

Fig. 2 Experimental setup for the volume holographic correlators with speckle modulation

The holographic diffuser is put in front of the spatial light modulator(SLM size: 1024 × 768, pixel size: 26 × 26μm^2), and it is illuminated by a collimated plane wave; the crystal is located at the Fourier plane of the SLM; the translating stage can change the incident angle of the reference beam along the horizontal and the vertical directions to implement multiplexing; the size of this system is: 400 × 400 × 150 mm^3

In this system, a holographic diffuser is put in front of the SLM, when it is illuminated by the collimated plane wave, because of the phase random distribution, a random speckle field is obtained which will speckle modulate the object plane. According to the experiments, we find that the closer the distance between the diffuser and the SLM, the higher the frequency of the speckle field, so the sharper the correlation peak can be obtained. Experiment result is coincident to the numerical simulation in Fig. 1.

In our experimental system in Fig. 2, the object beam is very weak compared with the reference beam especially after being scattered by diffuser. And the higher frequency of the speckle field, the weaker of the object beam. Comprehensive analyzing, we choose a high frequency band limited diffuser with the spatial spectrum width of 65cm^{-1} and the transmissivity of 90% in our experiment system, so that not only the correlation peak is sharp enough to suppress the crosstalk caused by the sidelobes, but also the light intensity of the object beam is strong enough to record.

Fig. 3 shows the autocorrelation patterns of the contrastive experiment between the correlators without and with speckle modulation under the same condition; Fig. 4 shows the experimental analysis. The speckle modulation can reduce the multiplexing spacing which is one-half along the horizontal direction and one-fifth along the vertical direction. We define channel density $\omega = 1/(\Delta x \times \Delta y)$, where Δx and Δy is the multiplexing spacing along the vertical and the horizontal directions, so the channel density can be increased 10 times compared with the correlation system without speckle modulation.

Furthermore, we present a contrastive experiment of 1000 human faces (50×20) between the experiment without and with speckle modulation in the same multiplexing spacing. In Fig. 5, the image sample is presented, and the wavelet transform is used to improve the quality of the correlated

Fig. 3 Autocorrelation pattern under the same experiment condition
(a) without speckle modulation;
(b) with speckle modulation

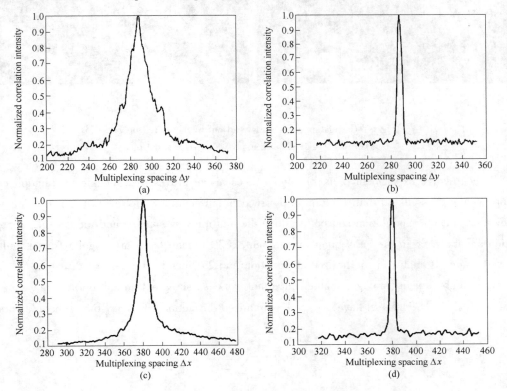

Fig. 4 Experimental interpretation
(a) the correlation peak along the vertical direction without speckle modulation; (b) the corresponding result with speckle modulation; (c), (d) the same interpretation along the horizontal direction

output. In Fig. 6, a "white" image (all the pixels of the SLM are set on) which is correlated to every image stored is input. In Fig. 6(a) without speckle modulation, we can see that the sidelobes of the correlation spots especially along the vertical direction are almost overlap so that the crosstalk is very notable and the accuracy of correlated recognition is approximate lower than 80%. In Fig. 6(b) with speckle modulation, the crosstalk caused by sidelobes is remarkably suppressed along vertical direction as well as horizontal direction, and a much clearer spots array is exhibited, so the correlation accuracy is notably increased to better than 95%.

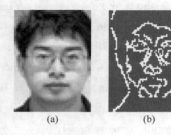

Fig. 5 Pretreatment of the human face
(a) original face pattern; (b) the binary edge character extracted with wavelet transform

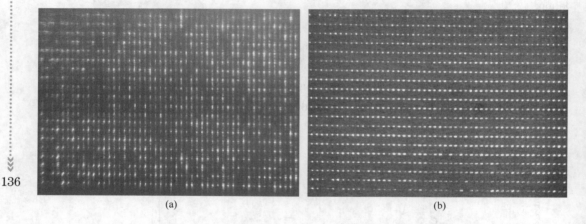

Fig. 6 Correlation spots array read out by a "white" image
(a) without speckle modulation; (b) with speckle modulation

The contrastive experiment of 1000 human faces above proves that speckle modulation can suppress the crosstalk so that to increase the recognition accuracy. However, it is not enough to show that speckle modulation can both reduce the multiplexing spacing and increase the recognition accuracy. So in the subsequent experiment, 2030 human faces are angular-fractal multiplexed using 70 angles in the horizontal direction and 29 lines in the vertical direction.

But in high-capacity storage, another important work is fishing out an appropriate time schedule. In 1991, F. H. Mok ets. have deduced a equalized diffraction efficiency time schedule equation[9]:

$$\Delta n_N \approx \frac{\tau_e}{\tau_r} \frac{\Delta n_{sat}}{N}, \quad \sum_{m=1}^{N} \Delta n_m \approx \frac{\tau_e}{\tau_r} \Delta n_{sat} \qquad (2)$$

According to this equation and the maximum exposure time of our Fe: $LiNbO_3$ crystal based on testing experiment, we found an appropriate erasing time constant τ_e, and then gained a time schedule curve in Fig. 7(a). However, this time schedule is not suitable for our high-capacity storage because that we immerses Fe: LiNbO$_3$ crystal in the NaCl solution to suppress the influence of a photovoltaic dc field[8]. During the storage process, when the crystal is illuminated for a long time, it will generate a surface electric field, though it will be counteracted by the NaCl solution, the photovoltaic electric current will form a circuit in the NaCl solution, so the electron in the conduction band of the crystal to increase, that means the photoconductivity increases. It causes the sensitivity of the crystal increase, so the recording and erasing time constant are changed in some sort. So we need to change the time constant τ_e in the lower half rows. According to the experimental result using the time schedule in Fig. 7(a), we found that the diffraction efficiencies of the correlation patterns in the left area are lower compared to the right area of the spots array especially focused in the lower half. So we amended the time schedule in the lower half rows by changing the time constant τ_e in different rows, and the time schedule curve is in Fig. 7(b), with this time schedule, we obtained a homogeneous correlation spots array.

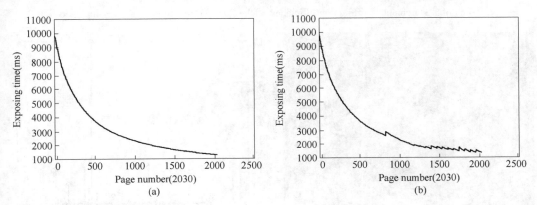

Fig. 7   (a) The original time schedule curve; (b) The amended time schedule curve

Fig. 8(a) shows the correlation spots array of the 2030 human faces. Compared with Fig. 6, the channel density is twice, and the correlation spots array is still clear because of the suppressing of the sidelobes. (In Fig. 6 and Fig. 8(a), the target of the CCD is in the same proportion). The numerical statement of diffraction efficiency of the correlation spots array is shown in Fig. 8(b), the diffraction efficiency is very homogeneous except a few images which are not recorded because of the inaccuracy of the shutters. Experimental result for the correlated recognition of the 1353th page and the corresponding numerical statement is shown in Fig. 9. Only one sharp peak corresponding to the 1353th page is focused on the output plane. So, if we choose a suitable threshold value, we can play correlated recognition. According to the recognitions of all the 2030 human faces, 37 pages are error recognition on average and the correlation accuracy is better than 95%. Furthermore, the correlated recognition is implemented by the object beam which is one-tenth intense relative to the reference beam, therefore the erasing effect during the recognition is much less than the image-readout process which is implemented by the reference beam. After 10000 times of recognition, the accuracy declines less than 1%. And the data can be reserved enduringly by the subsequent fixing.

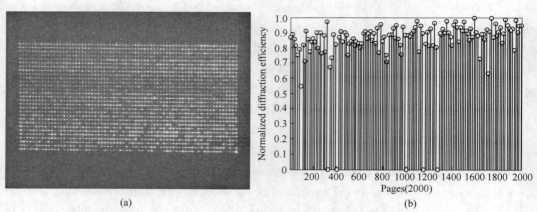

Fig. 8 (a) The correlation spots array of the 2030 human faces;
(b) Numerical statement of diffraction efficiency of the 2030 correlation spots array in Fig. 8(a)

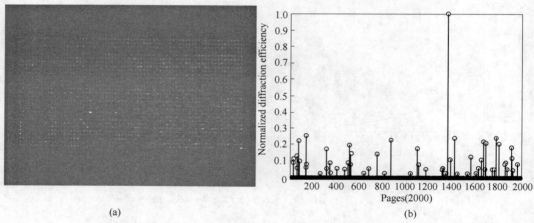

Fig. 9 Experimental result for the correlated recognition of the 1353th
page and the corresponding numerical statement
(a) the 1353th; (b) the numerical statement

4 Conclusion

A high parallel correlated storage and recognition of 2030 human faces in Fe: LiNbO$_3$ crystal is presented, a very homogeneous correlation spots array was obtained, so the recognition accuracy increases to approximate 95%. It has well proved that speckle modulation generated by a high-frequency diffuser on the object beam of correlation system can suppress the crosstalk, both the channel density and the recognition accuracy are increased compared with the traditional holographic correlators without speckle modulation.

References

[1] G. W. Burr, S. Kobras. H. Hanssen, et al. "Content-addressable data storage by use of volume holograms",

Appl. Opt. 38,6779-6784(1999).
[2] S. H. Lee, S. Y. Yi and E. S. Kim, "Fingerprint Identification by use of a Volume Holographic Optical Correlator", in Optical Pattern Recognition X, Orlando, Florida, April, Proc. SPIE 3715, 321-325(1999).
[3] T. H. Chao, H. Zhou and G. Reyes, "Compact 512 × 512 Grayscale Optical Correlator", in Optical Pattern Recognition XIII; D. P. Casasent, T. H. Chao, Eds., Proc. SPIE 4734, 9-12(2002).
[4] C. Gu, H. Fu and J. R. Lien, "Correlation patterns and cross-talk noise in volume holographic optical correlators", J. Opt. Soc. Am. A 12, 861-868(1995).
[5] F. Wenyi, Y. Yingbai, J. Guofan, W. Minxian, H. Qingsheng, "Volume holographic wavelet correlation processor", Opt. Eng. 39, 2444-2450(2000).
[6] Ouyang Chuan, Cao Liangcai, He Qingsheng, Liao Yi, Wu Minxian, Jin Guofan, "Sidelobe suppression in volume holographic optical correlators by use of speckle modulation", Opt. Lett. 28, 1972-1974(2003).
[7] J. Goodman, "Statistical properties of laser speckle patterns", in Laser Speckle and Related Phenomena, J. C. Dainty, ed. (Springer-Verlag, Berlin, 1975), pp. 9-76.
[8] Qingsheng He, Guodong Liu, Xiaochun Li, Jinangang Wang, Winxian Wu, and Guofan Jin, "Suppression of the influence of a photovoltaic dc field on volume holograms in $Fe:LiNbO_3$", Appl. Opt. 41, 4104-4107 (2002).
[9] F. H. Mok, M. C. Tackitt and D. Psaltis, "Storage of 500 high-resolution holograms in a $LiNbO_3$ crystal", Opt. Lett. 16, 605-607(1991).

Improving Accuracy of Multichannel Volume Holographic Correlators by Using a Two-dimensional Interleaving Method*

Abstract We introduce a two-dimensional interleaving method to improve the output accuracy of volume holographic correlators. The method redistributes the pattern of every input and stored image without changing their inner products, so it can eliminate the impact of correlation pattern difference on calculated inner product values. Experimental tests show that this method can achieve more accurate output intensity for every channel, which agrees much better with the theoretical value than without this method.

A volume holographic correlator(VHC) is a multichannel optical correlator based on high-density holographic storage. It inherently has the characteristics of high-speed, high parallelism, and multichannel processing, which give it potential in applications that need fast, real time, and high-capacity correlation calculations, such as associative retrieval[1], pattern recognition[2], target tracking, and navigation[3], among others. It has been reported that 10,000 data pages can be stored in each of 16 locations in a crystal[4] and 2030 correlation spots can be parallely retrieved from a single location in a crystal[5].

Due to Bragg condition and medium defocusing, the output correlation pattern of the VHC is a suppressed correlation function, whose central point represents the two-dimensional(2D) inner product[6,7]. Therefore, in most cases VHC is used to calculate the 2D inner products between input and stored images. It is crucial that the detected signal be determined only by the inner product, but usually that is not the case: as measuring the brightness of a "mathematical" point is impossible, the correlation function over a small area covering the center of the correlation pattern has to be integrated to approximate the inner product. However, because the output correlation patterns strongly depend on the input and stored image patterns, the distribution in the integration area is also variable, as shown in Fig. 1(a). As a result, the integration result depends not only on the inner product but also on the actual distribution of the correlation pattern. This implies that the results may be different even if the inner products are equal. This pattern-dependent behavior is serious when the images have differences in low spatial frequencies, because the integration results are notably impacted by the profiles(the general shape) of the correlation distributions, and the profile depends mainly on the product of the low-frequency components of the input and stored images. The problem is also mentioned as a quadratic-linear

---

* Copartner: Kai Ni, Zongyao Qu, Liangcai Cao, Ping Su, Qingsheng He. Reprinted from *Optics Letters*, 2007, 32(20): 2972-2974.

phenomenon in Refs. [8,9].

If the integration area is small enough, the impact of the correlation pattern difference can be ignored. Several methods can be used to achieve the small area[6,7,9,10]. A diffuser placed in the object beam path can distribute the spectra more evenly throughout the storage media[11] and suppress the sidelobes of the correlation pattern[10]. It is equivalent to multiplying the correlation pattern with a narrow window centered on it. In theory, diffusers with large scattering angles will generate sharp windows, making the integration area small enough to reduce the impact of the correlation pattern difference to an undetectable level. However, in practical systems, diffusers with large scattering angles are not applicable for throughput considerations. In other words, the window generated by an "applicable diffuser" is required to have a certain width. As a result, the impact of the correlation pattern difference still exists after using the diffuser.

In this Letter we introduce a 2D interleaving method to improve the calculation accuracy. Interleaving is widely used in digital communications and data storage systems for burst-error correction[12]. It is also used in the joint transform correlator to enhance the contrast between the autocorrelation and the cross-correlation peaks[13]. We use it here in the VHC as an orthogonal transformation to make the correlation patterns of different images similar, which can eliminate the impact of correlation pattern difference on calculated inner product values. The interleaving procedure can be described as redistributing the pixels inside the image: first, we need to determine an interleaving rule P that maps every pixel to its new position in the same image. Then, every input or stored image should be interleaved according to the same rule P before being uploaded onto the spatial light modulator (SLM). It is easy to prove that the inner product remains unaltered after interleaving, i. e.,

$$I_1 I_2 = P\{I_1\} P\{I_2\} \tag{1}$$

where I_1 and I_2 are uninterleaved images and $P\{I_1\}$ and $P\{I_2\}$ are interleaved ones. The interleaving rule should be designed to make the target image patterns similar. Here random interleaving is used, and its effect can be explained as follows. Some of the real images, such as photographs, often present various patterns that will generate different correlation patterns, causing the calculated inner products to be inaccurate as discussed above. However, for random images, as shown in Fig. 1(b), correlation patterns are similar to each other. When we integrate over the center area to approximate the inner product, the impact of the correlation distributions is nearly the same for different images and can be canceled by normalization. Therefore, the calculated inner products between random images are more accurate even with a bit larger integration area. As random interleaving can transform the real images to random ones without changing their inner products, the calculation accuracy of the VHC for those real images will then be improved. Note that generally the spatial bandwidth of the optical system of the VHC is designed larger than or equal to the bandwidth of the SLM, and interleaving does not increase the bandwidth of the SLM; thus the interleaved images can be processed without the lost of the resolution information.

The experimental setup is shown in Fig. 2. A diodepumped solid-state laser (DPSSL, λ = 532

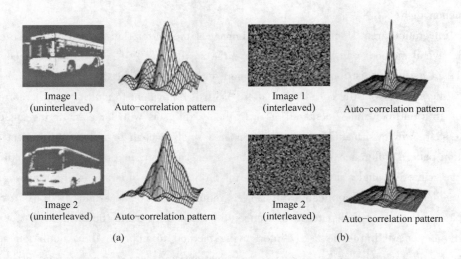

Fig. 1 (a) For uninterleaved images, correlation patterns are different;
(b) for interleaved images, correlation patterns are similar

nm) is the light source. A diffuser of 0.2° scattering angle is placed behind the SLM. The holograms are angle-fractal multiplexed in a Fe: LiNbO$_3$ crystal. A CCD camera (MINTRON MTV-1881EX) is used to read the correlation spots. Interleaving is performed by the computer before uploading images onto the SLM.

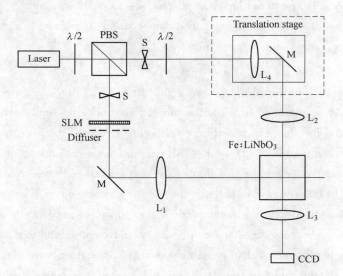

Fig. 2 Experimental setup for testing the effect of 2D interleaving.
PBS—polarizing beam splitter; SLM—spatial light modulator; S—shutter;
L$_1$-L$_4$—lenses; M—mirror; $\lambda/2$—half-wave-plate

Fig. 3 shows the two groups of images used to test the effect of the interleaving method. Each group consists of five pairs of complementary images. Fig. 3(b) is the random interleaved version of Fig. 3(a). For each image, the inner product is unity with itself, 0 with its complementary image, and 0.5 with the rest in the same group. For each group, the ten images were first stored in the crystal in the order $1p, 1n, 2p, 2n, \ldots, 5p, 5n$ by angle multiplexing, forming a ten-channel VHC. Then, the ten images were input to the VHC in the same order. For every input, the output consists of ten correlation spots. Ten inputs generate 10×10 spots.

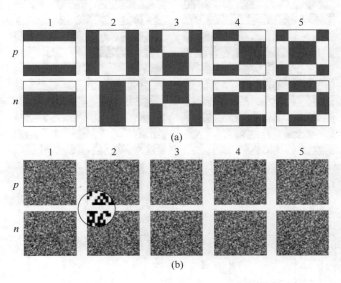

Fig. 3 Test images

(a) Uninterleaved; (b) interleaved by computer (A small area is magnified to show the fine structure of the interleaved images)

Fig. 4 gives the experimental results. Note that for both interleaved and uninterleaved images, the diffuser is always used. The ten spots in one row are generated by one input image, taken with one CCD shot, corresponding to the responses of ten channels to a single input. The ten spots in one column are generated by ten input images, corresponding to the responses of a single channel to ten different inputs. As shown in Fig. 4(a), with uninterleaved images, the spot brightness does not follow the theoretical values well. For rows, this means that the responses of different channels to a single input depend on the stored image patterns, indicating a stored-patterndependent behavior. For columns, this means that the responses of a single channel to different inputs depend on the input image patterns, indicating an input-pattern-dependent behavior. However, with random interleaved images, Fig. 4(b) gives a more reliable result. The spots' brightness agrees well with theoretical predictions in both rows and columns, meaning that the calculated result of every channel to every input is accurate. Figs. 5(a) and 5(b) give the plots of brightness of the spots in the fifth column in Figs. 4(a) and 4(b), respectively. The theoretical plot is given in Fig. 5(c). The spot brightness is calculated by summing up the values of all the pixels it covers. In both Figs. 5(a) and 5(b), the brightest (fifth) and darkest (sixth) spots can be correctly located, while the half-lighted ones vary from 0.22 to 0.56 in (a) and from 0.31 to 0.33 in (b), indicating that Fig. 5(b) gives a higher output accuracy (the expected spot brightness should be the square of the inner product). The excess of the half-lighted spots over the value of 0.25 was possibly caused by the nonlinearity and background noise of the CCD. The former could be corrected by a calibration of the CCD.

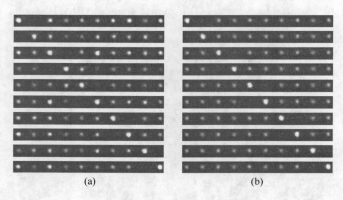

Fig. 4　Output correlation spots

(a) uninterleaved; (b) interleave

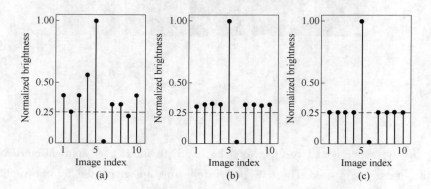

Fig. 5　Spot brightness of the fifth column in Figs. 4(a) and 4(b)

(a) uninterleaved; (b) interleaved; (c) theoretical

Fig. 6 gives a group of car images used to test the effect of the interleaving method. Fig. 6(a) was first stored in the crystal. Then the five images in Fig. 6(b) were input into the VHC. The detected correlation spots were normalized to the maximum value, and their square roots were taken to get the inner products. For uninterleaved images (Fig. 6) and their corresponding interleaved versions, the experiment was repeated four times. Fig. 7 gives the plots of calculation re-

Fig. 6　Real car test images

(a) the stored image; (b) five input images

sults for both versions. It clearly shows that the interleaving method, Fig. 7(b), gives better accuracy.

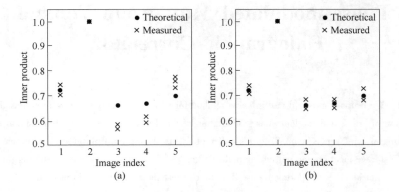

Fig. 7 Calculated inner products between Figs. 6(a) and 6(b)
(a) using the uninterleaved images in Fig. 6; (b) using the interleaved version of Fig. 6

In conclusion, by use of the 2D interleaving method, the correlation pattern difference in the VHC can be eliminated without changing the inner product, and the output accuracy of VHC is increased. This method adds a little time to the preprocessing of the correlation. However, in most cases, just as for digital correlators, preprocessing is also necessary for the VHC whether with or without this method. Since interleaving is quite simple compared with other preprocessing methods normally used in the VHC, the extra time added to the existing total time-consuming preprocessing is very limited. This method can also be combined with other existing coding schemes or preprocessing methods on the VHC.

References

[1] G. W. Burr, Proc. SPIE 5181, 70(2003).
[2] A. Heifetz, J. T. Shen, J. K. Lee, R. Tripathi, and M. S. Shahriar, Opt. Eng. 45, 025201(2006).
[3] A. Pu, R. Denkewalter, and D. Psaltis, Opt. Eng. 36, 2737(1997).
[4] G. W. Burr, F. H. Mok, and D. Psaltis, Proc. SPIE 2297, 402(1994).
[5] Y. Liao, Y. B. Guo, L. C. Cao, X. S. Ma, Q. S. He, and G. F. Jin, Opt. Express 12, 4047(2004).
[6] C. Gu, H. Fu, and J. R. Lien, J. Opt. Soc. Am. A 12, 861(1995).
[7] M. Levene, G. J. Steckman, and D. Psaltis, Appl. Opt. 38, 394(1999).
[8] F. Grawert, S. Kobras, G. W. Burr, H. Coufal, H. Hanssen, M. Riedel, C. Michael Jefferson, and M. Juric, Proc. SPIE 4109, 177(2000).
[9] S. Kobras, "Associative recall of digital data in volume holographic storage systems", Master's thesis (TechnischeUniversitiätMünchen, 1998).
[10] C. Ouyang, L. C. Cao, Q. S. He, Y. Liao, M. X. Wu, and G. F. Jin, Opt. Lett. 28, 1972(2003).
[11] D. K. McMillen, C. G. Zabel, P. S. Erbach, and D. A. Gregory, Opt. Eng. 34, 2232(1995).
[12] G. C. Clark, Jr., and J. Bibb Cain, Error-Correction Coding for Digital Communications(Plenum, 1981).
[13] Z. Zalevsky, A. Rubner, J. García, P. Garcia-Martinez, C. Ferreira, and E. Marom, Appl. Opt. 45, 7325 (2006).

Phase-modulated Multigroup Volume Holographic Correlator[*]

Abstract We propose a hybrid multiplexing method for volume holographic correlators based on the combination of orthogonal random phase modulation of an object beam and two-dimensional angular multiplexing of a reference beam. This method implements storing multiple groups of library images in a single storage location and outputting multiple groups of correlation spots by switching the modulation phase masks. Three experiments are designed to test the validity of this method.

A volume holographic correlator(VHC)[1,2] is a multichannel optical correlator based on high-density holographic storage. It is often used to calculate the two-dimensional(2D) inner products between input and library images. Compared with digital correlators, VHC inherently has the characteristics of high speed, high parallelism, and multichannel processing, which benefit from the capability of multiplexing volume holograms at a single storage location. High-performance VHCs provide new possibilities for applications that need fast, real time, and high-capacity correlation calculations, such as associative retrieval[3], pattern recognition[4], target tracking, and navigation[5].

In a VHC, the allowed number of library images depends mainly on two capacities: the high-density storage capacity and the correlation readout capacity. Currently the latter serves to be the primary limiting factor. In previous research, Burr et al. demonstrated the storage capacity of storing and reconstructing 10,000 images at a single location in a crystal[6]. The storage capacity can be even larger in a VHC: The maximum number of multiplexed holograms can be represented as $M \approx \tau_e A_0 / \tau_r A_m$[7], where T_r and T_e are the time constants for recording and erasing, respectively. A_0 and A_m are, respectively, the maximum possible and minimum detectable grating strength. Note that for a stored hologram, when used in reconstruction, the diffracted energy is distributed as an image on the whole sensitive area of the detector, covering thousands of pixels; when used in correlation, the diffracted energy is focused into a small correlation spot, covering only tens of pixels. Thus, for the same detector, the required A_m for correlation detection is much smaller than that for reconstruction detection, which allows storing more holograms. The correlation readout capacity is determined by the number of correlation spots that can be organized on the detector. When an image is sent into a VHC, each stored hologram will correspondingly output a correlation spot with certain size on the detector along the direction of the reference beam used to record it. The reference angle spacing between neighboring holograms must

---

[*] Copartner: Kai Ni, Wei Ren, Zongyao Qu, Liangcai Cao, Qingsheng He. Reprinted from *Optics Letters*, 2008, 33(10): 1144-1146.

be large enough to avoid overlaps of correlation spots, which significantly decreases the number of allowed angle positions. We used the speckle modulation technique[8] and the interleaving method[9] to obtain smaller correlation spot covering only 7 × 7 pixels on the detector, organizing up to 6000 correlation spots. However, this number is still less than $\tau_e A_0/\tau_r A_m$. Further organizing more library images by decreasing the correlation spot size or enlarging the panel of a CCD is difficult and costly.

In this Letter, we introduce a hybrid multiplexing method to implement storing multiple groups of library images, which would allow organizing more images in a single storage location in a VHC. Assuming n groups of library images to be stored into VHC, we first correspondingly generate n random phase masks $\varphi_i(x,y)$, $i = 1, 2, ..., n$, satisfying $\iint \exp[j\varphi_i(x,y)] \exp[-j\varphi_k(x,y)] \mathrm{d}x\mathrm{d}y \approx 0$ when $i \neq k$, meaning that they are approximately orthogonal to each other. During the storage stage, as shown in Fig. 1(a), the ith group of library images is stored by 2D angular multiplexing with the object beam modulated by the ith mask. Different groups repeat the same angular multiplexing, making correlation spots of different groups locating at the same positions on the CCD. During the correlation stage, as shown in Fig. 1(b), if we use the ith phase mask to modulate the object beam, only the ith group will output effective correlation spots. The approximate orthogonality between different random phase masks ensures the outputs of other groups to be nearly zero, which significantly decreases crosstalk among different groups, as shown in Eq. (1):

$$\iint \mathrm{d}x\mathrm{d}y f(x,y) \exp[j\varphi_i(x,y)] g^*(x,y) \exp[-j\varphi_k(x,y)]$$
$$= \begin{cases} \iint f(x,y) g^*(x,y) \mathrm{d}x\mathrm{d}y & (k = i) \\ \approx 0 & (k \neq i) \end{cases} \quad (1)$$

By switching the modulation phase masks, each group can be respectively accessed. There are several ways to generate the phase masks, such as using different zones of a large random phase plate(RPP) by shifting the plate or using a phase spatial light modulator(PSLM).

The experimental setup is shown in Fig. 2[10]. A CCD camera(MTV-1881EX, signal-to-noise ratio 48 dB) is used to detect the correlation spots. A commercial RPP(Luminit Co.) is placed behind the SLM. The plate is fixed on a translation stage parallel to the SLM surface. Three experiments are designed to test the validity of this method.

Generally, the phase distribution of a RPP can be regarded as a stationary random process. When the distances between different zones are larger than the half-width of their autocorrelation functions, those zones can be regarded as approximately orthogonal. The first experiment measured the required shift distance. A white image(all pixels ON) modulated by the RPP was first stored into the crystal. Then we sent this image into the VHC again and obtained a very bright correlation spot. Next we gradually shifted the plate and measured the brightness change of the spot. As shown in Fig. 3, the brightness decreases with the increase of the shift

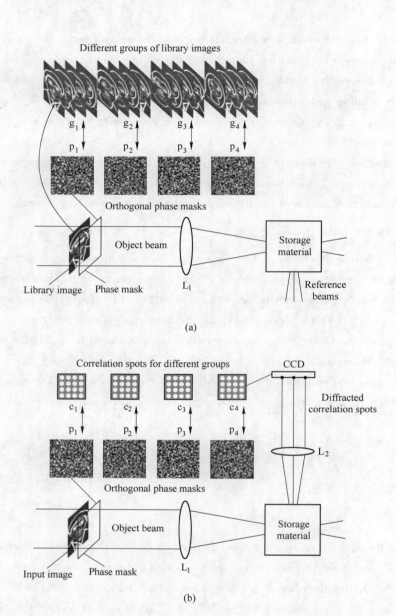

Fig. 1 Schematic diagram of (a) storing and (b) accessing multiple groups of library images in a VHC

distance. When the RPP was shifted about 200 μm away from the original position, the spot was undetectable, indicating that the new generated phase mask was approximately orthogonal to the original one and can be used to modulate a new group of library images.

The second experiment tested the crosstalk among different groups. We first set 50 stop positions for the RPP to generate 50 orthogonal phase masks. Then a binary image M was stored with the modulation of the ith mask, and its complementary image M_c with the rest of the masks. All images were stored at the same storage location with the same reference beam. This formed a VHC with 50 groups of library images (one for each group here) whose correlation

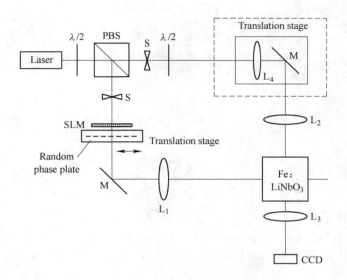

Fig. 2 Experimental setup

PBS—polarizing beam splitter; S—shutter; L_1-L_4—lenses; M—mirror; $\lambda/2$—half-wave plate

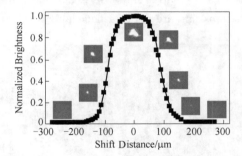

Fig. 3 Measured brightness of the correlation spot versus the shift distance of the RPP in Experiment 1

spots were located at the same position on the CCD. When we input M_c with the ith mask, the output of the ith group (M in that group) was zero, so the detected output represents the crosstalk from other 49 groups (M_c in each group). We repeated the experiment for $i = 1, 25, 50$, respectively. Fig. 4 gives the detected output of M and M_c, respectively. We can see that the output of M is a bright spot, while the output of M_c is just noise, indicating that the crosstalk within at least 50 groups was under a detectable level.

The third experiment demonstrated a VHC with four groups of library images, each containing about 1000 images. The first group contains 1024 real images, organized as 32×32 correlation spots, which is demonstrated as a content addressable database to find the most similar three images. The second group contains 512 pairs of complementary digital data pages, also organized as 32×32 spots. For each page, the inner product is unity with itself, 0 with its complementary page, and 0.5 with the rest. For the third and fourth groups, we specially designed the brightness of each correlation spot to make the output spot array form a predefined pattern. This

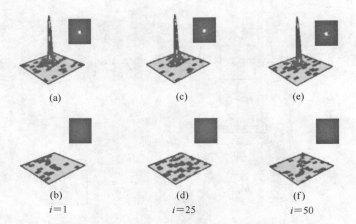

Fig. 4 Detected correlation spots in Experiment 2 with the input of (a), (c), (e) the original image M and (b), (d), (f) the complementary image M_c

was achieved by storing specific library images that have predetermined inner products with a common input image. The third group was organized as 40×30 spots and the fourth 23×36. The correlation result for each group is shown in Fig. 5. We can see that each group outputs the expected result and the crosstalk are under a detectable level.

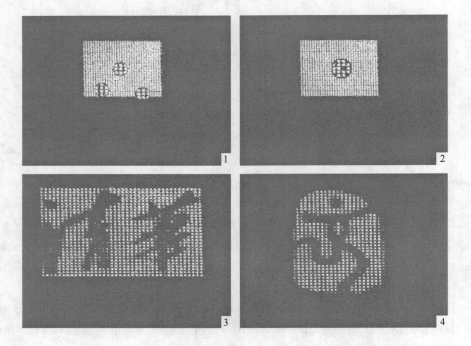

Fig. 5 Detected correlation spots for each group in Experiment 3. Areas in the black circles are magnified to clearly show the correlation spots

In conclusion, we proposed a hybrid multiplexing method for a VHC based on the modulation of the object beam and 2D angular multiplexing of the reference beam. It implements storing multiple groups of library images in a single storage location. By switching the modulation phase masks, the input image can be correlated with any group. This method would be a practical way to organize more library images at a single storage location in a VHC. If using a PSLM instead of the RPP, this method will not need movable mechanical components, making it much more promising.

References

[1] G. W. Burr, S. Kobras, H. Hanssen, and H. Coufal, Appl. Opt. 38, 6779(1999).
[2] X. C. Li, F. Dimov, W. Phillips, L. Hesselink, and R. McLeod, in 29th Applied Imagery and Pattern Recognition Workshop (AIPR2000), J. Aanstoos, ed. (IEEE, 2000), pp. 78-83.
[3] G. W. Burr, Proc. SPIE 5181, 70(2003).
[4] Heifetz, J. T. Shen, J. K. Lee, R. Tripathi, and M. S. Shahriar, Opt. Eng. 45, 025201(2006). A. Pu, R. Denkewalter, and D. Psaltis, Opt. Eng. 36, 2737(1997).
[5] G. W. Burr, F. H. Mok, and D. Psaltis, Proc. SPIE 2297, 402(1994).
[6] D. Psaltis, D. Brady, and K. Wagner, Appl. Opt. 27, 1752(1988).
[7] Ouyang, L. C. Cao, Q. S. He, Y. Liao, M. X. Wu, and G. F. Jin, Opt. Lett. 28, 1972(2003).
[8] K. Ni, Z. Y. Qu, L. C. Cao, P. Su, Q. S. He, and G. F. Jin, Opt. Lett. 32, 2972(2007).
[9] Q. S. He, G. D. Liu, X. C. Li, J. G. Wang, M. X. Wu, and G. F. Jin, Appl. Opt. 41, 4104(2002).
[10] J. W. Goodman, Statistical Optics (Wiley, 1985).

Fast Associative Filtering Based on Two-dimensional Discrete Walsh Transform by a Volume Holographic Correlator*

Abstract An optical fast associative filtering method based on a multichannel volume holographic correlator to search the database as a front-end filter is described. The features of searching query are parallelly extracted by a volume holographic correlator based on two-dimensional discrete Walsh transform, and are used to measure the similarity between the query and all the database records. The best matches are picked up for further searching. An experiment is carried out, and the experimental results prove the validity of the method.

1 Introduction

A volume holographic correlator (VHC) is a multi-channel optical correlator based on high density holographic storage. Its output correlation pattern of each channel is a suppressed correlation function[1]. By the use of speckle modulation[2] and 2D interleaving method[3], the output results can be more accuracy. It inherently has the characteristics of high-speed, high parallelism and multichannel processing, which give it potential in applications that need fast, real time calculations, such as database searching[4], associative retrieval[5], and pattern recognition[6].

In the traditional VHC-used database searching method[7], there are two problems. The first is that all the database records have to be stored into the VHC, thus the number of database records is restricted by the storage capacity of VHC. The second is that the searching accuracy is restricted because of the analog nature of the optical process.

The first problem can be solved thorough the 2D orthogonal transformation. Transform functions (TFs) are encoded and stored into VHC instead of all the database records. When a query is input, the inner products of the query and TFs are calculated by VHC and form the features of the query. The features are then used to measure similarity between the query and database records. In this case, the library images needed to be stored are reduced from all the database records to a few TFs. In other words, using the same storage capacity, this method can search a much larger database than the traditional one.

There are optical correlations have been reported to be used in performing 2D orthogonal transformations. 2D orthogonal transformation can be considered as a set of inner products of the data array to be transformed and the TFs. When an optical correlator is used to perform 2D

---

* Copartner: Qiang Ma, Kai Ni, Qingsheng He, Liangcai Cao. Reprinted from *Optics Express*, 2009, 17(2):838-843.

transformations, the calculation speed depends on two important points: the number of the inner product calculation channels and the resolution of the stored TFs. Vander Lugtcorrelator (VLC)[8] can be used to perform 2D transformation. But it is a serial process, thus the number of calculation channel is only one. Leger and Lee proposed an optical implementation of 2D transformation using a planar computer generated hologram(CGH)[9,10]. But the number of calculation channels and theresolution of TFs are restricted by the cross talk among the planar holograms and the practical resolution of the CGH. Lenslet array can also be used to perform 2D transformation[11-13]. But the resolution of TFs is restricted by the image section width of the lens[13]. Generally, the resolution is not better than 32×32[13]. There are obvious advantages to use VHC to perform 2D orthogonal transformations than the methods mentioned above. The number of calculation channels can be more than 2000. And the resolution of TFs is the same as the SLM, which can be better than 1024×1024 now.

The second problem can be solved by a two-step searching scheme consists of the frontend approximate search and the back-end accurate search. In the first step, the VHC is used to take an approximate search of the whole database very fast and pick up a small set of several nearly matching records. In the second step, the digital search engine is used to find out the best match record in the small picked up set with a much higher accuracy. By using this two-step searching scheme, the required accuracy of the first step is reduced and could be satisfied by VHC. This scheme would fully take advantage of the parallelism and high speed of VHC and the high accuracy of the digital processor[14].

In this paper, we suggest a fast associative filtering method based on 2D discrete Walsh transform(DWT) implemented by VHC to approximately search the database. The system scheme is shown in Fig. 1. In the front-end searching process, the VHC serves as a fast optical filter, which performs 2D DWT swiftly to get the features of the input query. The features are then used to approximately measure the similarity between the query and database records. As the query is compressed to the features with much less elements, the measuring time is tiny. Therefore the front-end searching can approach a very high process speed. Then the best matches are picked up and delivered to the back-end digital search engine for further accurate searching. There are some advantages of the method. First, the images needed to be stored are

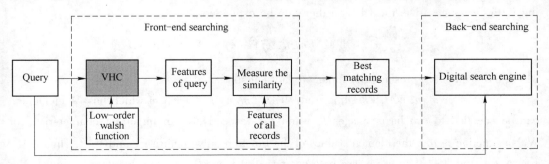

Fig. 1 System scheme

reduced from all the database records to a few Walsh functions. In other words, with the same storage capacity, this method can search a much larger database than the traditional one. Second, the two-step searching scheme would fully take the advantage of the parallelism and high processing speed of the multi-channel VHC and the high accuracy of the digital processor.

2 Feature extraction performed by VHC using 2D DWT

DWT is characterized by the energy packing capability. Furthermore, Walsh functions only have two states and are determined, thus possess many special properties. They are easy to be represented, and the calculation speed of DWT is high. Thereby it is widely used in feature extraction[15].

The 2D DWT of a data array $x_{i,j}$ of $N \times N$ points is[15]

$$X_{m,n} = \frac{1}{N_2} \sum_{i=0}^{N-1} \sum_{j=0}^{N-1} x_{i,j} WAL(m,n,i,j), \quad m,n = 0.1,\cdots,N-1 \quad (1)$$

where $WAL(m, n, i, j)$ is the 2D discrete Walsh function, and m, n is the order. Eq. (1) shows that 2D DWT is essentially a set of inner products of the data array and a series of 2D discrete Walsh functions.

VHC is often used to parallelly calculate the 2D inner products of the input image and all the stored images. The output of VHC can be expressed as

$$g(x_c = -x_m, y_c = -y_m) \propto \iint dx_0 dy_0 f'(x_0,y_0) f_m(x_0,y_0), \quad m = 1,2,3,\cdots \quad (2)$$

where $f'(x_0, y_0)$ is the input image, $f_m(x_0, y_0)$ is the mth stored image. Eq. (2) shows that the output of each channel is the inner product of $f'(x_0, y_0)$ and $f_m(x_0, y_0)$. Comparing Eq. (1) and Eq. (2), if x_{ij} and $WAL(m, n, i, j)$ are encoded to $f'(x_0, y_0)$ and $f_m(x_0, y_0)$ respectively, then VHC can be used to perform the 2D DWT.

The amplitude-modulation SLM which is used to upload images can only express nonnegative real quantities. In order to encode $WAL(m, n, i, j)$, it should be decomposed as

$$WAL(m,n,i,j) = WAL^0(m,n,i,j) + (-1)WAL^1(m,n,i,j) \quad (3)$$

where $WAL^0(m, n, i, j)$ and $WAL^1(m, n, i, j)$ are all nonnegative quantities. For simply, assuming $x_{i,j}$ to be nonnegative, Eq. (1) can therefore be expressed as

$$X_{m,n} = \frac{1}{N^2}[x_{i,j} \cdot WAL^0(m,n,i,j) - x_{i,j} \cdot WAL^1(m,n,i,j)] \quad (4)$$

where the symbol · represents the inner product calculation.

For each order m, n, $WAL^0(m, n, i, j)$ and $WAL^1(m, n, i, j)$, each of which has N^2 elements, can be encoded to two basis images $f^0_{m,n}$ and $f^1_{m,n}$ respectively. An important characteristic of Walsh function is that the element values of the function only comprise 1 and -1[14]. Thus $WAL^0(m, n, i, j)$ and $WAL^1(m, n, i, j)$ are both composed of 1 and 0, and can therefore be encoded to binary images: $f^0_{m,n}$ is divided equally by N^2 blocks $f^0_{m,n}(i,j)$, $i,j = 0, 1, \cdots, N-1$, which are used to represent the elements of $WAL^0(m, n, i, j)$. Completely white blocks are used to represent element value 1 and completely dark ones are used to represent 0. So does $f_{m,n}^1$. Here is an example: There are 16 Walsh functions when $N=4$. The 32 basis images obtained by encoding them are shown in Fig. 2.

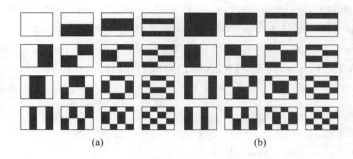

Fig. 2   32 basis images obtained by encoding walsh functions with $N=4$
(a) $f_{m,n}^0$; (b) $f_{m,n}^1$

$x_{i,j}$ can also be encoded to a data image f'. The filling ratio of each block is used to represent the normalized value of an element.

$WAL^0(m, n, i, j)$, $WAL^1(m, n, ii,)$ and $x_{i,j}$ are encoded to $f_{m,n}^0$, $f_{m,n}^1$ and f' respectively in the way described above. Thus

$$X_{i,j} \cdot WAL^0(m,n,i,j) \propto f' \cdot f_{m,n}^0$$
$$X_{i,j} \cdot WAL^1(m,n,i,j) \propto f' \cdot f_{m,n}^1 \quad (5)$$

Substitute Eq. (5) to Eq. (4)

$$X_{m,n} \propto f' \cdot f_{m,n}^0 - f' \cdot f_{m,n}^1 \quad (6)$$

$X_{m,n}$ are the transform coefficients. Eq. (6) shows that 2D DWT can be performed by VHC. Duo to the energy packing capability of DWT, it can be used to reduce information redundancy because only a subset of the transform coefficients is necessary to preserve the most important features. And the subset of coefficient is enough to approximately search the records in the front-end search. Means only the corresponding Walsh functions are needed to be stored into VHC instead of the whole database.

The method that uses 2D DWT performed by VHC to extract the features has several advantages. First, it can parallelly extract the features of the data array and the calculation time does not increase with enlarged data arrays, thus this method can reach a high processing speed. Assuming 2000 images are stored in VHC, with the resolution of 1024 × 1024, and the frame rate of SLM and CCD are both 500fps, then the equivalent computation speed of VHC is about 1×10^{12} MAC/s (MAC: Multiply-Accumulate Calculation), which is 2-3 orders of magnitude higher than that of a general microprocessor. Second, the images needed to be stored are reduced from all the database records to a few Walsh functions, means this method can search a much larger database than the traditional method using the same storage capacity. Furthermore, the method has a flexibility to extract features using other transformations if the transform functions are encoded properly and stored into VHC.

3 Front-end filtering experiment

The experiment setup is shown in Fig. 3. A diode-pumped solid-state laser (DPSSL, λ = 532nm) is the light source. A holographic diffuser of 0.2° scattering angle used as a speckle modulation device[2] is inserted behind the SLM. The holographic recording material is a Fe: LiNbO$_3$ crystal. A CCD camera (MINTRON MTV-1881EX) is used to read the output.

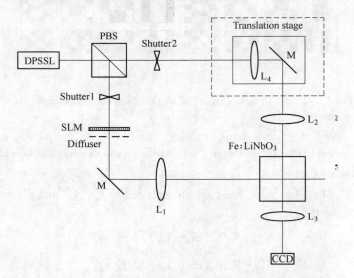

Fig. 3 Experimental setup for demonstrating the method

Fig. 4 One of the 2000 records

To demonstrate the fast front-end parallel filtering method, we use a database containing 2000 records which is also used in reference[7]. Fig. 4 shows one of the records. The energy of these records concentrates in low frequency, thus 16 low-order transform coefficients $X_{0,0}$, $X_{0,1}$, $\cdots X_{3,3}$ are used to approximately measure the similarity between the query and database records in front-end searching. Generally speaking, the larger the number of database records is, or the higher the spatial frequency of records, the more coefficients would be necessary.

The 16 Walsh functions $WAL(0,0,I,j)$, $WAL(0,1,i,j)$, $\cdots$, $WAL(3,3,i,j)$ are encoded to 32 basis images $f^0_{m,n}$ and $f^1_{m,n}$, $m, n = 0, 1, 2, 3$, which are then stored into a common location of the crystal in order $f^0_{0,0}, f^1_{0,0}, f^0_{0,1}, f^1_{0,1}, \cdots, f^0_{3,3}, f^1_{3,3}$ by angle multiplexing.

When a searching query is input to VHC, the output of VHC is detected by the CCD. The experimental output of the 1350th record input as the query is shown in Fig. 5. The 32 spots (The brightness of the second spot is zero) are the correlation results of the query and the 32 basis images respectively. Square root of each spot's brightness is taken to get the inner product of the query and the corresponding basis image. In the experiment, the error of the detected brightness is less than 5%.

Fig. 5　Experimental output of the 1350th record

The transform coefficients $X_{m,n}$ can be calculated by Eq. (6) from the inner products obtained above and form the features of the query. Then the features are used to measure the similarity between the query and each database record using the cross-correlation value R defined as

$$R = \frac{\sum_{m,n} X_{m,n} \cdot \overline{X}_{m,n}}{\sqrt{\sum_{m,n} X_{m,n}^2 \cdot \sum_{m,n} \overline{X}_{m,n}^2}} \tag{7}$$

where $X_{m,n}$ are the features of the query, and $\overline{X}_{m,n}$ are that of a database record. Fig. 6 shows the similarity between the 1350th record and all the database records. The best two matches are marked by red circle.

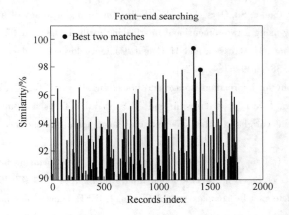

Fig. 6　Similarity between the 1350th record and all the database records in the front-end searching

The two best matching records (1350th, 1412th) are picked up and delivered to the back-end. Then the digital search engine can easily find that the query is the 1350th record from the two. As the delivered set is much smaller than the whole database (2 of 2000), the time consuming of back-end searching is tiny. The processing speed of the method mainly depends on the frame rate of SLM and CCD. In the experiment, the frame rate of VHC is 25 fps, which is mainly limited by the CCD we used.

To test the system performance, we search all the 2000 records one by one. If the most matching record picked up by front-end is directly used as the final searching result, there are 1991 records that can be searched out correctly and 9 errors. If the two or more best matching records picked up by front-end are delivered to back-end for further searching, all the records can be found out correctly. This searching result is better than the 37 errors given by the traditional method[7], while the library images needed to be stored are reduced from 2000 to 32.

4 Conclusion

A fast front-end associative filtering method based on 2D DWT implemented by VHC is presented. With necessary Walsh functions stored into VHC, the features of an input query are swiftly extracted by VHC and used to approximate measure the similarity between the query and database records. The best matching records are picked up and delivered to backend digital search engine for further accurate searching. Using the same storage capacity, the method can search a much larger database than the traditional one. The two-step searching scheme fully takes advantage of the parallelism and high speed of VHC and the high accuracy of the digital processor. In the experiment, all the record can be searched out correctly, which well proves the validity of the method.

References

[1] C. Gu, H. Fu, and J. R. Lien, Correlation patterns and cross-talk noise in volume holographic optical correlators, J. Opt. Soc. Am. A 12, 861-868(1995).

[2] C. Ouyang, L. C. Cao, Q. S. He, Y. Liao, M. X. Wu, and G. F. Jin, Sidelobe suppression in volume holographic optical correlators by use of speckle modulation, Opt. Lett. 28, 1972-1974(2003).

[3] K. Ni, Z. Y. Qu, L. C. Cao, P. Su, Q. S. He, and G. F. Jin, Improving accuracy of multichannel volume holographic correlators by using a two-dimensional interleaving method, Opt. Lett. 32, 2972-2974(2007).

[4] A. G. W. Burr, S. Kobras, H. Hanssen, and H. Coufal, Content-addressable data storage by use of volume holograms, Appl. Opt. 38, 6779-6784(1999).

[5] G. W. Burr. Holography for information storage and processing, Proc. SPIE 5181, 70-84(2003).

[6] A. Heifetz, J. T. Shen, J. K. Lee, R. Tripathi, and M. S. Shahriar. Translation-invariant object recognition system using an optical correlator and a super-parallel holographic random access memory, Opt. Eng. 45, 025201(2006).

[7] Y. Liao, Y. B. Guo, L. C. Cao, X. S. Ma, Q. S. He, and G. F. Jin. Experiment on parallel correlated recognition of 2030 human faces based on speckle modulation, Opt. Express 12, 4047-4052(2004).

[8] A. V. Lugt, "Signal detection by complex spatial filtering", IEEE Transactions on Information Theory 10, 139-146(1964).

[9] J. R. Leger and S. H. Lee. Coherent optical implementation of generalized two-dimensional transforms, Opt. Eng. 18, 518-523(1979).

[10] J. R. Leger and S. H. Lee. Hybrid optical processor for pattern recognition and classification using a generalized set of pattern functions, Appl. Opt. 21, 274-287(1982).

[11] Glaser. Noncoherent parallel optical processor for discrete two-dimensional linear transformations, Opt. Lett. 5, 449-451(1980).

[12] Glaser. Lenslet array processors, Appl. Opt. 21, 1271-1280(1982).

[13] Glaser. Compact lenslet-array-based holographic correlator/convolver, Opt. Lett. 20, 1565-1567(1995).

[14] F. Grawert, S. Kobras, G. W. Burr, H. Coufal, H. Hassen, M. Riedel, C. M. Jefferson, and M. Jurich. Content-addressable holographic database, Proc. SPIE 4109, 177-188(2000).

[15] K. G. Beauchamp. Walsh functions and their application, Academic Press, London 1975.

Multi-sample Aarallel Estimation in Volume Holographic Correlator for Remote Sensing Image Recognition[*]

Abstract Based on volume holographic correlator, a multi-sample parallel estimation method is proposed to implement remote sensing image recognition with high accuracy. The essential steps of the method including image preprocessing, estimation curves fitting, template images preparation and estimation equation establishing are discussed in detail. The experimental results show the validity of the multi-sample parallel estimation method, and the recognition accuracy is improved by increasing the sample numbers.

1 Introduction

Based on high-density holographic storage technology, multichannel volume holographic correlator(VHC) has the characteristics of high-speed, high parallelism and multichannel processing[1,2]. It may have the potential applications in the area requiring high speed, real time, high-capacity correlation calculations, such as associative retrieval[3,4], pattern recognition[5], target tracking[6], navigation[7], and so on. It has been reported that 4000 correlation spots can be parallelly retrieved from a single location in a crystal[8]. The phase modulated multi-group method has been used in the VHC to further increase the number of the retrieved correlation spots[9]. Random modulation[10] has been implemented to extract inner product of the correlation results, and interleaving technology[11] has been adopted to eliminate "pattern dependent behavior". Thus the accuracy of the multichannel VHC is greatly improved.

The remote sensing image recognition is widely used in space exploration, guided cruise, target tracking, and so on. The remote sensing image recognition is to locate of the target image in the reference remote sensing image[12,13]. In the template matching recognition, as shown in Fig. 1, the reference image with $K \times L$ pixels is divided vertically and horizontally into a set ($M \times N$) of template images, each of which has same $P \times Q$ pixels as the target image and same vertical and horizontal segmentation interval Δ_1, Δ_2. So the locations of template images in the reference image are exactly known. The matching recognition process is that the correlations between the target image and all of the template images are calculated, and the template image with the maximum inner product value corresponding to the brightest spot is determined to be the target image.

---

[*] Copartner: Shunli Wang, Qiaofeng Tan, Liangcai Cao, Qingsheng He. Reprinted from *Optics Express*, 2009, 17(24): 21739-21747.

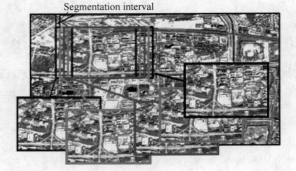

Fig. 1 The schematic diagram of the segmentation of the remote sensing image

VHC can be used to realize such correlation parallelly. When $\Delta_1 = \Delta_2 = 1$, and if $M \times N$ template images divided from the reference image can be all recorded in VHC, the remote sensing image can be perfectly recognized. However K and L are generally too large and then the VHC cannot record all template images with $\Delta_1 = \Delta_2 = 1$. Δ_1 and Δ_2 should be large enough to ensure $M \times N$ template images to be all recorded in VHC. Now if the brightest spot is only used to determine the location of the target image in reference image, the recognition accuracy is sharply decreased to Δ_1 and Δ_2. Furthermore, VHC should be precisely adjusted to ensure the correlation accuracy. However for remote sensing image recognition, the system may work under the real-time, high speed condition, and precise adjustment is very hard to maintain. The influence of noise should be considered and new technology should be adopted.

In statistics, the spatial grayscale distribution $g(x, y)$ of most remote sensing images is approximately a stationary random process[14,15]. A large number of project practices show that any point in the remote sensing image has some relevance with the points around it, and the correlation function R of the image can be expressed as[16]

$$R(\Delta x, \Delta y) = E[g(x,y)g(x + \Delta x, y + \Delta y)] \\ = a \times \exp(-\alpha|\Delta x| - \beta|\Delta y|) + b \quad (1)$$

where $E[\cdot]$ expresses the mathematic expectation, and a, b are the constant, determined by the variance and mean of the image grayscale respectively. And Δx, Δy represent horizontal and vertical coordinate difference respectively. And α, β are also constant, whose reciprocal values are called correlation lengths, determined by the characteristics of the horizontal and vertical spatial grayscale distributions respectively. A remote sensing image and its correlation function are shown in Fig. 2. According to the characteristic of the stationary random process, the correlation value of any two points is only determined by the coordinate difference Δx and Δy, but independent of the absolute coordinate of the points.

For a remote sensing image without any preprocessing, the typical one-dimensional correlation curve is shown in Fig. 3. The brightness of the correlation spots corresponding to the template images adjacent to each other have small difference. Misjudgement usually happens because of the noise effect if we use the brightest spot to estimate the location of the target and the

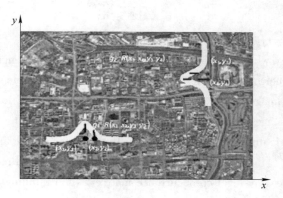

Fig. 2 A remote sensing image and its correlation function

recognition accuracy is decreased. Since the VHC is able to extract the inner product of the correlation results parallelly, more spots besides the brightest spot can be used to determine the target image location in reference image with higher accuracy.

In the VHC, thermal noise and shot noise of the detector and amplifier are Gaussian distributed, while the statistic property of optical signal is subject to Poisson distribution[17]. However, many experimental measurements show that the difference is very small to use the same Gaussian distribution to approximate the noise distribution of the light intensity of a single point[18]. According to the probability theory and multiple estimation theory[19,20], if the noise of each observed sample has the same distribution and is independent of each other, then multiple samples can be used to improve the accuracy. The multiple correlation spots in the VHC accord well with the request of the probability theory and multiple estimation theory. In this paper, we introduce a multi-sample parallel estimation method, which can be used to estimate the location of the target image with high accuracy, and the characteristics of high speed, high parallelism, multichannel processing can also be exploited sufficiently. Furthermore the system has higher tolerance and can work without precise adjustment.

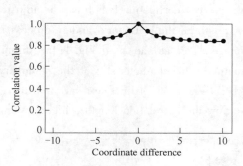

Fig. 3 A typical correlation curve of remote sensing image without preprocessing

2 Multi-sample parallel estimation method (MPE)

The MPE method with the VHC has the essential steps of image preprocessing, estimation curves fitting, template images preparation and estimation equation establishing.

Suppose the actual coordinate of the target image in the reference image is (x_1, y_1) and the recognized coordinate of the target image is (x_2, y_2), the recognition accuracy can be determined by error radius ρ, and

$$\rho = \sqrt{dx^2 + dy^2} \qquad (2)$$

where $dx = x_2 - x_1$, $dy = y_2 - y_1$.

Image preprocessing, which can be used to adjust the correlation length of the image and reduce the redundancy correlation (which is mean of the image grayscale) between the target image and the template images, is an important step to implement MPE for remote sensing image recognition. Without image preprocessing, which is shown in Fig. 3, the redundancy correlation is so much and the brightness of them will be mostly the same. There are two categories of the image preprocessing[21], including feature-based and gray-scale-based methods. The mathematical morphology[22], the gray-level co-occurrence matrix[23] and thinning algorithm[21] are actual examples of image preprocessing.

Thinning algorithm can narrow the width of the lines but maintain the basic skeleton of the image. The correlation length can be adjusted with the width of the lines changing and the characteristic of the stationary random process is determined by the skeleton of the remote sensing image. Then, as shown in Fig. 4, the thinning algorithm can adjust the correlation length but maintain the characteristic of the stationary random process, which well satisfies the need of MPE. When lots of lines are narrowed to 1-2 pixels, then more times thinning gives no help to narrow the correlation length, which is shown in Fig. 5.

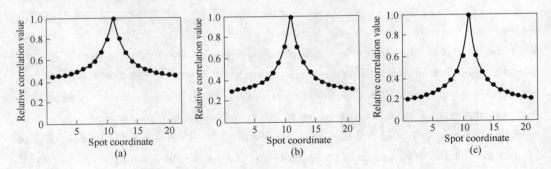

Fig. 4　Typical correlation curves of remote sensing image using different times of thinning algorithm
(a) once; (b) three times; (c) six times

The reference image needs to be segmented into a set of template images, which has been mentioned above. The segmentation intervals 1 and 2 should be in consideration. According to the characteristics of the stationary random process, for the remote sensing image, if the distance between two points exceeds the correlation length, the points will have little relevance to each other. Then to use multiple template images, the segmentation intervals must be less than the correlation length. The choice of the segmentation interval is not only determined by the accuracy requested, but also determined by the limited storage capacity of the holographic correlator, the scope of the recognized remote image and the noise of the system. The bigger size of the segmented template images, which gives better characteristic of the stationary random, would be benefit to the MPE, but it also should match the size of the SLM.

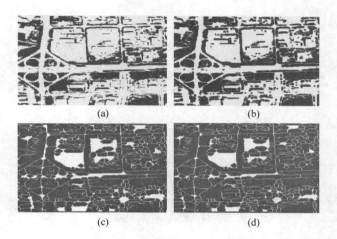

Fig. 5 Remote sensing image after different times of thinning algorithm
(a) without preprocessing; (b) once; (c) three times; (d) six times

The schematic diagram of one-dimensional MPE of remote sensing image recognition is shown in Fig. 6. For sake of convenience, three template images $n-1$, n and $n+1$ with the interval Δ being less than the correlation length, are used to determine the location of target image around nth template image. For a target image, between the template images $n-1$ and n, is correlated with the template images $n-1$, n and $n+1$ simultaneously, and the detected brightness of the correlation spots are respectively a_1 at $n-1$, a_2 at n and a_3 at $n+1$. To get the location of the target image, the estimation function f_{n-1}, f_n and f_{n+1} should be used to establish the estimation equations. According to the ergodic characteristic of the stationary random process and the relationship of the template images, the estimation function can be expressed by the same function form as

$$f_n(\Delta x, \Delta y) = R^2(\Delta x, \Delta y) = (a \times e^{-\alpha|\Delta x| - \beta|\Delta y|} + b)^2 \qquad (3)$$

Seen from the Eq. (3), the Eq. (3) is the square of the Eq. (1). The reason is that the $f_n(x, y)$ represents the brightness of the spots and the $R(x, y)$ represents the complex amplitude. Any image, whose size is the same as the target image, in the reference image is correlated with the reference image to fit Eq. (3). Then the estimation equations are given as

$$\begin{cases} f_n(x + \Delta) = a_1 \\ f_n(x) = a_2 \\ f_n(x - \Delta) = a_3 \end{cases} \qquad (4)$$

and the location x of the target can be obtained. Generally, the parameters in Eq. (3) should be determined according to the reference image.

According to the obtained correlation length and considering the storage capacity of the VHC, the scope of the remote sensing image and the noise of the system, the interval Δ_1 and Δ_2 can be chosen to segment the reference image. The template images are all stored in the VHC.

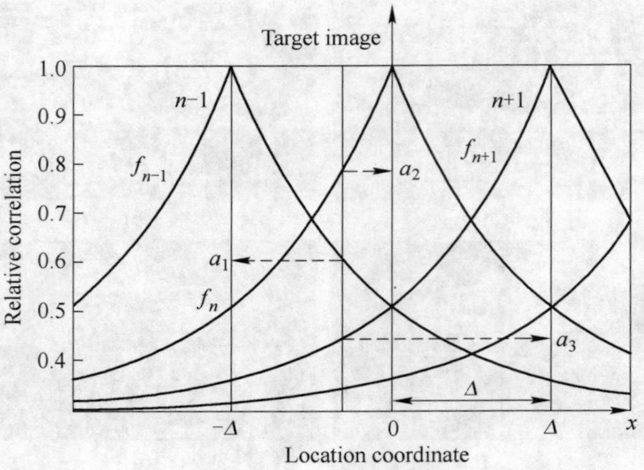

Fig. 6 The schematic diagram of one-dimensional MPE of remote sensing image recognition

Generally $\Delta_1 = \Delta_2 = \Delta$. When the target image is input into the VHC, the correlation spots can be parallelly detected. And the estimation equations can be established by using the brightness of different correlation spots. Now the number of the correlation spots used in the estimation should be decided. The number of the estimation spots is g(line p, row q, $g = p \times q$). As shown in Fig. 7, the location of the target image can be determined by the variable x, y. Then the estimation equation is

$$F(x,y) = \begin{cases} n_{11} = f_{11}(x,y) \\ m_{12} = f_{12}(x,y) \\ \vdots \\ m_{uv} = f_{uv}(x,y) \\ \vdots \\ m_{pq} = f_{pq}(x,y) \end{cases} \tag{5}$$

where $f_{11}(x, y)$, $f_{12}(x, y)$, $\cdots$, $f_{pq}(x, y)$ are the estimation functions, and $F(x, y)$ expresses the function group composed of the estimation functions above, and $m_{11}, m_{12}, \ldots, m_{pq}$ are the brightness of the used correlation spots. If the template image with sequences u and v is chosen as the benchmark image, according to the relationship of the template images, the other estimation functions can all be expressed by $f_{uv}(x, y)$, i.e.

$$F(x,y) = \begin{cases} m_{11} = f_{uv}[x + (u-1)\Delta, y + (v-1)\Delta] \\ m_{12} = f_{uv}[x + (u-2)\Delta, y + (v-1)\Delta] \\ \vdots \\ m_{uv} = f_{uv}(x,y) \\ \vdots \\ m_{pq} = f_{uv}[x + (u-p)\Delta, y + (v-q)\Delta] \end{cases} \tag{6}$$

Eq. (6) can be solved with g equations and 2 unknowns x and $y(g \geqslant 2)$, and the accuracy of x and y can be improved by increasing the number of the correlation spots, which is proved by the following experimental results.

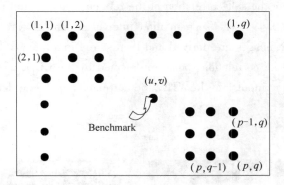

Fig. 7 The schematic diagram of estimation equation of correlation spots

3 Experimental results

The experimental setup is shown in Fig. 8, which is the same as the apparatus in literature[11]. The light source is a diode-pumped solid-state laser (DPSSL, λ = 532nm). A diffuser with 0.2° scattering angle is placed behind the SLM. The holograms are angle fractal multiplexed in a Fe: LiNbO$_3$ crystal. A CCD camera (MINTRON MTV-1881EX) is used to detect the correlation spots. The thickness of the recording medium is 15mm and the thickness of volume grating in the recording medium is about 6mm.

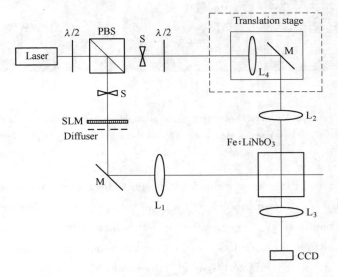

Fig. 8 Experiment setup for test the MPE method used in the VHC
PBS—polarizing beam splitter; SLM—spatial light modulator; S—shutter;
L$_1$-L$_4$—lenses; M—mirror; $\lambda/2$—half-wave-plate

A remote sensing reference image with size of 787 ×543 is used to test the MPE method. The template image has size of 640 ×480 limited by the SLM. The reference image after preprocessing(after three times thinning) is shown in Fig. 9. The parameters in estimation function Eq. (3) are derived by doing the self-correlation of the reference image, and we get $\alpha = 0.122, \beta = 0.160, a = 0.453, b = 0.491$. The segmentation interval Δ_1 and Δ_2 are chosen to be 3. And 1100(50 ×22) template images are derived and then stored into the VHC. When a white image is input into the VHC, the correlation spots are detected by the CCD, as shown in Fig. 10(a). When the target image is input into the VHC, the correlation spots are detected by the CCD, as shown in Fig. 10(b).

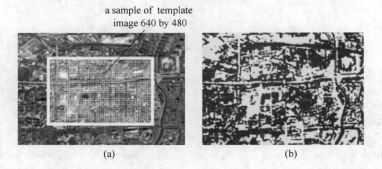

Fig. 9 An actual example of remote sensing image recognition
(a) reference image and a sample of the template image; (b) preprocessing of the reference image

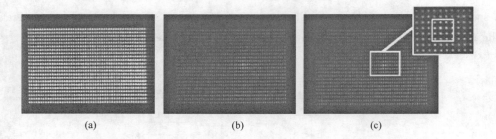

Fig. 10 The experiment of the multiple sample estimation
(a) detected correlation spots by inputting white image; (b) detected correlation spots by inputting the target image;
(c) 16(4 × 4) sample spots to be used

The top left corner of the reference image is chosen to be the coordinate origin(1,1) and a random target image with coordinate(83,33) is input into the VHC, which locates between the template images. Coordinate(X, Y) means the top left corner of the target image is X-1 pixels and Y-1 pixels apart from the coordinate origin along horizontal and vertical directions, respectively. The coordinate of the brightest spot is derived as (85,31) and the error radius ρ is 2.8. Around the brightest spot, we choose 4 × 4 spots(sampling number is 16) to estimate the location of the input target image, as shown in Fig. 10(c). The correlation result of the white image in Fig. 10(a) is used to be the base for the normalizing the brightness. And the normalized brightness of 4 × 4 spots is

$$\begin{bmatrix} 0.4816 & 0.5971 & 0.2231 & 0.4541 \\ 0.5607 & 0.6329 & 0.6221 & 0.4645 \\ 0.7736 & 0.7071 & 0.8141 & 0.5524 \\ 0.3122 & 0.4902 & 0.5751 & 0.3389 \end{bmatrix}$$

The image with brightness 0.3212 is chosen to be the benchmark image with coordinate (79, 28). The relationship of the estimation function can be inferred by the template images. And then the final equation can be written as

$$\begin{bmatrix} f_{11}(x,y) = 0.3212 & f_{11}(x-3,y) = 0.5324 & f_{11}(x-6,y) = 0.5429 & f_{11}(x-9,y) = 0.0198 \\ f_{11}(x,y-3) = 0.2135 & f_{11}(x-3,y-3) = 0.6035 & f_{11}(x-3,y-3) = 0.8789 & f_{11}(x-9,y-3) = 0.3869 \\ f_{11}(x,y-6) = 0.1206 & f_{11}(x-3,y-6) = 0.7878 & f_{11}(x-6,y-6) = 0.6057 & f_{11} = (x-9,y-6) = 0.4108 \\ f_{11}(x,y-9) = 0.4293 & f_{11}(x-3,y-6) = 0.6512 & f_{11}(x-6,y-6) = 0.1213 & f_{11}(x-9,y-9) = 0.3909 \end{bmatrix}$$

(7)

Usually, the location of the target image is determined by the real-time location of the remote sensing image system. Sometimes the location of the target image may coincide with the integer pixel of the reference image but sometimes may be inter-pixel. So we use the fraction coordinate to denote the location of the target image. After solving Eq. (7), $x = 4.3$ and $y = 4.4$. Since the coordinate of benchmark image is (79, 28), the location of the target image is (83.3, 32.4) and the error radius ρ is 0.7, one forth of that determined only by the brightest spot. In other words, using the MPE method, the recognition accuracy has been improved by about four times. For sixteen sampling correlation spots arbitrarily extracted around the brightest spot, the accuracy is approximately the same level.

Another random target image with the coordinate (34, 58), the same as one of the template images, is input. The coordinate of the brightest spot is derived as (37, 58) and the error radius ρ is 3.0. The normalized brightness of 4 × 4 spots around the brightest spot is

$$\begin{bmatrix} 0.3212 & 0.5324 & 0.5429 & 0.0198 \\ 0.2135 & 0.6035 & 0.8789 & 0.3869 \\ 0.1206 & 0.7878 & 0.6057 & 0.4108 \\ 0.4293 & 0.6512 & 0.1213 & 0.3909 \end{bmatrix}$$

In the similar manner, the image with brightness 0.4816 is chosen to be the benchmark image with coordinate (31, 52). Also we use the final equation and derive the result (34.5, 56.7). The error radius ρ is 1.4, which is half of that obtained only by the brightest spot.

217 random target images are input into the VHC to test the estimation error of different sample numbers in statistics. One part of data is shown in Table 1. The statistical error radius distribution can be obtained from the 217 groups of the correlation results, as shown in Fig. 11. As is shown in Table 1, for one group observation, the accuracy cannot be always improved by increasing the sample number because of the gross error of the individual spots. In statistics, the larger the sample number is, the smaller the maximum error radius and the higher the recogni-

tion accuracy are, which agrees with the probability theory and multiple estimation theory. As is shown in the Fig. 11, the sample number is from 1, 4, 16 to 36, and the maximum error radius is from 8, 6.2, 4.6 to 4. Then the curve of the relationship between the accuracy and the sample number is nearly as the reciprocal of the square root and it will approach the horizontal with the number increasing, which is shown in Fig. 12. Thus the accuracy can be improved by increasing the sample number, but it cannot be increased unlimitedly, which fits the theory simulation very well. In our experiment, the sample number 36(6 × 6) is nearly enough, and a larger sample number does not mean a higher accuracy.

Table 1　The error radius with different sample number

The coordinate of the input image	The error radius of the brightest spot	Sample number 4 (2 ×2)	Sample number 16 (4 ×4)	Sample number 36 (6 ×6)
[15,8]	2.7	1.0	1.9	1.2
[58,10]	3.5	3.3	1.3	0.6
[147,6]	1.3	1.0	0.8	0.5
[12,33]	4.4	4.2	2.1	1.2
[58,35]	3.1	2.9	1.2	0.8
[130,36]	4.2	3.7	2.6	1.0
[14,52]	1.5	1.6	1.1	0.9
[88,60]	1.8	1.2	0.9	0.8
[144,58]	5.3	2.9	3.1	1.7
[14,12]	6.2	5.3	3.8	2.1
[72,9]	7.8	5.6	4.2	2.8

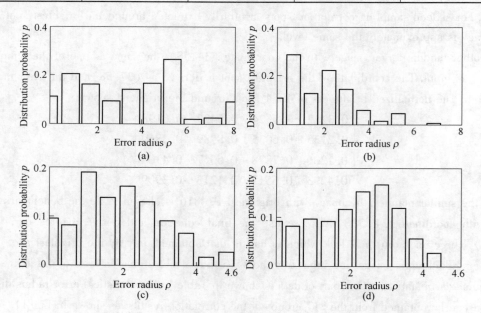

Fig. 11　The error radius distribution with different sample number
(a) only with the brightest spot; (b) sample number is 4(2 × 2); (c) sample number is 16(4 × 4); (d) sample number is 36(6 × 6)

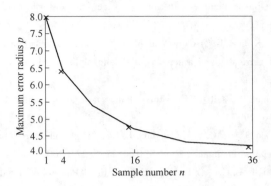

Fig. 12 Curve of relationship between the accuracy and increasing number

4 Conclusions

The MPE method is proposed to increase the remote sensing image recognition accuracy under the VHC, which makes good use of the characteristics of the high speed, high parallelism, multi-channel processing of the VHC and the characteristic of the stationary random of the remote sensing image. The image preprocessing, estimation curves fitting, template images preparation and the estimation equation establishing are important steps to accomplish the MPE method. The segmentation interval must be less than the correlation length. Most of these steps can be completed before the system begins to work. During the real-time correlation, only the pre-processing of target image and the simple estimation computation are needed, which cost little time and can meet the need of the high-speed application. The experimental results verify that the MPE method can improve the recognition accuracy of the remote sensing image. Furthermore the requirement of the storage capacity of the VHC is relaxed. Generally speaking, the larger the sample number is, the higher the recognition accuracy is. Because of the slope of the correlation curve and limited dynamic range of the CCD detector, more sampling spots can hardly narrow the error radius in a practical VHC system. The optimization of the MPE method should be in consideration in future.

References

[1] G. W. Burr, F. H. Mok, and D. Psaltis, "Large-scale volume holographic storage in the long interaction length architecture", Proc. SPIE 2297, 402-414(1994).

[2] Y. Takashima, and L. Hesselink, "Media tilt tolerance of bit-based and page-based holographic storage systems", Opt. Lett. 31(10), 1513-1515(2006).

[3] G. W. Burr, S. Kobras, H. Hanssen, and H. Coufal, "Content-addressable data storage by use of volume holograms", Appl. Opt. 38(32), 6779-6784(1999).

[4] B. J. Goertzen, and P. A. Mitkas, "Volume holographic storage for large relational databases", Opt. Eng. 35(7), 1847-1853(1996).

[5] L. Hesselink, S. S. Orlov, and M. C. Bashaw, "Holographic Data Storage Systems", in Proceedings of IEEE

Conference on Digital Object Identifier (Institute of Electrical and Electronics Engineers, New York, 2004), pp. 1231-1280.

[6] A. Heifetz, J. T. Shen, J. K. Lee, R. Tripathi, and M. S. Shahriar, "Translation-invariant object recognition system using an optical correlator and a superparallel holographic random access memory", Opt. Eng. 45(2), 1-5(2006).

[7] A. Pu, R. Denkewalter, and D. Psaltis, "Real-time vehicle navigation using a holographic memory", Opt. Eng. 36(10), 2737-2746(1997).

[8] K. Ni, Z. Y. Qu, L. C. Cao, P. Su, Q. S. He, and G. F. Jin, "High accurate volume holographic correlator with 4000 parallel correlation channels", Proc. SPIE 6827, 6827J(2007).

[9] K. Ni, W. Ren, Z. Y. Qu, L. C. Cao, Q. S. He, and G. F. Jin, "Phase-modulated multigroup volume holographic correlator", Opt. Lett. 33(10), 1144-1146(2008).

[10] C. Ouyang, L. C. Cao, Q. S. He, Y. Liao, M. X. Wu, and G. F. Jin, "Sidelobe suppression in volume holographic optical correlators by use of speckle modulation", Opt. Lett. 28(20), 1972-1974(2003).

[11] K. Ni, Z. Y. Qu, L. C. Cao, P. Su, Q. S. He, and G. F. Jin, "Improving accuracy of multichannel volume holographic correlators by using a two-dimensional interleaving method", Opt. Lett. 32(20), 2972-2975(2007).

[12] J. Capon, "A Probabilistic Mode for Run Length Coding of Picture", IEEE Trans. Inf. Theory 5(4), 157-163(1959).

[13] S. Fumihiko, "Image template matching based on edge-spin correlation", Electr. Eng. 153, 1592-1596(2005).

[14] S. D. Wei, and S. H. Lai, "Robust and efficient image alignment based on relative gradient matching", IEEE Trans. Image Process. 15(10), 2936-2943(2006).

[15] T. S. Huang, "PCM Picture Transmission", IEEE Spectr. 2, 57-63(1965).

[16] L. E. Franks, "A Mode for the Random Video Process", Bell Syst. Tech. J. 45, 609-630(1966).

[17] H. Andrew, Jazwinskl, *Stochastic process and filtering theory* (New York and London, 1970).

[18] P. M. Lundquist, C. Poga, R. G. Devoe, Y. Jia, W. E. Moerner, M.-P. Bernal, H. Coufal, R. K. Grygier, J. A. Hoffnagle, C. M. Jefferson, R. M. Macfarlane, R. M. Shelby, and G. T. Sincerbox, "Holographic digital data storage in a photorefractive polymer", Opt. Lett. 21(12), 890-892(1996).

[19] M.-P. Bernal, H. Coufal, R. K. Grygiel, J. A. Hoffnagle, C. M. Jefferson, R. M. Macfarlane, R. M. Shelby, G. T. Sincerbox, P. Wimmer, and G. Wittmann, "A precision tester for studies for holographic optical storage materials and recording physics", Appl. Opt. 35(14), 2360-2374(1996).

[20] R. V. Hogg, and A. T. Craig, Introduction to Mathematical Statistics(The Macmillan Company, 1959).

[21] A. Baraldi, and F. Paramiggiani, "An investigation of the textural characteristics associated withgray level cooccurrence matrix statistical parameters", IEEE Trans. Geosci. Rem. Sens. 3, 293-304(1993).

[22] C. Rafael, Gonzalez, Digital image processing(New York, 2005).

[23] R. M. Haralick, K. Shanmugan, and I. H. Dinstein, "Textural features for image classification", IEEE Trans. Syst. Man Cybern. 3(6), 610-621(1973).

Optical Fingerprint Recognition Based on Local Minutiae Structure Coding[*]

Abstract A parallel volume holographic optical fingerprint recognition system robust to fingerprint translation, rotation and nonlinear distortion is proposed. The optical fingerprint recognition measures the similarity by using the optical filters of multiplexed holograms recorded in the holographic media. A fingerprint is encoded into multiple template data pages based on the local minutiae structure coding method after it is adapted for the optical data channel. An improved filter recording time schedule and a post-filtering calibration technology are combined to suppress the calculating error from the large variations in data page filling ratio. Experimental results tested on FVC2002 DB1 and a forensic database comprising 270,216 fingerprints demonstrate the robustness and feasibility of the system.

1 Introduction

Automatic fingerprint identification system has been widely used in forensic and civilian applications. The recognition algorithm itself and how the algorithm should be implemented are the two critical issues attracting most of the research interests. While many algorithms and corresponding digital implementation systems are currently available, the time consumption and I/O traffic are becoming the bottlenecks of large scale digital systems[1]. Optical recognition has the merit of parallel processing and high data transfer rate. Specially, associative retrieval of volume holographic optical correlator offers a unique approach for fingerprint recognition because of its high storage density, integration of storing and computing, and multi-channel parallel filtering ability[2,3].

Previous studies have proposed several optical fingerprint recognition systems, including systems based on joint transform correlator(JTC) and volume holographic correlator(VHC). Fielding et al. built an optical fingerprint identification system using a binary JTC[4]. Grycewicz et al. utilized Fourier plane binarization and output peak intensity normalization techniques to improve the performance of a JTC system for fingerprint recognition[5]. Bal et al. used a dynamic neural-network based supervised filtering technique to enhance the fingerprint and the fringe-adjusted JTC algorithm for the identification process[6]. Yan et al. and Lee et al. demonstrated multi-channel matched correlators using photorefractive material that can output multiple correlation results with single input[7,8]. Watanabe et al. proposed an ultrahigh-speed compact optical correlation system using volume holographic disc that can be used for

---

[*] Copartner: Yao Yi, Liangcai Cao, Wei Guo, Yaping Luo, Jianjiang Feng, Qingsheng He. Reprinted from *Optics Express*, 2013,21(14):17108-17121.

fingerprint recognition[9].

The systems and methods described above show promising prospects of optical fingerprint recognition in the aspects of parallel processing and easy gallery data accessing. They perform well for pre-aligned, undeformed fingerprint images. However, the performances degrade greatly when fingerprint translation, rotation and nonlinear skin distortion occur, which is very common in practice. The advanced correlation filters combining multiple training images proposed by Kumar et al. [10] alleviated this problem, but it still needs multiple training templates and fingerprint alignment. The reason of performance degradation of optical systems mainly lies on the fact that most of these systems treat fingerprint images as general gray scale or binary pictures when constructing fingerprint templates.

The robustness problem of fingerprint recognition was addressed in digital algorithms by using local minutiae structure matching in the last decade[11-16]. Minutiae are the ridge ends and bifurcations of the fingerprint as shown in Fig. 2(a). Local minutiae structures encode the relationship between a minutia and its neighboring minutiae in terms of distances and angles. These attributes are invariant with respect to translation and rotation, and need no global alignment. Besides, since nonlinear distortions are trivial in local areas of a fingerprint, this method is also robust to global nonlinear skin distortion. However, the local structure encoding require for a very large memory size of fingerprint template[16], which drastically increases the time consumption and the demand for high data transfer rate during matching process.

This paper integrates a multi-page local minutiae structure coding method for fingerprint template construction in a volume holographic optical fingerprint recognition system, aiming at combining the advantages of both optical processing (parallel processing and high data transfer rate) and local minutiae structure matching (robustness to deformation).

The rest of this paper is organized as follows: Section 2 presents the algorithm selection criteria for optical recognition and the local minutiae structure coding method. Section 3 focuses on the optical considerations including optimizing the coding parameters, improving the filtering recording time schedule and compensating channel nonuniformity by postfiltering calibration. In Section 4, experiments conducted on two fingerprint databases are reported. Section 5 draws the conclusion.

2 Template data page construction with local minutiae structure coding

How the template data pages for a fingerprint are constructed is crucial for an optical fingerprint recognition system. Although many local minutiae structure coding and matching algorithms[11-16] have been proposed to cope with fingerprint deformation, not all of them are suitable for optical implementation.

2.1 Algorithm selection criteria for optical recognition

In a 90-degree geometry VHC system which is shown in Fig. 1, if M data pages $f_i(x_0, y_0)$ ($i = 1, 2, \cdots, M$) are stored as filters in the photorefractive crystal by angular multiplexing, the output intensity corresponding to the ith data page on the focal plane of L_1 can be expressed as:

$$I(f,f_i) = [g(f,f_i)]^2 \propto \left[\iint dx_0 dy_0 f(x_0,y_0) f_i^*(x_0,y_0) \right]^2 \quad (1)$$

when a query template(also called search argument)$f(x_0, y_0)$ is loaded to the spatial light modulator(SLM) in retrieval[17]. The correlation peak values between the search argument and all the stored gallery data pages are calculated in parallel. The intensity(grayscale) of each output correlation spot detected by the CCD stands for the square of the 2-D inner product.

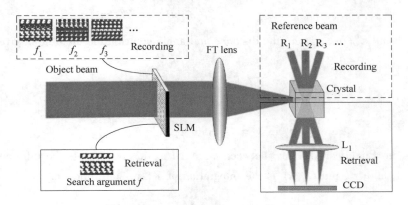

Fig. 1 Demonstration of the recording and retrieval processes in an angular multiplexing volume holographic correlator

SLM—spatial light modulator; FT lens—Fourier Transform Lens

The VHC architecture and working mode of optical correlation recognition exert several restrictions on the algorithms. The template encoding method and the corresponding matching algorithm should meet the following criteria: (1) The encoded template data page is a fixedlength representation which can be formatted into a 2-D image and modulated by the SLM. (2) The optical matching process has no further operation on the stored gallery templates such as translation and rotation once they are stored in the VHC. (3) The similarity between a gallery template and a query template can be evaluated based on the correlation between them.

2.2 Template data page construction with Minutiae Cylinder-Code(MCC)

After compared with many local minutiae matching algorithms[11-16], the MCC representation[16] could satisfy the previously described requirements. The basic idea of the MCC is to generate a minutia structure template for each minutia according to the spatial and directional relationships between the minutia and its fixed-radius(local) neighboring minutiae. Thus a fingerprint image is represented by N_m minutiae template data pages, where N_m is the minutiae number of that fingerprint. The fingerprint encoding method is named as multi-page local structure coding.

As is shown in Fig. 2, the neighborhood of minutia M is represented by a cylinder with radius R and height 2π. The base of the cylinder is centered on the minutia location (x_m, y_m), and aligned to the minutia direction θ_m. The cuboid enclosing the cylinder is discretized into $N_C = $

$N_A \times N_A \times N_H$ cells. Thus each cell(i, j, k) has a location $P_{i,j}$ and an angle φ_k. A numerical value is calculated according to the following sigmod equation:

$$\psi(i,j,k) = \psi(v) = Z(v,\mu,\tau) = \frac{1}{1 + \exp[\tau(v-\mu)]}, \quad (2)$$

where μ and τ are two parameters that limit the contribution of dense minutiae clusters, and v is calculated by accumulating distance and angle contributions from all the neighboring minutiae around the cell ($m_t \in N_{P_{i,j}}$). The accumulating contribution is defined as follows:

$$v(i,j,k) = \Sigma_{m_t \in N_{P_{i,j}}} [C_M^S(m_t, P_{i,j}) \cdot C_M^D(m_t, \varphi_k)] \quad (3)$$

$C_M^S(m_t, P_{i,j})$ and $C_M^D(m_t, \varphi_k)$ are the distance and angle contribution minutia m_t gives to cell(i, j, k):

$$C_M^S(m_t, P_{i,j}) = \frac{\exp(-d_S^2/2\sigma_S^2)}{\sigma_S \sqrt{2\pi}} \quad (4)$$

$$C_M^D(m_t, \varphi_k) = G_D\{d\phi[\varphi_k, d\phi(\theta_M, \theta_{m_t})]\} \quad (5)$$

where d_S is the Euclidean distance between m_t and $P_{i,j}$, $d\phi(\alpha_1, \alpha_2)$ is the difference between two angles α_1 and α_2, and $G_D(x)$ is the integration of a Gaussian function.

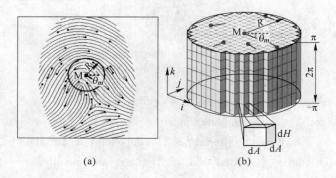

Fig. 2 A graphic representation of the local minutiae structure associated to a given minutia M
(a) the skeletonized fingerprint with its minutiae labeled; (b) the discretized cylinder of minutia M. The cylinder is rotated horizontally so that axis i is aligned to the corresponding minutia direction

It can be inferred from the definition of C_M^S, C_M^D and ψ that the value $\psi(i, j, k)$ attached to the cell(i, j, k) represents the likelihood of finding minutiae near the cell within a certain directional difference interval. The calculation of $\psi(i, j, k)$ only involves the relative distance and angle difference, both of which are irrelevant to fingerprint translation as well as rotation. Thus there is no need for fingerprint alignment and registration when using the MCC coding method. Besides, since each minutia cylinder is calculated locally (limited to the disk with radius R around that minutia), the MCC representation is also robust to global deformation brought by the elasticity of skin. Fig. 3 demonstrates the robustness of MCC representation. Although large variation occurs between the two fingerprints, the local structures and corresponding templates of minutia M are quite stable.

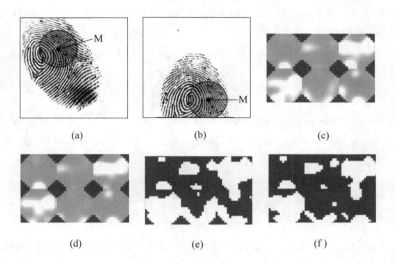

Fig. 3　Demonstration of MCC representation robust to fingerprint translation, rotation and elastic deformation
(a), (b) are two impressions from the same finger; (c)-(f) are the grayscale and binarized MCC templates for the corresponding minutia M labeled in the center of the circle in (a) and (b), respectively

After each minutia cylinder is calculated, a threshold can be chosen to binarize $\psi(i, j, k)$, which will be discussed in Section 3.2. The fingerprint image is now totally represented by its multiple binarized minutiae templates (cylinders). These templates are further selected and reshaped to form data pages to adapt to the two-dimensional optical data channel. The characteristics of the VHC system is carefully considered during these procedures.

3　Optical considerations

When it comes from the algorithm to optical implementation, several challenges have to be overcome besides the criteria described in section 2.1. These challenges include the relationship between the inner product and the similarity, large variation of the template data page filling ratio, and channel nonuniformity. This is because unlike traditional "hit-or-miss" method, we have to know not only which data page (representing a minutia) in the gallery is the most similar to the query data page, but also how similar they are.

3.1　Similarity between two minutiae using inner product

Since the intensities of correlation spots, which stand for the square of inner products, are employed to identify matching pairs of minutiae, it is important to establish the relationship between the inner product and template similarity. Traditional optical recognition uses inner product directly to represent similarity between the gallery and probe templates. However, this similarity computing method causes a problem that a probe template with higher filling ratio, which is defined as the number of pixels ON divided by the total number of pixels, tends to get higher inner products with all the gallery templates. The output cannot reflect the real similarity when the filling factor is not uniform for all the data pages. Here we propose the following approach for similarity calculation.

The similarity of two vectors $\mathbf{v}_1$ and $\mathbf{v}_2$ with the same length can be defined as

$$\gamma(\mathbf{v}_1,\mathbf{v}_2) = 1 - \frac{\|\mathbf{v}_1 - \mathbf{v}_2\|}{\|\mathbf{v}_1\|_p + \|\mathbf{v}_2\|_p} \tag{6}$$

where $\|\mathbf{v}\|_p$ denotes the p-norm of vector $\mathbf{v}$. It can easily be inferred that γ is always in the range $[0,1]$: One represents maximum similarity while zero denotes minimum similarity. If $\mathbf{v}_1$ and $\mathbf{v}_2$ are binarized vectors, and $p=1$, Eq. (6) can be simplified as

$$\gamma(\mathbf{v}_1,\mathbf{v}_2) = \frac{2 \times \sum_{i=1}^{N}[\mathbf{v}_1(i) \cdot \mathbf{v}_2(i)]}{N \times [fill(\mathbf{v}_1) + fill(\mathbf{v}_2)]} \tag{7}$$

where N is the length of vector $\mathbf{v}_1$ and $\mathbf{v}_2$, $fill(\mathbf{v}_1)$ and $fill(\mathbf{v}_2)$ are the filling ratios, $\mathbf{v}_1(i)$ and $\mathbf{v}_2(i)$ are the ith element of $\mathbf{v}_1$ and $\mathbf{v}_2$, respectively. This expression can easily be extended to 2-D case if we substitute the two vectors with two 2-D minutia template data pages:

$$\gamma(T,T_s) = \frac{2\sum_{i=1}^{Row}\sum_{j=1}^{Col}[T(i,j) \cdot T_s(i,j)]}{(Row \cdot Col)[fill(T) + fill(T_s)]} = \frac{2 \cdot \langle T,T_s \rangle}{(Row \cdot Col)[fill(T) + fill(T_s)]} \tag{8}$$

where Row and Col represent the numbers of rows and columns of a data page, and $\langle T, T_s \rangle$ is the inner product between T and T_s, $fill(T)$ and $fill(T_s)$ are the filling ratios.

In Eq. (8) the filling ratio $fill(T_s)$ of the search argument T_s affects the numerator and denominator simultaneously. The change of inner product caused by the change of filling ratio is counteracted by the denominator, thus $\gamma(T, T_s)$ can reflect the real similarity even when the filling ratios of different search arguments are not uniform.

3.2 Template data page binarization and filling ratio

Binarizing the template data pages provides better cross-correlation discrimination in the optical recognition system[4]. While choosing an appropriate filling ratio range, two factors are under consideration: optical energy passing through the SLM, and the discriminating power of minutiae templates. Low filling ratio may cause low signal-to-noise response both in the recording and retrieval process because of low optical energy pass. Meanwhile, if the filling ratio is too high, the differences between minutiae templates may be decreased because the inner products of both the matched and unmatched templates could be very high. Thus, the minutiae could be difficult to be discriminated and there is a trade-off between the two factors.

A number of experiments are conducted to determine the threshold. The results show that a threshold value of 0.1 could make most of the tested minutiae templates (from 800 different fingerprint impressions) fall into the interval $[0.1, 0.5]$, which is suitable both for storing and retrieval according to experiments in the VHC system. The filling ratio histogram of all the tested minutiae templates with binarization threshold $Thr = 0.1$ is shown in Fig. 4.

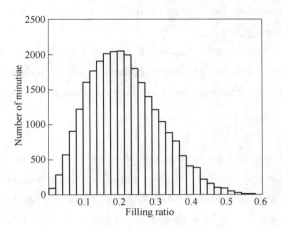

Fig. 4   The filling ratio histogram of all the tested minutiae templates with binarization threshold $Thr = 0.1$

After binarization, the minutiae templates with filling ratio less than 7.81% (corresponding to 120 cells out of a total of 1536 cells in a minutia cylinder) are labeled invalid and not used for recognition. The invalid templates are approximately 6.7% of all the minutiae templates (1798 /27073). The discard of these invalid templates would not degrade the recognition rate, yet it helps to avoid the long recording time and the low intensity response problems. The last step is to interleave each template[18] and reshape it to a 640 ×480 pixels image so as to match the pixel numbers of SLM. Four typical coded minutia data pages are shown in Fig. 5.

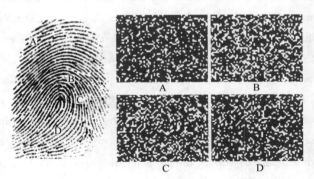

Fig. 5   Typical coded minutia data pages adapting to the SLM, with a size of 640 × 480 pixels

3.3   Channel nonuniformity and calibration

Now the challenge of the VHC system is how to capture diffraction spots that can truly reflect the inner products. The intensity of each spot on the image captured by CCD should be proportional to the inner product between the search argument and the gallery minutia data page in the corresponding channel. The main challenges come from large variation in filling ratio and channel-to-channel variations in diffraction efficiency. We adopt two approaches to solve these problems: improving filter recording time schedule in the recording process and post-filtering calibration in the retrieval process.

Recording time schedule is very important for the channel uniformity in multi-channel VHC. In order to achieve uniform diffraction efficiency for all channels, sequential exposure schedule was proposed[19-21]. The exposure time for each data page is:

$$t_{M-k} = \tau_r \ln\left[\frac{[k+1]T_M k}{kT_M - (k-1)}\right] \quad (k = 0,1,2,\cdots,M-1) \quad (9)$$

Where τ_r is the time constant of the crystal and M is the total number of stored data pages. As shown in Fig. 4, although the threshold of binarization is delicately selected, the filling ratios of the minutia data pages still vary from 7.81% to 59%. However, Eq. (9) ignores the influence of filling ratio variation and it can be regarded as the proto exposure time schedule based on the average filling ratio of all the data pages. Then each exposure time is modified by multiplying it with a factor, which is related to the ratio between the filling ratio of the corresponding data page and the average filling ratio of all the data pages. For a proto time schedule T_0, the modified time schedule T'_0 is experimentally determined as:

$$T'_0(k) = \begin{cases} t_k \cdot \dfrac{Ave\_fill}{fill(k)} & if \quad fill(k) \leq Ave\_fill \\ \max\left\{t_{\min}, t_k \cdot \sqrt{\dfrac{Ave\_fill}{fill(k)}}\right\} & if \quad fill(k) > Ave\_fill \end{cases} \quad (10)$$

where Ave_fill is the average filling ratio of all the data pages, $fill(k)$ is the filling ratio of the kth data page. This modification in recording exposure time schedule is intended to record correlation channels that are relatively uniform rather than rigorously uniform. It assures that all the data pages with a filling ratio between 7.81% and 59% could achieve a channel diffraction efficiency around that of a data page with the average filling ratio.

The residual differences are further compensated by post-filtering calibration in the retrieval process. The role of the calibration scheme is to ensure that all the residual nonuniformity of optical channels is compensated. This can be achieved by uploading a white image (all pixels ON) to obtain a base intensity, $I_B(k)$, of the response of each channel. Meanwhile, since the filling ratio of the white image is 100%, the expected response of each channel, $\bar{I}_B(k)$, is proportional to the square of the filling ratio of the stored minutia data page corresponding to that channel. This relationship is described as

$$\bar{I}_B(k) = A_B(k)^2 = \beta \cdot \langle T_k, W \rangle^2 = \beta \cdot [(Row \cdot Col) \cdot fill(T_k)]^2 \quad (11)$$

where $A_B(k)$ is the amplitude of the diffraction beam in the kth channel, $\langle T_k, W \rangle$ is the inner product of the kth minutia data page and the white page, $fill(T_k)$ is the filling ratio of the kth minutia data page, and β is the expected constant relation between inner product and diffraction intensity, epresenting the expected uniform diffraction efficiency of the VHC system. Define $\xi_{\text{calibration}}(k)$ as

$$\xi_{\text{calibration}}(k) = \frac{\overline{I_B(k)}}{I_B(k)} = \beta \cdot (Row \cdot Col)^2 \frac{fill(T_k)^2}{I_B(k)} \qquad (12)$$

then $\xi_{\text{calibration}}(k)$ can be taken as the intensity calibrating factor for the kth channel in the post-filtering to compensate the variations of channel diffraction efficiency. When a search argument T_s is input to the system, the calibrated inner product with the kth minutia data page can be calculated as

$$\langle T_s, T_k \rangle_{\text{calibration}} = \sqrt{I_s(k) \cdot \xi_{\text{calibration}}(k)/\beta} = (Row \cdot Col) \cdot fill(T_k) \cdot \sqrt{\frac{I_s(k)}{I_B(k)}} \qquad (13)$$

where $I_s(k)$ is the intensity of the kth correlation spot. Substitute Eq. (13) into Eq. (8),

$$\gamma(T_k, T_s) = \frac{2 \cdot fill(T_k)}{fill(T_k) + fill(T_s)} \cdot \sqrt{\frac{I_s(k)}{I_B(k)}} \qquad (14)$$

Thus, the calibrated expression of data page similarity is obtained from the intensity of correlation spot. $fill(T_s)$ requires only single calculation in the retrieval process. $I_B(k)$ and $fill(T_k)$ can be stored as a table which is easy to look up during post-processing. Fig. 6 illustrates the effect of calibration. The calibrated experimental similarities are more consistent with the theoretical similarities than the experimental similarities without calibration.

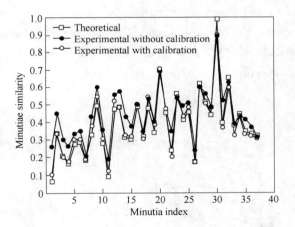

Fig. 6 Illustration of the similarities between one minutia in a fingerprint and all the 37 minutiae obtained both theoretically and experimentally

After minutiae similarities are calculated between all minutiae in two fingerprints, a single value is obtained to denote the overall matching score between the two fingerprints from their minutiae similarities. A simple technique called Local Similarity Sort (LSS)[14] is utilized for this purpose. The LSS technique sorts all the minutiae similarities between two fingerprints and selects the top n_p. The overall matching score is the average of the n_p similarities. The value of n_p is adaptively calculated based on the numbers of minutiae, n_A and n_B, of two fingerprints:

$$n_p = n_{\text{pmin}} + R\{Z(min(n_A, n_B), \mu_p, \tau_p) \cdot (n_{\text{pmax}} - n_{\text{pmin}})\} \qquad (15)$$

where Z is defined in Eq. (2), $R\{\cdot\}$ denotes the rounding operation, $min(\cdot)$ denotes selecting the minimum, and the parameters are experimentally set as $\mu_p = 20$, $\tau_p = 2/5$, $n_{pmin} = 4$ and $n_{pmax} = 12$.

4 Experimental demonstration

In this section, experiments aimed at evaluating the robustness and feasibility of the proposed VHC based optical recognition system are carried out and the matching results are reported.

The robustness to fingerprint translation, rotation and nonlinear distortion is demonstrated on a whole set of public testing database, and the feasibility for large scale database recognition is demonstrated on 270,216 fingerprints collected from a forensic department.

4.1 Optical setup and system working procedure

The experimental setup of the VHC system for testing is shown in Fig. 7. The light source is a 200mW diode-pumped solid-state laser ($\lambda = 532$nm). The SLM used to modulate the amplitude of the object beam is a transmissive twisted nematic liquid crystal display (TNLCD), with a resolution of 640×480 and pixel size of $9\mu m \times 9\mu m$. A diffuser of 0.2° scattering angle is placed behind the SLM to suppress the side-lobes and cross talk[22]. The binarized working mode simplifies the amplitude linearization calibrating of the SLM. The expanded beam is split by the PBS into reference beam and object beam during the recording process, and an Fe:LiNbO$_3$ crystal is utilized for storing the template images. Then shutter ST$_1$ is shut down during the retrieval process, and a CCD of 768×576 pixels, with pixel size of $14\mu m \times 14\mu m$ and 8-bit depth, is employed for detecting the correlation spots array.

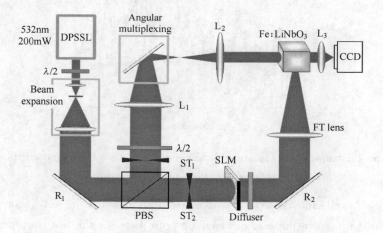

Fig. 7 Experimental setup of the volume holographic optical correlator system
PBS—polarizing beam splitter; SLM—spatial light modulator; ST$_1$, ST$_2$—shutters; L$_1$-L$_3$—lens; R$_1$, R$_2$—reflectors; $\lambda/2$—half-wave plate

The working procedure of the recognition system comprises the following four main steps (Fig. 8):

(1) Data page construction. The minutiae of each fingerprint are digitally extracted using a

commercial fingerprint recognition software, VeriFinger[23]. For each extracted minutia, an MCC data page is constructed and filtered using the modified coding method described in Section 2.2.

(2) Data page storing. The encoded data pages of the database are sequentially stored into the Fe: LiNbO$_3$ crystal by using 2-D angular multiplexing. The recorded holograms are filters for the fingerprint recognition. Hologram fixing is not used in the current demonstration since the stored data pages work well in the short-time experiments (less than 24 hours). In long-time practical applications, thermal fixing can be used to prevent the loss of fingerprint information[24].

(3) Fingerprint retrieval. All the minutia data pages of the query fingerprint are sequentially uploaded to the SLM as search arguments. For each search argument, an array of correlation spots can be obtained in parallel.

(4) Post-filtering. Minutiae similarities of all the minutiae pairs are digitally obtained from the spots array. Fingerprint matching scores are further calculated with the LSS technique described in Section 3.3.

Step 1 and step 4 require a computer and can be done in less than a second for a single search task. The database storing process of Step 2 is a time-consuming task. Fortunately it is pre-finished off-line and only needs to be done once. Compared to the digital fingerprint recognition, the real merits of the optical fingerprint recognition system is reflected in step 3 (the fingerprint retrieval step). This is because data transfer from the memorizer to the calculator is not needed and the inner products are optically calculated in parallel, which would otherwise be the main causes of time consumption and I/O traffic in a digital system if millions of fingerprint identification should be performed.

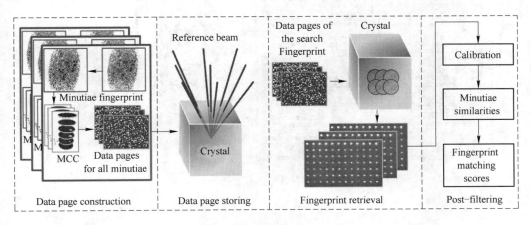

Fig. 8 Flow chart of the VHC fingerprint recognition system

4.2 Robustness test

The system robustness to fingerprint deformation is tested on a public collection of fingerprint images proposed in FVC2002[25]. The benchmark data is labeled DB1, which comprises 800 flat fingerprint images captured at a resolution of 500dpi, with an optical sensor, from 100 different fingers (eight impressions per finger). Remarkable deformations including translation (as large as 150 pixels), rotation (as large as ± 40 degrees) and nonlinear distortion exist in these impressions. Fig. 9 shows eight impressions of a sample finger in the database.

Fig. 9　Eight different impressions of the same sample finger(#1) in FVC2002 DB1, remarkable translation, rotation and nonlinear distortion exist

Fig. 10 demonstrates the minutiae matching results between two impressions of the same finger. A pair of minutiae whose similarity is above a predefined threshold is determined as a matching pair and linked. Although remarkable translation, rotation and nonlinear distortion exist and no pre-alignment is conducted before data page construction, only one pair of minutiae is wrongly linked (marked with blue line). A correct "Match" decision is successfully achieved with the matching score calculated from the matched minutiae.

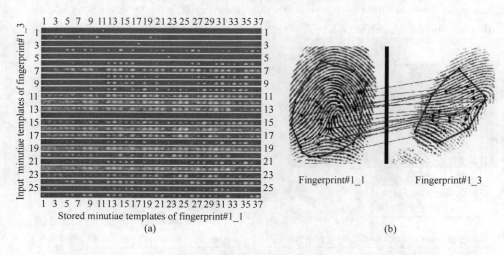

Fig. 10　Minutiae matching results of two impressions of the same finger
(a) optical correlation results between minutia data pages of fingerprint #1_1 (37 minutiae) and #1_3 (26 minutiae);
(b) corresponding minutiae pairs linked according to optical correlation results

The 800 fingerprints are divided into 8 subsets. Each subset has 100 different fingerprint impressions. A total of 3250 minutia template data pages of subset #1 are stored in a common volume of the crystal as gallery templates by angular multiplexing. Then all the 800 fingerprints are searched over the whole gallery to test the system robustness. Fig. 11 demonstrates part of the optical recognition results. The matching scores of the fingerprints with no deformation, small deformation and large deformation are all above the chosen threshold and genuine fingerprint pairs are correctly matched. It verifies the robustness of the proposed optical recognition system to fingerprint deformation.

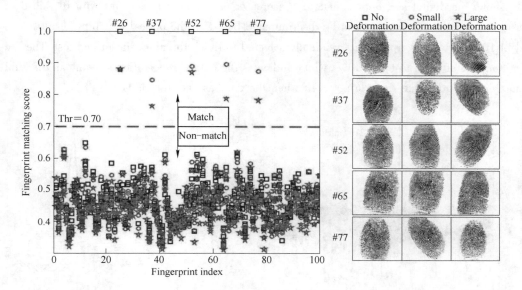

Fig. 11　Part of the matching results of the fingerprints with no deformation, small deformation and large deformation when searched over the database stored in the optical system

4.3 Feasibility test

The most concerned challenge for the optical recognition system to be used in large scale database (>100,000 fingerprints) recognition task is the system feasibility. Can the system discriminate the genuine fingerprint from its similar fake fingerprints in such a large database? The system feasibility requires both the matching algorithm and the optical system can discriminate two similar fingerprints. In this work, it is tested on a database of 270,216 rolled fingerprints collected from a forensic department both digitally and optically.

The testing protocol is as follows. A sample fingerprint is randomly chosen from the database to be the query fingerprint. The similarities between it and all the fingerprints in the database are digitally calculated. The 20 fingerprints with the highest matching scores are selected as the sub-database to represent the whole database in the feasibility test. Obviously, the query fingerprint is included in the sub-database and the other 19 fingerprints are the most similar ones with the query fingerprint. The sub-database represents the fingerprints which are the most difficult ones to discriminate from the query fingerprint in the whole database. It can be inferred that the genuine sample fingerprint can be recognized from the whole database if it can be recognized from the sub-database. Then the sub-database (their minutiae templates) are encoded and stored in the crystal as filter bank and the query fingerprint is optically retrieved.

Fig. 12(a) demonstrates the digitally calculated matching scores between the randomly selected 120,091th fingerprint and the whole database. The fingerprints with the highest scores are marked with triangular and selected as the sub-database. The matching score difference between the matched and unmatched fingerprints verifies the feasibility of the algorithm. Fig. 12 (b) shows the digital and optical matching scores of one sample fingerprint with its sub-database. The matched fingerprint can be discriminated based on the optical matching results. A total of 100 sample fingerprints are randomly selected to finish the above-described test. The recognition rate of these tests is 100%, which indicates the optical recognition system can discriminate genuine from fake fingerprints even when the database is quite large.

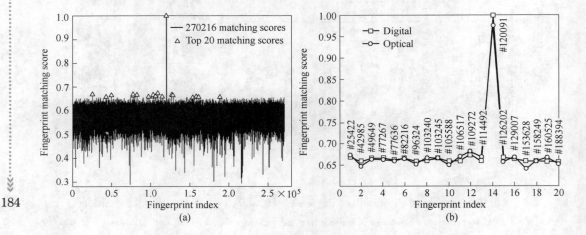

Fig. 12 (a) Matching scores of the 120,091th fingerprint with all the 270,216 fingerprints in the database;
(b) Comparison of digitally and optically calculated matching scores between the
sample fingerprint and its corresponding most similar 20 fingerprints

The discriminating power comes from the different template data pages and different multi-page combinations. Without fingerprint encoding, the real discriminative information may be submerged by the redundant correlation noise if all the information is retained, especially when fingerprint distortion occurs. The feature extraction and binarizationoperation cause some information lost compared with using the fingerprint image directly. However, these operations can retain most of the discriminative information while removing the redundant correlation noise between fingerprints, provided that the encoding parameters such as the MCC fixed-radius R and the binarization threshold are optimized. The experiments show that the recognition accuracy of the optical system after the multi-page fingerprint encoding is improved on the whole.

The proposed method could be more efficiently implemented in a holographic disc based configuration with shift multiplexing. Up to ten million data pages can be stored in a regular size photopolymer holographic disc with a multiplexing displacement of approximately 20μm. The stored database can be quickly retrieved with the high-speed rotation of the holographic disc.

5 Conclusions

We integrate a local minutiae structure coding method into the VHC based fingerprint recognition system in order to promote the robustness of the optical system against translation, rotation and nonlinear distortion. The hybrid system decomposes each fingerprint into multiple data pages by using a modified local minutiae structure coding method adapted for the optical data channel. The robustness of the system comes from the invariant properties of local structure coding and no pre-alignment of fingerprint is needed. The modified similarity calculation method, binarization and post-filtering calibration are combined to suppress the similarity calculating error. Experiments conducted on a public database FVC2002 DB1 including 800 deformed flat fingerprints and a forensic database including 270,216 rolled fingerprints verify the robustness and feasibility of the optical system.

References

[1] D. Maltoni, D. Maio, A. K. Jain, and S. Prabhakar, Handbook of Fingerprint Recognition (Springer, 2009), Chap. 4, pp. 167-233.

[2] P. A. Mitkas and G. W. Burr, "Volume holographic optical correlators", in Holographic Data Storage, H. J. Coufal, D. Psaltis, and G. T. Sicebox, eds. (Springer-Verlag, 2000), pp. 429-446.

[3] B. V. K. Vijaya Kumar, A. Mahalanobis, and R. D. Juday, Correlation Pattern Recognition (Cambridge University, 2005), Chap. 8, pp. 295-356.

[4] K. H. Fielding, J. L. Horner, and C. K. Makekau, "Optical fingerprint identification by binary joint transform correlation", Opt. Eng. 30(12), 1958-1961(1991).

[5] T. J. Grycewicz, "Techniques to improve binary joint transform correlator performance for fingerprint recognition", Opt. Eng. 38(1), 114-119(1999).

[6] A. Bal, A. M. El-Saba, and M. S. Alam, "Improved fingerprint identification with supervised filtering enhancement", Appl. Opt. 44(5), 647-654(2005).

[7] Y. Yan, G. Huang, W. Feng, G. Jin, and M. Wu, "Multichannel wavelet correlators for fingerprint identification by the use of associative storage in a photorefractive material", Proc. SPIE 3458, 259-266(1998).

[8] S. H. Lee, S. Y. Yi, and E. S. Kim, "Fingerprint identification by use of a volume holographic optical correlator", Proc. SPIE 3715, 321-325(1999).

[9] E. Watanabe, A. Naito, and K. Kodate, "Ultrahigh-speed compact opticalcorrelation system using holographic disc", Proc. SPIE 7442, 74420X(2009).

[10] B. V. K. Vijaya Kumar, M. Savvides, C. Xie, K. Venkataramani, J. Thornton, and A. Mahalanobis, "Biometric verification with correlation filters", Appl. Opt. 43(2), 391-402(2004).

[11] X. Jiang and W. Y. Yau, "Fingerprint minutiae matching based on the local and global structures", in Proceedings of the 15th International Conference on Pattern Recognition2 (Institute of Electrical and Electronics Engineers, New York, 2000), pp. 1038-1041.

[12] N. K. Ratha, R. M. Bolle, V. D. Pandit, and V. Vaish, "Robust fingerprint authentication using local structural similarity", in Proceedings of the 5th IEEE Workshop on Applications of Computer Vision (Institute of Electrical and Electronics Engineers, New York, 2000), pp. 29-34.

[13] J. Feng, "Combining minutiae descriptors for fingerprint matching", Pattern Recognit. 41(1), 342-352

(2008).

[14] A. A. Paulino, J. Feng, and A. K. Jain, "Latent fingerprint matching using descriptor-based hough transform", IEEE Trans. Inf. Foren. Sec. 8(1), 31-45(2013).

[15] J. Dai, J. Feng, and J. Zhou, "Robust and efficient ridge-based palmprint matching", IEEE Trans. Pattern Anal. Mach. Intell. 34(8), 1618-1632(2012).

[16] R. Cappelli, M. Ferrara, and D. Maltoni, "Minutia Cylinder-Code: a new representation and matching technique for fingerprint recognition", IEEE Trans. Pattern Anal. Mach. Intell. 32(12), 2128-2141(2010).

[17] G. W. Burr, S. Kobras, H. Hanssen, and H. Coufal, "Content-addressable data storage by use of volume holograms", Appl. Opt. 38(32), 6779-6784(1999).

[18] K. Ni, Z. Qu, L. Cao, P. Su, Q. He, and G. Jin, "Improving accuracy of multichannel volume holographic correlators by using a two-dimensional interleaving method", Opt. Lett. 32(20), 2972-2974(2007).

[19] F. H. Mok, M. C. Tackitt, and H. M. Stoll, "Storage of 500 high-resolution holograms in a $LiNbO_3$ crystal", Opt. Lett. 16(8), 605-607(1991).

[20] A. Adibi, K. Buse, and D. Psaltis, "Multiplexing holograms in $LiNbO_3$: Fe: Mn crystals", Opt. Lett. 24(10), 652-654(1999).

[21] K. Curtis, K. Anderson, and M. R. Ayres, "M/# requirements for holographic data storage", in Proceedings of the Optical Data Storage Topical Meeting, IEEE, 9-11(2006).

[22] L. Cao, Q. He, C. Ouyang, Y. Liao, and G. Jin, "Improvement to human-face recognition in a volume holographic correlator by use of speckle modulation", Appl. Opt. 44(4), 538-545(2005).

[23] http://www.neurotechnology.com/vf_sdk.html

[24] X. An, D. Psaltis, and G. W. Burr, "Thermal fixing of 10,000 holograms in $LiNbO_3$: Fe", Appl. Opt. 38(2), 386-393(1999).

[25] FVC2002, http://bias.csr.unibo.it/fvc2002/default.asp

二元光学

计算机源生的全息光学元件（COHOE）的合成及优化设计*

摘 要 本文讨论了计算机源生的全息光学元件（COHOE）的合成及优化设计问题，文后给出了一个实际的结果，得到了特殊曲面的波面。

1 引言

1966年，Brown 和 Lohmann 首次提出了使用计算机绘制全息图的方法[1]，并将这种全息图定名为计算机产生的全息图，一般称之为计算全息图（Computer Generated Holograms，简称 CGH）。计算全息图原则上可以产生任意波面，但由于计算机容量、绘图密度等客观因素的限制，一般只能用来产生空间带宽积较小的波面，且衍射效率很低（约5%），使得这种方法灵活性的优点只能在很小的范围内得以发挥。因此，有必要发展一种组合的技术，把一般光学全息和计算全息及体积全息的优点结合起来，使其名副其实。

本文将对这种组合技术的方法及一般设计原则作进一步的讨论，文后给出了一个初步实验的结果。

2 HOE 的合成

广义而论，HOE 的功能是波面变换，其变换作用可用 $\varphi_H(x,y)$ 来描述：

$$\varphi_H(x,y) = \varphi_{out}(x,y) - \varphi_{in}(x,y) \tag{1}$$

式中，φ_{out}、φ_{in} 分别是输出及输入波面的相位分布。

众所周知，合成一个 HOE 需要有物光和参考光，因此有：

$$\varphi_H = \varphi_{obj} + \varphi_{ref} \tag{2}$$

式中，φ_{obj}、φ_{ref} 分别是物光和参考光的相位分布，以下为简明起见，将各个 $\varphi(x,y)$ 的变量略去，记作 φ。这里所说的物光和参考光仅仅是为了区别两个不同波面而借用的全息术习惯用语，并无全息术中本来的物理意义，在以后的公式中物光、参考光均可互易。

显然，φ_{obj}、φ_{ref} 的复杂程度决定了 φ_H 所能实现的变换的复杂性。根据 φ_{obj}、φ_{ref} 的产生方式，我们把 HOE 的合成分为下述三种类型：

2.1 光学+光学方式（O+O）[2]

这种方法只适宜产生与基准球面或柱面偏差不大的波面，有很大的局限性。

---

* 本文合作者：陆达。原发表于《光学学报》，1985，5（7）：594～599。

2.2 计算全息+光学方式（C+O）[3,4]

C+O方式中引进CGH调制，大大突破了球面波的局限，原则上可得到任意波面，但CGH空间带宽积小的缺点并没得到克服。

2.3 改进的计算全息+光学合成方式[(C+O)+O]

1980年，J. R. Fienup等提出了用计算机全息图产生光学元件的方法，并将这种元件定名为计算机源生的全息光学元件（Computer Originated Holographic Optical Elements，简称COHOE）[5]。本文主要讨论它。

Fienup的基本思想是将φ_{obj}分解为常规和非常规两部分。常规的部分由普通光学元件承担，记作φ_1；非常规的部分由CGH产生，记作φ_2。这种合成方式既保留了HOE具有较高空间带宽积的优点，又具有较高的灵活性。因此，它比一般的HOE有更大的潜在应用价值。Fienup曾用这种方式合成了一个旋转对称的非球面傅氏透镜，目前国内外也仅此一例报道。

图1（a）、(b)给出了COHOE的两种可能的合成光路。COHOE的相位表达式为：

$$\varphi_H = (\varphi_{1obj} + \varphi_{ref}) + \varphi_{2obj} \tag{3}$$

φ_1和φ_2的分配可以有许多方式，视不同需要而定，但作为一般原则，应使φ_2负担最小。将式（3）与式（1）相联，可以得到COHOE的输出与输入的关系：

$$\varphi_{out} = \varphi_{in} + (\varphi_{1obj} + \varphi_{ref}) + \varphi_{2obj} \tag{4}$$

图1（a）中的φ_1由参考光源产生，而在图1（b）中则由L_3产生，φ_1可以是球面，也可以是柱面波。

这种技术不仅可以合成旋转对称的波面，原则上可以产生任意的波面，但要再次强调指出的是，如果(C+O)+O的合成方式不与厚膜全息记录结合起来，则衍射效率很低，没有太大的实际意义。因此，在实际制作COHOE时，必须采用厚膜全息的记录材料。

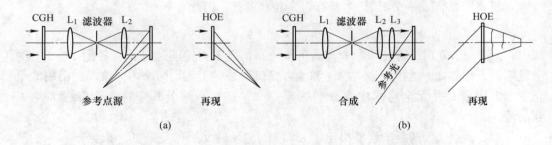

图1　(C+O)+O合成方式（a）和（b）

3　COHOE的优化设计问题

由式（4）可知，当φ_{obj}确定以后，COHOE的波面变换性质即已确定。但在COHOE设计中，确定φ_{obj}、φ_{ref}往往是很困难的。困难在于许多情况下不能用一个统一的

解析函数来描述 φ_{in} 和 φ_{out}，而只能把它们写成若干解析函数集合的形式，即：$\{\varphi_{in}^{(k)}, \varphi_{out}^{(k)}\}$，$k=1,\cdots,N$。例如傅氏变换透镜中不同空间频率的入射光及在频谱面会聚的位置就只能用一一对应的集合写出，对每个 $\varphi_{in}^{(k)}$ 和 $\varphi_{out}^{(k)}$ 都有：

$$\varphi_H^{(k)} = \varphi_{out}^{(k)} - \varphi_{in}^{(k)} \tag{5}$$

但往往找不到一个 φ_H，使它对所有的 k 都满足式（5），而只能得到一个误差最小的 $\hat{\varphi}_H$。这就是所谓的优化问题，也是 COHOE 设计中要引入优化设计的原因。

定义入射波面 $\varphi_{in}^{(k)}(x,y)$ 的波矢量为 $I^{(k)}(x,y)$；出射波面 $\varphi_{out}^{(k)}(x,y)$ 的波矢量为 $O^{(k)}(x,y)$；COHOE 优化波面 $\hat{\varphi}_H(x,y)$ 的波矢量为 $H(x,y)$。对绝大多数实际应用的光场而言，$\varphi_{in}^{(k)}$、$\varphi_{out}^{(k)}$、φ_H 均满足连续可微、光滑平顺的条件，因而可按下述准则进行优化：

$$\min\left\{\sum_{k=1}^{N}\left|H(x,y) - O^{(k)}(x,y) + I^{(k)}(x,y)\right|^2\right\} \tag{6}$$

其中

$$H(x,y) \triangleq (1/2\pi)[(i\partial\varphi_H/\partial x) + (j\partial\varphi_H/\partial y)]$$
$$O^{(k)}(x,y) \triangleq (1/2\pi)[(i\partial\varphi_{out}^{(k)}/\partial x) + (j\partial\varphi_{out}^{(k)}/\partial y)]$$
$$I^{(k)}(x,y) \triangleq (1/2\pi)[(i\partial\varphi_{in}^{(k)}/\partial x) + (j\partial\varphi_{in}^{(k)}/\partial y)]$$

这样得到的 $\hat{\varphi}_H$ 称为最小二乘意义下的优化解。

在许多情况下，COHOE 是光学系统中的一个元件（见图2），因此 $I^{(k)}$ 要通过前光学系统追迹而得，而 $O^{(k)}$ 可通过后光学系统反向追迹求出。

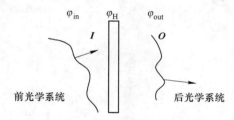

图2 光学系统中的 COHOE

利用式（3），我们可以得到优化目标函数新的表达式：

$$R \triangleq \sum_{k=1}^{N}\{[(\partial\hat{\varphi}_{obj}/\partial x) + (\partial\varphi_{ref}/\partial x) - (\partial\varphi_{out}^{(k)}/\partial x) + (\partial\varphi_{in}^{(k)}/\partial x)]^2 +$$
$$[(\partial\hat{\varphi}_{obj}/\partial y) + (\partial\varphi_{ref}/\partial y) - (\partial\varphi_{out}^{(k)}/\partial y) + (\partial\varphi_{in}^{(k)}/\partial y)]^2\} \tag{7}$$

式（6）的问题转化为求 $\min\{R\}$。令：

$$\hat{\varphi}_{obj}(x,y) \triangleq \sum_{i=0}^{M_1}\sum_{j=0}^{M_2} c_{ij} x^i y^j \tag{8}$$

$$\hat{\varphi}_{obj}(x,y) \triangleq \sum_{i=0}^{M_2}\sum_{j=0}^{M_1} c_{ij} L_i(x) L_j(y) \tag{9}$$

上式中的 c_{ij} 是待定的系数，$L_i(x)$ 是 i 阶关于 x 的正交多项式，如 Legendre 多项式。同样的说明也适用于 $L_i(y)$。从计算的角度出发，式（9）更为实用。但为直观起见，本文在做理论分析时采用了式（8）。

由

$$\partial R/\partial c_{ij} = 0 \quad i=0,1,\cdots,M_1; j=0,1,\cdots,M_2 \quad (i,j \text{ 不同时为 } 0) \tag{10}$$

可得到 $(M_1+1)(M_2+1) - 1$ 个关于 c_{ij} 的线性方程组，解之可得到 c_{ij}。

4 COHOE 合成实例

我们按 (C + O) + O 的思想实际制作了一块 COHOE，用于处理合成孔径雷达（SAR）的图像。根据系统的设计结果，COHOE 所产生的波面函数为：

$$\varphi_H(x,y) = (2\pi/\lambda)\{[x^2/f_{xx}(y)] + [y^2/f_y]\} - (2\pi/\lambda)\sin\theta_x \quad (11)$$

式中，$(2\pi/\lambda)\sin\theta_x$ 为 φ_{ref}，是斜入射的平面参考波；$f_{xx}(y)$ 是 y 的函数，其随 y 变化的规律见图 3。由式（11）可知，$\varphi_H(x,y)$ 在 y 方向是一个柱面波，焦距 f_y；在 x 方向是一个锥面波，焦距 $f_{xx}(y)$。图 3 中用"△"标出的是 $df_{xx}(y)/dy$ 和 y 的函数关系，显然各 y 值的导数不等，即锥面波的母线是弯曲的。式（11）所描述的波面的立体示意图见图 4。这样复杂的波面用一般的方法是不易产生的。

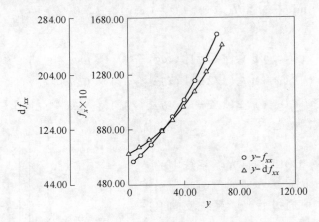

图 3　$y - f_{xx}(y)$，$y - df_{xx}(y)/dy$ 曲线 HOE

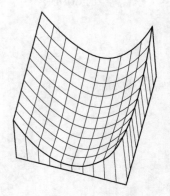

图 4　COHOE 波面立体图

经分析，我们在 φ_{obj} 中引入 CGH 调制，其中柱面波的部分由光学元件承担，CGH 只描述曲母线的锥面波，即

$$\varphi_{1obj} = 2\pi y^2/\lambda f_y \quad (12)$$

$$\varphi_{2obj} = 2\pi x^2/\lambda f_{xx}(y) \quad (13)$$

使用的合成装置原理图见图 5。我们可以把图中的球面镜设想成两个正交的柱面透镜组合，CGH 的波面经过两个柱面透镜后仍保留了 φ_{2obj} 的形式，只是 $f_{xx}(y)$ 有所变化；而球面镜中的另一维柱面透镜则用来产生 φ_{1obj}，这两个波面是自然正交的。

实际使用的光路见图 6。合成所用的干板是重铬酸盐明胶干板，由河北化工学院提供，处理工艺与一般重铬酸盐明胶干板无异。

图 7 是合成光路中的计算机全息图照片。图 8 为使用我们合成的 COHOE 所处理出的合成孔径雷达（SAR）图像。图像中的接缝痕迹是由于机械运动的误差造成的，与 COHOE 的光学质量无关。

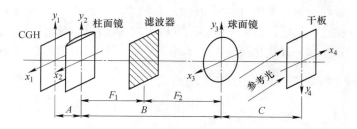

图 5　COHOE 合成原理图

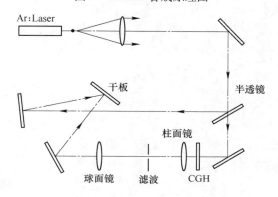

图 6　实际使用的 COHOE 合成光路

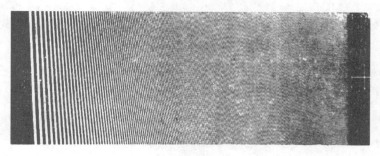

图 7　合成光路中使用的 CGH

图 8　COHOE 处理的 SAR 图像

在我们的实际例子中，由于优化对象的特殊性，需要取较多的抽样点（$N=405$），因此整个优化运算需要较大的计算机贮存，约 600×10^3 字节。这对刻制计算全息图造成了很大的障碍。一般刻图机所配的计算机内存只有几万字节左右，由大计算机到小计算机的数据交换也存在一些问题，因而使得第三节讨论的优化设计未能从物理上实现，以下仅给出计算结果。图9（a）是优化前的像点点列图分布，图9（b）是优化后校正的结果。

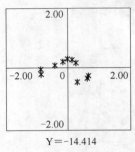 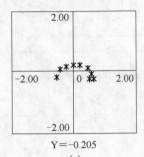 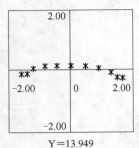

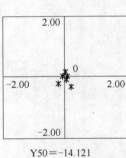 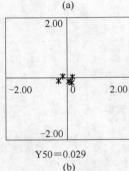 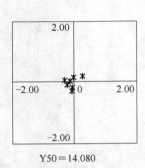

图9　像点分布
（a）优化前；（b）优化后

在本工作过程中，曾得到北京邮电学院、北京师范学院、北京光电技术研究所、清华大学材料力学实验室及河北化工学院等有关同志的大力支持和卓有成效的合作，谨此表示谢意。

参 考 文 献

[1] B. R. Brown, A. W. Lohmann. Appl. Opt., 1966, 5(6):967.
[2] W. C. Sweatt. Appl. Opt., 1978, 17(8):1220.
[3] S. K. Case. Appl. Opt., 1978, 17(16):2537.
[4] Chang-yuan Han, et al. Appl. Opt., 1983, 22(22):3644.
[5] J. R. Fienup. AD-A085219, 1980.

Composition and Optimum Design of Computer Originated Holographic Optical Elements（COHOE）

Abstract　This paper discusses the composition and optimum design of the computer originated holographic optical elements （COHOE）. The preliminary experimental result with a special wavefront is presented.

电子束计算全息图的制作*

摘　要　本文论述了利用电子束曝光机制作计算全息图的方法及其优点。并得到了制作 Lohmann 型计算全息图和计算全息微分滤波器的初步实验结果。

1　制作方法及其优点

自 A. W. Lohmann 1966 年发明计算全息术以来，由于数字计算机和绘图仪器的普及，这个领域得到了迅速发展，并推动了全息术的发展和应用。但由于绘图精度和分辨率的原因，计算全息图的空间频率受到限制。而使用电子束曝光机制作全息图就有可能较好地解决这一问题。

电子束计算全息图是指用电子束曝光机扫描获得的计算全息图。电子束曝光技术不仅为集成电路工业提供了一种全新的超微细图形加工手段，而且促进了声表面波、磁泡、超导等新型半导体器件的进展。同样，也在光学工程中获得应用。利用它制作计算全息图的工作，在国外已有一些开展，我们也开始作了尝试。与传统的计算全息相比，电子束计算全息图有其独特的优点。(1) 线条细，空间频率高；(2) 误差源少，制作精度高；(3) 工序简单，制作方便，效率高。

本实验使用的 DB-3 型电子束曝光机（长沙半导体设备研究所设计研制[2]），它由电子束曝光机和电子计算机控制系统两部分组成。主机包括电子光学镜筒、激光干涉定位工作台、电气控制及真空系统等部分组成（如图 1 所示）。

DB-3 型机利用内存 4K 的专用计算机进行自动控制，使光闸将电子束通断，磁偏转放大以 $0.3\mu m$ 束斑在单元面积内逐点扫描。用激光干涉仪定位提供单元图形之间的步进重复和拼结。电子束计算全息图的制作工艺包括数据处理、扫描曝光、显影定影和坚膜腐蚀等。可以根据图形的要求选用不同的电子束感光胶。

电子束计算全息图为计算全息术开辟了新的前景。用传统方法难以制作或不能制作的计算全息图，如散射板计算全息图，用于校正大波差的非球面计算全息图和计算全息扫描器等，都将可以用电子束计算全息来解决。

2　实验结果

我们利用电子束曝光机制作了 Lohmann 型"字符"的计算全息图，微分滤波器，取得了初步成功。

---

* 本文合作者：严瑛白，余东校。原发表于《仪器仪表学报》，1986，7(1)：83～86。

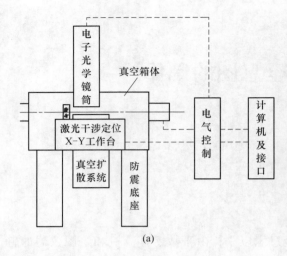

图 1　DB-3 型电子束曝光机图
(a) 示意图；(b) 外形图

2.1　Lohmann 型计算全息图

电子束曝光机配有扫描矩形的软件，特别适合于制作 Lohmann 型计算全息图，只要给定矩形孔径的起点和增量，就能自动扫描出所需要的各矩形图形。

对于傅里叶变换全息图中各单元的复值函数 u_{jk}，即

$$u_{jk} = C \sum_{m=-M/2}^{M/2} \sum_{n=-M/2}^{M/2} f(n,m) \exp\{-2\pi i[(mj/M) + (nk/N)]\} \tag{1}$$

按照 Lohmann 提出的迂回相位方法进行编码，每一抽样单元的矩形孔径中有

$$W_{nm} = A_{nm}, \quad P_{nm} = \varphi_{nm}/2\pi \tag{2}$$

式中，A_{nm}、φ_{nm} 分别为 (n, m) 单元的振幅和相位；W_{nm}、P_{nm} 分别为矩孔的高度（或宽度）、中心距，如图 2 所示。

使用电子束曝光机制作了 Lohmann Ⅰ型和Ⅱ型傅里叶变换全息图。图 3 为 Lohmann Ⅱ型的再现字符"巾"。曝光条件：加速电压 30 kV，电子束束斑约 0.3 μm，采用 2×2 扫描场，数模阶梯 13 位，扫描频率 100 kc，采用 COP 电子束感光胶制作于铬版上。计算机控制电子束曝光机按图 2 的几何尺寸扫描图形，考虑到机器的系统误差，编制了误差修正程序，以消除机器的系统误差对全息图的影响，如电子束的

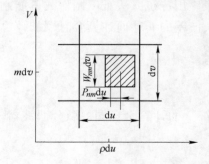

图 2　矩形单元图

图 3　Lohmann Ⅱ型再现象

邻近效应，工作台位移的过冲量等。

傅里叶变换全息图的再现象噪声与抽样单元数有关。增加抽样数可以增加信噪比。由于 DB-3 型电子束曝光机的内存有限，本全息图计算单元仅为 32×32，影响了再现象的质量。

2.2 计算全息微分滤波器[4]

微分滤波器是一种用来突出图像边缘信息的滤波器，一维微分滤波器的滤波函数为[3]：

$$H(u,v) = 2\pi i v \Delta x \tag{3}$$

其分量形式

$$A(u,v) = 2\pi |v| \Delta x$$

$$\varphi(u,v) = \begin{cases} \pi/2 & v \geq 0 \\ -\pi/2 & v < 0 \end{cases} \tag{4}$$

滤波器 $H(u,v)$ 可以用全息照相或计算全息制取，而计算全息微分滤波器更有吸引力，因为计算全息容易按照数学公式制作各种理想的滤波器。与全息微分滤波器相比，计算全息微分滤波器的微分阈值低、噪声小、制作方便。且因载频高，能够处理大面积图像，而得到清晰的各级衍射图像。

根据公式（3），可以有多种编码方法制取计算全息微分滤波器。我们制作了两种不同编码方式的一维微分滤波器，得到了相同的结果。一维微分滤波器的最细线条为 $1\mu m$，面积 $5.3mm \times 5.3mm$，400 条线，制作在铬版上。

微分图像的实验在 4f 处理系统进行。图 4 为一维微分图像实例，原始物体为"±"。经过微分滤波后，中央零级像保持原样，两边 ±1 级衍射象即为物体的一维微分图像，如图 5 所示。

图 4　原始物体"±"　　　　　　　　图 5　一维微分图像

依据一维微分滤波器，可以把它推广到二维的结果。其滤波函数为

$$H(u,v) = u + iv \tag{5}$$

使用电子束制作二维微分滤波器，最小扫描单元 $1\mu m \times 1\mu m$。为了保证 45° 斜线的微分效果，沿 v 轴上半平面图形与下半平面图形相错 $\pi/2$ 位相。图 6 为原始物体"G"，物体面积 $15mm \times 15mm$，图 7 为其微分图像。

 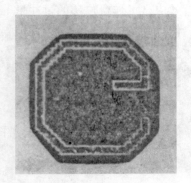

图 6 原始物体 "G"　　　　　　　　　图 7 二维微分图像

3　结束语

电子束计算全息特别适合于制作空间频率高的计算全息图。空间频率比绘图缩版法可以提高一个数量级，与光学全息相近。

参 考 文 献

［1］B. R. Brown, and A. W. Lohmann. Appl. Opt., 5(1966).
［2］长沙半导体设备研究所四室. 半导体工艺设备，1(1982).
［3］Wan-Hon Lee. Progress in Optics, 16(1978).
［4］A. W. Lohmann. Appl. Opt., 7(1968).
［5］R. C. Gonazolex, P. Wintz. Digital image Processing.
［6］R. M. Scott. Appl. Opt., 8(1969).

A Computer-generated Electron-beam Holograms

Abstract　A new approach for generating holograms by electron-beams is described. Primary results of Lohmann-type CGH (computer-generated holograms) and differential filters are presented.

Optimization of Grating Multi-beamsplitters*

Abstract We derive a numerical method of calculating binary grating structures, which have predetermined spatial spectra with a typical accuracy of less than one percent. As an experimental verification, we present test results of a grating that converts a plane wave into a central block of 11 × 11 uniformly diffracted waves.

1 Introduction

Modulation of the periodic structure of a binary diffraction grating to shape the distribution of intensities of diffracted orders has acquired considerable interest in recent years. This kind of generalized (holographic) gratings can be conveniently used as image multipliers and multi-beam beamsplitters in optical information processing and fiber-optic communication systems. Several authors have considered the problem of calculating structures, which give a central block of equally intense diffracted beams[1-5]. Perhaps the most successful of the various approaches is the one developed by Killat et al. in Ref. [5], in which several one-dimensional structures are presented, with diffraction efficiencies of 70% ~ 85% and nonuniformities of 0.3% ~ 1.7%.

In this investigation, we derive a considerably simple gradient-type optimization algorithm, which gives results comparable to those of Ref. [5]. Our method can, in addition, be used to design grating structures with nonuniform, predetermined spectra. We also analyze the accuracy required in the fabrication process of the grating structure, and fabricate a 11 × 11 beam two dimensional uniform beamsplitter.

2 Spectra of binary diffraction gratings

A general binary amplitude grating is a periodic structure, which has only two different values of amplitude transmission $f(x)$, as illustrated in Fig. (1). Clearly, the grating and its spectrum are completely determined by the set $\{a_l, b_l\}$, $l = 1, \cdots, L$, of transition point positions of $f(x)$. Since $f(x)$ is periodic, it can be represented in the form of Fourier series,

$$f(x) = \sum_{m=-\infty}^{\infty} F(m) \exp\{2\pi i m x\} \tag{1}$$

By straightforward calculation, the Fourier coefficients of the grating structure are found in terms of the transition point positions

---

* 本文合作者:Jari Turunen,卞新高。原发表于《光学学报》,1988,8(10):946~953.

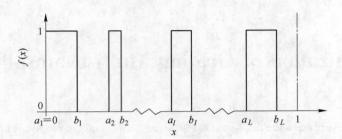

Fig. 1 Amplitude transmittance $f(x)$ of a general binary amplitude grating over one period

Length of the period is normalized to unity

$$F(0) = \sum_{l=1}^{L} (b_l - a_l) \tag{2}$$

for order $m = 0$ and

$$F(m) = \frac{1}{2\pi m}[F_R(m) + iF_I(m)] \tag{3}$$

for higher orders. In Eq. (3),

$$\begin{cases} F_R(m) = \sum_{l=1}^{L} [\sin(2\pi m b_l) - \sin(2\pi m a_l)] \\ F_I(m) = \sum_{l=1}^{L} [\cos(2\pi m b_l) - \cos(2\pi m a_l)] \end{cases} \tag{4}$$

The power spectrum $\{P(m)\}, m = -\infty, \cdots, \infty$, of the structure, defined as

$$P(m) = |F(m)|^2 \tag{5}$$

gives the intensities to various diffraction orders. It can now be calculated from the known transition point positions with the aid of Eqs. (2) ~ (4). Note that $P(-m) = P(m)$, which indicates that the spectrum of a binary diffraction grating in symmetrical with respect to the undiffracted beam.

A binary phase grating is a periodic structure, which has a uniform amplitude transmission and only two different values, say $\pi/2 + \theta$ and $\pi/2 - \theta$, of phase retardation. The complex amplitude transmittance function $g(x) = f(x)\exp\{-i\theta(x)\}$ is then of the form

$$g(x) = \pm \sin\theta + i\cos\theta \tag{6}$$

for all $x \in [0,1]$. The power spectrum of this phase grating is simply related to the power spectrum of the corresponding amplitude grating (which has the same transition point positions). Straightforward calculation, where use is made of the identity

$$\int_0^1 \exp\{-2\pi i m x\} dx = 0$$

and linearity of sum and integral operators, gives the Fourier coefficients $\{G(m)\}$ of a general binary phase grating. We get

$$G(0) = [2F(0) - 1]\sin\theta + i\cos\theta \tag{7}$$

for order $m = 0$ and

$$G(m) = 2\sin\theta F(m) \tag{8}$$

for higher orders. Now the power spectrum $\{P(m)\}$ of the binary phase grating, defined as $P(m) = |G(m)|^2$, can be calculated from the transition point positions with the aid of Eqs. (2) ~ (4), (7) and (8).

3 Optimization of the grating structure

Our aim is to find a set $\{a_l, b_l\}, l = 1, \cdots, L$, of transition point positions (and a phase angle θ if it is a free variable) in such a way that the power spectrum $\{P(m)\}$, $|m| = 0, \cdots, M$, closely approximates some desired power spectrum $\{\hat{P}(m)\}$, with good diffraction efficiency

$$P_E = P(0) + 2\sum_{m=1}^{M} P(m) \tag{9}$$

This optimization problem can be analytically solved only in the case of a three-beam ($m = 0$, ± 1) phase grating. For example, a uniform set of three diffracted beams is given by a structure $a_1 = 0$, $b_1 = 1/2$ and $\theta = \arctan(\pi/2) \approx 57.52°$. Diffraction efficiency into three central orders is $P_E = 12/(\pi^2 + 4) \approx 86.5\%$.

In more complicated cases, numerical optimization techniques have to be used. As a measure of the progress of optimization, we define a merit or error function by the expression

$$E^2 = \alpha\{[P(0) - P_E\hat{P}(0)]^2 + 2\sum_{m=1}^{M}[P(m) - P_E\hat{P}(m)]\} + (1 - \alpha)(1 - P_E)^2 \tag{10}$$

where α is a parameter within the range $[0, 1]$. The first term in the merit function is a measure of the similarity of distributions $\{P(m)\}$ and $\{P_E\hat{P}(m)\}$, and the second term is proportional to the diffraction efficiency of the structure.

Due to the considerable complexity of the dependence of the merit function on the transition point positions $\{a_l, b_l\}$ and on phase angle θ, it generally has large number local minima. Starting from arbitrary initial transition point positions, one of these minima can be found by the following gradient-type iteration technique.

Let the transition points in the beginning of the nth iteration step be $\{a_{l,n}, b_{l,n}\}$ and the value of the error function be E_m. A small trial change is made in each $a_{l,n}$ and $b_{l,n}$, and corresponding change of the error function, $\Delta E_{a_{l,n}}$ and $\Delta E_{b_{l,n}}$ are calculated by using the formalism of the previous section. A new gradient in the $2L$-dimensional space of transition point positions is determined by equations

$$\begin{cases} \Delta a_{l,n} = -\gamma\Delta E_{a_{l,n}}/|\Delta E_{\max,n}| \\ \Delta b_{l,n} = -\gamma\Delta E_{b_{l,n}}/|\Delta E_{\max,n}| \end{cases} \tag{11}$$

where γ is a free parameter and

$$\Delta E_{\max,n} = \max\{\Delta E_{a_{l,n}}, \Delta E_{b_{l,n}}\}$$

With the aid of these equations, the error E_n can be interpreted as a function of γ. As illustrated in Fig. 2, the function $E_n(\gamma)$ is well behaved around the minimum, and only a few simple evaluations of the merit function are required to find $\gamma_{\min,n}$ with a sufficient degree of accuracy. Once $\gamma_{\min,n}$ is found, new values for the transition point positions are found by equations

$$\begin{cases} a_{l,n+1} = a_{l,n} - \gamma_{\min,n}\Delta E_{a_{l,n}}/|\Delta E_{\max,n}| \\ b_{l,n+1} = b_{l,n} - \gamma_{\min,n}\Delta E_{b_{l,n}}/|\Delta E_{\max,n}| \end{cases} \quad (12)$$

If the set $\{a_{l,n+1}, b_{l,n+1}\}$ is not an acceptable solution, a new iteration step is performed.

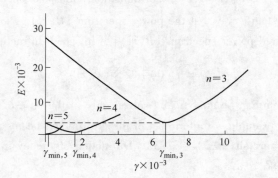

Fig. 2 Illustration of the progress of iteration in a realistic case of a uniform five-beam grating. Starting from randomly chosen transition point positions, five iteration steps were needed to obtain an acceptable optimum; the calculated dependence $E_n(\gamma)$ is shown for the last three steps

If the phase angle is a free variable, it can be treated in the iteration process just as any one of the transition point positions. However, change of phase angle does not affect the relative intensities of the diffraction orders $|m|>0$, as can be deduced from Eq. (8). Hence optimization of phase angle can be done algebraically. For example, if the aim distribution is a central block of uniformly diffracted beams, the intensity of the zero-order beam can be equated to the average intensity of the higher-order beams by the formula

$$\sin\theta_{m+1} = \sin\theta_n[1 + P_{\max,n} - P_n(0)]^{1/2} \quad (13)$$

in the end of each iteration step.

Free phase angle adds an additional degree of freedom in optimization. This is valuable especially when the number of diffracted orders in the aim distribution is low. For example, if the phase angle of a uniform three-beam grating is fixed to $\pi/2$, the maximum efficiency that can be obtained is $P_E = 66\%$, compared with 87% if the phase angle is a free variable. For a large number of diffracted orders, the optimal phase angle is often close to $\pi/2$.

The iteration procedure outlined in this section gives a local minimum of error function from almost any starting distribution in less than ten iteration steps. If the number of free parameters in the grating structure is adequately large ($L \geqslant M$, see Ref. [2]) and $\alpha = 1$ is chosen in merit function, the maximum relative deviation from the normalized aim distribution

$$\eta = \max_m \left\{ 1 - \left| \frac{P(m)}{P_E \hat{P}(m)} \right| \right\} \tag{14}$$

is often less than 1%. However, when the desired number of diffraction orders becomes large, many randomly chosen starting distributions may be required to find a structure with good efficiency, and there seems to be no simple method to ascertain that the minimum obtained is the global minimum. If the local minimum closest to the initial transition point positions yields low efficiency, it is often possible to force the iteration toward another, better local minimum by choosing $\alpha < 1$ for a while. The phase retrieval algorithm of Ref. [4], which as such does not generally give solutions with $\eta < 10\% \sim 20\%$, may also be employed to find suitable starting systems.

4 Spectrum shaping examples

In this section, we give some examples of the use of the optimization algorithm to design grating structures with specific spectra. We begin by defining some practically important aim distributions. An ideal uniform distribution is defined as

$$\hat{P}_U(m) = \frac{1}{2M+1}, \quad |m| = 0, \cdots, M \tag{15}$$

and $\hat{P}_U(m) = 0$ if $|m| > M$. Triangle distribution is defined by the requirement that the difference between intensities to two adjacent diffraction orders remains constant up to order M, after which intensities are equal to zero. This condition is equivalent with $\hat{P}_T(m) - \hat{P}_T(m+1) = \hat{P}_T(0)/(M+1)$ or $\hat{P}_T(m) = \hat{P}_T(0)[1 - m/(M+1)]$. By requiring that the efficiency to M central orders is unity, and using the properties of arithmetic series, we obtain for the triangle distribution an expression

$$\hat{P}_T(m) = \frac{1}{M+1}\left(1 - \frac{m}{M+1}\right), \quad |m| = 0, \cdots, M \tag{16}$$

and $\hat{P}_T(m) = 0$ if $|m| > M$. Another interesting nonuniform distribution, which we call power distribution, is characterized by the requirement that the ratio of two adjacent orders is a constant ε. Algebraically, $\hat{P}_P(m) = \hat{P}_P(0)\varepsilon^m$. Taking the efficiency to $2M+1$ central orders to be unity, and using properties of geometric series, we get

$$\hat{P}_P(m) = \frac{1-\varepsilon}{1+\varepsilon - 2\varepsilon^{M+1}}\varepsilon^m, \quad |m| = 0, \cdots, M \tag{17}$$

and $\hat{P}_P(m) = 0$ if $|m| > M$.

A five-beam power distribution with $\varepsilon = 1/2$, $P_E \approx 75.9\%$ and $\eta \approx 0.18\%$ is given by the structure

$$a_1 = 0, \quad b_1 = 0.3540$$

$$a_2 = 0.73900, \quad b_2 = 0.81125$$

and phase angle $\theta \approx 57.55°$. A five-beam triangle distribution with $P_E \approx 72.1\%$ and $\eta \approx 0.08\%$ is given by the structure

$$a_1 = 0, \qquad b_1 = 0.05794$$
$$a_2 = 0.52433, \quad b_2 = 0.81299$$

and phase angle $\theta \approx 66.30°$.

As an example of uniform distribution, we design a two-dimensional 11×11-beam beamsplitter. Fabrication problems strongly suggest that a two-dimensional phase grating should have only two levels of phase modulation, which sets the restriction that the phase angle has to equal $\pi/2$. A solution

$$a_1 = 0, \qquad b_1 = 0.06857$$
$$a_2 = 0.20885, \quad b_2 = 0.44467$$
$$a_3 = 0.52930, \quad b_3 = 0.72101$$
$$a_4 = 0.72854, \quad b_4 = 0.86437$$

was found. The maximum deviation from uniformity was $\eta \approx 0.47\%$. Diffraction efficiency to 11 central orders was $P_E \approx 76.4\%$ for one-dimensional structure and $P_E^2 \approx 58.49$ for two-dimensional structure. We also analyzed the accuracy required in the fabrication process of the grating. If the transition points were rounded to four decimals, uniformity was $\eta \approx 0.54\%$. Rounding to three decimals gave $\eta \approx 2.69\%$, and two decimals gave $\eta \approx 16.8\%$.

5 Experiment

As an experimental verification, we fabricated the uniform 11×11 beamsplitter designed in previous section. The structure shown in Fig. 3(a) was written, photoreduced and multiplied; the grating consisted of 10×10 periods, length of one period was $633\mu m$. A phase grating corresponding to the structure shown in Fig. 3(a) was prepared by ion beam etching. The far field diffraction pattern of the grating, photographed in the focal plane of a Fourier-transform lens, is shown in Fig. 3(b).

Because of an error in phase angle, the zero-order spot is more intense than the others, which show reasonable uniformity. This error is partly due to the finite number of grating periods, which causes noticeable structure (see Fig. 3(b)) especially around the zero-order spot. Maximum deviation from the uniform aim distribution in the upper row of uniformly diffracted beams in Fig. 3(b) was measured to be 9.5%. The difference between measured and theoretical uniformities is due to manufacturing inaccuracies in locations of the phase transition points. These inaccuracies were, according to the discussion in the end of the previous section, estimated to be a little less than 1% of the grating period, or of the order of $5\mu m$. Better uniformity could naturally be obtained by more sophisticated fabrication techniques, like electron beam lithography.

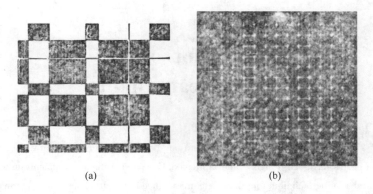

Fig. 3 Experiment results
(a) a grating structure yielding a uniform central block of 11 × 11 diffraction orders; (b) the far field diffraction pattern of the grating, photographed in the focal plane of a Fourier-transform lens

6 Conclusions

The optimization method presented in this paper was found to be successful in design of holographic beamsplitter structures, which give arbitrary predetermined spectra. The optimization program we prepared is efficient enough to be run in a personal computer, and it gives the transition point positions with an accuracy that is beyond most practical fabrication techniques, which therefore set the limit to the accuracy of the spatial spectra.

References

[1] H. Damman and K. Görtler, "High-efficiency in-line multiple imaging by means of multiple phase holograms", *Optics Communications*, 3:312, 1971.

[2] H. Dammann and E. Klotz, "Coherent optical generation and inspection of two-dimensional periodic structures", *Optica Acta*, 24:505, 1977.

[3] Wai-Hon Lee, "High-efficiency multiple beam gratings", *Applied Optics*, 18:2153, 1979.

[4] Wu Zhouling, Lu Da, and Jin Guofan, "Application of an iterative method in the design of high-efficiency binary diffraction gratings", *Acta Optica Sinica*, 6:567, 1986.

[5] U. Killat, G. Rabe, and W. Rave, "Binary phase gratings for star couplers with high splitting ratio", *Fiber and Integrated Optics*, 4:159, 1982.

多光束分光光栅的优化设计

摘　要　本文描述了一种设计计算二元光栅结构参数的数值方法，这种方法可以把光栅衍射谱的分布误差控制在1%以内。作为实验验证，本文给出了一个二元光栅谱分布的测试结果，该光栅将入射平面波转换成在中央11×11级次内均匀分布的衍射波。

The Fabrication of a 25 × 25 Multiple Beam Splitter[*]

Abstract A kind of beam splitter which can split one beam of a laser into an array of 25 × 25 uniform intense beams has been designed and fabricated. Compared with the commonly used designing method and fabricating technique (ion beam etching method), we introduce a new improved numerical algorithm and an economical method to design and manufacture this kind of binary phase grating beam-splitters. The design and manufacture is done by nonlinear equations method and chemical etching technique, respectively. We further present the actual results of these gratings which show uniformity better than 96% and diffraction efficiency (splitting ratio) over 60%.

1 Introduction

In many fields such as image multiplication systems, fiber-optic communications, and optical digital processing systems, an important optical element is usually used which can convert one incident beam into a one-or two-dimensional array of equally intense laser beams. So far several fabricating methods have been reported, such as the use of volume holographs, waveguides, fibers, and gratings. However, when the requirement of uniformity and spots number increases, the last method seems to be the most efficient one. This kind of grating, which can equally split a laser beam into multiple beams, was first designed by Dammann in 1971 as an image multiplier[1], so we also call it Dammann gratings. Nowadays, this kind of grating is developed and used as a multichannel element in parallel processing systems, so with the laser source we can simultaneously process a lot of digital signals. The development shows that an optical logic element has been fabricated in a density of 10^7 devices in one square centimeter[2], but elements which can access the logic gate array can only reach a density of 100 channels per square centimeter by using 10 × 10 holographs. In our approach a Dammann grating is used to meet the needs and it shows that higher density and splitting ratio can be reached.

In order to search for the suitable binary phase grating structures which can give out the feature of equal-splitting, several methods have been reported in Refs. [1,3,4]. But the most successful one seems to be the method stated in Ref. [4] which is developed by Killat et al. with the augmented Lagrangian function. Other improvements were also presented in Ref. [5] by Turunen et al. in which the problem is treated as a nonlinear optimization. But usually iterations do not converge.

In this paper we derive another considerably simple algorithm in which we treat the problem

---

[*] Copartner: Xuenong Lu, Ying Wang, Minxian Wu. Reprinted from *Optics Communications*, 1989, 72(3,4):157-162.

as nonlinear equations and try to find out the optimized root. Several solutions including some actual results will be presented. A beam-splitter, which can convert one laser beam into a central block of 25 × 25 beam array with the same intensity distribution, is chemically etched and gives a good verification.

2 The spectrum distribution of a binary phase grating

The structure of a general binary phase grating is showed in Fig. 1. Note that this is only in one dimension, actually the grating is extended into two dimensions and this periodic structure is photoreduced and multiplied in both the X and Y axis. A description of the outline is

$$f(x) = \sum_{i=1}^{(M+1)/2} \frac{x - (x_{2i} + x_{2i-1})/2}{x_{2i} - x_{2i-1}} \tag{1}$$

here $x_0 = 0$, $x_{M+1} = 1$, $x \in [0,1]$, M is an odd number. Then the complex transmittance function would be

$$g(x) = [2f(x) - 1]\sin\varphi + i\cos\varphi \tag{2}$$

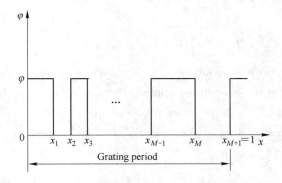

Fig. 1 The outline of the binary phase grating ($\{x_i\}$ are the relative positions of the transition points, φ is the phase shift)

We then perform the Fourier transformation to it and get the spectrum distribution of a general binary phase grating. Suppose that

$$G(k) = F\{g(x)\}, \quad (k = 0, \pm 1, \pm 2, \pm 3, \cdots)$$

then for zero order

$$G(0) = (1 - \sum_{m=1}^{M+1}(-1)^m x_m)\exp(i\varphi) + \sum_{m=1}^{M+1}(-1)^m x_m \tag{3}$$

and for non-zero orders ($k = \pm 1, \pm 2, \pm 3, \ldots$)

$$G(k) = \frac{\sin(\varphi/2)}{i2\pi k} \times \sum_{i=1}^{(M+1)/2} [(\sin 2\pi k x_{2i} - \sin 2\pi k x_{2i-1})] + i(\cos 2\pi k x_{2i} - \cos 2\pi k x_{2i-1}) \tag{4}$$

so the power spectrum of the grating, defined as

$$I_k(x,\varphi) = |G(k)|^2 \quad (k = 0, \pm 1, \pm 2, \cdots) \tag{5}$$

gives the intensities to various diffraction orders,

$$I_0(x,\varphi) = 1 + 4\sin^2(\varphi/2) \times \left[\sum_{m=1}^{M+1}(-1)^m x_m\right]\left[\sum_{m=1}^{M}(-1)^m x_m\right] \tag{6}$$

and

$$I_k(x,\varphi) = \frac{2\sin^2(\varphi/2)}{\pi^2 k^2} \times \sum_{m=2}^{M+1}\sum_{j=1}^{m-1}(-1)^{m+j}\cos[2\pi k(x_m - x_j)] \tag{7}$$

3 The derivation of suitable grating structures

Our purpose is to try to find some binary structures which can diffract equal intense orders in the range of -Nth to +Nth far field diffracting orders. The problem then becomes the derivation of a set of transition sudden-changing points stated as $\{x_i\}$, where $i = 0, 1, 2, \cdots, M+1$, and a proper phase shift φ if it is treated as a free variable[6,7]. Mathematically the requirement can be described as

$$I_{-N} = I_{-N+1} = \cdots = I_{-1} = I_0 = I_1 = \cdots = I_N \tag{8}$$

As the spectrum is symmetrical about the non-diffracted beam, zero order we get $I_k(x,\varphi) = I_{-k}(x,\varphi)$, so the number of equations in relation (8) should reduce by half. Meanwhile, in order to take advantage of all the laser power we also hope to make the diffraction efficiency as high as possible. That means most of the energy should be concentrated in the central orders from $-N$ to $+N$. Eqs. (8) become

$$I_0(x,\varphi) = I_1(x,\varphi) = \cdots = I_{N-1}(x,\varphi) = I_N(x,\varphi) = \text{maximum} \tag{9}$$

From Eq. (9) we establish nonlinear equations like

$$I_k(x,\varphi) = p\hat{I}_k(x,\varphi), \quad (k = 0, \pm 1, \cdots, \pm N) \tag{10}$$

Here p stands for the diffraction efficiency and $\hat{I}_k$ stands for the ideal spectrum distribution, $\hat{I}_k = 1/(2N+1)$, and it means all the power is being distributed in 0 to $\pm N$ orders. The standard nonlinear equations can be written as

$$F_k(x,\varphi) = I_k(x,\varphi) - P/(2N+1) = 0 \tag{11}$$

Notice that the phase shift φ is usually not a free variable since the extension of one to two dimensions requires it to be π radians. Otherwise the two-dimensional grating will not be a binary one because three different phase angles exist: $0, \varphi$ and 2φ. This would mean much complicated fabricating problems when we use the photolithographic technique.

From Eqs. (5) and (6) it is easy to know that the analytical solution of Eq. (11) is almost impossible. Only when $N = 1$ have we found a mathematical root, that is $x_0 = 0$, $x_1 = 0.5$, $x_2 = 1$

and $\varphi = \arctan(\pi/2) = 0.639\pi$, efficiency is about 87% (see also Fig. 2). For $N \geqslant 2$ numerical technique has to be used.

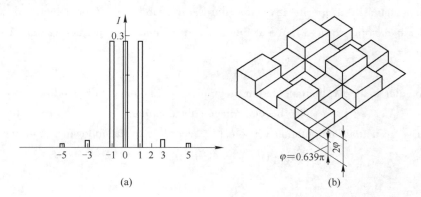

Fig. 2 Analytical solution for a 3×3 binary phase grating
(a) the spectrum distribution; (b) the structure of a two-dimensional binary phase grating

Recently the development of computer science has been considered a key to solving these kinds of problems. Applied to such nonlinear equations a lot of numerical methods have also been discovered. However, the basic or the most significant is Newton's method. But, in general it will only converge when the iteration is started near a root of the equations. So much attention has therefore been given to suitable modifications of Newton's method in order to enlarge the region of convergence, or say find a proper starting point for Newton's method. Here, based on Newton's method, another new improvement is introduced to solve Eq. (11), we call it automatic homotopy changing method.

Our approach will be based on Newton's homotopy (imbedding) or convex homotopy, stated as

$$H(X,t,X^0) \equiv F(X) + (1-t)F(X^0) \tag{12}$$

We then extended it to be

$$H(X,t,A,X^0) \equiv F(X) - (1-t)F(X^0) - (1-t^3)A(X-X^0) \tag{13}$$

here $A \in IR^{n \times n}$ is an n-dimensional diagonal matrix, $A = \mathrm{diag}(a_1, a_2, \cdots, a_n)$. Obviously when $A = 0$, or say $a_1 = a_2 = \cdots = a_n = 0$, Eq. (13) then becomes a convex homotopy like (12). For arbitrary $X^0 \in IR^n$ and A, Eq. (13) will give out the following conditions

$$H(X^0,0,A,X^0) \equiv 0 \tag{14}$$
$$H(X,1,A,X^0) \equiv F(X) \tag{15}$$

So the solution will be found by continuing the solution curve $X(t)$ of

$$H(X,t,A,X^0) = 0 \tag{16}$$

from $t_0 = 0$ to $t_1 = 1$. According to Eq. (14) the starting point will be the initial one given, and the end of the tracing will be the root of Eq. (11). However, in the procedure of the iteration the routine of tracing along $X(t)$ should always keep

$$(\partial/\partial x)H(X,t,A,X^0) = H'_x(X,t,A,X^0)$$

nonsingular. But normally H'_x will easily change its feature and become singular. The way to keep the nonsingularity in the iterative procedure is to search for a proper matrix A. For example, consider homotopy Eq. (13). We at first make $A = 0$ and start interacting from X^0. Assume that the matrix

$$\{H'_x(X^k,t_k,A_i,X^0)\}$$

becomes singular in step k; the calculation program should automatically modify A in order to let the matrix $\{F(X^k) + (1-t^3)A\}$ become diagonally dominant in order to meet the needs of nonsingularity. In actual computation we select a_i according to the following conditions,

$$\text{sign}(a_i) = \text{sign}(\partial f_i/\partial x_i) \tag{17}$$

and

$$\left|(1-t_k^3)a_i + \partial f_i/\partial x_i\right| > \sum_{j\neq i}\left|\partial f_i/\partial x_i\right| \tag{18}$$

here $i = 1, 2, \cdots, n$.

In this way the iteration can be performed continuously, and when $t=1$ we then get the approximate root $X_{t=1}$ of Eq. (11). More accurate solutions will be obtained with general Newton's method by starting our iteration at the point of $X_{t=1}$.

4 Solution examples

Generally we will search the structure by giving a high efficiency range of p and making sure the best solution is found. For example the following conditions

$$p > 70\%, \quad N = 4, \quad \text{and error } e < 0.0001$$

are given, we may get several structures, but the best is

$$x_0 = 0.00000, \quad x_1 = 0.06668, \quad x_2 = 0.12871, \quad x_3 = 0.28589,$$
$$x_4 = 0.45666, \quad x_5 = 0.59090, \quad x_6 = 1.00000$$

The efficiency is about 73%, so the relative intensity of the nine beams is about $I_i = 0.081$ ($i = -4, -3, \cdots, 0, \cdots, +3, +4$), see also Fig. 3.

Another example is a structure which can give out a 25 × 25 uniform intensity beam array with a high efficiency of 83%. The structure is given in Fig. 4 with the power distribution. The parameter of transition points is presented as

$x_0 = 0.00000, \quad x_1 = 0.15387, \quad x_2 = 0.16452,$
$x_3 = 0.27298, \quad x_4 = 0.33211, \quad x_5 = 0.38120,$
$x_6 = 0.44909, \quad x_7 = 0.52349, \quad x_8 = 0.64396,$
$x_9 = 0.76823, \quad x_{10} = 0.82082, \quad x_{11} = 0.85980,$
$x_{12} = 0.91639, \quad x_{13} = 0.95832, \quad x_{14} = 1.00000$

Fig. 3 The structure for a 9 × 9 binary phase grating (black for the phase shift of "π", white for "0")

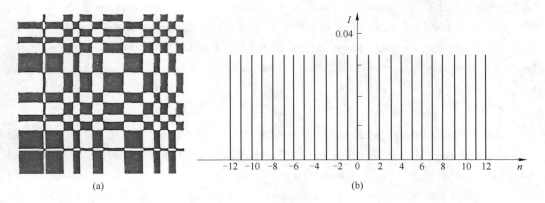

Fig. 4 (a) The structure for a 25 ×25 phase grating;
(b) The spectrum distribution of the 25 ×25 grating

5 Fabrication technique

There are several techniques which can be applied in the fabrication of such binary phase gratings. The most significant one seems to be the laser beam etching technique, which is developed in these few years for the VLSI. The binary structure showed in Fig. 4 is firstly photoreduced and multiplied on a mask by photolithographic technique, the ion beam will etch the nonmasked part until reaching the predetermined depth, as in Fig. 4 the white blocks will be etched. Then the structure is copied in the glass substrate. After removing the mask the phase grating is formed. Fig. 5 shows the far field distribution of a 9 × 9 grating which is fabricated with ion beam etching technique.

For the 25 ×25 structure we use glass corrodent ($HF : H_3PO_4 : H_2SO_4 = 1 : 5 : 1$) to etch the masked substrate. By controlling the etching time we can reach the depth predetermined. The depth d is proportional to the etching time T, experimental curve is obtained and showed in Fig. 6.

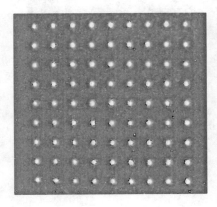

Fig. 5 Experimental results of a 9 ×9 beam-splitter which has the structure of Fig. 3

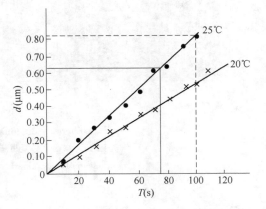

Fig. 6 Experimental results for etching depth versus etching time

According to the calculation the error between different diffracting beams should be less than 10^{-4}, however the fabricating error is involved here and influences the feature of the splitting. An approximate formula can be derived from Eqs. (6) and (7) for estimating the error between the zero order and non-zero orders if we suppose that when $\varphi = \pi$, $I_0 = I_k$, here $k \neq 0$, that is

$$\Delta = I_0 - I_k = 1 - \sin^2(\varphi/2) = \cos^2(\varphi/2) \qquad (19)$$

The actual error also depends on the accuracy of transition point positions and the correctness of the grating shape. This part of error can be calculated by a program. For instance in our experiment the absolute position error is $0.3\mu m$ and the smallest block is $3 \times 3\mu m^2$, it then induces a relative error of 1% in non-uniformity. Consider (19), if the phase error $\Delta\varphi > 0.02\pi$, the absolute error between zero order and non-zero orders will be 0.001, and the relative error for a 25×25 grating is as poor as 3%. So accurate controlling to the depth of etching is necessary, especially when you want to split out many beams. In this case the dominant error will be induced by the incorrect phase angle. Fig. 7 shows the power spectra of two gratings which have slightly different phase angles. In Fig. 7(a) the central point is much brighter than the other though the incorrectness of phase angle is only 0.04π (in depth it is about $0.002\mu m$, when the wavelength is $0.5145\mu m$). The satisfied result is in Fig. 7(b), in which the phase angle is about 1.015π. The outline of the grating is also measured by a Taylor-V, see also Fig. 8. The testing result of the grating shows that the uniformity is about 96% in all 625 beams, efficiency better than 60%. The far field diffraction pattern is showed in Fig. 7(b).

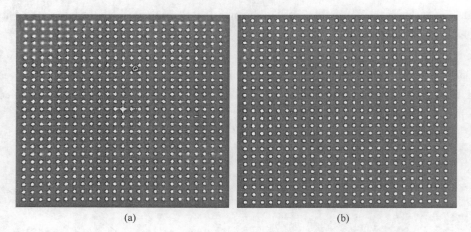

Fig. 7 Uniformity difference and the phase shift error
(a) a 25×25 grating with the phase shift of $\varphi = 1.04\pi$;
(b) an uniform beam splitter with the phase shift of $\varphi = 1.015\pi$

6 Discussion

The new method to design a grating by solving nonlinear equations proved to be very power-

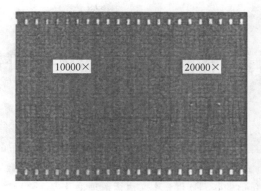

Fig. 8 The outline of the 25 ×25 beam splitter
(the etching depth here is about 12mm/20000 =0. 6μm)

ful. Even by a microcomputer a lot of structures can be obtained. More complicated structures which can split as large as 100 × 100 beams will be able to be found by some more powerful computers. This also means much accurate depth controlling in real time, which may be achieved in an interferometer. The chemical etching method also has disadvantages like temperature dependence and the density sensitivity of corrodent liquid. Studies for enhancing the designing program and the fabricating technique are still continuing.

Acknowledgements

We want to thank Mr. L. W. Ren and Professor Q. Y. Li for their helpful work in solving the nonlinear equations, and Mr. J. Turunen for helpful discussions in letters.

References

[1] H. Damman and K. Görtler, Optics Comm. ,3:312(1971).
[2] J. L. Jewell, A. Scherer, S. L. McCall, A. C. Gossard and J. H. English. Appl. Phys. Lett. ,51:94(1987).
[3] Wai-Hon Lee. Appl. Opt. ,18:2152(1979).
[4] U. Killat, G. Rabe and W. Rave. Fiber and Integrated Optics,4:159(1982).
[5] J. Turunen, Bian Xingao and Jin Guofan. Acta Optica Sinica,8:946(1988).
[6] G. H. Meyer. SIAM J. Number. Anal. ,5:739(1968).
[7] L. W. Ren and Q. Y. Li. Scientific Reports in Tsinghua University,1987.

平行传输阵列光斑器件的研制*

摘　要　本文基于周期物体的自成像效应，研制了二元位相光栅型平行传输阵列光斑器件。16×16、32×32 等二维二元 $\pi/2$ 及 $2\pi/3$ 位相光栅阵列器件，在 500mm 的传输距离内，边长为 0.3mm 及 0.4mm 的方形单元光斑，其光束扩展率 <1%。为光数字处理及光计算等系统研制了一种有实用价值的阵列器件。本文由衍射理论及数值分析对该器件的设计原理、技术特点等进行了理论分析，并给出了实验结果。

1　概述

阵列光斑器件的作用是将一均匀连续光波分割成离散的阵列光束，用以照明光学微阵列元件。它是多重成像、光纤通讯、光数字处理及光计算等系统中的重要器件。照明光束或阵列光束可以是平行的、发散（会聚）的或与某种固有模式的微结构相谐调的。阵列光斑器件有其广泛的应用，例如功率或时钟信号的分配、阵列微像的再现，以及某些光学网络（如理想洗牌）中，实现图像或光斑尺寸的压缩等。

利用光栅的衍射现象来产生等光强阵列光斑，是一种很有效的方法。Dammann 光栅[1,2]是利用位相光栅的衍射频谱来获得阵列光斑。光斑必须产生在透镜的焦面上。本文研制的基于自成像效应的位相光栅阵列器件，是利用光栅的菲涅耳衍射原理来产生阵列光斑[3]。在使用中勿需加透镜。系统简单，而且只要改变照明光束的发散（会聚）度，就可以改变阵列光斑的位置、大小和扩展率。

2　二元位相光栅的自成像效应

当一周期物体（如光栅）被一单色的和空间相干光照明时，在光栅后面的某些确定位置上可以重现光栅的清晰像。这种现象称为自成像效应或泰伯（Talbot）自成像效应。

设以单位振幅的单色平面波照明一维光栅，光栅的复振幅透过率函数为

$$E(x,0) = \sum_{n=-\infty}^{\infty} C_n \exp[\mathrm{i}2\pi(n/d)x] \tag{1}$$

作傅氏变换

$$T(f_x,0) = F\{E(x,0)\} = \sum_{n=-\infty}^{\infty} C_n \delta\left(f_x - \frac{n}{d}\right) \tag{2}$$

将 $T(f_x,0)$ 乘以菲涅耳近似下的传递函数 $H(f_x,z)$ 即得任一 z 平面上的复振幅分布的傅氏变换

---

* 本文合作者：严瑛白，姚长坤，邬敏贤。原发表于《仪器仪表学报》，1992，13(3)：263~271。

$$T(f_x,z) = T(f_x,0)H(f_x,z) \tag{3}$$

其中

$$H(f_x,z) = \exp(ikz)\exp(-i\pi z f_x^2) \tag{4}$$

再作 $T(f_x,z)$ 的逆傅氏变换,则得任一 z 平面上的复振幅分布。由式(2)~(4),得

$$E(x,z) = \sum_{n=-\infty}^{\infty} C_n \delta\left(f_x - \frac{n}{d}\right)\exp(ikz)\exp(-i\pi z f_x^2)$$

$$= \exp(ikz)\sum_{n=-\infty}^{\infty} C_n \exp[i2\pi(n/d)x]\exp[-i\pi\lambda z(n/d)^2] \tag{5}$$

对于振幅光栅(图1(a)),有

$$E_a(x,0) = \sum_{n=-\infty}^{\infty} \text{rect}\left(\frac{x-nd}{a}\right) \tag{6}$$

二元 $\pi/2$,$a = b = d/2$ 位相光栅(图1(b)),有

$$E_p(x,0) = \sum_{n=-\infty}^{\infty}\left[\text{rect}\left(\frac{x-nd}{a}\right)\exp(-i\pi/4) + \text{rect}\left(\frac{x-nd-d/2}{a}\right)\exp(-i\pi/4)\right] \tag{7}$$

由式(6)、式(7)分别求得

$$C_n = \frac{a}{d}\text{sinc}\left(n\frac{a}{d}\right) \quad (振幅光栅) \tag{8}$$

$$C_n = \frac{\sqrt{2}}{2}\text{sinc}\left(\frac{n}{2}\right)\exp(-i\pi n^2/2) \quad (位相光栅) \tag{9}$$

令 $Z = P\dfrac{d^2}{\lambda}$,$P$ 为一常数,由式(5)~式(7),对于振幅光栅

$$E(x,z) = E_a\left(x - \frac{P}{2}d\right) \quad (P 为整数) \tag{10}$$

$$E(x,z) = \sqrt{2}/2 E_p\left[x - \left(\frac{P}{2} + \frac{1}{4}\right)d\right] \quad (P 为 1/2 的奇数倍) \tag{11}$$

对于位相光栅,

$$E(x,z) = E_p\left(x - \frac{P}{2}d\right) \quad (P 为整数) \tag{12}$$

$$E(x,z) = \sqrt{2}/2 E_a\left[x - \left(\frac{P}{2} + \frac{1}{4}\right)d\right] \quad (P 为 1/2 的奇数倍) \tag{13}$$

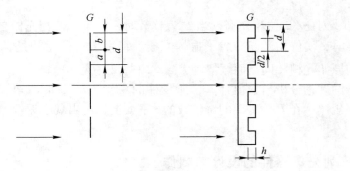

图1　(a)振幅光栅;(b) $\pi/2$ 位相光栅

由计算机编程进行数值计算，结果如图 2 及图 3 所示。由图 2 知，对振幅光栅，当 P 为偶数及奇数时，分别产生正自成像和负自成像，而当 P 为 1/2 的整数倍时，振幅几乎均匀分布，即变成了位相光栅。

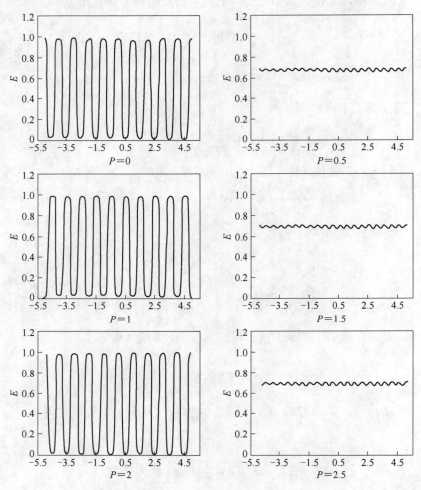

图 2　振幅光栅的菲涅耳衍射像

由图 3 可知，对 $\pi/2$ 位相光栅，当 P 为整数时，产生了位相光栅自成像，而当 P 为 1/2 的奇数倍时，使位相光栅变成了振幅光栅。特别值得注意的是，这些位置的振幅光栅的周期与原始光栅相同，而振幅增大到 $\sqrt{2}$（图 2 中是 1）。

通过进一步的理论分析与数值计算，所设计的二元 $2\pi/3$，$a/d=1/3$ 的位相光栅，有如图 4 所示的规律。在 P 为 1/3 的奇倍数的 z 平面上，也出现了振幅像，而且振幅增至 $\sqrt{3}$。

3　位相光栅阵列光斑器件的设计与制作

基于上述原理，将两个一维二元位相光栅正交重叠，构成二维光栅。设 N 为一维

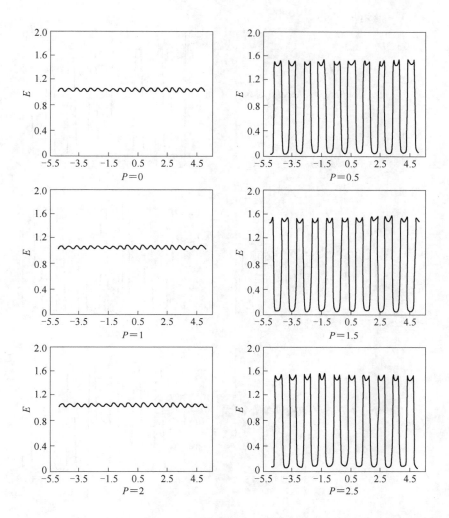

图 3 π/2 位相光栅的菲涅耳衍射像

光栅的总线对数,则在光栅后的一些确定位置上,将产生 $N \times N$ 清晰的阵列光斑。位相光栅产生的光斑比振幅光栅有更大的光强 (π/2 及 2π/3 位相光栅分别是振幅光栅的 2 倍和 3 倍)。所设计的两种位相光栅参数见表 1。

表 1 位相光栅参数

光栅	周期 d/mm	a/mm	b/mm	位相 δ/π	槽深 $h/\mu m$
1	0.8	0.4	0.4	1/2	0.316
2	0.9	0.3	0.6	2/3	0.422

设位相光栅的槽深为 h (图 1 (b)),则位相延迟与深度的关系为

$$\delta = \frac{2\pi}{\lambda}(n-1)h$$

即

$$h = \frac{\lambda \delta}{2\pi(n-1)} \tag{14}$$

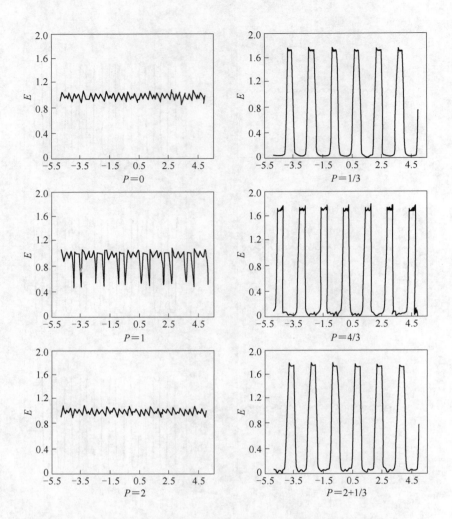

图 4 2π/3 位相光栅的菲涅耳衍射像

设 $\lambda = 0.6328\mu m$，$n = 1.5$，则 π/2 及 2π/3 位相光栅的槽深 h 分别为 $0.316\mu m$ 及 $0.422\mu m$。

在加工中，要求高透过率且均匀性好；光栅轮廓形状准确（包括线宽精度及线形的规整性）及槽深准确。我们采用了大规模集成电路的制作工艺流程。掩模板经初缩 $40^\times$ 及精缩 $10^\times$，并分步重复，线宽精度 $1.25\mu m$。用涂有 AZ-350 型胶的感光铬板及接触式曝光。并用 HF 腐蚀基片。通过电视干涉显微镜检测槽深及 CCD 摄像机测量光斑均匀性。结果表明，其质量满足了设计要求。

4 几个问题的讨论及实验结果

4.1 传输光束扩展率的变化

由平行光照明 π/2 位相光栅时，分别在 $P = 0.5$、1.5 处得到的阵列光斑，如图 5 所示。实测宽度为 0.4mm 的单元光斑，在 500mm 的传输距离内，其扩展率小于 1%。

图 5　平行传输阵列光斑

由衍射理论证明，当用发散或会聚球面波照明时，也产生类似平行光照明的自成像现象。只是自成像的位置、周期及传输光束的扩展率发生变化[4]。图 6 表示在三种不同光束照明下的阵列光斑照片，在不同距离 Z 处得到同一 P 级、但大小不同的阵列光斑。因此，利用这一特性可以方便地改变光斑的位置、大小及传输光束的扩展率。特别是在用会聚光照明时，可以用较粗的光栅获得较小的阵列光斑。

图 6　不同光照明时的自成像
（a）发散光；（b）平行光；（c）会聚光

4.2　光栅疵病的影响

基于自成像效应的光栅器件，其光斑质量受光栅本身质量的影响较大。但随着使用级次 P 的增加，其影响减小。图 7 表示一组实验结果。到图 7（d）时，缺陷的影响基本消失。如果选用光栅周期 d 较小，则在一个较小的距离 Z 之内，可以使用较高级次 P，因此，光栅质量的影响较容易减小或消失。

4.3　产生自成像的最大距离及光斑均匀性

对一块有一定口径大小的光栅，离散衍射波随着传播距离越大，所能重叠的部分就越小。在一定距离之后，自成像现象不再存在。

若仅考虑 $n=0$，± 1 级 3 个衍射波的叠加，设光栅宽度为 $2b$，周期为 d（图 8），

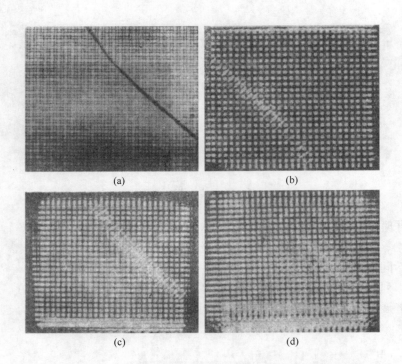

图 7　光栅疵病的影响随 Z 的增大（从（a）到（d））而减小

则产生自成像的最大距离

$$Z_{max} = \frac{b}{\tan\alpha} \approx \frac{b}{\sin\alpha}$$

而

$$\frac{\sin\alpha}{\lambda} = \frac{1}{d}$$

所以

$$Z_{max} = \frac{b}{\sin\alpha} \frac{bd}{\lambda} \tag{15}$$

从图 8 可知，产生自成像的横向范围在阴影区 $2b'$ 内，$2b'$ 随 Z 的增大而减小。在 $2b$ 范围内虽均可相干重叠而产生光斑，但非阴影部分重叠光束减少而使光强减弱，因此造成阵列光斑光强不均匀。但因高级次衍射波的贡献较小，其影响不大。此外，入射波的不均匀性也直接影响光斑的均匀性，图 9 是 CCD 摄像机扫描的阵列光斑光强分布，其不均匀性在 20% 以内。

4.4　光栅占宽比（a/b）及位相误差对阵列光强分布光斑的影响

由于加工误差，占宽比及位相均要偏离设计值。尤其位相（槽深）的控制受到环境条件的影响较大。经理论计算表明，对于光栅的加工精度要求并不十分严格。图 10 表示 $2\pi/3$ 位相光栅，占宽比误差为 5% 及位相误差为 10% 的计算结果。与图 4 的理想值比较，主要是暗背景略微变亮，但对使用影响不大。我们的工艺方法及测试手段可以保证在以上误差之内。

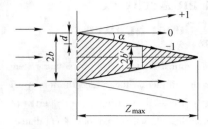 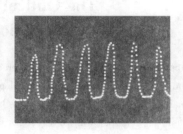

图 8　光栅自成像的最大距离　　　　　图 9　阵列光斑的光强分布

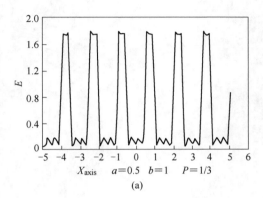 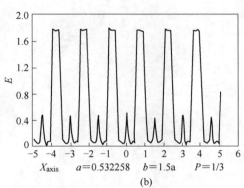

图 10　误差影响

（a）占宽比；（b）位相误差的影响

5　结束语

利用位相光栅的菲涅耳衍射及自成像效应制作平行或发散（会聚）阵列光斑器件，具有许多优点：设计、制造简单；使用方便，系统中勿需加透镜，阵列光斑的位置、大小及扩展率可以通过改变入射光束的发散度而方便地加以控制。作为平行传输阵列器件，它可以在图形变换或坐标变换等系统中作级联之用，还可以利用在不同 P 处光栅像周期不同的特点[5]，进行像的合成，利用光栅的负自成像现象可能作"非"门逻辑处理。这些尚待进一步研究。

参 考 文 献

[1] H. Dammann and K. Gortler. Optics Communications，3(5)：312(1971).

[2] Jari Turunen，Bian Xingao and Jin Guofan. 光学学报，8(10)：946(1988).

[3] A. W. Lohmann. Optik，79(1)：41～45(1988).

[4] 严瑛白，余官正. 仪器仪表学报，3(2)：149～158(1982).

[5] J. R. Leger and G. J. Swanson. Optics Letters，15(5)：233(1990).

The Optical Element of an Array of Two Dimensional Parallel Beam Splitter

Abstract A phase grating array interconnection element for optical digital image processing system and optical computing system is developed. On the basis of the diffraction theory and the numerical analysis, the self-image property of the binary phase grating is discussed. By using these property, $\pi/2$ and $2\pi/3$ grating with $16 \times 16, 32 \times 32$ arrays have been fabricated. A light spot of 0.4mm diameter has a divergence less than 1% in the transmission distance of 500mm. The main technical features are analyzed and some experimental results are demonstrated.

实现 ICF 均匀照明的二元光学器件的混合优化设计*

摘　要　结合模拟退火和遗传算法各自的特点，提出了一种混合算法，用以设计阵列型惯性约束核聚变（ICF）均匀照明系统中二元阵列单元的位相分布。混合算法充分利用了遗传算法的并行性及保留一定历史信息的特性，并用模拟退火的温度参数控制收敛性。模拟运算表明，混合算法具有较高的效率及寻优可靠性。用本算法设计的均匀照明阵列，可以得到顶部均匀性为 3.2%（rms）、能量利用率接近 90% 的焦斑。

关键词　模拟退火；遗传算法；混合算法；ICF 均匀照明；均匀性

1　引言

在激光驱动惯性约束核聚变（ICF）中，要求激光束有极高的均匀性，即所谓的"平顶"光束。而大功率激光器的输出光束往往有波面畸变，其振幅与位相都带有噪声。为提高聚焦系统的抗近场噪声能力，实现均匀照明，一般采用阵列型结构的聚焦系统（见图 1）。入射光束被分成多个子束分别处理，在每个小单元中近似认为入射波面是均匀的，最后由主聚焦透镜将各子束在焦面叠加。典型的技术方案有随机位相板[1]、透镜阵列[2]等。这些方法都大大提高了焦斑的均匀性，但由于硬边衍射的影响，并不能得到令人满意的强度包络。为消除硬边衍射对焦斑均匀性的影响，阵列单元需具有特殊的非球面位相分布[3]，而用传统的光学器件是很难实现这种特殊功能的。

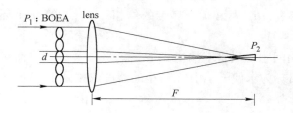

图 1　阵列型 ICF 均匀照明系统原理图

Fig. 1　The schedule diagram of the ICF uniform illumination array system

近年来，二元光学技术迅速发展，尤其在波面整形方面表现出极大的优越性，它被认为是很有希望实现 ICF 均匀照明的新技术途径。由于均匀照明问题与位相恢复问题的相似性，一般采用迭代优化的方法来设计二元光学器件的位相结构，如较早提出的 KPP（Kinoform Phase Plate）[4]，DPP（Distributed Phase Plate）[5]，用 GS 及其改进算法[6,7]进行设计，可以达到接近 8 阶的超高斯分布；考虑离散计算时采样造成的影响，

---

* 本文合作者：郑学哲，王凌，严瑛白。原发表于《中国激光》，1998，A25(3)：265~269。

采用 Y-G（杨-顾）算法[8]进行优化设计，可以进一步提高焦斑边缘的陡度[9]，但其顶部的均匀性还不能满足物理实验的要求。这些优化算法都由目标函数的极值点处导数为零而得到基本迭代公式，优化过程只接受目标函数下降的自变量取值，普遍存在的问题是易陷于局部最优解，因而优化结果依赖于初值。尽管结合几何变换等方法提供的优化初始解，可以改善优化结果，但尚未能找到产生足够均匀焦斑的器件位相分布。

模拟退火算法（Simulated Annealing，简称 SA）和遗传算法（Genetic Algorithm，简称 GA）是两种近似求解的启发式算法，大大提高了求解问题的全局最优解的可能性。它们已被成功地用于衍射位相光学元件的设计[10,11]。为尽可能地得到近"平顶"的焦斑强度包络，本文结合 SA 和 GA 的特点，提出以 SA&GA 混合算法设计 ICF 均匀照明系统中二元阵列单元的位相分布，分析了如何控制优化过程并给出了混合算法实施的基本规则，取得了较好的效果。

2 设计模型

为简化分析，取图 1 中中心阵列单元，考虑圆对称情况，由菲涅耳积分公式[12]可得焦面的输出光场为

$$U_2(r) = \frac{ik_0}{f}\exp\left[-ik_0\left(f+\frac{r^2}{2f}\right)\right]\int_0^R J_0\left(\frac{k_0 dr}{f}\right)\exp[ih(d)]ddd \tag{1}$$

式中，f 为透镜焦距；k_0 为 $2\pi/\lambda$；R 为阵列单元半径；J_0 为零阶 Bessel 函数；h 为阵列单元位相。

由于数值计算必须进行离散采样处理，因而最后设计得到由不同常数位相环带组成的器件。上述积分可以改写为各个环带积分的和，

$$U_2(r) = \frac{ik_0}{f}\exp\left[-ik_0\left(f+\frac{r^2}{2f}\right)\right]\sum_{n=1}^{N}e^{ih_n}\int_{d_{n-1}}^{d_n} J_0\left(\frac{k_0 dr}{f}\right)ddd \tag{2}$$

式中，N 为器件环带数；h_n 为第 n 个环带的位相；d_{n-1} 为第 n 个环带的最小半径；d_n 为第 n 个环带的最大半径。

对 ICF 均匀照明问题，只要求输出光强呈"平顶"分布，而对焦面位相分布无特殊要求，所以设计目标为求使下式中 F 最小的器件位相分布 h_n

$$F = \left\| |U_2|^2 - |U_I|^2 \right\|^2 \tag{3}$$

其中，F 为实际输出焦斑复振幅分布 U_2 和理想焦斑分布 U_I 间的距离。也就是说，希望通过优化设计过程，寻求特殊的位相结构，重新分配焦斑的强度分布，来抑制由阵列单元硬边衍射引起的焦斑强度调制。

3 SA&GA 混合算法

SA 算法借鉴不可逆动力学的思想，是基于 Monte Carlo 迭代求解法的一种启发式随机优化方法。算法在某一温度下，经不断降温，在全局解空间中随机搜索最优解，同时具有概率突跳特点，即在局部极小以一定概率跳出并最终趋于全局最优。SA 算法通用性强，不依赖问题的特殊信息，求解全局最优的可靠性高，对初始解选择的鲁棒（Robust）性好，同时算法简单而易于实现。

GA 算法借鉴生物进化论的思想,是一种启发式群体概率性迭代优化方法。GA 将问题的求解表示成"染色体"的适者生存过程,其中包括选择、交叉和变异操作,通过"染色体"群的一代代不断进化,最终收敛到"最适应环境"的个体,从而求得问题的最优解。GA 算法具有并行搜索能力,同样不依赖问题的特殊信息,但存在进化缓慢和"早熟"现象。

结合 SA、GA 各自的优点,利用 GA 的并行搜索及保留一定历史信息的能力,为 SA 提供初始解;同时利用 SA 的温度参数控制 GA 的收敛性,可以构造出一种具有更强搜索能力的混合算法。混合算法过程如下:

(1) 选择初始解群,给定初温;
(2) 确定每一个体的适配值;
(3) 重复以下步骤直至算法收敛准则满足:
(3.1) 按适配值进行比例复制操作(Proportional Reproduction);
(3.2) 进行交叉操作(Crossover)并采用保优原则;
(3.3) 进行变异操作(Mutation)并采用保优原则;
(3.4) 对种群中每一个体进行模拟退火操作(Simulated Annealing)直至抽样稳定。
(3.4.1) 由状态产生函数产生新个体(Generation);
(3.4.2) 利用状态接受函数以一定概率接受新状态(Acception)。
(3.5) 退温操作。
(4) 输出最优结果。

由算法结构可见,混合算法中模拟退火步骤的初始解来自遗传的结果,而模拟退火步骤得到的解又作为遗传的初始解,遗传算法具有并行处理能力,同时能保留一定历史信息,模拟退火算法具有概率突跳性,当温度较小时就成为趋化性寻优,如此将两种算法混合,一定程度上会提高算法的效率和寻优的可靠性。

4 模拟运算及设计结果

根据上述设计模型,利用所提出的混合算法设计二元阵列单元的位相分布。设计参数为:单元口径 $d=30\text{mm}$;聚焦透镜焦距 $f=800\text{mm}$;波长 $\lambda=0.6328\mu\text{m}$;焦斑半径 $r_0=200\mu\text{m}$。由式(2)知,迭代过程中被改变的量只有 h_n,因而可以得到一个由不变量形成的系数矩阵,它与更新后位相的乘积就是新的输出光场分布。这可以大大缩短求适配值的时间,从而提高算法的效率。

考虑设计条件及计算效率,初始解群只选取了两个个体,一个是几何变换的结果,另一个取常数位相分布。优化所得器件连续位相分布如图 2 所示;焦斑强度分布如图 3 曲线 2 所示。焦斑光强分布接近 16 阶超高斯分布:$I_0\exp\left[-2\left(\frac{x}{w}\right)^{16}\right]$,顶部均匀性为 $rms=3.2\%$;能量利用率大于 90%。rms 定义为

$$rms = \sqrt{\sum_{i=1}^{n}\left[\frac{I_i - \bar{I}}{\bar{I}}\right]^2 \bigg/ (n-1)}, \quad \bar{I} = \sum_{i=1}^{n} I_i/n \tag{4}$$

其中,I_i 表示焦斑顶部各点光强。

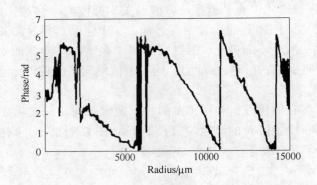

图 2　混合算法所得阵列单元位相分布

Fig. 2　Phase distribution of the array cell obtained by the hybrid algorithm

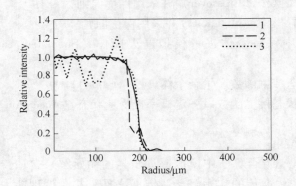

图 3　优化设计所得焦斑光场分布

（曲线 1 表示理想 16 阶超高斯分布；2 表示混合算法优化所得焦斑光强分布；
3 表示 Y-G 算法优化所得焦斑光强分布）

Fig. 3　Results of the optimization

(Curve 1 is the ideal 16th order super-Gaussian function; curve 2 is the focus intensity
distribution obtained by the hybrid algorithm while curve 3 is the result of Y-G algorithm.)

同样条件下，用 Y-G 算法的优化结果如图 3 曲线 3 所示。虽然焦斑边缘更陡，但顶部均匀性显然不如混合算法结果。

同时还进行了混合算法与模拟退火算法的性能比较。由于不是在线问题，算法的时间特性并不是关键问题，所以只给出优化结果所得焦斑光强分布的比较示于图 4，混合算法所得的焦斑比 SA 结果有更高的顶部均匀性，更小的旁瓣。与 GA 的有机结合确实提高了 SA 的寻优性能。可见，混合算法是很有效的。

5　小结

本文利用 SA&GA 混合算法优化设计了 ICF 均匀照明二元光学器件，得到了较有希望的结果。

分析混合算法的特点可知，当 GA 的并行化全空间搜索能力和 SA 的概率突跳特点

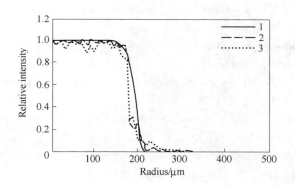

图 4　混合算法与 SA 结果的比较

（曲线 1 表示理想 16 阶超高斯分布；2 表示混合算法优化所得焦斑光强分布；
3 表示 SA 算法优化所得焦斑光强分布）

Fig. 4　The comparison of the hybrid algorithm and SA

(Curve 1 is the ideal 16th order super-Gaussian function; curve 2 is the focus intensity
distribution obtained by SA&GA algorithm while curve 3 is the result of SA algorithm)

结合后，混合算法具有更好的收敛性能，并在较大程度上避免陷入局部极小解。种群中各个体完全相同时，复制和交叉操作不能改变个体，而只能通过变异和模拟退火操作使得种群发生转移，当变异概率较小且不引入模拟退火操作，算法将长时间在旧状态"徘徊"，搜索效率很低且易出现"早熟收敛"现象。SA 算法的引入使这种僵局打破的概率大大加大，从而提高搜索效率；同时，SA 算法在局部极小解具有概率突跳性，从而避免"早熟"现象而加大趋于全局最优的概率。当温度降到较低时，SA 算法就成为概率为 1 的保优变异操作（即趋化性算法），从而提高变异的效率和性能。另一方面，为提高全局收敛的效率，自然希望适配值高的状态得以较大概率生存，复制操作在这方面起较大作用，但过强的复制操作会将状态过分地吸引到局部极小解而出现"早熟"现象，因此 SA 算法的引入很有必要。混合策略归纳为遗传算法利用模拟退火算法得到的解作为初始种群，通过并行化遗传操作使种群得以进化，模拟退火算法对遗传算法得到的进化种群进行进一步优化（温度较高使具有较强的概率突跳性以免陷入局部极小解，相对于对种群的"粗搜索"；温度较低时演化为趋化性算法，相对于对种群的"细搜索"）。这种不仅是算法上的混合，而且是进化思想上的混合策略，对解决优化问题提供了很好的途径。

对用二元光学方法实现 ICF 均匀照明问题，如果增加初始解群中的个体数，使其含有更多的优良基因，会大大提高得到全局最优解的可能性，从而得到近"平顶"分布的焦斑。

参 考 文 献

[1] Y. Kato, K. Mima, N. Miyanaga, et al.. Random phasing of high-power lasers for uniform target acceleration and plasma-instability suppression. *Phys. Rev. Lett.*, 1984, 53(11): 1057.

[2] Ximing Deng, Xiangchun Liang, Zezun Chen, et al.. Uniform illumination of large targets using a lens array. *Appl. Opt.*, 1986, 25(3): 377～381.

[3] N. Nishi, T. Jitsuno, K. Tsubakimoto, et al.. Aspherical multilens array for uniform target irradiation. *SPIE*, 1993, 1870: 105~111.

[4] S. N. Dixit, J. K. Lawson, K. R. Manes, et al. Kinoform phase plates for focal plane irradiance profile control. *Opt. Lett.*, 1994, 19(6): 417~419.

[5] Y. Lin, T. J. Kessler, G. N. Lawrence. Distributed phase plates for super-Gaussian focal plane irradiance profiles. *Opt. Lett.*, 1995, 20(7): 764~766.

[6] R. W. Gerchberg, W. O. Saxton. A practical algorithm for the determination of phase from image and diffraction plane pictures. *Optik*, 1972, 35: 237~246.

[7] J. R. Fienup. Reconstruction and synthesis applications of an iterative algorithm. In *Transformations in Optical Signal Processing*, W. T. Rhodes, ed., Proc. Soc. Photo-Opt. Instrum. Eng., 1981, 373: 147~160.

[8] G. Z. Yang, B. Y. Gu, B. Z. Dong. Theory of the amplitude-phase retrieval in any linear transform system and its applications. *Int. J. Mod. Phys.*, 1993, B7: 3152~3224.

[9] X. Z. Zheng, Y. B. Yan, G. F. Jin, et al.. Diffractive optical elements for inertial confinement fusion (ICF). *SPIE*, 1996, 2866: 99~103.

[10] Yang Guoguang. Genetic algorithm to the optimal design of diffractive optical elements and its comparison with simulated annealing algorithm. *Acta Optica Sinica* (光学学报), 1993, 13(7): 577~584(in Chinese).

[11] M. P. Dames, R. J. Dowling, P. Mchee, et al.. Efficient optical elements to generate intensity weighted spot arrays design and fabrication. *Appl. Opt.*, 1991, 30(19): 2685~2691.

[12] J. W. Goodman. Introduction to Fourier Optics. New York: McGraw-Hill, 1968, Chap. 4, 57~61.

Design Binary Optical Elements for ICF Uniform Illumination with Hybrid Optimization Method

Abstract In this paper, a novel hybrid algorithm which combines SA (simulated annealing) algorithm with GA (genetic algorithm) is used to design binary optics elements for ICF (inertial confinement fusion) uniform illumination. GA, which is a parallel algorithm, reserves historical information for SA, while SA's parameter, temperature, is used to control GA's convergence. Simulation calculation shows that the hybrid algorithm is effective and robust. A focal spot with a uniformity 3.2% (*rms*) and an efficiency approximating to 90% is obtained by this method.

Key words simulated annealing algorithm (SA); genetic algorithm (GA); hybrid algorithm; ICF uniform illumination; uniformity

Aberration Theory of Arrayed Waveguide Grating*

Abstract The general aberration theory of arrayed waveguide grating (AWG) is proposed in this paper. The derived conclusions can be applied to any symmetrical AWGs. With the aberration theory, the spectral response can be optimized. Some performances of an AWG, for example, crosstalk and channel number, are expected to be improved by reducing the aberrations. A common AWG with the Rowland geometry has low aberrations for the absence of the second-order aberration. To further reduce the aberrations, nonstandard AWGs are considered. The AWG based on three stigmatic points is proved to have a perfect performance.

Key words aberration theory; arrayed waveguide grating (AWG); rowland circle; stigmatic point

1 Introduction

Arrayed waveguide gratings (AWGs), also referred to as phased-array gratings (PHASARs) and waveguide grating routers (WGRs), are the most promising devices for filters or multi/demultiplexers in WDM systems because of their low insertion loss, high stability, and low cost. A lot of theoretical analysis and realization of AWG have been reported[1-9].

The aberration of an AWG strongly affects the spectral response[5,8,9]. Low aberration results in narrow spectrum widening. Thus, the channel number and the crosstalk of an AWG can be improved by reducing its aberration. In the previous works, the star coupler of AWG commonly consists of the geometry of Rowland circle. In this design, the aberration is quite low because the second-order aberration is cancelled. However, if nonstandard structure is employed, even lower aberration can be obtained, just like the case of planar spectrograph[10].

The general aberration theory of AWG is presented in this paper. In our model, the star coupler can be of any geometry. The arms of arrayed waveguides are distributed along the end-face according to an arbitrary function, and the length of each arm is also a free parameter for design.

This paper is arranged as follows. Section 2 discusses the general principles based on the optical path function (OPF). A series of iterative formulas for aberration analysis with the accuracy of any desired order is derived. Section 3 discusses the question of the second-order imaging. Under this condition, the first two aberration coefficients are cancelled. Section 4 concerns the Rowland-type AWG, which is widely used in previous works. AWGs based on stigmatic points are discussed in Section 5 as a special example of applying the developed aberration the-

---

* Copartner: Daoyi Wang, Yingbai Yan, Minxian Wu. Reprinted from *Journal of Lightwave Technology*, 2001, 19(2):279-284.

ory. Some comparisons between the traditional AWG and the AWG with stigmatic points are carried out in Section 6, which contains some numerical examples. Finally, in Section 7, the whole work is summarized.

2 General theory

As shown in Fig. 1, an $N \times N$ AWG consists of N input waveguides, two $N \times M$ star couplers jointed by M arrayed waveguides, and N output waveguides. The end-face of arrayed waveguides is called as grating curve in our convention. A rectangle coordinate system, XOY, is set up by letting the convex of the grating curve be the origin and letting the X axis be the normal of the grating curve. Another coordinate system $X'O'Y'$ is symmetrical to XOY. In other words, the corresponding coordinates in the two coordinate systems are equal to each other. Thus, in the following analysis, all the coordinates are assumed to all be those of in XOY. The grating curve can be expressed as a power series

$$u(w) = u(0) + u'(0)w + u''(0)w^2/2 + \cdots + u^{(n)}(0)w^2/n! + \cdots \quad (1)$$

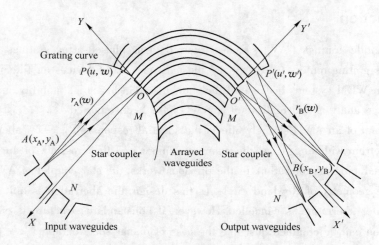

Fig. 1 Illustrative diagram of an $N \times N$ AWG

Under the coordinate system, $u(0) = 0$ and $u'(0) = 0$. For clearness, the coordinate of a common point P along the grating curve is assumed to be (u, w).

For the ray emitted from the input waveguide at $A(x_A, y_A)$, coupled into the arrayed waveguides at $P(u, w)$, then transferred through one single waveguide with length of $L(w)$, and diffracted by the output end-face of the arrayed waveguides at $P'(u', w')$, finally coupled into the output waveguide at $B(x_B, y_B)$, the OPF can be written as

$$F(w) = N_s r_A(w) + N_w L(w) + G(w)m\lambda + N_s r_B(w) \quad (2)$$

where N_s and N_w are effective refractive index of the star coupler and the arrayed waveguide, respectively; m is spectral order; $L(w)$ is geometrical length of the waveguide between P and P'; $G(w)$ is number of waveguides counted from the origin O to the point P.

$$G(w) = G'(0)w + G''(0)w^2/2 + \cdots + G^{(n)}(0)w^n/n! + \cdots \quad (3)$$

To discuss the aberration analytically, the number of waveguides is extended to a continuous function having the proper integer value at the corresponding position of the waveguide. In the numerical calculations, the coordinate of the ith arm of the arrayed waveguides (u_i, w_i) will be determined by letting $G(w_i) = i$.

$r_A(w)$ and $r_B(w)$ in Eq. (2) are the geometrical length in the star coupler

$$r_A(w) = \langle AP \rangle = \sqrt{(u - x_A)^2 + (w - y_A)^2}$$
$$r_B(w) = \langle P'B \rangle = \sqrt{(u - x_B)^2 + (w - y_B)^2} \quad (4)$$

Obviously, the OPF can be divided into four independent parts. The term $r_A(w)$ and the term $r_B(w)$ correspond to the propagation distances in the star coupler. The term $L(w)$ is the contribution of the arrayed waveguide. These three terms mainly determine the focusing characteristics of the AWG. While the term $G(w)m\lambda$ indicates the diffraction of the output end-face of the arrayed waveguides (like a grating), which mainly determines the dispersion of the AWG.

In the following sections, the OPF described in Eq. (2) will be expanded into a power series

$$F(w) = F(0) + F'(0)w + F''(0)w^2/2 + \cdots + F^{(n)}(0)w^2/n! + \cdots \quad (5)$$

where

$$F^{(n)}(0) = N_s[r_A^{(n)}(0) + r_B^{(n)}(0)] + N_w L^{(n)}(0) + G^{(n)}(0)m\lambda \quad (6)$$

is commonly called as the nth aberration coefficient. $F(0)$ is a constant term. Letting $F'(0)$ be zero, one can obtain the grating equation at O. $F^{(2)}(0)$, $F^{(3)}(0)$ and $F^{(4)}(0)$ correspond to defocus, coma, and spherical aberration, respectively.

Due to the intrinsic relation between the differential coefficients of a specialized function, such iterative expressions for Eq. (6) can be obtained[11]

$$r_A^{(1)}(0) = -y_A/r_A$$
$$r_A^{(n)}(0) = \frac{1}{2}\sum_{k=1}^{n-1} C_n^k [u^{(k)}(0)u^{(n-k)}(0) - r_A^{(k)}(0)r_A^{(n-k)}(0)]r_A -$$
$$u^{(n)}(0)x_A/r_A + \delta(n-2)/r_A \quad (n > 1) \quad (7)$$

where $r_A = \langle AO \rangle = \sqrt{x_A^2 + y_A^2}$ is the optical path, $\delta(q) = \begin{cases} 1 & q = 0 \\ 0 & \text{else} \end{cases}$ is the Kronecker function, $C_n^k = \dfrac{n!}{k!(n-k)!}$ is the combination number. The iterative formula for $r_B^{(n)}(0)$ is similar to that of $r_A^{(n)}(0)$.

As to the length $L(w)$, it is a free parameter for design. Once the $L(w)$ is determined, the geometrical layout of the arrayed waveguides can be carried out by the analysis of performance [5,12]. In the traditional schemes [4-6], $L(w)$ is specialized so that the difference of length between two adjacent arms of the arrayed waveguides keeps constant.

3 Second-order imaging

The low-order aberrations of a grating are the main sources of aberration[13]. Therefore, most of the efforts to reduce aberrations are to eliminate or to reduce the low-order aberration coefficients.

For a planar grating just as AWG, only two coordinates are necessary for the focal point of a wavelength. Therefore, only the first two aberration coefficients can be cancelled for all the wavelengths of operation. The imaging under this condition is called as the second-order imaging.

For an AWG, the formulas of the second-order imaging are as follows

$$-N_s(y_A/r_A + y_B/r_B) + N_w L^{(1)}(0) + G^{(1)}(0)m\lambda = 0$$

$$N_s[x_A^2/r_A^3 + x_B^2/r_B^3 - u^{(2)}(0)(x_A/r_A + x_B/r_B)] + N_w L^{(2)}(0) + G^{(2)}(0)m\lambda = 0 \quad (8)$$

The first formula is the general grating equation for an AWG. Compared with common gratings, an additional term is introduced by the arrayed waveguides. The second formula is the focusing equation. With these two equations, the focal point of a specialized wavelength can be determined. As will be shown in Section 4, Rowland circle is a special geometry meeting the condition of the second-order imaging.

Once the first two aberration coefficients vanish, the main task to design a low-aberration AWG is to reduce the aberrations upper than the second-order. An important method based on stigmatic points will be discussed in Section 5.

4 Rowland-type AWG

Rowland-type star coupler is widely used in the reported AWGs [2-9]. This kind of geometry suffers low aberrations for the absence of the second-order aberration. In this section, a Rowland-type AWG will be studied as a special example to apply the general aberration theory derived in Section 2.

The geometrical functions for this scheme are as follows

$$u(w) = R - \sqrt{R^2 - w^2}$$

$$G(w) = \frac{R}{d}\sin^{-1}\left(\frac{w}{R}\right) \quad (9)$$

$$L(w) = L_0 + G(w)\Delta L$$

where R is radius of the grating curve; d is period of the arrayed waveguides measured along the grating curve; L_0 is geometrical length of the central waveguide; ΔL is difference of lengths between two adjacent arms, a constant.

Usually, ΔL is determined by the spectral order m and the central wavelength λ_c

$$\Delta L = -m\lambda_c/N_w \quad (10)$$

After some calculations and induction, the differential coefficients of these functions can be expressed as

$$u^{(n)}(0) = \begin{cases} \dfrac{1}{(n-1) \cdot 2^n \cdot R^{n-1}} \dfrac{(n!)^2}{[(n/2)!]^2} & n = \text{even} \\ 0 & \text{else} \end{cases}$$

$$G^{(n)}(0) = \begin{cases} \dfrac{1}{n^{n-1}R^{n-1}d} \left\{ \dfrac{(n-1)!}{[(n-1)/2]!} \right\}^2 & n = \text{odd} \\ 0 & \text{else} \end{cases}$$

$$L^{(n)}(0) = G^{(n)} \Delta L \tag{11}$$

Introducing them into the general expressions described in Eq. (6) and Eq. (7), one can obtain the detailed OPF for the Rowland-type AWG. The grating equation can be derived by letting $F'(0) = 0$

$$F'(0) = -N_s \left(\frac{y_A}{r_A} + \frac{y_B}{r_B} \right) + N_w \frac{\Delta L}{d} + \frac{m\lambda}{d} = 0 \tag{12}$$

This equation concords with that reported in [6] and [7]. To get the focusing relation, the second-order aberration coefficient must be considered, i.e.,

$$F''(0) = N_s \left(-\frac{y_A^2}{r_A^3} - \frac{y_B^2}{r_B^3} - \frac{1}{R}\frac{x_A}{r_A} - \frac{1}{R}\frac{x_B}{r_B} + \frac{1}{r_A} + \frac{1}{r_B} \right) = 0 \tag{13}$$

The geometry of Rowland circle is a special solution of this focusing equation [13].

Now all the necessary formulas to calculate the aberration of the Rowland-type AWG have been obtained. A numerical example will be given in Section 6.

5 AWG with stigmatic points

As shown in Section 4, the Rowland-type AWG cancels the second-order aberration for all wavelengths of operation. Therefore, the overall aberrations are quite low. However, the Rowland geometry is not the optimum one. For example, the period can be varied to get better results. In fact, all the geometrical parameters can be considered as variables to be optimized by applying the general aberration theory derived in Section 2. Obviously, this is a complex process.

However, there are some analytical methods, among which introducing stigmatic points (i.e., aberration-free focal points) into gratings is the most effective one[10]. As all the aberrations at the stigmatic wavelength vanish, the aberrations at the adjacent wavelengths can be low. In fact, the focal point of the Rowland-type AWG corresponding to the central wavelength defined by Eq. (10) is a stigmatic point. In other words, the Rowland-type AWG has one stigmatic point. It should be notified that the AWG based on more than one stigmatic point is no longer the standard one with Rowland geometry. In the remaining part of this section the structure of such an AWG will be studied.

For a stigmatic point $S_1(x_1, y_1)$ at wavelength λ_1, all the aberration coefficients described in Eq. (6) are equal to zero

$$N_s [r_A^{(n)}(0) + r_1^{(n)}(0)] + N_w L^{(n)}(0) + G^{(n)}(0) m\lambda_1 = 0 \tag{14}$$

Thus, a restraint is imposed on $G^{(n)}(0), L^{(n)}(0)$ and $u^{(n)}(0)$. In other words, only two of these three parameters are independent. Therefore, an AWG can have no more than three stigmatic points. Introducing Eq. (14) into Eq. (6), the aberration coefficient for AWG with one stigmatic point is

$$F^{(n)}(0) = N_s[r_B^{(n)}(0) - r_1^{(n)}(0)] + G^{(n)}(0)m(\lambda - \lambda_1) \quad (15)$$

Adding another stigmatic point $S_2(x_2, y_2)$ at wavelength λ_2, another restraint equation similar with Eq. (14) comes into effect. Consequently, there is only one free parameter among $G^{(n)}(0), L^{(n)}(0)$, and $u^{(n)}(0)$. Once one additional condition besides the stigmatic imaging is given, the geometry will be completely determined. $F^{(n)}(0)$ for the AWG based on two stigmatic points is

$$F^{(n)}(0) = N_s[r_B^{(n)}(0) - r_1^{(n)}(0)] - N_s[r_2^{(n)}(0) - r_1^{(n)}(0)](\lambda - \lambda_1)/(\lambda_2 - \lambda_1) \quad (16)$$

Adding the third stigmatic point $S_3(x_3, y_3)$ at wavelength λ_3 results in the third constrain equation similar with Eq. (14). Therefore, the geometry of the AWG is fixed. The differential coefficients of the grating curve can be given in an iterative manner

$$u^{(1)}(0) = 0$$

$$u^{(n)}(0) = [(\psi_1 - \psi_3)(\lambda_1 - \lambda_2) - (\psi_1 - \psi_2)(\lambda_1 - \lambda_3)]/$$

$$\left[\left(\frac{x_2}{r_2} - \frac{x_1}{r_1}\right)(\lambda_1 - \lambda_3) - \left(\frac{x_3}{r_3} - \frac{x_1}{r_1}\right)(\lambda_1 - \lambda_2)\right] \quad (17)$$

where

$$\psi_i = \frac{1}{2} \sum_{k=1}^{n-1} C_n^k [u^{(k)}(0) u^{(n-k)}(0) - r_i^{(k)}(0) r_i^{(n-k)}(0)]/r_i + \delta(n-2)/r_i \quad (i = 1,2,3)$$

The OPF can be calculated by introducing Eq. (17) into Eq. (16).

6 Numerical examples

Some illustrative examples for AWGs in SiO_2/Si material system will be given in this section. In total, two 1×8 AWGs with the central wavelength of 1.550 μm and the channel spacing of 2 nm will be evaluated. Device I is the traditional Rowland-type AWG. Device II is based on three stigmatic points corresponding to the wavelength $\lambda_1 = 1.536$μm, $\lambda_2 = 1.550$μm, and $\lambda_3 = 1.564$μm. The output waveguides of all these devices are located along the second-order imaging curves. For the calculations, the beam emitted from the input waveguide is approximated by a Gaussian beam. The mode size u_0 is defined as its half width at $1/e^2$ of the maximum optical intensity. The following device specifications are assumed: the input port $A(x_A, y_A) =$

$(10\text{mm}, 0)$, $w_0 = 1.2\mu\text{m}$, the Rowland circle radius $R = 10\text{mm}$, $N_s = 1.442$, $N_w = 1.454$, the period $d = 12\mu\text{m}$, the spectral order $m = -48$.

Fig. 2 demonstrates the aberration curves of two devices, given in units of wavelength. Here the maximum variation of OPF on the grating curve of operation, i.e., $\max\{F(w) - F(0)\}$, is adopted as the merit function of geometrical aberration. The Rowland-type AWG suffers serious aberration ($\Delta f \gg \lambda/4$). It has only one stigmatic point corresponding to $1.550\mu\text{m}$. The geometric aberration will be significantly reduced if more stigmatic points are introduced. Compared with Device I, the aberration of Device II is several orders better. In the narrow spectral range, AWG with three stigmatic points suffers nearly zero aberration (less than $10^{-4}\lambda$).

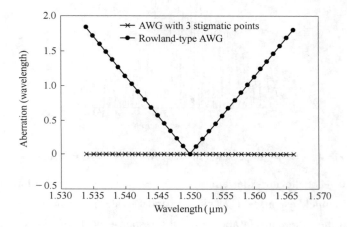

Fig. 2　Aberration curves of Device I and Device II

The dependence of the aberration on the finite aperture for $\lambda = 1.536\mu\text{m}$ is also given in Fig. 3. The Rowland-type AWG is affected by the increase of the pupil coordinate more seriously than the AWG based on three stigmatic points. However, the realistic aperture of an AWG is determined by the divergence of the beam (corresponding to the mode size w_0) emitted from the input port and the structure of the star coupler. In our design the total aperture is almost 6

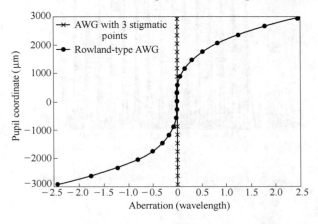

Fig. 3　Aberration curves for $\lambda = 1.536\mu\text{m}$ with respect to the pupil coordinate

mm. Consequently near 450 arrayed waveguides are laid out in the devices.

To show the influence of aberration, a method based on the scalar diffraction theory is utilized to simulate the spectral responses of two devices[9]. As shown in Fig. 4, for the output channel (corresponding to $\lambda = 1.536\mu m$), the AWG based on three stigmatic points exhibits smaller spectral widening than Rowland-type AWG. As a result, low crosstalk and larger channel number can be obtained.

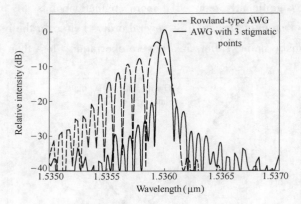

Fig. 4 Channel responses of the AWG based on three stigmatic points

Another advantage of the low-aberration AWG lies in the channel uniformity. Figs. 5 and 6 demonstrate the channel responses of Device I and Device II, respectively. For the Rowland-type AWG, due to the difference in aberration, the response of the channel far from the central channel is worse than that of the channel near the central channel. For the AWG based on three stigmatic points, however, the difference in channel responses is indistinctive. In addition, better loss uniformity is obtained in the low-aberration AWG.

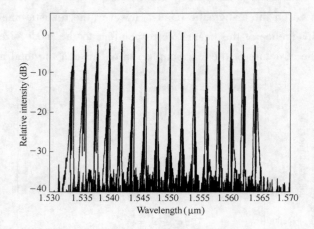

Fig. 5 Channel responses of the Rowland-type AWG

7 Conclusion

In this paper, the aberration theory of AWG has been proposed. Our model generalized the traditional geometry of AWG. The derived conclusions can be applied to any symmetrical AWG. With the aberration theory, the spectral response can be optimized.

The Rowland-type AWG suffers low aberration for absence of the second-order aberration. To obtain even lower aberrations, stigmatic points are introduced to get a nonstandard AWG. Such an AWG is promising to improve such performances as crosstalk, channel number, etc.

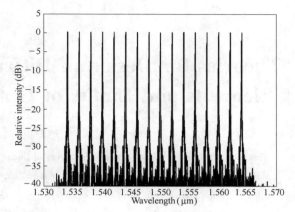

Fig. 6 Channel responses of the AWG based on three stigmatic points

References

[1] M. K. Smit, "New focusing and dispersive planar component based on an optical phased array", *Electron. Lett.*, 24:385-386, 1988.

[2] C. Dragone, "An N × N optical multiplexer using a planar arrangement of two star couplers", *IEEE Photon. Technol. Lett.*, 3:812-815, 1991.

[3] K. Okamoto, K. Syuto, H. Takahashi, and Y. Ohmori, "Fabrication of 128-channel arrayed-waveguide grating multiplexer with 25 GHz channel spacing", *Electron. Lett.*, 32:1474-1476, 1996.

[4] H. Takahashi, K. Okamoto, and Y. Inoue, "Arrayed-waveguide grating wavelength multiplexers for WDM systems", *NTT Rev.*, 10:37-44, 1998.

[5] M. K. Smit and C. Van Dam, "PHASAR-based WDM-devices: Principles, design and applications", *IEEE J. Select. Topics Quantum Electron.*, 2:236-249, 1996.

[6] H. Takahashi, K. Oda, H. Toba, and Y. Inoue, "Transmission characters of arrayed waveguide N × N wavelength multiplexer", *J. Lightwave Technol.*, 13:447-455, 1995.

[7] K. A. McGreer, "Arrayed waveguide gratings for wavelength routing", *IEEE Commun. Mag.*, 36:62-68, 1998.

[8] R. Radar, C. H. Henry, C. Dragone, R. C. Kistler, and M. A. Milbrodt, "Broad-band array multiplexers made with silica waveguides on silicon", *J. Lightwave Technol.*, 11:212-219, 1993.

[9] Y. P. Ho, H. Li, and Y. J. Chen, "Flat channel-pass band-wavelength multiplexing and demultiplexing devices by multiple-Rowland-circle design", *IEEE Photon. Technol. Lett.*, 9:342-344, 1997.

[10] R. Mars and C. Cremer, "On the theory of planar spectrographs", *J. Lightwave Technol.*, 10:2017-2022, 1992.

[11] Wang Daoyi, Jin Guofan, Yan Yingbai, and Wu Minxian, Iterative Analysis of Planar Spectrograph, Optik, submitted for publication.

[12] T. Goh, S. Suzuki, and A. Sugita, "Estimation of waveguide phase error in silica-based waveguides", *J. Lightwave Technol.*, 15:2107-2113, 1997.

[13] C. H. F. Velzel, "A general theory of the aberrations of diffraction gratings and grating-like optical instruments", *J. Opt. Soc. Amer.*, 66:346-353, 1976.

Theories for Design of Diffractive Superresolution Elements and Limits of Optical Superresolution[*]

Abstract This paper suggests using the theory of linear programming to design diffractive superresolution elements if the upper bound of the intensity distribution on the input plane is restricted, and using variation theory of functional or wide-sense eigenvalue theory of matrix if the upper bound of the radiation flux through the input plane is restricted. Globally optimal solutions can be obtained by each of these theories. Several rules of the structure and the superresolution performance of diffractive superresolution elements are provided, which certify the validity of these theories and set some limits of optical superresolution.

1 Introduction

Superresolution[1] owns considerable significance in many cases, such as optical data storage[2,3], confocal scanning microscopy[4,5] and laser lithography[6]. A filter of amplitude-only[3], phase-only[7,8] or hybrid[9,10] type is often placed at the exit pupil of an optical system as a diffractive superresolution element[7] (DSE) to change the spread function whose first zero position determines the resolving power, as is shown in Fig. 1(a). Design of a hybrid type filter can be regarded as a design of the field distribution on an input plane with the upper bound of the intensity distribution on the input plane restricted, which will be discussed in this paper considering it has more freedom than the design of the other two types. Here the input plane is defined as a plane just against to the exit surface of DSE and an output plane as the image plane. In order to achieve superresolution with little energy loss, a double cone-shaped lens [11] or a phase-only grating[12] is fitted in front of the exit pupil to generate an annular intensity distribution. This method is extended in this paper to a design of both the intensity distribution of the light incident into the exit pupil and the complex transmittance of the DSE at the exit pupil with a beam shaping system in front of the exit pupil to generate the designed intensity distribution, as is shown in Fig. 1(b). Design of this type of DSE may be treated as a design of the field distribution on the input plane with the upper bound of the radiation flux through the input plane restricted, which is another content of this paper.

The performance of superresolution is described in this paper by a normalized spot size G [1], a Strehl ratio S[1] and a normalized adjacent side lobe intensity K. G determines the resolving power, which is defined as the ratio between the radius of the zero position of the main

---

[*] Copartner: Haitao Liu, Yingbai Yan, Qiaofeng Tan. Reprinted from *Journal of Optical Society of America A*, 2002, 19 (11):2185-2193.

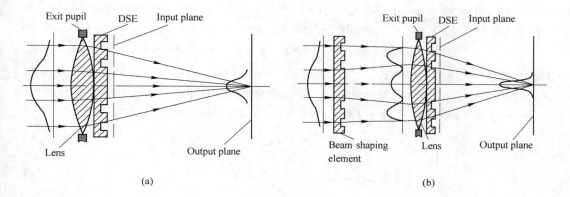

Fig. 1 (a) DSE with the upper bound of the intensity distribution on the input plane restricted;
(b) DSE with the upper bound of the radiation flux through the input plane restricted

lobe of superresolved pattern and the radius of the zero position of the main lobe of the Airy disk pattern. S is the ratio between the central intensity of superresolved pattern and that of the Airy disk pattern. The effects of side lobes are characterized by K, which is defined as the maximum intensity of the side lobe adjacent to the main lobe relative to the central intensity. S^{eu} is defined as the maximum value of S under certain constraints. An upper bound of S^{eu} is provided as S^u by Ref. [1] published in *Optics Letters* so that the physical content of some limits of optical superresolution are revealed. However, the accurate value of S^{eu} is obtained in this paper, which is much more improved than S^u in term of the mathematics. The discussion goes along in the realms of scalar diffraction theory satisfying the Fresnel approximation with the system being of rotational symmetry.

In this paper, the theory of linear programming is abbreviated to TLP, variation theory of functional to VTF, wide-sense eigenvalue theory of matrix to WETM, the upper bound of the intensity distribution on the input plane to UBIDIP and the upper bound of the radiation flux through the input plane to UBRFIP.

2 Theories of design

2.1 Theory of Linear Programming

Design of a DSE with UBIDIP restricted is depicted as a problem of

$$\max_{u_1(r_1)} i_2(0) \tag{1a}$$

such that

$$\begin{cases} i_2(r_2^{G_i}) = 0, \quad i = 1, \cdots, N & (1b) \\ |u_1(r_1)|^2 \leq i_1^u(r_1) & (1c) \end{cases}$$

where

$$i_2(r_2) = \left(\frac{2\pi}{\lambda f}\right)^2 \left| \int_0^R u_1(r_1) J_0\left(\frac{2\pi r_1 r_2}{\lambda f}\right) r_1 \mathrm{d}r_1 \right|^2 \tag{2}$$

is the intensity on the output plane, $u_1(r_1)$ is the complex field on the input plane, $i_1^u(r_1)$ is UBIDIP, r_1 and r_2 are correspondingly the radial coordinate on the input plane and that on the output plane, r_2^{Gi} is the radial coordinate of the position of a zero intensity constraint on the output plane which satisfies $0 < r_2^{G_{i-1}} < r_2^{G_i}$, $i = 2, \cdots, N$, R is the radius of the exit pupil, λ is the wavelength and f is the distance from the input plane to the output plane.

To normalize problem (1a) to (1c), it is convenient to define

$$E_1^u = 2\pi \int_0^R i_1^u(r_1) r_1 dr_1 \tag{3}$$

$$I_1^u(\eta_1) = i_1^u(r_1)/[E_1^u/(\pi R^2)] \tag{4}$$

$$U_1(\eta_1) = u_1(r_1)/\sqrt{E_1^u/(\pi R^2)} \tag{5}$$

$$I_2(\eta_2) = 4(r_a/R)^2 i_2(r_2)/[E_1^u/(\pi R^2)] \tag{6}$$

where $\eta_1 = r_1/R$ and $\eta_2 = r_2/r_a$ are respectively the normalized radial coordinate on the input plane and that on the output plane, $r_a = \lambda f/(2\pi R)$, E_1^u is UBRFIP, $U_1(\eta_1)$ is the normalized complex field on the input plane, $I_2(\eta_2)$ is the normalized intensity on the output plane. Then the normalized intensity on the input plane is $I_1(\eta_1) = |U_1(\eta_1)|^2$ and the phase on the input plane is $\phi_1(\eta_1) = \arg[U_1(\eta_1)] + \phi_0$, where ϕ_0 is an arbitrary constant. $I_1^u(\eta_1)$ is thus the upper bound of $I_1(\eta_1)$ and $I_1^u(\eta_1) = 1$ for uniform distribution. $G = \eta_2^{G1}/\eta_2^A$, where $\eta_2^{Gi} = r_2^{Gi}/r_a$ is the normalized radial coordinate of the position of a zero intensity constraint on the output plane, and $\eta_2^A \approx 3.8325$ is the normalized radial coordinate of the first zero position of the Airy disk pattern, scilicet the minimum positive zero point of $J_1(x)$. Eq. (2), Eq. (5) and Eq. (6) may lead to

$$I_2(\eta_2) = 4 \left| \int_0^1 U_1(\eta_1) J_0(\eta_1 \eta_2) \eta_1 d\eta_1 \right|^2 \tag{7}$$

which ensures that

$$S = I_2(0) \tag{8}$$

considering that $U_1(\eta_1) = 1$ for the Airy disk pattern. Then problem (1a) to (1c) can be normalized to be

$$\max_{U_1(r_1)} I_2(0) \tag{9a}$$

such that

$$\begin{cases} I_2(\eta_2^{G_i}) = 0 & (9b) \\ |U_1(\eta_1)|^2 \leq I_1^u(\eta_1) & (9c) \end{cases}$$

Let $A_1(\eta_1)$ and $B_1(\eta_1)$ denote $\mathrm{Re}[U_1(\eta_1)]$ and $\mathrm{Im}[U_1(\eta_1)]$ correspondingly, then problem (9a) to (9c) comes to the form of

$$\max_{A_1(\eta_1), B_1(\eta_1)} \left\{ \left[\int_0^1 A_1(\eta_1) \eta_1 d\eta_1 \right]^2 + \left[\int_0^1 B_1(\eta_1) \eta_1 d\eta_1 \right]^2 \right\} \tag{10a}$$

such that

$$\begin{cases} \int_0^1 A_1(\eta_1) J_0(\eta_1 \eta_2^{G_i}) \eta_1 \mathrm{d}\eta_1 = 0 & (10\mathrm{b}) \\ \int_0^1 B_1(\eta_1) J_0(\eta_1 \eta_2^{G_i}) \eta_1 \mathrm{d}\eta_1 = 0 & (10\mathrm{c}) \\ A_1(\eta_1)^2 + B_1(\eta_1)^2 \leqslant I_1^u(\eta_1) & (10\mathrm{d}) \end{cases}$$

If

$$U_1(\eta_1) = U_{1m}(\eta_1) \tag{11}$$

is a globally optimal solution of problem (10a) to (10d), it can be easily proved that

$$U_1(\eta_1) = U_{1m}(\eta_1) \exp(i\phi_0) \tag{12}$$

is also a globally optimal solution of problem (10a) to (10d), where ϕ_0 is an arbitrary constant phase. To linearize objective function (10a), a particular ϕ_0 is selected to enable Eq. (12) to meet

$$\int_0^1 \mathrm{Re}[U_1(\eta_1)] \eta_1 \mathrm{d}\eta_1 = \int_0^1 \mathrm{Im}[U_1(\eta_1)] \eta_1 \mathrm{d}\eta_1 \tag{13}$$

Eq. (12) and Eq. (13) give ϕ_0 as

$$\phi_0 = n\pi + \arctan\left(\frac{\sigma_m - \mu_m}{\sigma_m + \mu_m}\right) \tag{14}$$

where

$$\sigma_m = \int_0^1 \mathrm{Re}[U_{1m}(\eta_1)] \eta_1 \mathrm{d}\eta_1 \tag{15}$$

$$\mu_m = \int_0^1 \mathrm{Im}[U_{1m}(\eta_1)] \eta_1 \mathrm{d}\eta_1 \tag{16}$$

and n is an arbitrary integer. So no less than one of the globally optimal solutions of problem (10a) to (10d) satisfies a constraint

$$\int_0^1 A_1(\eta_1) \eta_1 \mathrm{d}\eta_1 = \int_0^1 B_1(\eta_1) \eta_1 \mathrm{d}\eta_1 \tag{17}$$

Then objective function (10a) can be linearized to be

$$\min_{A_1(\eta_1), B_1(\eta_1)} \left[\min\left(\int_0^1 A_1(\eta_1) \eta_1 \mathrm{d}\eta_1, -\int_0^1 A_1(\eta_1) \eta_1 \mathrm{d}\eta_1 \right) \right] \tag{18}$$

where function $\min(a, b)$ is equal to the less one between the real number a and the real number b. So problem (10a) to (10d) can be converted into another one named as $P_\pm$ with (18) as its objective function and with (17) and (10b) to (10d) as its constraints. By problem $P_\pm$, a problem P_+ is defined as

$$\min_{A_1(\eta_1), B_1(\eta_1)} \int_0^1 A_1(\eta_1) \eta_1 \mathrm{d}\eta_1 \tag{19a}$$

such that

$$\int_0^1 A_1(\eta_1) J_0(\eta_1 \eta_2^{G_i}) \eta_1 d\eta_1 = 0 \tag{19b}$$

$$\int_0^1 B_1(\eta_1) J_0(\eta_1 \eta_2^{G_i}) \eta_1 d\eta_1 = 0 \tag{19c}$$

$$\int_0^1 A_1(\eta_1) \eta_1 d\eta_1 = \int_0^1 B_1(\eta_1) \eta_1 d\eta_1 \tag{19d}$$

$$A_1(\eta_1)^2 + B_1(\eta_1)^2 \leq I_1^u(\eta_1) \tag{19e}$$

and another problem P_- is defined as

$$\min_{-A_1(\eta_1), -B_1(\eta_1)} \int_0^1 [-A_1(\eta_1)] \eta_1 d\eta_1 \tag{20a}$$

such that

$$\int_0^1 [-A_1(\eta_1)] J_0(\eta_1 \eta_2^{G_i}) \eta_1 d\eta_1 = 0 \tag{20b}$$

$$\int_0^1 [-B_1(\eta_1)] J_0(\eta_1 \eta_2^{G_i}) \eta_1 d\eta_1 = 0 \tag{20c}$$

$$\int_0^1 [-A_1(\eta_1)] \eta_1 d\eta_1 = \int_0^1 [-B_1(\eta_1)] \eta_1 d\eta_1 \tag{20d}$$

$$[-A_1(\eta_1)]^2 + [-B_1(\eta_1)]^2 \leq I_1^u(\eta_1) \tag{20e}$$

Let $U_1(\eta_1) = U_{1m\pm}(\eta_1)$ denote a globally optimal solution of problem $P_\pm$, $U_1(\eta_1) = U_{1m+}(\eta_1)$ denote a globally optimal solution of problem P_+, $U_1(\eta_1) = U_{1m-}(\eta_1)$ denote a globally optimal solution of problem P_-, F_{m+} denote the value of objective function (19a) corresponding to $U_{1m+}(\eta_1)$ and F_{m-} denote the value of objective function (20a) corresponding to $U_{1m-}(\eta_1)$. Then it can be easily proved that

$$U_{1m\pm}(\eta_1) = \begin{cases} U_{1m+}(\eta_1), & \text{if } F_{m+} \leq F_{m-} \\ U_{1m-}(\eta_1), & \text{if } F_{m-} \leq F_{m+} \end{cases} \tag{21}$$

It is obvious that P_- can be turned into P_+ with $A_1(\eta_1)$ substituted for $-A_1(\eta_1)$ and $B_1(\eta_1)$ for $-B_1(\eta_1)$, so P_- and P_+ are the same problems. Thus, problem $P_\pm$ can be transformed into problem P_+ by Eq. (21).

But the constraint (19e) is still nonlinear. Fortunately, this constraint can be linearized geometrically. As is shown in Fig. 2, the constraint (19e) corresponds to an area surrounded by a circle with original point as its center and $I_1^u(\eta_1)^{1/2}$ as its radius, so this constraint can be approximated to some constraints corresponding to an area surrounded by a regular polygon with $4P$ sides. The bigger P is, the more exact this approximation is. The equation of line $l_{++}(p)$ shown in Fig. 2 can be easily attained as

$$A_1(\eta_1) \cos\gamma(p) + B_1(\eta_1) \sin\gamma(p) = \sqrt{I_1^u(\eta_1)} \cos[\pi/(4P)] \tag{22}$$

where $\gamma(p) = (p - 1/2)\pi/(2P), p = 1, \cdots, P$. Similarly, the equation of line $l_{-+}(p)$, that of $l_{--}(p)$ and that of $l_{+-}(p)$ are respectively

$$-A_1(\eta_1)\cos\gamma(p) + B_1(\eta_1)\sin\gamma(p) = \sqrt{I_1^u(\eta_1)}\cos[\pi/(4P)] \qquad (23)$$

$$-A_1(\eta_1)\cos\gamma(p) - B_1(\eta_1)\sin\gamma(p) = \sqrt{I_1^u(\eta_1)}\cos[\pi/(4P)] \qquad (24)$$

$$A_1(\eta_1)\cos\gamma(p) - B_1(\eta_1)\sin\gamma(p) = \sqrt{I_1^u(\eta_1)}\cos[\pi/(4P)] \qquad (25)$$

So the constraints corresponding to the area surrounded by the regular polygon are

$$(-1)^m A_1(\eta_1)\cos\gamma(p) + (-1)^n B_1(\eta_1)\sin\gamma(p) \leq \sqrt{I_1^u(\eta_1)}\cos[\pi/(4P)] \qquad (26)$$

where $m, n \in \{1, 2\}$ and $p = 1, \cdots, P$. Thus, the nonlinear constraint (19e) is converted into the linear constraints (26) with the number of $4P$. So problem (10a) to (10d) is now transformed into a linear one named as P_+^a with (19a) as its objective function and with (19b), (19c), (19d) and (26) as its constraints.

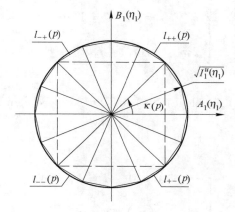

Fig. 2 Constraint (19e) corresponding to an area surrounded by a circle can be approximated to constraints (26) corresponding to an area surrounded by a regular polygon with $4P$ sides. $\kappa(p) = (p-1)\pi/(2P), p = 1, \cdots, P$

To discretize problem P_+^a, let $A_1(\eta_1) = A_k$, $B_1(\eta_1) = B_k$ and $I_1^u(\eta_1) = I_k^u$ if $\eta_1 \in [\eta_{k-1}, \eta_k]$, where $k = 1, \cdots, k_m$, $\eta_0 = 0$, $\eta_{k_m} = 1$ and $(\eta_k - \eta_{k-1})$ is little enough to enable $(A_k + iB_k)$ to approach $[A_1(\eta_1) + iB_1(\eta_1)]$ sufficiently. Then the discretization form of problem P_+^a is

$$\min_{\{A_k, B_k\}} \sum_{k=1}^{k_m} A_k \alpha_k(0) \qquad (27a)$$

such that

$$\begin{cases} \sum_{k=1}^{k_m} A_k \alpha_k(\eta_2^{G_i}) = 0 & (27b) \\ \sum_{k=1}^{k_m} B_k \alpha_k(\eta_2^{G_i}) = 0 & (27c) \\ \sum_{k=1}^{k_m} (A_k - B_k) \alpha_k(0) = 0 & (27d) \\ (-1)^m A_k \cos\gamma(p) + (-1)^n B_k \sin\gamma(p) \\ \leq \sqrt{I_k^u}\cos[\pi/(4P)], \quad k = 1, \cdots, k_m & (27e) \end{cases}$$

where

$$\alpha_k(\eta_2) = \begin{cases} [J_1(\eta_k \eta_2)\eta_k - J_1(\eta_{k-1}\eta_2)\eta_{k-1}]/\eta_2, & \text{if } \eta_2 \neq 0 \\ (\eta_k^2 - \eta_{k-1}^2)/2, & \text{if } \eta_2 = 0 \end{cases} \quad (28)$$

Problem (27a) to (27e) belongs to linear programming problems, whose globally optimal solutions can be obtained with TLP [13].

2.2 Variation Theory of Functional

Design of a DSE with UBRFIP restricted is narrated as a problem of

$$\max_{u_1(r_1)} i_2(0) \quad (29a)$$

such that

$$\begin{cases} i_2(r_2^{G_i}) = 0, \quad i = 1, \cdots, N & (29b) \\ 2\pi \int_0^R |u_1(r_1)|^2 r_1 dr_1 \leq E_1^u & (29c) \end{cases}$$

where E_1^u is UBRFIP.

With Eq. (4) to Eq. (6), problem (29a) to (29c) can be normalized to be

$$\max_{U_1(\eta_1)} I_2(0) \quad (30a)$$

such that

$$\begin{cases} I_2(\eta_2^{G_i}) = 0, \quad i = 1, \cdots, N & (30b) \\ 2\int_0^1 |U_1(\eta_1)|^2 \eta_1 d\eta_1 \leq 1 & (30c) \end{cases}$$

which can be rewritten to be

$$\max_{A_1(\eta_1), B_1(\eta_1)} \left\{ \left[\int_0^1 A_1(\eta_1) \eta_1 d\eta_1 \right]^2 + \left[\int_0^1 B_1(\eta_1) \eta_1 d\eta_1 \right]^2 \right\} \quad (31a)$$

such that

$$\begin{cases} \int_0^1 A_1(\eta_1) J_0(\eta_1 \eta_2^{G_i}) \eta_1 d\eta_1 = 0 & (31b) \\ \int_0^1 B_1(\eta_1) J_0(\eta_1 \eta_2^{G_i}) \eta_1 d\eta_1 = 0 & (31c) \\ 2\int_0^1 [A_1(\eta_1)^2 + B_1(\eta_1)^2] \eta_1 d\eta_1 \leq 1 & (31d) \end{cases}$$

where $A_1(\eta_1) = \text{Re}[U_1(\eta_1)]$ and $B_1(\eta_1) = \text{Im}[U_1(\eta_1)]$.

$A_1(\eta_1)$ and $B_1(\eta_1)$ are both assumed to be continuous functions so that VTF is available to solve problem (31a) to (31d).

Let

$$\begin{cases} A_1(\eta_1) = A_{1m}(\eta_1) \\ B_1(\eta_1) = B_{1m}(\eta_1) \end{cases} \quad (32)$$

denote any of the globally optimal solutions of problem (31a) to (31d).

If Eq. (32) satisfy the inequality of constraint (31d), the constructed functional is

$$F_{ie}[A_1(\eta_1), B_1(\eta_1), \tau_i, \omega_i] = \left[\int_0^1 A_1(\eta_1)\eta_1 d\eta_1\right]^2 + \left[\int_0^1 B_1(\eta_1)\eta_1 d\eta_1\right]^2 +$$
$$\sum_{i=1}^N \tau_i \int_0^1 A_1(\eta_1) J_0(\eta_1 \eta_2^{G_i}) \eta_1 d\eta_1 +$$
$$\sum_{i=1}^N \omega_i \int_0^1 B_1(\eta_1) J_0(\eta_1 \eta_2^{G_i}) \eta_1 d\eta_1 \quad (33)$$

where τ_i and ω_i are both Lagrange multipliers. By VTF, Eq. (32) meet

$$\begin{cases} \delta_{A_1(\eta_1)} F_{ie} = \delta_{B_1(\eta_1)} F_{ie} = 0 & (34a) \\ \partial F/\partial \tau_i = \partial F/\partial \omega_i = 0 & (34b) \end{cases}$$

where

$$\delta_{A_1(\eta_1)} F_{ie} = \int_0^1 \left[2c_A + \sum_{i=1}^N \tau_i J_0(\eta_1 \eta_2^{G_i})\right] \eta_1 \delta A_1(\eta_1) d\eta_1 \quad (35)$$

$$\delta_{B_1(\eta_1)} F_{ie} = \int_0^1 \left[2c_B + \sum_{i=1}^N \omega_i J_0(\eta_1 \eta_2^{G_i})\right] \eta_1 \delta B_1(\eta_1) d\eta_1 \quad (36)$$

$$\partial F_{ie}/\partial \tau_i = \int_0^1 A_1(\eta_1) J_0(\eta_1 \eta_2^{G_i}) \eta_1 d\eta_1 \quad (37)$$

$$\partial F_{ie}/\partial \omega_i = \int_0^1 B_1(\eta_1) J_0(\eta_1 \eta_2^{G_i}) \eta_1 d\eta_1 \quad (38)$$

with

$$c_A = \int_0^1 A_1(\eta_1) \eta_1 d\eta_1 \quad (39)$$

$$c_B = \int_0^1 B_1(\eta_1) \eta_1 d\eta_1 \quad (40)$$

both being constant numbers independent of η_1. Since $\delta A_1(\eta_1)$ and $\delta B_1(\eta_1)$ are both arbitrary real continuous functions satisfying $\delta A_1(0) = \delta A_1(1) = \delta B_1(0) = \delta B_1(1) = 0$[14], Eq. (34a) can deduce

$$\begin{cases} 2c_A J_0(\eta_1 \eta_2^{G_0}) + \sum_{i=1}^N \tau_i J_0(\eta_1 \eta_2^{G_i}) = 0 & (41a) \\ 2c_B J_0(\eta_1 \eta_2^{G_0}) + \sum_{i=1}^N \omega_i J_0(\eta_1 \eta_2^{G_i}) = 0 & (41b) \end{cases}$$

where $\eta_2^{G_0} = 0$. Considering that a function group of $\{J_0(\eta_1 \eta_2^{G_i}) \mid i = 0, 1, \cdots, N\}$ with η_1 as an independent variable is linearly independent due to $\eta_2^{G_{i-1}} < \eta_2^{G_i}$, $i = 1, \cdots, N$, Eq. (41a) and Eq. (41b) may lead to

$$\begin{cases} c_A = c_B = 0 & (42a) \\ \tau_i = \omega_i = 0 & (42b) \end{cases}$$

Eq. (42a) will induce $I_2(0)$ of the globally optimal solutions of problem (31a) to (31d) to be zero, which is considered to be impossible. So Eq. (32) must satisfy the equality of constraint (31d).

Upon that, a new functional in view of constraint (31b), constraint (31c) and the equality of constraint (31d) is constructed by VTF to be

$$F_e[A_1(\eta_1), B_1(\eta_1), \tau_i, \omega_i, \theta] = [\int_0^1 A_1(\eta_1)\eta_1 d\eta_1]^2 + [\int_0^1 B_1(\eta_1)\eta_1 d\eta_1]^2 +$$

$$\sum_{i=1}^N \tau_i \int_0^1 A_1(\eta_1) J_0(\eta_1 \eta_2^{G_i}) \eta_1 d\eta_1 +$$

$$\sum_{i=1}^N \omega_i \int_0^1 B_1(\eta_1) J_0(\eta_1 \eta_2^{G_i}) \eta_1 d\eta_1 +$$

$$\theta \{ 2\int_0^1 [A_1(\eta_1)^2 + B_1(\eta_1)^2] \eta_1 d\eta_1 - 1 \} \quad (43)$$

where θ is another Lagrange multiplier. And VTF indicates that Eqs. (32) satisfy

$$\begin{cases} \delta_{A_1(\eta_1)} F_e = \delta_{B_1(\eta_1)} F_e = 0 & (44a) \\ \partial F_e / \partial \tau_i = \partial F_e / \partial \omega_i = 0 & (44b) \\ \partial F_e / \partial \theta = 0 & (44c) \end{cases}$$

where

$$\delta_{A_1(\eta_1)} F_e = \int_0^1 [2c_A + \sum_{i=1}^N \tau_i J_0(\eta_1 \eta_2^{G_i}) + 4\theta A_1(\eta_1)] \eta_1 \delta A_1(\eta_1) d\eta_1 \quad (45)$$

$$\delta_{B_1(\eta_1)} F_e = \int_0^1 [2c_B + \sum_{i=1}^N \omega_i J_0(\eta_1 \eta_2^{G_i}) + 4\theta B_1(\eta_1)] \eta_1 \delta B_1(\eta_1) d\eta_1 \quad (46)$$

$$\partial F_e / \partial \tau_i = \int_0^1 A_1(\eta_1) J_0(\eta_1 \eta_2^{G_i}) \eta_1 d\eta_1 \quad (47)$$

$$\partial F_e / \partial \omega_i = \int_0^1 B_1(\eta_1) J_0(\eta_1 \eta_2^{G_i}) \eta_1 d\eta_1 \quad (48)$$

$$\partial F_e / \partial \theta = 2\int_0^1 [A_1(\eta_1)^2 + B_1(\eta_1)^2] \eta_1 d\eta_1 - 1 \quad (49)$$

Because $A_1(\eta_1)$ and $B_1(\eta_1)$ are both assumed to be continuous functions with $\delta A_1(\eta_1)$ and $\delta B_1(\eta_1)$ both being arbitrary real continuous functions satisfying $\delta A_1(0) = \delta A_1(1) = \delta B_1(0) = \delta B_1(1) = 0$[14], Eq. (44a) can deduce

$$A_1(\eta_1) = c_0^A [1 + \sum_{i=1}^N d_i^A J_0(\eta_1 \eta_2^{G_i})] \quad (50)$$

$$B_1(\eta_1) = c_0^B [1 + \sum_{i=1}^N d_i^B J_0(\eta_1 \eta_2^{G_i})] \quad (51)$$

where

$$c_0^A = -c_A/(2\theta) \quad (52)$$

$$c_0^B = -c_B/(2\theta) \quad (53)$$

$$d_i^A = \tau_i/(2c_A) \quad (54)$$

$$d_i^B = \omega_i/(2c_B) \quad (55)$$

No less than one of the globally optimal solutions of problem (31a) to (31d) can meet
$$c_A = c_B \tag{56}$$
due to a reason similar to that for Eq. (17). In view of $I_2(0)$ of the globally optimal solutions of problem (31a) to (31d) not being zero, an important deduction of Eq. (52), Eq. (53) and Eq. (56) is
$$c_0^A = c_0^B = c_0 \neq 0 \tag{57}$$
with which, insertion of Eq. (50) and Eq. (51) into Eq. (44b) yields
$$\int_0^1 [1 + \sum_{i=1}^N d_i^A J_0(\eta_1 \eta_2^{G_i})] J_0(\eta_1 \eta_2^{G_j}) \eta_1 d\eta_1 = 0 \tag{58}$$
$$\int_0^1 [1 + \sum_{i=1}^N d_i^B J_0(\eta_1 \eta_2^{G_i})] J_0(\eta_1 \eta_2^{G_j}) \eta_1 d\eta_1 = 0 \tag{59}$$
Eq. (58) and Eq. (59) can determine d_i^A and d_i^B as
$$[d_i^A] = [d_i^B] = [d_i] = -[T_{ji}]^{-1}[q_j] \tag{60}$$
where
$$T_{ji} = \int_0^1 J_0(\eta_1 \eta_2^{G_i}) J_0(\eta_1 \eta_2^{G_j}) \eta_1 d\eta_1$$
$$= \begin{cases} \dfrac{\eta_2^{G_j} J_0(\eta_2^{G_i}) J_1(\eta_2^{G_j}) - \eta_2^{G_i} J_0(\eta_2^{G_j}) J_1(\eta_2^{G_i})}{(\eta_2^{G_j})^2 - (\eta_2^{G_i})^2}, & \text{if } j \neq i \\ [J_0(\eta_2^{G_j})^2 + J_1(\eta_2^{G_j})^2]/2, & \text{if } j = i \end{cases} \tag{61}$$
$$q_j = \int_0^1 J_0(\eta_1 \eta_2^{G_j}) \eta_1 d\eta_1 = J_1(\eta_2^{G_j})/\eta_2^{G_j} \tag{62}$$
In Eq. (60), symmetric matrix $[T_{ji}]$ is of full rank since a function group of $\{J_0(\delta_1 \delta_2^{G_i}) \mid i = 1, \cdots, N\}$ with η_1 as an independent variable is linearly independent. Eq. (50), Eq. (51), Eq. (57) and Eq. (60) can derive
$$A_1(\eta_1) = B_1(\eta_1) = c_0[1 + \sum_{i=1}^N d_i J_0(\eta_1 \eta_2^{G_i})] \tag{63}$$
Insertion of Eq. (63) into Eq. (44c) yields
$$c_0 = \pm 0.5 \{\int_0^1 [1 + \sum_{i=1}^N d_i J_0(\eta_1 \eta_2^{G_i})]^2 \eta_1 d\eta_1\}^{-1/2} \tag{64}$$
The integral in Eq. (64) can be simplified by Eq. (60), Eq. (61) and Eq. (62) to be
$$\int_0^1 [1 + \sum_{i=1}^N d_i J_0(\eta_1 \eta_2^{G_i})]^2 \eta_1 d\eta_1 = 0.5 + 2\sum_{i=1}^N d_i \int_0^1 J_0(\eta_1 \eta_2^{G_i}) \eta_1 d\eta_1 +$$
$$\sum_{i=1}^N \sum_{j=1}^N d_i d_j \int_0^1 J_0(\eta_1 \eta_2^{G_i}) J_0(\eta_1 \eta_2^{G_j}) \eta_1 d\eta_1$$
$$= 0.5 + 2[d_i]^T[q_i] + [d_i]^T[T_{ij}][d_j]$$
$$= 0.5 - [q_j]^T[T_{ij}]^{-1}[q_i] \tag{65}$$
Insertion of Eq. (65) into Eq. (64) gives c_0 as

$$c_0 = \pm 0.5(0.5 - [q_j][T_{ij}]^{-1}[q_i])^{-1/2} \tag{66}$$

Upon that, Eq. (7), Eq. (8), Eq. (60), Eq. (62), Eq. (63) and Eq. (66) present an analytical expression of S^{eu} as

$$S^{eu} = 1 - 2[q_j]^T[T_{ij}]^{-1}[q_i] \tag{67}$$

If $N = 1$, Eq. (61) and Eq. (62) can get

$$[T_{ij}] = [J_0(\eta_2^G)^2 + J_1(\eta_2^G)^2]/2 \tag{68}$$

$$[q_i] = J_1(\eta_2^G)/\eta_2^G \tag{69}$$

with $\eta_2^G = \eta_2^{G_1} = \eta_2^A G$. So a distinct form of S^{eu} for $N = 1$ can be obtained by Eq. (67), Eq. (68) and Eq. (69) as

$$S^{eu} = 1 - \frac{4}{J_0(\eta_2^G)^2 + J_1(\eta_2^G)^2}\left[\frac{J_1(\eta_2^G)}{\eta_2^G}\right]^2 \tag{70}$$

which satisfies

$$S^{eu}(G = 0) = 0 \tag{71}$$

$$S^{eu}(G = 1) = 1 \tag{72}$$

2.3 Wide-sense eigenvalue theory of matrix

The assumption of $A_1(\eta_1)$ and $B_1(\eta_1)$ both being continuous functions enables VTF to solve problem (31a) to (31d), but the assumption is not necessary for a design of DSEs. So VTF cannot obtain any of the globally optimal solutions of problem (31a) to (31d) if none of these solutions satisfies the assumption. To get rid of this assumption and testify the validity of VTF, WETM is presented to solve problem (30a) to (30c).

With a technique of penalty functions, problem (30a) to (30c) can be transformed into

$$\min_{A_1(\eta_1), B_1(\eta_1)} [-I_2(0) + \sum_{i=1}^{N} v_i I_2(\eta_2^{G_i})]/4 \tag{73a}$$

such that

$$2\int_0^1 [A(\eta_1)^2 + B(\eta_1)^2]\eta_1 d\eta_1 \leq 1 \tag{73b}$$

where v_i is a penalty factor satisfying $v_i \gg 1$.

After the same discretization procedure as that of problem P_+^a, problem (73a) and (73b) is discretized to be

$$\min_x x^T H x \tag{74a}$$

such that

$$x^T L x \leq 1 \tag{74b}$$

where

$$x^T = (x_1^T, x_2^T) \tag{75}$$

$$H = \text{diag}(h, h) \tag{76}$$

$$L = \text{diag}(\beta, \beta) \tag{77}$$

with definitions

$$[x_1]_{k,1} = A_k \tag{78}$$

$$[x_2]_{k,1} = B_k \tag{79}$$

$$h = -\alpha(0)\alpha(0)^T + \sum_{i=1}^{N} v_i \alpha(\eta_2^{G_i})\alpha(\eta_2^{G_i})^T \tag{80}$$

$$\alpha(\eta_2)_{k,1} = \alpha_k(\eta_2) \tag{81}$$

$$\beta_{k,l} = \begin{cases} 0, & \text{if } k \neq l \\ (\eta_k^2 - \eta_{k-1}^2), & \text{if } k = l \end{cases} \tag{82}$$

$\alpha_k(\eta_2)$ in Eq. (81) is defined by Eq. (28).

Let

$$x = x_m \tag{83}$$

denote any of the globally optimal solutions of problem (74a) and (74b).

If Eq. (83) reaches the equality of constraint (74b), the method of Lagrangian multiplier presents an augmented function as

$$F(x,\lambda) = x^T H x - \lambda(x^T L x - 1) \tag{84}$$

and indicates that Eq. (83) satisfies

$$\partial F/\partial x = \partial F/\partial \lambda = 0 \tag{85}$$

where

$$(\partial F/\partial x)^T = 2(Hx - \lambda Lx) \tag{86}$$

$$\partial F/\partial \lambda = -(x^T L x - 1) \tag{87}$$

Then Eq. (85) can be rewritten as

$$\begin{cases} Hx = \lambda Lx & (88a) \\ x^T L x = 1 & (88b) \end{cases}$$

Inserting Eq. (88a) and Eq. (88b) into objective function (74a) may simplify it to be

$$\min_{x} \lambda \tag{88c}$$

with which, Eq. (88a) and Eq. (88b) indicate that Eq. (83) is the solution of simultaneous Eqs.

$$\begin{cases} Hx = \lambda_{\min} Lx & (89a) \\ x^T L x = 1 & (89b) \end{cases}$$

where $\lambda_{\min}$ is the minimum wide-sense eigenvalue of the real symmetric matrix H and the real positive definite matrix L and x is a normalized wide-sense eigenvector belonging to the wide-sense eigenvalue of $\lambda_{\min}$, since all of the wide-sense eigenvalues of a real symmetric matrix and a real positive definite matrix are real by WETM.

Objective Eq. (73a) shows that the positive index of inertia of matrix H is $2N$ and the negative index of inertia of matrix H is 2. So objective Eq. (88c) indicates that the value of objective function (74a) corresponding to the globally optimal solutions must be negative because the number of negative wide-sense eigenvalues of matrix H and matrix L is equal to the negative index of inertia of matrix H by WETM.

If Eq. (83) keeps the inequality of constraint (74b), it is satisfied by Eq. (83) that

$$[\partial(x^T Hx)/\partial x]^T = 2Hx = 0 \qquad (90)$$

After inserting Eq. (90) into objective function (74a), it is deduced that the value of objective function (74a) corresponding to the globally optimal solutions is zero, which contradicts a proved conclusion that the value of objective function (74a) of the globally optimal solutions is negative. So Eq. (83) must obey the equality of constraint (74b).

Upon that, problem (74a) and (74b) can be transformed into simultaneous Eqs. (89a) and (89b), which own ripe algorithms to achieve a solution [15].

3 Simulation results

The parameters for simulation are selected to be $I_1^u(\eta_1) = 1$, $\eta_2^{G_2} - \eta_2^{G_1} = \eta_2^A/2$, $v_1 = v_2 = 10000$, $\eta_k = k/k_m, k = 0, 1, \cdots, k_m$. TLP and WETM can be actually implemented with software for scientific computation. To reduce the amount of numerical calculation while ensuring enough precision, an appropriate value for P is 10 and one for k_m is 50, which is determined by actual numerical computation.

As is shown in Fig. 3 to Fig. 7, the simulation results of VTF and those of WETM appear to be identical. So the validity of VTF is testified by WETM, as is mentioned at the beginning of subsection 2.3.

When $\eta_2^{G_1} = \eta_2^A/2$ and $N = 1$ or 2, two DSEs with UBIDIP restricted and another two with UBRFIP restricted are designed, $I_1(\eta_1)$ and $I_2(\eta_2)$ of which are respectively shown in Fig. 3 and Fig. 4.

The DSE attained through designing is a phase-only element, and the value of $\phi_1(\eta_1)$ of the globally optimal solutions is 0 or π. $\phi_1(\eta_1)$ of globally optimal solutions has at most one dis-

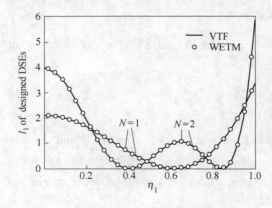

Fig. 3 Curves of $I_1(\eta_1)$ of designed DSEs with $\eta_2^{G_1} = \eta_2^A/2$ and with UBRFIP restricted

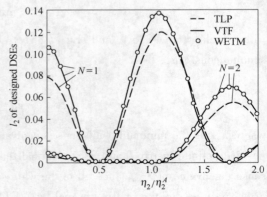

Fig. 4 Curves of $I_2(\eta_2)$ of designed DSEs with $\eta_2^{G_1} = \eta_2^A/2$

continuous point $\eta_1 = \eta_1^b$ if $N = 1$ and has at most two discontinuous points $\eta_1 = \eta_1^{b_1}$, $\eta_1^{b_2}$ if $N = 2$. The dependence of η_1^b, $\eta_1^{b_1}$ or $\eta_1^{b_2}$ upon G is shown in Fig. 5.

The curves of $S^{eu}(G)$ are shown in Fig. 6 and those of $K(G)$ of globally optimal solutions in Fig. 7, which indicate that S^{eu} decreases rapidly and K of the globally optimal solutions increases fast with G reduced, so that some limits of optical superresolution are found. Here $K(G)$ of globally optimal solutions is defined as the normalized adjacent side lobe intensity corresponding to globally optimal solutions under certain constraints

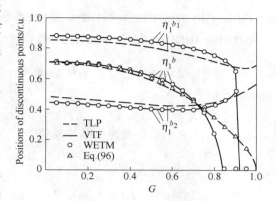

Fig. 5 Curves of the dependence of η_1^b, $\eta_1^{b_1}$ or $\eta_1^{b_2}$ upon G, where $\eta_1^b = 0$ or $\eta_1^{b_1} = \eta_1^{b_2} = 0$ means no discontinuous point

with a resolving power described by G. A curve of $S^u(G)$ similar to that of $S^{eu}(G)$ is provided by Ref. [1], which reveals the physical content of some limits of optical superresolution, but it bears the following drawbacks in term of the mathematics: $S^u(G)$ is not equal to $S^{eu}(G)$ but larger than it; the value given by the analytic formula of $S^u(G)$ is close to that of $S^{eu}(G)$ only when $G \in [0, 0.46]$, and if $G \in (0.46, 1]$, the curve of $S^u(G)$ is obtained by extrapolation which is not rigorous. The curves shown in Fig. 6 not only get rid of these drawbacks, but also can be adapted to more cases due to more freedom of design with UBRFIP restricted than that with UBIDIP restricted and the arbitrary number of the zero intensity constraints on the output plane. It can also be seen that S^{eu} and K of the globally optimal solutions decrease if N increases and the superresolution performance with UBRFIP restricted is better than that with UBIDIP restricted because the design freedom corresponding to the former is more than that of the latter.

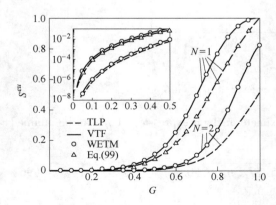

Fig. 6 Curves of $S^{eu}(G)$. The insert is the curves of logarithmic coordinate to magnify the curves of homogeneous coordinate

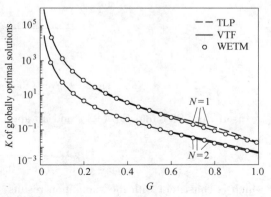

Fig. 7 Curves of $K(G)$ of globally optimal solutions

4 Analytic results with the upper bound of the intensity distribution on the input plane restricted

If $N = 1$ and $I_1^u(\eta_1) = 1$, the DSE attained through above designing is a phase-only element with

$$U_1(\eta_1) = \begin{cases} 1, & \text{if } \eta_1 \in [0, \eta_1^b] \\ -1, & \text{if } \eta_1 \in (\eta_1^b, 1] \end{cases} \quad (91)$$

Eq. (7) and Eq. (91) lead to

$$I_2(\eta_2) = 4[2J_1(\eta_1^b \eta_2)\eta_1^b - J_1(\eta_2)]^2 / \eta_2^2 \quad (92)$$

Eq. (9b) and Eq. (92) show that η_1^b meets

$$2J_1(\eta_1^b \eta_2^G)\eta_1^b - J_1(\eta_2^G) = 0 \quad (93)$$

where $\eta_2^G = \eta_2^{G_1} = \eta_2^A G$. If $G \in [0,1]$, Fig. 8 shows that

$$\eta_1^b \eta_2^G \in [0, 1.2356] \quad (94)$$

so that

$$J_1(\eta_1^b \eta_2^G) \approx \eta_1^b \eta_2^G / 2 - (\eta_1^b \eta_2^G)^3 / 16 \quad (95)$$

bears a little error not more than 7.2658×10^{-3}.

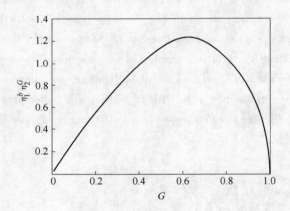

Fig. 8 Curve of the dependence of $\eta_1^b \eta_2^G$ on G

Insert Eq. (95) into Eq. (93) and an approximate analytic expression of η_1^b can be attained as

$$\eta_1^b \approx \sqrt{1 - \sqrt{1 - 0.5\eta_2^G J_1(\eta_2^G)}} / (0.5\eta_2^G) \quad (96)$$

which is consistent with the simulation results

$$\eta_1^b(G = 0) = \sqrt{2}/2 \quad (97)$$

$$\eta_1^b(G = 1) = 0 \quad (98)$$

Upon that, an analytic formula of S^{eu} is given by Eq. (8), Eq. (92) and Eq. (96) as

$$S^{eu} = [1 - 2(\eta_1^b)^2]^2 \tag{99}$$

which corresponds to the simulation results

$$S^{eu}(G = 0) = 0 \tag{100}$$
$$S^{eu}(G = 1) = 1 \tag{101}$$

in view of Eq. (97) and Eq. (98).

Eq. (96) and Eq. (99) are both considerably accurate, as are respectively shown in Fig. 5 and Fig. 6.

5 Conclusions

The globally optimal solutions of a DSE can be obtained with TLP if UBIDIP is restricted and can be calculated out with VTF or WETM if UBRFIP is restricted. The simulation results of VTF and those of WETM appear to be identical, which enables the validity of VTF to be certified by WETM. As examples of design, DSEs with specific parameters are devised. The DSE attained through designing is a phase-only element and the value of $\phi_1(\eta_1)$ of the globally optimal solutions is 0 or π. The curves showing the relations between η_1^b, $\eta_1^{b_1}$ or $\eta_1^{b_2}$ and G are provided. To indicate some limits of optical superresolution, the curves of $S^{eu}(G)$ and $K(G)$ of globally optimal solutions are figured, which are much more improved than the corresponding results in Ref. [1] in term of the mathematics. If $N = 1$ and $I_1^u(\eta_1) = 1$, the rules of DSE's structure mentioned above can lead to approximate analytic formulas of η_1^b and S^{eu} with considerable accuracy.

However, vector properties of an actual electromagnetic field become significant in many cases, such as an optical system of high numerical aperture [16]. In these cases, theories considering the vector properties are necessary for a design of DSEs, which will be a further work for us.

Acknowledgements

This work is sponsored by the National Research Fund for Fundamental Key Project of China, No. 973 (G19990330), and the Fund of High Technology of China (863-804-2).

References

[1] T. R. M. Sales and G. M. Morris, "Fundamental limits of optical superresolution", Opt. Lett. 22, 582-584 (1997).

[2] I. J. Cox, "Increasing the bit packing densities of optical disk systems", Appl. Opt. 23, 3260-3261 (1984).

[3] Masahisa SHINODA and Kenjiro KIME, "Focusing characteristics of an optical head with superresolution using a high-aspect-ratio red laser diode", Jpn. J. Appl. Phys. 35, 380-383 (1996).

[4] T. Wilson and Barry R. Masters, "Confocal microscopy", Appl. Opt. 33, 565-566 (1994).

[5] K. Carlsson, P. E. Danielsson, R. Lenz, A. Liljeborg, L. Majlof and N. Aslund, "Three-dimensional microscopy using a confocal laser scanning microscope", Opt. Lett. 10, 53-55 (1985).

[6] Shen Yibing, Yang Guoguang, Hou Xiyun, "Research on phenomenon of the super-resolution in laser lithography", Acta Opt. Sin. 19, 1512-1517(1999).

[7] Tasso R. M. Sales and G. Michael Morris, "Diffractive superresolution elements", J. Opt. Soc. Am. A 14, 1637-1646(1997).

[8] Michael R. Wang and Xu Guang Huang, "Subwavelength-resolvable focused non-Gaussian beam shaped with a binary diffractive optical element", Appl. Opt. 38, 2171-2176(1999).

[9] R. Boivin and A. Boivin, "Optimized amplitude filtering for superresolution over a restricted field", Opt. Acta. 27, 587-610(1980).

[10] I. J. Cox, C. J. R. Sheppard, and T. Wilson, "Reappraisal of arrays of concentric annuli as superresolving filters", J. Opt. Soc. Am. 72, 1287-1291(1982).

[11] Ding Zhihua, Tian Weijian, Bao Zhengkang, "Superresolution with high throughput via irradiance redistribution element", Acta Opt. Sin. 20, 701-706(2000).

[12] Han Anyun, "Wavefront engineering and optical superresolution", Semicond. Inf. 35, 1-17(1998).

[13] J. K. Strayer, *Linear Programming and Its Applications* (Springer-Verlag, New York, 1989), Chap. 2.

[14] L. E. Elsgolc, *Calculus of Variations* (Pergamon Press, Oxford, 1961), Chap. I-2.

[15] J. H. Wilkinson, *The Algebraic Eigenvalue Problem* (Clarendon Pr., Oxford, 1965), Chap. 5.

[16] M. Born and E. Wolf, *Principles of Optics*, 5th ed. (Pergamon, New York, 1975), Chap. 3, Chap. 8.

Theories for the Design of a Hybrid Refractive-diffractive Superresolution Lens with High Numerical Aperture*

Abstract   By the geometrical optics and the Rayleigh-Sommerfeld diffraction formula, theories for the design of a hybrid refractive-diffractive superresolution lens (HRDSL) with high numerical aperture are constructed in this paper. Differences between the profile of the diffractive superresolution element (DSE) with high numerical aperture and that with low numerical aperture are indicated. The optimization theory can obtain a globally optimal solution through a linear programming much more simplified than the corresponding one in Ref. [9]. The rules of the structure of the designed DSE are both theoretically proved and numerically verified. Comparison of the optimization theory in this paper with the other design theories and examples of designing the HRDSL with high numerical aperture are provided. Lastly, some limits of optical superresolution with high numerical aperture are set and compared with those of low numerical aperture.

1 Introduction

A focal spot of a micrometer or less is required by optical data storage[1], laser lithography[2] or confocal scanning microscopy[3,4]. A laser source with short wavelength or an objective lens with high numerical aperture are typical approaches[5], but the former is limited by the level of manufacture and the latter is limited by many factors, such as a high energy loss due to the reflection on the surfaces, large mass and volume, and a strong sensitivity to aberrations and misalignment. An alternative method is to place a DSE[6-10] behind an objective lens to reduce the radius of the first zero position of the transverse pattern on the focal plane. Here a normalized spot size G is defined as the ratio between the radius of the first zero position of a superresolved pattern and the radius of the first zero position of the Airy disk pattern, a Strehl ratio S is defined as the ratio between the central intensity of a superresolved pattern and the central intensity of the Airy disk pattern, and a normalized adjacent side lobe intensity K is defined as the ratio of the maximum intensity of the sidelobe adjacent to the main lobe to the central intensity of a superresolved pattern. G determines the resolving power, and S and K describe the performance of the main lobe and that of the side lobes of a superresolved pattern, respectively. It has been well known that S decreases rapidly and K increases fast with a reduction of G[9,10]. Upon that, an eclectic approach to obtain a focal spot of a micrometer or less is to apply both an objective lens with practicable high numerical aperture and a DSE with appropriate resolving power.

---

* Copartner: Haitao Liu, Yingbai Yan, Deer Yi. Reprinted from *Journal of Optical Society of America A*, 2003, 20(5):913-924.

Compared with a design of a DSE with an objective lens of low numerical aperture[6-10], a design of a DSE with an objective lens of high numerical aperture bears many particular problems. First, a lens with high numerical aperture can no longer be treated as a thin one[11] because both the propagation direction and the intensity distribution of a normally incident wave are distinctly changed after it passes through the lens. Since the change of the slow varying amplitude in the scale of a wavelength is far smaller than the slow varying amplitude for the field within the lens, both the transmissive wave on the two surfaces of the lens and the propagating wave within the lens can be calculated by the geometrical optics[12,13]. Second, the Fresnel approximation no longer applies to the diffraction of the wave in the image space of the lens. However, the Rayleigh-Sommerfeld diffraction formula[14,15] of the scalar diffraction theory is still available to compute the field distribution near the focus, because both the focal length and the radius of the lens are much larger than a wavelength. Lastly, a DSE behind the lens is not perpendicularly illuminated so that the relation[16] $h = [\mod(\psi, 2\pi)/(2\pi)]\lambda/(n-1)$ applicable to a thin phase-only element vertically illuminated does not work as well, where λ is the wavelength, h is the thickness of the element, ψ is the phase delay introduced by the element, and n is the refractive index of the element. Some approaches to deal with a thick diffractive object illuminated arbitrarily have been presented, such as the Born approximation[17,18], the perturbation theory[19] or the ray tracing method[20]. In view of the practicability of these theories for the design of the DSE, the ray tracing method is developed in this paper to obtain a set of transformation equations describing the transformation of the field near the focus resulting from the DSE.

For a design of a HRDSL with high numerical aperture described in Section 2, a dependence of the field on the angular coordinate is deduced by the rotational symmetry of the system in Section 3, which simplifies a double integral of diffraction of the AAPL into a definite integral in Section 4. Based on above work, a transformation of the field near the focus of the AAPL resulting from a phase-only DSE is described by a set of transformation equations in Section 5, which actualizes a construction of an optimization theory to design the DSE of the HRDSL in Section 6. Lastly in Section 7, comparison of this optimization theory with the other design theories and examples of designing HRDSLs with high numerical aperture are presented, and some limits of optical superresolution with high numerical aperture are revealed.

2 Description of the hybrid refractive-diffractive superresolution lens

As shown in Fig. 1, a HRDSL of rotational symmetry is composed of an AAPL and a step-profiled hybrid type DSE introducing both phase delay and energy attenuation in view of a hybrid type DSE owning more design freedom than a phase-only DSE or an amplitude-only DSE. However, it will be theoretically proved in Section 6 and numerically verified in Section 7 that the designed DSE must be a phase-only element, so that an implementation of a hybrid type DSE is not necessary. The optical axis is selected as z axis and the center of the aspheric surface is selected as the original point. An incident wave in the object space of the HRDSL is set to be

$$\mathbf{E}^{(i)}(\rho,\theta,z) = \mathbf{y}E^{(i)}(\rho,\theta,z) = \mathbf{y}e^{(i)}\exp(-\mathrm{i}k_0 z) \tag{1}$$

with **x**, **y** and **z** separately defined as a unit vector along x axis, one along y axis and one along z axis, and $e^{(i)}$ being homogeneous in the object space of the HRDSL. And an incident wave of elliptical polarization can be decomposed into two orthogonal components of linear polarization. The dependence of a field on time is selected to be $\exp(\mathrm{i}\omega\tau)$ in this paper, where ω denotes the angular frequency of the field and τ is time. ρ, θ and z in Eq. (1) are respectively the radial coordinate, the angular coordinate and the axial coordinate, $k_0 = 2\pi/\lambda$, and λ is the wavelength. The refractive index of the HRDSL is denoted by n, the thickness of the AAPL is denoted by t, the axial distance between the diaphragm and the original point is denoted by βt with $\beta \in (0,1)$, the distance between the planar surface of the AAPL and the focal plane is denoted by f_a named as the focal length of the HRDSL, and the maximum radial coordinate of the field on the planar surface of the AAPL determined by the geometrical optics is denoted by $\rho_{1,M}$. Then define $NA_s = \rho_{1,M}/(\rho_{1,M}^2 + f_a^2)^{1/2}$ as the sinusoidal numerical aperture of the HRDSL and $NA_t = \rho_{1,M}/f_a$ as the tangential numerical aperture of the HRDSL. $NA_t \ll 1$ for low numerical aperture and $NA_t \sim 1$ for high numerical aperture, where " $\ll$ " means "far smaller than" (" $\gg$ " means "far larger than", contrarily) and " $\sim$ " means "close in order of magnitude".

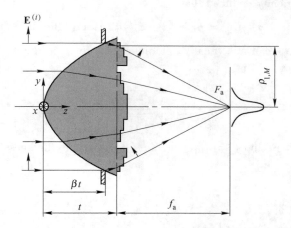

Fig. 1 Scheme of the HRDSL

As shown in Fig. 2, let $z_a(\rho_1)$ and $\rho_a(\rho_1)$ respectively denote the axial coordinate and the radial coordinate of any point B on the aspheric surface, let ρ_1 denote the radial coordinate on the planar surface of the AAPL, and let $\alpha^{(i)}(\rho_1)$ denote the incident angle of the field on the planar surface of the AAPL expressed by

$$\alpha^{(i)}(\rho_1) = \arcsin\left(\frac{\rho_1}{n\sqrt{\rho_1^2 + f_a^2}}\right) \tag{2}$$

which can be easily obtained by the refraction law of a ray through the planar surface of the AAPL. By the principle of equal optical path[12], the profile of the aspheric surface shown in

Fig. 2 satisfies an optical path equation of

$$z_a(\rho_1) + n[t - z_a(\rho_1)]/\cos\alpha^{(i)}(\rho_1) + \sqrt{\rho_1^2 + f_a^2} = [ABCF_a] = [ODF_a] = nt + f_a \tag{3}$$

which leads to an expression of $z_a(\rho_1)$ as

$$z_a(\rho_1) = \frac{nt[1/\cos\alpha^{(i)}(\rho_1) - 1] + \sqrt{\rho_1^2 + f_a^2} - f_a}{n/\cos\alpha^{(i)}(\rho_1) - 1} \tag{4}$$

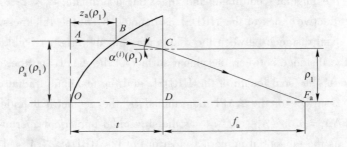

Fig. 2 Computation of the profile of the aspheric surface by the principle of equal optical path

And it can be geometrically deduced that

$$\rho_a(\rho_1) = \rho_1 + [t - z_a(\rho_1)]\tan\alpha^{(i)}(\rho_1) \tag{5}$$

Eqs. (2),(4) and (5) can analytically determine the profile of the aspheric surface. If $\rho_1 = \rho_{1,M}$, Eqs. (2) and (4) may lead to

$$z_a(\rho_{1,M}) = \frac{nt\{\sqrt{(1+NA_t^2)/[1+(1-1/n^2)NA_t^2]} - 1\} + \sqrt{\rho_{1,M}^2 + f_a^2} - f_a}{n\sqrt{(1+NA_t^2)/[1+(1-1/n^2)NA_t^2]} - 1} \tag{6}$$

Considering $z_a(\rho_{1,M}) = \beta t$ and $f_a = \rho_{1,M}/NA_t$, Eq. (6) gives the thickness t expressed by specified $n, \beta, \rho_{1,M}$ and NA_t as

$$t = \frac{\rho_{1,M}(\sqrt{1+1/NA_t^2} - 1/NA_t)}{(n-\beta) - n(1-\beta)\sqrt{(1+NA_t^2)/[1+(1-1/n^2)NA_t^2]}} \tag{7}$$

We select $\lambda = 550$nm, $n = 1.5, \beta = 0.8$ and $\rho_{1,M} = 5$mm for all simulations in this paper. The profile of the aspheric surface of an AAPL with $NA_t = 1$ is shown in Fig. 3.

3 Calculation of the transmissive field on the planar surface of the anaberrational aspheric-planar lens without the diffractive super-resolution element

A field of angular coordinate 0 or $\pi/2$ excited by the incident wave $\mathbf{y}E^{(i)}$ is

$$\mathbf{E}(\rho,0,z) = \mathbf{y}E_y(\rho,0,z) \tag{8}$$

$$\mathbf{E}(\rho,\pi/2,z) = \mathbf{y}E_y(\rho,\pi/2,z) + \mathbf{z}E_z(\rho,\pi/2,z) \tag{9}$$

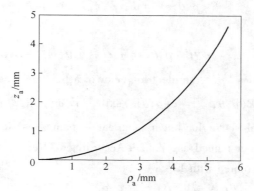

Fig. 3 The profile of the aspheric surface of an AAPL with $NA_t = 1$

As shown in Fig. 4, $x' - y'$ coordinate system comes from a rotation of $x - y$ coordinate system round z axis by θ angle, and denote a unit vector along x' axis and one along y' axis severally by $\mathbf{x}'$ and $\mathbf{y}'$. Considering the rotational symmetry of the system, a field of angular coordinate θ excited by an incident wave $\mathbf{y}'E^{(i)}$ or $\mathbf{x}'E^{(i)}$ can be obtained by Eq. (8) or Eq. (9) as

$$\mathbf{E}_\perp(\rho,\theta,z) = \mathbf{y}'E_y(\rho,0,z) \tag{10}$$

$$\mathbf{E}_\parallel(\rho,\theta,z) = \mathbf{x}'E_y(\rho,\pi/2,z) + \mathbf{z}E_z(\rho,\pi/2,z) \tag{11}$$

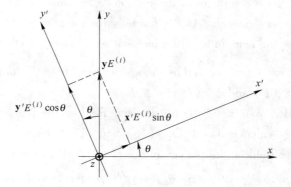

Fig. 4 Coordinate systems to get the dependence of the field on the angular coordinate with the rotational symmetry of the system

By the linearity of electromagnetic fields, the field of angular coordinate θ excited by the incident wave $\mathbf{y}E^{(i)}$ is

$$\mathbf{E}(\rho,\theta,z) = \mathbf{E}_\perp(\rho,\theta,z)\cos\theta + \mathbf{E}_\parallel(\rho,\theta,z)\sin\theta \tag{12}$$

Insertion of Eqs. (10) and (11) into Eq. (12) can deduce

$$E_x(\rho,\theta,z) = -0.5[E_y(\rho,0,z) - E_y(\rho,\pi/2,z)]\sin2\theta \tag{13a}$$

$$E_y(\rho,\theta,z) = 0.5[E_y(\rho,0,z) + E_y(\rho,\pi/2,z)] + 0.5[E_y(\rho,0,z) - E_y(\rho,\pi/2,z)]\cos2\theta \tag{13b}$$

$$E_z(\rho,\theta,z) = E_z(\rho,\pi/2,z)\sin\theta \qquad (13c)$$

where

$$\mathbf{x}E_x(\rho,\theta,z) + \mathbf{y}E_y(\rho,\theta,z) + \mathbf{z}E_z(\rho,\theta,z) = \mathbf{E}(\rho,\theta,z) \qquad (14)$$

By the geometrical optics[12,13], the electric vector of a light field can be expressed as

$$\mathbf{E}(\rho,\theta,z) = \mathbf{e}(\rho,\theta,z)\exp[-ik_0\phi(\rho,\theta,z)] \qquad (15)$$

where $\mathbf{e}(\rho,\theta,z)$ is a real vector for linear polarization named as slow varying amplitude and $\phi(\rho,\theta,z)$ is a real number named as eikonal. According to Ref. [12], the propagating wave within a lens can be calculated with Eqs.

$$\phi(P_2) = \phi(P_1) + [P_1 P_2] \qquad (16a)$$

$$\mathbf{e}(P_2) = \mathbf{e}(P_1)\exp\left[-\frac{1}{2n}\int_{P_1}^{P_2}\nabla^2\phi(P)\,\mathrm{d}s\right] \qquad (16b)$$

where P_1 and P_2 are two points on a ray L, $[P_1 P_2]$ is the optical path from P_1 to P_2, $P\in L$, and n is the refractive index of the lens. By Ref. [13], the transmissive wave on the two surfaces of a lens can be computed with Eqs.

$$\phi^{(t)}(P) = \phi^{(i)}(P) \qquad (17a)$$

$$\mathbf{e}^{(t)}(P) = \overset{\leftrightarrow}{\mathbf{T}}(P)\cdot\mathbf{e}^{(i)}(P) \qquad (17b)$$

where P is a point on the two surfaces of the lens, $\phi^{(i)}(P)$ and $\phi^{(t)}(P)$ are respectively the eikonal of the incident wave and that of the transmissive wave, $\mathbf{e}^{(i)}(P)$ and $\mathbf{e}^{(t)}(P)$ are respectively the slow varying amplitude of the incident wave and that of the transmissive wave, and $\overset{\leftrightarrow}{\mathbf{T}}(P)$ is a transmissive tensor given by the well known Fresnel's formulas. So it is convenient to denote both $\phi^{(i)}(P)$ and $\phi^{(t)}(P)$ by $\phi(P)$.

Then $\mathbf{e}(\rho,\theta,z)$ and $\phi(\rho,\theta,z)$ of the transmissive field on the planar surface of the AAPL without the DSE can be calculated with Eqs. (16) and (17). Denote the slow varying amplitude of the transmissive field on the planar surface of the AAPL without the DSE by

$$\mathbf{e}^{(t)}(\rho_1,\theta_1) = \mathbf{x}e_x^{(t)}(\rho_1,\theta_1) + \mathbf{y}e_y^{(t)}(\rho_1,\theta_1) + \mathbf{z}e_z^{(t)}(\rho_1,\theta_1) \qquad (18)$$

where θ_1 is the angular coordinate on the planar surface of the AAPL. $\mathbf{e}^{(t)}(\rho_1,\theta_1)$ with $NA_t = 1$ is computed as shown in Fig. 5. By the principle of equal optical path[12], the eikonal of the field on the planar surface of the AAPL without the DSE is given as

$$\phi(\rho_1) = -\sqrt{\rho_1^2 + f_a^2} + \phi(F_a) \qquad (19)$$

where $\phi(F_a) = nt + f_a$ is the eikonal of the field at focus F_a of the AAPL without the DSE.

4 Computation of the field near the focus of the anaberrational aspheric-planar lens without the diffractive superresolution element

Let

$$\mathbf{E}_k(\rho',\theta',f) = \mathbf{x}E_{x,k}(\rho',\theta',f) + \mathbf{y}E_{y,k}(\rho',\theta',f) + \mathbf{z}E_{z,k}(\rho',\theta',f) \qquad (20)$$

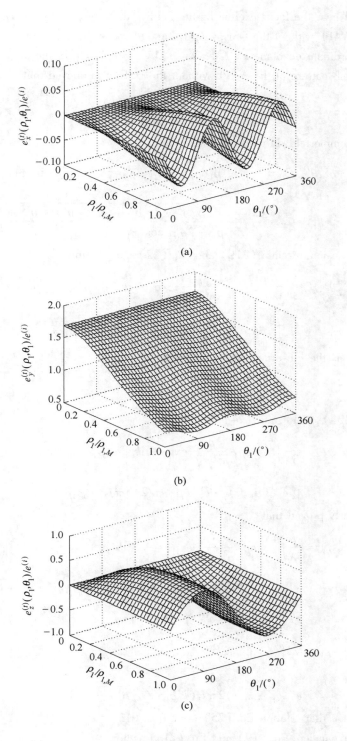

Fig. 5 (a)-(c) respectively show the x component, the y component and the z component of the slow varying amplitude of the transmissive field on the planar surface of the AAPL with $NA_t = 1$ without the DSE denote the field near the focus excited by the field in the zone $\rho_1 \in [\rho_{1,k-1}, \rho_{1,k}]$ on the planar surface of the AAPL without the DSE, where ρ' and θ' are respectively the radial coordinate and the angular coordinate on plane $z = f$, $k = 1, \cdots, M$, and $\rho_{1,0} = 0$. In the area of the focal depth and the Airy disk pattern of the AAPL without the DSE, it is obeyed that

$$|f - f_a| \sim \lambda/NA_t^2 \quad \text{or} \quad |f - f_a| < \lambda/NA_t^2 \tag{21a}$$

$$\rho' \sim \lambda/NA_t \quad \text{or} \quad \rho' < \lambda/NA_t \tag{21b}$$

The Rayleigh-Sommerfeld diffraction formula yields

$$E_{x,k}(\rho', \theta', f) = \frac{f}{i\lambda} \int_{\rho_{1,k-1}}^{\rho_{1,k}} \rho_1 d\rho_1 \int_0^{2\pi} d\theta_1 e_x^{(t)}(\rho_1, \theta_1) \exp[-ik_0\phi(\rho_1)] \times$$

$$\frac{\exp[-ik_0\sqrt{\rho_1^2 - 2\rho_1\rho'\cos(\theta_1 - \theta') + \rho'^2 + f^2}]}{\rho_1^2 - 2\rho_1\rho'\cos(\theta_1 - \theta') + \rho'^2 + f^2} \tag{22}$$

With series expansion, formula (21b) and Eq. (22) can obtain

$$E_{x,k}(\rho', \theta', f) \approx \frac{f}{i\lambda} \int_{\rho_{1,k-1}}^{\rho_{1,k}} d\rho_1 \frac{\rho_1}{\rho_1^2 + f^2} \exp\{-ik_0[\phi(\rho_1) + \sqrt{\rho_1^2 + f^2}]\} \times$$

$$\int_0^{2\pi} d\theta_1 e_x^{(t)}(\rho_1, \theta_1) \exp[-ik_0\rho_1\rho'/\sqrt{\rho_1^2 + f^2}\cos(\theta_1 - \theta')] \tag{23}$$

with J_n defined as the nth-order Bessel function,

$$\exp(ix\cos\theta) = \sum_{n=-\infty}^{+\infty} J_n(x)i^n\exp(-in\theta) \tag{24}$$

Relation Eq. (13a), (23), and (24) yield

$$E_{x,k}(\rho', \theta', f) \approx \frac{f}{i\lambda} \int_{\rho_{1,k-1}}^{\rho_{1,k}} d\rho_1 \frac{\rho_1}{\rho_1^2 + f^2} \exp\{-ik_0[\phi(\rho_1) + \sqrt{\rho_1^2 + f^2}]\} \times$$

$$[e_y^{(t)}(\rho_1, 0) - e_y^{(t)}(\rho_1, \pi/2)]\pi J_2(k_0\rho_1\rho'/\sqrt{\rho_1^2 + f^2})\sin 2\theta' \tag{25a}$$

It can be similarly proved that

$$E_{y,k}(\rho', \theta', f) \approx \frac{f}{i\lambda} \int_{\rho_{1,k-1}}^{\rho_{1,k}} d\rho_1 \frac{\rho_1}{\rho_1^2 + f^2} \exp\{-ik_0[\phi(\rho_1) + \sqrt{\rho_1^2 + f^2}]\} \times$$

$$\{[e_y^{(t)}(\rho_1, 0) + e_y^{(t)}(\rho_1, \pi/2)]\pi J_0(k_0\rho_1\rho'/\sqrt{\rho_1^2 + f^2}) -$$

$$[e_y^{(t)}(\rho_1, 0) - e_y^{(t)}(\rho_1, \pi/2)]\pi J_2(k_0\rho_1\rho'/\sqrt{\rho_1^2 + f^2})\cos 2\theta'\} \tag{25b}$$

$$E_{z,k}(\rho', \theta', f) \approx \frac{f}{\lambda} \int_{\rho_{1,k-1}}^{\rho_{1,k}} d\rho_1 \frac{\rho_1}{\rho_1^2 + f^2} \exp\{-ik_0[\phi(\rho_1) + \sqrt{\rho_1^2 + f^2}]\} \times$$

$$e_z^{(t)}(\rho_1, \pi/2) 2\pi J_1(k_0\rho_1\rho'/\sqrt{\rho_1^2 + f^2})\sin\theta' \tag{25c}$$

It is easy to prove that relation Eq. (25) obey Eqs. (13).

With series expansion, Eq. (19) and Eq. (21a) yield

$$\exp\{-ik_0[\phi(\rho_1) + \sqrt{\rho_1^2 + f^2}]\} \approx \exp[-ik_0\phi(F_a)]\exp[-ik_0f_a(f-f_a)/\sqrt{\rho_1^2 + f_a^2}]$$

$$\tag{26}$$

Inserting relation (26) into relation (25) and applying formulas (21) lead to

$$E_{x,k}(\rho',\theta',f) \approx \frac{f_a}{i\lambda}\exp[-ik_0\phi(F_a)]\int_{\rho_{1,k-1}}^{\rho_{1,k}}d\rho_1\frac{\rho_1}{\rho_1^2+f_a^2}\exp[-ik_0f_a(f-f_a)/\sqrt{\rho_1^2+f_a^2}] \times$$

$$[e_y^{(t)}(\rho_1,0)-e_y^{(t)}(\rho_1,\pi/2)]\pi J_2(k_0\rho_1\rho'/\sqrt{\rho_1^2+f_a^2})\sin2\theta' \qquad (27a)$$

$$E_{y,k}(\rho',\theta',f) \approx \frac{f_a}{i\lambda}\exp[-ik_0\phi(F_a)]\int_{\rho_{1,k-1}}^{\rho_{1,k}}d\rho_1\frac{\rho_1}{\rho_1^2+f_a^2}\exp[-ik_0f_a(f-f_a)/\sqrt{\rho_1^2+f_a^2}] \times$$

$$\{[e_y^{(t)}(\rho_1,0)+e_y^{(t)}(\rho_1,\pi/2)]\pi J_0(k_0\rho_1\rho'/\sqrt{\rho_1^2+f_a^2}) -$$

$$[e_y^{(t)}(\rho_1,0)-e_y^{(t)}(\rho_1,\pi/2)]\pi J_2(k_0\rho_1\rho'/\sqrt{\rho_1^2+f_a^2})\cos2\theta'\} \qquad (27b)$$

$$E_{z,k}(\rho',\theta',f) \approx \frac{f_a}{\lambda}\exp[-ik_0\phi(F_a)]\int_{\rho_{1,k-1}}^{\rho_{1,k}}d\rho_1\frac{\rho_1}{\rho_1^2+f_a^2}\exp[-ik_0f_a(f-f_a)/\sqrt{\rho_1^2+f_a^2}] \times$$

$$e_z^{(t)}(\rho_1,\pi/2)2\pi J_1(k_0\rho_1\rho'/\sqrt{\rho_1^2+f_a^2})\sin\theta' \qquad (27c)$$

5 Transformation of the field near the focus of the anaberrational aspheric-planar lens resulting from a phase-only diffractive superresolution element

As shown in Fig. 6, relation Eq. (25a) yields

$$E_{x,k}(\rho',\theta',f;h_k) \approx \frac{f-h_k}{i\lambda}\int_{\rho_{2,k-1}^+}^{\rho_{2,k}^-}d\rho_2\frac{\rho_2}{\rho_2^2+(f-h_k)^2}\exp\{-ik_0[\phi(\rho_2;h_k) +$$

$$\sqrt{\rho_2^2+(f-h_k)^2}]\} \times [e_y^{(t)}(\rho_2,0;h_k)-e_y^{(t)}(\rho_2,\pi/2;h_k)]$$

$$\pi J_2[k_0\rho_2\rho'/\sqrt{\rho_2^2+(f-h_k)^2}]\sin2\theta' \qquad (28a)$$

$$E_{x,k}(\rho',\theta',f;0) \approx \frac{f}{i\lambda}\int_{\rho_{1,k-1}^+}^{\rho_{1,k}^-}d\rho_1\frac{\rho_1}{\rho_1^2+f^2}\exp\{-ik_0[\phi(\rho_1;0)+\sqrt{\rho_1^2+f^2}]\} \times$$

$$[e_y^{(t)}(\rho_1,0;0)-e_y^{(t)}(\rho_1,\pi/2;0)]\pi J_2(k_0\rho_1\rho'/\sqrt{\rho_1^2+f^2})\sin2\theta'$$

$$(28b)$$

where $E_{x,k}(\rho',\theta',f;h_k)$ is the contribution from the field on plane 1 with radial coordinate $\rho_1 \in [\rho_{1,k-1}^+, \rho_{1,k}^-]$ transformed by the kth step of a phase-only DSE with refractive index n and thickness h_k to the x component of the field on plane $z=f$, ρ_2 is the radial coordinate on plane 2, $e_y^{(t)}(\rho_2,0;h_k)$ and $e_y^{(t)}(\rho_2,\pi/2;h_k)$ are both the y components of the slow varying amplitude of the transmissive field on the surface of the step, $\phi(\rho_2;h_k)$ is the eikonal of the field on the surface of the step, $k=1,\cdots,M$, $\rho_{1,0}^+=\rho_{1,0}=0$, and $\rho_{1,M}^-=\rho_{1,M}$ is the maximum radial coordinate of the field on plane 1 determined by the geometrical optics. Here the planar surface of the AAPL is defined as plane 1, and the planar surface of the kth step of the phase-only DSE is defined as plane 2. $E_{x,k}(\rho',\theta',f;0)$ is the contribution from the field on plane 1 with $\rho_1 \in [\rho_{1,k-1}^+, \rho_{1,k}^-]$ not transformed by the DSE to the x component of the field on plane $z=f$, $e_y^{(t)}$

$(\rho_1, 0; 0)$ and $e_y^{(t)}(\rho_1, \pi/2; 0)$ are both the y components of the slow varying amplitude of the transmissive field on the planar surface of the AAPL without the DSE, and $\phi(\rho_1; 0)$ is the eikonal of the field on the planar surface of the AAPL without the DSE. Ray tracing in the step gives

$$\rho_2 = \rho_1 - h_k \tan\alpha^{(i)}(\rho_1) \tag{29a}$$

$$\rho_{2,k-1}^+ = \rho_{1,k-1}^+ - h_k \tan\alpha^{(i)}(\rho_{1,k-1}^+) \tag{29b}$$

$$\rho_{2,k}^- = \rho_{1,k}^- - h_k \tan\alpha^{(i)}(\rho_{1,k}^-) \tag{29c}$$

Select $\rho_{1,k-1}^+$ and $\rho_{1,k}^-$ to satisfy

$$\rho_{1,k}^- - \rho_{1,k-1}^+ \gg \lambda \tag{30a}$$

$$\rho_{1,k}^- - \rho_{1,k-1}^+ \ll \rho_{1,M}^- \tag{30b}$$

As a component of the DSE, the step satisfies

$$h_k \sim \lambda \quad \text{or} \quad h_k < \lambda \tag{31}$$

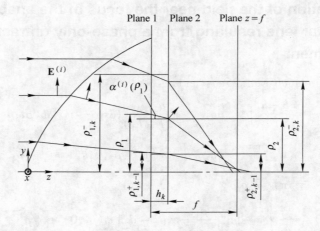

Fig. 6 Transformation of the field near the focus of the AAPL resulting from a phase-only DSE

Equations (29b) and (29c), inequality (30a) and formula (31) yield

$$\rho_{1,k}^- \approx \rho_{1,k}^+, \quad k = 0,1,\cdots,M \tag{32}$$

with prescription of $\rho_{1,0}^- = \rho_{1,0}^+$ and $\rho_{1,M}^+ = \rho_{1,M}^-$, so it is convenient to denote both $\rho_{1,k}^-$ and $\rho_{1,k}^+$ by $\rho_{1,k}$ with $k = 0, 1, \cdots, M$.

With Eq. (17a) and Eqs. (29), Eq. (16a) becomes

$$\phi(\rho_2; h_k) = \phi(\rho_1; 0) + nh_k/\cos\alpha^{(i)}(\rho_1) \tag{33}$$

With Eqs. (29) and formula (31), we can deduce from Eqs. (16b) and (17b) that

$$e_y^{(t)}(\rho_2, 0; h_k) \approx e_y^{(t)}(\rho_1, 0; 0) \tag{34a}$$

$$e_y^{(t)}(\rho_2, \pi/2; h_k) \approx e_y^{(t)}(\rho_1, \pi/2; 0) \tag{34b}$$

Replacing integration variable ρ_2 in Eq. (28a) with ρ_1 by Eqs. (29) and applying formulas

(21b) and (31) and expressions (32)-(34) yields

$$E_{x,k}(\rho',\theta',f;h_k) \approx \frac{f}{i\lambda}\int_{\rho_{1,k-1}}^{\rho_{1,k}} d\rho_1 \frac{\rho_1}{\rho_1^2+f^2}\exp\{-ik_0[\phi(\rho_1;0)+nh_k/\cos\alpha^{(i)}(\rho_1)+$$

$$\sqrt{\rho_2^2+(f-h_k)^2}]\}\times[e_y^{(t)}(\rho_1,0;0)-e_y^{(t)}(\rho_1,\pi/2;0)]$$

$$\pi J_2(k_0\rho_1\rho'/\sqrt{\rho_1^2+f^2})\sin 2\theta' \qquad (35)$$

With series expansion, formula (21a), Eqs. (29), inequality (30b) and formula (31) lead to

$$\exp\{-ik_0[\phi(\rho_1;0)+nh_k/\cos\alpha^{(i)}(\rho_1)+\sqrt{\rho_2^2+(f-h_k)^2}]\}$$

$$\approx \exp\{-ik_0[\phi(\rho_1;0)+\sqrt{\rho_1^2+f^2}]\}$$

$$\exp\left\{-ik_0 h_k\left[\frac{n}{\cos\alpha^{(i)}(\bar{\rho}_{1,k})}-\frac{\bar{\rho}_{1,k}\tan\alpha^{(i)}(\bar{\rho}_{1,k})+f_a}{\sqrt{\bar{\rho}_{1,k}^2+f_a^2}}\right]\right\} \qquad (36)$$

where $\bar{\rho}_{1,k} \in [\rho_{1,k-1},\rho_{1,k}]$ can be arbitrarily selected. Inserting relation (36) into relation (35) and applying relation (28b), we can derive that

$$\frac{E_{x,k}(\rho',\theta',f;h_k)}{E_{x,k}(\rho',\theta',f;0)} \approx \exp\left\{-ik_0 h_k\left[\frac{n}{\cos\alpha^{(i)}(\bar{\rho}_{1,k})}-\frac{\bar{\rho}_{1,k}\tan\alpha^{(i)}(\bar{\rho}_{1,k})+f_a}{\sqrt{\bar{\rho}_{1,k}^2+f_a^2}}\right]\right\} \qquad (37)$$

It can be similarly proved that

$$\frac{E_{y,k}(\rho',\theta',f;h_k)}{E_{y,k}(\rho',\theta',f;0)} \approx \frac{E_{z,k}(\rho',\theta',f;h_k)}{E_{z,k}(\rho',\theta',f;0)} \approx \frac{E_{x,k}(\rho',\theta',f;h_k)}{E_{x,k}(\rho',\theta',f;0)} \qquad (38)$$

By Eqs. (37) and (38), the phase delay ψ_k introduced by h_k is

$$\psi_k = -k_0 h_k\left[\frac{n}{\cos\alpha^{(i)}(\bar{\rho}_{1,k})}-\frac{\bar{\rho}_{1,k}\tan\alpha^{(i)}(\bar{\rho}_{1,k})+f_a}{\sqrt{\bar{\rho}_{1,k}^2+f_a^2}}\right], \forall \bar{\rho}_{1,k} \in [\rho_{1,k-1},\rho_{1,k}] \qquad (39)$$

In view of inequality (30b), two sectionally continuous functions $\psi(\rho_1)$ and $h(\rho_1)$ can be defined as

$$\psi(\rho_1) \approx \psi_k, \quad h(\rho_1) \approx h_k, \quad \text{if } \rho_1 \in [\rho_{1,k-1},\rho_{1,k}], \quad k=1,\cdots,M \qquad (40)$$

Then Eq. (39) becomes

$$\psi(\rho_1) = -k_0 h(\rho_1)\left[\frac{n}{\cos\alpha^{(i)}(\rho_1)}-\frac{\rho_1\tan\alpha^{(i)}(\rho_1)+f_a}{\sqrt{\rho_1^2+f_a^2}}\right] \qquad (41)$$

If $NA_t \ll 1$, Eq. (41) becomes

$$\psi_L(\rho_1) = -k_0 h_L(\rho_1)(n-1) \qquad (42)$$

with $\psi(\rho_1)$ and $h(\rho_1)$ becoming $\psi_L(\rho_1)$ and $h_L(\rho_1)$, respectively. Let $\psi(\rho_1) = \psi_L(\rho_1)$, then from Eqs. (41) and (42) we can deduce that

$$h(\rho_1)/h_L(\rho_1) = \frac{n-1}{\dfrac{n}{\cos\alpha^{(i)}(\rho_1)}-\dfrac{\rho_1\tan\alpha^{(i)}(\rho_1)+f_a}{\sqrt{\rho_1^2+f_a^2}}} \qquad (43)$$

For a HRDSL with $NA_t = 1$, Fig. 7 shows that $h(\rho_1)/h_L(\rho_1)$ with $\psi(\rho_1) = \psi_L(\rho_1)$ will decrease from 1 if ρ_1 increases from 0. This result is consistent with the geometrical optics since $\alpha^{(i)}(\rho_1)$ will increase from 0 if ρ_1 increases from 0.

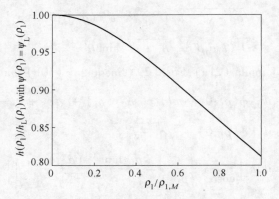

Fig. 7 Curve of $h(\rho_1)/h_L(\rho_1)$ with $\psi(\rho_1) = \psi_L(\rho_1)$ for a HRDSL with $NA_t = 1$

6 Optimization theory for the design of the diffractive superresolution element

Define the attenuation coefficient of the kth step of the hybrid type DSE as $T_k \in [0,1]$. Then the intensity on the focal plane of the HRDSL is[12]

$$I(\rho',\theta',f_a) = \left| \sum_{k=1}^{M} T_k E_{x,k}(\rho',\theta',f_a;h_k) \right|^2 + \left| \sum_{k=1}^{M} T_k E_{y,k}(\rho',\theta',f_a;h_k) \right|^2 + \left| \sum_{k=1}^{M} T_k E_{z,k}(\rho',\theta',f_a;h_k) \right|^2 \tag{44}$$

The design of the DSE may be narrated as

$$\max_{\{T_k,h_k\}} I(0,0,f_a) \tag{45a}$$

subject to

$$I(\rho_j^{(G)},\theta',f_a) = 0, \quad \forall \theta' \in [0,2\pi), \quad j = 1,\cdots,N \tag{45b}$$

where $\rho_j^{(G)}$ is the radial coordinate of a zero intensity constraint on the focal plane satisfying $0 < \rho_{j-1}^{(G)} < \rho_j^{(G)}$ ($j = 2,\cdots,N$). The first zero point $\rho_1^{(G)}$ determines the resolving power, while the other zero points determines the utilizable field of view[6] (FOV). It is kept on Eqs. (27) in the major area of ρ_1 that

$$|e_y^{(t)}(\rho_1,0) + e_y^{(t)}(\rho_1,\pi/2)| \gg |e_y^{(t)}(\rho_1,0) - e_y^{(t)}(\rho_1,\pi/2)| \tag{46a}$$

$$|e_y^{(t)}(\rho_1,0) + e_y^{(t)}(\rho_1,\pi/2)| \gg |2e_z^{(t)}(\rho_1,\pi/2)| \tag{46b}$$

with which, Eqs. (27),(37)-(39) and (44) can approximate constraints (45b) to

$$\sum_{k=1}^{M} T_k \exp(i\psi_k) g_k(\rho_j^{(G)}) = 0 \tag{47}$$

where

$$g_k(\rho_j^{(G)}) = \int_{\rho_{1,k-1}}^{\rho_{1,k}} d\rho_1 \frac{\rho_1}{\rho_1^2 + f_a^2} [e_y^{(t)}(\rho_1,0;0) + e_y^{(t)}(\rho_1,\pi/2;0)] J_0(k_0\rho_1\rho_j^{(G)}/\sqrt{\rho_1^2 + f_a^2})$$

(48)

Let

$$A_k = \text{Re}[T_k \exp(i\psi_k)] \tag{49a}$$

$$B_k = \text{Im}[T_k \exp(i\psi_k)] \tag{49b}$$

then Eqs. (45) can be approximated to

$$\max_{\{A_k,B_k\}} \left\{ \left[\sum_{k=1}^{M} A_k g_k(0)\right]^2 + \left[\sum_{k=1}^{M} B_k g_k(0)\right]^2 \right\} \tag{50a}$$

subject to

$$\sum_{k=1}^{M} A_k g_k(\rho_j^{(G)}) = 0, \quad j = 1,\cdots,N \tag{50b}$$

$$\sum_{k=1}^{M} B_k g_k(\rho_j^{(G)}) = 0, \quad j = 1,\cdots,N \tag{50c}$$

$$A_k^2 + B_k^2 \leq 1, \quad k = 1,\cdots,M \tag{50d}$$

After almost the same procedure as is presented in Section 2A of Ref. [9], Eqs. (50) can be transformed into

$$\min_{\{A_k,B_k\}} \sum_{k=1}^{M} A_k g_k(0) \tag{51a}$$

subject to

$$\sum_{k=1}^{M} A_k g_k(\rho_j^{(G)}) = 0, \quad j = 1,\cdots,N \tag{51b}$$

$$\sum_{k=1}^{M} B_k g_k(\rho_j^{(G)}) = 0, \quad j = 1,\cdots,N \tag{51c}$$

$$\sum_{k=1}^{M} A_k g_k(0) = \sum_{k=1}^{M} B_k g_k(0) \tag{51d}$$

$$A_k^2 + B_k^2 \leq 1, \quad k = 1,\cdots,M \tag{51e}$$

Let

$$A_k = A_k^* \quad \text{and} \quad B_k = B_k^* \tag{52}$$

denote a globally optimal solution of Eqs. (51), which means that

$$\sum_{k=1}^{M} A_k^* g_k(\rho_j^{(G)}) = 0, \quad j = 1,\cdots,N \tag{53a}$$

$$\sum_{k=1}^{M} B_k^* g_k(\rho_j^{(G)}) = 0, \quad j = 1,\cdots,N \tag{53b}$$

$$\sum_{k=1}^{M} A_k^* g_k(0) = \sum_{k=1}^{M} B_k^* g_k(0) \tag{53c}$$

$$A_k^{*2} + B_k^{*2} \leq 1, \quad k = 1,\cdots,M \tag{53d}$$

and the value of objective function (51a) corresponding to a globally optimal solution is

$$F_{\min} = \sum_{k=1}^{M} A_k^* g_k(0) \tag{54}$$

Then we intend to prove in the following that

$$A_k = B_k = (A_k^* + B_k^*)/2 \tag{55}$$

is also a globally optimal solution of Eqs. (51).

Insertion of Eq. (55) into objective function (51a) shows that

$$\sum_{k=1}^{M} g_k(0)(A_k^* + B_k^*)/2 = F_{\min} \tag{56}$$

where Eqs. (53c) and (54) are used. Eqs. (53a) and (53b) yields that Eq. (55) satisfies constraints (51b) and (51c), and it is obvious that Eq. (55) meets constraint (51d). With the Cauchy-Schwarz inequality[21], Eq. (55) can meet constraint (51e) as

$$\left(\frac{A_k^* + B_k^*}{2}\right)^2 + \left(\frac{A_k^* + B_k^*}{2}\right)^2 = \frac{(A_k^* \cdot 1 + B_k^* \cdot 1)^2}{2} \leq \frac{(A_k^{*2} + B_k^{*2}) \cdot (1^2 + 1^2)}{2}$$

$$= A_k^{*2} + B_k^{*2} \leq 1 \tag{57}$$

where inequation (53d) is applied.

So Eq. (55) is also a globally optimal solution of Eqs. (51). Then Eqs. (51) can be simplified to be

$$\min_{\{A_k\}} \sum_{k=1}^{M} A_k g_k(0) \tag{58a}$$

subject to

$$\sum_{k=1}^{M} A_k g_k(\rho_j^{(G)}) = 0, \quad j = 1,\cdots,N \tag{58b}$$

$$-1/\sqrt{2} \leq A_k \leq 1/\sqrt{2}, \quad k = 1,\cdots,M \tag{58c}$$

$$B_k = A_k, \quad k = 1,\cdots,M \tag{58d}$$

Eqs. (58) need much less amount of numerical computation than the corresponding Eqs. (27) in Ref. [9]. Moreover, Eqs. (58) prove theoretically that the phase delay introduced by the designed DSE must be 0 or π. A globally optimal solution of Eqs. (58a)-(58c) can be obtained by the linear programming theory[22].

Finally, we try to prove theoretically that any of the globally optimal solutions of Eqs. (58a)-(58c) must reach the equality of constraint (58c), so that the designed DSE must be a phase-only element introducing a phase delay of 0 or π, which can be produced with high quality. Therefore an implementation of a hybrid type DSE is not necessary.

In view of inequality (30b), a sectionally continuous function $A(\rho_1)$ of real value can be defined as

$$A(\rho_1) \approx A_k, \quad \text{if} \quad \rho_1 \in [\rho_{1,k-1}, \rho_{1,k}], \quad k = 1, \cdots, M \tag{59}$$

Then Eqs. (58a)-(58c) become a functional optimization problem of

$$\min_{A(\rho_1)} \int_0^{\rho_{1,M}} d\rho_1 A(\rho_1) \frac{\rho_1}{\rho_1^2 + f_a^2} [e_y^{(t)}(\rho_1, 0; 0) + e_y^{(t)}(\rho_1, \pi/2; 0)] \tag{60a}$$

subject to

$$\int_0^{\rho_{1,M}} d\rho_1 A(\rho_1) \frac{\rho_1}{\rho_1^2 + f_a^2} [e_y^{(t)}(\rho_1, 0; 0) + e_y^{(t)}(\rho_1, \pi/2; 0)] J_0(k_0 \rho_1 \rho_j^{(G)}/\sqrt{\rho_1^2 + f_a^2}) = 0 \tag{60b}$$

$$-1/\sqrt{2} \leq A(\rho_1) \leq 1/\sqrt{2} \tag{60c}$$

Let

$$A(\rho_1) = [\cos\theta(\rho_1)]/\sqrt{2} \tag{61}$$

to meet constraint (60c), where $\theta(\rho_1)$ is a sectionally continuous function of real value. Thus, Eqs. (60) take the form of

$$\min_{\theta(\rho_1)} \int_0^{\rho_{1,M}} d\rho_1 \cos\theta(\rho_1) \frac{\rho_1}{\rho_1^2 + f_a^2} [e_y^{(t)}(\rho_1, 0; 0) + e_y^{(t)}(\rho_1, \pi/2; 0)] \tag{62a}$$

subject to

$$\int_0^{\rho_{1,M}} d\rho_1 \cos\theta(\rho_1) \frac{\rho_1}{\rho_1^2 + f_a^2} [e_y^{(t)}(\rho_1, 0; 0) + e_y^{(t)}(\rho_1, \pi/2; 0)] J_0(k_0 \rho_1 \rho_j^{(G)}/\sqrt{\rho_1^2 + f_a^2}) = 0 \tag{62b}$$

By the variation theory of a functional [23], an augmented objective functional is constructed as

$$F[\theta(\rho_1), \lambda_j] = \int_0^{\rho_{1,M}} d\rho_1 \cos\theta(\rho_1) \frac{\rho_1}{\rho_1^2 + f_a^2} [e_y^{(t)}(\rho_1, 0; 0) + e_y^{(t)}(\rho_1, \pi/2; 0)] +$$

$$\sum_{j=1}^N \lambda_j \int_0^{\rho_{1,M}} d\rho_1 \cos\theta(\rho_1) \frac{\rho_1}{\rho_1^2 + f_a^2} [e_y^{(t)}(\rho_1, 0; 0) +$$

$$e_y^{(t)}(\rho_1, \pi/2; 0)] J_0(k_0 \rho_1 \rho_j^{(G)}/\sqrt{\rho_1^2 + f_a^2})$$

$$= \int_0^{\rho_{1,M}} d\rho_1 \cos\theta(\rho_1) \frac{\rho_1}{\rho_1^2 + f_a^2} [e_y^{(t)}(\rho_1, 0; 0) + e_y^{(t)}(\rho_1, \pi/2; 0)]$$

$$\sum_{j=0}^N \lambda_j J_0(k_0 \rho_1 \rho_j^{(G)}/\sqrt{\rho_1^2 + f_a^2}) \tag{63}$$

where $\lambda_j (j = 1, \cdots, N)$ is a Lagrangian multiplier, $\lambda_0 = 1$, and $\rho_0^{(G)} = 0$. And any of the globally optimal solutions of Eqs. (62) must obey

$$\delta_{\theta(\rho_1)} F[\theta(\rho_1), \lambda_j] = -\int_0^{\rho_{1,M}} d\rho_1 \delta\theta(\rho_1) \sin\theta(\rho_1) \frac{\rho_1}{\rho_1^2 + f_a^2}$$

$$[e_y^{(t)}(\rho_1,0;0) + e_y^{(t)}(\rho_1,\pi/2;0)] \sum_{j=0}^{N} \lambda_j J_0(k_0\rho_1\rho_j^{(G)}/\sqrt{\rho_1^2+f_a^2}) = 0 \quad (64)$$

where $\delta\theta(\rho_1)$ is an arbitrary continuous function of real value satisfying $\delta\theta(0) = \delta\theta(\rho_{1,M}) = 0$. Then Eq. (64) can lead to

$$\sin\theta(\rho_1) \frac{\rho_1}{\rho_1^2+f_a^2} [e_y^{(t)}(\rho_1,0;0) + e_y^{(t)}(\rho_1,\pi/2;0)]$$

$$\sum_{j=0}^{N} \lambda_j J_0(k_0\rho_1\rho_j^{(G)}/\sqrt{\rho_1^2+f_a^2}) = 0, \quad \forall \rho_1 \in [0,\rho_{1,M}] \quad (65)$$

Since $\lambda_0 \neq 0$ and the function group $\{J_0[k_0\rho_1\rho_j^{(G)}/(\rho_1^2+f_a^2)^{1/2}] \mid j=0,1,\cdots,N\}$ with ρ_1 as the independent variable is linearly independent owing to $\rho_{j-1}^{(G)} < \rho_j^{(G)}$ ($j=1,\cdots,N$), it is kept that

$$\sum_{j=0}^{N} \lambda_j J_0(k_0\rho_1\rho_j^{(G)}/\sqrt{\rho_1^2+f_a^2}) \neq 0, \quad \exists \rho_1 \in [a,b], \quad \forall [a,b] \subseteq [0,\rho_{1,M}] \quad (66)$$

Eqs. (65) and (66) can derive

$$\sin\theta(\rho_1) = 0, \quad \forall \rho_1 \in [0,\rho_{1,M}] \quad (67)$$

By Eqs. (61) and (67), any of the globally optimal solutions of Eqs. (60) must satisfy

$$A(\rho_1) \in \{-1/\sqrt{2}, 1/\sqrt{2}\}, \quad \forall \rho_1 \in [0,\rho_{1,M}] \quad (68)$$

Inview of Eqs. (59) and (68), we can prove theoretically that any of the globally optimal solutions of Eqs. (58a)-(58c) must obey

$$A_k \in \{-1/\sqrt{2}, 1/\sqrt{2}\}, \quad k = 1,\cdots,M \quad (69)$$

7 Simulation results

In this paper, the Airy disk pattern is defined as the transverse pattern on the focal plane of the AAPL without the DSE, and the radius of the first zero position of the Airy disk pattern is approximated to the minimum zero point of $\sum_{k=1}^{M} g_k(\rho')$ in view of Eqs. (27) and inequalities (46). The ratio between the radius of the first zero position of the superresolved pattern denoted by $\rho_1^{(G)}$ and the radius of the first zero position of the Airy disk pattern is defined as the normalized spot size G, the ratio of the central intensity of the superresolved pattern to the central intensity of the Airy disk pattern is defined as the Strehl ratio S, and the maximum value of S under certain constraints is defined as S^{eu}. Since the intensity distribution on the focal plane is of no rotational symmetry, $[K(\theta'=0) + K(\theta'=\pi/2)]/2$ is defined as an average normalized adjacent side lobe intensity denoted by $K^{(a)}$, and a $K^{(a)}$ corresponding to S^{eu} is denoted by $K_m^{(a)}$. $S^{eu} = S^{eu}(G)$ and $K_m^{(a)} = K_m^{(a)}(G)$ if $N=1$. To describe the side lobes of a superresolved pattern more sufficiently, a maximum normalized side lobe intensity W is defined as the ratio of the maximum intensity of all side lobes to the central intensity of a superresolved pattern, and ρ_W is defined as the radius where W is reached. Then define $W^{(a)} = [W(\theta'=0) + W(\theta'=\pi/$

2)]/2 and $\rho_W^{(a)} = [\rho_W(\theta'=0) + \rho_W(\theta'=\pi/2)]/2$.

For an AAPL with low numerical aperture, $\rho_L = 0.61\lambda/NA_t$ is the radius of the first zero position of the field on the focal plane, $HFD_L = 2\lambda/NA_t^2$ is the distance between the focus and a zero position of the field on the optical axis closest to the focus, and $I_L = [e_y^{(t)}(0,0)0.61\pi\rho_{1,M}/\rho_L]^2$ is the focal intensity. Let $I(\rho', \theta', f)$ denote the intensity of the field near the focus. So in Figs. 8 and 9, it is convenient to normalize ρ', $(f-f_a)$ and $I(\rho', \theta', f)$ by ρ_L, HFD_L and I_L, respectively.

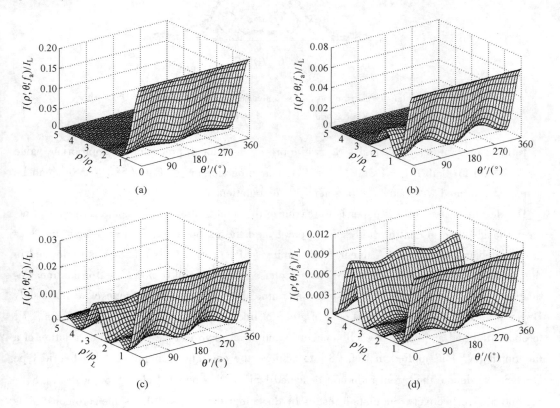

Fig. 8 The intensity distribution on the focal plane of the designed HRDSLs with
$G = 0.68966$ (for $N = 1, 2$ and 3) and $NA_t = 1$. Here $N = 0$ means no DSE
(a) $N = 0$; (b) $N = 1$; (c) $N = 2$; (d) $N = 3$

To quantize the profile of the DSE into a set of steps, we select $\rho_{1,k} = (k/M)\rho_{1,M}$, $k = 0, 1, \cdots, M$. Here M should be selected carefully to meet inequalities (30) and validate the scalar diffraction theory. Firstly, inequalities (30) become

$$\lambda \ll \rho_{1,M}/M \ll \rho_{1,M} \Leftrightarrow 1 \ll M \ll \rho_{1,M}/\lambda \tag{70}$$

Secondly, the conditions to validate the scalar diffraction theory [24] become

$$\rho_{1,M}/M \geq P\lambda \Leftrightarrow M \leq \rho_{1,M}/(P\lambda) \tag{71}$$

where $P = 30$ if $\alpha^{(i)}(\rho_{1,M}) \approx 30°$ and $P = 20$ if $\alpha^{(i)}(\rho_{1,M}) \approx 0°$. For the parameters selected at the end of Section 2, $\alpha^{(i)}(\rho_{1,M}) = 28.1255° \approx 30°$ if $NA_t = 1$ and $\alpha^{(i)}(\rho_{1,M}) \approx 0°$ if $NA_t \ll$

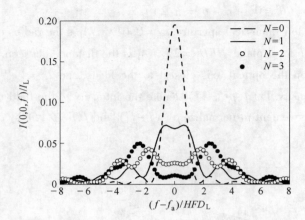

Fig. 9 The intensity distribution on the optical axis of the designed HRDSLs with $G = 0.68966$ (for $N = 1, 2$ and 3) and $NA_t = 1$. Here $N = 0$ means no DSE

1. To meet inequalities (70) and (71) while preserving enough precision, an appropriate value for M is 200 for both $NA_t = 1$ and $NA_t \ll 1$. Linear programming of Eqs. (58a)-(58c) can be actually performed with software for scientific computation.

The design strategy of this paper is to maximize the Strehl ratio S under some zero constraints on the focal plane to control the resolving power and the usable FOV. Similar design strategies are adopted in Refs. [6-8]. A linear programming (LP) of Eqs. (58a)-(58c) is performed in this paper to design a phase-only DSE with a phase modulation of 0 or π, a set of nonlinear equations (NE) is solved in Ref. [6] to determine the position of each zone of a phase-only DSE introducing a phase shift of 0 or π, a group of linear equations (LE) is solved in Ref. [7] to obtain the real amplitude of each concentric annuli of a hybrid type DSE, and a variation of a functional (VF) is applied in Ref. [8] to achieve the continuous amplitude of a hybrid type DSE. So we denote the Strehl ratio of a designed DSE by this paper, Refs. [6-8] with S_{LP}, S_{NE}, S_{LE} and S_{VF}, respectively. Simulation results by these four theories with the same zero constraints as in Ref. [7] are shown in Table 1, where $z_J = 3.8325$ is the first zero point of the first order Bessel function, and η_i^b is the ratio of the ith break point of $\psi(\rho_1)$ defined on Eq. (40) to $\rho_{1,M}$. Since the theories in Refs. [6-8] are all limited to low numerical aperture, we select $NA_t \ll 1$ while solving Eqs. (58a)-(58c). Table 1 shows a relation of

$$S_{LP} \approx S_{NE} > S_{LE} > S_{VF} \qquad (72)$$

except for example 1 not keeping the second " $>$ ". Relation (72) shows that the theory in Ref. [6] may also reach a globally optimal solution of Eqs. (45) because S_{LP} just corresponds to this globally optimal solution obtained through a linear programming of Eqs. (58a)-(58c). This conclusion, however, cannot be proved by the theory in Ref. [6]. Moreover, linear programming of Eqs. (58a)-(58c) needs much less amount of numerical computation than solving the set of nonlinear equations in Ref. [6].

Table 1 Simulation results by the four theories with the same zero constraints as in Ref. [7]

Examples	1	2	3	4	5
$z_J \rho_1^{(G)}/\rho_L$	1.91585	1.91585	3.19308	2.79401	2.79401
$z_J \rho_2^{(G)}/\rho_L$	3.5078	4.1	5.55	4.92388	5.35
$z_J \rho_3^{(G)}/\rho_L$	5.08675	5.9	8.42	6.52682	8.1
$z_J \rho_4^{(G)}/\rho_L$	6.66185	—	11.1031	—	—
$z_J \rho_5^{(G)}/\rho_L$	8.23532	—	—	—	—
G	0.5	0.5	0.833	0.73	0.73
η_1^b	0.2275	0.3225	0.2475	0.2975	0.2500
η_2^b	0.4525	0.6275	0.4125	0.6125	0.5575
η_3^b	0.6575	0.8825	0.6625	0.8525	0.7875
η_4^b	0.8325	—	0.8325	—	—
η_5^b	0.9525	—	—	—	—
S_{LP}	1.98549×10^{-7}	6.62295×10^{-4}	7.33787×10^{-2}	1.28813×10^{-2}	6.48272×10^{-2}
S_{NE}	1.98865×10^{-7}	6.62762×10^{-4}	7.34043×10^{-2}	1.28854×10^{-2}	6.48444×10^{-2}
S_{LE}	1.25830×10^{-8}	1.46721×10^{-4}	2.64231×10^{-2}	3.46088×10^{-3}	2.65224×10^{-2}
S_{VF}	2.4826084×10^{-8}	1.2530898×10^{-4}	1.5574213×10^{-2}	2.5044974×10^{-3}	1.5165610×10^{-2}

Three HRDSLs with $NA_t = 1$ and $N = 1, 2$ and 3 are designed. $\rho_1^{(G)}, \rho_2^{(G)}$ and $\rho_3^{(G)}$ are selected to be $\rho_L, 2\rho_L$ and $3\rho_L$, respectively. Since the radius of the first zero position of the Airy disk pattern is about $\rho^{(A)} = 1.45\rho_L$, the normalized spot size $G = \rho_1^{(G)}/\rho^{(A)} = 0.68966$. The parameters of the designed HRDSLs are shown in Table 2. The designed DSEs are all phase-only elements introducing a phase shift of 0 or π, so that Eq. (69) theoretically proved in Section 6 are now numerically verified. The profile of the designed DSEs can be determined by Eq. (43) or Fig. 7. Table 2 and Fig. 8 show that if N increases with a constant G, S and $K^{(a)}$ will decrease while $W^{(a)}$ and $\rho_W^{(a)}$ will increase. A reduction of $K^{(a)}$ and an increment of $\rho_W^{(a)}$ can enlarge the utilizable FOV effectively, while an excessive reduction of S and an immoderate increment of $W^{(a)}$ will result in a high energy loss. So N should be selected eclectically. Fig. 9 shows that the axial intensity of the designed HRDSLs keeps almost constant in a range symmetrical about the focus with an approximate width of $2HFD_L$. This property can enhance the axial tolerance of focalization for optical data storage or laser lithography.

Table 2 The parameters of the designed HRDSLs with $NA_t = 1$. Here $N = 0$ means no DSE

N	η_i^b	S	$K^{(a)}$	$W^{(a)}$	$\rho_W^{(a)}/\rho_L$
0	0	1	0.013854	0.015464	1.65
1	0.3225	0.34634	0.31772	0.33775	1.75
2	0.2725, 0.5325	0.13273	0.26129	0.47189	2.90
3	0.1825, 0.4150, 0.6575	0.055843	0.076482	0.82101	4.00

To reveal some limits of optical superresolution, an upper bound of S under resolving power of G was provided as $S^u(G)$ in Ref. [10], and the exact upper bound of S under resolving power of G was later presented as $S^{eu}(G)$ in Ref. [9]. Comparison between $S^u(G)$ and $S^{eu}(G)$ has been performed in Ref. [9]. However, both the discussions in Ref. [9] and those in Ref. [10] are limited to low numerical aperture, so a calculation of $S^{eu}(G)$ with high numerical is necessary for both applications and theoretical discussions of optical superresolution. This calculation can be easily carried out through linear programming of Eqs. (58a)-(58c) with $N = 1$. Fig. 10 shows the curves of $S^u(G)$, $S^{eu}(G)$ with $NA_t \ll 1$ and $S^{eu}(G)$ with $NA_t = 1$, where $S^u(G)$ corresponds to the dashed curve if $G \in [0, 0.46]$ and corresponds to the dotted curve if $G \in [0.46, 1]$. The curve of $S^{eu}(G)$ with $NA_t = 1$ is almost identical with the curve of $S^{eu}(G)$ with $NA_t \ll 1$, so the approximate analytic expression of $S^{eu}(G)$ for low numerical aperture deduced in Ref. [9] applies to high numerical aperture as well. As shown in Fig. 11, the curve of $K_m^{(a)}(G)$ with $NA_t = 1$ is almost the same as that with $NA_t \ll 1$ provided in Ref. [9]. Upon that, the optical superresolution with high numerical aperture obeys almost the same limits as those with low numerical aperture.

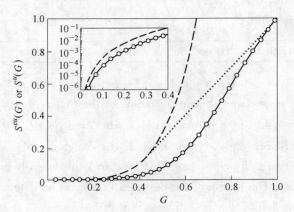

Fig. 10 The dashed curve and the dotted curve correspond to $S^u(G)$, the solid curve corresponds to $S^{eu}(G)$ with $NA_t \ll 1$, and the "o" curve corresponds to $S^{eu}(G)$ with $NA_t = 1$. The inserted curves are of logarithmic coordinate to magnify those of linear coordinate

8 Conclusions

To design a hybrid refractive-diffractive superresolution lens (HRDSL) with high numerical aperture, a set of transformation equations to describe the diffractive superresolution element (DSE) of the HRDSL is ultimately derived in Section 5, which shows the differences between the profile of the DSE with low numerical aperture and that with high numerical aperture while introducing the same phase shift. Based on this set of transformation equations, an optimization theory is constructed in Section 6. This optimization theory can reach a globally optimal solution through a linear programming of Eqs. (58a)-(58c), and needs much less amount of numerical

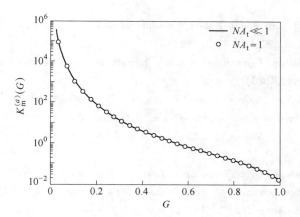

Fig. 11 Curves of $K_m^{(a)}(G)$

computation than the corresponding Eqs. (27) in Ref. [9]. It is theoretically proved in Section 6 and numerically verified in Section 7 that the designed DSE must be a phase-only element introducing a phase delay of 0 or π which can be produced with high quality, so that an implementation of a hybrid type DSE is not necessary. In Section 7, comparison of the optimization theory in this paper with the other three design theories under the same zero constraints shows the validity and the advantages of the former. Three design examples of the HRDSL with $NA_t = 1$ reveal the capability of the theories in this paper to maximize the Strehl ratio, compress the side lobe adjacent to the main lobe, and keep the highest side lobe away from the main lobe for high numerical aperture. Curves of $S^{eu}(G)$ and $K_m^{(a)}(G)$ with $NA_t = 1$ are figured to show some limits of optical superresolution with high numerical aperture. These curves are almost identical with those of $NA_t \ll 1$, so the optical superresolution with high numerical aperture obeys almost the same limits as those with low numerical aperture.

Acknowledgements

This work is sponsored by the National Research Fund for Fundamental Key Project of China, No. 973 (G19990330), and the Fund of High Technology of China (863-804-9-2).

References

[1] I. J. Cox, "Increasing the bit packing densities of optical disk systems", Appl. Opt. 23, 3260-3261(1984).
[2] Yibing Shen, Guoguang Yang, Xiyun Hou, "Research on phenomenon of the super-resolution in laser lithography", Acta Opt. Sin. 19, 1512-1517(1999).
[3] T. Wilson and B. R. Masters, "Confocal microscopy", Appl. Opt. 33, 565-566(1994).
[4] K. Carlsson, P. E. Danielsson, R. Lenz, A. Liljeborg, L. Majlof and N. Aslund, "Three-dimensional microscopy using a confocal laser scanning microscope", Opt. Lett. 10, 53-55(1985).
[5] Duanyi Xu, *Principle and Design of Optical Storage Systems* (National Defence Industry, Beijing, 2000), Chap. 1.
[6] T. R. M. Sales and G. M. Morris, "Diffractive superresolution elements", J. Opt. Soc. Am. A 14, 1637-1646

(1997).

[7] I. J. Cox, C. J. R. Sheppard, and T. Wilson, "Reappraisal of arrays of concentric annuli as superresolving filters", J. Opt. Soc. Am. 72, 1287-1291(1982).

[8] R. Boivin and A. Boivin, "Optimized amplitude filtering for superresolution over a restricted field", Opt. Acta. 27, 587-610(1980).

[9] Haitao Liu, Yingbai Yan, Qiaofeng Tan, and Guofan Jin, "Theories for the design of diffractive superresolution elements and limits of optical superresolution", J. Opt. Soc. Am. A 19, 2185-2193(2002).

[10] T. R. M. Sales and G. M. Morris, "Fundamental limits of optical superresolution", Opt. Lett. 22, 582-584 (1997).

[11] J. W. Goodman, *Introduction to Fourier Optics* (McGraw-Hill, New York, 1968), Chap. 5.

[12] M. Born and E. Wolf, *Principles of Optics* (Pergamon, New York, 1975), Chap. 3.

[13] S. Solimeno, B. Crosignani and P. DiPorto, *Guiding, Diffraction, and Confinement of Optical Radiation* (Academic, San Diego, Calif., 1986), Chap. II.

[14] J. W. Goodman, *Introduction to Fourier Optics* (McGraw-Hill, New York, 1968), Chap. 3.

[15] J. D. Jackson, *Classical Electrodynamics* (John Wiley & Sons, New York, 1975), Chap. 9.

[16] Y. Han, L. N. Hazra, and C. A. Delisle, "Exact surface-relief profile of a kinoform lens from its phase function", J. Opt. Soc. Am. A 12, 524-529(1995).

[17] W. Singer and K. H. Brenner, "Transition of the scalar field at a refracting surface in the generalized Kirchhoff diffraction theory", J. Opt. Soc. Am. A 12, 1913-1919(1995).

[18] W. Singer and H. Tiziani, "Born approximation for the nonparaxial scalar treatment of thick phase gratings", Appl. Opt. 37, 1249-1255(1998).

[19] M. Testorf, "Perturbation theory as a unified approach to describe diffractive optical elements", J. Opt. Soc. Am. A 16, 1115-1123(1999).

[20] Y. Chugui, V. P. Koronkevitch, B. E. Krivenkov, and S. V. Mikhlyaev, "Quasi-geometrical method for Fraunhofer diffraction calculations for three-dimensional bodies", J. Opt. Soc. Am. 71, 483-489(1981).

[21] T. S. Blyth and E. F. Robertson, *Further linear algebra* (Springer, London; New York, 2002), Chap. 1.

[22] J. K. Strayer, *Linear Programming and Its Applications* (Springer-Verlag, New York, 1989), Chap. 2.

[23] L. E. Elsgolc, *Calculus of Variations* (Pergamon, Oxford, UK, 1961), Chap. I.

[24] D. A. Pommet, M. G. Moharam, and E. B. Grann, "Limits of scalar diffraction theory for diffractive phase element", J. Opt. Soc. Am. A 11, 1827-1834(1994).

Broadband Polarizing Beam Splitter Based on the Form Birefringence of a Subwavelength Grating in the Quasi-static Domain*

Abstract We propose a novel broadband polarizing beam splitter with a compact sandwich structure that has a sub-wavelength grating in the quasi-static domain as the filling. The design is based on effective-medium theory and anisotropic thin-film theory, and the performance is investigated with rigorous coupled-wave theory. The design results show that the structure can provide a high polarization extinction ratio in a broad spectral range.

Polarizing beam splitters (PBSs) have numerous applications, such as magneto-optic data storage in optical information processing and optical switching in optical communication[1,2]. Conventional PBSs, such as a Wollaston prism and a PBS cube, are either bulky and heavy or applicable in only a narrow wavelength range. With the development of microfabrication technologies, subwavelength gratings (SWGs) have attracted more and more attention. SWGs are expected to realize special optical functions based on their form-birefringence effect. Moreover, their compact size and light weight are advantageous to the miniaturization and integration of optical systems. However, the PBSs based on SWGs that have been developed so far have suffered from narrow bandwidth or a low extinction ratio (ER)[3-5]. In this Letter we propose a novel compact PBS with a high ER in a broad spectral range based on the form birefringence of a SWG in the quasi-static domain. A combination of effective-medium theory (EMT) and anisotropic thin-film theory, along with rigorous coupled-wave theory, is applied to the design and analysis of this PBS[6,7].

As shown in Fig. 1, the broadband PBS has a sandwich structure, with a rectangular-groove SWG as the filling and substrates with refractive indices n_0 and n_3 as the top and bottom layers, respectively. We assume that $n_0 = n_3$. The rectangular-groove SWG is specified by groove depth h, period d, and duty cycle $\alpha = d_1/d$, with low and high refractive indices n_1 and n_2, respectively. The grating is illu-

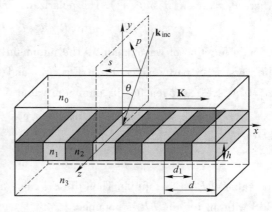

Fig. 1 Schematic diagram of the broadband PBS

---

* Copartner: Deer Yi, Yingbai Yan, Haitao Liu, Si Lu. Reprinted from *Optics Letters*, 2004, 29(7):754-756.

minated by an incident wave with an azimuth angle of 90°.

From EMT, in the quasi-static domain, where the period-to-wavelength ratio d/λ is much smaller than unity (roughly $d/\lambda < 1/10$), the rectangular-groove SWG is equivalent to an anisotropic homogeneous layer of thickness h, as shown in Fig. 2. The layer can be characterized by two effective refractive indices, n_p and n_s, depending on whether the electric field of the incident wave is oriented to be perpendicular or parallel to the grating vector **K**, i.e., the p and s components as shown in Fig. 1. The effective refractive indices are determined from EMT with the following first-order equations:

$$n_p = [n_1^2(1-\alpha) + n_2^2\alpha]^{1/2}$$
$$n_s = \{[(1-\alpha)/n_1^2] + (\alpha/n_2^2)\}^{-1/2} \quad (1)$$

We assume that $n_0 > n_p, n_s$. Because $n_p > n_s$ ($0 < \alpha < 1$), the angles of total reflection at the top surface of the grating layer have the relation $\theta_c^p > \theta_c^s$, where θ_c^p and θ_c^s are the angles of total reflection of the p and s components, respectively.

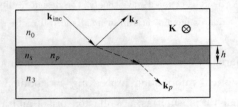

Fig. 2 Equivalent anisotropic homogeneous layer of the rectangular-groove SWG

For the incident wave with incident angle $\theta \in [\theta_c^s, \theta_c^p]$, the s component is totally reflected with reflectance $R_s = 1$ (i.e., the transmittance $T_s = 0$). From anisotropic thin-film theory the reflectance of the p component is

$$R_p = \left(\frac{n_0^2\cos\theta_1}{n_p\cos^2\theta} - \frac{n_p}{\cos\theta_1}\right)^2 \bigg/ \left[\frac{4n_0^2}{\cos^2\theta}\cot^2\left(\frac{2\pi}{\lambda}n_p h\cos\theta_1\right) + \left(\frac{n_0^2\cos\theta_1}{n_p\cos^2\theta} - \frac{n_p}{\cos\theta_1}\right)^2\right] \quad (2)$$

where $\theta_1 = \arcsin(n_0\sin\theta/n_p)$. The polarization ER is

$$ER = \min(R_s/R_p, T_p/T_s) = R_s/R_p = 1/R_p \quad (3)$$

ER is a function of wavelength λ. The minimum value of ER corresponds to the wavelength of the worst PBS performance. From Eqs. (2) and (3), when $\lambda = 2n_p h\cos\theta_1/(m + 1/2)$, where $m = 0, 1, 2, \cdots$, ER reaches its minimum value

$$ER_{\min} = [(n_0^2 n_p^2 - n_0^4\sin^2\theta + n_p^4\cos^2\theta)/(n_0^2 n_p^2 - n_0^4\sin^2\theta - n_p^4\cos^2\theta)]^2 \quad (4)$$

When the denominator equals zero, i.e., when

$$n_p = n_0 \quad \text{or} \quad n_p = n_0\tan\theta \quad (5)$$

the value of $ER_{\min}$ is infinite. Then the ER value at any wavelength is infinite; i.e., a broadband PBS with an infinite ER is obtained.

For the case of $n_p = n_0$ the effective refractive index of the SWG for the p component is equal to the indices of the top and bottom substrates. The p component transmits through the grating with $T_p = 1$ (omitting the absorption of the material). When the incident angle satisfies $\theta \in [\theta_c^s, \theta_c^p]$, the s component is totally reflected at the top surface of the grating layer with $R_s = 1$. Therefore we obtained an infinite ER independent of the incident wavelength. For the case of

$n_p = n_0\tan(\theta)$ the incident angles at both the top and the bottom surfaces of the grating layer are equal to the Brewster's angle of the p component. Therefore the p component is totally transmitted at these two interfaces, resulting in $T_p = 1$. When $\theta \in [\theta_c^s, \theta_c^p]$, $R_s = 1$. An infinite ER independent of the incident wavelength is obtained. We can see that high ER in a broad spectral range can be achieved in both cases.

The design principles of structure parameters d, h, and α are listed as follows:

(1) In both cases the structure of the rectangular-groove SWG is related to the refractive indices of the materials by the duty cycle. For the case of $n_p = n_0$, α and the refractive indices are related as follows

$$\begin{cases} n_p = n_0 \\ n_p = [n_1^2(1-\alpha) + n_2^2\alpha]^{1/2} \end{cases} \Rightarrow \alpha = \frac{n_0^2 - n_1^2}{n_2^2 - n_1^2} \tag{6}$$

For the case of $n_p = n_0\tan(\theta)$ the incident angle should satisfy

$$\begin{cases} n_p = n_0\tan\theta \\ \theta > \theta_c^s \end{cases} \Rightarrow \begin{cases} \theta = \arctan\{[n_1^2(1-\alpha) + n_2^2\alpha]^{1/2}/n_0\} \\ \theta = \arcsin\{[(1-\alpha)/n_1^2 + \alpha/n_2^2]^{-1/2}/n_0\} \end{cases} \tag{7}$$

Eq. (7) can be equally transformed into

$$n_0^2(n_1^2 - n_2^2)^2\alpha^2 - (n_1^2 - n_2^2)[n_0^2(n_1^2 - n_2^2) + n_1^2 n_2^2]\alpha + n_1^4 n_2^2 < 0 \tag{8}$$

if and only if n_1 satisfies

$$n_1 < \frac{n_0 n_2}{n_0 + n_2} \quad \text{or} \quad n_1 > \frac{n_0 n_2}{|n_0 - n_2|} \quad (n_0 \neq n_2) \tag{9}$$

Expression (8) has a real solution $\begin{cases} \alpha \in [\alpha_{\min}, \alpha_{\max}] \\ \alpha \in [0,1] \end{cases}$, where

$$\alpha_{\min} = \frac{n_0^2 n_1^2 + n_1^2 n_2^2 - n_0^2 n_2^2 - (n_0^4 n_1^4 - 2n_0^2 n_1^4 n_2^2 - 2n_0^4 n_1^2 n_2^2 + n_1^4 n_2^4 - 2n_0^2 n_1^2 n_2^4 + n_0^4 n_2^4)^{1/2}}{2n_0^2(n_1 + n_2)(n_1 - n_2)}$$

$$\alpha_{\max} = \frac{n_0^2 n_1^2 + n_1^2 n_2^2 - n_0^2 n_2^2 + (n_0^4 n_1^4 - 2n_0^2 n_1^4 n_2^2 - 2n_0^4 n_1^2 n_2^2 + n_1^4 n_2^4 - 2n_0^2 n_1^2 n_2^4 + n_0^4 n_2^4)^{1/2}}{2n_0^2(n_1 + n_2)(n_1 - n_2)}$$

(10)

The ranges of n_1 versus those of n_0 and n_2 are shown in Fig. 3.

(2) To ensure the validity of the EMT analysis, the grating must be in the quasi-static domain, i.e., $d/\lambda < 1/10$. If the grating period meets $d/\lambda > 1/10$, the first-order EMT analysis loses accuracy. However, the broadband PBS still works with a lower ER.

(3) When the groove depth is too small, the evanescent wave at the top surface of the grating layer will take energy and the R_s will be less than 1.

Therefore we can optimize the structure parameters according to the applications, available materials, and fabrication technologies. The ER might decrease from infinity to an acceptable value.

We give an example for the case of $n_p = n_0\tan(\theta)$. The refractive indices of the materials are

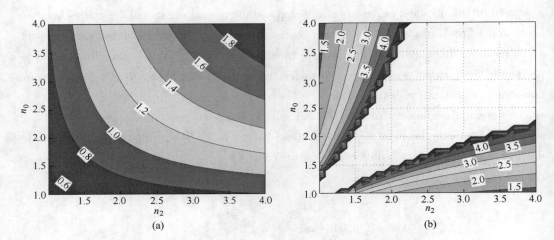

Fig. 3 Ranges of n_1 for the case of $n_p = n_0 \tan\theta$

(a) for the case of $n_1 < n_0 n_2/(n_0 + n_2)$, n_1 should be less than the values marked on the solid curves; (b) for the case of $n_1 > n_0 n_2/(|n_0 - n_2|)$, with $n_0 \neq n_2$, n_1 should be greater than the values marked on the solid curves

$n_1 = 1.44 (\mathrm{SiO_2})$, $n_2 = 3.45 (\mathrm{Si})$, and $n_0 = n_3 = 3.45 (\mathrm{Si})$. The optimized parameters are $d = 0.10 \mu\mathrm{m}$, $h = 1.5 \mu\mathrm{m}$, $\alpha = 0.50$, and $\theta = 37.7°$. The performance of the device is calculated based on rigorous coupled-wave theory. Fig. 4 shows the polarization ER plotted as a function of wavelength. Throughout the broad spectral range, from 1.3 to 2.3 μm, an excellent ER of better than 1000 is obtained. Fig. 5 shows the polarization ER plotted as a function of the incident angle at $\lambda = 1.55 \mu$m.

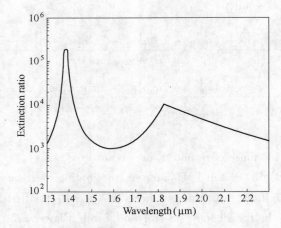

Fig. 4 ER versus wavelength

In both Figs. 4 and 5 there are very sharp peaks caused by multiple-beam interference that reinforces the zero transmission of the p component and increases the ER for special wavelengths. The peaks can be moved around within the spectral range of interest by optimization. Moreover, simulations show that the dispersions of materials have a slight effect on

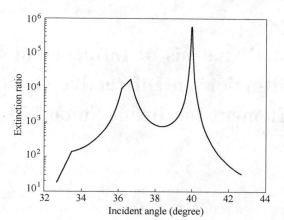

Fig. 5 ER versus incident angle ($\lambda = 1.55 \mu m$)

the performance of the broadband PBS.

In conclusion, we have proposed a novel broadband PBS with a compact sandwich structure with a rectangular-groove SWG in the quasi-static domain as the filling. The PBS can provide a super high ER without any limitation on the wave band or bandwidth. The structure offers a universal and powerful method for broadband polarizing beam splitting.

The authors are grateful to L. F. Li for providing KAPPA, a computer program for modeling planar, one-dimensionally periodic diffraction gratings.

References

[1] M. Ojima, A. Saito, T. Kaku, M. Ito, Y. Tsunoda, S. Takayama, and Y. Sugita, Appl. Opt. 25, 483(1986).
[2] Q. W. Song, M. C. Lee, and P. J. Talbot, Appl. Opt. 31, 6240(1992).
[3] R. C. Tyan, P. C. Sun, A. Scherer, and Y. Fainman, Opt. Lett. 21, 761(1996).
[4] D. Dias, S. Stankovic, H. Haidner, L. L. Wang, T. Tschudi, M. Ferstl, and R. Steingrüber, J. Opt. A 3, 164 (2001).
[5] M. Schmitz, R. Brauer, and O. Bryngdahl, Opt. Lett. 20, 1830(1995).
[6] S. M. Rytov, Sov. Phys. JETP 2, 466(1956).
[7] L. F. Li, J. Chandezon, G. Granet, and J. -P. Plumey, Appl. Opt. 38, 304(1999).

Statistic Analysis of Influence of Phase Distortion on Diffractive Optical Element for Beam Smoothing[*]

Abstract In this paper, the diffractive optical element (DOE) is used to realize beam smoothing with the smoothing by spectral dispersion (SSD) technique. The influences of the high and low frequency phase distortions on the DOE for beam smoothing are statistically analyzed based on the spatial frequency spectrum method. The amplitude and initial phase spectra of the far field intensity distribution are modulated by the characteristic of the phase distortion. The relationship between the performance parameters of the beam smoothing and the characteristic of the phase distortion is obtained. This can afford a theoretical tool to determine whether the characteristic of the phase distortion and the performance of the designed DOE for beam smoothing are able to satisfy the requirement of the light efficiency and the non-uniformity simultaneously or not.

1 Introduction

Beam smoothing, with a prescribed intensity distribution on the focal plane, is required in many kinds of application, for example, inertial confinement fusion (ICF)[1]. Diffractive optical element (DOE) is a good option for realizing such kind of beam smoothing, due to its high light efficiency, high flexibility in phase design, matured fabricating technology and broad selectivity of materials. The phase design of the DOE is usually considered as an optimization problem. In fact, many kinds of optimization algorithms have been applied for designing DOE, such as Gerchberg-Saxton (GS) algorithm[2,3] and its related modified algorithms[4,5], Yang-Gu (YG) algorithm[6] and Global/local united search algorithm (GLUSA)[7]. The near field phase distribution of the DOE is generally discretized with some same intervals, and the sampling interval on the focal plane is generally chosen as $\Delta = \lambda f/D$, where f, λ and D are the focal length of the Fourier transform lens, the incident wavelength and the aperture size of the DOE respectively. With this sampling interval on the focal plane, good performances of DOE for beam smoothing have been obtained by using any algorithms mentioned above. The intensities of the chosen sampling points on the focal plane are consistent with the required beam smoothing, but the other points are hardly fulfilled. Therefore, the sampling interval on the focal plane should be reduced, i.e. $\Delta \leq \lambda f/D/2$. To save the optimization time, let $\Delta = \lambda f/D/2$. Then the intensity of any point on the focal plane will be fully filled with the required beam smoothing, not only the used sampling points in the optimization[8]. However, the performance parameters of the DOE

---

[*] Copartner: Qiaofeng Tan, Yingbai Yan. Reprinted from *Optics Express*, 2004, 12(14):3270~3278.

for beam smoothing are generally calculated with a group of discrete sampling points on the focal plane, and the continuous far field intensity distribution of the DOE is not truly reflected in practice. To reflect the continuous far field intensity distribution and actually evaluate the performance of the beam smoothing, based on the spatial-frequency spectrum of the far field intensity distribution of the DOE, a novel evaluation criterion has been developed[9].

However, in practical applications, the designed DOE is very difficult to realize the expected beam smoothing on the focal plane, no matter what the sampling interval on the focal plane is chosen in the optimization. The reason is that the DOE is very sensitive to the incident phase distortion. In a realistic laser system, in particular the high power laser system, the output laser beam is not an ideal plane wave and usually it contains a rather large phase distortion, which exceeds the tolerance of the DOE for beam smoothing. To decrease the influence of the phase distortion and improve the performance of the DOE for beam smoothing, some techniques to realize smoothing have been adopted in time domain, for example, smoothing by spectral dispersion(SSD) in the ICF system[10,11]. For a laser pulse, the effective phase distortion is different at different time, therefore the obtained intensity distribution on the focal plane is a statistic result with numerous phase distortions. And the performance of the DOE for beam smoothing is improved. To explain this phenomenon theoretically, in this paper, based on the spatial-frequency spectrum of the far field intensity distribution of the DOE for beam smoothing, the relationship between the characteristic of the phase distortion and the performance parameters of the DOE for beam smoothing is quantitatively analyzed.

2 Theory of spatial-frequency spectrum analysis

For simplicity, only one-dimensional DOE is considered. In the optimization, the near field phase distribution of the DOE is discretized with same intervals. However, the continuous phase DOE cannot be obtained by interpolation with sinc function from the discrete phase data, as after interpolation, the continuous DOE has not only a phase modulation but also an amplitude modulation. In this case, the designed DOE is a multi-phase level element. Supposing the size of the DOE is D, the number of the phase cells is K, then the transmittance of the DOE is

$$T(x) = \sum_{p=1}^{K} \exp(i\varphi_p) \mathrm{rect}\left(\frac{x - pD/K}{D/K}\right) \tag{1}$$

where x is the coordinate of the DOE, φ_p is the phase value of the pth cell, and

$$\mathrm{rect}(x) = \begin{cases} 1 & \text{when } |x| \leq \dfrac{1}{2} \\ 0 & \text{else} \end{cases} \tag{2}$$

When an ideal plane wave is incident on the DOE, according to Kirchhoff theory, the far field intensity distribution can be written as

$$I(x') = \left| \sum_{p=1}^{K} \exp(i\varphi_p) \exp\left(-i2\pi \frac{pD}{k\lambda f} x'\right) \right|^2 \tag{3}$$

where x' is the coordinate of the focal plane, f is the focal length of Fourier transform lens, λ is the incident wavelength. In Eq. (3), the sinc function and the phase factor are ignored.

If the phase of the DOE is symmetrical, expanding Eq. (3), then

$$I(y) = \sum_{p=1}^{K}\sum_{q=1}^{K} \cos[\varphi_p - \varphi_q + (p-q)y] = K + 2\sum_{m=1}^{K-1} A_m \cos(my + B_m) \quad (4)$$

where $y = -\dfrac{2\pi D}{K\lambda f}x'$, $A_m = \sqrt{[\sum_{p=m+1}^{K}\cos(\varphi_p - \varphi_{p-m})]^2 + [\sum_{p=m+1}^{K}\sin(\varphi_p - \varphi_{p-m})]^2}$ and $B_m = \arctan[\sum_{p=m+1}^{K}\sin(\varphi_p - \varphi_{p-m}) / \sum_{p=m+1}^{K}\cos(\varphi_p - \varphi_{p-m})]$ are the amplitude and initial-phase spectra of the intensity distribution respectively.

Based on Eq. (4), the total intensity in the required area of the main lobe of the far field intensity distribution of the DOE for beam smoothing can be calculated. Assuming the required area is $d_1 \leq x' \leq d_2$, the total intensity is

$$I_{\text{total}} = \int_{y_1}^{y_2} I(y)\,dy = K(y_2 - y_1) + 2\sum_{m=1}^{K-1}\frac{A_m}{m}[\sin(my_2 + B_m) - \sin(my_1 + B_m)] \quad (5)$$

where $y_i = -\dfrac{2\pi D}{K\lambda f}d_i$, $i = 1, 2$.

The input intensity can be calculated from Eq. (5) with boundaries $y_1 = -\pi$ and $y_2 = \pi$. The value is $2\pi K$, and the light efficiency η can be defined as

$$\eta = \frac{I_{\text{total}}}{2\pi K} \quad (6)$$

The non-uniformity at the top, rms, can be defined as

$$rms = \sqrt{\frac{\int_{y_1}^{y_2}(I(y) - \bar{I})^2 dy}{(y_2 - y_1)\bar{I}^2}} \quad (7)$$

where $\bar{I} = \dfrac{I_{\text{total}}}{y_2 - y_1}$ is the average intensity in the required area.

3 Statistic analysis of influence of phase distortion

If the incident wave is not an ideal plane wave and the phase distortion of the pth cell is $\Delta\phi_p$, then the far field intensity distribution is

$$I(y) = \left|\sum_{p=1}^{K}\exp[i(\varphi_p + \Delta\phi_p)]\exp(ipy)\right|^2 \quad (8)$$

In the ICF system, the SSD technique is used to improve the performance of the DOE for beam smoothing. The SSD technique does not reduce the phase distortion. It creates an entirely different phase distortion, which changes in time[10]. Supposing the realized number of phase distortions in one laser pulse is L, the intensity distribution on the focal plane can be considered as a sum of L intensity distributions with the corresponding phase distortion. Supposing the pulse shape in time domain is rectangular, and, for convenience, the normalized far field intensity distribution is here concerned,

$$\bar{I}_L(y) = \sum_{l=1}^{L} \left| \sum_{p=1}^{K} \exp[i(\varphi_p + \Delta\phi_p^l)] \exp(ipy) \right|^2 / L \qquad (9)$$

where $\Delta\phi_p^l$ is the phase distortion of the pth cell at the lth moment.

Expanding Eq. (9), and

$$\bar{I}_L(y) = \frac{1}{L} \sum_{l=1}^{L} \left[\sum_{p=1}^{K} \sum_{q=1}^{K} \exp[i(\varphi_p - \varphi_q + (p-q)y)] \exp[i(\Delta\phi_p^l - \Delta\phi_q^l)] \right]$$

$$= K + \sum_{p=1}^{K} \sum_{q=1,q\neq p}^{K} \exp[i(\varphi_p - \varphi_q + (p-q)y)] \frac{1}{L} \sum_{l=1}^{L} \left[\exp[i(\Delta\phi_p^l - \Delta\phi_q^l)] \right]$$

$$(10)$$

The key problem is how to evaluate $\frac{1}{L}\sum_{l=1}^{L}\{\exp[i(\Delta\phi_p^l - \Delta\phi_q^l)]\}$. For simplicity, let L be infinity, so the key problem is to obtain the expected value of $\frac{1}{L}\sum_{l=1}^{L}\{\exp[i(\Delta\phi_p^l - \Delta\phi_q^l)]\}$, i.e. $E\{\exp[i(\Delta\phi_p - \Delta\phi_q)]\}$.

In the ICF system, the output laser beam contains high frequency and low frequency phase distortions. The remained high frequency phase distortion after phase correction is approximately a Gaussian distribution[12]. Here, the Gaussian distributed high frequency phase distortion is analyzed, and it can be expressed as

$$\Delta\phi_H = N[0,\sigma] \qquad (11)$$

where $N[0,\sigma]$ takes place the Gaussian distribution with zero mean and σ variance.

The low frequency phase distortion can be simplified as a random phase plate, with a phase distribution written as[13]

$$\Delta\phi_L = a \cdot \text{rand}(-1,1) * \exp[-(x/x_s)^2] \qquad (12)$$

where $\text{rand}(-1,1)$ and $*$ take place the uniform distribution over $(-1,1)$ and convolution respectively. x_s and a reflect the spatial characteristic size and the magnitude of the phase distortion respectively.

Hence

$$\Delta\phi_p^l = \Delta\phi_{p,H}^l + \Delta\phi_{p,L}^l \qquad (13)$$

where $\Delta\phi_{p,H}^l$ and $\Delta\phi_{p,L}^l$ are the high and low frequency phase distortions of the pth cell at the lth moment respectively.

Assume that the high and low frequency phase distortions are mutual independent, then

$$E\{\exp[i(\Delta\phi_p - \Delta\phi_q)]\} = E\{\exp[i(\Delta\phi_{p,H} - \Delta\phi_{q,H})]\} \cdot E\{\exp[i(\Delta\phi_{p,L} - \Delta\phi_{q,L})]\}$$

$$(14)$$

For the high frequency phase distortion, when $q \neq p$, $\Delta\phi_{p,H}^l$ and $\Delta\phi_{q,H}^l$ are mutually independent, then

$$\Delta\phi_{p,H}^l - \Delta\phi_{q,H}^l = N[0,\sqrt{2}\sigma] \tag{15}$$

$$E\{\exp[i(\Delta\phi_{p,H} - \Delta\phi_{q,H})]\} = \int_{-\infty}^{+\infty} \exp[i\phi] \frac{1}{2\sigma\sqrt{\pi}} \exp\left(-\frac{\phi^2}{4\sigma^2}\right) d\phi = \exp(-\sigma^2) \tag{16}$$

For the low frequency, $q \neq p$, $\Delta\phi_{p,L}^l$ and $\Delta\phi_{q,L}^l$ are not mutually independent. According to Eq. (12), for the lth moment,

$$\Delta\phi_{p,L}^l - \Delta\phi_{q,L}^l = a \cdot \sum_{k=1}^{K} R_k^l \cdot \left\{ \exp\left\{ -\left[(k-p)\frac{D}{K}\bigg/x_s\right]^2 \right\} - \exp\left\{ -\left[(k-q)\frac{D}{K}\bigg/x_s\right]^2 \right\} \right\}$$

$$= \sum_{k=1}^{K} F(k,p,q) R_k^l \tag{17}$$

where R_k^l is the kth random number obeyed with uniform distribution rand$(-1,1)$ at the lth moment, and $F(k,p,q) = a \cdot \left\{ \exp\left\{ -\left[(k-p)\frac{D}{K}\bigg/x_s\right]^2 \right\} - \exp\left\{ -\left[(k-q)\frac{D}{K}\bigg/x_s\right]^2 \right\} \right\}$.

Supposing R_k^l is mutual independent at different k or different l, then

$$E\{\exp[i(\Delta\phi_{p,L} - \Delta\phi_{q,L})]\} = E\{\exp[i\sum_{k=1}^{K} F(k,p,q) R_k]\} = \prod_{k=1}^{K} E\{\exp[iF(k,p,q) R_k]\} \tag{18}$$

For each k,

$$E\{\exp[iF(k,p,q) R_k]\} = \int_{-1}^{1} \exp[iF(k,p,q)\phi] \frac{1}{2} d\phi = \text{sinc}[F(k,p,q)] \tag{19}$$

where $\text{sinc}(x) = \sin x/x$.

Therefore, Eq. (10) can be rewritten as

$$I(y) = K + \exp(-\sigma^2) \sum_{p=1}^{K} \sum_{q=1,q\neq p}^{K} \prod_{k=1}^{K} \text{sinc}[F(k,p,q)] \exp[i(\varphi_p - \varphi_q + (p-q)y)]$$

$$= K + 2\exp(-\sigma^2) \sum_{m=1}^{K-1} A_m \cos(my + B_m) \tag{20}$$

where $A_m = \sqrt{C_m^2 + D_m^2}$ and $B_m = \arctan[D_m/C_m]$ are the amplitude and initial-phase spectra of the intensity distribution after using the SSD technique respectively.

$$C_m = \sum_{p=m+1}^{K} \prod_{k=1}^{K} \text{sinc}[F(k,p,p-m)] \cos(\varphi_p - \varphi_{p-m}) \tag{21}$$

$$D_m = \sum_{p=m+1}^{K} \prod_{k=1}^{K} \text{sinc}[F(k,p,p-m)] \sin(\varphi_p - \varphi_{p-m}) \tag{22}$$

The forms of Eq. (20) and Eq. (4) are nearly the same, but the amplitude and initial-phase spectra in Eq. (20) are modulated by the characteristic of the phase distortion. According to Eq. (20), the high frequency phase distortion has the same influence on each $m(m \geq 1)$, while the low frequency phase distortion has different influence on each $m(m \geq 1)$.

4 Simulated results

For showing an example, the values for f, D, and λ are taken as 600mm, 100mm, and 1.053μm respectively. The size of the main lobe of the far field intensity distribution of the DOE for beam smoothing is 100μm and K is 256. The sampling interval on the focal plane is chosen as $\Delta = \lambda f/D/2$ to realize a true beam smoothing[8]. The near field phase and the far field intensity distributions of the designed DOE are shown in Fig. 1. Its complex amplitude spatial-frequency spectrum is shown in Fig. 2. When $\Delta = \lambda f/D/2$, the GS algorithm or its related modified algorithms and YG algorithm are not suitable to design the DOE for beam smoothing. The adopted optimization algorithm in this paper is a kind of hybrid algorithm which merges hill-climbing (HC) with simulated annealing(SA). The basic structure of this hybrid algorithm is described as follow. This hybrid algorithm has sufficient ability of strong convergence of HC and the global optimization potential of SA.

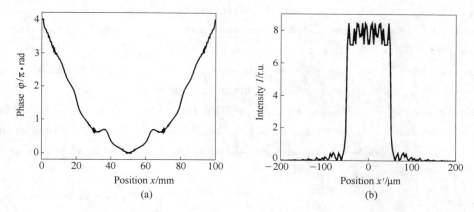

Fig. 1 The design results of the DOE for beam smoothing
(a) near field phase distribution; (b) far field intensity distribution

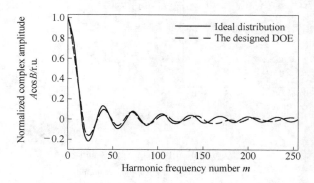

Fig. 2 Spatial-frequency spectrum of the designed DOE for beam smoothing

Step1: let phase distribution be a sine function with random period, amplitude and initial phase;

Step2: calculate the value cost 1 of the cost function;

Step3: choose a sine function with random period and initial phase as the searching direction, and optimize the amplitude of the sine function along this direction (hill-climbing optimization);

Step4: calculate the new cost function value cost 2 after every hill-climbing optimization step;

Step5: if cost 1 > cost 2, accept the optimization step, let cost 1 = cost 2, and go to step9;

Step6: if cost 1 < cost 2, calculate the accepted and refused probability p1 and p2, respectively;

Step7: if p1 > p2, accept the optimization step, and go to step9;

Step8: if p1 < p2, refuse the optimization step, and go to step3;

Step9: determine whether the algorithm can stop or not, if can, go to step1, else, go to step3;

Step10: stop the algorithm.

The optimized phase distribution is the sum of many sine functions with different period, amplitude and initial phase, which the continuity of the phase distribution can be assured.

The values of η and rms, calculated with Eq. (6) and Eq. (7) are 92.7% and 7.5% respectively. The ideal distribution in Fig. 2 is the corresponding complex amplitude spatial-frequency spectrum of the rectangular function with the same size.

Now, let us have a discussion. Firstly, in this simulation, only the high frequency phase distortion is considered, i.e. $a = 0$ and $\text{sinc}[F(k, p, q)] = 1$ for any k, p and q. Then the difference between Eq. (20) and Eq. (4) is the factor $\exp(-\sigma^2)$. According to the definition of the top non-uniformity, the larger the variance σ, the smaller the factor $\exp(-\sigma^2)$ is and the better the top non-uniformity. However, according to the definition of the light efficiency, the larger the variance σ, the smaller the light efficiency is. Calculating with the designed DOE as shown in Fig. 1, the relation between η, rms and σ is shown in Fig. 3, and the variations of η and rms are both monotone.

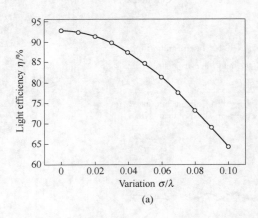
(a)

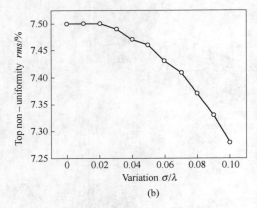
(b)

Fig. 3 Relation between η, rms and σ
(a) light efficiency; (b) top non-uniformity

Secondly, only the low frequency phase distortion is considered, i.e. $\sigma = 0$ and $\exp(-\sigma^2) =$

1. From $\mathrm{sinc}[F(k,p,q)]$, the qualitative relation between η, rms and a, x_s is not immediately obvious. When x_s is chosen as 60mm or 80mm, calculating with the designed DOE as shown in Fig. 1, the relation between η, rms and a is shown in Fig. 4. The variation of η is monotone. When x_s is relatively large, i. e. the spatial frequency of the low frequency phase distortion is relatively low, in this case, the light efficiency decreases relatively slowly. The variation of rms is not monotone but a little bit complicated. rms improves firstly, then goes to bad, and finally improves again.

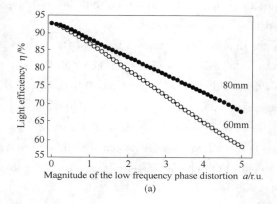 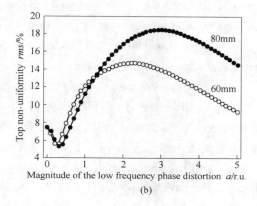

Fig. 4 Relation between η, rms and a
(a) light efficiency; (b) top non-uniformity

Finally, the high and low frequency phase distortions are both considered, when x_s is chosen as 60mm, the relation between η, rms and a, σ is shown in Fig. 5. σ varies from 0.01λ to 0.10λ step by step.

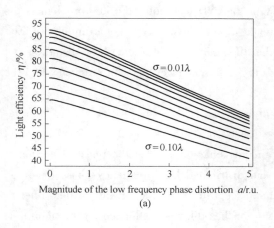 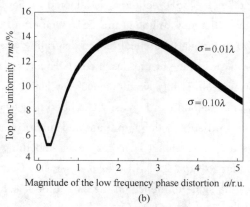

Fig. 5 Relation between η, rms and a
(a) light efficiency; (b) top non-uniformity

5 Conclusion

In this paper, considering the use of the SSD technique in the ICF system, the influence of the high and low frequency phase distortions on DOE for beam smoothing is statistically analyzed based on the spatial frequency spectrum method. The amplitude and initial phase spectra of the far field intensity distribution are modulated by the characteristic of the phase distortion. The relationship between the performance parameters of the beam smoothing and the characteristic of the phase distortion is obtained. For the Gaussian distributed high frequency phase distortion, the light efficiency decreases but the top non-uniformity improves with the increase of the variance of the phase distortion. For the low frequency phase distortion, the light efficiency decreases with the increase of the magnitude of the low frequency phase distortion, but the top non-uniformity improves firstly, then goes to bad, and finally improves again.

With the SSD technique, the *rms* value is better than the one realized by the DOE with an ideal plane wave. This is very useful and important in a realistic laser system. For high frequency phase distortion, though the value of *rms* changes slightly when σ varies from 0 to 0.10λ, the *rms* value will be less than 1% when σ is larger than 0.25λ. With the SSD technique, the top non-uniformity of the far field intensity distribution is correspondingly easy to satisfy the requirement of the beam smoothing. Even if the top non-uniformity of the designed DOE for beam smoothing is not good enough, we can choose or control the characteristic of phase distortion to improve the top non-uniformity according to Eq.(20).

With the SSD technique, the light efficiency is basically determined by the variance σ of the Gaussian distributed high frequency phase distortion and the magnitude a of the low frequency phase distortion. The higher the frequency of the phase distortion, the quicker decrease of the light efficiency is. For high frequency phase distortion, when σ is larger than 0.25λ, the value of η will be less than 30%.

Therefore, the phase distortion should be controlled to be as small as possible, no matter whether the SSD technique is adopted or not. With the SSD technique, the key problem is the light efficiency. Without the SSD technique, the key problem is the top non-uniformity. Eq.(20) affords a theoretical tool to determine whether the characteristic of the phase distortion and the performance of the designed DOE can satisfy the requirement of the light efficiency and the top non-uniformity simultaneously or not.

This method can be easily extended to a two-dimensional DOE for beam smoothing. The relationship between the amplitude and initial phase spectra of the far field intensity distribution of the two-dimensional DOE for beam smoothing and the characteristic of the phase distortion is similar to Eq.(20). In like manner, the Gaussian distributed high frequency phase distortion has the same influence on the amplitude spectrum, except the zero order, of the far field intensity distribution, and the low frequency phase distortion has different influence on the amplitude and initial-phase spectra. The qualitative variation rules of η and *rms* in two-dimensional case are similar to those in one-dimensional case.

Acknowledgements

This work is supported by the Fund of High Technology of China 2004AA849027.

References

[1] S. N. Dixit, J. K. Lawson, K. R. Manes, H. T. Powell and K. A. Nugent, "Kinoform phase plates for focal plane irradiance profile control", Opt. Lett. 19, 417-419(1994).

[2] R. W. Gerchberg, W. O. Saxton, "Phase determination for image and diffraction plane pictures in the electron microscope", Optik 34, 275-284(1971).

[3] R. W. Gerchberg, W. O. Saxton, "A practical algorithm for the determination of phase from image and diffraction plane pictures", Optik 35, 237-246(1972).

[4] J. R. Fienup, "Phase retrieval algorithms: a comparison", Appl. Opt. 21, 2758-2769(1982).

[5] J. S. Liu, M. R. Taghizadeh, "Iterative algorithm for the design of diffractive phase elements for laser beam shaping", Opt. Lett. 27, 1463-1465(2002).

[6] B. Y. Gu, G. Z. Yang, and B. Z. Dong, "General theory for performing an optical transform", Appl. Opt. 25, 3197-3206(1986).

[7] J. H. Zhai, Y. B. Yan, G. F. Jin, and M. X. Wu, "Design of continuous phase screens by Global/local united search algorithm for focal-plane irradiance profile control", Chinese Journal of Lasers B7, 235-240 (1998).

[8] Q. F. Tan, Y. B. Yan, G. F. Jin, and M. X. Wu, "Design of diffractive optical element for true beam smoothing", Opt. Commun. 189, 167-173(2001).

[9] Q. F. Tan, Q. S. He, Y. B. Yan, G. F. Jin and D. Y. Xu, "Spatial-frequency spectrum analysis of the performance of diffractive optical element for beam smoothing", Optik 113, 163-166(2002).

[10] "Two-dimensional SSD on OMEGA", LLE Review 69, 1-10(1996), http://www.lle.rochester.edu/pub/review/v69/1-two.pdf.

[11] G. Miyaji, N. Miyanaga, S. Urushihara, K. Suzuki, S. Matsuoka, M. Nakatsuka, A. Morimoto, and T. Kobayashi, "Three-dimensional spectral dispersion for smoothing a laser irradiance profile", Opt. Lett. 27, 725-727(2002).

[12] K. R. Manes, R. A. London, S. B. Sutton, L. E. Zapata, "Shot rate-thermal recovery", in NIF Laser System Performance Ratings, J. M. Auerbach, E. S. Bliss, S. N. Dixit, M. D. Feit, D. M.

[13] Gold, S. W. Haan, M. A. Henesian, O. S. Jones, J. K. Lawson, R. A. London, K. R. Manes.

[14] D. Munro, J. R. Murray, S. M. Pollaine, P. A. Renard, J. E. Rothenberg, R. A. Sacks, D. R. Speck, eds., Proc. SPIE 3492(supplement), 136-149(1999).

[15] Q. H. Deng, X. M. Zhang, F. Jing, L. Q. Liu, "Research on the rule of laser beam's low-frequency phase aberration superimposition", High Power Laser and Particle Beams 14, 81-84(2002).

High Quality Light Guide Plates that can Control the Illumination Angle Based on Microprism Structures*

Abstract In order to make the backlight system thinner and brighter, we propose a high quality poly methylmethacrylate light guide plate(LGP) based on microprism structures, which can be designed to control the illumination angle, and to get enough uniformity of intensity, so the backlight system will be simplified to use only one LGP and requires no other optical sheets. Design results reveal that our LGP can achieve a uniformity of intensity better than 86%, and get an illumination angle between ±25°, without using any optical sheets.

Liquid crystal displays (LCDs) offer several advantages, such as low energy consumption, low weight, and high uniformity of intensity, which make them ideal for many applications including monitors in notebook personal computers, screens for TV, and many portable information terminals, such as mobile phones, personal digital assistants, etc. Research and development on LCDs have made displays thinner, brighter, and more lightweight recently, meanwhile, many researchers have devoted a tremendous amount of effort to improve the qualities of backlight systems which convert a linear light source (such as a cold cathode fluorescent lamp) or some point sources [such as light emitting diodes(LEDs)] into an area of uniform light. A conventional LCD backlight system is composed of light sources, a light guide plate (LGP), and optical sheets, such as reflection sheet, diffusion sheet, and prism sheets, etc. In this system, the light rays from the source are incident on one of the sides of the LGP and are guided inside it based on the principle of total internal reflection, and then the rays are reflected and refracted by an array of etched or ink printed white spots at the bottom of the LGP. The emanated light from the top surface of the LGP disperses by the diffusion sheet that also weakens bright or dark fringes made by spots. Usually, two cross prism sheets are used to collimate and enhance the light transmitted by the diffusion sheet, and one reflection sheet under the LGP reduces the optical loss from the LGP's bottom surface.

To satisfy market requirements for mobile and personal display panels, it is more and more necessary to modify the backlight system and make it thinner, lighter, and brighter all at once. Considerable effort has been extended on the design of a backlight system with high qualities for LCDs. Recently, Käläntär, Matsumoto, and Onishi[1] have proposed a backlight system with a modified LGP made by poly methylmethacrylate(PMMA), in which only one prism sheet

---

* Copartner: Di Feng, Yingbai Yan, Shoushan Fan. Reprinted from *Applied Physics Letters*, 2004, 85(24):6016-6018.

is used instead of using one diffusion sheet and two collimating prism sheets in the conventional type, and Koike et al. [2,3] have reported a backlight using a highly scattering optical transmission(HSOT) polymer with high quality. But LGPs in these systems do not have the ability to control the illumination angle. More recently, Okumura et al. [4] have reported a highly efficient HSOT backlight system that has no optical sheets and can control the light directly into the front direction. But the backlight system's LGP is made of a HSOT polymer, not a traditional acrylic material, polycarbonate(PC) material, or PMMA material; moreover, it is not easy designing a LGP to control the illumination angle according to customers' requirements. In order to overcome these limitations, and motivated by the increasing demand on the backlight system, we have proposed a class of LGPs made by PMMA that can control the illumination angle easily based on microprism structures, and, meanwhile, achieve a high uniformity of intensity.

In this letter, we report a backlight system that is composed of light sources and a LGP based on microprism structures without any other optical sheets. A conventional LGP that is not designed to control the illumination angle cannot make the light emit toward a direction perpendicular to the top surface of LGP. The light is usually scattered to the direction lapsing from light sources, as shown in Fig. 1, so it is necessary to adopt diffusion sheet and prism sheets to convert the light direction into the front direction. In order to remove these optical sheets, it is necessary to design the LGP that can control the illumination angle (angle $i = 0$, as shown in Fig. 1) and can limit the angular extent of this illumination cone over the LGP also. Meanwhile, in a conventional backlight system, a reflection sheet is set at the bottom of a LGP and brings back the leakage light into the LGP, but the reflection sheet is not useful for the LGP to control illumination angle [Fig. 2(a)], so in our designed LGP, a highly reflective material(such as aluminum) is coated on the back surface of the LGP, as shown in Fig. 2(b), which can easily make most of light emitting toward the front direction, and the conventional reflection sheet will be omitted in our backlight system.

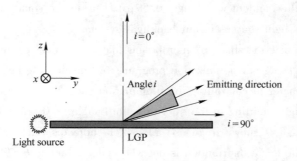

Fig. 1　The principle of light emitting characteristic for a conventional LGP

In order to calculate and analyze the scattering and diffraction of light, the Monte Carlo nonsequential ray tracing method based on Mie theory is employed in our study[5]. The Monte Carlo approach is based upon the law of large numbers, which says that the probability of an event occurring may be determined by observing the number of times the event occurs out of a large

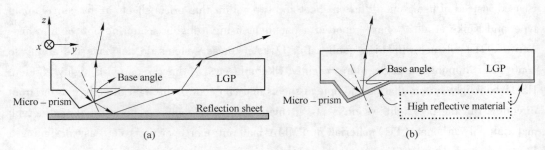

Fig. 2 Schematic diagram of the ray emitting trajectory for a LGP based on microprism structures (a) using a reflection sheet (b) no reflection sheet, but coating a high reflective material at the bottom of the LGP

number of trials, so in our study, scattering and diffraction are treated as random processed, and instead of propagating a distribution of light, discrete samples of the distribution, or rays, are propagated.

Our designed LGP is made of PMMA with a refractive index of $n_1 = 1.49$ at the wavelength 0.59 μm, and its bottom surface is convex microprisms that are the same shape, but their density and sizes are varied with distance from light sources. The width and length of the LGP are 40mm and 30mm, respectively. The LGP is a rectangular shape with a thickness of 0.75mm. As a real material, there must be some absorption when light rays propagate in the LGP. So rays that enter a material with nonzero absorption coefficient are attenuated according to Lambert's law of absorption as follows

$$\phi_T = \phi_0 \exp(-\alpha S) \quad (1)$$

where ϕ_T and ϕ_0 are transmitted and incident flux, α is absorption coefficient, and S is the distance of material through which the ray travels. Here in our design, the absorption coefficient α is 0.026 (mm^{-1}) at the incident wavelength 0.59 μm. Due to the small size of LGP, three distinct LEDs are set as light sources in our backlight system.

First, we must optimize the shape of microprisms that is a critical parameter to control the illumination angle. By using our optimization program, we find that the base angle of microprism is an important parameter in controlling the illumination cone. When the base angle is 35°, our LGP can make the light emitting in the front direction (angle $i = 0$), and the illumination angle distribution on the LGP is shown in Fig. 3(a), which indicates little light is leaked from the front direction. But the angle distribution is not well collected along $\theta = 90° - 270°$, so in order to compress the angle distribution, an array of microcompressor elements are designed on the top of the LGP, as shown in Fig. 3(b). After detailed optical design, the LGP can get a good illumination angle distribution along both $\theta = 90° - 270°$ and $\theta = 0° - 180°$, as shown in Fig. 3 (c). Fig. 3(d) shows the normalized angle i distributions along $\theta = 0° - 180°$ and $90° - 270°$ and we can get an angle i between ±25° with our designed LGP without using any optical sheets.

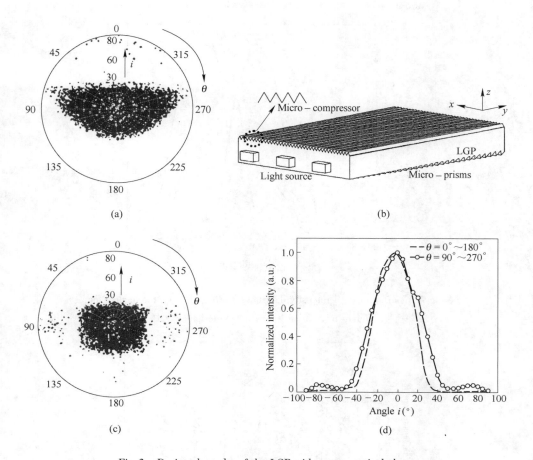

Fig. 3 Designed results of the LGP without any optical sheets
(a) illumination angular light distribution on the designed LGP; (b) structures of the LGP with a microcompressor; (c) illumination angular light distribution on the LGP with a microcompressor; (d) the normalized intensity distributions dependent on the angle i, getting an illumination angle between $\pm 25°$ with our designed LGP

Moreover, the geometry of the microprism can be optimized to direct the output light at a desired angle other than perpendicular, so in order to narrate our LGP's ability to control the illumination angle, we design the angle of microprisms to make the peak brightness shift from about $i = 40°$ to $i = -20°$, as shown in Fig. 4.

Second we must design our LGP to get sufficient intensity uniformity through optimizing the microprisms' distribution and sizes, respectively. Against the base angle of $35°$, we optimize the LGP that can get the normalized intensity distribution versus the distance from light sources (LEDs), as shown in Fig. 5, which expresses that our LGP can achieve a uniformity of intensity better than 86%.

In this letter, a high quality LGP made by PMMA is designed to control the illumination angle, and to get enough uniformity of intensity for thin backlight system applications. The LGP has a microcompressor element on the top surface and a microprism element at the bottom surface. By adjusting the microprism's shape, the LGP can control the illumination angle, and by

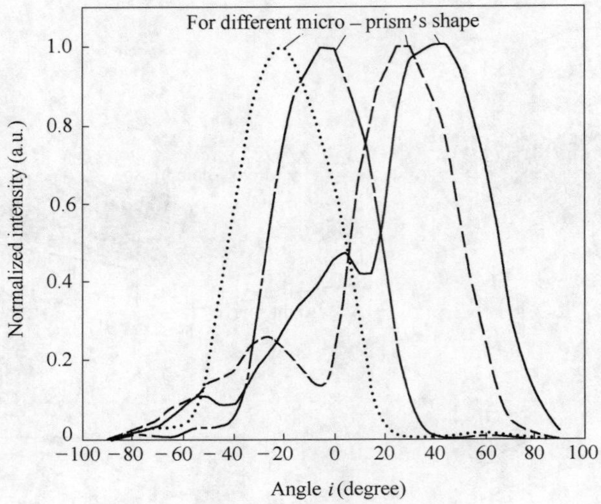

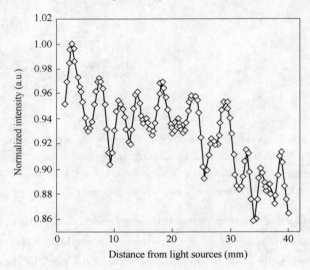

Fig. 4 The normalized intensity distributions dependent on the angle i for different microprism shape of the designed LGP

Fig. 5 The normalized intensity distribution dependent on distance from light sources (three LEDs); the designed LGP can achieve a uniformity of intensity better than 86%

adjusting microprisms' distribution and size, the LGP can achieve a uniformity of intensity. With our optimizing program, the designed LGP can achieve a uniformity of intensity better than 86%, and get an illumination angle between ±25°, without using any optical sheets. Our LGP will make the backlight system thinner, brighter, and lighter, because no any optical sheets will be used in the system. We believe our results can be expanded to other traditional material such as polycarbonate, acrylic material, and are a benefit to the design and the production of LGP for thin LCD applications.

References

[1] K. Käläntär, S. Matsumoto, and T. Onishi, IEICE Trans. Electron. E84-C, 1637(2001).
[2] A. Horibe, M. Baba, E. Nihei, and Y. Koike, IEICE Trans. Electron. E81-C, 1697(1998).
[3] A. Tagaya, M. Nagai, Y. Koike, and K. Yokoyama, Appl. Opt. 40, 6274(2001).
[4] T. Okumura, A. Tagaya, Y. Koike, M. Horiguchi, and H. Suzuki, Appl. Phys. Lett. 83, 2515(2003).
[5] I. Lux and L. Koblinger, *Monte Carlo Particle Transport Methods: Neutron and Photon Calculations* (CRC, Boca Raton, FL, 1991), p. 1.

Polarized Light-guide Plate for Liquid Crystal Display*

Abstract A novel polarized light-guide plate(LGP) for the illumination of liquid crystal display is proposed in this paper. For the substrate of the LGP, stress-induced birefringence is introduced to achieve the polarization state conversion. An aluminum sub-wavelength grating(SWG) is designed on the top surface as a polarizing beam-splitter(PBS). The structure of the novel LGP is optimized for three wavelengths of LEDs: 625nm, 533nm and 452nm, and high efficiencies of polarization conversion are obtained. The backlight system with the designed LGP does not require prism sheets and quarter wavelength plate. The backlight with the novel LGP achieved large gain in energy efficiency, and the peak luminous intensity is about 2 times higher than that of a conventional backlight.

1 Introduction

Liquid crystal displays(LCDs) are widely used in a range of electronic productions. The LCD panel has no function of emitting light spontaneously. And there are three LCD panel illumination technologies: transmissive LCD, transflective LCD and reflective LCD. The transmissive LCD panel for portable multimedia devices always needs edge-lit backlight system(BLS) for illumination. The conventional BLS always consists of a light source, a LGP and some optical sheets, such as a reflection sheet, a diffusion sheet and two crossed prism sheets. In order to make the backlight thinner and lighter, Käläntär et al.[1] have proposed a BLS, in which only one prism sheet is used instead of one diffusion sheet and two crossed prism sheets. Based on highly scattering optical transmission, Okumura et al.[2] have reported a highly-efficient BLS without any optical sheet. In these un-polarized BLSs, the optical efficiency is low due to the lack of polarization conversion. In the generation of polarized light, over 50% of the energy is absorbed by the rear polarizer of the LCD panel. For high efficiency, in recent years, lots of researches focus on the polarized BLS which recycles the undesired polarized light and emits linearly polarized light directly. Z. Pang and L. Li have proposed a high-efficiency polarized BLS based on the thin film PBS on the bottom surface of the LGP[3]. However, because there are no patterns on the bottom surface of the LGP, the backlight uniformity is difficult to control. For the blue light, the extinction ratio is only 20. Henri J. B. Jagt, Hugo J. Cornelissen et al. have reported a polarized BLS, in which the s-polarized light is extracted owing to the selective total internal reflection at microgrooves in the anisotropic layer[4]. However, the smoothness requirements of microgroove surfaces are stringent for reducing scattered light. Ko-Wei Chien et al. have re-

---

* Copartner: Xingpeng Yang, Yingbai Yan. Reprinted from *Optics Express*, 2005, 13(21):8349-8356.

ported an integrated polarized LGP and a 1.7 gain factor of polarization efficiency is achieved[5]. But the slot structures on the bottom surface of the LGP cannot control the illumination angle easily. Moreover, all these polarized BLSs require achromatic quarter wavelength plates to achieve the polarization state conversion.

In the polarized BLS, the two essential issues are the reflecting PBS and the polarized light conversion. The reflecting PBS separates polarized light, such as transmitting the p-polarized light and reflecting the s-polarized light. Then, the reflected s light is turned into p light by the polarization conversion device, and then emitted by the PBS. Finally both p and s light are utilized.

In this paper, a novel LGP is developed for three wavelengths of 625nm, 533nm and 452nm which emitted by red, green and blue(RGB) LEDs. The LGP substrate with the stress induced birefringence realizes the conversion of polarized light and this is described in detail in Section 2. In Section 3, an aluminum SWG is designed on the top surface as a PBS to extract linearly polarized light. Compared with the conventional BLS, the polarized BLS with the novel LGP is simulated in Section 4.

2 Polarized light conversion

In the published polarized BLSs, quarter wavelength plates are used to realize p-s polarized light conversion[3,5]. In this Section, the stress-induced birefringence is introduced instead of the quarter wavelength plate, and the stress is optimized. The stress-optical law of the plane photoelasticity can be expressed as

$$\Delta n = n_{\sigma y} - n_{\sigma x} = C\Delta\sigma \quad (1)$$

The amount of produced birefringence (Δn) is proportional to the stress difference ($\Delta\sigma = \sigma_y - \sigma_x$) provided the stress is not too large. C indicates the stress-optical coefficient.

As shown in Fig. 1, σ_x axis is perpendicular to σ_y axis and the angle between σ_x axis and x axis is denoted as β. The stress is applied along σ_x axis and σ_y axis directions respectively. The thickness of the LGP substrate denotes as a. The phase retardation value δ of passing through the substrate twice can be written as

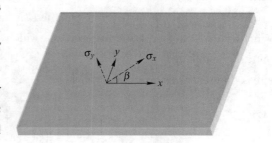

Fig. 1　Coordinate system of the LGP substrate with the applied stress

$$\delta = 2\pi C\Delta\sigma L/\lambda \quad (2)$$

where λ denotes the wavelength of the incident light, and $L = 2a$.

In x-y coordinates, the Jones matrix T of the substrate can be expressed as

$$T = R(\beta)T_\delta R(-\beta) = \begin{bmatrix}\cos\beta & -\sin\beta\\ \sin\beta & \cos\beta\end{bmatrix}\begin{bmatrix}1 & 0\\ 0 & e^{j\delta}\end{bmatrix}\begin{bmatrix}\cos\beta & \sin\beta\\ -\sin\beta & \cos\beta\end{bmatrix} \quad (3)$$

The incident y-directional polarized light can be written as $E_i = A \begin{bmatrix} 0 \\ 1 \end{bmatrix}$, here A denotes the light wave amplitude. The light passed the substrate twice can be written as

$$E_o = \begin{bmatrix} E_{ox} \\ E_{oy} \end{bmatrix} = TE_i = A \begin{bmatrix} \sin\beta\cos\beta - \sin\beta\cos\beta e^{j\delta} \\ \sin^2\beta + \cos^2\beta e^{j\delta} \end{bmatrix} \quad (4)$$

The intensity of the x-directional polarized light transmitted from the PBS is

$$I_{ox} = A^2 \sin^2 2\beta \sin^2(\delta/2) = A^2 \sin^2 2\beta \sin^2(\pi C \Delta\sigma L/\lambda) \quad (5)$$

Under the conditions of

$$\begin{cases} \beta = \pi/4 \\ \delta = 2k\pi + \pi, \quad k = 0,1,2,3,4,\cdots \end{cases} \quad (6)$$

the intensity achieves the maximum value A^2, and the efficiency of polarization conversion is 100%. Eq. (6) means that the LGP with applied stress is similar to the quarter wavelength retardation plate. As the achromatic BLS, the phase retardation value should be close to $2k\pi + \pi$ for the multiple wavelengths of $\lambda_R(625\text{nm})$, $\lambda_G(533\text{nm})$ and $\lambda_B(452\text{nm})$. Hence, the stress difference $\Delta\sigma$ should be optimized. The optimization problem can be expressed as

$$y = \min_{\Delta\sigma} \{ w_R \times \text{abs}[\text{mod}(\delta_R, 2\pi) - \pi] + w_G \times \text{abs}[\text{mod}(\delta_G, 2\pi) - \pi] +$$
$$w_B \times \text{abs}[\text{mod}(\delta_B, 2\pi) - \pi] \} \quad (7)$$

where mod denotes modulus after division, abs returns absolute value, and δ_R, δ_G and δ_B denote the phase retardation values for wavelengths of λ_R, λ_G and λ_B respectively. w_R, w_G and w_B denote weight factors of the light of red, green and blue respectively. In our design, all the weight factors are set to 1.0.

The LGP substrate of 0.8mm thickness is made of Bisphenol-A Polycarbonate (BAPC), which is a traditional plastic material and widely used in BLSs. The value of the BAPC's stress optical coefficient C is $8.9 \times 10^{-12} \text{ Pa}^{-1}$[6]. The objective function values with respect to the stress difference $\Delta\sigma$ are plotted in Fig. 2.

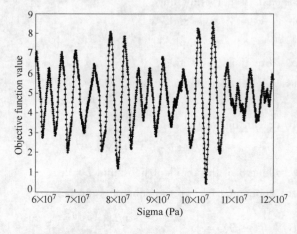

Fig. 2 The objective function value with respect to the stress difference

When $\Delta\sigma = 10.31 \times 10^7 \text{Pa}$, the local minimum value of Eq. (7) is obtained. The phase retardations of the red, green and blue light are

$$\begin{cases} \delta_R = 46\pi + 0.97\pi \\ \delta_G = 54\pi + 1.07\pi \\ \delta_B = 64\pi + 0.94\pi \end{cases} \quad (8)$$

Compared with the ideal achromatic wave plate, the maximum error is only 7%. As $\beta = \pi/4$, Eq. (5) indicates that the polarization conversion efficiency of this substrate is more than 99%. The 7% error of phase retardation only leads to a 1% decrease in conversion efficiency. The LGP substrate with applied stress can realize the polarization conversion, and the quarter wavelength plate can be left out. The stress-induced birefringence can remain in LGP by using the stress-freezing techniques[7]. Be similar, the strain-induced birefringence can be applied to achieve the polarization conversion too.

3 Polarizing beam splitter

The PBS is the other key issue in the polarized BLS. As a PBS, the wire-grid SWG is studied for visible light in both classical and conical mounting by M. Xu, et al[8]. Based on the rigorous coupled-wave theory[9], we design a SWG on the top surface of the LGP as shown in Fig. 3. The SWG as a reflecting PBS transmits p-polarized light and reflects s-polarized light. Aluminum is chosen as the material of the grating, and the material of the substrate is BAPC as mentioned in Section 1. The period of the SWG d is chosen as $0.14\mu\text{m}$. In order to achieve achromatism, the SWG should have the same performance for the red, green, and blue light. The desired goals are gain the maximum transmission for the p light, the minimum transmission for the s light, the maximum extinction ratio and the achromatism.

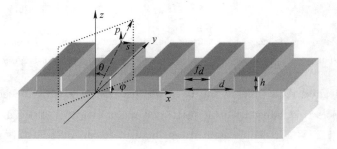

Fig. 3 The coordinate system and the structure of the SWG on the top surface of the LGP. Here, θ denotes the incident angle, φ denotes the azimuth angle, and f denotes the grating duty cycle

As shown in Fig. 4, the transmission of the p-and s-polarized light depends on the duty cycle f of the grating. Fig. 4 shows that the duty cycle of 0.5 is appropriate.

As shown in Fig. 5, the transmission of the red, green and blue light depends on the depth of the aluminum h respectively too.

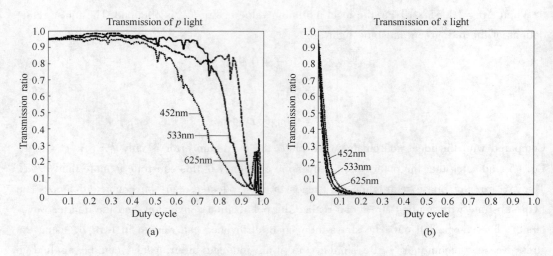

Fig. 4 The calculated dependences of the transmission on the duty cycle for (a) p-polarized light and (b) s-polarized light of red, green and blue. Here, incident angle $\theta = 0°$

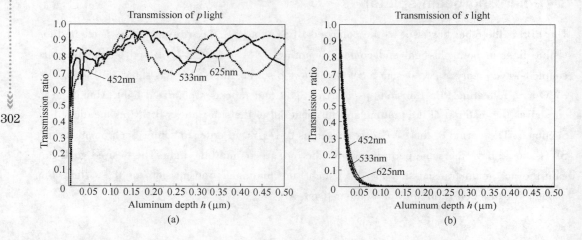

Fig. 5 The calculated dependences of the transmission on the grating depth for (a) p-polarized light and (b) s-polarized light of the red, green and blue light. Here, incident angle $\theta = 0°$

Based on the desired requirements, the depth of the grating h is optimized as 0.16 μm. At this depth, the p-polarized transmission ratios of the red, green and blue light are all equal to 0.91 as shown in Fig. 5(a). The calculated results of the designed SWG are shown in Fig. 6.

When $\theta < 40°$, the transmission of the p-polarized light is high. The s light transmission is low, and the minimum extinction ratio is as high as 11000. Because of the high extinction ratio of the SWG, the rear absorbing polarizer of the liquid crystal display panel is not required anymore.

4 Polarized light-guide plate

The substrate of the LGP acts as an achromatic quarter wavelength plate, and the SWG on the

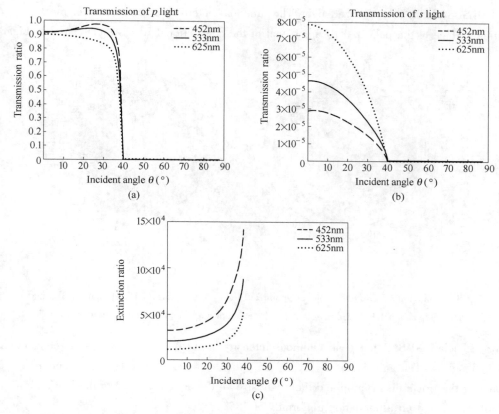

Fig. 6　Calculated results of the designed SWG with different wavelengths
(a) transmission of the *p*-polarized light; (b) transmission of *s*-polarized light;
(c) extinction ratio as a function of the incident angle

top surface acts as a reflecting PBS. The patterns on the bottom surface of the LGP are v-type grooves as shown in Fig. 7, the base angle and the apex angle are optimized as 53 and 74 degrees respectively. They are used to change the incident angle (to $\theta < 40°$) and control the backlight illumination angle instead of the prism sheets. There is a reflection sheet under the LGP, and the source is the LED array. The 2-inch polarized BLS with the novel LGP is simulated by the Monte Carlo non-sequential ray tracing program. An absorb polarizer is used in the conven-

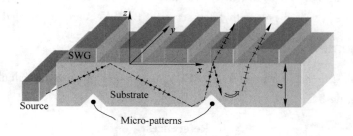

Fig. 7　Ray tracing in the proposed LGP. The *s* light is converted into the *p* light by the stress induced birefringence of the LGP substrate

tional BLS to obtain the polarized light. The luminous intensity at the center of the top light-emitting surface of the polarized BLS and that of the conventional BLS are compared in Fig. 8.

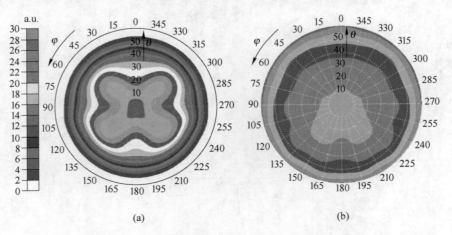

Fig. 8 Polar luminous intensity plots of polarized light emitting form (a) the proposed BLS and (b) the conventional BLS. Here, φ denotes the azimuth angle, and θ denotes inclination angle

For the polarized BLS, the peak luminous intensity is about 28, and that of the conventional BLS is about 14. The gain factor is 2, and the optical efficiency is substantially improved. By optimizing the prism distribution and the prism size on the bottom surface of the LGP, the polarized BLS achieved an illumination uniformity of 78%.

5 Conclusion

In this paper, the stress-induced birefringence is introduced to the substrate of the LGP. The SWG is designed on the top surface of the LGP to emit linearly polarized light. Both the stress and the structure of the SWG are optimized for the red, green and blue light to realize achromatism. The polarization conversion and separation are realized in one LGP. For the BLS with this LGP, the prism sheets, the diffusion sheet and the quarter wavelength plate are not needed anymore. The simulation results show that the luminous intensity greatly increased. We believe it will make the LCD thinner, lighter and brighter.

In our design, the material of the LGP substrate is polycarbonate, however, other polymers, such as polystyrene(PS) and polymethylmethacrylate(PMMA), can be selected too.

Acknowledgements

The authors thank Lifeng Li for providing KAPPA, a computer program for calculation of periodic diffraction gratings.

References

[1] K. Käläntär, S. Matsumoto, and T. Onishi, "Functional light-guide plate characterized by optical micro-deflector and micro-reflector for LCD backlight", IEICE Trans. Electron. E84-C, 1637-1646(2001).

[2] T. Okumura, A. Tagaya, Y. Koike, M. Horiguchi, and H. Suzuki, "Highly-efficient backlight for liquid crystal display having no optical films", Appl. Phys. Lett. 83, 2515(2003).
[3] Z. Pang and L. Li, "Novel high-efficiency polarizing backlighting system with a polarizing beam splitter", *SID 99 Technical DIGEST*(Society for Information Display, San Jose, Calif. , 1999), 916(1999).
[4] Henri J. B. Jagt, Hugo J. Cornelissen and Dirk J. Broer, "Polarized light LCD backlight based on liquid crystalline polymer film: a new manufacturing process", *SID 99 Technical DIGEST*(Society for Information Display, Boston, Mass. , Calif. , 1999), 1236(2002).
[5] Ko-Wei Chien and Han-Ping D. Shieh, "Design and fabrication of an integrated polarized light guide for liquid-crystal-display illumination", Appl. Opt. 43, 1830(2004).
[6] R. Winberger-Friedl, J. G. de Bruin, and H. F. M. Schoo, "Residual birefringence in modified polycarbonates", Polym. Eng. Sci. 43, 62(2003).
[7] G. D. Shyu, A. I. Isayev, and C. T. Li, J. Polym. "Residual Thermal Birefringence in Freely Quenched Plates of Amorphous Polymers: Simulation and Experiment", Sci. Part B: Polym. Phys. 41, 1850(2003).
[8] M. Xu, H. P. Urbach, D. K. G deBoer and H. J. Cornelissen, "Wire-grid diffraction gratings used a polarizing beam splitter for visible light and applied in liquid crystal on silicon", Opt. Express 13, 2303-2320 (2005), http://www.opticsexpress.org/abstract.cfm? URI = OPEX-13-7-2303.
[9] Lifeng Li, "A modal analysis of lamellar diffraction grating in conical mountings", J. Mod. Opt. 40, 553-573 (1993).

基于 PWC 方法的折衍混合红外物镜设计*

摘 要 折衍混合设计为选用廉价材料设计像质优良的红外物镜提供了新的途径。基于衍射结构的高折射率模型和传统的 PWC 方法,分析了折衍混合红外单透镜的光焦度分配和二级光谱;在 8~12μm 波段,采用 GASIR2 和 AMTIR1 红外玻璃,分别设计了可模压生产的折衍混合单片型和 Petzval 型红外物镜。结果表明:采用新型红外玻璃的折衍混合设计可以代替锗材料设计出具有大相对孔径,且像质优良而廉价的红外物镜。

关键词 光学设计;红外物镜;折-衍混合系统;PWC 方法

Design of Hybrid Diffractive-refractive Infrared Objectives Based on PWC Method

Abstract Hybrid diffractive-refractive (HDR) design offers a new approach to design infrared objectives having excellent image quality and low cost. Based on the ultrahigh index model of diffractive structure and the conventional PWC method, the power assigning and the secondary spectrum in the HDR infrared singlet were analyzed and the HDR singlet type and Petzval type infrared objectives that could be molded were designed by using GASIR2 and AMTIR1 infrared glass in 8~12μm waveband respectively. The result shows that the low cost far infrared objectives having large relative aperture and excellent image quality can be obtained by using new infrared glass to replace of germanium and HDR design.

Key words optical design; infrared objective; hybrid refractive-diffractive system; PWC method

1 引言

具有良好光学性能和机械性能的红外材料较少,给红外物镜的设计带来了一定的困难。而随着非致冷红外相机的广泛应用,又提出了具有大相对孔径廉价的 8~12μm 波段红外物镜的要求。锗为 8~12μm 波段内广泛采用的材料[1],由于该材料色散较小,折射率很高,吸收很小,可以采用较为简单的结构形式达到优良的成像质量。但锗的价格昂贵,对温度较敏感,且不能模压生产,很难应用于诸如非致冷红外相机等低价位的相机中。目前新型红外玻璃(AMTIR 和 GASIR 系列)的光学性能和机械性能都得到了很大改善,与锗材料相比较,这类玻璃材料价格低廉,对温度较不敏感,透镜可模压生产,因此较适合于低价位的非致冷相机应用。虽然这类红外玻璃材料色散较大,折射率较低,但采用折衍混合设计[2,3],仍可以设计出具有优良像质的红外物镜[4]。

本文采用 PWC 表示的折衍混合光学系统初级像差理论和衍射透镜高折射率设计方

---

* 本文合作者:曾吉勇,王民强,严瑛白。原发表于《红外与毫米波学报》,2006,25(3):213~216。

法，研究了折衍混合红外单透镜的光焦度分配和二级光谱；在 8～12μm 波段，采用 GASIR2 和 AMTIR1 红外玻璃，分别设计了折衍混合单片型和 Petzval 型红外物镜。

2 基本理论

衍射结构可以视为折射率无限大的薄透镜（衍射透镜）[5]，因此可以建立折衍混合单透镜的衍射透镜与折射透镜的双胶合模型[6]。

2.1 含衍射透镜的薄透镜系统的初级像差[6]

光阑与透镜组非密接的情况，含非球面衍射透镜的初级像差和数为

球差：
$$S_1 = \sum h(P + \Delta P) \tag{1}$$

彗差：
$$S_2 = \sum h_z(P + \Delta P) - J\sum W \tag{2}$$

像散：
$$S_3 = \sum \frac{h_z^2}{h}(P + \Delta P) - 2J\sum \frac{h_z}{h}W + J^2\sum \varphi \tag{3}$$

弧矢场曲：
$$S_4 = J^2 \sum \mu_r \varphi_r \tag{4}$$

畸变：
$$S_5 = \sum \frac{h_z^3}{h^2}(P + \Delta P) - 3J\sum \frac{h_z^2}{h^2}W + J^2\sum \frac{h_z}{h}(3\varphi + \mu_r\varphi_r) \tag{5}$$

轴向色差：
$$S_{1C} = \sum h^2 C \tag{6}$$

垂轴色差：
$$S_{2C} = \sum hh_z C \tag{7}$$

式中，P、W 为光学系统内部参数；h 为轴上点发出经过孔径边缘的第一辅助光线在各透镜组上的投射高；h_z 为视场边缘发出经过孔径光阑中心的第二辅助光线在各透镜组上的投射高；φ 为各透镜组的光焦度；J 为拉格朗日不变量。在本文中下标为 r 对应于折射透镜的量，下标为 d 对应于衍射透镜的量。$\Delta P = (n-1)(c^3 k + 8A_4)h^3$ 为衍射透镜非球面项附加光程差产生的 Seidel 和数增量。

2.2 折衍混合单透镜的光焦度分配和二级光谱

衍射透镜的等效折射率可以写为[6] $n(\lambda) = \lambda \times 10^s + 1$，设 λ_C、λ_S 和 λ_L 分别为设计的闪耀波长、光谱区的短波长和长波长，对应的折射率分别为 n_C、n_S 和 n_L，则阿贝数和相对部分色散分别为

$$v = \frac{n_C - 1}{n_S - n_L} = \frac{\lambda_C}{\lambda_S - \lambda_L} \qquad P = \frac{n_S - n_C}{n_S - n_L} = \frac{\lambda_S - \lambda_C}{\lambda_S - \lambda_L} \tag{8}$$

令 $S = 7$，在 ZEMAX 软件的玻璃库中建立了名为 DOE 的材料，在 8～12μm 光谱区内，对应的 $n_C = 100001$、$n_S = 80001$、$n_L = 120001$，阿贝数 $v_d = -2.5$，相对部分色散为 $P_d = 0.5$。

设折衍混合单透镜中折射透镜的光焦度为 φ_r、阿贝数为 v_r、相对部分色散为 P_r，衍射透镜的光焦度为 φ_d、阿贝数为 v_d、相对部分色散为 P_d。折衍混合单透镜的焦距规一化，在 8～12μm 波段内，折射透镜和衍射透镜的消色差光焦度分配和二级光谱详见表 1。计算二级光谱时，物在无穷远，折衍混合透镜的焦距为 f'。

表1 消色差折衍混合单透镜的光焦度分配和二级光谱
Table 1 The power assigning and secondary spectrum of HDR achromatic infrared singlet

$v_d = -2.5 \quad P_d = 0.5$

Material	n_C	v_r	P_r	φ_d	φ_r	Δ_L
AMTIR1	2.49749	113.58	0.461	0.022	0.978	$0.00033f'$
GASIR2	2.58416	100.51	0.448	0.024	0.976	$0.00051f'$
GE	4.00438	783.21	0.618	0.003	0.997	$-0.00015f'$
Z_nS_e	2.40644	57.47	0.443	0.042	0.958	$0.00095f'$
Z_nS	2.19991	22.76	0.432	0.099	0.901	$0.00270f'$

2.3 $\overline{P}_\infty$、$\overline{W}_\infty$ 与折衍混合单透镜结构参量的函数关系

折衍混合单透镜结构参量包括衍射透镜和折射透镜的折射率 n_d 和 n_r,透镜曲率半径 r_1、r_2 和 r_3。令 $C_2 = 1/r_2$,则透镜弯曲系数 $Q = C_2 - \varphi_1$,已知 n_d、n_r、φ_1、Q,就能计算 r_1、r_2 和 r_3。

$$\overline{P}_\infty = a(Q - Q_0)^2 + P_0 \tag{9}$$

$$\overline{W}_\infty = -\frac{a+1}{2}(Q - Q_0) + W_0 \tag{10}$$

式中,$Q_0 = -\frac{b}{2a}$;$P_0 = c - \frac{b^2}{4a}$;$W_0 = \frac{1-\varphi_1}{3} - \frac{3-a}{6}Q_0$。当衍射面在透镜的前表面时,$a = 1 + 2\frac{\varphi_r}{n_r}$,$b = -\frac{3}{n_r - 1}\varphi_r^2 - 2\varphi_r$,$c = \frac{n_r}{(n_r-1)^2}\varphi_r^3 + \frac{n_r}{n_r - 1}\varphi_r^2$。

3 折衍混合单片型红外物镜设计

设计要求为焦距 $f = 80$ mm,相对孔径为 $F/2$,视场角为 $\pm 2°$。

3.1 初始结构确定

由于相对孔径和视场均较小,采用折射混合单透镜结构,可以校正球差、彗差和色差以满足设计要求。由 2.1 节的初级像差公式得到校正初级球差、彗差和色差的条件分别为 $P + \Delta P = 0$,$\overline{W}_\infty = 0$,$\overline{C} = 0$。设计以 GASIR2 玻璃材料为基底,衍射结构置于透镜前表面。由 2.3 节的公式计算得到 $a = 1.7552$,$P_0 = 0.5019$,$Q_0 = -1.0695$,$W_0 = 0.1033$,$\varphi_d = 0.9757$。

由式(10)得到透镜弯曲系数,由此计算透镜的曲率半径。由式(9)得到 $\overline{P}_\infty = 0.5118$,衍射透镜的非球面系数 $A_4 = \frac{-\overline{P}_\infty (h\varphi)^3}{8(n-1)h^3}$。表 2 为折衍混合单片型物镜的初始结构,初级像差和数为 $S_1 = 0.000252$、$S_2 = -0.000055$、$S_{1C} = 0.000067$、$S_{2C} = -0.000001$,可见初始结构的初级球差、彗差、色差均已良好校正。

表2 折衍混合单片型物镜的初始结构
Table 2 Original design of HDR singlet type objective

Surf	r/mm	d/mm	Glass
1STOP*	67.44710456	0	DOE
2	68.4471188	0.1	GASIR2
3	144.6980	79.86	

* Coefficients of EVENASPH: $A_4 = -1.2494 \times 10^{-12}$。

3.2 与常规单片型锗红外物镜像质比较

图1为优化设计的折衍混合单片型玻璃物镜的传递函数曲线。图2为常规单片型锗物镜的传递函数曲线,该物镜结构取自文献[1],做了进一步非球面优化。可见在0.75视场内折衍混合设计达到了衍射受限成像质量,而常规设计接近衍射受限成像质量。常规设计较折衍混合设计成像质量差的原因在于前者色差没有校正,而后者色差已良好校正。

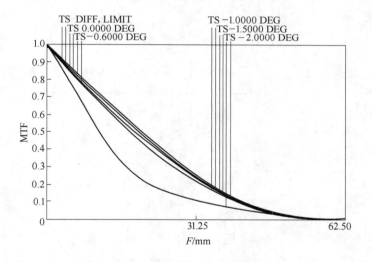

图1 折衍混合单片型物镜的传递函数曲线
Fig. 1 MTF curve of HDR singlet type objective

4 折衍混合 Petzval 红外物镜设计

光学特性要求焦距 $f = 25$ mm,相对孔径为 $F/1$,视场角为 $\pm 5°$。Petzval物镜要求校正球差、彗差、像散和色差。在常规设计中,Petzval物镜由间隔一定距离的前后两透镜组组成(图3),每一透镜组至少由两片透镜组成以实现每一透镜组单独消色差。本文设计中前后透镜组均采用单片折衍混合玻璃透镜。

根据设计要求,物镜总偏角 $\Delta u = 0.5$,偏角分配为前组0.2,后组0.3。取前后透镜组间距 $d = 0.7f$,设光阑与前组密接,表3为物镜的外部参量:前后组偏角分配 Δu,光焦度 φ 和焦距 f;第一辅助光线的入射高 h 和入射角 u;第二辅助光线的入射高 h_z 和

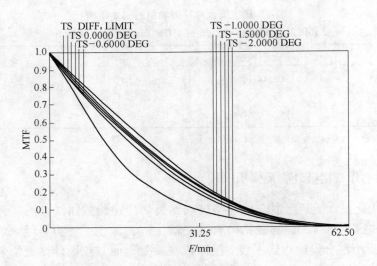

图 2 常规锗单片型物镜的传递函数曲线
Fig. 2 MTF curve of conventional sing let-type objective

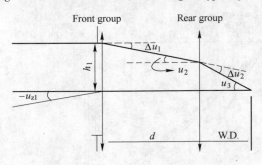

图 3 Petzval 物镜
Fig. 3 Petzval objective

入射角 u_z。根据像差校正要求和物镜的外部参量得到初级像差方程

$$\overline{P}_{1\infty t} + 2.43\,\overline{P}_{2\infty t} + 6.4803\,\overline{W}_{2\infty} + 2.48434 = 0 \tag{11}$$

$$\overline{P}_{2\infty t} + 0.2858\,\overline{W}_{2\infty} - 1.0582\,\overline{W}_{1\infty} - 2.78724 = 0 \tag{12}$$

$$\overline{P}_{2\infty t} - 2.0953\,\overline{W}_{2\infty} + 1.793 = 0 \tag{13}$$

后组满足方程（13）以校正像散，令 $\overline{W}_{2\infty} = 0.6$，则 $\overline{P}_{2\infty t} = -0.5358$。由方程（11）、方程（12），前组提供 $\overline{P}_{1\infty t} = -5.0705$，$\overline{W}_{1\infty} = -2.9782$ 以校正球差和彗差。

前后组均采用 AMTIR1 玻璃为基底，衍射结构置于透镜前表面，由表 1 衍射透镜和折射透镜的光焦度分配 $\varphi_d = 0.022$、$\varphi_r = 0.978$，由 2.3 节的公式计算得到透镜参量 $a = 1.7836$，$P_0 = 0.5354$，$Q_0 = -1.0863$，$W_0 = 0.1059$。透镜组 1、2 各参量的计算与上一节中折衍混合单片型红外物镜计算方法相同。表 4 为物镜的初始结构，其初级像差和数 $S_1 = 0.003104$、$S_2 = 0.003629$、$S_3 = 0.000079$、$S_{1C} = -0.000373$、$S_{2C} = -0.000005$，可见初始结构的初级球差、彗差、像散和色差均已良好校正。图 4 为优化设计的物镜的传递函数曲线，在0.8视场内达到了衍射受限的成像质量。

表3 外部参量
Table 3 External parameter

No.	$h\varphi(\Delta u)$	φ	f	h	u	h_z	u_z
Group1	0.2	0.016	62.5	12.5	0	0	−0.0872
Group2	0.3	0.033	30	9	0.2	1.526	−0.0872

表4 折衍混合Petzval红外物镜初始结构
Table 4 Original design of HDR Petzval infrared objective

Surf	Radius/mm	Thickness/mm	Glass	A_4
STOP	18.80433239	0	DOE	-7.35424×10^{-10}
2	18.80434456	0.1	AMTIR1	
3	23.40586	17.5		
4	39.85153179	0	DOE	-6×10^{-10}
5	39.85164561	0.1	AMTIR1	
6	301.923	17.79234		

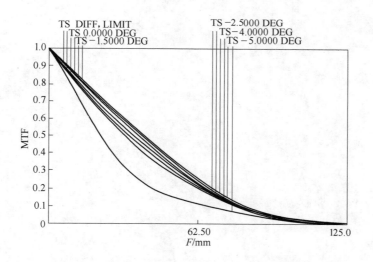

图4 折衍混合Petzval红外物镜的传递函数曲线
Fig.4 MTF curve of HDR Petzval infrared objective

5 结论

采用传统PWC表示的折衍混合光学系统初级像差理论和折衍混合光学系统设计的高折射率方法，研究了折衍混合红外单透镜的光焦度分配和二级光谱；在8~12μm波段，采用GASIR2和AMTIR1红外玻璃，分别设计了可模压生产的折衍混合单透镜红外物镜和Petzval红外物镜。设计结果表明：采用新型红外玻璃和折衍混合设计可以代替锗材料设计出具有优良成像质量的红外物镜，从而为设计具有大相对孔径廉价的红外物镜提供了新的途径。

References

[1] Zhao Xiuli. *Design of Infrared Optical System*. Beijing: Mechanical Industry Publishing House（赵秀丽. 红外光学系统设计. 北京：机械工业出版社），1984.

[2] Wood A P. Design of infrared hybrid refractive-diffractive lenses. Appl. Opt., 1992, 31(13):2253~2258.

[3] Liu Liping, Wang Yongtian, Li Ronggang, et al. Infrared diffractive optical element fabricated on aspheric substrate. J. Infrared Millim. Waves（刘莉萍，王涌天，李荣刚，等. 制作在非球面基底上的红外衍射光学元件. 红外与毫米波学报），2004, 23(4):308~312.

[4] Graham A, LeBlanc RA, Hilton R. Low cost infrared glass for IR imaging applications. Proc. SPIE, 2003, 5078: 216~224.

[5] Sweat W C. Describing holographic optical elements as lenses. J. Opt. Soc. Am., 1977, 67(6):803~808.

[6] Zeng Jiyong, Jin Guofan, Wang Minqiang, et al. PWC primary aberration expression of thin lens system including diffractive optical element. Acta Optica Sinica（曾吉勇，金国藩，王民强，等. 含衍射结构薄透镜系统初级像差的 PWC 表示. 光学学报），2006, 26(2):96~100.

Enhancement of the Light Output of Light-emitting Diode with Double Photonic Crystals[*]

Abstract A light-emitting diode(LED) with double photonic crystals(PhCs) is designed to enhance the light output. Based on the configuration of the PhC assisted LED with a single PhC(SPC-LED), a second PhC is added on the bottom surface of the active layer to improve the light output. The optical properties of this double PhCs assisted LED are simulated using the three-dimensional(3D) finite-difference time-domain(FDTD) method. The calculation results show that its light output can be 3.2 times higher than that of LED without PhC, and 1.39 times higher than that of SPC-LED.

1 Introduction

Photonic crystals(PhCs) have been introduced into the light-emitting diodes(LEDs) to obtain the high light output, and the relevant LEDs mainly involve the GaN-based LEDs[1-3], the GaAs-based LEDs[4,5], and the organic LEDs(OLEDs)[6-8]. By the effect of photonic band gap (PBG) or coherent diffraction of PhCs, the light output of this type of PhC assisted LED(PC-LED) can ideally reach 10-20-fold of that of the conventional planar LED(C-LED)[9,10]. But actually, only about 2-fold enhancement has been realized under current injection due to some realistic factors, such as the degradation of internal quantum efficiency and the absorption of semiconductor material[1-5]. Therefore, high extraction efficiency LED with two PhCs structure has been demonstrated[11], where the second PhC serves as a reflector, and the light output is enhanced significantly on the basis of the LED with a single PhC. However, the material absorption is not concerned. What is worse, the aspect ratio of the PhC in sapphire needs to be chosen larger than 3.5, which is somewhat difficult for the actual fabrication.

Meanwhile, laser lift-off and wafer-bonding technologies have been employed to add some microstructures on the bottom surface of the active layer(e. g., GaN) of LED one is on the top surface as usual is 2.77 times brighter than the C-LED. In this work, we present a high light output LED with two PhCs: the first PhC is in the indium-tin-oxide(ITO) layer(or the transparent electrode), and the second PhC is on the bottom surface of the active layer. Compared with the two microstructures fabricated by the surface-roughening[12,13], the PhC structure is more predictable and more convenient for the parameter optimization. Moreover, the aspect ratio of our designed PhC structure is low enough for the actual fabrication. A three-dimensional(3D) finite-

---

[*] Copartner: Zhenfeng Xu, Liangcai Cao, Qiaofeng Tan, Qingsheng He. Reprinted from *Optics Communications*, 2007, 278: 211-214.

difference time-domain (FDTD) method with perfectly matched layer (PML) and periodic boundary condition (PBC) is carried out to simulate the optical properties of the presented LED, and the material absorption is involved during the calculation. Based on this simulation method, the influences of the hole depth, hole diameter, and lattice constant of PhC are investigated, respectively.

2 Design and analysis

The schematic diagram of the LED with single PhC (SPC-LED) is shown in Fig. 1. The PhC_1 is only etched in the ITO layer in order to avoid the degradation of the electrical properties[14]. A metal layer is added between the active layer and substrate so as to reflect back the downward emission, which can be implemented using the laser lift-off and wafer-bonding technologies[12]. For the convenience of comparison, we designate the LED with the metal layer but without the PhC_1 as "M-LED". Theoretically, once the emitted light from the active layer enters the ITO layer, it can totally radiate out by several times of diffraction of the PhC structure according to the multi-diffraction model[15]. However, for the realistic LED structure, the absorptions of semiconductor material and metal layer cannot be neglected. Thus, the "extraction speed" of the PhC structure is quite critical, that is, the emitted light should be extracted out as "soon" as possible before it is absorbed eventually. As depicted in Fig. 1, when the light with large incident angle is diffracted by PhC_1, only a little portion will be coupled out, thus a large portion of the light will be reflected between the PhC_1 and metal layer and then absorbed by the active layer and metal material. Thereby, the SPC-LED still has a relatively low "extraction speed", and its light output can be increased by enhancing the "extraction speed".

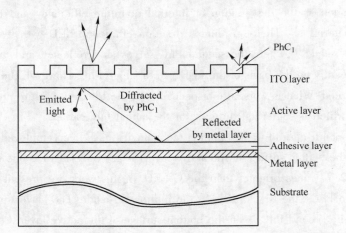

Fig. 1 The schematic diagram of SPC-LED structure, where the metal layer works as a mirror to reflect the downward light

In order to improve the "extraction speed", another PhC (PhC_2) is introduced on the bottom surface of active layer, as shown in Fig. 2. The PhC_2 can be fabricated between the laser lift-off and the wafer-bonding, which is familiar with the process demonstrated by Peng et al. [13]. In the presence of PhC_2, the light with otherwise large incident angle can be redirected to have a small incident angle, and then more portion of the light can be coupled out, as shown in Fig. 2. Consequently, this PC-LED with double PhCs(DPC-LED) will have a faster "extraction speed" compared with the SPC-LED. Therefore, its light output can be improved with the combination of PhC_1 and PhC_2.

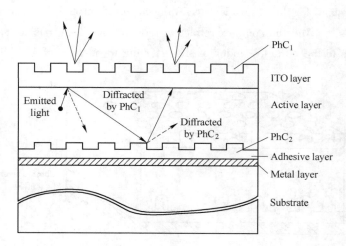

Fig. 2 The schematic diagram of DPC-LED structure, where PhC_2 is added on the bottom surface of active layer to improve the "extraction speed"

3 Simulation and results

To verify the analysis above, the light output of the DPC-LED(e. g. , GaN-based) is calculated by the 3D FDTD method, which is widely used in the computation of PC-LED structure[9-11,16]. The light output is represented by the vertical component of the Poynting vector integrated over the top surface of simulation domain. Totally 625(25 × 25)TE-polarized point dipoles distributed evenly in the multi-quantum-wells (MQWs) plane, which locates at the center of the GaN active layer, are used to simulate the emitted field. The phase and electric field direction of each dipole are set randomly to generate the incoherent light, and the emitted wavelength is chosen as λ = 0. 45 μm. The PhC_1 and PhC_2 are both designed as square lattices of air holes in the background materials due to their larger density of leaky modes[10]. The lateral simulation domain is set to be a 5 × 5 square-lattice air-hole array to reduce the size of calculation, and the PBCs are applied in the lateral directions to reduce the finite-size effect[16]. In the vertical directions, the PMLs are employed as the numerically absorbing boundaries. We have tested our numerical results with those published in Refs. [11,16] and found good agreements.

The 3D model of DPC-LED is shown in Fig. 3. In the following simulation, PhC_1 and PhC_2 have the same lattice constant a and hole diameter d, which are taken as 0. 8μm and 0. 8a, respectively. The thicknesses of ITO t_1, GaN layer t_2, adhesive layer t_3 and metal layer t_4 are chosen as 0.3μm, 2.5μm, 0.1μm and 0.1μm, respectively. The hole depth of PhC_1 h_1 is set as 0.2μm, which means the PhCs is only drilled partly in the ITO layer. The holes of PhC_2 are filled with the adhesive material, and the depth h_2 will be varied to study its influence. The refractive index of adhesive layer is around 1.5. The complex refractive indices of GaN and ITO layer are assumed as 2.5 + i0.002 and 2.0 + i0.006, respectively, where the imaginary parts of refractive indices stand for the absorption according to the absorption coefficients of the corresponding materials[17]. And the complex refractive index of metal layer is set as 0.04 + i2.66, which corresponds to the refractive index of silver for the wavelength of 450 nm[18].

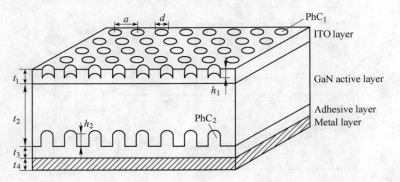

Fig. 3 The 3D model of DPC-LED structure, where $t_1 - t_4$ and h_1 are set invariable, and the effects of a, d and h_2 will be studied

The hole depth of PhC_2 h_2 is changed from 0.0μm (i.e., the SPC-LED) to 0.5μm with an interval of 0.1μm. The corresponding normalized light output as functions of the simulation time are calculated and plotted in Fig. 4, where the normalized light output is defined as the ratio of steady light output of PC-LED to that of M-LED (i.e., $h_1 = 0$ and $h_2 = 0$). It can be seen that (1) in less than 500 fs, the light output of PC-LED asymptotically reaches its steady-state while that of M-LED saturates within 100 fs; (2) the enhancement of light output is ladderlike, and each step comes from a diffraction of the PhC structure, which agrees well with the multi-diffraction model[15]; (3) the normalized light output of SPC-LED is nearly 2.1; (4) when the depth h_2 increases to 0.3μm, the normalized light output of DPC-LED reaches the value of 3.2, which is around 1.52 times higher than that of the SPC-LED; the higher output implies that the DPC-LED has a faster "extraction speed"; and (5) if the depth h_2 increases from 0.3μm, the light output of DPC-LED will not increase correspondingly, but decrease slightly; thus the optimum aspect ratio of PhC_2 is about 0.47, which is easy for the actual fabrication.

The normalized light outputs of SPC-LED and DPC-LED with different hole diameter d and lattice constant a are also simulated and plotted in Figs. 5(a) and (b), respectively. Since the light output is enhanced by the diffraction effect of PhCs, the lattice constant a should satisfy $a > \lambda_0/\sqrt{2} \approx 0.32$μm to realize fully diffraction[4].

Compared with the SPC-LED, the DPC-LED is a little more sensitive to those parameters, but

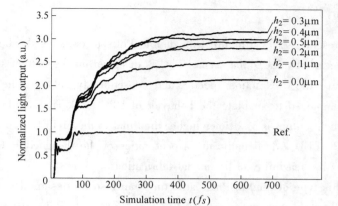

Fig. 4 The FDTD simulation results of normalized light outputs of the conventional planar LED (as the reference), SPC-LED (i.e., $h_2 = 0$), and the DPC-LED with different h_2

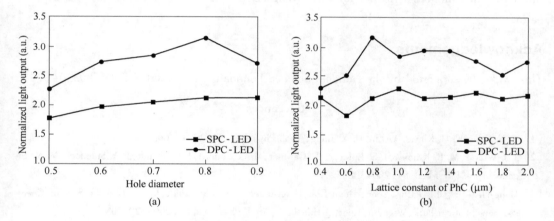

Fig. 5 Normalized light outputs as functions of (a) hole diameter and (b) lattice constant

its light output is always higher than that of SPC-LED. When the hole diameter is in the range of $0.7a$-$0.8a$ and the lattice constant is in the range of 0.8-1.4 μm, the normalized output is relatively high. As shown in Fig. 5(b), the output of SPC-LED reaches its maximum (i.e., ~2.3) when the lattice constant is around 1.0 μm (i.e., $a/\lambda_0 \approx 2$), which coincides with the previous results[5,8,19]; the maximum output of DPC-LED is nearly 3.2 when the lattice constant shifts to 0.8 μm, and the light output is still improved by 39%. The optimum value of the lattice constant of PhC is nearly 2-fold of the emitted wavelength in vacuum, which still needs further study to fully understand the mechanism. Furthermore, PhC_1 and PhC_2 can be optimized with different lattice constants, but it will be somewhat difficult for the presented FDTD calculation because the PBC cannot be suitable for the different lattice constants simultaneously. The appropriate calculation method is under investigation.

4 Conclusion

In summary, we present a LED structure with double PhCs to enhance the light output. Based on the configuration of SPC-LED, the second PhC is designed on the bottom surface of the active layer to improve the "extraction speed", and consequently enhance the light out. A 3D FDTD method is employed to simulate the behavior of DPC-LED. The simulation results show that its optimized light output is 3.2 times higher than that of the M-LED, and 1.39 times higher than that of SPC-LED. Additionally, the aspect ratios of the designed PhC are lower than 0.5, which ensures the feasibility of the actual fabrication.

We note that this type of double PhCs structure can also be designed in the OLEDs. For OLEDs, no crystal growth process is involved[6-8], so the double PhCs structure can be added without the laser lift-off and wafer-bonding techniques during the fabrication of OLEDs. It will reduce the fabrication cost while realizing relatively high enhancement. In that case, the calculation model for the 3D FDTD and the corresponding analysis are still valid.

Acknowledgements

This work was supported by the National Science Foundation of China (No. 60678033).

References

[1] T. N. Oder, K. H. Kim, J. Y. Lin, H. X. Jiang, Appl. Phys. Lett. 84(2004)466.

[2] J. J. Wierer, M. R. Krames, J. E. Epler, N. F. Gardner, M. G. Craford, J. R. Wendt, J. A. Simmons, M. M. Sigalas, Appl. Phys. Lett. 84(2004)3885.

[3] Dong-Ho Kim, Chi-O Cho, Yeong-Geun Roh, Heonsu Jeon, Yoon Soo Park, Jaehee Cho, Jin Seo Im, Cheolsoo Sone, Yongjo Park, Won Jun Choi, Q-Han Park, Appl. Phys. Lett. 87(2005)203508.

[4] H. Ichikawa, T. Baba, Appl. Phys. Lett. 84(2004)457.

[5] T. Kim, P. O. Leisher, A. J. Danner, R. Wirth, K. Streubel, K. D. Choquette, IEEE Photon. Technol. Lett. 18(2006)1876.

[6] Yong-Jae Lee, Se-Heon Kim, Joon Huh, Guk-Hyun Kim, Yong-Hee Lee, Sang-Hwan Cho, Yoon-Chang Kim, Young Rag Do, Appl. Phys. Lett. 82(2003)3779.

[7] M. Fujita, T. Ueno, T. Asano, S. Noda, H. Ohhata, T. Tsuji, H. Nakada, N. Shimoji, Electron. Lett. 39(2003).

[8] A. M. Adawi, R. Kullock, J. L. Turner, C. Vasilev, D. G. Lidzey, A. Tahraoui, P. W. Fry, D. Gibson, E. Smith, C. Foden, M. Roberts, F. Qureshi, N. Athanassopoulou, Org. Electron. 7(2006)222.

[9] Shanhui Fan, Pierre R. Villeneuve, J. D. Joannopoulos, E. F. Schubert, Phys. Rev. Lett. 78(1997)3294.

[10] Han-Youl Ryu, Jeong-Ki Hwang, Yong-Jae Lee, Yong-Hee Lee, IEEE J. Sel. Top. Quantum Electron. 8(2002)231.

[11] Chia-Hsin Chao, S. L. Chuang, Tzong-Lin Wu, Appl. Phys. Lett. 89(2006)091116.

[12] Wei Chih Peng, YewChuang Sermon Wu, Appl. Phys. Lett. 88(2006)181117.

[13] Wei Chih Peng, YewChuang Sermon Wu, Appl. Phys. Lett. 89(2006)041116.

[14] Shyi-Ming Pan, Ru-Chin Tu, Yu-Mei Fan, Ruey-Chyn Yeh, JungTsung Hsu, IEEE Photon. Technol.

Lett. 15(2003)649.
[15] Hisao Kikuta, Shunsuke Hino, Akira Maruyama, Akio Mizutani, J. Opt. Soc. Am. A 23(2006)1207.
[16] W. J. Choi, Q-Han Park, Dongho Kim, Heonsu Jeon, Cheolsoo Sone, Yongjo Park, J. Korean Phys. Soc. 49(2006)877.
[17] E. F. Schubert, Light Emitting Diodes, Cambridge University Press, Cambridge, 2003.
[18] P. B. Johnson, R. W. Christy, Phys. Rev. B. 6(1972)4370.
[19] H. K. Cho, J. Jang, J. H. Choi, J. Choi, J. Kim, J. S. Lee, B. Lee, Y. H. Choe, K. D. Lee, S. H. Kim, K. Lee, S. K. Kim, Y. H. Lee, Opt. Exp. 14(2006)8654.

Achromatic Generation of Radially Polarized Beams in Visible Range Using Segmented Subwavelength Metal Wire Gratings[*]

Abstract We propose an efficient method to achromatically transform a circularly polarized beam to a radially polarized beam in visible range using segmented subwavelength metal wire gratings. We present a theoretical analysis of the relationship between the polarization purity of transmitted beams and the number of segments. To verify our analysis, we fabricate a device composed of four quadrant sectors of subwavelength metal wire gratings, and measure the transformation properties of the device at visible wavelengths of 488nm, 532nm and 633nm, which show a good agreement with theoretical results and the broadband achromatic property of the generation method.

Radially polarized beams have attracted much attention due to their unique properties and many potential applications. And various active and passive methods have been developed to generate such beams, such as liquid-crystal polarization converters[1], space-variant subwavelength metallic gratings[2] and segmented retardation plates with properly oriented half-wave plates[3,4]. However most methods are operated at a certain wavelength in near-infrared or far-infrared range, the achromatic generation of radially polarized beams in visible range is rarely studied. The polarization converters[1] show an achromatic effect if tuned carefully for the whole visible range, but such devices suffer from temporal instability of the orientation of liquid-crystal molecules and can not be used for high powers. In the letter, we present a simple and efficient method to achromatically generate the nearly radially polarized beams in visible range using segmented subwavelength metal wire gratings(SMWGs).

Just as shown in Fig. 1(a), the device is composed of n segments of SMWG, and each one has radial orientation of the grating vector. The SMWG acts as a polarizing beam splitter, which reflects most of the beam polarized perpendicular to the grating vector(TE polarized beam) and transmits most of the beam polarized parallel to the grating vector(TM polarized beam)[5]. When passing a circularly polarized beam through the segmented structure, most TM polarized beam component transmits and the polarization distribution will be nearly radial. So with the method the maximum possible value of conversion efficiency is 50%, this is 50% of the incident total power can be transformed to radial polarization[4]. The extent of radial approximation increases with the number of segments, so does the radial polarization purity of transmitted beams. The radial polarization purity describes the overall degree of radial polarization of transmitted beams, which is defined as the ratio of the power with pure radial polarization to the total

---

[*] Copartner: Zhehai Zhou, Qiaofeng Tan, Qunqing Li. Reprinted from *Optics Letters*, 2009, 34(21):3361-3363.

beam power[6]. Fig. 1(b) shows the theoretical analysis of the relationship between the polarization purity of transmitted beams and the number of segments n for different extinction ratio (ER) of transmitted TM polarized beams. If n is 4, the polarization purity of transmitted beams approaches about 90% even for low ER 15dB, and exceeds 95% for n larger than 16. The analysis indicates the segmented device can efficiently transform a circularly polarized beam to a radially polarized beam without large number of segments, which lowers the fabrication difficulty and the cost as well. Moreover, the polarization purity is not susceptible to the ER, which means a broadband achromatic property of the generation method. During the research, we use the software KAPPA, a computer program for modeling one-dimensionally periodic diffraction gratings based on rigorous coupled-wave analysis[7,8] and obtain the transmitted efficiencies of TE and TM polarized beams, then calculate the ER of transmitted TM polarized beams as ER = 10log τ_{TM}/τ_{TE}, where τ_{TM} and τ_{TE} are transmission efficiencies of TM and TE polarized beams respectively.

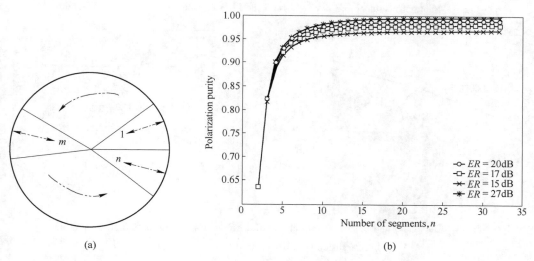

Fig. 1 Schematic device structure composed of segmented SMWGs(a), where the arrows schematically show directions of the grating vector of SMWG, and n is the total number of segments. Calculated results of the relationship between polarization purity of transmitted beams and the number of segments(b)

To verify our analysis, we fabricate a device with four segments, and Figs. 2(a), (b) respectively show the photo of manufactured device and the SEM of a cross section of the grating stripes. The device is fabricated on a quartz substrate, and the aluminum grating stripes are realized using e-beam lithography and RIE etching technology, with the period of 200nm, the duty cycle of 0.5 and the etched depth of 100nm. Fig. 3 shows the calculated and measured ER and transmission efficiency(η) of transmitted TM polarized beams of SMWGs at visible wavelengths of 488nm, 532nm and 633nm. The measured ER results are close to the calculated values, and they are all larger than 16dB, resulting in a high polarization purity of transmitted beams. While the measured η values are about 0.45, which deviate from the calculated results of about 0.66, and the deviation mainly attributes to the fabrication flaws and no antireflection coating on the backside of the wafer.

 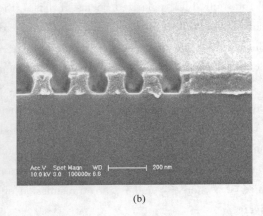

Fig. 2 Photo of manufactured device(a) and the SEM of a cross section of SMWGs(b)

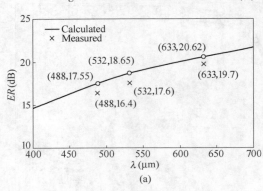 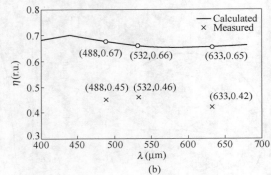

Fig. 3 Calculated and measured extinction ratio(ER)(a) and transmission efficiency(η)(b) of transmitted TM polarized beams of SMWGs

To test the polarization purity of the transmitted beam, the circularly polarized beams at wavelengths of 488nm, 532nm and 633nm were separately introduced into the transformation system similar to that of Ref. [3], and made four measurements of the transmitted intensity at each wavelength. Figs. 4(a)-(d) show the measured intensity distributions at 633nm after passing the beam through an analyzing polarizer in vertical(a), horizontal(b), 45°(c), and through a quarter-wave plate oriented horizontally and a polarizer oriented at 45°(d). Then the four measurements were used to calculate the Stokes parameters S_0, S_1, S_2 and S_3[6] at each point of the beam, from which the local ellipticity and azimuthal angle were obtained as $\chi = \frac{1}{2}\arcsin\left(\frac{S_3}{S_0}\right)$ and $\psi = \frac{1}{2}\arctan\left(\frac{S_2}{S_1}\right)$. Figs. 5(a), (b) show the distribution of the absolute value of the ellipticity angle χ and the deviation of the azimuthal angle ψ from the direction of local grating vector in each transverse point, measured in degrees. The deviations of χ and ψ from theoretical results are small enough in whole beam cross-section. We also obtained the similar results at 488nm and 532nm based on the same processing. The polarization purities at wavelengths of 488nm, 532nm and 633nm, calculated from the measured data, are 84.5%, 87.8% and 89.3% respectively, which agree quite well with the corresponding theoretical results of 89.1%, 90.5% and 91.2%, indicating a good performance of the segmented device and the broadband achromatic property in visible range.

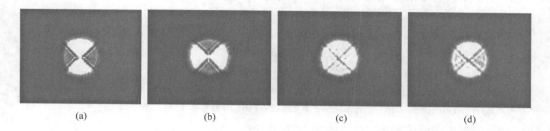

Fig. 4 Measured intensity distributions with spatial filter after passing the beam through an analyzing polarizer in vertical(a), horizontal(b), 45°(c), and through a quarter-wave plate oriented horizontally and a polarizer oriented at 45°(d)

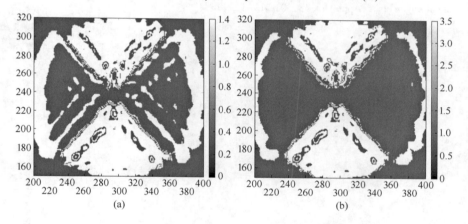

Fig. 5 Measured local polarization results at 633nm, the distribution of(a) the absolute value of the ellipticity angle χ and(b) the deviation of the azimuthal angle ψ from the direction of local grating vector in each transverse point, measured in degrees

In conclusion, we demonstrated achromatic generation of nearly radially polarized beams using segmented SMWGs and presented the experiment measurement for a device with four segments, which showed a good performance and a broadband achromatic property in visible range. The proposed method has relatively low conversion efficiency but it can be improved by applying high-transmission-efficiency SMWG and more segments. The achromatic generation method may find many applications in some visible multi-wavelength optical systems due to its simple structure, low fabrication cost as well as high polarization purity of transformed beams.

The research was supported by National Basic Research Program of China under the Grants No. 2007CB935303 and 2007CB935301. The authors are grateful to Prof. Lifeng Li for providing KAPPA.

References

[1] Stalder M and Schadt M, "Linearly polarized light with axial symmetry generated by liquid crystal polarization converters", Opt. Lett. 21, 1948-1950(1996).

[2] Ze'ev Bomzon, Vladimir Kleiner, Erez Hasman, "Formation of radially and azimuthally polarized light using space-variant subwavelength metal stripe gratings", Appl. Phys. Lett. 79, 1587-1589(2001).

[3] G. Machavariani, Y. Lumer, I. Moshe, A. Meir, et al., "Efficient generation of radially and azimuthally polarized beams", Opt. Lett. 32, 1468-1470(2007).

[4] S. Quabis, R. Dorn, G. Leuchs, "Generation of a radially polarized doughnut mode of high quality", Appl. Phys. B 81, 597-600(2005).

[5] D. Zhang, P. Wang, X. Jiao, C. Ming, G. Yuan, et al., "Polarization properties of subwavelength metallic gratings in visible light band", Appl. Phys. B 85, 139-146(2006).

[6] M. Born and E. Wolf, *Principles of Optics* (Pergamon Press, 1999).

[7] M. G. Moharam and T. K. Gaylord, "Rigorous coupled-wave analysis of metallic surface-relief gratings", J. Opt. Soc. Am. A 3, 1780-1787(1986).

[8] L. Li, J. Chandezon, G. Granet, J. -P. Plumey, "Rigorous and efficient grating-analysis method made easy for optical engineers", Appl. Opt. 38, 304-313(1999).

Security Enhanced Optical Encryption System by Random Phase Key and Permutation Key*

Abstract Conventional double random phase encoding (DRPE) encrypts plaintext to white noise-like ciphertext which may attract attention of eavesdroppers, and recent research reported that DRPE is vulnerable to various attacks. Here we propose a security enhanced optical encryption system that can hide the existence of secret information by watermarking. The plaintext is encrypted using iterative fractional Fourier transform with random phase key, and ciphertext is randomly permuted with permutation key before watermarking. Cryptanalysis shows that linearity of the security system has been broken and the permutation key prevent the attacker from accessing the ciphertext in various attacks. A series of simulations have shown the effectiveness of this system and the security strength is enhanced for invisibility, nonlinearity and resistance against attacks.

1 Introduction

With the rapid development of networked multimedia techniques, the information security is facing more and more challenges nowadays. Optical systems are of growing interests for image encryption because of their distinct advantages of processing two-dimensional complex data in parallel and at high speed. Since the pioneer work of optical encryption based on the concept of double random phase encoding (DRPE) proposed by Réfrégier and Javidi[1], the research area in optical cryptography has been enlightened and various optical encryption schemes have been proposed during the past decades[2-8]. These methods convert the plaintext to stationary white noise by use of two random phase keys. However, the characteristics of ciphertext may expose the existence of secret information in the data transmission and attract attention of eavesdroppers. The security of optical cryptographic system has become a great concern in recent years[9-16]. According to cryptanalysis, any optical security system could not be claimed secure unless it was able to endure various attacks. Chosen-ciphertext attack[9], known-plaintext attack[10-12] and chosen-plaintext attack[13] were proposed to explore the security strength of DRPE. Those attacks have demonstrated that the security flaws originate from the linearity of the DRPE[14-16].

To conceal the existence of secret information, various optical information hiding[17-19] and watermarking[20-22] methods have been proposed. However these methods conceal the secret information without considering the security strength. Therefore the embedded secret information is vulnerable at various attacks. Here we propose an optical security system combines the high

---

* Copartner: Mingzhao He, Qiaofeng Tan, Liangcai Cao, Qingsheng He. Reprinted from *Optics Express*, 2009, 17(25): 22462-22473.

security strength of encryption system and invisibility of watermarking technique. The proposed method is based on cascaded fractional Fourier transform system. The plaintext image is first encoded to ciphertext by iterative fractional Fourier transform with use of random phase key. Then the ciphertext is random permuted and embedded in an overt image. The watermarked image for transmission is not white noise-like ciphertext as in DRPE, but similar to overt image with subtle changes that eavesdroppers may not notice. After authorized user receives the watermarked image, ciphertext can be extracted and then restored by inverse permutation. Decryption can be simply performed by fractional Fourier transform with the correct phase key. Cryptanalysis indicates that system linearity has been broken. Furthermore, the security system has good resistance against various attacks because permutation key prevent attacker from accessing the ciphertext directly. A series of numerical simulations have verified the effectiveness of the method and its resistance against attacks.

This paper is organized as follows: In Section 2 the security system is proposed and the encryption and watermarking principle is presented in detail. In Section 3 numerical simulations have verified the effectiveness of the method and phase key quantization, and robustness of watermarked image has also been discussed. In Section 4, we have analyzed the system linearity and demonstrate the resistances against known-plaintext attack and chosen-plaintext attack respectively. Conclusions are presented in Section 5.

2 Security scheme

The proposed optical encryption and hiding system consists of two fractional Fourier transform (FrFT) systems with fractional order P_1 and P_2 cascaded together as illustrated in Fig. 1. We consider three planes of this optical system, referred to as input plane, encryption plane, and output plane. Two Fourier transform lenses are sandwiched in three planes. The input and the encryption planes are two planes located symmetrically with respect to lens 1. Similarly, the encryption plane and the output plane located symmetrically with respect to lens 2. The plaintext for encryption is placed on input plane, and a random phase mask is placed on encryption plane as secret key. The phase distributions on the input plane and output plane can be regarded as two virtual phase masks(VPMs). The VPM in the output plane is defined as ciphertext. The encryption is carried out using iterative FrFT with plaintext and overt image as magnitude constraint conditions. After detailed encryption process that given below, the ciphertext is obtained and then embedded in the overt image as watermark after permutation. The watermarked image has no hint of white noise. Eavesdropper may not notice the secret image in the watermarked image. Since the permutation operation interchange pixel positions of the ciphertext, it is impossible to extract the ciphertext without knowledge of permutation key even if the eavesdropper perceives the existence of secret image. In decryption, the authorized user can extract and restore the ciphertext from watermarked image with inverse permutation operation. The ciphertext is first multiplied with overt image and then transformed by lens 2. Multiplied with the complex conjugation of phase key at encryption plane, the complex amplitude is transformed to input plane by lens 1. Considering the error in the iterative algorithm, the amplitude at input plane can be obtained as an approximation of the plaintext. The decryption can be expressed mathematically as

$$p(x,y) = |F^{-P_1}\{F^{-P_2}[o(x,y) \cdot e^{ic(x,y)}] \cdot k^*(u,v)\}| \tag{1}$$

where $p(x,y)$, $c(x,y)$ and $o(x,y)$ denote plaintext, ciphertext and overt image respectively. $F^{-P_1}[\cdot]$ and $F^{-P_2}[\cdot]$ are the inverse FrFT operators with the fractional order $-P_1$ and $-P_2$ respectively. $k^*(u,v)$ is the complex conjugation of phase key at encryption plane.

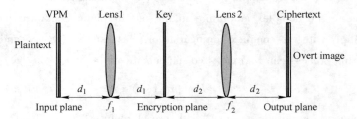

Fig. 1 System structure of optical encryption and hiding system. VPM, virtual phase mask

The objective of encryption algorithm is to calculate phase distribution in the output plane with the known amplitudes of input and output planes which can be referred to a phase retrieval problem. Gerchberg-Saxton(G-S) algorithm[23] is an efficient and extensively used phase retrieval algorithm in Fourier transform domain. Y. Shi et al. have extended the phase retrieval algorithm to Fresnel domain for optical image hiding[18] and multiple-image hiding[19]. In this paper, we utilize iterative phase retrieval in fractional Fourier transform domain for encryption. As the flowchart shown in Fig. 2, the algorithm can be described as follows:

Step 1. Initialize $\theta(x,y)$ and phase key $k(u,v)$. Here $\exp[i\theta(x,y)]$ is VPM of the input plane and phase key $k(u,v)$ can be expressed as $\exp[i\varphi(u,v)]$. We assign $\theta(x,y)$ and $\varphi(u,$

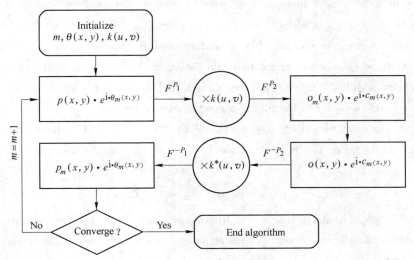

Fig. 2 Flow chart of iterative phase retrieval algorithm for encryption

v) with random phase uniformly distributed in $[0, 2\pi)$. The complex amplitude on the input plane is $p(x, y) \cdot \exp[i\theta(x, y)]$, where $p(x, y)$ is input plaintext image to be encrypted.

Step 2. Transform complex amplitude in input plane to the encryption plane by FrFT with fractional order P_1, then multiply with the phase key $k(u, v)$, and then transform to output plane by FrFT with fractional order P_2. This process can be expressed mathematically as

$$o_m(x,y) \cdot e^{ic_m(x,y)} = F^{P_2}\{F^{P_1}[p(x,y) \cdot e^{i\theta_m(x,y)}] \cdot k(u,v)\} \quad (2)$$

Step 3. Impose magnitude constraint in the output plane by replacing the amplitude $o_m(x, y)$ with overt image $o(x, y)$ while keeping the phase $c_m(x, y)$ unchanged.

Step 4. Transform the complex amplitude in the output plane backward to the encryption plane. Then multiply with the complex conjugation of phase key $k^*(u, v)$, and then transform to input plane. This process can be expressed mathematically as

$$p_m(x,y) \cdot e^{i\theta_m(x,y)} = F^{-P_1}\{F^{-P_2}[o(x,y) \cdot e^{ic_m(x,y)}] \cdot k^*(u,v)\} \quad (3)$$

Step 5. Evaluate the amplitude $p_m(x, y)$ with preset plaintext $p_0(x, y)$. One common used criterion is mean square error(MSE) which can be defined as

$$MSE = \frac{1}{M \times N} \sum_{i=1}^{M} \sum_{j=1}^{N} [p_m(i,j) - p(i,j)]^2 \quad (4)$$

where $p_m(i, j)$ is the amplitude at input plane after mth iteration, M and N are the numbers of pixels in x and y directions. The iteration process is carried out until the convergence condition is satisfied or the maximum loop number is reached.

Step 6. If the convergence condition is not satisfied, replace the amplitude of input plane using $p(x, y)$, while retaining the phase component unchanged, and then return to Step 2.

After iteration, $c_m(x, y)$ is extracted as the ciphertext. Then each pixel of ciphertext is permuted according to a permutation key which controls the random interchange of pixels. Permutation operation is widely used in cryptography[24] and optical security system[7,25]. The permuted ciphertext is generally like a random noise distribution, and the original ciphertext can be restored by inverse permutation with the same permutation key. The permutation operation provides additional freedom for encryption, and furthermore the permutation key enhances the security level by preventing phase retrieval in attacks.

To hide the existence of ciphertext, the ciphertext is then embedded in the overt image with a proper weighting factor α. The permutation and spatial domain watermarking can be expressed as

$$o'(x,y) = o(x,y) + \alpha \cdot P[c_m(x,y)] \quad (5)$$

where $o'(x, y)$ is watermarked image for transmission and $P[\ ]$ denotes permutation operation. The watermarked image is similar to overt image except for some subtle changes.

When the authorized user received the watermarked image, ciphertext can be extracted and restored by inverse permutation. Then plaintext can be easily decrypted using Eq. (1).

3 Numerical simulations

In order to verify the feasibility of proposed optical encryption and hiding method, a series of computer simulations have been carried out. In all simulations the images are 256×256 pixels in size, and 256 gray levels for grayscale images. Considering the cascaded FrFT system shown in Fig. 1, the parameters are $d_1 = 80$mm, $d_2 = 90$mm, and $f_1 = 100$mm, $f_2 = 120$mm, and the corresponding fractional orders are $P_1 = 0.8718$, and $P_2 = 0.7836^{[5]}$.

In the simulation, we encrypt a binary image shown in Fig. 3(a) to demonstrate the information encryption and hiding ability of the security scheme. The overt image shown in Fig. 3(b) is used as magnitude constraint in the output plane and also can be used as host image in the watermarking. Without lose of generality, we use random phase distributions, which are uniform distribution between $[0, 2\pi)$, as initial values of $\theta(x, y)$ and $\varphi(u, v)$. After each loop, the amplitude in the input plane is measured by MSE criterion. From the MSE evolution curve shown in Fig. 4, we can find that MSE is converging fast at first, and then slow down after 20 loops. So we can set maximum loop number as 20 to save computation time. The ciphertext shown in Fig. 3(c) is obtained by extracting the phase component at output plane. Then the ciphertext is permuted according to the permutation key. The permuted ciphertext shown in Fig. 3(d) looks like random white noise. Next, we utilize spatial domain watermarking to hide the ciphertext with weighting factor 0.02. The value range of weighting factor has been studied intensively[20,22]. In Ref. [20], the cases of constant-level weighting and image-dependent weighting are discussed, and Ref. [22] exhibits the effect of different weighting factors on retrieved hidden image. Here we take a small value as weighting factor to hide the ciphertext invisibly. From the watermarked image shown in Fig. 3(e), it is impossible to perceive the secret information with

Fig. 3 (a) plaintext; (b) overt image; (c) ciphertext; (d) permuted ciphertext; (e) watermarked image; (f) decryption result with correct keys

human vision. Even eavesdroppers find the hidden information, it is hard to restore the ciphertext without the knowledge of permutation key. The authorized user can restore the ciphertext with permutation key and decrypted it with phase key. The decrypted image shown in Fig. 3(f) contains some noises because after encryption the amplitude of input plane $p_m(x, y)$ is an approximation of plaintext $p(x, y)$. The *MSE* for decrypted image is 0.014 as given in Table 1. Considering the maximum loop number is 20, a smaller *MSE* may be achieved with a large loop number. Besides, noises in Fig. 3(f) can be suppressed by filtering.

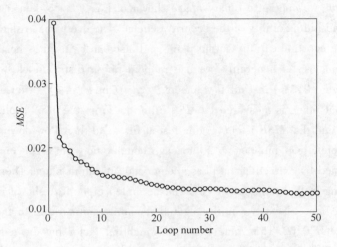

Fig. 4　Convergence curves of *MSE* in encryption

In above simulation the plaintext and overt images have 256 gray levels. In the phase key $k(u, v)$ which can be expressed as $\exp[i\varphi(u, v)]$, $\varphi(u, v)$ is float point of 32 bit for calculating ciphertext at maximum accuracy. Taking the state-of-the-art fabricating technology into account, it is reasonable to represent the phase key in discrete form. By uniform quantization[26], we quantize phase key to 2 level, 4 level and 8 level phase keys respectively. From the decrypted results (using 2 level, 4 level and 8 level phase keys) shown in Figs. 5(a)-(c), we notice that the noises in the decrypted images decrease with more quantization levels, and 4 level quantization may obtain a recognizable decryption result for binary plaintext images.

Fig. 5　(a)-(c) are decryption results using 2, 4 and 8 quantization level phase keys, respectively

Quantization noise in the decryption results can be evaluated using *MSE* criterion. The *MSE*s of decrypted images after 20th loop are calculated using different quantized phase keys. From the *MSE* evolution curves shown in Fig. 6, it can be concluded that *MSE*s are in the same level for a certain phase quantization level. The reason for the same *MSE* level is the same quantization error level of quantized phase key. In a desire to recover plaintext image with high fidelity, more quantization level should be adopted.

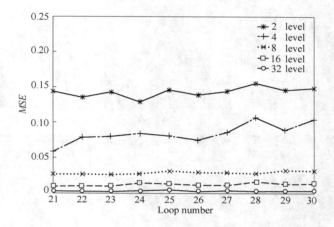

Fig. 6 *MSE* evolution curves of 2, 4, 8, 16 and 32 quantization level phase keys

The watermarked image can be sent to the receiver through public communication channel. In data transmission, JPEG compression[27,28] is widely used to reduce the image size. We have investigated the effect of JPEG compression on the watermarked image as well as the noise and occlusion robustness of this security system. The watermarked image shown in Fig. 3(e) is compressed in Matlab R2008b with 8 bit-depth and quality factor 80[29]. In the JPEG compression, some original image information is lost and cannot be restored. From the compressed image shown in Fig. 7(a), it has little perceptible difference with Fig. 3(e) because the eliminated high frequency details are beyond human visual sense. However, as shown in Fig. 7(d), the decryption result using the compressed image is degenerated seriously.

As we can obverse from Table 1, the *MSE* value of Fig. 7(d) is much larger than the *MSE* of Fig. 3(f). The watermarked image with Gaussian noise of mean 0 and variance 0.01 is shown in Fig. 7(b), and corresponding decryption result is shown in Fig. 7(e). The occlusion is applied by cutting of 25% of watermarked image as shown in Fig. 7(c). Corresponding decryption result exhibited in Fig. 7(f) reveals that occlusion has less effect on decryption because only 25% of ciphertext is lost. As can be concluded from simulation results, this security technique is vulnerable to JPEG compression and noise attack. The main reason is that the ciphertext embedded in high frequency components of watermarked image which is compressed or disturbed in the JPEG compression and noise attack. To achieve better JPEG compression and noise resistance, we can employ some watermarking techniques with high frequency attack robustness. We will investigate it in our future work.

Table 1 The *MSE* values of decryption results

Decryption results	Fig. 3(f)	Fig. 7(d)	Fig. 7(e)	Fig. 7(f)
Attack type	No attack	JPEG	noise	occlusion
MSE values	0.0141	0.17	0.18	0.052

Fig. 7 Watermarked images attacks and corresponding decryption results
(a) JPEG compression; (b) noise; (c) occlusion; (d), (e) and
(f) are decryption results using (a), (b) and (c)

4 Cryptanalysis

Cracking a security system means finding the value of keys with some knowledge about the input and corresponding output of the system. According to the Kerckhoffs' principle[24,30], the security system is publicly-known except the keys. As illustrated in the pervious work[1-6], the benefit of two dimensional phase key is that the key space is extraordinarily large so that brute force attack is computationally intractable. In this section, we first analysis the security system linearity, and then testify the resistance against known-plaintext attack, chosen-ciphertext attack and chosen-plaintext attack.

4.1 System linearity analysis

The optical security system based on DRPE and its extensions have been found vulnerable under various attacks[9-16]. The flaws originate from the linearity of the encryption algorithm. For a linear system, the output of a weighted sum of two (or more) functions is simply the identically weighted sum of their individual outputs[31]. This relation can be expressed mathematically as

$$S\{\alpha g + \beta h\} = \alpha S\{g\} + \beta S\{h\} \tag{6}$$

As the permutation is linear operation obviously, we discuss the system linearity without considering permutation in this subsection. In the encryption system, the linearity is broken for two reasons. First, the initial value of VPM in the input plane is randomly generated before iteration and updated in the iteration. Second, the substitutions in the input plane and output plane are nonlinear operations obviously. We have examined the relation of input and output using Eq. (2) after iteration. Since $p_m(x,y)$ and $o_m(x,y)$ are reaching $p(x,y)$ and $o(x,y)$ respectively, Eq. (2) can be rewritten as

$$o(x,y)e^{ic_m(x,y)} = F^{P_2}\{F^{P_1}[p(x,y) \cdot e^{i\theta_m(x,y)}] \cdot k(u,v)\} \tag{7}$$

Then, the ciphertext can be expressed as function of plaintext as

$$c_m(x,y) = i\ln[o(x,y)] - i\ln\{F^{P_2}\{F^{P_1}[p(x,y) \cdot e^{i\theta_m(x,y)}] \cdot k(u,v)\}\} \tag{8}$$

Obviously, Eq. (8) violates this definition because $\ln\{g+h\}$ may not always equal to $\ln\{g\}$ + $\ln\{h\}$, and besides there is a phase variable $e^{i\theta_m(x,y)}$ that taking random value in every iteration. And similarly, it is not possible to solve plaintext as a linear function of ciphertext from Eq. (1). Therefore the linearity between input plaintext and output ciphertext is broken in the security system.

The system nonlinearity can be further exhibited by the simulations. To show evident results, the decryption result of sum of two ciphertext images is compared with sum of their individual decryption results. First, we encrypted two plaintexts using the same phase key, and the corresponding ciphertext images are shown in Figs. 8(a) and (b) respectively. Their individual decryption results are shown in Figs. 8(c) and (d). Fig. 8(e) shows retrieved plaintext with the sum of two ciphertext images using the same phase key. Obviously, it is completely different from the sum of individual decryption results shown in Fig. 8(f). The simulation results confirm that the system linearity has been broken by random VPM and substitution operations.

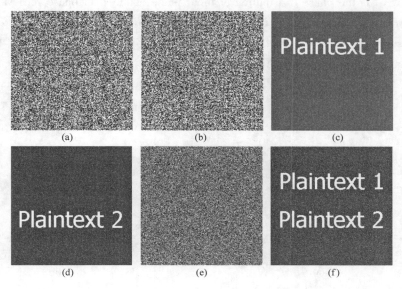

Fig. 8 (a), (b) ciphertext 1 and 2; (c), (d) individual decryption results 1 and 2; (e) decryption result with sum of linear combination of two ciphertext; (f) sum of individual decryption results 1 and 2

4.2 Known-plaintext attack

The optical security system based on DRPE and its extensions have been found vulnerable under various attacks[9-15]. Chosen-ciphertext attack[9], known-plaintext attack[10,11] and chosen-plaintext attack[13] have been proposed to explore the security strength of optical system based on DRPE. The security system based on Projection-Onto-Constraint-Sets has also been cracked by Peng et al[12]. using known-plaintext attack. The attack principle is to convert the known-plaintext attack to phase retrieval problem with some prior knowledge. To resist this type of attack, we can permute the pixel position of ciphertext randomly before watermarking. The two dimensional permutation can be realized by row permutation and column permutation respectively. The key space of permutation key is $256! \times 256! \approx 7 \times 10^{1013}$ which is large enough to exhaust attackers. Therefore it is impossible to retrieve correct plaintext without the permutation key.

Similarly, it is also impossible to implement known-plaintext attack without the permutation key. Suppose an attacker intercepts the watermarked image $o'(x,y)$ as shown in Fig. 3(e) and extracts the secret image $P[c_m(x,y)]$ as shown in Fig. 3(d). Because the attacker does not have the permutation key or even does not know the application of permutation, he uses $P[c_m(x,y)]$ to crack the phase key using KPA method in Ref. [12]. The KPA includes three steps: calculate the amplitude in encryption plane; obtain phase distribution of input plane using phase retrieval; crack the phase key with the knowledge of complex amplitudes of input plane and output plane. The cracked phase key is given by

$$k_{\text{crack}}(u,v) = F^{P_1}[p(x,y)e^{i\theta_{\text{trial}}(x,y)}]/F^{-P_2}[o(x,y)e^{ic_m(x,y)}] \quad (9)$$

where $\theta_{\text{trial}}(x,y)$ is the phase distribution of input plane obtained in phase retrieval. The *MSE* evolution curve of amplitude in the input plane is shown in Fig. 9(a). The phase key evolution can be measured by correlation coefficient (CC)[7,8] which shown the difference of the cracked phase key and original phase key. From the *MSE* curve shown in Fig. 9(a), we can find that

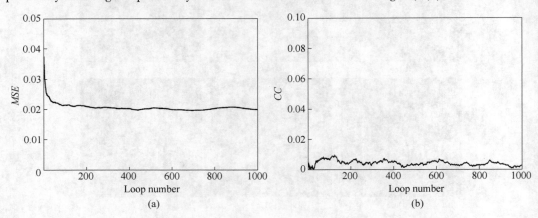

Fig. 9 Amplitude and phase evolutions in KPA
(a) *MSE* curve; (b) *CC* curve

the MSE curve is converging, while the *CC* shown in Fig. 9(b) won't reach 1 as the attacker expected. When *MSE* reach a preset threshold value, the attacker gets the cracked phase key. However, the cracked key can be used for this plaintext and ciphertext pair only. Fig. 10 (a) shows the decryption result for the original ciphertext and Fig. 10(b) shows another plaintext for phase key test. This plaintext is encrypted with $k(u,v)$ and the corresponding ciphertext is attacked using $k_{crack}(u,v)$. From the attack result shown in Fig. 10(c), obviously the attack result with cracked phase key can not reveal any information of plaintext. Therefore the introduction of permutation key enhanced the resistance against known-plaintext attack.

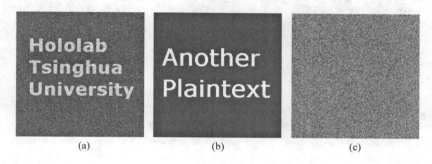

Fig. 10　Simulation results for known-plaintext attack
(a) attack result for original ciphertext; (b) another plaintext for phase key test; (c) attack result with cracked phase key

4.3　Chosen-plaintext attack

The optical security system based on DRPE has been cracked by chosen-plaintext attack[13] and chosen-ciphertext attack[9]. As we know, chosen-plaintext attack and chosen-ciphertext attack on iterative encryption system such have not been reported. Here we investigate the resistance of proposed security system against this chosen-plaintext attack.

In chosen-plaintext attack, we assume the attacker has the ability to trick a legitimate user of the system into encrypting particular images. As reported in Ref. [14], a simple and yet effective attack can be mounted by obtaining the ciphertext corresponding to a Dirac delta function as plaintext. The corresponding ciphertext is $c_\delta(x,y)$, and the complex amplitude of encryption plane after phase key $\beta(u,v)$ is known as $\beta_\delta(u,v) = F^{-P_2}[o(x,y)e^{ic_\delta(x,y)}]$. From Eq. (2), we can deduce that

$$\beta_\delta(u,v) = F^{P_1}[\delta(x,y) \cdot e^{i\theta_m(x,y)}] \cdot k(u,v) = e^{i\theta_m(0,0)} \cdot k(u,v) \qquad (10)$$

Then phase key can be expressed as

$$k(u,v) = e^{-i\theta_m(0,0)} \cdot \beta_\delta(u,v) = e^{-i\theta_m(0,0)} \cdot F^{-P_2}[o(x,y)e^{ic_\delta(x,y)}] \qquad (11)$$

where phase constant $e^{-i\theta_m(0,0)}$ can be ignored because it does not affect the decryption[13]. Using above cracked phase key, the ciphertext of Fig. 10(b) can be cracked successfully as shown in Fig. 11(a). As a conclusion, this security system is vulnerable to chosen-plaintext attack in the absence of permutation key because there is only one phase key. When permutation key is applied, the attacker can only obtain the permuted ciphertext $P[c_\delta(x,y)]$. Another phase key can be cracked as shown in Fig. 11(b), however it is completely different with correct phase key. The ciphertext of Fig. 10(b) is attacked using this phase key, but the decryption result shown in Fig. 11(c) reveals any information of original plaintext, Therefore, the resistance against chosen-ciphertext attack is greatly enhanced by permutation key.

Fig. 11 Simulation results for chosen-plaintext attack
(a) attack result with cracked phase key in the absence of permutation; (b) another phase key obtained when permutation is applied; (c) attack result using Fig. 8(b)

In chosen-ciphertext attack, the attacker can decrypt any ciphertext and obtain corresponding plaintext. Thus the phase of output plane can be arbitrarily designed, and then $\beta(u,v)$ is arbitrary. From Eq. (1), we can deduce that

$$p(x,y) = |F^{-P_1}\{\beta(u,v) \cdot k^*(u,v)\}| \tag{12}$$

Then phase retrieval algorithm can be utilized to solve the phase key from above equation. The security analysis is similar to the case of known-plaintext attack and omitted here.

From above analysis, we can conclude that permutation key play an important role in the security system. The resistances against known-plaintext attack and chosen-plaintext attack are greatly enhanced by denying direct accessing to the ciphertext.

5 Conclusions

In this paper, we have proposed a security enhanced optical encryption system by random phase key and permutation key. The encryption is realized by phase retrieval in fractional Fourier transform domain with the random phase key, and after random permutation the ciphertext is hidden in the overt image by watermarking. The ciphertext is imperceptible in watermarked image for transmission. The security strength of this system is then enhanced for the attacker may not perceive the existence of secret image. Even if the watermarked image is intercepted, the attacker cannot extract the ciphertext without the permutation key. Numerical simulations have verified the performance of the encryption system, and the robustness of watermarked image against compression, noise and occlusion have been discussed. Cryptanalysis indicates that the system linearity has been broken for random VPM and substitution operations. In known-plaintext attack and chosen-plaintext attack, the permutation key prevents the attacker from accessing the ciphertext directly. Hence the security strength is significantly enhanced for the introduction of random permutation key. In conclusion, security enhanced encryption with invisibility, nonlinearity and attack resistance has been achieved with the use of random phase key and permutation key.

Acknowledgements

This work is supported by National Basic Research Program of China(2009CB724007), National Natural Science Foundation of China (60807005), and 863 High Technology (2009AA01Z112).

References

[1] Ph. Réfrégier and B. Javidi, "Optical image encryption based on input plane and Fourier plane random encoding", Opt. Lett. 20, 767-769(1995).

[2] T. Nomura, B. Javidi, "Optical encryption using a joint transform correlator architecture", Opt. Eng. 39, 2031-2035(2000).

[3] T. Nomura, B. Javidi, "Securing information by use of digital holography", Opt. Lett. 25, 28-30(2000).

[4] P. C. Mogensen and J. Glückstad, "Phase-only optical encryption", Opt. Lett. 25, 566-568(2000).

[5] G. Unnikrishnan, J. Joseph, and K. Singh, "Optical encryption by double-random phase encoding in the fractional Fourier domain", Opt. Lett. 25, 887-889(2000).

[6] G. Situ and J. Zhang, "Double random-phase encoding in the Fresnel domain", Opt. Lett. 29, 1584-1586 (2004).

[7] X. F. Meng, L. Z. Cai, X. L. Yang, X. X. Shen, and G. Y. Dong, "Information security system by iterative multiple-phase retrieval and pixel random permutation", Appl. Opt. 45, 3289-3297(2006).

[8] X. C. Cheng, L. Z. Cai, Y. R. Wang, X. F. Meng, H. Zhang, X. F. Xu, X. X. Shen, and G. Y. Dong, "Security enhancement of double-random phase encryption by amplitude modulation", Opt. Lett. 33, 1575-1577 (2008).

[9] A. Carnicer, M. Montes-Usategui, S. Arcos, and I. Juvells, "Vulnerability to chosen-cyphertext attacks of optical encryption schemes based on double random phase keys", Opt. Lett. 30, 1644-1646(2005).

[10] U. Gopinathan, D. S. Monaghan, T. J. Naughton, and J. T. Sheridan, "A known-plaintext heuristic attack on the Fourier plane encryption algorithm", Opt. Express 14, 3181-3186(2006). http://www.opticsinfobase.org/abstract.cfm?URI=oe-14-8-3181.

[11] X. Peng, P. Zhang, H. Wei, and B. Yu, "Known-plaintext attack on optical encryption based on double random phase keys", Opt. Lett. 31, 1044-1046(2006).

[12] H. Wei, X. Peng, "Known-Plaintext Attack on Optical Cryptosystem Based on Projection-Onto-Constraint-Sets Algorithm and a 4f Correlator", Acta Optica Sinica, 28, 429-434(2008).

[13] X. Peng, H. Wei, and P. Zhang, "Chosen-plaintext attack on lensless double-random phase encoding in the Fresnel domain", Opt. Lett. 31, 3261-3263(2006).

[14] Y. Frauel, A. Castro, T. J. Naughton, and B. Javidi, "Resistance of the double random phase encryption against various attacks", Opt. Express 15, 10253-10265(2007). http://www.opticsinfobase.org/oe/abstract.cfm?URI=oe-15-16-10253.

[15] T. J. Naughton, B. M. Hennelly, and T. Dowling, "Introducing secure modes of operation for optical en-

cryption", J. Opt. Soc. Am. A 25,2608-2617(2008).
[16] D. S. Monaghan, U. Gopinathan, G. Situ, T. J. Naughton, and J. T. Sheridan, "Statistical investigation of the double random phase encoding technique", J. Opt. Soc. Am. A 26,2033-2042(2009).
[17] S. Kishk, B. Javidi, "Information hiding technique with double random phase encoding", Appl. Opt. 41, 5462-5470(2002).
[18] Y. Shi, G. Situ, J. Zhang, "Optical image hiding in the Fresnel domain", J. Opt. A:Pure Appl. Opt. 8,569-577(2006).
[19] Y. Shi, G. Situ, J. Zhang, "Multiple-image hiding in the Fresnel domain", Opt. Lett. 32,1914-1916(2006).
[20] N. Takai, Y. Mifune, "Digital watermarking by a holographic techniuqe", Appl. Opt. 41,865-873(2002).
[21] S. Kishk, B. Javidi, "Watermarking of three-dimensional objects by digital holography", Opt. Lett. 28, 167-169(2003).
[22] H. Zhang, L. Z. Cai, X. F. Meng, X. F. Xu, X. L. Yang, X. X. Shen, G. Y. Dong, " Image watermarking based on an iterative phase retrieval algorithm and sine-cosine modulation in the discrete-cosine-transform domain", Opt. Commun. 278,257-263(2007).
[23] J. R. Fienup, "Phase retrieval algorithms:a comparison", Appl. Opt. 21,2758-2769(1982).
[24] B. Schneier, *Applied Cryptography*, 2nd ed. (John Wiley & Sons, 1996).
[25] B. Hennelly and J. T. Sheridan, "Optical image encryption by random shifting in fractional Fourier domains", Opt. Lett. 28,269-271(2003).
[26] Joseph N. Mait, "Understanding diffractive optic design in the scalar domain", J. Opt. Soc. Am. A 12, 2145-2158(1995).
[27] G. K. Wallace, "The JPEG still picture compression standard", IEEE Transactions on Consumer Electronics,38,18-34(1992).
[28] R. C. Gonzalez and R. E. Woods, *Digital Image Processing*, 2nd ed. (Prentice Hall, 2002).
[29] http://www.mathworks.com.
[30] http://en.wikipedia.org/wiki/Kerckhoffs'_principle.
[31] J. W. Goodman, *Introduction to Fourier Optics*, 2nd ed. (McGraw-Hill, 1996).

Achromatic Phase Retarder Applied to MWIR & LWIR Dual-band[*]

Abstract The development of the dual-band IR imaging polarimetry creates the need for achromatic phase retarder used in dual-band. Dielectric grating with the period smaller than the illuminating wavelength presents a strong form-birefringence. With this feature, the combination of several subwavelength gratings can be used as achromatic phase retarders. We proposed a combination of 4 subwavelength structured gratings(SWGs) used as an achromatic quarter-wave plate(QWP) applied to MWIR & LWIR bandwidths. Design method using effective medium theory and optimization algorithms is described in detail. The simulation results led to the possibility of an dual-band achromatic QWP whose retardance deviates from 90° by $< \pm 0.75°$ with the fast axis unfixed and by $< \pm 1.35°$ with the fast axis fixed over MWIR(3-5μm)& LWIR(8-12μm) bandwidths.

1 Introduction

The development of dual-band focal plane arrays[1] in which each pixel consists of superimposed mid-wave and long-wave photo detectors enables the combination of MWIR and LWIR polarimetric imagers into a single unit. Due to the increased target information obtained from the dual-band sensor, MWIR & LWIR polarimetric imager may have better image contrast, longer detecting range and higher spatial target discrimination than the single-band ones. In view of this, some key dual-band polarization elements such as polarizers and phase retarders are in need. An IR polarizer[2] called metal wire grating whose period is much smaller than the illuminating wavelength can easily provide extinction ratio of 25dB over MWIR & LWIR dual-band. But to our knowledge, dual-band achromatic phase retarder has not yet been reported.

Standard phase retarders usually can be used only at a single wavelength, since the phase retardation strongly depends on the frequency. Achromatic phase retarders can be realized on the basis of several physical principles. Achromatic prism retarders[3], which are based on the phenomenon of phase shift at total reflection, are voluminous and often result in an output beam displaced from the input beam which may disturb the imaging process. Another type of achromatic phase retarder is the combination of several crystal waveplates[4] with different kinds of birefringent materials. The design process is similar to making an achromatic lens, but there are far fewer IR birefringent materials to choose from. Besides, the weak natural birefringence of crystals usually makes a thick combination of multi-order waveplates, which results in a relatively higher absorption of IR radiation. A single piece of dielectric subwavelength grating[5]

---

[*] Copartner: Guoguo Kang, Qiaofeng Tan, Xiaoling Wang. Reprinted from *Optics Express*, 2010, 18(2):1695-1703.

(SWG) with period comparable to the illuminating wavelength can also be designed as an achromatic retarder, due to its strong birefringence dispersion proportional to the wavelength over a certain spectra bandwidth. However, the grating of this kind designed in the resonance region has nonzero-order diffractions and is sensitive to the variance of incident angle. Further more, the birefringence of a single piece of grating monotonously changes with the wavelength[6]. Its birefringence dispersion curve can't keep the same proportionality factor over two different wavebands at the same time, indicating that a single piece of SWG is unable to achieve the dual-band achromatism.

In this paper, a combination of four SWGs designed as IR dual-band achromatic quarter-wave plate is proposed for the first time. The period of each grating is chosen small enough to exclude non-zero order diffractions. The depth and the orientation angle of each grating are carefully optimized to achieve the best achromatic performance. In order to reduce volume and increase IR transmittance, the grating structures are supposed to be fabricated on both sides of the substrate.

2 Subwavelength grating

Grating in quasi-static domain has a period much smaller than the illuminating wavelength ($d/\lambda \leqslant 1/10$). Its birefringence is analyzed by the zero-order approximation effective medium theory (EMT)[7]. A surface-relief grating of a rectangular-groove profile, as shown in Fig. 1, is equivalent to an anisotropic thin film and the effective refractive indices n_{TE} and n_{TM} can be written as

$$n_{TE}^{(0)} = [fn_2^2 + (1-f)n_1^2]^{1/2} \quad (1)$$

$$n_{TM}^{(0)} = [fn_2^{-2} + (1-f)n_1^{-2}]^{-\frac{1}{2}} \quad (2)$$

where superscript 0 indicates the zero-order approximation, f is the filling factor, n_1 and n_2 are the refractive indices composing the grating. Like crystalline waveplate, the modulation of the polarization state is achieved by use of the form-birefringent properties of the SWG: the TE and the TM components of the incoming light undergo different phase shifts, so the relative phase difference between TE and TM polarizations changes. Compared with IR crystalline materials, dielectric SWGs exhibit much stronger birefringence. The birefringence dispersion curves of Cadmium Sulfide (CdS), Cadmium Selenide (CdSe) and silicon SWG within IR range are plotted in Fig. 2 respectively. The birefringence of SWG is calculated by using Eq. (1) and Eq. (2) with $f = 0.5$, $n_1 = 1$ and $n_2 = n_{Si}(\lambda)$[8]. The value of form-birefringence provided by SWG ($\Delta \bar{n} \approx 1.16$) is two orders of magnitude larger than that of natural birefringence ($\Delta \bar{n} \approx 0.015$). Consequently, by controlling the relatively shallow grating depth, a single piece of

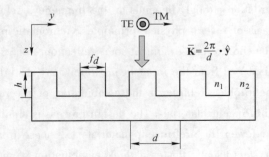

Fig. 1 Schematic diagram of a dielectric grating. It has a period of d, groove depth of h and a filling factor of f. The width of the ridge is fd in case $n_1 < n_2$

SWG with a small period-to-wavelength ratio could achieve any phase retardance with fast axis parallel to the grating vector $\overline{K}$ (shown in Fig. 1).

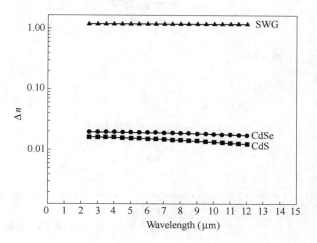

Fig. 2   Birefringence of CdS, CdSe and SWG

Grating in resonance domain has a period comparable to the illuminating wavelength. When the grating's period approaches the illuminating wavelength, the zero-order EMT becomes inaccurate. The second-order EMT, which includes a second-order correction of the finite ratio d/λ[9], has improved accuracy for $1/10 < d/\lambda < 3/2$. Although the second-order EMT is not a rigorous analysis for gratings in resonance domain, it leads to an approximate value of the phase of light waves that have passed through the grating. We can use it to estimate the dependence of birefringence on wavelength. The second-order EMT has the form,

$$n_{\text{TE}}^{(2)} = \left[(n_{\text{TE}}^{(0)})^2 + \frac{1}{3}\left(\frac{d}{\lambda}\right)^2 \pi^2 f^2 (1-f)^2 (n_2^2 - n_1^2)^2 \right]^{\frac{1}{2}} \quad (3)$$

$$n_{\text{TM}}^{(2)} = \left[(n_{\text{TM}}^{(0)})^2 + \frac{1}{3}\left(\frac{d}{\lambda}\right)^2 \pi^2 f^2 (1-f)^2 (n_2^{-2} - n_1^{-2})^2 \cdot (n_{\text{TM}}^{(0)})^6 (n_{\text{TE}}^{(0)})^2 \right]^{\frac{1}{2}} \quad (4)$$

where superscript 2 indicates the second-order approximation and d is the grating period. The effective refractive indices $n_{\text{TM}}^{(2)}$ and $n_{\text{TE}}^{(2)}$ change with the wavelength, so the birefringence of the grating ($\Delta n = n_{\text{TE}}^{(2)} - n_{\text{TM}}^{(2)}$) has spectra dependence. Fig. 3 depicts some birefringence spectra curves with respect to different grating periods. The birefringence is calculated by using Eq. (3) and Eq. (4) with $f = 0.7$, $n_1 = 1$ and $n_2 = n_{\text{Si}}(\lambda)$. With the increase of the illuminating wavelength, the decrease of the value of radio d/λ makes the dispersion curve go flat. The flat part of the curve is similar to that plotted in Fig. 2, indicating that the second-order EMT approaches the zero-order EMT when d/λ becomes smaller.

The phase retardance introduced by an SWG can be expressed as

$$\phi(\lambda) = \frac{2\pi}{\lambda} \cdot \Delta n(\lambda) \cdot h \quad (5)$$

where $\Delta n(\lambda)$ is the birefringence with respect to the wavelength and h is the groove depth. According to EMT, the wave propagating through a grating region is equivalent to passing

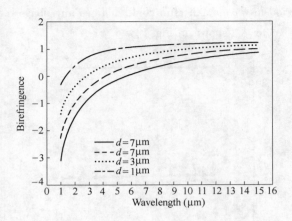

Fig. 3 Birefringence spectra curves with respect to different grating periods

through an anisotropic film. So the phase retardance is determined by the effective film thickness-groove depth h and the form-birefringence $\Delta n(\lambda)$. To make an ideal achromatic phase retarder, i. e. , the phase retardance does not change with the wavelength, the birefringence $\Delta n(\lambda)$ must be proportional to the wavelength, that is,

$$\Delta n(\lambda) = \frac{\phi}{2\pi h} \cdot \lambda \qquad (6)$$

Thus, the ideal birefringence dispersion curve should be an oblique line.

As shown in Fig. 2, the birefringence dispersion curve for a single piece of grating in quasistatic domain goes like a horizontal line indicating that birefringence has no spectra dependence. So, just like a standard waveplate, the phase retardance introduced by the grating varies with the incident wavelength.

As shown in Fig. 3, for the grating in resonance domain, its birefringence has spectra dependence. Although the actual curve does not agree with the ideal oblique line for all wavelengths, it is possible for these two curves to coincide with each other in a specified wavelength range. We can appropriately adjust the gradient of the ideal line by varying the groove depth, making the oblique line tangent to the actual curve within the target waveband. So the grating of this kind can be used as the achromatic phase retarder for a single-band. But it fails for dual-band, because we can't make the oblique line tangent to the birefringence spectra curve within the two separate wavebands simultaneously.

3 Achromatic design

Enlightened by the idea from Destriau and Prouteau, i. e. , combination of a standard half-wave plate and a standard quarter-wave plate creates a new achromatic quarter-wave plate[10,11], we utilized SWGs with different structural parameters to take place crystalline waveplates and extended the design to a combination of up to four SWGs(schematically drawn in Fig. 4). The principle of achromatism is based on the partial cancellation of the change of retardance from each grating with respect to the wavelength. We performed simulations to combinations of 2 to 4 gratings and their retardance spectra curves are shown in Fig. 5. As expected, the retardance gets closer to $\pi/2$ with the number of gratings.

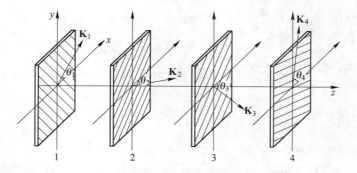

Fig. 4 Schematic diagram for the proposed combination of 4 SWGs
K is the grating vector and θ is the orientation angle

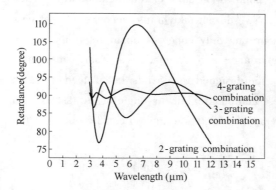

Fig. 5 Retardance spectra curves with respect to different number of gratings

A single piece of SWG is equivalent to a standard phase plate which can be described by its corresponding Jones matrix J:

$$J(\phi,\theta) = \begin{bmatrix} \cos\phi/2 + i\cos2\theta\sin\phi/2 & i\sin2\theta\sin\phi/2 \\ i\sin2\theta\sin\phi/2 & \cos\phi/2 - i\cos2\theta\sin\phi/2 \end{bmatrix} \quad (7)$$

where ϕ is the phase retardance determined by the birefringence and the groove depth of the SWG, θ is the orientation angle of the fast axis (grating vector). J is a unitary matrix with a form of:

$$J = \begin{bmatrix} a & b \\ -b^* & a^* \end{bmatrix} \quad (8)$$

where matrix elements satisfies $a \cdot a^* + b \cdot b^* = 1$. It has been shown by Jones[12] that any combination of retardation plates is optically equal to a system containing only two elements: a retarder and a rotator. The product of two or more of the J matrices is still unitary. That is

$$\prod_{i=1}^{n} J(\phi_i, \theta_i) = R(\overline{\omega}) J(\overline{\phi}, \overline{\theta}) = \begin{bmatrix} A & B \\ -B^* & A^* \end{bmatrix} \quad (9)$$

where

$$R(\overline{\omega}) = \begin{bmatrix} \cos\overline{\omega} & -\sin\overline{\omega} \\ \sin\overline{\omega} & \cos\overline{\omega} \end{bmatrix} \quad (10)$$

$$A = \cos\overline{\omega}\cos\frac{\overline{\phi}}{2} + i\sin\frac{\overline{\phi}}{2} \cdot [\cos\overline{\omega}\cos 2\overline{\theta} - \sin\overline{\omega}\sin 2\overline{\theta}] \quad (11)$$

$$B = -\sin\overline{\omega}\cos\frac{\overline{\phi}}{2} + i\sin\frac{\overline{\phi}}{2} \cdot [\cos\overline{\omega}\sin 2\overline{\theta} + \sin\overline{\omega}\cos 2\overline{\theta}] \quad (12)$$

ϕ_i and θ_i are the phase retardance and the orientation angle of the ith SWG, $\overline{\omega}$ is the amount of rotation, $\overline{\phi}$ and $\overline{\theta}$ are the phase retardance and the orientation angle of the combination. Since the ϕ_i are functions of wavelength, $\overline{\phi}$, $\overline{\theta}$ and $\overline{\omega}$ will in general vary with wavelength. For a rotator[13], it only rotates the polarization state of the incident wave at an angle of $\overline{\omega}$ but does not change it essentially. Thus, we do not take $\overline{\omega}$ into account in most cases. Usually, there are two types of achromatic design: one only concerns about retardance achromatism, while the other one aims to achieve retardance and orientation achromatisms simultaneously. The former is highly achromatic within the designed wavebands, but it needs to rotate when target wavelength changes. The latter is less achromatic, but the angle of the fast axis remains nearly fixed.

The phase retardance $\overline{\phi}$ and orientation angle $\overline{\theta}$ can be obtained from Eqs. (9) to (12), that is,

$$\tan^2\frac{\overline{\phi}}{2} = \frac{|\mathrm{Im}A|^2 + |\mathrm{Im}B|^2}{|\mathrm{Re}A|^2 + |\mathrm{Re}B|^2} \quad (13)$$

$$\tan 2\overline{\theta} = \frac{\mathrm{Im}B + \frac{\mathrm{Re}B}{\mathrm{Re}A} \cdot \mathrm{Im}A}{\mathrm{Im}A - \frac{\mathrm{Re}B}{\mathrm{Re}A} \cdot \mathrm{Im}B} \quad (14)$$

For the combination of SWGs used as achromatic QWP within the MWIR&LWIR wavebands with the fast axis unfixed, the design of achromatism for phase retardance is described by a two-objective optimization model[14] which can be expressed as

$$\mathrm{Min} \begin{cases} \mathrm{function1} = \Sigma_\lambda \left| \overline{\phi_\lambda}(h_i, \theta_i) - \frac{\pi}{2} \right| \\ \mathrm{function2} = \max |\overline{\phi_\lambda}(h_i, \theta_i)| \end{cases} \quad \lambda \in (3-5\mu m) \cup (8-12\mu m)$$

$$\mathrm{s.t.} \ 0 < h_i < \mathrm{fabrication\ limits} \quad (15)$$

where $\overline{\phi_\lambda}$ is the phase retardance for the wavelength of λ introduced by the combination, h_i is the groove depth and θ_i is the orientation angle of the ith piece of SWG. Objective function1 is used to minimize the total phase retardance error, while function2 aims to even the phase retardance error over MWIR&LWIR wavebands. We performed simulation for the combination of 4

SWGs. By use of the optimization algorithm, the parameters of each SWG are calculated (listed in Table 1). The result obtained with the combination of these 4 SWGs is depicted in Fig. 6.

Table 1 Parameters of QWP with its fast axis unfixed

Parameters	1	2	3	4
Thickness h_i/nm	2956	2112	2296	1880
Orientation angle θ_i/(°)	38.0	-22.3	49.9	87.9

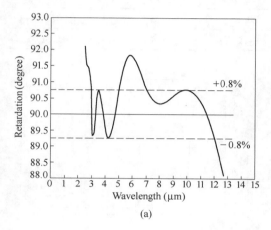
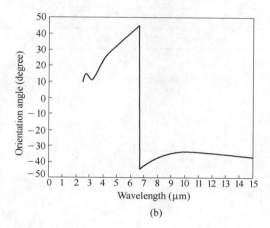

Fig. 6 Calculated results of achromatic QWP with the fast axis unfixed (a) phase retardance dispersion curve and (b) orientation angle dispersion curve

The designed achromatic QWP exhibits perfect retardance stability. The phase retardance deviates very slightly from ideal $\pi/2$ (the maximal relative deviation $\delta = |\overline{\phi_\lambda} - \pi/2|/\pi/2$ is 0.8%) over both MWIR and LWIR wavebands. Since the fast axis is not involved in the design process, the orientation angle of the fast axis depicted in Fig. 6(b) changes rapidly with different wavebands. The orientation angle is positive for MWIR and negative for LWIR.

For the archromatic QWP with the fast axis fixed, orientation achromatism is as important as retardance achromatism. The value of the standard deviation (STD) for orientation angles is selected as the evaluation criteria for the degree of orientation achromatism. The design of achromatic QWP of this kind is described by a three-objective optimization model, that is,

$$\text{Min} \begin{cases} \text{function1} = \Sigma_\lambda \left| \overline{\phi_\lambda}(h_i, \theta_i) - \frac{\pi}{2} \right| \\ \text{function2} = \max \left| \overline{\phi_\lambda}(h_i, \theta_i) \right| \quad \lambda \in (3-5\mu m) \cup (8-12\mu m) \\ \text{function3} = \text{STD} \left[\overline{\theta_\lambda}(h_i, \theta_i) \right] \end{cases}$$

$$\text{s. t. } 0 < h_i < \text{fabrication limits} \qquad (16)$$

Still, function1 and function2 are used to achieve retardance achromatism. The added function3 aims to control the uniformity of the orientation angles. For a multi-objective optimization problem, it is impossible to get a solution that minimizes the value of each objective function. What we can do is to seek for a compromising solution that satisfies the three objective functions simultaneously. Based on this thought, the optimized parameters of QWP are listed in Table 2. The calculated retardance and orientation achromatisms with these parameters are demonstrated in Fig. 7.

Table 2 Parameters of QWP with its fast axis fixed

Parameters	1	2	3	4
Thickness h_i/nm	2059	2051	2045	3072
Orientation angle θ_i/°	58.0	34.2	68.8	−11.1

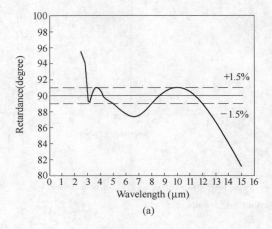 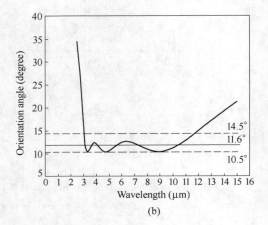

(a)   (b)

Fig. 7 Calculated results of achromatic QWP with the fast axis fixed (a) phase retardance dispersion curve and (b) orientation angle dispersion curve

As expected, the relative retardance deviation increases from 0.8% to 1.5% with the involvement of function3. But the orientation angle of the fast axis varies from 10.5° to 14.5° over both MWIR and LWIR wavebands. The nominal value of the orientation angle is 11.6° with its standard deviation of 1.27°.

4 Fabrication consideration

The QWP is supposed to be used in imaging polarimetry system. Since higher-order diffractions may sneak into the imaging system and form "ghost images", the gratings should be zero-order elements. In other words, the period should be small enough to exclude nonzero diffraction orders. For any dielectric grating under normal incidence, as depicted in Fig. 1, the threshold period[15] under which only the transmitted and the reflected zero-order diffractions are nonevanescent is

$$d_{th} = \frac{\lambda}{n_1 + n_2} \qquad (17)$$

The threshold period is proportional to the illuminating wavelength and the value of the grating period is specified by the smallest wavelength in the waveband. For MWIR&LWIR dual-band, the minimal wavelength is 3μm and the refractive index of silicon for the wavelength of 3μm is 3.4307. By inserting $n_1 = 1$ and $n_2 = 3.4307$ into Eq. (17), the threshold period of the grating should be 677nm, which guarantees a zero-order grating. The linewidth of the grating should be 338nm with the filling factor of 0.5 ($f = 0.5$ helps to ease the fabrication). As seen from Table 1 and Table 2, the largest groove depth is 3072nm, which corresponds to an aspect-ratio of 1:9. This aspect-ratio is a little bit challenging but still in the range of fabrication capabilities[16] of electron-beam writing and ion-beam etching machines.

In order to increase the IR transmittance and ease the adjustment, the combination of 4 SWGs are supposed to be fabricated on 2 silicon wafers, that is, 4 gratings are orderly on the front and back sides of the two substrates. A 200μm-thick silicon wafer has MWIR transmittance of 55% and LWIR transmittance of 45%[17]. Then the total thickness of the QWP is 400μm and the total transmittance would be around 25%. Of course, other IR materials with higher transmittance, such as ZnSe and BaF_2, may also be selected. But compared with Si, other IR materials have smaller refractive indices which brings a larger aspect-ratio to the grating, making the fabrication even more difficult.

Each grating is designed with a groove depth precision of better than 10nm, and the relative angular adjustment between the two silicon wafers is better than 1°. With these fabrication tolerances, the relative retardance deviation will be limited in the range below 3%.

5 Conclusion

In this paper, we discussed the limitations of the currently existing achromatic waveplates applied to dual-band. A combination of 4 SWGs used as MWIR & LWIR dual-band achromatic QWP is proposed. The grating period is specified to exclude nonzero order diffractions, while the groove depth and the orientation of the grating vector are optimized to achieve retardance achromatism. The designed dual-band achromatic QWP has potential application in MWIR & LWIR imaging polarimetry where crystalline retarders are not available.

Acknowledgements

This work was supported by the National Basic Research Program of China under Grant No 2007CB935303 and by Beijing Municipal Commission of Education Project under Grant No KM200910772005.

References

[1] J. F. Scholl, E. L. Dereniak, M. R. Descour, C. P. Tebow, and C. E. Volin, "Phase grating design for a dual-band snapshot imaging spectrometer", Appl. Opt. 42, 18-29 (2003).

[2] I. Yamada, K. Takano, M. Hangyo, M. Saito, and W. Watanabe1, "Terahertz wire-grid polarizers with micrometer-pitch Al gratings", Opt. Lett. 34, 274-276 (2009).

[3] R. M. A. Azzam and C. L. Spinu, "Achromatic angle-insensitive infraredquarter-wave retarder based on to-

tal internal reflection at the Si-SiO$_2$ interface", J. Opt. Soc. Am. A. 21,2019-2022(2004).

[4] J. B. Masson and G. Gallot, "Terahertz achromatic quarter-wave plate", Opt. Lett. 31,265-267(2006).

[5] H. Kikuta, Y. Ohira, and K. Iwata, "Achromatic quarter-wave plates using the dispersion of form birefringence", Appl. Opt. 36,1566-1572(1997).

[6] G. G. Kang, Q. F. Tan, and G. F. Jin, "Optimal Design of an Achromatic Angle-Insensitive Phase Retarder Used in MWIR Imaging Polarimetry", Chin. Phys. Lett. 26,074218:1-4(2009).

[7] I. Richter, P. C. Sun, F. Xu, and Y. Fainman, "Design considerations of form birefringent microstructures", Appl. Opt. 34,2421-2429(1995).

[8] B. J. Frey, D. B. Leviton, and T. J. Madison, "Temperature-dependent refractive index of silicon and germanium", Proc. SPIE 6273,62732J(2006).

[9] P. Lalanne and J. P. Hugonin, "High-order effective-medium theory of subwavelength gratings in classical mounting: application to volume holograms", J. Opt. Soc. Am. A 15,1843-1851(1998).

[10] A. M. Title, "Improvement of Birefringent Filters. 2: Achromatic Waveplates", Appl. Opt. 14,229-237(1975).

[11] G. Destriau and J. Prouteau, "Realisation d'um quart d'onde quasi acromatique par juxtaposition de deux lames cristallines de meme nature", J. Phys. Radium. 2,53-55(1949).

[12] R. C. Jones, "A New Calculus for the Treatment of Optical Systems I. Description and Discussion of the Calculus", J. Opt. Soc. Am. A. 31,488-493(1941).

[13] R. C. Jones, "A New Calculus for the Treatment of Optical Systems II. Proof of Three General Equivalence Theorems", J. Opt. Soc. Am. A. 31,493-499(1941).

[14] D. Kalyanmoy, *Muiti-Objective Optimization Using Evolutionary Algorithms* (John Wiley & Sons, 2009).

[15] N. Bokor, R. Shechter, N. Davidson, A. A. Friesem, and E. Hasman, "Achromatic phase retarder by slanted illumination of a dielectric grating with period comparable with the wavelength", Appl. Opt. 40,2076-2080(2001).

[16] M. Okano, H. Kikuta, Y. Hirai, K. Yamamoto, and T. Yotsuya, "Optimization of diffraction grating profiles in fabrication by electron-beam lithography", Appl. Opt. 43,5137-5142(2004).

[17] Instrument networks discussion, "Infrared Spectrum transmittance" (Instrument networks,2009) http://bbs.instrument.com.cn/shtml/20090906/2098092/.

Experimental Demonstration of Tunable Directional Excitation of Surface Plasmon Polaritons with a Subwavelength Metallic Double Slit[*]

Abstract We demonstrate experimentally the directional excitation of surface plasmon polaritons (SPPs) on a metal film by a subwavelength double slit under backside illumination, based on the interference of SPPs generated by the two slits. By varying the incident angle, the SPPs can be tunably directed into two opposite propagating directions with a predetermined splitting ratio. Under certain incident angle, unidirectional SPP excitation can be achieved. This compact directional SPP coupler is potentially useful for many on-chip applications. As an example, we show the integration of the double-slit couplers with SPP Bragg mirrors, which can effectively realize selective coupling of SPPs into different ports in an integrated plasmonic chip.

Plasmonic circuits utilizing surface plasmon polaritons (SPPs) as optical information carriers offer opportunities for integrating compact nanoelectronic and nanophotonic devices onto a single chip due to their remarkable ability of breaking the diffraction limit[1]. The components for effective excitation and flexible manipulation of SPPs are of fundamental importance in such circuits[2-15]. For SPP launchers, which couple incident light to SPPs, the directional excitation of SPPs is often required[8]. A symmetrical excitation of SPPs can be achieved at oblique[2-4] or normal[5-7] incidence in prism or grating couplers. For miniaturization and high density integration purposes, SPP couplers with more compact geometries are preferred, for which subwavelength slits and ridges are good candidates. However, it is difficult to realize efficient directional (especially unidirectional) excitation of SPPs with such spatially symmetric structures.

To improve the performance of nanoslit couplers, several previous designs have been proposed[8-15], in which the key is to introduce either structural or illumination asymmetry to the system. For example, by adding a periodic array of grooves on one side of a slit, SPPS can be unidirectional launched efficiently to the other side of the slit by Bragg reflection in the grooves[8,9]. Nevertheless, the inclusion of the groove array increases the size and complexity of the device. In another way, by manipulating the constructive or destructive interference of SPPs generated by multislits, asymmetrical excitation of SPPs in nanoslit couplers with backside illumination were also presented[11-15]. Most of these previous works were numerical demonstrations because the proposed structures are difficult to fabricate with challenging design parameters[11-15]. Meanwhile, the designs usually did not take into account the tunability of SPP launching that is highly demanded in various applications.

---

[*] Copartner: Xiaowei Li, Qiaofeng Tan, Benfeng Bai. Reprinted from *Applied Physics Letters*, 2011, 98: 251109.

In this letter, we report on an experimental demonstration of asymmetrical and unidirectional excitation of SPPs using a compact double-slit coupler under backside illumination, in which the asymmetrical excitation ratio of SPPs can be flexibly tuned by varying the incident angle. We further show that the double-slit coupler can be integrated with other plasmonic components, such as SPP Bragg mirrors to realize selective coupling of SPPs into different ports, in an integrated plasmonic chip.

The proposed structure is depicted in the upper-left inset of Fig. 1, in which two subwavelength slits spaced by d are perforated into a gold film deposited on a silica substrate with a 15 nm thick titanium adhesion layer. When a TM polarized light (whose magnetic field vector is parallel with the slits) impinges on the structure from backside, the scattered light from each slit may excite SPPs (with wave vector k_{SPP}) on the gold-air interface[16]. At normal incidence, equal strength of SPPs are excited toward the left and right sides of the double slit due to symmetry. However, at oblique incidence, we may achieve asymmetrical excitation of SPPs by modulating the phase difference (and thereby the interference) of SPPs launched separately from the two slits. Denoting the initial phases of SPPs generated by the right and left slits as φ_R and φ_L, respectively, we can write their phase difference, which is solely induced by the oblique incidence, as

$$\varphi_R - \varphi_L = \frac{2\pi}{\lambda_0} d\sin\theta \qquad (1)$$

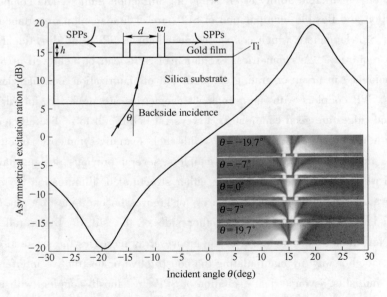

Fig. 1 Numerically calculated SPP asymmetrical excitation ratio r with respect to the incident angle θ in a double-slit coupler with $d = 790$nm. The insets show the geometry of the coupler (upper left) and the calculated time-averaged power flow distribution on the metal surface at incident angles $\theta = 0°$, $\pm 7°$, $\pm 19.7°$ (lower right)

By introducing this asymmetry, the left and right SPPs can be controlled to interfere constructively to one side and destructively to the other side, by which to realize unidirectional excitation of SPPs. For example, if we want to excite SPPs only to the left side, the following phase matching conditions should be satisfied

$$\varphi_L - (\varphi_R + k_{SPP}d) = 2M\pi \quad (2)$$

$$\varphi_L - (\varphi_R + k_{SPP}d) = (2N+1)\pi \quad (3)$$

where M and N are arbitrary integers. k_{SPP} and the associated SPP wavelength λ_{SPP} are obtained by $k_{SPP} = 2\pi/\lambda_0 \sqrt{(\varepsilon_m\varepsilon_d)/(\varepsilon_m + \varepsilon_d)}$ and $\lambda_{SPP} = 2\pi/\text{real}(k_{SPP})$, where ε_d and ε_m are the permittivities of air and gold, respectively.

The performance of the double-slit coupler was numerically investigated using a commercial finite-element-method software COMSOL MULTIPHYSICS 3.5a. The coupler is designed to operate at wavelength $\lambda_0 = 1064$ nm, for which the permittivities of gold and silica are $-52.02 + 3.87i$ and 1.45^2, respectively. The SPP wavelength is calculated to be $\lambda_{SPP} = 1054$ nm. Then according to Eqs. (1)-(3), the structural parameters are determined as $d = 790$ nm and $\theta = 19.7°$ by selecting $M = -1$ and $N = -1$ so as to keep a compact size of the device. Similarly, SPPs can be unidirectionally excited to the right side with $d = 790$ nm and $\theta = -19.7°$. The gold film is chosen thick enough ($h = 250$ nm) to prevent direct penetration of light through the film. The line in Fig. 1 shows the calculated asymmetrical SPP excitation ratio $r = 10\lg(P_L/P_R)$, where P_L and P_R represent the power flows of left-and right-propagating SPPs, respectively. It is seen that around $\theta = \pm 19.7°$ unidirectional excitation of SPPs can be realized with the largest ratio $|r| = 19.55$ dB. To illustrate the dependence of SPP excitation on the incident angle more intuitively, the time-averaged power flow distribution on the metal surface was simulated at different incident angles $\theta = 0°, \pm 7°, \pm 19.7°$, as shown in the lower-right inset of Fig. 1. Obviously, with the increase of incident angle θ, the SPP excitation can be tuned from symmetrical (at normal incidence) to asymmetrical (at $\theta = \pm 7°$) and to unidirectional ($\theta = \pm 19.7°$), as predicted.

To verify this effect experimentally, a couple of double-slit and single-slit samples were fabricated and their SPP excitation performances were characterized. The nanoslits of 50 μm length and 200 nm width were perforated by focused ion beam milling (with FEI Nova 200 Nanolab) in the gold film. Fig. 2(a) shows the scanning electronic microscope (SEM) image of a double-slit sample, where two shallow groove gratings with groove depth 80 nm and period $\Lambda = \lambda_{SPP} = 1054$ nm were located symmetrically (90 μm away from the double slit) on the two sides. The gratings convert SPPs to radiative light so that the excitation of SPPs can be characterized by far-field imaging[10]. The samples were illuminated from backside by a TM polarized laser beam. The scattered light emanating from the slits and grooves on the upper gold surface was collected by a long-working-distance objective (Nikon, 40×, NA = 0.60) and imaged by a charge coupled device (CCD). Then the detected CCD signal intensity on the grating area is proportional to the intensity of SPPs propagating to the grating, which is a direct evaluation of the SPP intensity.

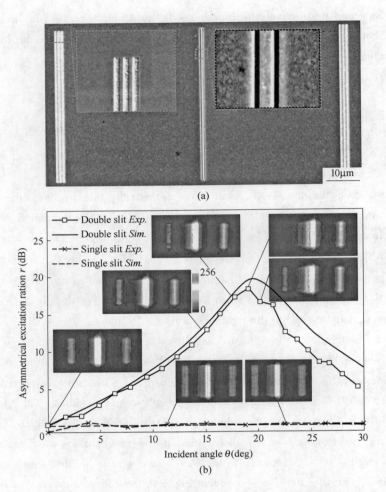

Fig. 2 (a) SEM image of a double-slit sample with $d = 790$ nm, (b) Experimentally measured and numerically simulated asymmetrical excitation ratio r with respect to the incident angle θ for the double-slit sample and a single-slit sample. The insets show the CCD images (false color images) at different incident angles. All the CCD images have the same color scale

Fig. 2(b) shows the CCD images of a double-slit and a single-slit samples measured at different incident angles, from which the asymmetrical excitation ratio r was derived and compared with simulation results. The bright speckles in the middle are the directly transmitted light from the slits; while the left and right speckles are radiation caused by SPPs propagating to the respective gratings. It is seen that the variation of r with respect to the change of θ matches the theoretical prediction nicely. Especially, at an angle of $\theta = 19.08°$, the right-side SPP intensity reaches the maximum while the left-side SPP intensity is nearly zero, with a peak r of 18.32 dB, indicating an efficient unidirectional excitation of SPPs to the right side. The left-side unidirectional excitation of SPPs can be realized simply by switching the incident angle to $-19.08°$ (not shown here). The small discrepancies between experiment and simulation are mainly caused by the deviations of material properties and structural parameters in fabrication.

As a comparison, we also characterized the SPP excitation in a single slit with all the other parameters the same as the double-slit one. It is seen in Fig. 2(b) that the left and right SPPs are excited almost symmetrically, irrespective of the incident angle, and the ratio r is always close to 0. This shows that the directional excitation of SPPs can only be achieved in the double-slit coupler.

Finally, to show the potential of the double-slit coupler integrated with other plasmonic components for the application in plasmonic circuitry, we designed a sample integrating the double-slit couplers with Bragg grating mirrors to realize selective coupling of SPPs into different ports, as shown in Fig. 3(a). The double slit launches SPPs to Bragg mirrors composed of 20 periodic grooves of depth 150nm, period 745 nm, and tilted by 45° with respect to the double slit. The SPPs reflected by the Bragg mirrors are outcoupled to radiation by two shallow groove gratings. Fig. 3(b) shows the CCD image at normal TM incidence, in which both the directly transmitted light in the double slit and the outcoupled SPPs via the gratings are visible. To distinguish the SPP signal more clearly, we employed a cross-polarization detection technique[17], by adding a TE analyzer between the microscope objective and the CCD to block the direct transmitted light from the double slit. Figs. 3(c)-3(e) show the detected CCD images at $\theta = 0°, 7°, 19.7°$ with the analyzer. It is clearly observed that at $\theta = 19.7°$ the SPPs unidirectionally excited to the right is reflected by the Bragg mirror and then directed to port 2.

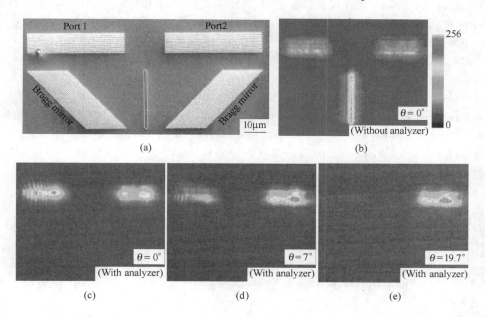

Fig. 3 (a) SEM image of a sample integrating a double-slit coupler, Bragg grating mirrors, and outcoupling gratings. (b)-(e) CCD images (false color images) of the sample under illumination at different incident angles, with or without using the analyzer. All the CCD images have the same color scale

In summary, we have demonstrated experimentally the tunable directional excitation of SPPs by a compact double-slit coupler based on the interference of SPPs generated by the slits and its capability of integration with other plasmonic components such as Bragg mirrors. By controlling the incident angle, the asymmetrical excitation ratio can be tuned effectively and flexibly and unidirectional SPP excitation can be realized. The coupling efficiency of the coupler may be further improved by adopting multislit design. The same principle can be extended to other wavelength ranges such as mid-infrared and THz. The coupler is potentially useful in applications such as integrated plasmonic circuits, pickup head for high density data storage, and plasmonic biosensors.

Acknowledgements

We acknowledge the support by the National Basic Research Program of China (Grant No. 2007CB935303), the National Natural Science Foundation of China (Project No. 11004119), and the Academy of Finland (Project No. 128420).

References

[1] W. L. Barnes, A. Dereux, and T. W. Ebbesen, Nature (London) 424, 824 (2003).

[2] D. Egorov, B. S. Dennis, G. Blumberg, and M. I. Haftel, Phys. Rev. B 70, 033404 (2004).

[3] J. Y. Laluet, E. Devaux, C. Genet, T. W. Ebbesen, J. -C. Weeber, and A. Dereux, Opt. Express 15, 3488 (2007).

[4] I. P. Radko, S. I. Bozhevolnyi, G. Brucoli L. Martín-Moreno, F. J. García-Vidal, and A. Boltasseva, Opt. Express 17, 7228 (2009).

[5] N. Bonod, E. Popov, L. Li, and B. Chernov, Opt. Express 15, 11427 (2007).

[6] B. Bai, X. Meng, J. Laukkanen, T. Sfez, L. Yu, W. Nakagawa, H. P. Herzig, L. Li, and J. Turunen, Phys. Rev. B 80, 035407 (2009).

[7] A. Roszkiewicz and W. Nasalski, J. Phys B. 43, 185401 (2010).

[8] F. López-Tejeira, S. G. Rodrigo, L. Martín-Moreno, F. J. García-Vidal, E. Devaux, T. W. Ebbesen, J. R. Krenn, I. P. Radko, S. I. Bozhevolnyi, M. U. González, J. C. Weeber, and A. Dereux, Nat. Phys. 3, 324 (2007).

[9] S. B. Choi, D. J. Park, Y. K. Jeong, Y. C. Yun, M. S. Jeong, C. C. Byeon, J. H. Kang, Q. H. Park, and D. S. Kim, Appl. Phys. Lett. 94, 063115 (2009).

[10] J. Chen, Z. Li, S. Yue, and Q. Gong, Appl. Phys. Lett. 97, 041113 (2010).

[11] H. Kim and B. Lee, Plasmonics 4, 153 (2009).

[12] G. Lerosey, D. F. P. Pile, P. Matheu, G. Bartal, and X. Zhang, Nano Lett. 9, 327 (2009).

[13] T. Xu, Y. Zhao, D. Gan, C. Wang, C. Du, and X. Luo, Appl. Phys. Lett. 92, 101501 (2008).

[14] J. Wang, Y. Wang, X. Zhang, K. Yang Y. Wang, S. Liu, and Y. Song, J. Mod. Opt. 57, 1630 (2010).

[15] Y. Wang, X. Zhang, H. Tang, K. Yang, Y. Wang, Y. Song, T. Wei, and C. H. Wang, Opt. Express 17, 20457 (2009).

[16] L. Salomon, G. Bassou, H. Aourag, J. P. Dufour F. de Fornel, F. Carcenac, A. V. Zayats, Phys. Rev. B 65, 125409 (2002).

[17] K. A. Tetz, L. Pang, and Y. Fainman, Opt. Lett. 31, 1528 (2006).

Design Method of Surface Contour for a Freeform Lens with Wide Linear Field-of-view*

Abstract In this paper, a design method of surface contour for a freeform imaging lens with a wide linear field-of-view (FOV) is developed. During the calculation of the data points on the unknown freeform surfaces, the aperture size and different field angles of the system are both considered. Meanwhile, two special constraints are employed to find the appropriate points that can generate a smooth and accurate surface contour. The surfaces obtained can be taken as the starting point for further optimization. An $f\text{-}\theta$ single lens with a $\pm 60°$ linear FOV has been designed as an example of the proposed method. After optimization with optical design software, the MTF of the lens is close to the diffraction limit and the scanning error is less than 1 μm. This result proves that good image quality and scanning linearity were achieved.

1 Introduction

Compared with conventional rotationally symmetric surfaces, freeform optical surfaces have more degrees of freedom, therefore reduce the aberrations and simplify the structure of the system in optical design. In recent years, with the development of the advancing manufacture technologies, freeform surfaces have been successfully used in the imaging field, such as head-mounted-display[1-3], reflective systems[4-9], varifocal panoramic optical system[10] and microlens array[11].

Traditional freeform imaging system design uses spherical or aspherical system as the starting point. Then, some surfaces in the system are replaced by freeform surfaces to obtain satisfactory results[1,2,10]. However, with the increasing novelty and complexity of optical systems, this method is difficult to satisfy the design goal. A possible solution to this problem is to directly design freeform surfaces based on the object-image relations. One common method is to establish the partial differential equations based on incident and exit rays which determine the shape of the surfaces[12-14]. The points on the surfaces can be calculated next and the freeform surfaces are obtained after surface fitting. This method is simple and effective in imaging optics, especially for designing systems with a small field-of-view (FOV). Another ingenious method is the Simultaneous Multiple Surface (SMS) design method[15,16]. Multiple freeform surfaces can be generated simultaneously, and several input and output tangential-ray bundles become fully coupled by the optical system.

For freeform imaging system design, a wide FOV is difficult to be achieved. Furthermore, the

---

* Copartner: Jun Zhu, Tong Yang. Reprinted from *Optics Express*, 2013, 21(22):26080-26092.

real size of aperture is expected to be considered while designing a wide FOV system. So, the light beams of different fields possibly have overlap area on the unknown surface during the design. The position and shape of the overlap area should meet the imaging requirement of different fields. It is both an interest and a challenge to directly design freeform surfaces under these conditions.

In this paper, a novel method to design a freeform imaging lens with a wide FOV has been developed. The proposed method has two key contents. Firstly, the aperture size and different field angles in a wide FOV system are both considered during the calculation of the data points on the freeform surfaces. Secondly, two special constraints are employed during the calculation. With these two key contents, a series of data points can be obtained, and a smooth and accurate surface contour can be generated after curve fitting with these data points. More importantly, the coordinates and normal vectors of these original data points can be approximately maintained, and the expected imaging relationship is ensured. The surfaces obtained can be taken as the starting point for further optimization in optical design software. Here, only the design method of the two-dimensional surface contour for tangential rays is covered in this paper. An f-θ single lens with a $\pm 60°$ linear FOV is designed as an example. A good starting point is obtained with the proposed method, and it is then optimized in optical design software to achieve good image quality and scanning linearity. The design method of three-dimensional freeform surfaces will be discussed in the future study.

2 Method

A freeform single lens is shown schematically in Fig. 1(a). Parallel light beams of different fields from the entrance pupil are refracted by the lens and focus on the image plane. The two-dimensional contour of the front surface in the tangential plane is firstly generated to control the tangential rays without the back surface, as shown in Fig. 1(b). This single surface is taken as the starting point for further optimization with optical design software. The back surface is then added in the next step. Assume that the center of the entrance pupil is located at the origin of an orthogonal coordinates system and the optical axis is along with the z-axis. The FOV of the system $2\omega(\pm\omega)$ is divided into $2k+1$ fields with equal interval $\Delta\omega$ between each two neighboring fields during the design process. So $\Delta\omega$ can be expressed as

$$\Delta\omega = \frac{\omega}{k} \tag{1}$$

The design is improved with the increasing number of sampling fields. When the field angle is small, the beams of neighboring fields with fixed intervals generally have overlap area on the front surface, as shown in Fig. 2. When the field angle is increasing, the overlap area of neighboring fields is getting smaller, and finally disappears. In addition, three rays corresponding to three different pupil coordinates in each field (the chief ray, the top marginal ray and the bottom marginal ray) are specified in Fig. 3.

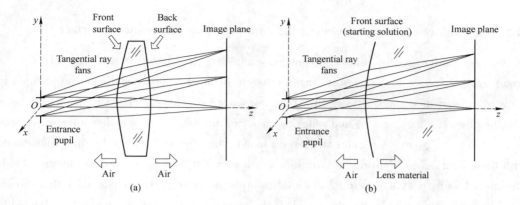

Fig. 1 Layout of a freeform single lens and its starting point
(a) layout of a freeform single lens; (b) the front surface (the surface contour in the tangential plane) is taken as the starting point for further optimization

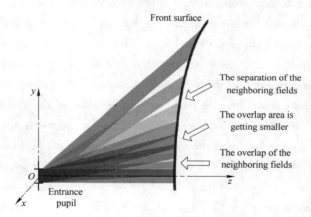

Fig. 2 The change of the overlap area of neighboring fields with increasing field angle. When the field angle is small, the beams of neighboring fields with fixed intervals generally have overlap area on the front surface. When the field angle is increasing, the overlap area of neighboring fields is getting smaller, and finally disappears

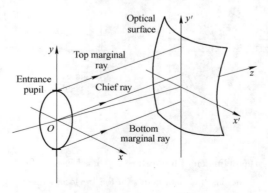

Fig. 3 The rays of different pupil coordinates specified in each field. The three rays are the chief ray and two marginal rays from the top and the bottom of the entrance pupil

2.1 The feature rays for calculating the data points on the unknown surface

During the design of the freeform contour, each two neighboring fields are taken as a group. Several feature rays corresponding to different fields and pupil coordinates are defined in each group, and their intersections with the front surface can be calculated based on the relationships between the incident and outgoing rays. So, the data points on the unknown surface can be obtained group after group and the contour of the front surface can be then constructed with these data points. Consider an arbitrary group shown in Fig. 4. The two neighboring fields are labeled as field #1 and field #2 and the feature rays are respectively marked with a circled number and plotted in bold. Feature ray ① is the chief ray of field #1 and it will be refracted to its ideal image point P_{f1} by the front surface at data point 1. Feature ray ② is the bottom marginal ray of field #2 and it will be refracted to its ideal image point P_{f2} by the front surface at data point 2. When the beams of neighboring fields have overlap area on the unknown surface, the beams from two different fields have to be simultaneously controlled at this area on the surface. It means that two rays from field #1 and #2 which hit on the same point in the overlap area should be refracted to P_{f1} and P_{f2} respectively. This problem generally does not have an exact solution due to the over-determined problem. So in this paper, the calculation of the data points is taken as an optimization problem. As only the starting point of the system is concerned, an approximate solution obtained by a mathematical optimization process is adequate. Here in particular, at data point 2 (the bottom of the overlap area), feature ray ② from field #2 as well as another feature ray ③ from field #1 will be refracted to their ideal image point P_{f2} and P_{f1} respectively. In short, two data points (point 1 and 2) on the surface are calculated in one field group using three feature rays (①②③) when the light beams have overlap area on the unknown surface. When the field angle is increasing and the light beams are separated, as shown

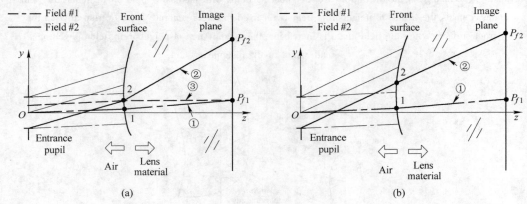

Fig. 4 The definition of the feature rays used in each field group to calculate the data points on the freeform surface

(a) When the two neighboring fields have overlap area on the unknown surface, two data points (point 1 and 2) are calculated in one field group using three feature rays (①②③); (b) When the light beams are separated, two data points (point 1 and 2) on the surface are calculated in one field group with only two feature rays (①②)

in Fig. 4(b), there is no feature ray ③ and two data points (point 1 and 2) on the surface are calculated in one field group using only two feature rays (①②).

2.2 Establishing the constraints to generate a smooth link line of the points

After the feature rays used in each group are defined, the data points on the front surface can be calculated based on the relationships between the incident and outgoing light rays. However, there are still some problems. As there are no geometric relationships between different groups, the two data points calculated in each field group are the optimum solution of a single problem specific to two fields. Therefore, the data points from different groups distribute irregularly in the tangential plane. A smooth and accurate surface contour is difficult to be obtained.

Another problem is related to the realization of the expected imaging relationship. The outgoing direction of a light ray which goes through a data point is determined by both the coordinates and the normal vector of this point. In this paper, the surface contour is obtained by curve fitting with the data points. In the ideal case, all the data points are on the fitted contour, and the original normal vector at each data point which determines ray direction is ensured as well. In this way, the light rays can be shifted into the expected direction by the surface. However, if no constraints are added when calculating the data points, the points obtained generally have considerable deviations from the surface contour after curve fitting, in other words, the fitting error is big. More seriously, the normal vector of the surface contour after curve fitting is not consistent with the normal at each original data point. As a consequence, the outgoing light rays are deviated from the expected direction and the expected imaging relationship will be not ensured.

To solve these problems, two special constraints are employed during the calculation of the data points. One constraint is used to establish the geometric relationships between neighboring field groups using the surface normal. The other one called stairs-distribution elimination constraint is used to improve the smoothness of the link line. With these constraints, the data points distribute regularly and form a smooth link line in the tangential plane. In this way, an accurate fitted contour is achieved and the deviation of the original data points from the fitted contour is very small. The way to maintain the consistency of the normal vectors after curve fitting is also involved in these constraints. Detailed analyses of these constraints are depicted in the following.

The data points calculated in the previous neighboring group are used during the calculation of the data points in each field group. The first constraint uses the surface normal vector at each data point to establish the geometric relationships between neighboring field groups. As shown in Fig. 5, P_3 and P_4 are the data points to be calculated in the current field group, P_1 and P_2 are the data points already calculated in the previous field group. The direction vector $\mathbf{e}_{23}$ from P_2 to P_3 is constrained to be perpendicular to the unit normal $\mathbf{N}_3$ at P_3, and the direction vector $\mathbf{e}_{34}$ from P_3 to P_4 is constrained to be perpendicular to the unit normal $\mathbf{N}_4$ at P_4. So, the constraints can be written as

$$\mathbf{N}_3 \cdot \mathbf{e}_{23} = 0 \tag{2}$$

$$\mathbf{N}_4 \cdot \mathbf{e}_{34} = 0 \tag{3}$$

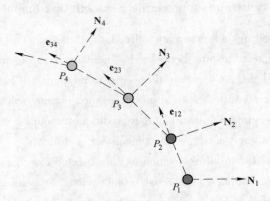

Fig. 5 The constraint to establish the geometric relationships between neighboring field groups using the normal vector at each data point

(P_3 and P_4 are the data points to be calculated in the current field group, P_1 and P_2 are the data points already calculated in the previous field group. The direction vector $\mathbf{e}_{23}$ from P_2 to P_3 is constrained to be perpendicular to the unit normal $\mathbf{N}_3$ at P_3, and the direction vector $\mathbf{e}_{34}$ from P_3 to P_4 is constrained to be perpendicular to the unit normal $\mathbf{N}_4$ at P_4)

Eqs. (2) and (3) establish the geometric relationships between neighboring field groups. The data points no longer distribute irregularly in the tangential plane after the constraint is added. Moreover, in this constraint, the original normal vector at each data point which determines the outgoing direction of light ray is perpendicular to the line connecting the neighboring point. As a consequence, the consistency of the normal vectors after curve fitting is approximately ensured, and the light rays can be shifted in the expected directions.

However, the link line of the data points may be not smooth enough. A typical case is shown in Fig. 6. The line connecting the two data points in each group is approximately parallel to the one in the neighboring group (Figs. 6(a) and 6(b)), which yields a distribution of points like stairs, as shown in Fig. 6(c). So the fitting accuracy of the contour is low and the data points have considerable deviations from the fitted contour. To solve this problem, the stairs-distribution elimination constraint is proposed. The intersection P_i of the two lines which connect the two data points in each group is expected to be between P_2 and P_3, as shown in Fig. 7. So the y coordinate P_{iy} of P_i is constrained to be between the y coordinates of P_2 and P_3, and the z coordinate P_{iz} of P_i is constrained to be between the z coordinates of P_2 and P_3. The constraint can be written as

$$(P_{2y} - P_{iy})(P_{3y} - P_{iy}) < 0 \tag{4}$$

$$(P_{2z} - P_{iz})(P_{3z} - P_{iz}) < 0 \tag{5}$$

Using this constraint, the stairs-distribution can be eliminated and a smooth link line of the data points is obtained. In this way, an accurate fitted surface contour is achieved and the deviation of the original data points from the fitted contour is further reduced, which contributes to maintaining the expected imaging relationship.

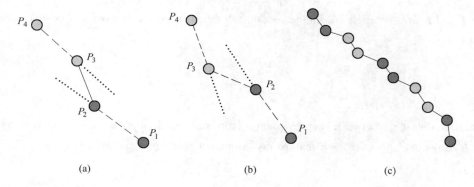

Fig. 6 The stairs-distribution of data points. The connecting the two data points in each group is approximately parallel to the one in the neighboring group, which is shown in (a) and (b). It causes the stair-distribution of data points shown in (c)

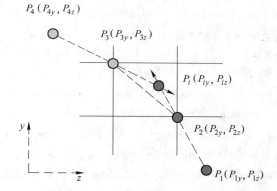

Fig. 7 The stairs-distribution elimination constraint. The intersection P_i of the two lines which connect the two data points in each group is expected to be between P_2 and P_3. The y coordinate P_{iy} of P_i is constrained to be between the y coordinates of P_2 and P_3, and the z coordinate P_{iz} of P_i is constrained to be between the z coordinates of P_2 and P_3

2.3 Calculating the data points on the unknown freeform surface

With the defined feature rays and the special constraints depicted above, the next step is to calculate all the data points and obtain the freeform surface contour. In the ideal case, the feature rays used in each field group are refracted by the surface to their ideal image points respectively based on the Snell's law. The vector form of the Snell's law can be written as

$$n'(\mathbf{r}' \times \mathbf{N}) = n(\mathbf{r} \times \mathbf{N}) \tag{6}$$

where $\mathbf{r} = (\alpha, \beta, \gamma)$, $\mathbf{r}' = (\alpha', \beta', \gamma')$ are the unit vectors along the directions of the incident and exit ray. n is the refractive index of the medium around the lens. n' is the refractive index of the lens material. $\mathbf{N} = (i, j, k)$ represents the unit normal vector at the data point. In the YOZ plane, Eq. (6) can be written in the scalar form[17]

$$n'\beta' - n\beta = j(n'\cos I' - n\cos I) \quad (7)$$

$$n'\gamma' - n\gamma = k(n'\cos I' - n\cos I) \quad (8)$$

where I and I' are the angles of incidence and refraction respectively. $\cos I$ and $\cos I'$ can be obtained by

$$\cos I = \beta j + \gamma k \quad (9)$$

$$\cos I' = \frac{1}{n}\sqrt{n'^2 - n^2 + n^2\cos^2 I} \quad (10)$$

In addition, the angle between each incident feature ray and the optical axis is equal to its field angle θ respectively. So, for each feature ray, a tangent relation is required

$$\tan\theta = \frac{\beta}{\gamma} \quad (11)$$

The components of $\mathbf{r}$ and $\mathbf{r}'$ of each feature ray used in Eqs. (7)-(11) can be easily written out with the coordinates of its intersections with the entrance pupil, the unknown surface and the ideal image point. The calculation of the two data points in each field group is taken as a mathematical optimization problem. Two special constraints used in the optimization to obtain the data points that can generate a smooth link line have been depicted in section 2.2. As an exact solution may be not achievable to satisfy the Snell's law for all the feature rays in each field group, Eqs. (7), (8) and (11) are also taken as constraints to control the direction of each feature ray in the optimization process. So, the constraints used to get the corresponding optimum solution in each group are Eqs. (2)-(5), (7), (8) and (11). Note that the constraints Eqs. (7), (8) and (11) will be used several times as there are more than one feature ray in a field group. The y and z coordinates (y_1, z_1), (y_2, z_2) as well as the y and z component (j_1, k_1), (j_2, k_2) of surface normal vector of the two data points in each field group are set as unknown variables. So, all the constraints can be expressed in terms of $(y_1, z_1, y_2, z_2, j_1, k_1, j_2, k_2)$. A merit function $\Phi(y_1, z_1, y_2, z_2, j_1, k_1, j_2, k_2)$ is formed by the sum of residual squares of the constraints. The optimization process is to minimum Φ and to get the corresponding (y_1, z_1) (y_2, z_2). In this paper, the optimization is done by the commercial optimization software 1stOpt ®[18]. Other commercial optimization software such as MATLAB ® and Lingo ® are also recommended.

The whole algorithm starts from the group of the first two fields containing the marginal field of the system. As shown in Fig. 8, when the coordinates (y_1, z_1), (y_2, z_2) of data point 1 and 2 in the first group are obtained, field #2 and the next neighboring field #3 are taken as the next group, and data point 3 and 4 can be then calculated with the same method. The above mentioned process is repeated until the all the fields are calculated. It should be emphasized that when calculating data point 1 and point 2, as no previous neighboring group exists, they can be obtained with the constraints Eqs. (3), (7), (8) and (11). The freeform contour in the tangential plane is finally obtained after curve fitting with all the data points, and it is taken as the starting point for subsequent optimization in optical design software, which will be depicted later in section 3.2.

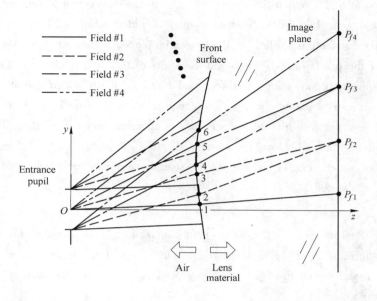

Fig. 8 The calculation of the data points on the contour of the front surface (the starting point)
(The whole algorithm starts from the group of the first two fields containing the marginal field of the system. When the coordinates of data point 1 and 2 in this group are obtained, field #2 and the next neighboring field #3 are taken as the next group, and data point 3 and 4 can be then obtained. The same procedure is done for the group containing field #3 and #4 to obtain data point 5 and 6. The mentioned method is repeated until the all the fields are calculated. The black areas on the front surface stand for the overlap areas of the neighboring fields)

3 Design example: A freeform f-θ single lens

3.1 Designing the starting point of the system

As an example of the proposed method, an f-θ single lens with a wide linear FOV has been designed. The f-θ lens is used for a scanning range of ±210mm in y direction. The system has a linear FOV of ±60°, and it is divided equally into 61 fields with a 2° interval during the design process. As the scanning width y(mm) has a linear relationship with the scanning angle θ (°) for an f-θ lens, the f-θ property can be written as

$$y = \frac{210}{60} \cdot \theta = 3.5 \cdot \theta \tag{12}$$

The system has a circular entrance pupil with 3mm diameter. The scanning light is 780nm infrared laser. The material of the lens is PMMA.

Next, the starting point of the system was designed with the proposed method. Note that only half of the contour for 0° to 60° fields is needed to be generated because of the plane-symmetrical structure. If the data points are calculated only based on the equations to control the ray direction (Eqs. (7),(8) and (11)), these points are irregularly distributed in the tangential plane, as shown in Fig. 9. When the constraint to establish the geometric relationships between neighboring field groups using the surface normal is added (Eqs. (2) and (3)), the data points no longer distribute irregularly in the tangential plane, as shown in Fig. 10(a). However, the link line of the points is not smooth and the stairs-distribution is obvious. When the stairs-distribution elimination constraint (Eqs. (4) and (5)) is finally added, the stairs-distribution is removed and the data points obtained can generate a smooth link line, as shown in Fig. 10 (b). A surface contour with high fitting accuracy is obtained after curve fitting. Then this starting point is entered into optical design software, in this paper, Code V ®[19]. Fig. 11(a) shows the layout of the system. Fig. 11(b) shows the scanning error of each field. In this paper, the scanning error is defined as

$$\Delta h = h' - h \tag{13}$$

where h is the ideal image height, h' is the actual image height. For most of the sampling fields (0° to 50°), the error is within ± 0.4mm. For some larger field angles, the error is no more than ± 1mm. It shows that the light beams are well controlled by the starting point designed with the proposed method.

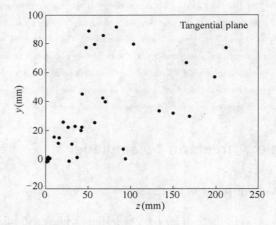

Fig. 9 The irregularly data points calculated by the method when the geometric relationships between neighboring groups are not established

3.2 Optimization of the starting point

The optimization of the starting point was conducted in Code V. In this paper, the surface type of the front surface chooses to be XY polynomials. XY polynomial surface is a commonly used non-rotationally symmetric freeform surface[1,3,6,10,14,19]. The general expression for XY polynomials is shown in Eq. (14):

$$z(x,y) = \frac{c(x^2 + y^2)}{1 + \sqrt{1 - (1 + k)c^2(x^2 + y^2)}} + \sum_{i=1}^{N} A_i x^m y^n \tag{14}$$

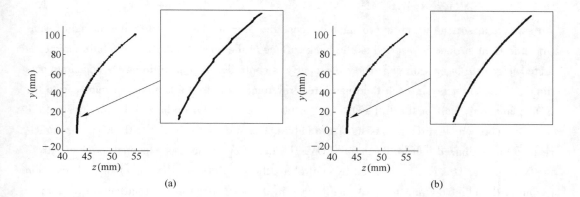

Fig. 10 The effect of the two constraints used during the calculation of the
data points that can generate a smooth link line

(a) The point distribution after adding the constraint to establish the geometric relationships between neighboring field groups using surface normal. The data points do not distribute irregularly in the tangential plane, but the stairs-distribution is obvious; (b) The point distribution when the stairs-distribution elimination constraint is finally added. The stairs-distribution is removed and the data points can generate a smooth link line

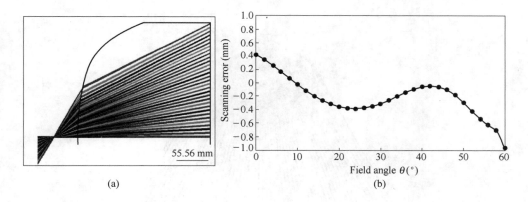

Fig. 11 Design result of the starting point of the system (a) layout of the starting point (front surface); (b) the scanning error of each field of the starting point

where c is the curvature of the surface, k is the conic constant, and A_i is the coefficient of the x-y terms. It has more parameters and thus offers more degrees of freedom to the optical design process. Generally, the XY polynomial surface can achieve a significantly lower error function and better image quality than other surface types such as anamorphic and even aspheric surface[1,3]. Since the optical system is symmetric about the YOZ plane, only the even items of x in XY polynomials are used. Moreover, the higher order polynomial terms only slightly improve the optimization result, and they lower the ray tracing speed and increase the difficulty in manufacture. Therefore, an 11 terms XY polynomial surface up to the 5th order is used in the design.

$$z(x,y) = \frac{c(x^2 + y^2)}{1 + \sqrt{1 - (1 + k)c^2(x^2 + y^2)}} + A_2 y + A_3 x^2 + A_5 y^2 + A_7 x^2 y +$$

$$A_9 y^3 + A_{10} x^4 + A_{12} x^2 y^2 + A_{14} y^4 + A_{16} x^4 y + A_{18} x^2 y^3 + A_{20} y^5 \tag{15}$$

An aspherical surface which approximately keeps the previous outgoing direction of light beams from the front surface is inserted as the back surface of the lens. In addition, a rotating mirror is added to realize laser scanning. The $f\text{-}\theta$ property is controlled by constraining the imaging coordinate of the chief ray in each field using real ray trace data. As the system is symmetric to the XOZ plane, only half of the full FOV (0° to 60°) needs to be optimized. In this design, the front XY polynomial surface consists of two halves. The surface for $-60°$ to 0° FOV is symmetrical to the optimized half for 0° to 60° FOV. The final front surface is expected to be smooth in the XOZ plane. This can be realized by controlling the local $M(+y$ direction) optical direction cosine of the chief ray and the $+X$ sagittal ray in 0° field after the front surface to be zero in real ray trace data. The layout of the final design is shown in Fig. 12. The design result of the system is summarized in Table1. The MTF of each field in the final design is close to the diffraction limit, as shown in Fig. 13(a). Fig. 13(b) shows the spot diagram. Fig. 14 shows the scanning error of each field. For most of the sampling fields, the error is within ± 0.2μm. For some larger field angles, the error is no more than ±1μm. This result proves that good scanning linearity was achieved.

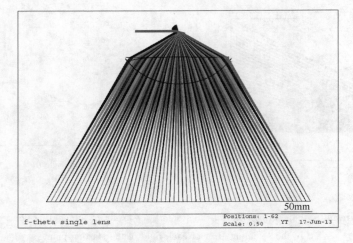

Fig. 12　Layout of the $f\text{-}\theta$ single lens

Table 1　Design result of the single $f\text{-}\theta$ lens

parameter	result
FOV/(°)	±60
Scanning width/mm	±210
Entrance pupil diameter/mm	3
Number of freeform surfaces	1
Total length of the system/mm	280
Back focal length/mm	193.02
Image quality	MTF at diffraction limit @ all fields
Scanning error	Within ±1μm @ all fields

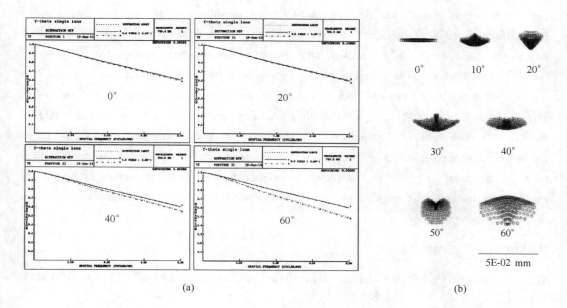

Fig. 13 Image quality analysis of different fields
(a) MTF plot; (b) Spot diagram

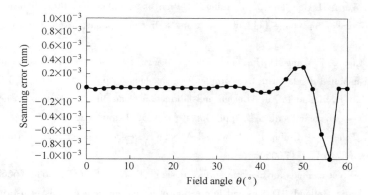

Fig. 14 Scanning error of different fields

4 Conclusion

A freeform single lens design method to achieve a wide linear FOV is depicted in detail in this paper. The aperture size in a wide FOV system is considered during the calculation of the data points on the unknown freeform surfaces, and the calculation of the data points on the freeform surface is a mathematical optimization problem. Two special constraints are employed to find the appropriate data points which can generate a smooth link line. The constraint using surface normal vector at each data point establishes the geometric relationships between neighboring field groups. Moreover, the consistency of the normal vectors after curve fitting can be maintained. Then the smoothness of the link line is improved effectively by adding the stairs-distribution elimination constraint. With these constraints, a smooth and accurate surface contour can be finally obtained after curve fitting. The coordinates and normal vectors of the original data points can be approximately satisfied, and the expected imaging relationship can be ensured. The freeform surfaces are taken as the starting point for further optimization in Code V. A freeform f-θ single lens has been designed as an example of the proposed method. The overall system achieves a wide FOV of $\pm 60°$ with a scanning range of ± 210 mm, and the MTF of each field is close to the diffraction limit. The proposed method to calculate the data points on the unknown surface is effective. Here, only the design of the two-dimensional surface contour for tangential rays is covered in this paper. The design method of the freeform contour can be extended to designing a three-dimensional freeform surface for imaging or illumination optics in the future study.

Acknowledgements

This work is supported by the National Basic Research Program of China (973, No. 2011CB706701).

References

[1] D. Cheng, Y. Wang, H. Hua, and M. M. Talha, "Design of an optical see-through head-mounted display with a low f-number and large field of view using a freeform prism", Appl. Opt. 48 (14), 2655-2668 (2009).

[2] Q. Wang, D. Cheng, Y. Wang, H. Hua, and G. Jin, "Design, tolerance, and fabrication of an optical see-through head-mounted display with free-form surface elements", Appl. Opt. 52(7), C88-C99 (2013).

[3] Z. Zheng, X. Liu, H. Li, and L. Xu, "Design and fabrication of an off-axis see-through head-mounted display with an x-y polynomial surface", Appl. Opt. 49(19), 3661-3668 (2010).

[4] K. Garrard, T. Bruegge, J. Hoffman, T. Dow, and A. Sohn, "Design tools for free form optics", Proc. SPIE 5874, 58740A, 58740A-11 (2005).

[5] R. A. Hicks, "Direct methods for freeform surface design", Proc. SPIE 6668, 666802, 666802-10 (2007).

[6] O. Cakmakci and J. Rolland, "Design and fabrication of a dual-element off-axis near-eye optical magnifier", Opt. Lett. 32(11), 1363-1365 (2007).

[7] L. Xu, K. Chen, Q. He, and G. Jin, "Design of freeform mirrors in Czerny-Turner spectrometers to suppress astigmatism", Appl. Opt. 48(15), 2871-2879 (2009).

[8] X. Zhang, L. Zheng, X. He, L. Wang, F. Zhang, S. Yu, G. Shi, B. Zhang, Q. Liu, and T. Wang, "Design and fabrication of imaging optical systems with freeform surfaces", Proc. SPIE 8486, 848607, 848607-10 (2012).

[9] T. Hisada, K. Hirata, and M. Yatsu, "Projection type image display apparatus", U. S. Patent, 7,701,639 (April 20, 2010).

[10] T. Ma, J. Yu, P. Liang, and C. Wang, "Design of a freeform varifocal panoramic optical system with specified annular center of field of view", Opt. Express 19(5), 3843-3853 (2011).

[11] L. Li and A. Y. Yi, "Design and fabrication of a freeform microlens array for a compact large-field-of-view compound-eye camera", Appl. Opt. 51(12), 1843-1852 (2012).

[12] G. D. Wassermann and E. Wolf, "On the Theory of Aplanatic Aspheric Systems", Proc. Phys. Soc. B 62

(1),2-8 (1949).

[13] D. Knapp,"Conformal Optical Design",Ph. D. Thesis,University of Arizona (2002).

[14] D. Cheng,Y. Wang,and H. Hua,"Free form optical system design with differential equations",Proc. SPIE 7849,78490Q,78490Q-8 (2010).

[15] J. C. Miñano, P. Benítez, W. Lin, J. Infante, F. Muñoz, and A. Santamaría, "An application of the SMS method for imaging designs",Opt. Express 17(26),24036-24044 (2009).

[16] F. Duerr,P. Benítez,J. C. Miñano,Y. Meuret,and H. Thienpont,"Analytic design method for optimal imaging:coupling three ray sets using two free-form lens profiles",Opt. Express 20(5),5576-5585 (2012).

[17] Y. Wang and H. H. Hopkins, "Ray-tracing and aberration formulae for a general optical system", J. Mod. Opt. 39(9),1897-1938 (1992).

[18] 1st Opt Manual,7D-Soft High Technology Inc. (2012).

[19] Code V Reference Manual,Synopsys Inc. (2012).

体全息存储

1000 幅数字图像的晶体体全息存储与恢复*

摘 要 利用 Fe：LiNbO$_3$ 光折变晶体和角度多重方法，在晶体内的一个公共体积内实现了 1000 多幅数字图像的存储与重建。由于同时采用差分编码与纠错编码，因而系统抗噪能力强，误码率低。整个系统由计算机控制，自动化程度高。

关键词 全息存储器；纠错码；角度多重

在体全息光学存储技术[1]中，待存数据按"页"方式并行读写，与传统存储技术中按"位"操作的串行方式显著不同；同时，由于体全息图严格的布拉格选择性，体全息存储还可很容易地实现大量信息的多重存储。因此，晶体全息存储器不但存储容量巨大（理论极限为 $1/\lambda^3$，λ 为光波波长，即约 1012bits/cm^3），而且传输速率极高（可达 1Gbyte/s）。若对读出参考光采用声光、电光等非机械式寻址方式，全息存储器的寻址速度也将极快。另外，更为重要的一点是，由于晶体存储器是对二维图像直接读写，因而它天然地具有快速的内容相关寻址功能，该特性对基于图像运算的军事目标快速识别、医用图像处理、模板匹配等应用十分重要。这种新的存储技术的其他应用领域还包括：并行计算、光学神经网络、光通讯中的光互联开关等。

体全息信息存储概念的提出始于 20 世纪 60 年代[2]，在经过 30 多年的徘徊以后，目前，随着新型优良体全息记录材料（如光折变晶体、光聚合物等）的研制以及相关光电子元器件（如高密度高速率空间光调制器和 CCD 光电探测阵列）制造技术的不断进步，实用化、商品化的大规模体全息存储系统正逐步成为现实。由于全息存储技术的上述诸多优点，美国、日本、英国等发达国家都投入了巨大人力、财力竞相研究。1995 年在美国国家存储工业联合会（NSIC）组织下，组成了一个包括大学、工业部门及政府机构共 12 个单位参加的联合研究体，实施"光折变信息存储材料（PRISM）"和"体全息数据存储系统（HDSS）"两个研究项目，共投资 7000 万美元，从存储材料、关键器件、光学系统结构、信号处理等各方面对全息存储技术展开有分工、有组织的全方位研究，同时组织专家对全息存储技术可能的应用领域进行需求分析，力争抢占这一新兴高技术的战略制高点[3]。联合采用空间与角度等几种多重方法，目前美国已实现上万幅全息图的记录存储[4]。

实现大容量体全息存储的首要关键是要求记录材料有足够的动态范围，对此本文作者以前已进行过较深入的理论研究[5]。本文报道在体全息大容量数据存储研究中的最新实验结果。Fig.1 为实验装置简图。激光波长为 514.5nm，物光与参考光夹角近似为 90°，物光中的空间光调制器（SLM）为— Megashow MP-1830 液晶投影板，像素数

---

* 本文合作者：李晓春，何庆声，邬敏贤，严瑛白，宋修宇。原发表于《光学学报》，1997, 18(6):722~725。

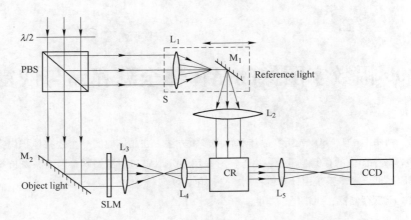

Fig. 1　Schematic illustration of the experimental arrangements
PBS—polarizing beam splitter; CR—crystal; M—mirror; L—lens

720×480，参考光通过透镜 L 反射镜 M_1 及透镜 L_2 到达记录晶体，其中 L_1 与 M_1 被固定在同一精密平移台 S 上，通过计算机控制 S 左右移动，可精密调整参考光入射角度。记录晶体为 $1.5cm \times 1.5cm \times 2.0cm$ 的掺铁铌酸锂晶体。参考光强度约 $5mW/cm^2$，物光的最大光强约 $2.3mW/cm^2$。对应相邻两幅全息图的参考光在空气中的夹角约 $0.01°$。

利用上述实验装置，在晶体中的一个公共体积内已实现 1000 多幅数字图像的存储与恢复，图像均为由计算机产生的随机数页。Fig. 2 为记录的第一幅图像在不同存储容量时的读出效果。当存储 1000 幅以后，记录的第一幅全息图的衍射效率约为 10^{-5}，Fig. 3 为第一幅存储图像的读出信号强度 I 随存储容量 C 的变化情况。

Fig. 2　Reconstructions of the first stored hologram under different storage capacities
（a）Just after its storage；（b）After 150 hs. storage；（c）After 250 hs. storage；（d）After 300 hs. storage；（e）After 400 hs. storage；（f）After 550 hs. storage；（g）After 700 hs. storage；（h）After 800hs. storage；（i）After 1000 hs. storage. hs. Stands for holograms

为降低误码率，原始数据先经过 Hamming 纠错编码；为了在数据判读时可采用自适应阈值，在将编码信息映射为空间光调制器像素的亮暗状态时，又采用了差分编码技术。实验表明，经过这两重编码之后，系统抗噪能力大大增强，可有效地防止各种大尺度噪声及随机噪声[6]。最后取得的误比特率约为 10^{-6}。Fig. 4 为读出的第一幅存储图像中的错误比特数随存储容量 C 的变化情况。

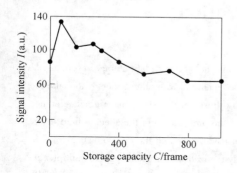

Fig. 3 The signal intensity of the reconstructed image of the first stored hologram varies with the storage capacity C

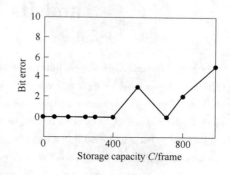

Fig. 4 Variation of the number of bit errors in the reconstructed image of the first stored hologram versus the storage capacity C

有关详细实验情况将另文发表。

参 考 文 献

[1] 金国藩，李晓春，光学体全息数据存储新技术. 光子学报，1997, 26(Z1):14~20.
[2] P. J. vanHeerden, Theory of optical information storage in solids. Appl. Opt., 1963, 2(4):393~400.
[3] L. Hesselink. Digital holographic data storage looks ahead. Photonics Spectra, 1996, 30(3):44~46.
[4] G. W. Burr, F. H. Mok, D. Psaltis, Large scale volume holographic storage in the long inferaction length architecture. Proc. SPIE, 1994, 2 297:402~414.
[5] Xiaochun Li, Minxian Wu, Yingbai Yan et al., Dynamic range metric for a photorefractive crystal in a volume holographic memory. Opt. Commun., 1997, 138(3):143~150.
[6] 李晓春，邹敏贤，成罡等，数字体全息数据存储中的两种信道编码方法. 光子学报，1997, 26(4):330~335.

Volume Holographic Storage and Retrieval of 1000 Digital Images in Lithium Niobate

Abstract As many as 1000 digital images have been recorded in a common volume of an Fe: LiNbO$_3$ photorefractive crystal using angular multiplexing. The combination of the error-correcting encoding and the differential encoding results in a significant noise-tolerant feature and thus a lower bit-error rate is achieved. As a highly automated digital holographic memory, the overall system is under the control of a computer. Experimental results are presented.

Key words holographic memory; error-correcting encoding; angular multiplexing

Dynamic Speckle Multiplexing Scheme in Volume Holographic Data Storage and Its Realization*

Abstract Dynamic speckle multiplexing scheme in volume holographic data storage is proposed, since it offers a novel multiplexing geometry, and could be combined with other schemes to make the full use of the dynamic ranges. In this scheme, a random diffuser is added in the original reference path of the classical 90° setup. In this paper, we analyzed the propagation of the speckle field in the holographic system and established the related theoretical model based on the dynamic speckle autocorrelation function and diffraction theory. We successfully realized the dynamic speckle multiplexing in our experimental system and reached a storage density of 4.6 Gigapixels/cm^3 based on the DPL laser source.

1 Introduction

Volume holographic data storage is more and more attractive because of its promising high storage density and fast data transfer rate[1]. Holographic multiplexing methods, such as wavelength[2], angular[3], shift[4], and phase encoding[5] have been widely used. These multiplexing methods of volume holograms are based on the selective reconstruction of specific holograms out of an entire ensemble of holograms. However, due to the limitation of the multiplexing methods themselves, none of the methods could exhaust the dynamic range of the storage media entirely. Recently, V. Markov proposed the static speckle multiplexing scheme[6,7], which is based on the spatial autocorrelation character of the speckle field. This scheme permits the real three-dimensional storage and offers a much smaller multiplexing shift. Compared with the conventional plane wave reference beam, it may enhance the storage density greatly.

Unlike V. Markov, we proposed a novel dynamic speckle multiplexing scheme, which generates different speckle fields by moving the random phase diffuser itself, instead of moving the storage crystal. As Fig. 1 shows, we put a random phase diffuser in the reference beam, where, along with the variation of the diffuser, different parts of the diffuser are employed to generate the speckle field. Therefore, the speckle distribution on the surface of the crystal will change. Consequently, the interference of the reference beam and the signal beam will change, finally realizing the optical data storage.

Compared with static speckle multiplexing, in the dynamic speckle multiplexing scheme the storage media will not move during the storage process; therefore, all information is stored in a

---

* Copartner: Qingsheng He, Jinnan Wang, Peikun Zhang, Jiangang Wang, Minxian Wu. Reprinted from *Optics Express*, 2003, 11(4):366-370.

common volume. This will definitely benefit the storage density of the holographic system. In addition, the dynamic speckle multiplexing scheme can be combined with other schemes, such as angular and spatial multiplexing, to form novel hybrid multiplexing schemes. Later in this paper, we will discuss angular speckle multiplexing and its realization.

In this paper, based on the auto-correlation character of the speckle field and diffraction theory, we establish the theoretical model of the dynamic speckle multiplexing, and discuss the relationship between the correlation length and multiplexing shift. At the end of the paper, we present the experimental result of a high density storage based on dynamic speckle multiplexing, which reaches a storage density of 4.6 Gigapixels/cm^3.

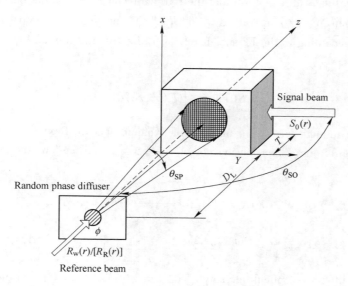

Fig. 1 Geometry of the hologram recording by signal plane wave $S_0(r)$ and dynamical speckle reference wave $R_W(r)$

(Here T is the thickness of volume hologram. D_L is the distance from the hologram front surface to random-phase diffuser. Δ is the shift of the diffuser)

2 Theoretical analysis

As Fig. 1 shows, in the dynamic speckle multiplexing scheme, the speckle-coded reference wave $R_W(r)$ interfere with the plane object wave $S_0(r) = A\exp(ikS_0 \cdot r)$, and form the holographic gratings in the storage media. After the exposure, the permittivity of the storage media $\varepsilon(r)$ will change to $\varepsilon(r) = \varepsilon_0 + \delta\varepsilon(r)$, where $\delta\varepsilon(r)$ is the modulated component of the permittivity, direct proportion to the square of the electric field of the interacting waves, i.e.,

$$\delta\varepsilon(\vec{r}) \propto |E|^2 = |\vec{S}_0(\vec{r}) + \vec{R}_W(\vec{r})|^2 \propto S_0(\vec{r}) \vec{R}_W^*(\vec{r}) \tag{1}$$

$\vec{R}_W^*(\vec{r})$ is the conjugate of $R_W(\vec{r})$.

Now, we use the read out speckle encoded field $R_R(r)$ to retrieve the recorded hologram. The retrieve beam $R_R(r)$ and the diffractive beam $S(r)$ could be described by the Maxwell equations. Considering the monochromatic waves of identical polarization in an isotropic media, the Maxwell equation could be reduced to a scalar wave function. With the first Born approximation and perturbation theory method, the diffractive optical field $S(r)$ should be:

$$S(\vec{r}) = k_0^2 \int_{-\infty}^{\infty} \delta\varepsilon(\vec{r}')R_R(\vec{r}')G(\vec{r},\vec{r}')\mathrm{d}V' \tag{2}$$

$G(\vec{r},\vec{r}') = \dfrac{\exp(ik_0|\vec{r}-\vec{r}'|)}{4\pi|\vec{r}-\vec{r}'|}$ is the Green function.

In the dynamic speckle multiplexing system, when we are multiplexing holograms, the diffuser will move Δ each time. Now, suppose that the retrieve beam $R_R(r)$ is the beam that formed after recording beam $R_W(r)$ moved Δ. And, $R_R(r)$ and $R_W(r)$ propagate in the same direction. Then, according to the Fresnel-Kirchhoff diffractive integration, Eq. (2) could be expressed as,

$$S(\vec{r}) \propto \int_{-\infty}^{\infty} \Gamma(\vec{r},\vec{r}')\mathrm{d}V' \tag{3}$$

$\Gamma(\vec{r},\vec{r}')$ is the mutual correlation function of dynamic speckle field. Generally, we consider only the condition of one dimension movement, i. e., $\Delta = \Delta_y Y$. Therefore, the mutual correlation function is,

$$\Gamma(\vec{r},\vec{r}') = [R_W^*(\vec{r})R_R(\vec{r}')] = \frac{\sqrt{\pi}\omega_0^2}{2\sqrt{2}\omega(Z_0)\lambda^2 Z^2}\exp\left(-\frac{\Delta_y^2}{2\omega^2(z_0)}\right) \times$$
$$\exp\left\{-\frac{\pi^2\omega^2(z_0)\Delta_y^2}{2\lambda^2 z^2}\left[1+\frac{z}{\rho(z_0)}\right]^2\right\} \times \exp\left(i\frac{2\pi\Delta_y}{\lambda z}y'\right) \tag{4}$$

Use normalized diffractive optical intensity $I_{DN}(\Delta) = I_D(\Delta)/I_D(\Delta=0)$ and the shift of diffuser Δ to describe the multiplexing selectivity of dynamic speckle, where, $I_D(\Delta=0)$ is the diffractive intensity when the diffuser shift is zero, we will get,

$$I_{DN}(\Delta_y) = \exp\left[-\frac{\Delta_y^2}{\omega^2(z_0)}\right] \frac{\left|\iiint_{v'} \frac{1}{z^2}\exp\left\{-\frac{\pi^2\omega^2(z_0)\Delta_y^2}{2\lambda^2 z^2}\left[1+\frac{z}{\rho(z_0)}\right]^2\right\}\exp\left(i\frac{2\pi\Delta_y}{\lambda_z}y'\right)\mathrm{d}x'\mathrm{d}y'\mathrm{d}z\right|^2}{\left|\iiint_{v'} \frac{1}{z^2}\mathrm{d}x'\mathrm{d}y'\mathrm{d}z\right|^2}$$

(5)

$\omega(z_0)$ and $\rho(z_0)$ are the radius and the curve radius of the illumination beam respectively.

The correlation length of the speckle field on the surface of the storage media is $\delta = 0.66 \cdot \dfrac{\lambda D_L}{\omega(z_0)}$, D_L is the distance from the diffuser to the surface of the storage media.

3 Computational and experimental results and some discussions

3.1 The relationship between the angular selectivity and the correlation length δ

From the numerical computation Eq. (5), it is possible to discuss the relationship between the diffractive intensity $I_{DN}(\Delta_y)$ and the shift of diffuser Δ_y. The correlation length δ could be adjusted by changing the distance D_L. From the computational results Fig. 2, we could find that, the diffraction intensity will drop monotonically along with the increase of Δ_y. Furthermore, the smaller of the correlation length δ, the faster $I_{DN}(\Delta_y)$ drops.

In the holographic data storage system, we put a random diffuser before the storage media along the reference optical path.

As we move the motor controller, different parts of the diffuser will be illuminated, and thus, different speckle fields will be generated on the surface of the storage

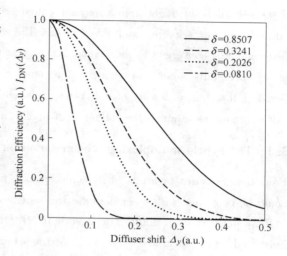

Fig. 2 Calculated dependence of the normalized diffratcted beam intensity $I_{DN}(\Delta_y)$ on shift Δ_y at reconstruction with different speckle sizes δ

media. We adjust the distance of D_L to get different speckle field with different correlation length δ. We have known that, the smaller of D_L, the smaller of δ. From Fig. 3, it is easy to find that, the smaller of correlation length δ, the better of the multiplexing shift of the dynamic speckle multiplexing system, which bears well accordance with the theoretical analysis.

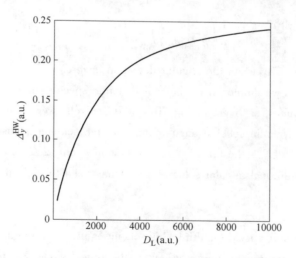

Fig. 3 The shifting selectivity Δ_y^{HW} as a function of the distance D_L from the hologram front surface to random phase diffuser

3.2 High density storage experimental results

The storage system was designed for high-density dynamic speckle multiplexed holographic storage using the 90° geometry. The storage material is Iron Doped Lithium Niobate (LiNbO$_3$: Fe,

0.03% Fe-doped). Light from a frequency-doubled diode pumped Nd: YAG laser was expanded and split into reference and object beams. The reference beam passed the random phase diffuser and was converged to the crystal. 500 digital frames, each curtaining 1024×768 pixels, were successfully stored and retrieved in a common volume of 0.086 cm^3 of the storage crystal, reaching the storage density of 4.6 Gagapixels/cm^3. Besides, the crystal was immersed in the NaCl solution to suppress the influence of photovoltaic field and the possible degradation[8].

3.3 The hybrid multiplexing scheme based on dynamic speckle multiplexing

From the theoretical analysis of dynamic speckle multiplexing scheme, we could know that, it could be combined with other multiplexing schemes to get new hybrid multiplexing schemes.

If we put a random diffuser in the original 90° angular multiplexing geometry, then it turns into the dynamic speckle angular multiplexing scheme. Suppose that the retrieve beam $R_R(r)$ is a different speckle field that forms after the recording beam $R_W(r)$ turns a tiny angle $\delta\theta_A$ ($\delta\theta_A \ll 1$), and they have the same propagation direction, then, the diffraction optical field could be expressed as[9],

$$S(\delta\theta_A, q) = \exp(\vec{k}_0 \sin\theta_s) t_0^2 \int_0^T \frac{1}{z'\delta\theta_A} \exp\left[\frac{-ik_0\delta\theta_A}{d_L}(z'\delta\theta_A + 2y)\right] J_1\left(\frac{k_0\phi_L\delta\theta_A}{2d_L}\right) dz' \quad (6)$$

After normalization, we can get the relation between the multiplexing shift and the deviation angle $\delta\theta_A$[9],

$$\frac{I_D(\delta\theta_A)}{I_{Dmax}} = \frac{1}{\pi}\left(\frac{4d_L}{k_0 D_H \phi_L T}\right) \iint_{0 \leq q^2 \leq D_H^2/4} |S(\delta\theta_A, \vec{q}')|^2 d^2 q' \quad (7)$$

I_{Dmax} is the readout intensity when the angular deviation is zero.

To prove its validity, we employ it in the real dynamic speckle angular multiplexing storage system. 300 digital frames, each curtaining 1024×768 pixels, were stored in a common volume. Because of the dynamic speckle angular multiplexing scheme is insensitive to the multi-longitudinal mode of DPL[10], the full angle of the system is only 2.28°, which is much smaller compared with the traditional angular scheme which may need the full angle of 13.6°.

4 Conclusion

In this paper, the dynamic speckle multiplexing scheme is proposed to enhance the storage density of a volume holographic data storage system. A theoretical model was proposed, and numerical and experimental results correspond very well with each other, proving the correlation length of the speckle field in the reference beam determines the multiplexing shift, where, the smaller of the length δ, the better the multiplexing selectivity. In addition, a very high storage density of 4.6 Gigapixels/cm^3 was reached based on the dynamic speckle multiplexing in our holographic system. We also realized the hybrid dynamic speckle angular multiplexing scheme experimentally, which has the best potential to realize super-high storage density.

References

[1] J. F. heanue, M. C. Bashaw, and L. Hesselink, "Volume holographic storage and rettieval of digital date", Science 256, 749-752(1994).

[2] G. A. Rakuljic, V. Leyva, and A. Yariv, "Optical data storage by using orthogonal wavelength-multiplexed volume holograms", Opt. Lett. 17, 1471-1473(1992).

[3] F. H. Mok, "Angule-multiplexed storage of 5000 holograms in lithium niobate", Opt. Lett. 18, 915-917 (1993).

[4] D. Psaltis, M. Levene, A. Pu and G. Barbastathis, "Holographic stroage using shift multiplexing", Opt. Lett. 20, 782-784(1995).

[5] J. E. Ford, Y. Fainman, and S. H. Lee, "Array interconnection by phase-coded optical correlation", Opt. Lett. 15, 1088-1090(1990).

[6] A. Darsky, V. Markov, "Information capacity of holograms with reference speckle wave", Proc. SPIE 1509, 36-46(1991).

[7] Darskii A. M. Markov V. B, "Some properties of 3D holograms with a reference speckle-wave and their application to information storage", Proc. SPIE 1600, 318-332(1992).

[8] Qingsheng He, Guodong Liu, Xiaochun Li, Jiangang Wang, Minxian Wu, Guofan Jin, "Suppression of the influence of photovoltaic dc field on volume hologram in $Fe:LiNbO_3$", Appl. Opt. 41, 4104-4107(2002).

[9] V. Markov, "Spatial-angular selectivity of 3-D speckle-wave holograms and information storage", J. Imaging Sci. Technol. 41, 383-388(1997).

[10] Jinnan Wang, Qingsheng He, Dong Huang, "High density Volume holography data storage based on Speckle Angular Multiplexing", OSA Annual Meeting, (2002).

[11] Darskii A. M. and Markov, V., "Shift selectivity of the holograms with a reference speckle wave", Opt. Spectroscopy 65, 392-395(1988).

[12] Peikun Zhang, Qingsheng He, Guofan Jin, "A novel speckle angular-shift multiplexing for high-density holographic storage", Proceedings of SPIE 4081, 236-241(2000).

Exposure-schedule Study of Uniform Diffraction Efficiency for DSSM Holographic Storage[*]

Abstract An exposure-schedule theory of uniform diffraction efficiency for a dynamic-static speckle multiplexing (DSSM) volume holographic storage system is proposed. The overlap-factor $\gamma_{overlap}$ is introduced into the system to compensate for the erasure effect of the static speckle multiplexing scheme. The exposure-schedule which is an inverse recursion formula is determined. Experimental results are obtained in a $LiNbO_3$: Fe crystal and 400 holograms with uniform diffraction efficiency are achieved by the use of the new exposure-schedule.

1 Introduction

Recently, holographic storage has received increasing attention owing to its potentially high storage capacity and fast data access rate. Its large storage capacity is realized by means of multiplexing schemes, such as angular multiplexing, wavelength multiplexing, shift multiplexing and phase-code multiplexing[1,2]. Most of them relate to the Bragg selective reconstruction of specific holograms out of an entire ensemble of holograms except phase encoding techniques.

In the last decade, V. Markov proposed the speckle multiplexing scheme[3,4], which is based on the spatial autocorrelation character of speckle field and realized by using the same speckle reference beam and shifting the storage material. It can be called the static speckle multiplexing scheme[5,6]. In our last paper, Q. S He ets. proposed a novel dynamic speckle multiplexing scheme[7,8], which generates different speckle fields by moving the random phase diffuser, instead of the storage crystal as implemented in the static speckle multiplexing scheme. One advantage of dynamic speckle multiplexing is that it can be combined with other multiplexing schemes easily. Subsequently, we incorporated both the dynamic speckle multiplexing and static speckle multiplexing schemes for high capacity holographic storage. As can be seen from Fig. 1, the idea of this hybrid multiplexing scheme is to record some holograms by first shifting a holographic diffuser in the reference beam as a random-phase modulator and then shifting the $LiNbO_3$: Fe crystal to record another serious of holograms. We call this multiplexing scheme "Dynamic-Static speckle multiplexing (DSSM)", a holographic data storage method chosen for its simple setup and the potential to offer further data storage capability. Each reference beam interferes in the crystal with object beam carrying information. To retrieve the data, the object beam is blocked and the phase modulator reproduces the specific phase code to which the re-

---

[*] Copartner: Xiaosu Ma, Qingsheng He, Jinnan Wang, Minxian Wu. Reprinted from *Optics Express*, 2004, 12(6): 984-989.

quired data has been addressed.

In this paper we propose a recording schedule for DSSM holographic storage. The theoretical analysis is applied to the recording of multiplexed holograms that reconstruct with equal diffraction efficiencies. Experimental results demonstrating the validity of this schedule are presented for holograms recorded in both $LiNbO_3$: Fe and $LiNbO_3$: Fe: In crystals.

As is well known, each newly recorded hologram has tiny effect on the retrieval, but partially erases all previously stored images.

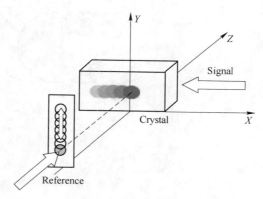

Fig. 1   Geometry of DSSM holographic storage

Therefore, if uniform recording time and uniform exposure intensity are applied to record all the holograms, each hologram recorded previously must have a lower diffraction efficiency than that recorded at a later time[9,10]. Burke and Psaltis proposed a recording schedule called sequential exposure that effectively compensates the erasure of the later-recorded holograms[11,12]. But all these conclusions were obtained by using a collimated light as the reference beam, so some basic experiments in our DSSM system were implemented to demonstrate that it is still valid when we use a speckle reference beam. According to the experimental results, the recording and erasing process can also be drawn as an exponential curve like using a collimated light. So the recording schedule can be applied directly in the dynamic speckle multiplexing process.

In the next section, we propose a novel exposure schedule that can optimize DSSM holographic storage scheme. Before our analysis, two assumptions should be considered for DSSM: one is that all the holograms in one position of the crystal are governed by the dynamic speckle multiplexing; the other is that erasing effect between every two positions is proportional to the superposed area of every two recording faculae. When these two positions are completely overlapped, the schedule is the same as that of the dynamic multiplexing storage. Along this way, our overriding goal will be to explore: for a sequence of DSSM holograms, how long should each frame be recorded to achieve uniform image playback intensity?

2 Principle

As stated earlier in the paper, the DSSM recording process records M frames of holograms in one position by shifting the holographic diffuser in the reference beam and then to record another M frames of holograms by shifting the storage material. Finally, $M \times N$ holograms can be recorded in the crystal as shown in Fig. 2.

According to the two assumptions above, if the facular point radius of the reference beam is r, and the shift distance between two points is $1(1 < 2r)$, the superposition area can be calculated as follows.

The shaded area showed in Fig. 3 can be calculated as:

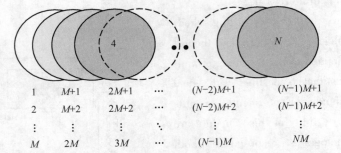

Fig. 2 Schematic diagram of the DSSM holograms

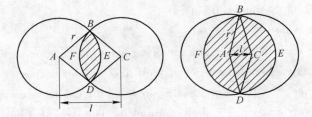

Fig. 3 Relationship diagram of the overlapped holograms

$$S_{\text{overlap}} = 2r^2\cos^{-1}\left(\frac{l}{2r}\right) - l\sqrt{r^2 - l^2/4} \tag{1}$$

We can now define an overlap-factor γ_{overlap}, which represents the proportion of overlapped areas

$$\gamma_{\text{overlap}} = \frac{S_{\text{overlap}}}{S_{\text{whole}}} = \frac{2}{\pi}\cos^{-1}\left(\frac{l}{2r}\right) - \frac{l}{\pi r^2}\sqrt{r^2 - l^2/4} \tag{2}$$

This factor can be regarded as a constant to compensate for the erasure effect of the static speckle multiplexing scheme. So we can adopt the sequential exposure schedule here in our DSSM system by introducing a factor γ_{overlap}. Since the diffraction efficiency of each hologram is roughly proportional to the square of the refractive-index modulation depth, the formulae can be derived. First, the recording and erasing process can be respectively written as:

$$\begin{aligned}\text{recording process} \quad & \Delta n = \Delta n_s[1 - \exp(-t/\tau_r)] \\ \text{erasing process} \quad & \Delta n' = \Delta n \exp[-t'/(\tau_e/\gamma_{uv})]\end{aligned} \tag{3}$$

where Δn_s is the saturated steady-state index of refraction modulation, Δn and $\Delta n'$ are the refractive-index modulation after being exposed for time t and erased for time t' respectively and τ_r, τ_e, is the recording and erasure-time constants. γ_{uv} is the overlap-factor between the uth hologram and the vth hologram. The function of the overlap-factor is to increase the erasure-time constant and weaken the erasure effect.

When M holograms are recorded in one recording position, and then N positions are implemented by shifting the crystal in the DSSM storage system. Every hologram must be partially erased by all other holograms recorded after it. Here Δn_g is defined as the refractive-index modulation of the gth ($g = 1, 2, \cdots, NM$) hologram.

$$\Delta n_1 = \Delta n_s [1 - \exp(-t_1/\tau_r)] \exp\left[-\left(\sum_{j=2}^{MN} \gamma_{j,1} t_j\right)/\tau_e\right]$$

$$\Delta n_{(i-1)M+i'} = \Delta n_s \left[1 - \exp\left(-\frac{t_{(i-1)M+i'}}{\tau_r}\right)\right] \exp\left[-\left(\sum_{j=(i-1)M+i'+1}^{MN} \gamma_{j,[(i-1)]M+i'} t_j\right)/\tau_e\right] \quad (4)$$

$$\Delta n_{NM} = \Delta n_s [1 - \exp(-t_{NM}/\tau_r)]$$

Generally, the recording-and erasure-time constants are not the same, and we equate the index of refraction modulation of every hologram,

$$\Delta n_s = \left(\frac{\tau_r}{\tau_e}\right) \sum_{i=1}^{N} \sum_{j=1}^{M} \Delta n_{ij} \quad (5)$$

In this model, it is reasonable to expect uniform diffraction efficiency if a recording sequence satisfies the criterion that $\Delta n_{ij} = \Delta n_{i'j'}$,

$$\Delta n_{ij} = \left(\frac{\tau_r}{\tau_e}\right) \frac{\Delta n_s}{MN} = \frac{\alpha \Delta n_s}{MN} \quad \text{and} \quad \alpha = \frac{\tau_r}{\tau_e} \quad (6)$$

To simplify the recursion formula of the exposure schedule, the numbers of the two-dimension holograms here are converted to one-dimension $1, 2, \cdots, M \times N$. Substitute Eq. (6) into Eq. (4), then the exposure schedule can be obtained as an inverse recursion formula,

$$t_{N \times M} = -\tau_r \ln\left(1 - \frac{\Delta n_{N \times M}}{\Delta n_s}\right) = -\tau_r \ln\left(1 - \frac{\alpha}{M \times N}\right) \quad (7)$$

where $t_{N \times M}$ is the exposure time of the last hologram in the last point.

$$t_{(i-1) \times M+k} = -\tau_r \ln\left\{1 - \frac{\alpha}{M \times N} \exp\left[\left(\sum_{j=(i-1) \times M+k+1}^{N \times M} \gamma_{j,(i-1) \times M+1} t_j\right)/\tau_e\right]\right\} \quad (8)$$

Where $t_{(i-1) \times M+k}$ represent the exposure time of the kth hologram in the ith point. Finally, the exposure time of the first hologram can be expressed by,

$$t_1 = -\tau_r \ln\left\{1 - \frac{\alpha}{M \times N} \exp\left[\left(\sum_{j=2}^{N \times M} \gamma_{j,1} t_j\right)/\tau_e\right]\right\} \quad (9)$$

Now, we deduced the exposure schedule, which is an inverse recursion formula. It is similar with the schedule of the dynamic speckle multiplexing, but with an erasure factor γ_{overlap} to compensate for the erasure effects of the DSSM scheme. Some experiments are executed by using an iron-doped lithium niobate. The validity of the exposure schedule of the DSSM has been validated and the diffraction efficiencies of the recorded holograms are comparatively uniform.

3 Experiment

We used a diode pumped laser at 532 nm as the light source. The hologram recording medium is a $LiNbO_3$: Fe crystal of 45° cut, and its dimensions are 10mm × 10mm × 10mm. The laser was collimated and split into two parts. One was the signal beam modulated by a resolution chart incident on the crystal. The other beam was incident on a diffuser, with the passing wave as the reference beam. The signal and the reference beams were incident on adjacent surfaces of the crystal. Thus the incident angle is approximately 90°. Two accurate motion stages are used in this system to execute the DSSM process. Stage. 1 is used to control the vertically lateral movement of the diffuser, and stage 2 is used to control the horizontal movement of the crystal. The experimental setup is shown in Fig. 4.

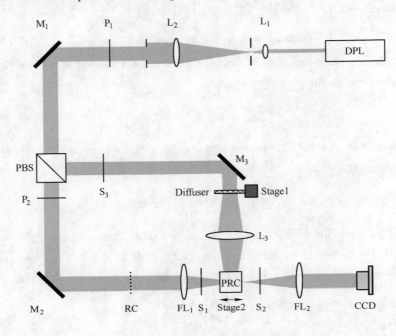

Fig. 4 Experimental setup

It should be emphasized that the result is valid only if τ_w and τ_e are all constants for every frame in the recording sequence. This requires precise survey of the writing-and the erasure-time constants of the crystal. Another practical problem arises from the fact that we cannot directly measure τ_w and τ_e, but must deduce these constants (under our exposure conditions) from experimental observations.

We adopted Eq. (8) to generate different series of recording time schedules. By using this schedule, we obtained DSSM storage with 10×10 holograms. This means that 10 holograms are recorded in one position and 10 positions are recorded in a $LiNbO_3$: Fe crystal doped with 0.03wt% Fe. Generally, the intensity of the gray scale is proportional to the diffraction efficiency of the hologram. The experiment result is shown in Fig. 5(a). To reduce the scattering noise in the reconstructed figures and increase the recording rate, a second type of crystal ($LiNbO_3$: Fe: In which is doped with 0.03wt% Fe and 1.0mol% In) is employed to record 10×10 holograms with the same scheme. The experiment result has a less than ±6.2% fluctuation and is as shown in Fig. 5(b). The result also demonstrates the validity of the schedule.

The demonstrative experiment with 20×20 holograms was executed in a $LiNbO_3$: Fe: In crystal. The expected uniformity in diffraction efficiency with ±7.67% fluctuation can be observed, as shown in Figs. 6(a) and (b).

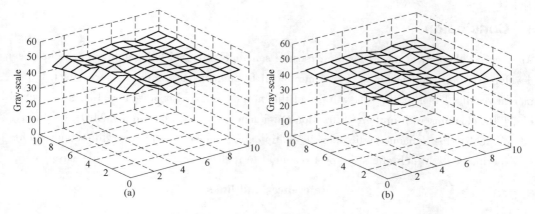

Fig. 5 Experimental result
(a) in a LiNbO$_3$: Fe crystal; (b) in LiNbO$_3$: Fe: In crystal

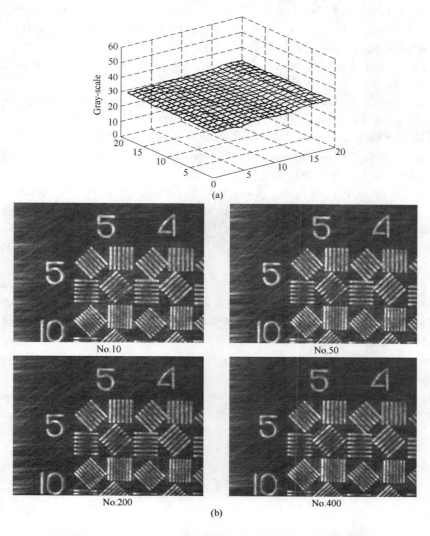

Fig. 6 Experimental result in a LiNbO$_3$: Fe crystal with 20 × 20 holograms

4 Conclusion

An accurate theoretical model for the exposure-schedule in DSSM holographic storage has been developed. An erasure factor $\gamma_{overlap}$ is introduced to compensate for the erasure effect. Experimental results demonstrating the validity of this approach are also presented. Using this schedule, 400 holograms with uniform diffraction efficiencies were recorded in a $LiNbO_3$: Fe: In crystal. We realize the dynamic-static speckle multiplexing scheme with uniform diffraction efficiency experimentally, which offers the best potential to realize super-high storage density.

References and links

[1] H. J. Coufal, D. Psaltis, and G. T. Sincerbox, Holographic data storage, Springer Series in Optical Sciences Vol. 76.

[2] K. Curtis, W. L. Wilson, M. Tackitt, A. J. Hill, and S. Campbell, "High Density, High Performance Data Storage via Volume Holography: The Lucent Technologies Hardware Platform", Optical Data Storage 8, OSA Technical Digest Series, 1998:168-170.

[3] A. Darsky and V. B. Markov, "Information capacity of holograms with reference speckle wave", Proc. SPIE 1509, 1991:34-36.

[4] A. M. Darskii and V. B. Markov, "Some properties of 3D holograms with a reference speckle-wave and their application to information storage", Proc. SPIE 1600, 1992:318-332.

[5] V. B. Markov, "Holographic memory with speckle-wave volume hologram", SPIE 3486, 1997:68-79.

[6] C. C. Sun, W. C. Su, "Three-dimensional shifting selectivity of random phase encoding in volume holograms", Appl. Opt. 40, 2001:1253-1260.

[7] Q. S. He, J. N. Wang, P. K. Zhang, J. G. Wang, M. X. Wu, and G. F. Jin, "Dynamic speckle multiplexing scheme in volume holographic data storage and its realization", Opt. Express 11, 366-370 (2003), http://www.opticsexpress.org/abstract.cfm? URI = OPEX-11-4-366.

[8] J. N. Wang, S. R. He, Q. S. He, D. Huang, G. F. Jin, "Insensitivity of Speckle Multiplexing to Multi-Longitudinal Modes of Laser in Volume Holographic Storage", Chin. Phys. Lett. 20, 2003:1047.

[9] M. L. Delong, B. D. Duncan, and J. H. Parker, "Parametric extension of the classical exposure-schedule theory for angle-multiplexed photorefractive recording over wide angles", Applied Optics 37, 1998.

[10] E. S. Maniloff and K. M. Johnson, "Maximized photorefractive holographic storage", J. Appl. Phys. 70, 1991:4702-4707.

[11] D. Psaltis, D. Brady, and K. Wagner, "Adaptive optical networks using photorefractive crystals", Appl. Opt. 27, 1988:1752-1759.

[12] W. J. Burke, and P Sheng, "Crosstalk noise from multiple thick-phase holograms", J. App. Phys. 48, 1977:681.

10Gb/cm³ 小型化体全息数据存储及相关识别系统*

摘　要　简述了体全息数据存储的最近研究进展，讨论了体全息存储系统小型化、实用化的关键技术，并对全息存储系统的结构、复用及寻址方式进行了优化，在此基础上研制了小型化体全息存储及相关识别系统，其存储密度达 10Gb/cm³，兼有多页面高速并行相关识别处理功能，识别速率大于 2000 幅/s；提出的动态散斑复用的新概念及动态散斑—角度混合复用技术，其复用灵敏度比单纯角度复用灵敏度提高了 8 倍，为进一步提高存储密度提供了有效途径，展示了体全息存储系统在海量存储及光学并行处理领域具有很大的应用潜力。

关键词　光学数据存储；光折变材料；相关识别；体全息；信息处理

如何有效地存储越来越多的数据是当前信息社会所面临的一个重要课题，尽管目前已经有了大容量的计算机磁盘，光盘 CD（Compaet Disk），DVD（Digital Versatile Disk）甚至磁光盘 MO（Magnetooptieal Disk）[1]，但是伴随计算机网络和多媒体技术的发展，信息的容量以爆炸的形式增长着，人们需要进一步寻求性能更优异的存储技术。在众多的新型存储中，体全息存储技术格外受到人们的青睐。

光学体全息存储是超高密度存储中最重要的研究领域之一早在 1963 年美国科学家 Heerden 就曾提出利用全息术进行数据存储的概念[2]，当时由于缺少合适的记录材料，光电元器件技术还不成熟，因而体全息存储的研究工作进展很小。20 世纪 90 年代后，随着计算机科学和现代信息处理技术的不断发展，对于具有大容量、高传输率、可快速存取的数据存储系统提出了日益迫切的要求；同时，随着新型优良全息记录材料（如光折变晶体和光聚合物）的研制出现以及相关元器件，如高速空间光调制器 SLM（Spatial Light Modulator）和电荷耦合光电探测阵列 CCD（Charge Coupled Device）制造技术的不断进步，可满足各种实际应用要求的体全息数据存储系统正逐步成为可能，人们对体全息存储的兴趣又重新高涨起来。

在体全息光存储研究 40 多年的历史中，许多研究机构都对体全息存储的研究作出了很大贡献，特别是在最近几年有了很大突破。1995 年美国实施 DARPA 计划，由美国国家存储工业联合会主持，组织 IBM 公司的 Almaden 研究中心，Rockwell 科学中心，加州理工学院，斯坦福大学，卡耐基梅隆大学等共 12 家单位联合协作，实施了光折变存储材料（PRISM）项目和全息数据存储系统（HDSS）项目，预期在 5 年内开发出具有容量为 10^{12}b、存取速率为 1Gb/s 的一次写入或可重复写入的全息数据存取系统。在这两个项目的支持与带动下，许多体全息存储系统先后问世[3~8]。

---

* 本文合作者：曹良才、何庆声、尉昊赟、刘国栋、欧阳川、赵健、邬敏贤。原发表于《科学通报》，2004，49(27)：2495~2500。

一般的光学体全息数据存取机理可简单描述为：待存储的数据（数字或模拟图像）经 SLM 调制到信号光上，形成二维信息页，然后与参考光在记录介质中发生干涉，利用光折变晶体或光聚合物的光折变效应形成体全息图，从而完成信息的记录；读出时使用和原来相同的参考光寻址，可以再现存储在晶体或聚合物中的全息图，然后使用光信号探测器件如 CCD 或 CMOS（Complementary Metal Oxide Semiconductor Transistor）读出图像并传送给计算机。根据体全息图的布拉格角度选择性或者波长选择性，改变参考光的入射角度或波长以实现多重存储。因为布拉格选择性非常高，所以体全息存储可以在一个单位体积内复用多幅图像，达到超高密度存储的目的。

与已成熟的磁存储技术和光盘存储技术相比，体全息数据存储系统具有存储容量大、数据传输速度高、寻址速度快和存储冗余度高等优点，并具有并行快速内容寻址功能。此外，利用体全息存储器中数据以"页"为存储单位的特性，通过非相干-相干实时转换器（如液晶光阀），还可实现对大量卫星侦察图像的近实时直接存储。由于体全息存储器的读出为原存储图像数据的光学重建，因而在其转化为电信号之前可进行光信息处理，这样存储系统的整体功能可大大增强。尽管体全息存储技术取得了很大的进展，但是距离实用化还有一定的距离。其中一个主要的问题是目前还很难获得具有光学质量高、动态范围高、灵敏度高和非易失性记录的廉价记录材料。由于一般的体全息存储器使用激光器和元器件体积比较大，不便于集成和移动，系统的小型化面临许多障碍。另外，目前体全息存储器的误码率指标还很难达到实用化工业标准。为了提高数字体全息存储系统的综合性能，本文对一些关键技术进行了深入研究，提出了一些新的结构与方法。并对全息存储器在相关识别中的应用进行了讨论，在此基础上设计并构建了小型化存储系统，存储密度达到了 $10Gb/cm^3$，同时可实现快速并行人脸识别。

1 小型化体全息存储系统的设计

1.1 离焦90° Fourier 谱面全息记录

全息存储的光学系统通常有像面全息、Fourier 谱面全息、菲涅耳全息和范德拉格全息等4种结构。其中 Fourier 谱面全息光学系统结构简单，冗余度大，存储密度高，在谱面全息中只需记录图像的±1级以内的空间频率成分，读出图像就能够完全复现二值数字图像信息。和像面全息相比，谱面存储的有以下几个优点：（1）光斑尺寸小，再现图像的像质均匀；而像面全息中若要缩小记录的光斑尺寸就会受到图像大小和光学系统参数的影响，并且会受到材料吸收的影响，图像均匀性严重下降。（2）冗余度大，对材料表面灰尘或内部缺陷不敏感。（3）具有平移不变性，可以在位置复用中实现记录光束相对材料移动，而在读出时，能够保证 CCD 上的图像不移动。（4）易于实现分维复用，提高了存储密度，谱面全息进行分维复用时不需要对光路进行任何改动；而要实现像面全息的分维复用必须在材料和 CCD 之间放置孔径光阑或渐晕光阑。因此，本文在系统设计中将采用90°Fourier 谱面全息记录光路，如图1所示。

在谱面全息存储中，零级谱的能量与一级谱的能量相差是很大的、如果材料的线性区域很小，将出现这种情况：当能量很大的零级谱过饱和时，其他级次的谱线有的

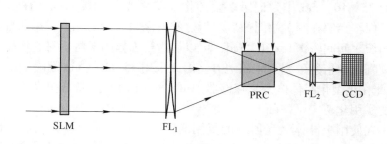

图 1　离焦 90°Fourier 谱面全息记录结构示意图

没有被记录,有的还没有达到饱和。因为全息系统实际上是一个线性系统,如果引入这个非线性因素,显然会给全息存储的过程带来很大的误差,常常会导致非期望的光折变效应。采用离焦记录可以克服由于零级过强引起的像质蜕变及材料损伤等现象。根据谱面强度分布随离焦量变化的理论分析,在我们的系统中离焦量取值在 8~15mm 之间时,谱面强度分布比较均匀,一级谱的能量得到加强,零级谱强度受到明显抑制,从而达到改善像质和充分利用材料的动态范围的目的。

为了提高光学系统的成像质量,傅里叶变换镜头组经过精密设计,大大简化了系统的光学结构,避免了一般存储系统中即使采用高质量的照相物镜仍然存在严重渐晕现象的问题。考虑到所使用的 SLM 为 1024×768 像素,像素尺寸大小为 26μm,CCD 像素尺寸大小为 9μm,经过设计得到的 Fourier 镜头组的焦距分别为 154.6mm 和 53.5mm,全视场畸变为 0.001704%,无渐晕,分辨率为 135 线/mm,调制传递函数 MTF 大于 0.15[9]。

1.2　复用与寻址方式

为了在全息材料的同一体积内存储尽量多的信息,实现超高密度存储,就要进行复用。复用一般是指在记录材料同一体积内存储多幅全息图的方法,复用的基础是体全息光栅的布拉格选择性,即只有当入射参考光束满足 $2\Lambda\sin\theta = N(\lambda/n)$ 时(Λ 是折射率光栅的空间周期,λ 是光波波长),才能重现信号光所加载的信息。复用通过改变参考光的参数(包括入射角度、波长、波面)和载有信息的物光光束相干涉记录数据,分别对应于角度复用[10,11]、波长复用[12]、位相编码复用[13]。如果在材料中选择不同的记录点,可以实现移位复用[14,15],当参考光束为一散斑调制的光束时又可利用散斑场的相关性质实现散斑复用[16]。如果将上述几种复用方式适当组合,可成为混合复用。

全息图之间的复用间距越小,存储密度越高,但是图像之间的串扰越大。根据布拉格选择性公式得到全息图的角度选择性可表示为 Sinc 函数的形式,所以为了避免串扰,需要全息图的中心位置正好在位于相邻的全息图的 Sinc 函数的零点。如果使用价格低廉,体积小的半导体泵浦固体(DPL)激光器,还需要考虑激光器的多纵模对全息图串扰的影响。这是因为 DPL 激光器谐振腔短且增益曲线较宽,存在多个频率间隔(间隔大小不至于被彼此的主瓣淹没)的纵模。实验发现,DPL 激光器作为系统光源记录和读出时,所记录的全息图明显具有多个旁瓣。单幅图像的旁瓣扩展将近 300×10^{-5} rad,如果采用 150×10^{-5} rad 作为间隔,虽然串扰比较小,但是将严重影响存储容量;

可以考虑采用 75×10^{-5} rad 作为记录多幅图像时参考光的角度间隔。在这个角度间隔下，第 2 幅图像将存储在第 1 幅图像后的衍射效率函数为零的地方；第 3 幅存储图像和第 1 幅存储图像之间间隔 150×10^{-5} rad，可以认为第 1 幅图像将不对它产生影响，且它正好落在第 2 幅图像衍射效率为零的地方，因此采用 75×10^{-5} rad 间隔能够有效地避免串扰[17]。由此在小型化晶体体全息存储器中可以采用 DPL 激光器作为光源，这也为全息存储技术的实用化奠定了基础。

为了缩小在使用半导体激光器时的复用间隔，提高存储密度，我们提出了基于动态散斑复用的新技术。动态散斑复用是指在体全息存储系统中加入移动的随机相位板，使得参考光由原来的平面波或者球面波成为散斑场调制的光场。利用散斑场的随机性和相关长度小的特性来提高体全息存储的选择性，从而提高体全息存储的存储密度和实现空间三维存储。在图 2（a）中，使用散斑参考光 $R_w(r)$ 和平面物光 $S_0(r)$ 进行相干在存储介质中形成相位全息，在多重全息复用时随机相位板每次移动 Δ。在图 2（b）中，当使用散斑参考光读出时，仅当漫射板的位置及入射平面光波角度和 $R_w(r)$ 非常接近时，才能读出全息图。实验结果表明，散斑参考光的引入大大提高了角度选择性，动态散斑—角度复用比单纯角度复用的选择性灵敏度提高了 8 倍，使用散斑复用装置能够很大程度上抑制 DPL 激光器多纵模模式对体全息系统角度选择性的影响，使得其与在单纵模模式情况下的选择性基本相同，只要存储介质的动态范围足够大，在存储介质的单一位置上即可最大限度的存储全息图，从而提高了存储密度。另外，动态散斑复用的角度选择性与记录材料的厚度无关，特别适合于在光聚合物存储材料上进行高密度存储[18,19]。

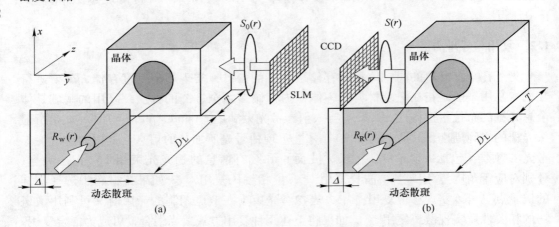

图 2　动态散斑复用示意图
（a）使用物光 $S_0(r)$ 和散斑参考光 $R_w(r)$ 记录；（b）使用散斑参考光 $R_R(r)$ 读出得到信号光 $S(r)$

1.3　存储数据的可靠性

作为一种存储手段，必须保证存储数据的可信度，即误码率要控制在一定数值。目前数字体全息系统的误码率水平还比较低。因此降低误码率仍然是体全息存储技术中的一个核心问题。要得到高质量，低误码率的重建图像，必须保证输入器件 SLM 和探测器件 CCD 像素之间的一一匹配。对于较大的像素阵列，这要求非常细致的光学设计及对准、输入输出相对倾斜角度及位移的精确调节和非常低的光学失真。能否实现一对一像素匹配直接关系到能否充分发挥已有设备的性能，因而成为决定存储系统性能的重要因素。此外为了减弱体全息存储中的散斑噪声，在 SLM 前面使用旋转的漫射器可以得到均匀的部分相干照明，降低了物光的相干性，这样可消除体全息存储中部分散斑噪声，同时在记录平面还可得到均化的强度分布。

误码率的降低不仅要求存储系统硬件的完备，同时对一些硬件无法避免的固有噪声，如系统有限空间响应带宽引起的页内像素间干涉噪声和空气中尘埃引起的突发性错误，应采用软件方法加以抑制。针对体全息存储系统中常见的几种噪声源，即大尺度噪声、像素间干涉噪声以及突发性随机噪声的特点[20]，分别采用减背景、页内均衡化和所罗门编码三种软件方法予以补偿[21]。在实验中，通过将三种方法有机地结合在一起，可获得明显的降噪效果，误码率由初始的 10^{-2} 降低到 10^{-4} 以下。进一步采用动态补偿技术和编码技术有望将误码率降低以达到实用化水平[22]。

1.4 相关识别系统

体全息存储的一个突出优点是具有关联寻址功能，如果在读出时不用参考光而改由物光中的某幅图像照射公共体积内由角度复用存储的多重全息图，那么将会读出一系列不同方向的"参考光"，各光的强度大小代表对应存储图像与输入图像之间的相似程度，利用此关联特性，可以实现内容寻址操作和基于图像相关运算的快速目标识别。体全息相关器基于超高密度体全息存储和光学相关识别系统，具有容量大，通道多，速度快，准确率高等优点，能广泛应用于如电子门卫，自动导航，导弹防卫等各种图像与模式识别领域。

体全息相关输出既不是参考光的简单再现，也不同于平面匹配滤波器中输入图像与各模板图像间相关结果的简单叠加，它受到 Sinc 函数的调制，调制的程度取决于介质的厚度。Sinc 函数的引入给识别结果带来了旁瓣噪声[23]，旁瓣噪声是高密度体全息相关识别系统的主要噪声，也是影响和限制提高体全息相关系统并行性和识别率的主要因素。传统抑制旁瓣噪声的办法是通过增加记录介质的厚度来实现，但这是一种各向异性的抑制方法，即在水平方向抑制后，竖直方向仍保持较强的旁瓣噪声，进一步的抑制又受到厚度的限制。

采用随机函数调制输入信号的新方法可以很好地抑制旁瓣噪声，利用散斑函数的自相关函数为 δ 函数的性质，可对物函数引入随机因子，对体全息相关公式进行修正，修正后的公式表明该方法可以完全从各个方向抑制旁瓣，得到非常突出的相关峰。通过数值模拟计算，发现 δ 函数的抑制作用与记录材料的厚度无关，相关峰近似为一个点，不带有任何旁瓣。为了实现随机调制，在物光光路放置低通有限带宽透射式全息漫射板器件，该器件产生的散斑对物光信号进行调制，可以实现抑制旁瓣的作用。漫射板的透射率大于 90%，可以获得较大的光能量利用率，同时低通有限带宽保证了在记录谱面具有较强的强度，使得引入散斑调制后物光光强仍然能够满足记录和复用要求。采用散斑调制技术抑制相关旁瓣可以使得角度分维复用技术的复用间隔在水平减小为原来的 1/2，在竖直方向上减小为原来的 1/4～1/5；使相关识别的并行度提高为原来的8~10倍，并大大提高了识别的准确率[24~26]。

2 小型化存储及相关识别系统

小型化系统的光路图如图3所示，系统的尺寸为400mm×400mm×150mm，使用的光源为半导体泵浦Nd：YAG倍频连续激光器，其波长为532nm，功率为150mW。在物光光路，物光透过SLM经M_2反射后通过Fourier透镜FL_1，然后经过反Fourier透镜成像到CCD1上，记录的介质为掺铁铌酸锂晶体，放置在谱面离焦位置。在参考光路，为了实现角度分维复用，使用一个4-f成像系统和两个精密移动台控制，在水平方向和竖直方向实现二维扫描。另外在光路中还加入快门和波片等相应的元器件，用于控制记录光的偏振态，参物比，曝光时序等参数。存储过程由计算机控制，计算机软件采用多线程机制，具有很好的人机界面，实现了全息存储过程快速存储、准确读出的自动化。

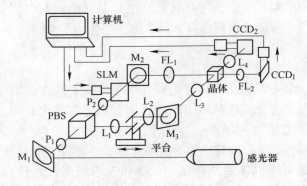

图3 小型化系统的光路图

$M_1 \sim M_3$—反射镜；P_1，P_2—半波片；$L_1 \sim L_4$—透镜；FL_1，FL_2—Fourier透镜；PBS—偏振分光棱镜

在记录工作模式中，数据可以通过二值化数据页或灰度数据页的方式存入晶体中，通过控制精密移动台，在水平方向记录了50个角度，竖直方向记录了20维，即在单一公共体积内使用角度分维复用的方法存储了1000幅图像。记录点的孔径大小为0.3mm，每一个数据页的数据量为1024×768，经计算存储密度大于$10Gb/cm^3$。同时，为了保证所有图像具有均衡的衍射效率，需要根据晶体的记录和擦除时间常数确定合适的递减曝光时序，在本次实验中，第1幅图的记录时间为6s，最后一幅图的记录时间约为1s。图4所显示的再现读出的第1、500和1000幅图，可以看到它们具有相同的强度。

  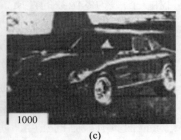

(a)   (b)   (c)

图4 存储后再现读出的第1、500和1000幅图

在图像复现工作模式中，使用参考光照明晶体即可在 CCD_1 中得到所存储的图像，改变参考光的角度，可依次复现所有的存储图像。在相关识别工作模式中，使用物光照明晶体，在 CCD_2 上所得到的是输入物图像和所有 1000 幅图像的相关结果，为 20×50 点阵，这是一个并行处理的过程。如果使用白图作为物光照明，由于白图和每一幅图的相关均有值，那么所得到的点阵为 1000 个亮点，如图 5(a)所示。如果使用某一幅人脸图像作为输入，那么点阵中只有一个点非常亮，该亮点所对应的图像与输入图像最相关。如图 5(b)所示，使用第 528 人脸作为输入，系统经过判别后给出了和该图最相近的前三幅图，依次为第 528、740 和 828 幅。对系统识别进行重复实验，测得准确率大于 95%。考虑快门，SLM，CCD 和计算机判别等所消耗的时间，整个识别过程大约需要 0.5s 的时间，识别速率达到 2000 幅/s。另外引入子波变换特征提取等图像处理技术，该系统还具有图像相关的抗畸变能力。

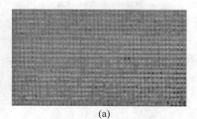

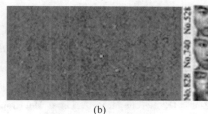

图 5　全息相关器的人脸输出结果
(a) 使用"白图"读出的 20×50 相关点阵；
(b) 输入第 528 幅图后所得到的最相关的前三幅图

3　总结与展望

本文对体全息存储技术的最新进展和关键技术进行了论述，并介绍了 $10Gb/cm^3$ 小型化体全息存储及人脸相关识别系统样机。目前性能优异的商品化光致聚合物已经问世，以及高速高对比度反射式 SLM 和高速 CMOS 等光电元器件的研究进展迅速，这对体全息存储系统综合性能的提高将起到至关重要的推动作用，可以预见，体全息存储技术将会成为下一代主流数据存储技术之一。

参 考 文 献

[1] TerLris B D, Mamin H J, Rugar D. Near-field optical data storage. APPl Phy Lett., 1996, 68(2)：141~143.

[2] Van Heerden P J, Theory of optical information storage in solids. APPlOPt, 1963, 2：393~400.

[3] Burr G W, Jefferson C M, Coufal H, et al. Volume holograPhie data storage at an areal density of 250 gigapixels/in². Opt Lett, 2001, 26(7)：444~446.

[4] Coufal H J, Psaltis D, Sincerbox G T, Holographic Data Storage. New York：Springer-VerlagBurlin Heidelberg, 2000：3~20.

[5] Heanue J F, Bashaw M C, Hesselink L. Volume holographic storage and retrieval of digital data. Science, 1994, 16：605~607.

[6] Psaltis D, Mork F. Holograpphic memories. Science America, 1995, 273：70~76.

[7] Ashley J, Bernal M P, Burr G W, et al. Holographic data storage. IBM J Res and Dev, 2000, 44(3)：341~368.

[8] Kincade K. Holographic data storage prepares for the real word. Laser Focus World, 2003. 39(10)68~73.
[9] Yang Y P, Adibi A, Psaltis D. et al. Comparison of transmission and the 90-degree holographic recording geometry. Appl Opt, 2003. 42(17):3418~3427.
[10] Leith E N, Kozma A, Upatnieks J. et al. Holographic data storage in three-dimensional media. Appl Opt, 1966, 5(8):1303~1311.
[11] Curtis K, Pu A, Psaltis D. et al. Method for holographic storage using peristrophic multiplexing. Opt Lett. 1994, 19(13):993~994.
[12] Rakuljic G A, Levya A. Optical data storage by using orthogonal wavelength-multiplexed volume holograms. Opt Lett, 1992. 17(20):1471~1473.
[13] Denz C, Pauliat G. Roosen G. Volume hologram multiplexing using a deterministic phase encoding method. Opt Comm, 1991, 85:171~176.
[14] Psaltis D, Levene M. Pu A, et al. Holographic storage using shift multiplexing. Opt Lett, 1995. 7(20):782~784.
[15] Tao S Q, Song Z H, Selviah D R, et al. Spationangular multiplexing scheme for dense holographic storage. Appl Opt, 1996, 35(14):2380~2388.
[16] Markov V, Millerd J. et al. Multilayer volume holographic optical memory. Opt Lett. 1999, 24(4):265~267.
[17] 宋修宇,何庆声,邹敏贤,等. DPL 在晶体体全息存储的串扰问题研究. 光电子激光, 2002, 11(3):258~261.
[18] Wang J N. He S R, He Q S, et al. Insensitivity of speckle multiplexing to multy-longitudinal modes of laser in volume holographic storage. Chin Phy Lett, 2003, 20(7):1047~1050.
[19] He Q S, Wang J N, Zhang P K, et al. Dynamic speckle multiplexing scheme in volume holographic data storage and its realization. Opt Expk, 2003, 11(4):366~370.
[20] Vadde V, Vijaya Kumar V V K, Channel modeling &estimation for intrapage equalization in pixel-matched volume holographic data storage. Appl Opt, 1999, 38:4368~4374.
[21] 黄雄斌,何庆声,商未雄,等. 体全息存储页内串扰抑制的研究中国激光(增刊), 2003, 30:79~80.
[22] Burr B W, Ashley J, Coufal H J, et al., Modulation coding for pixel-matched holographic data storage. Opt Lett. 1997, 22(9):639~641.
[23] Gu C, Fu H, Lien J R. Correlation patterns and cross-talk noise in volume holographic optical correlators. J Opt Soc of Am A, 1995, 12(5):861~868.
[24] Feng W Y, Yan Y B, Jin G F, et al. Volume holographic wavelet correlation processor. Opt Eng, 2000, 39(9):2444~2450.
[25] Ouyang C, Can L C. He Q S, etal. Sidelobe suppression in volume holographic optical correlators by use of speckle modulation. Opt Lett. 2003, 28(20):1972~1974.
[26] Su W C, Chen Y W, Ouyang Y, et al. Optical identification using a random phase mask. Opt Comm, 2003, 219:117~123.

Orthogonal Polarization Dual-channel Holographic Memory in Cationic Ring-opening Photopolymer*

Abstract A dual-channel holographic recording technique and its corresponding memory scheme in the cationic ring-opening photopolymer are presented. In the dual-channel technique, a pair of holograms are recorded simultaneously with two orthogonal polarization channels in the common volume of the material, and are reconstructed concurrently with negligible inter channel crosstalk. The grating strengths of these two channels are investigated and the relevant parameters for equal diffraction intensity readout are optimized. Combining the dual-channel technique with speckle shift multiplexing, a high-density holographic memory is realized. This dual-channel scheme enables the users to interact with the storage medium from an additional channel. The simultaneous nature of the two channels also offers a faster data transfer rate in both the recording and reading processes.

1 Introduction

Volume holographic memory has attracted great attention due to its potential of high storage density and fast data transfer rate[1]. To increase the storage density, many multiplexing methods, such as angle multiplexing[2,3], shift multiplexing[4-6], wavelength multiplexing[7,8], and polarization multiplexing[9-11], etc. have been proposed and investigated previously. Most of these methods require a sequential recording of the holograms. Although wavelength multiplexing can potentially record more than one hologram in each exposure, it will require more than one laser source and a complicated alignment system. Practically, polarization multiplexing is the only mechanism that allows simultaneous recording and retrieval of two holograms. Several methods of implementing polarization multiplexing have been studied in $LiNbO_3$ crystals and other photo-induced anisotropic materials. Su et al[9]. recorded two holograms sequentially in the same position of a $LiNbO_3$ crystal using a polarization multiplexing method based on the photovoltaic effect and the photorefractive effect. Todorov et al[10]. demonstrated another polarization multiplexing method in photo-induced anisotropic materials where the spatial modulations of polarization resulted in spatially modulated anisotropy and, consequently, in refractive index modulation. In Ref. [10], the reference and signal beams had orthogonal circular polarizations (left-hand and right-hand), and the two polarization holograms were superimposed sequentially in the common volume of a methyl orange/PVA film. Another kind of photo-induced anisotropic film, bacteriorhodopsin film, was also used to record two polarization multiplexed holograms by

---

* Copartner: Haoyun Wei, Liangcai Cao, Zhenfeng Xu, Qingsheng He, Claire Gu. Reprinted from *Optics Express*, 2006, 14: 5135-5142.

Koek et al[11]. In their system, the polarization of the reference beam remained unchanged during the two exposures, while the polarizations of the object beams of the two holograms were orthogonal. However, all of these polarization multiplexing schemes demonstrated only the simultaneous readout of the two polarization multiplexed holograms using the reference beam with a specific polarization. And no simultaneous recording and high density memory systems have been reported due to the following limitations: (1) the hologram generated via the photovoltaic effect in Ref. [9] is much weaker than that generated via the general photorefractive effect in $LiNbO_3$ crystals, and (2) the recorded holograms are volatile due to dark decay[12] in $LiNbO_3$ crystals and thermal relaxation[11] in photo-induced anisotropic materials.

In this paper we demonstrate a novel dual-channel holographic recording technique, in which the holograms of the two channels can be operated simultaneously in both the recording and the retrieving processes. The recording material employed in this work is the isotropic cationic ring-opening photopolymer (CROP polymer)[13] developed by Aprilis Inc. The grating evolution during the dual-channel holographic recording process is evaluated both theoretically and experimentally. The simultaneous recording and concurrent readout with equal diffraction intensities and negligible inter-channel crosstalk are demonstrated in a high-density holographic memory system based on the dual-channel holographic recording technique and a speckle shift multiplexing method. The simultaneous nature of new scheme provides the ability for the users to interact with the storage medium from an additional channel, and offers a faster data transfer rate in both the recording and reading processes.

2 Grating evolutions in dual-channel holographic recording

2.1 Theoretical analysis

Fig. 1 shows the basic geometry of our dual-channel holographic recording in a CROP photopolymer with thickness d in the z direction. In this scheme, the recording beams are separated into the p-and s-polarization channels so that the polarization states of the two channels are mutually orthogonal. Specifically, the p-polarization channel consists of the recording beam pair, $E_{rp}(r)$ and $E_{op}(r)$, with the p-polarization; and the s-polarization channel consists of the other recording beam pair, $E_{rs}(r)$ and $E_{os}(r)$, with the s-polarization. During the holographic recording exposure, the photo-induced polymerization profile of the CROP polymer is directly proportional to the spatial modulation of the exposure intensity, which leads to the corresponding modulation of refractive index[15]. Therefore, the holographic exposure in this scheme simultaneously records two (and only two) volume holograms which are produced by the two interferences of the recording beam pairs in the s-and p-polarization channels, respectively. Moreover, these two holograms are distinguished by the Bragg selectivity, where a separation angle between the two reference beams much larger than the Bragg angular selectivity ($\Delta\theta_B$) is applied. The incidence angles of the reference beams, $E_{rp}(r)$ and $E_{rs}(r)$, in the recording material, are θ_p and θ_s, respectively. For simplicity, $\theta_p = -\theta_s$ is adopted, thus the separation angles is $|2\theta_p|$

($|2\theta_p| \gg \Delta\theta_B$). In addition, the two object beams are both propagating along the direction of the z-axis that is normal to the surface of the recording material.

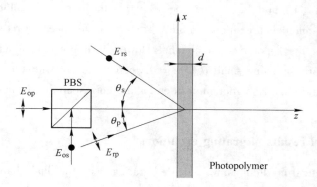

Fig. 1　Geometry of the dual-channel holographic recording by plane waves in a photopolymer: d, thickness of the photopolymer medium; PBS, polarizing beam splitter; E_{rp}, E_{op}, E_{rs} and E_{os}, electric fields of the recording beams; θ_p and θ_s, incidence angles of the reference beams in the medium. Here the subscript symbols: o, object beam; r, reference beam; p, p-polarization (in the x-z plane) recording channel (shows in blue); s, s-polarization (along the direction of the y-axis) recording channel (shows in red)

Under the diffusion free condition with a low exposure intensity in the CROP polymer, where the diffusion of the monomer is much faster than the polymerization, high values of index modulation can be recorded with high fidelity and the dark reaction effect can be neglected[16]. In this case, the diffusion-reaction model[17,18] is applicable to the CROP polymer[16]. Thus we can obtain the temporal evolutions of diffraction index modulations[18] of p-and s-polarization channels in the proposed scheme respectively, as:

$$\Delta n_p(t) = 2\alpha_M \frac{E_{rp}E_{op}\cos(\theta_p)}{I_0} \cdot \{1 - \exp\{\gamma[1 - \exp(t/\tau)]\}\} \tag{1a}$$

$$\Delta n_s(t) = 2\alpha_M \frac{E_{rs}E_{os}}{I_0} \cdot \{1 - \exp\{\gamma[1 - \exp(t/\tau)]\}\} \tag{1b}$$

where, $\Delta n_p(t)$ and $\Delta n_s(t)$ are the temporal modulations of the refraction index; E_{rp}, E_{rs}, E_{op}, and E_{os} are the amplitudes of the reference and object beams of the p-and s-polarization channels, respectively. $I_0 = \sum_{i=r,o}\sum_{j=s,p} E_{ij}^2$ is the average recording intensity, α_M is the maximum available modulation index, γ is a positive constant responsible for the time evolution, and τ ($\tau \propto I_0^{-1}$) is the time constant.

The temporal evolutions of the grating strengths (ν_p and ν_s) of these two gratings can be described as[19]:

$$\nu_p(t) = \{\pi\cos(\theta_{bp})d/[\lambda\sqrt{\cos(\theta_{bp})}]\} \cdot \Delta n_p(t) \tag{2a}$$

$$\nu_s(t) = \{\pi d/[\lambda\sqrt{\cos(\theta_{bs})}]\} \cdot \Delta n_s(t) \tag{2b}$$

where, λ is the wavelength of the probe wave, θ_{bp}, θ_{bs} are the Bragg diffraction angles of the probe waves for the p-and s-polarization channels respectively, and $\theta_{bp} = -\theta_{sp}$, due to $\theta_p = -\theta_s$.

According to Eqs. (1) and (2), it can be seen that the temporal evolutions of the grating strengths of the two channels are similar in form. The only difference is a reductive coefficient of $\cos(\theta_p)\cos(\theta_{bp})$ in the p-polarization channel. However, in practice, equal grating strengths can be achieved easily by using small θ_p ($\theta_{bp} = \theta_p$ when the beam with the same wavelength is used for reading) and slightly higher intensities for the corresponding beams of the p-polarization channel.

2.2 Experimental results of grating evolutions

To verify the theoretical predictions, we recorded gratings with the dual-channel holographic recording scheme shown in Fig. 1 in the CROP polymer. A diode pumped laser (wavelength $\lambda = 532$nm) is used to generate laser beams for the dual-channel holographic recording. The polarization of each beam is carefully adjusted as in Fig. 1. The recording intensity of each beam is about 0.8mW/cm^2. The incidence angle of the reference beam measured in the air is $25°$ for the s-polarization channel, and $-25°$ for the p-polarization channel. Using the CROP polymer index $n = 1.545$, we can calculate the corresponding incidence angle in the material, which is $\theta_s = 15.87°$ and $\theta_p = -15.87°$ respectively. Since the material has low sensitivity and absorption at wavelengths longer than 600nm, a diode laser at the wavelength of 650nm is used to monitor the evolution of gratings in real time by Bragg-matched probing: one exposure process for the s-polarization channel using an s-polarized probe beam at the probe angle of $\theta_{bs} = 17.62°$, and the other for the p-polarization channel using a p-polarized probe beam at the probe angle of $\theta_{bp} = -17.62°$. The temporal evolution of each grating is

$$\nu(t) = \arcsin(\sqrt{\eta(t)}) = \arcsin(\sqrt{I_d(t)/(I_d(t) + I_t(t))}) \quad (3)$$

where $I_d(t)$ and $I_t(t)$ are the intensities of the diffracted and transmitted beams, respectively.

The dotted curves in Fig. 2 illustrate the grating evolutions of the s-and p-polarization channels. We extracted $\nu_{p.sat} = 0.471$, $\gamma = 2.10$ and $\tau = 13.5$s from the experimental data of the p-

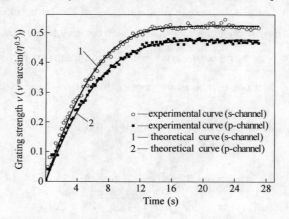

Fig. 2 Grating evolutions during dual-channel holographic recording

polarization channel, and $\nu_{s.sat} = 0.523$, $\gamma = 2.14$ and $\tau = 13.5$s from that of s-polarization channel by minimizing the sum of the squares of the difference between experimental data points and theoretical predictions. The values of γ and τ of the two gratings are similar, just as the theoretical predictions. The ratio of $\nu_{p.sat}$ and $\nu_{s.sat}$ is 0.90, which agrees with the theoretical value ($r = \nu_{p.sat}/\nu_{p.sat} = \cos(\theta_p)\cos(\theta_{bp}) = 0.91$).

3 Holographic memory with dual data-channel

3.1 System setup

The optical setup of the dual-channel holographic memory system is shown in Fig. 3. The linearly polarized light at the wavelength of 532nm generated from a diode-pumped laser source passes through a half-wave plate HP_1 and a polarized beam splitter PBS_1, the light is then divided into the reference and object arms. In the object arm, the PBS_3 divides the light into two parts to illuminate the two spatial light modulators, SLM_1 and SLM_2 (1280 × 768 arrays of 13.2μm × 13.2μm), respectively. One is the p-polarization data-channel. And the other is the s-polarization data-channel. In the reference arm, the PBS_2 divides the light into reference beams of the p- and s-polarization channels. A quarter-wave plate QP and an adjustable mirror are added to compensate the path-length difference between the object and reference beams in the p-polarization data-channel. The half-wave plates HP_1, HP_2 and HP_3 are employed to adjust the intensi-

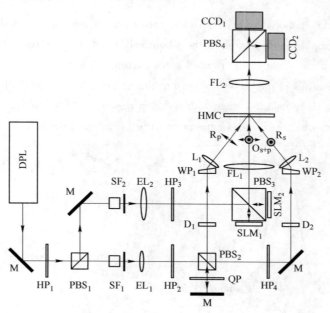

Fig. 3 Experimental setup for dual data-channel holographic memory: DPL, diode-pumped solid-state laser; HP, half-wave plate; PBS, polarizing beam splitter; SF, spatial filter; EL, beam-expanding lens; QP, quarter-wave plate; D, diffuser; SLM, spatial light modulator; FL, Fourier transfer Lens; WP, wedge prism; L, lens; M, mirror; HMC, holographic media card; CCD, charge coupled device

ties of the four recording beams. And the HP_4 is used to adjust the reference beam polarization. The diffusers, D_1 and D_2, with FHWM = 5°, generate random speckle modulation in reference beams for speckle shift multiplexing. The corresponding effective NA (numerical aperture) of each reference beam is about 0.1. The recording material is Aprilis HMC-series polymer with the thickness of 300μm, which features a high dynamic range, high recording sensitivity, and very low volumetric shrinkage[20], and is regarded as one of the most promising WORM material for nonvolatile holographic memory[14]. This material is mounted onto a positioning system composed of two linear translation stages which can perform addressing in sub-micrometer scale along the x- and y directions at the Fourier plane of the SLMs, respectively. The reconstructed beams are separated by the PBS_4 and then captured by CCD_1 and CCD_2 which receive the data from p-and s-polarization data-channels.

3.2 Experimental results

According to the analysis of grating evolutions in our dual-channel holographic recording scheme, in order to achieve equal diffraction intensity for the two channels, the reference (readout) power of the p-polarization channel is adjusted to a slightly higher level than that of the s-polarization channel. In the experiment, the powers of the s-and p-reference beams are 81μW and 100μW, and those of the signal beams are 38μW and 34μW, respectively. The diameter of the recorded spot is around 5mm. And the incidence angles of the reference beams are 25° and −25°, respectively.

Shift multiplexing method based on speckle correlation is adopted here, in order to perform high density holographic multiplexing in the whole volume of the material. The one dimensional shift selectivity of a hologram in the dual-channel system is plotted in Fig.4(a), by measuring the diffraction intensities of the reconstructed holograms every 0.425μm shift interval (50 shifting steps of the linear translation stage with the step resolution of 0.0085μm). The shift selectivity is about 6μm, which is in good agreement with theoretical results[5] in single-channel sys-

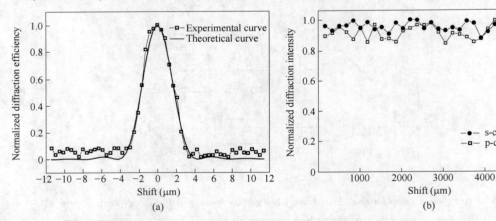

Fig. 4 (a) shift selectivity of a hologram recorded using speckle shift multiplexing in dual-channel system; (b) readout of 30 holograms per channel superimposed by speckle shift multiplexing

tems. The deviation of the experimental curve from the theoretical one in the tails of the diffraction efficiency curve is mainly caused by the noises of speckle field and the I/O devices. Moreover, the speckle noise intensity will increase linearly with the increasing number of multiplexed holograms, since the previously recorded holograms are still reconstructed randomly under an optical speckle field[14] in the speckle shift multiplexing. Thus a shift interval of 170μm is chosen in both the x and y directions to minimize the crosstalk and the speckle noise. 1800 (30 × 30 ×2) holograms with 0.96-Mpixels data each are recorded in a local square area by using the dual-channel technique and the speckle shift multiplexing method. The resulting density is about 20pixels/μm^2. The diffraction intensity of each hologram in the first row is plotted in Fig. 4(b). It shows that the average diffraction intensities of the p-and s-channels are similar, and the diffraction intensities of sequentially recorded holograms in the same channel are almost uniform with a fluctuation of ±7%. This is mainly caused by the fluctuation of the laser power and the tiny non-uniformity of polymer composition system. One pair of simultaneously reconstructed data pages is shown in Fig. 5. It can be seen that a good separation of the data pages of the two channels is achieved, and no evidence of inter-channel crosstalk is observed, as in Fig. 5 (a) and Fig. 5 (b). The overlapped data page captured by CCD_1 without PBS_4 is shown in Fig. 5 (c). It shows the diffraction intensities of two channels are almost equal. This implies that the parameters chosen for dual data-channel holographic memory are feasible.

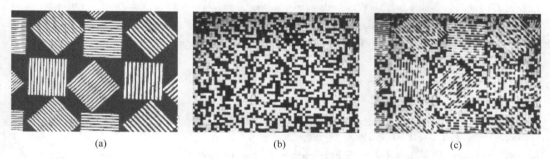

Fig. 5 Readout pages in dual data-channel holographic memory
(a) p-channel data page; (b) s-channel data page; (c) the overlapped
p-and s-channel data pages without PBS_4

4 Conclusions

A dual-channel holographic recording technique that allows simultaneously access of two orthogonally polarized holograms, in both recording and reading processes, with negligible inter-channel crosstalk is proposed. The grating evolution in the dual-channel holographic recording scheme is theoretically analyzed, and experimentally validated. A holographic memory system with equal diffraction intensities is demonstrated by combining the proposed technique with a speckle shift multiplexing method. In addition, our system uses only one 4-f optical head to transfer the pair of data pages with orthogonal polarization. This simultaneous nature of the two channels offers a compact system with a faster data transfer rate in both the recording and reading processes.

The result reported here is obtained with Aprilis CROP polymer but it is valid for most isotropic photopolymerizable recording media without photo-induced anisotropy in general. And in this kind of dual-channel system, the pair of pages recorded simultaneously can be independent data from different data sources, e. g. , one from an analog device and the other from a digital source; or they can be related data from correlated data sources. For example, one potential application of the dual-channel holographic memory system is the holographic projector[21] for stereoscopic video.

Acknowledgements

The authors thank Dr. Glenn Horner for valuable technical supports of the recording materials. This work was supported by National Natural Science Foundation of China(No. 60277011) and National Research Fund for Fundamental Key Projects No. 973 (G19990330). Claire Gu would like to acknowledge partial support by the National Science Foundation (ECS-0401206) and by the Special Research Grant of UC Santa Cruz.

References

[1] H. J. Coufal, D. Psaltis, and G. T. Sincerbox, eds. , *Holographic Data Storage*, Vol. 76 of Springer Series in Optical Sciences(Springer-Verlag, Berlin, 2000).

[2] L. d' Auria, J. P. Huignard, and E. Epitz, "holographic read-write memory and capacity enhancement by 3-D storage", IEEE Trans. Mag. 9, 83-94(1973).

[3] F. H. Mok, " Angle-multiplexed storage of 5000 holograms in lithium niobate", Opt. Lett. 18, 915-917 (1993).

[4] D. Psaltis, M. Levene, A. Pu, G. Barbastathis, and K. Curtis, " Holographic storage using shift multiplexing", Opt. Lett. 20, 782-784(1995).

[5] A. M. Darskii and V. B. Markov, "Shift selectivity of holograms with a reference speckle wave", Opt. Spectrosc. 65, 392-395(1988).

[6] L. Dhar, K. Curtis, M. Tackitt, M. Schilling, S. Campbell, W. Wilson, A. Hill, C. Boyd, N. Levinos, and A. Harris, "Holographic storage of multiplex high-capacity digital data pages in thick photopolymer systems", Opt. Lett. 23, 1710-1712(1998).

[7] G. A. Rakuljic, V. Levya, and A. Yariv, " Optical data storage by using orthogonal wavelength-multiplexed tunable diode-laser", Opt. Lett. 17, 1471-1473(1992).

[8] S. Yin, H. Zhou, F. Zhao, M. Wen, Y. Zang, J. Zhang, and F. T. S. Yu, "Wavelength-multiplexed holographic storage in a sensitive photorefractive crystal using a visible-light tunable diode-laser", Opt. Commun. 101, 317-321(1993).

[9] W. Su, C. Sun, N. Kukhtarev, and A. E. T. Chiou, " polarization-multiplexed volume holograms in $LiNbO_3$ with 90-deg geometry", Opt. Eng. 42, 9-10(2003).

[10] T. Todorov, L. Nikolova, K. Stoyanova, and N. Tomova, " Polarization holography. 3: Some applications of polarization holographic recording", Appl. Opt. 24, 785-788(1985).

[11] W. D. Koek, N. Bhattacharya, and J. J. M. Braat, "Holographic simultaneous readout polarization multiple-

xing based on photoinduced anisotropy in bacteriorhodopsin", Opt. Lett. 29, 101-103 (2004).

[12] YP. Yang, I. Nee, K. Buse, and D. Psaltis, "Ionic and electronic dark decay of holograms in $LiNbO_3$:Fe crystals", Appl. Phys. Lett. 78, 4076-4078 (2001).

[13] D. A. Waldman, R. T. Ingwall, P. K. Dhal, M. G. Horner, E. S. Kolb, H. -Y. S. Li, R. A. Minns, and H. G. Schild, "Cationic ring-opening photopolymerimization methods for volume hologram recording", in *Diffractive and Holographic Optical Technology III*, I. Cindrich, and S. H. Lee, Eds., Proc. SPIE 2689, 127-141 (1996).

[14] S. S. Orlov, W. Phillips, E. Bjornson, Y. Takashima, P. Sundaram, L. Hesselink, R. Okas, D. Kwan, and R. Snyder, "High-transfer-rate high-capacity holographic disk data-storage system", Appl. Opt. 43, 4902-4914 (2004).

[15] L. Paraschis and L. Hesselink, "Properties of compositional volume grating recording in photopolymers", in *International Symposium on Nonlinear Optics*. IEEE, 72-74 (1998).

[16] L. Paraschis, Y. Sugiyama, and L. Hesselink, "Physical properties of volume holographic recording utilizing photo-initiated polymerization for nonvolatile digital data storage", in *Advanced Optical Data Storage: Materials, Systems, and Interfaces to Computers*, P. A. Mitkas, Z. U. Hasan, H. J. Coufal, and G. T. Sincerbox, Eds., Proc. SPIE 3802, 72-83 (1999).

[17] G. Zhao and P. Mouroulis, "Diffusion model of hologram formation in dry photopolymer materials", J. Mod. Opt. 41, 1929-1939 (1994).

[18] S. Piazzolla and B. K. Jenkins, "Holographic grating formation in photopolymer", Opt. Lett. 21, 1075-1077 (1996).

[19] H. Kogelnik, "Coupled wave theory for thick hologram gratings", Bell Syst. Tech. J. 48, 2909-2947 (1969).

[20] For the detail information of Aprilis media properties, http://www.aprilisinc.com/Aprilils media product sheet.pdf.

[21] D. Papazoglou, M. Loulakis, G. Siganakis, and N. Vainos, "Holographic read-write projector of video images", Opt. Express 10, 280-285 (2002).

Improving Signal-to-noise Ratio by Use of a Cross-shaped Aperture in the Holographic Data Storage System*

Abstract A cross-shaped aperture is proposed to improve signal-to-noise ratio (SNR) in the holographic data storage system (HDSS). Both simulated and experimental results show that higher SNR can be achieved by the cross-shaped aperture than traditional square or circular apertures with the same area. A maximum gain of 20% in SNR is obtained for the optimized cross-shaped aperture. The sensitivities to pixel misalignment and magnification error are also numerically compared.

1 Introduction

Holographic data storage is a page-oriented data storage scheme and has high storage density and high data transfer rate. The next generation of digital storage might use a form of holographic data storage[1-3]. Recent progress in materials, multiplexing techniques, components and coding methods are finally making this vision a reality[4,5]. Besides storage density and data transfer rate, signal-to-noise ratio (SNR) of the holographic data storage system (HDSS) is one of the most important indices to evaluate its performance. Influencing factors of SNR of the HDSS include inter-pixel cross-talk, inter-page cross-talk, background noise, limited contrast of spatial light modulator (SLM), scattering noise, noise of electronics from detectors, pixel misalignment, optical aberration and so on. Researchers have made efforts to get rid of these factors mainly in two ways: (1) improvements of the optical system, which contain pixel matching[6], optimization of the size of the aperture[7], defocusing of the aperture and medium[8], making use of phase masks[9], et al. (2) post-processing and coding methods, including equalization of output image[10,11], compensation for pixel misregistration[12,13], modulation codes[14], error-correction codes[15], and so on. In this paper, we focus on one kind of improvements of the optical system, i. e. optimizing the aperture of the HDSS.

Impulse response of the system is determined by the size and geometrical shape of the aperture[16]. Limited size of the aperture causes the inter-pixel cross-talk which reduces the SNR. Obviously larger size of the aperture results in weaker inter-pixel cross-talk and higher SNR, whereas lower storage density. The storage density of correlation multiplexing in thin photopolymer disk media depends on the area of the aperture and the shift selectivity[3]. The shift selectivity is only determined by the autocorrelation function of the speckled beam, and is inde-

---

* Copartner: Huarong Gu, Songfeng Yin, Qiaofeng Tan, Liangcai Cao, Qingsheng He. Reprinted from *Appl. Opt*, 2009, 48: 6234-6240.

pendent of the shape of the aperture. If the area of the aperture is unchanged, the storage density of correlation multiplexing in thin photopolymer disk media will be maintained. Furthermore the geometrical shape of the aperture can be optimized to improve performances of the HDSS while maintaining the area of the aperture. In this paper, a cross-shaped aperture is proposed to improve the SNR. Simulated and experimental results show higher SNR can be achieved by the cross-shaped aperture than traditional square or circular apertures with the same area.

2 Theoretical Analysis

A schematic diagram of a typical holographic data storage system (HDSS) is shown in Fig. 1. The SLM has a pixel pitch Γ_1, a fill factor g_1 in both x and y directions and a contrast ratio c. f_1 and f_2 are focal lengths of Fourier Transform (FT) lenses L_1 and L_2, respectively. A complementary metal oxide semiconductor (CMOS) detector has a pixel pitch Γ_2, a fill factor g_{2x} in x direction and a fill factor g_{2y} in y direction. An aperture P is located in the common focal plane of lenses L_1 and L_2. We assume that the system images the SLM exactly onto the CMOS detector.

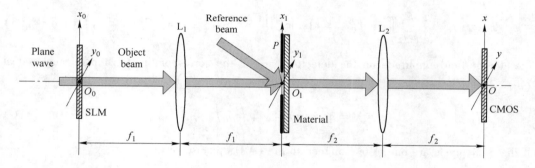

Fig. 1 Schematic of a typical holographic data storage system

For uniform plane-wave illumination, the amplitude of a single SLM pixel centered at $(m\Gamma_1, n\Gamma_1)$ is given by

$$U(x_0, y_0) = a(m,n) \text{rect}\left(\frac{x_0 - m\Gamma_1}{g_1 \Gamma_1}\right) \text{rect}\left(\frac{y_0 - n\Gamma_1}{g_1 \Gamma_1}\right) \quad (1)$$

where $a(m, n) = 1$ when this pixel is ON and $a(m, n) = 1/\sqrt{c}$ when this pixel is OFF. The electric-field amplitude right before the aperture plane is the two-dimensional spatial Fourier transform of Eq. (1)

$$\begin{aligned}
U(x_1, y_1) &= \frac{1}{i\lambda f_1} \mathcal{F}\{U(x_0, y_0)\} \\
&= \frac{a(m,n)}{i\lambda f_1} g_1^2 \Gamma_1^2 \text{sinc}\left(\frac{g_1 \Gamma_1 x_1}{\lambda f_1}\right) \text{sinc}\left(\frac{g_1 \Gamma_1 y_1}{\lambda f_1}\right) \exp\left[-i2\pi\left(\frac{m\Gamma_1 x_1}{\lambda f_1} + \frac{n\Gamma_1 y_1}{\lambda f_1}\right)\right]
\end{aligned} \quad (2)$$

where λ is the laser wavelength and F denotes Fourier transform. After the aperture, the electric-field amplitude becomes

$$U'(x_1,y_1) = U(x_1,y_1)P(x_1,y_1) \tag{3}$$

where $P(x_1,y_1)$ is the transmittance function of the aperture. The electric-field amplitude on the CMOS plane can be obtained by the Fourier transform of $U'(x_1,y_1)$ while ignoring the minus sign,

$$\begin{aligned}U(x,y) &= \frac{1}{i\lambda f_2}\mathcal{F}\{U'(x_1,y_1)\} \\ &= \frac{a(m,n)g_1^2\Gamma_1^2}{\lambda^2 f_1 f_2}\iint_P \mathrm{sinc}\left(\frac{g_1\Gamma_1 x_1}{\lambda f_1}\right)\mathrm{sinc}\left(\frac{g_1\Gamma_1 y_1}{\lambda f_1}\right) \\ &\quad \exp\left\{-i2\pi\left[\frac{x_1}{\lambda f_2}\left(x+\frac{f_2}{f_1}m\Gamma_1\right)+\frac{y_1}{\lambda f_2}\left(y+\frac{f_2}{f_1}n\Gamma_1\right)\right]\right\}\mathrm{d}x_1\mathrm{d}y_1\end{aligned} \tag{4}$$

Let

$$E(x,y) = \frac{g_1^2\Gamma_1^2}{\lambda^2 f_1 f_2}\iint_P \mathrm{sinc}\left(\frac{g_1\Gamma_1 x_1}{\lambda f_1}\right)\mathrm{sinc}\left(\frac{g_1\Gamma_1 y_1}{\lambda f_1}\right)\exp\left[-i2\pi\left(\frac{x_1}{\lambda f_2}x+\frac{y_1}{\lambda f_2}y\right)\right]\mathrm{d}x_1\mathrm{d}y_1 \tag{5}$$

the electric-field amplitude on the detector plane of the corresponding SLM pixel centered at $(m\Gamma_1, n\Gamma_1)$ is

$$U(x,y) = a(m,n)E\left(x+\frac{f_2}{f_1}m\Gamma_1, y+\frac{f_2}{f_1}n\Gamma_1\right) \tag{6}$$

If the SLM pixels are one to one aligned to the CMOS pixels, we have

$$\frac{f_2}{f_1} = \frac{\Gamma_2}{\Gamma_1} \tag{7}$$

Then Eq. (6) can be rewritten as

$$U(x,y) = a(m,n)E(x+m\Gamma_2, y+n\Gamma_2) \tag{8}$$

The pixel centered at $(-m\Gamma_2, -n\Gamma_2)$ on the CMOS plane receives greater than 98 percent of the energy from the SLM pixel centered at $(m\Gamma_1, n\Gamma_1)$ and its 12 surrounding pixels (Fig. 2). Therefore the contribution of the pixels beyond these 13 pixels to the central pixel can be ignored[7]. Then the electric-field amplitude of the pixel centered at $(-m\Gamma_2, -n\Gamma_2)$ can be expressed as

$$U_T(x,y) = \sum_{|j|+|k|\leq 2} a(m+j,n+k)E(x+(m+j)\Gamma_2, y+(n+k)\Gamma_2) \tag{9}$$

where j and k are integers. The signal received by the target CMOS pixel is obtained as

$$I(-m,-n) = \int_{-g_{2y}\Gamma_2/2}^{g_{2y}\Gamma_2/2}\int_{-g_{2x}\Gamma_2/2}^{g_{2x}\Gamma_2/2} |U_T(x,y)|^2 \mathrm{d}x\mathrm{d}y \tag{10}$$

We evaluate the intensity $I(-m,-n)$ of all the 2^{13} just equally likely combinations. Those in-

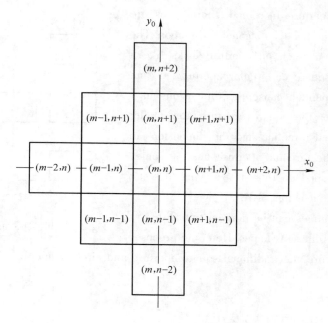

Fig. 2 Thirteen-pixel pattern used to study inter-pixel cross-talk

tensity corresponded to the "0" (OFF) SLM pixel is referred as $I_{p,0}$ and "1" (ON) as $I_{p,1}$, where $p = 1 \sim 2^{12}$. Assuming that the status of the SLM pixel under consideration has an equal possibility of ON and OFF, the SNR of the received signal is given by[17]

$$\text{SNR} = \frac{\mu_1 - \mu_0}{\sqrt{\sigma_0^2 + \sigma_1^2}} \tag{11}$$

where μ_1 is the average of $I_{p,1}$, μ_0 is the average of $I_{p,0}$, σ_1 is the standard deviation of $I_{p,1}$ and σ_0 is the standard deviation of $I_{p,0}$.

2.1 Consideration of the geometrical shape of the aperture

In order to improve the SNR of the received signal, the cross-talk between adjacent pixels must be reduced. A CMOS pixel (P_{00}) and its neighboring three pixels (P_{10}, P_{01} and P_{11}) shown in Fig. 3 are considered, and other neighboring pixels are not shown because of the biaxial symmetry. When an aperture with limited size is placed in the common focal plane (Fourier spectrum plane), the pixel P_{00} receives most of the energy transmitted by the corresponding SLM pixel, whereas the neighboring pixels P_{10}, P_{01} and P_{11} also receive a small amount of the energy called cross-talk. Assuming that the electric-field amplitude distribution transmitted by the aperture on the spectrum plane is uniform, we can increase the dimension of the aperture in one direction to weaken the cross-talk in the same direction. For example, increasing the dimension of the aperture in x direction will reduce the energy received by the pixel P_{10}. Since the pixels P_{10} and P_{01} are closer to the center pixel P_{00} than the pixel P_{11} (the distance between P_{00} and P_{11} is $\sqrt{2}$ times the distance between P_{00} and P_{10}), they might suffer severer cross-talk. Therefore, greater dimensions of the aperture in x and y directions are required than in the diagonal direction. The geometrical shape of such an aperture looks like a cross. Traditional square or circular apertures do not meet this requirement, because the dimensions in x and y directions of the square aperture are smaller than in the diagonal direction and the dimensions in all directions of the circular aperture are the same. Thus, the cross-shaped aperture is expected to result in less cross-talk and better SNR.

Calculation equations can be derived from Eq. (5) for apertures with different shapes. Here, the proposed cross-shaped aperture, the traditional square aperture and circular aperture (Fig. 4) are taken into account.

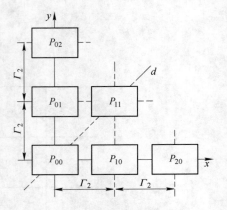

Fig. 3 Pixel distribution on the CMOS

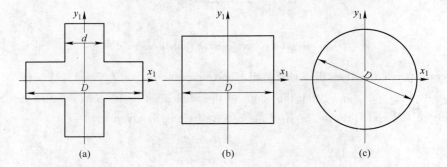

Fig. 4 Cross-shaped, square and circular apertures taken into account
(a) cross-shaped aperture; (b) square aperture; (c) circular aperture

2.2 Cross-shaped aperture

The cross-shaped aperture (Fig. 4(a)) can be described by its transmittance function

$$P(x_1, y_1) = \text{rect}\left(\frac{x_1}{D}\right)\text{rect}\left(\frac{y_1}{d}\right) + \text{rect}\left(\frac{x_1}{d}\right)\left[\text{rect}\left(\frac{y_1}{D}\right) - \text{rect}\left(\frac{y_1}{d}\right)\right] \tag{12}$$

where D is the length of the long side and d is the length of the short side. Eq. (5) is rewritten as

$$E(x, y) = 4\frac{f_1}{f_2}[\phi_1(x)\phi_2(y) + \phi_2(x)\phi_3(y)] \tag{13}$$

where

$$\phi_1(t) = \int_0^\alpha \text{sinc}(s)\cos\left(2\pi\frac{f_1}{f_2}\frac{t}{g_1\Gamma_1}s\right)ds \tag{14}$$

$$\phi_2(t) = \int_0^\beta \text{sinc}(s)\cos\left(2\pi\frac{f_1}{f_2}\frac{t}{g_1\Gamma_1}s\right)ds \tag{15}$$

$$\phi_3(t) = \int_\beta^\alpha \mathrm{sinc}(s)\cos\left(2\pi\frac{f_1}{f_2}\frac{t}{g_1\Gamma_1}s\right)ds \tag{16}$$

$$\alpha = \frac{k_1 g_1}{2},\ \beta = \frac{k_2 g_1}{2},\ k_1 = \frac{D}{D_N},\ k_2 = \frac{d}{D_N},\ D_N = \frac{\lambda f_1}{\Gamma_1} \tag{17}$$

k_1 and k_2 are normalized apertures and D_N is the size of Nyquist aperture.

2.3 Square aperture

When $d = D$ in Eq. (12), the cross-shaped aperture turns into a square aperture (Fig. 4(b)). Eq. (5) is rewritten as

$$E(x,y) = 4\frac{f_1}{f_2}\phi(x)\phi(y) \tag{18}$$

where

$$\phi(t) = \int_0^\alpha \mathrm{sinc}(s)\cos\left(2\pi\frac{f_1}{f_2}\frac{t}{g_1\Gamma_1}s\right)ds \tag{19}$$

$$\alpha = \frac{kg_1}{2},\quad k = \frac{D}{D_N},\quad D_N = \frac{\lambda f_1}{\Gamma_1} \tag{20}$$

2.4 Circular aperture

The transmittance function for the circular aperture (Fig. 4(c)) can be expressed as

$$P(x_1, y_1) = \begin{cases} 1 & \sqrt{x_1^2 + y_1^2} \leq D/2 \\ 0 & \sqrt{x_1^2 + y_1^2} > D/2 \end{cases} \tag{21}$$

where D is the diameter. Eq. (5) is rewritten as

$$E(x,y) = \frac{f_1}{f_2}\phi(x,y) \tag{22}$$

where

$$\phi(x,y) = \int_0^\alpha \int_0^{2\pi} \mathrm{sinc}(s\cos\theta)\mathrm{sinc}(s\sin\theta)\cos\left[2\pi\frac{f_1}{f_2}\frac{s}{g_1\Gamma_1}(x\cos\theta + y\sin\theta)\right]sdsd\theta \tag{23}$$

$$\alpha = \frac{kg_1}{2},\quad k = \frac{D}{D_N},\quad D_N = \frac{\lambda f_1}{\Gamma_1} \tag{24}$$

Substitute $E(x, y)$ into Eqs. (9), (10) and (11), we can obtain the SNR of the received signal.

3 Numerical Simulation

Simulation parameters are based on the equipments in our laboratory: $\Gamma_1 = 13.2\mu m$, $g_1 = 0.95$, $c = 22$, $f_1 = 44mm$, $f_2 = 40mm$, $\Gamma_2 = 12\mu m$, $g_{2x} = 0.87$, $g_{2y} = 0.54$. In the numerical simulation, each pixel is divided into 20×20 sub-pixels to provide an approximation. First we choose appropriate parameters for the cross-shaped aperture; then we compare the simulated SNR of the cross-shaped aperture with that of the square and circular apertures; finally we compare their sensitivities to pixel misalignment and magnification error.

3.1 Parameters for the cross-shaped aperture

The cross-shaped aperture is determined by D and d, or normalized apertures k_1 and k_2 in units of the Nyquist aperture. To determine their best value, we calculate SNRs of the cross-shaped apertures with different areas, assuming that SLM pixels and CMOS pixels are accurately one to one aligned. The results are depicted in Fig. 5. Each curve represents the SNRs of the apertures which have the same area s but vary in k_1. On the beginning of each curve, the cross-shaped aperture is actually a square one. For $1.8 \leqslant s \leqslant 3.8$, increasing k_1 to a certain extent will result in higher SNR. We choose the value of k_1 on each curve which leads to the maximum SNR. The value of k_2 for the cross-shaped aperture can be solved from s and k_1.

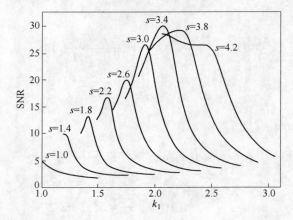

Fig. 5 Simulated SNRs of the cross-shaped apertures with different areas

3.2 Comparison of the simulated SNR

The linear size of the square aperture for a common HDSS is usually $1.0 \sim 2.0$, in units of D_N[7]. Corresponding area of the aperture is $1.0 \sim 4.0$, in units of D_N^2. Fig. 6 gives the simulated SNR versus area for the cross-shaped, square and circular apertures. Generally as expected in the consideration of the geometrical shape of the aperture in section 2, the cross-shaped aperture obtains higher SNR, the circular aperture obtains median SNR and the square aperture obtains lower SNR. When the area of the aperture is small, for example, in the interval [1,1.6], the cross-talk is so serious that distant pixels (P_{20} and P_{02} in Fig. 3) are also affected. To get similar cross-talk for the pixels P_{20}, P_{02} and P_{11}, the dimensions in x and y directions should be smaller than the dimension in the diagonal direction. The square aperture meets this requirement but the circular aperture does not, so the SNR of the square aperture is higher than that of the circular aperture with the same area. When the area is in the interval [2.6,4], the cross-shaped aperture obtains higher SNR than the square or circular apertures, because the dimension in the diagonal direction of the cross-shaped aperture is smaller than the dimensions in x

and y directions, and the pixels P_{10}, P_{01} and P_{11} are affected by similar cross-talk. When the area is around 3.5, the SNR of the cross-shaped aperture achieves its maximum, and is about 20% higher than the others. The area of the aperture in our system is designed to be 3.5 in units of D_N^2 to maximize the SNR and to eliminate the influences of noise sources.

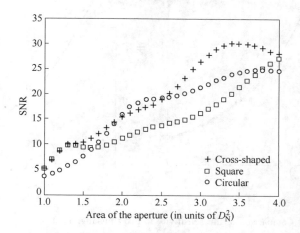

Fig. 6 Simulated SNR versus area for cross-shaped, square and circular apertures

3.3 Sensitivity to pixel misalignment

In an actual optical system, pixel misalignment can be hardly avoided due to limited machining and adjusting accuracy. Without compensation, the SNR of the received signal will decrease while the pixel misalignment increases as shown in Fig. 7. The pixel misalignment exists in x direction in Fig. 7(a) and in both x and y directions in Fig. 7(b). In both figures, the SNR of the cross-shaped aperture drops quicker than the other two apertures. The cross-shaped aperture is more sensitive to pixel misalignment. However, when the amount of pixel misalignment is less than about 1 μm, the proposed cross-shaped aperture gains higher SNR than the square or circular apertures with the same area (3.5, in units of D_N^2). When the pixel misalignment grows large, the SNRs of these three apertures become almost the same. Position accuracy of 1 μm can be easily obtained using our adjusting devices, so that better performance may be achieved with the proposed cross-shaped aperture.

3.4 Sensitivity to magnification error

When f_2/f_1 is not equal to Γ_2/Γ_1, magnification error occurs. We define the magnification error $\delta = (f_2/f_1 - \Gamma_2/\Gamma_1)/(\Gamma_2/\Gamma_1)$ and calculate the SNRs of the said three types of apertures with the same area (3.5, in units of D_N^2) plotted in Fig. 8. It is clear that SNR is very sensitive to magnification error without compensation. A magnification error of $\pm 6 \times 10^{-4}$ will cause the SNR to decrease half of the value. With carefully designed and adjusted optics, the magnifica-

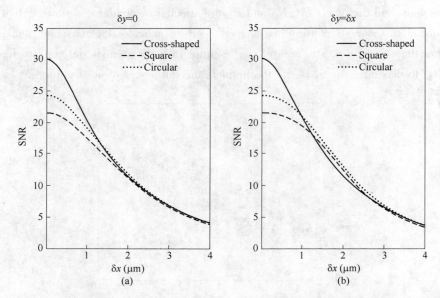

Fig. 7 Comparison of the sensitivity to pixel misalignment of the proposed cross-shaped aperture with that of the square and circular apertures
(a) SNR when pixel misalignment exists in x direction (denoted by δx); (b) SNR when pixel misalignment exists in both x and y directions (denoted by δx and δy respectively)

tion error can be kept within $\pm 2 \times 10^{-4}$, so the SNR of the cross-shaped aperture can be greater than that of the square or circular apertures.

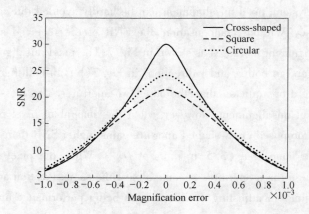

Fig. 8 Comparison of the sensitivity to magnification error of the proposed cross-shaped aperture with that of the square and circular apertures

4 Experimental Results

A test HDSS configuration is built to verify the performance of the proposed cross-shaped aperture. The parameters are identical to the numerical simulation. A frequency-doubled diode-pumped Nd: YAG laser (λ = 532nm) is used. The CMOS camera (Mikrotron MC1310) is mounted to a multi-dimensional stage to carry out pixel alignment. A series of the cross-shaped, square and circular apertures with different areas are placed in the common focal plane respectively. Randomly generated binary images (512×512 pixels) are used to calculate the SNR of the received signal, and the results are shown in Fig. 9. It can be seen that the proposed cross-shaped aperture gives higher SNR than the square or circular apertures when the area of the aperture lies in the interval [2.6, 4]. The circular aperture performs better than the square aperture. A maximum gain of 20% in SNR is obtained for the cross-shaped aperture whose area is 3.0 in units of D_N^2, which is basically identical to the simulation results. However, the overall SNR is relatively low compared to simulation results, because of the background noise and the interference pattern caused by multi-surface reflection. The SNR curves are not as smooth as simulated ones due to the discreteness of the areas and inaccurate positioning of the apertures in the spectrum plane.

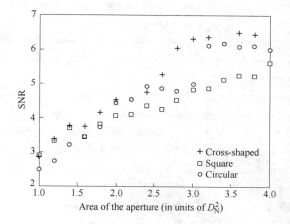

Fig. 9 SNR of the experimentally received signal of the proposed cross-shaped, square and circular apertures

5 Conclusions

We have considered the geometrical shape of the aperture in the HDSS to improve the SNR while maintaining the area of the aperture. Both simulated and experimental results showed that the proposed cross-shaped aperture could obtain higher SNR than the traditional square or circular apertures. Numerical simulations indicated that the cross-shaped aperture was a bit more sensitive to pixel misalignment and magnification error, but in our actual HDSS, the pixel misalignment and magnification error can be kept small enough to allow the cross-shaped aperture to perform better.

Acknowledgements

This work is supported by National Basic Research Program of China (2009CB724007), 863 High Technology (2009AA01Z112) and National Natural Science Foundation of China (60807005).

References

[1] F. H. Mok, "Angle-multiplexed storage of 5000 holograms in lithium niobate", Opt. Lett. 18, 915-917 (1993).

[2] G. W. Burr, C. M. Jefferson, H. Coufal, M. Jurich, J. A. Hoffnagle, R. M. Macfarlane, and R. M. Shelby, "Volume holographic data storage at areal density of 250 gigapixels/in. (2)", Opt. Lett. 26, 444-446 (2001).

[3] S. S. Orlov, W. Phillips, E. Bjornson, Y. Takashima, P. Sundaram, L. Hesselink, R. Okas, D. Kwan, and R. Snyder, "High-transfer-rate high-capacity holographic disk data-storage system", Applied Optics 43, 4902-4914(2004).

[4] W. L. Wilson, L. Dhar, and K. R. Curtis, "Progress toward the commercial realization of high performance holographic data storage: Architecture and function of the InPhase Technologies holographic drive", in *Organic Holographic Materials and Applications IV*, (SPIE, 2006), 63350G-63356.

[5] H. Horimai and X. D. Tan, "Holographic information storage system: Today and future", Ieee T Magn 43, 943-947(2007).

[6] R. M. Shelby, J. A. Hoffnagle, G. W. Burr, C. M. Jefferson, M. P. Bernal, H. Coufal, R. K. Grygier, H. Gunther, R. M. Macfarlane, and G. T. Sincerbox, "Pixel-matched holographic data storage with megabit pages", Opt. Lett. 22, 1509-1511(1997).

[7] M. P. Bernal, G. W. Burr, H. Coufal, and M. Quintanilla, "Balancing interpixel cross talk and detector noise to optimize areal density in holographic storage systems", Applied Optics 37, 5377-5385(1998).

[8] Z. Wang, G. F. Jin, Q. S. He, and M. X. Wu, "Simultaneous defocusing of the aperture and medium on a spectroholographic storage system", Applied Optics 46, 5770-5778(2007).

[9] J. W. Yang, L. M. Bernardo, and Y. S. Bae, "Improving holographic data storage by use of an optimized phase mask", Applied Optics 38, 5641-5645(1999).

[10] W. X. Shang, Q. S. He, and G. F. Jin, "Nonlinear blind equalization for volume holographic data storage", Chin. Phys. Lett. 21, 1741-1744(2004).

[11] A. He and G. Mathew, "Nonlinear equalization for holographic data storage systems", Applied Optics 45, 2731-2741(2006).

[12] G. W. Burr and T. Weiss, "Compensation for pixel misregistration in volume holographic data storage", Opt. Lett. 26, 542-544(2001).

[13] L. Menetrier and G. W. Burr, "Density implications of shift compensation postprocessing in holographic storage systems", Applied Optics 42, 845-860(2003).

[14] G. W. Burr, J. Ashley, H. Coufal, R. K. Grygier, J. A. Hoffnagle, C. M. Jefferson, and B. Marcus, "Modulation coding for pixel-matched holographic data storage", Opt. Lett. 22, 639-641(1997).

[15] W. C. Chou and M. A. Neifeld, "Interleaving and error correction in volume holographic memory systems", Applied Optics 37, 6951-6968(1998).

[16] J. W. Goodman, *Introduction to Fourier Optics* (McGraw-Hill, 1968).

[17] G. W. Burr, H. Coufal, R. K. Grygier, J. A. Hoffnagle, and C. M. Jefferson, "Noise reduction of page-oriented data storage by inverse filtering during recording", Opt. Lett. 23, 289-291(1998).

Orthogonal-reference-pattern-modulated Shift Multiplexing for Collinear Holographic Data Storage[*]

Abstract A novel hybrid shift multiplexing method for collinear holographic data storage (CHDS) by using orthogonal reference patterns (RPs) is proposed, analyzed, and demonstrated. For this method, holograms are multiplexed by not only shifting the media but also using different RPs. Compared with the traditional method, the shift pitch for the hybrid method is substantially reduced because of the selectivity introduced by different RPs. The interpage cross talk due to Bragg mismatch and degeneracy for multiplexing holograms in the same volume by using orthogonal RPs is also attenuated by utilizing the shift selectivity of the hologram. A 1.5 μm shift pitch is experimentally achieved by using three amplitude RPs in a system that would be 4.5 μm with only one RP. This new method offers an alternative to significantly increase the data density and transfer rate of the CHDS system given that the media has ideal properties.

Collinear (Coaxial) holographic data storage (CHDS)[1,2] has drawn much attention during recent years[3-6] since it entails several advantages such as being tolerant to environmental disturbances and having relatively large wavelength and tilt margins[7]. In addition, this configuration offers a unique degree of freedom in optimizing a reference pattern (RP) to further improve the performance of the system because the RP is also implemented by a spatial light modulator (SLM). Radial lines, multiple rings, and many other kinds of pattern shave been proposed as RPs to achieve a narrow point (pixel) spread function for the increasing of the signal-to-noise ratio[4,6,8]. However, analyses have shown that advanced design of these different kinds of RPs together with increasing media thickness has no significant influence on the shift selectivity for the traditional shift multiplexing method[9,10] and finally make no obvious benefit on storage capacity. Multiplexing holograms in the same volume by using orthogonal RPs[11] offers an alternative to increase data density. However, the interpage cross talk due to Bragg degeneracy and Bragg mismatch readout for this multiplexing method cannot be eliminated because the RP is along a two-dimensional plane and the angle between the reference beam and the signal beam is relatively small[12]. To circumvent these roadblocks and achieve the highest data density for ideal media, an orthogonal-RP modulated-shift multiplexing method by hybridization of these two multiplexing methods is proposed. Holograms are multiplexed by not only using orthogonal RPs but also shifting the media with a small distance separately in the recording

---

[*] Copartner: Jianhua Li, Liangcai Cao, HuarongGu, Xiaodi Tan, Qingsheng He. Reprinted from *Optics Letters*, 2012, 37 (5):936-938.

process. The interpage cross talk due to Bragg degeneracy and Bragg mismatch readout can be substantially attenuated by utilizing the shift selectivity of the hologram. The shift pitch for this hybrid shift multiplexing method can also be substantially reduced by utilizing shift selectivity in conjunction with Bragg selectivity, which offers a potential to significantly increase the data density and transfer rate of the CHDS system when the recording media has ideal properties.

The schematic of the orthogonal-RP modulated-shift multiplexing is shown in Fig. 1. N orthogonal RPs are used for this method. For an amplitude SLM, orthogonality means the positions of the ON pixels for RPs are totally different from each other. The objective lens in the Fourier geometry converts a pixel on the SLM into a plane wave with a unique angle corresponding to the pixel's coordinate. A recorded hologram can be treated as the accumulation of the gratings written by every pair of the plane waves from the ON pixels on the SLM[8]. For the recording process, as shown in the dashed line rectangle, both the RP and data page are simultaneously uploaded on the SLM to encode the reference and signal beams. Both beams will interfere with each other and form a hologram inside the holographic disk.

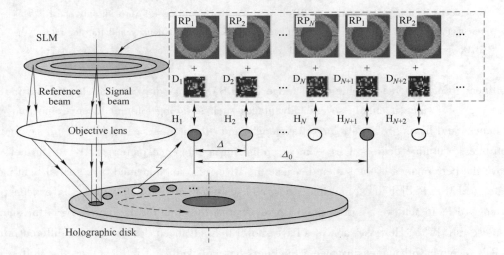

Fig. 1 Schematic of the orthogonal-RP modulated-shift multiplexing method by using N RPs

SLM—spatial light modulator; RP_i—the ith RP; D_i—the ith date page; H_i—the ith hologram; Δ_0—traditional shift pitch; Δ—new shift pitch. data page can be read out by using the corresponding RP. For this method, the shift pitch can be set to $1/N$ of that of the traditional shift multiplexing method with a single RP since holograms can even be multiplexed in the same volume of the media by using orthogonal RPs. However, the shift selectivity of the hologram imposes a limit on the maximum of N because the interpage cross talk will increase with the decrease of the shift pitch

On recording, the first data page (D_1) is recorded as the first hologram (H_1) by using the first RP (RP_1) Then the disk is shifted by a distance Δ, and RP_2 is used instead of RP_1 for the recording of D_2. D_3 is also recorded in this way, changing the RP and shifting the disk at the same time. If N RPs are employed for this method, the $(N+1)$th data page is then recorded again by using RP1. By repeating these two operations, all the data pages can be multiplexed. On reconstruction, the desired

For an amplitude SLM, the upper limit of N is also determined by the filling rate (η) of the RP and $N_{max} < 1/\eta$. Fig. 2 shows an example of three RPs for this method ($N = 3$). The radial line pattern shown in Fig. 2(a) is used as RP_1. The filling rate of this RP is 1/6. Fig. 2(b) is an enlarged version of the rectangle in Fig. 2(a). RP_2 and RP_3 shown in Fig. 2(c) are generated by rotating RP_1 along the center of the pattern clockwise by $\varphi/3$ and $2\varphi/3$, respectively, where φ represents the angle between the neighbor radial lines of RP_1.

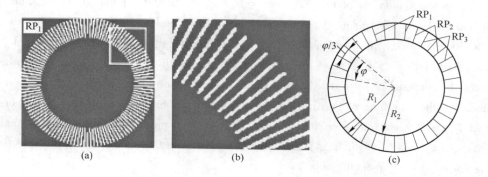

Fig. 2 Schematic and parameters of the three RPs
(a) RP_1; (b) the rectangular part in (a); (c) RP_1, RP_2, and RP_3 are uniformly spaced in the ring and orthogonal with each other

In order to analyze the shift selectivity and interpage cross talk for this hybrid method, we set the signal to be a point source in the center of the SLM and consider the intensity of the signal on the output plane reconstructed by using an RP. Based on Eq. (15) in Ref. [4], when the hologram is recorded by the ith RP and shifted with a displacement of Δu and Δv, the optical field on the output plane readout by the jth RP can be written as

$$PSF_{j \to i}(\varepsilon, \eta, \Delta u, \Delta v)$$
$$= \iint RP_j(x_0, y_0) RP_i^*(x_0 + \varepsilon, y_0 + \eta) \times tsinc\left[\frac{t}{\lambda f^2}(\varepsilon^2 + \eta^2 + \varepsilon x_0 + \eta y_0)\right] \times$$
$$\exp i \frac{2\pi}{\lambda f}(\Delta u \cdot x_0 + \Delta v \cdot y_0) dx_0 dy_0 \quad (1)$$

where (ε, η) and (x_0, y_0) denote the coordinates on the output plane and the SLM plane, respectively. RP_i and RP_j stand for the recording RP and reading RP uploaded on the SLM, respectively, t is the media thickness, f is the focus length of the objective lens, and λ is the wavelength. The sinc function in Eq. (1) is caused by Bragg mismatch, and the exponential function is introduced by the disk shift.

If the disk is kept at the same place for reading and only considering the image position of the point source on the output plane, $\varepsilon = \eta = \Delta u = \Delta v = 0$ and $PSF_{j \to i}(0,0,0,0)$ is proportional to the cross correlation between RP_i and RP_j. When the RP_i is orthogonal with RP_j ($i \neq j$),

$PSF_{j\to i}(0,0,0,0) = 0$, which means multiple holograms can be multiplexed without shifting of the disk by using orthogonal RPs. However, due to Bragg mismatch and Bragg degeneracy, $PSF_{j\to i}(\varepsilon, \eta, 0, 0) \neq 0$ when $\varepsilon \neq 0$ or $\eta \neq 0$, which will introduce interpage crosstalk for this scheme.

Shift selectivity can be defined as the variation of the integrated intensity of the reconstructed signal on the output plane when the disk shifts. It can be expressed as

$$I_{j\to i}(\Delta u, \Delta v) = \iint \left| PSF_{j\to i}(\varepsilon, \eta, \Delta u, \Delta v) \right|^2 d\varepsilon d\eta \tag{2}$$

Fig. 3 shows the simulation of Eq. (2) for $i = j$ and $i \neq j$. In the simulation, λ is 532nm, f is 4mm, and t is 1 mm. The pixel size of the SLM is 10.8μm, and the total pixel number used is 300 × 300. The outer radius R_1 of the RPs is 150 pixels, the inner radius R_2 is 105 pixels, and ϕ is 3°. Twenty-five (5 ×5) pixels are used for the integral in Eq. (2). It shows that, when a different RP is used ($i \neq j$) to read the hologram, the diffraction intensity has been substantially attenuated without shifting of the disk, and this intensity (interpage cross talk) will decrease fluctuately with the shift distance. It also shows the new shift pitch Δ can be almost as small as 1/3 of the traditional shift pitch Δ_0 with similar cross talk with these parameter, which are based on the following experimental setup.

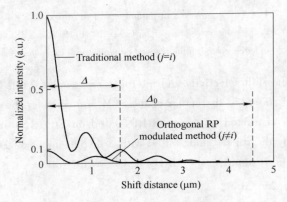

Fig. 3 Shift selectivity for the traditional and orthogonal-RP modulated-shift multiplexing method

The optical setup for the demonstration of the proposed method is shown in Fig. 4. A diode-pumped solid-state laser (DPSSL) at the wavelength of 532nm is employed as the light source. After passing through the beam expander, the laser beam is collimated into a plane wave. The intensity of the plane wave is adjusted to 97 mW/cm^2 by the neutral density filter (NDF). A digital micromirror device (DMD) (Texas Instruments, DLP0.55 XGA) is used as the SLM. The relay lens associated with iris$_1$ filters out the higher orders of the DMD. The objective lens (Nikon, NA = 0.55) Fourier transforms the DMD pattern into a PQ: PMMA polymer media, which is mounted on a precise microtranslation stage (Physik Instrumente, M-112). The resolution of the stage is 70nm. On reconstruction, iris$_2$ is used to prevent the reference beam from entering the complementary metaloxide semiconductor (CMOS) image sensor (Mikrotron, MC1310), which is employed to capture the reconstructed data pages.

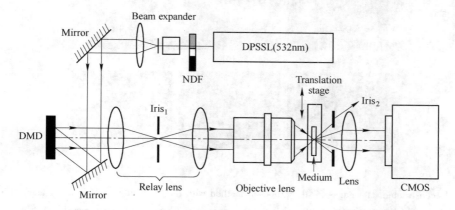

Fig. 4　Experimental setup of the CHDS system

The required shift pitch for the traditional shift multiplexing method was measured. A data page was stored into the media by using RP_1. Then the media was shifted, and the intensity of the reconstructed data page was monitored. As shown in Fig. 5, the normalized diffraction intensity decreases fluctuately with the shift distance. When the media was shifted 4.5 μm away from the original position, the diffraction intensity reached the minimum, and a new data page can be recorded in this position for the traditional shift multiplexing method. This value agrees well with the theoretical value in Fig. 3.

Three amplitude RPs in Fig. 2 were used for the orthogonal-RP modulated-shift multiplexing. After using RP_1 recorded the first hologram, the media was shifted only 1.5 μm away from the original position. Then RP_2 was used to record the second hologram. Six data pages were recorded by this method. Fig. 6 shows the reconstructed data pages. The varying location of the numbers in the data pages enables us to detect the undesired cross talk easily. It shows the reconstructed data pages exhibit a very low interpage cross talk, which has verified the feasibility of the proposed method.

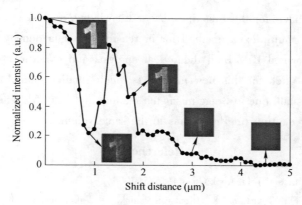

Fig. 5　Measured intensity of the reconstructed data page versus the shift distance

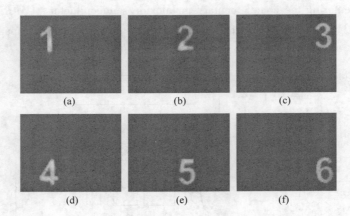

Fig. 6 Reconstructed data pages for the proposed method with the shift pitch of 1.5μm by using three RPs
(a),(d) reconstructed by RP_1; (b),(e) reconstructed by RP_2; (c),(f) reconstructed by RP_3

In order to compare the cross talk of this hybrid method with that of the method that multiplexes holograms in the same volume of the media, three data pages were multiplexed in a single spot by using the same three amplitude RPs used for Fig. 6. As shown in Fig. 7, the reconstructed data pages exhibit a relatively higher interpage cross talk than that in Fig. 6, which shows good agreement with the theoretical analysis.

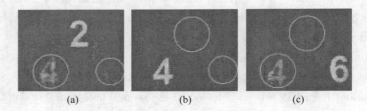

Fig. 7 Reconstructed data pages after multiplexing three holograms
in the same volume by using three RPs
(a)-(c) Reconstructed by RP_1 RP_2, and RP_3 respectively. The interpage cross talk is indicated by the circles

In conclusion, we have, for the first time to our knowledge, proposed, analyzed, and demonstrated a hybrid shift multiplexing method for increasing the data density and transfer rate of CHDS by using orthogonal RPs. Both theoretical analysis and experimental results have confirmed that the shift pitch for this new method is much smaller than that for the traditional method, and the cross talk due to Bragg mismatch readout and Bragg degeneracy for this method is lower than that for multiplexing holograms in the same volume.

References

[1] H. Horimai, X. D. Tan, and J. Li, Appl. Opt. 44, 2575(2005).
[2] K. Tanaka, M. Hara, K. Tokuyama, K. Hirooka K. Ishioka, A. Fukumoto, and K. Watanabe, Opt. Express 15, 16196(2007).

[3] S. Yasuda, J. Minabe, and K. Kawano, Opt. Lett. 32, 160(2007).
[4] C. C. Sun, Y. W. Yu, S. C. Hsieh, T. C. Teng, and M. F. Tsai, Opt. Express 15, 18111(2007).
[5] S. Yasuda, Opt. Lett. 34, 3101(2009).
[6] Y. W. Yu, C. Y. Chen, and C. C. Sun, Opt. Lett. 35, 1130(2010).
[7] X. Tan, H. Horimai, T. Shimura, S. Ichimura R. Fujimura, and K. Kuroda, Proc. SHE 6343, 63432W (2006).
[8] T. Shimura, S. Ichimura, R. Fujimura "K. Kuroda" X. D. Tan, and H. Horimai, Opt. Lett. 31, 1208(2006).
[9] T. Shimura, S. Ichimura, Y. Ashizuka, R. Fujimura K. Kuroda, X. Tan, and H. Horimai, Proc. SPIE 6282, 62820S(2007).
[10] Y. W. Yu, T. C. Teng, S. C. Hsieh, C. Y. Cheng, and C. C. Sun, Opt. Commun. 283, 3895(2010).
[11] M. Toishi, M. Hara, K. Tanaka, T. Tanaka. and K. Watanabe, Jpn. J. Appl. Phys. 46, 3775(2007).
[12] L. Hesselink, S. S. Orlov, and M. C. Bashaw, Proc. IEEE 92, 1231(2004).

Improvement of Volume Holographic Performance by Plasmon-induced Holographic Absorption Grating[*]

Abstract We report on the enhanced holographic performance by employing a strong volume holographic absorption grating induced by localized surface plasmon resonance effect in a bulk gold nanoparticles doped photopolymer. The contributions of plasmon-induced volume holographic absorption grating is characterized through the Kogelnik's coupled wave model and demonstrated experimentally by using two-beam interference technology. At the 0.05 vol. % concentration of the gold nanoparticles in the bulk photopolymer, 101.8% increase in the diffraction efficiency and more than four times suppression of the first side lobe in angular selectivity have been achieved.

Polymeric nanocomposite has attracted immense attention due to their advanced properties relative to conventional polymeric materials for the application of optical data storage[1], sensing[2-3], bioengineering[4], and organic solar cell[5,6]. These advanced optical and electrical properties are not only attributed to the characteristics of nanoparticles (NPs) but also sensitive to their spatial distribution in the matrix[7,8]. As a one-step, simple, and flexible technique, holographic photopolymerization has been used to create defect-free, sub-micrometer patterns of NPs over large dimensions in a three-dimensional polymeric matrix[9,10]. This also opens up practical possibilities to improve the coupling between the light wave and sub-micrometer patterns for designing functional polymeric nanocomposites[11-13].

In recent years, volume holographic gratings with a modulation of both the absorption coefficient and the refractive index have been investigated to improve the holographic performances in various materials such as silver halide emulsions, photorefractive crystals, or polymer[14-16]. Limited success has been reported on enhanced holographic properties with both refractive index and absorption modulation in the photopolymer material. This may be attributed to the unpredictable phase shift between refractive index and absorption grating by using organic dye[17]. Based on one-step holographic method, another grating formed by the spatially distribution of NPs has a fixed 0 or π phase shift to the polymeric refractive index grating[18]. An absorption grating in phase with the refractive index grating can theoretically increase holographic diffraction efficiency. Under the condition of weak absorption modulation, the maximum increase of the diffraction efficiency is only 3.7%[19]. Gold NPs exhibit strong absorption in the visible spectral range due to the localized surface plasmon resonance (LSPR) effect[20]. The absorp-

---

[*] Copartner: Chengmingyue Li, Liangcai Cao, Jingming Li, Qingsheng He, Shiman Zhang, and Fushi Zhang. Reprinted from *Aplied Physics Letters* 2013, 102: 061108.

tion of spatially distributed gold NPs two times greater than the polymer matrix was observed in a thin photopolymer film[21]. In this letter, we present the characteristics of strong holographic absorption grating in phase with the refractive index grating through Kogelnik's coupled wave model. By doping gold NPs into a bulk photopolymer, experimental results are also provided to demonstrate the improvement of plasmon-induced holographic absorption grating to holographic performances.

Holographic photopolymer contains one monomer or a mixture of monomers, a photosensitizer, and some nonreactive components as polymeric binder[22]. During holographic exposure, a few photons initiate a chain reaction of photopolymerizable monomer molecules. This photopolymerization process lowers the chemical potential of monomers in the bright regions, breaks the thermo-dynamical equilibrium balance, and leads to the diffusion of monomers from the dark to the bright regions[23]. The difference in refractive index between the reactive monomers in the bright region and the polymeric binder in the dark region results in the formation of holographic refractive index grating, as shown in Fig. 1(a). In the NPs dispersed photopolymer, NPs experience counter diffusion from the bright to the dark regions, driven by the photopolymerization process of the monomers since the NPs are not consumed and their chemical potential increases in the bright regions as a result of consumption of monomers[8]. Consequently, the spatially periodic distributions of NPs and photoproduct of monomers are formed under the inhomogeneous illumination in the NPs doped photopolymer. At the resonance wavelength of gold NPs, a strong absorption grating can be activated in phase with the refractive index grating as a result of density differences of NPs between the bright and the dark regions. Fig. 1(b) shows the schematic profile of the plasmon-induced absorption grating and the refractive index grating in the gold NPs doped photopolymer.

According to Kogelnik's coupled wave theory[19], the fringes of the mixed gratings can be re-

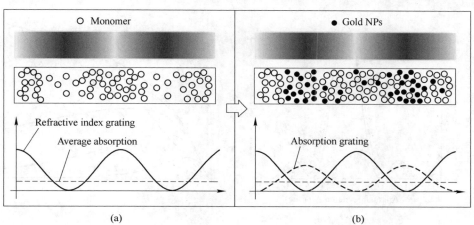

Fig. 1 (a) Formation of holographic refractive index grating in the pure photopolymer material; (b) Construction of holographic refractive index grating and plasmon-induced absorption grating in the gold nanoparticles doped photopolymer

presented by the spatial modulation of the dielectric constant $\varepsilon = \varepsilon_0 + \varepsilon_1 \cos(K \cdot x)$ and the conductivity $\sigma = \sigma_0 + \sigma_1 \cos(K \cdot x + \pi)$, where h_1 and σ_1 are the amplitudes of the spatial modulation, ε_0 is the average dielectric constant, and σ_0 is the average conductivity. Then the coupling constant k can be described as

$$k = \frac{\pi n_1}{\lambda} + j \frac{\alpha_1}{2} \tag{1}$$

where n_1 and α_1 are the refractive index modulation and absorption modulation, respectively. For the unslanted transmission hologram without wavelength deviation, the amplitude of the diffracted wave can be obtained as

$$S = -je^{-\alpha d/\cos\theta} e^{-j\varepsilon} \frac{1}{v} \mathrm{sinc}(\sqrt{v^2 + \varepsilon^2})$$

$$\varepsilon = \Delta\theta \cdot k_0 d \sin\theta_0 \tag{2}$$

$$v = \frac{\pi n_1 d}{\lambda \cos\theta} + j \frac{\alpha_1 d}{2\cos\theta}$$

where α is the average absorption constant, θ is the direction of the reading wave in the medium, and ε is the dephasing parameter related to the wave vector k_0, effective thickness of hologram d, Bragg angle θ_0, and angular derivation $\Delta\theta_0$. Then the diffraction efficiency of the mixed gratings is obtained by $h = SS^*$. For convenience, the loss parameters are described as the average absorption constant $D_0 = \alpha d/\cos\theta$, the absorption modulation $D_1 = \alpha_1 d/\cos\theta$, and the depth of absorption modulation $D_m = \alpha_1/\alpha_0$.

Fig. 2(a) shows the angular selectivity curves for several values of D_1 in Eq. (2) when $\pi n_1 d/\lambda \cos\theta = \pi/2$ and $D_0 = 0.5$. The thickness of the grating and the Bragg angle for the simulation are 1mm and 15° in air, respectively. There is a marked suppression of the side lobes

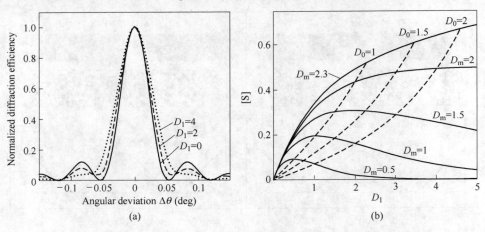

Fig. 2 (a) Angular selectivity curves of the mixed gratings for different values of absorption modulation D_1; (b) The diffracted amplitude as a function of absorption modulation D_1 for various values of average absorption constant D_0 and the depth of absorption modulation D_m

with some broadening in the sensitivity curves for the increasing D_1. This can be explained as the superimposing effect of absorption grating with the refractive index grating, which makes the prefect sinusoidal profile of the refractive index grating to near-Gaussian shape of the mixed gratings. It should be noted that the angular sensitivity of the mixed gratings is neither sensitive to the parameter D_0 nor D_m in such a thick holographic grating. Fig. 2(b) shows the diffracted amplitude of absorption grating as a function of D_1 for various values of D_m. It can be seen that the diffracted amplitude increases as the increased depth of absorption modulation. The black dashed curves for different values of D_0 show the grating behavior for constant background absorption. The average absorption constant could affect the diffracted amplitude, but the depth of absorption modulation is the major factor, which shows the great potential to enhance the volume holographic performance by increasing the spatial modulation of absorption grating.

To demonstrate the significantly enhanced holographic performance of strong absorption grating, photopolymer with different concentrations of gold NPs for volume holographic recording were fabricated by thermo-polymerization method. The photopolymer syrup consists of methyl methacrylate (MMA, Alfa Aesar), 2,2-azobis (2-methlpropionitrile) (AIBN, J&K Scientific Ltd.), and phenanthrenquinone (PQ, Alfa Aesar) with the optimized weight ratio of 98.3 : 0.7 : 1 [24]. The gold NPs were synthesized following a facile organic phase synthesis method via a burst nucleation of gold upon injection of a t-butylamine-borane complex[25]. The size of gold NPs was about 6-8nm in diameter, which are controlled below 10nm to avoid scattering loss, as shown in Fig. 3(a) inset. Different concentrations of gold nanoparticles were dispersed into the prepared polymer syrup. The mixture was then sonicated for 5min to make them uniform and stirred in an oil bath at 40℃ for an appropriate period of time until it became viscous. The viscous liquid was poured into a glass mold with a 1mm thick spacer and baked at 50℃ for 24h to thermally solidify the mixture. The resulting solid samples were removed from the molds and cut into 1 × 1 in. squares for holographic exposure. The absorption spectrum of the prepared solid bulk nanocomposites are shown in Fig. 3. The value of absorption in the phenanthrenequinone doped poly (methyl methacrylate) (PQ-PMMA) with gold NPs is about 2.4 times than that of pure PQ-PMMA at the wavelength of 532nm.

To demonstrate the contributions of strong absorption grating to holographic performances, the diffraction efficiency and angular selectivity of prepared bulk samples were measured by using two-beam interference technology at the wavelength of 532nm near the plasmon resonance peak of the gold NPs. Two split s-polarized beams with the incident angles of 15° in air were conducted to record unslanted hologram in the prepared samples. The intensity of each split beam was 5mW/cm^2, and the beam diameter was 6mm. The absolute diffraction efficiency of volume gratings were obtained by $\eta = I_d/I_{inc}$, which I_d and I_{inc} are the intensity of diffracted beam and incident beam, respectively. The temporal evolution of diffraction efficiency as the increasing exposure energy for pure PQ-PMMA, PQ-PMMA with 0.05 vol.% and 0.24 vol.% gold NPs, is shown in Fig. 3(b). The slight drop-off in temporal evolution of diffraction efficiency is attributed to the backward coupling of the diffracted beam with the transmitted beam and the fanning

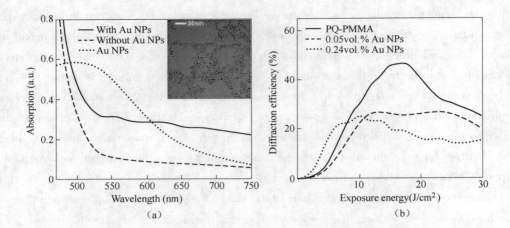

Fig. 3 (a) Absorption spectrum of the gold NPs soluble in hexane solvent, the solid PQ-PMMA films without gold NPs, and the PQ-PMMA films with gold NPs. The inset shows a TEM image of the gold NPs; (b) Temporal evolution of diffraction efficiency for PQ-PMMA, PQ-PMMA with 0.05 vol.%, and 0.24 vol.% gold NPs

of noise gratings[23].

Fig. 4(a) shows the measured angular selectivity curves of pure PQ-PMMA, PQ-PMMA with 0.05 vol.% and 0.24 vol.% gold NPs. The diffraction efficiency of the first side lobe is suppressed more than four times from 0.14% in the pure photopolymer to 0.03% in the photopolymer with 0.05 vol.% gold nanoparticles. The full width at half maximum (FWHM) of the angular selectivity is only broadened about 0.04° from 0.06° to 0.10°, which can be neglected in the practical application. The suppression of the side lobes is the direct result of the increasing absorption modulation induced by the LSPR effect of the doped gold NPs in the PQ-PMMA photopolymer corresponding to the characterization of holographic mixed gratings based on Kogelnik's coupled wave model.

Fig. 4(b) shows the maximum value in the temporal evolution of the diffraction efficiency for the PQ-PMMA with different doped concentration of gold NPs. It can be seen that the diffraction efficiency of gold NPs doped PQ-PMMA is higher than that of pure PQ-PMMA as a result of strong plasmon-induced holographic absorption grating. The maximum diffraction efficiency with 101.8% enhancement from 23.3% to 47.1% is observed in the PQ-PMMA containing 0.05 vol.% gold NPs. The dramatic decreased diffraction efficiency as the concentration of NPs higher than 0.05 vol.% attributes to the decreased depth of absorption modulation. The decreased density difference of gold NPs between bright and dark regions as the increasing concentration of gold NPs may attribute to the decrease in the initial photopolymerization driven process with higher concentration of gold NPs[21]. The reduced suppression of the side lobes in the 0.24 vol. % gold NPs doped PQ-PMMA [Fig. 4(a)] also confirms the decreased absorption modulation corresponding to the decreased depth of absorption modulation in the nanocomposite with higher concentration of gold NPs.

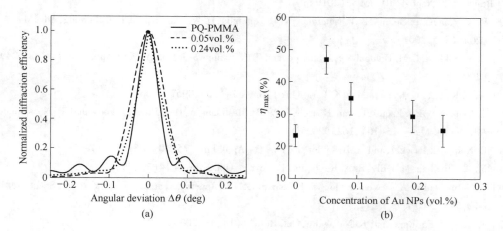

Fig. 4 (a) The angular selectivity curves of PQ-PMMA, PQ-PMMA with 0.05 vol. %, and 0.24 vol. % gold NPs; (b) Maximum value in volume holographic diffraction efficiency(max of the prepared samples as a function of gold NPs concentration)

In conclusion, we have demonstrated that the strong absorption grating can lead to the significant improvement of holographic performances through Kogelnik's coupled wave model. By dispersing gold NPs in a bulk PQ-PMMA photopolymer, more than four times improved side lobe suppression in angular selectivity and 101.8% increase in diffraction efficiency have been achieved at the 0.05 vol. % concentration of gold NPs. The quantitative model for the density of NPs after redistribution in gold NPs doped photopolymer is under progress. The mixed volume holographic grating mechanism may open up a way to understand the photophysical and photochemical process in the NPs doped photopolymer.

References

[1] R. A. Vaia and J. F. Maguire, Chem. Mater. 19(11), 2736(2007).

[2] E. Leite, I. Naydenova, S. Mintova, L. Leclercq, and V. Toal, Appl. Opt. 49(19), 3652 (2010).

[3] K. S. Lee and M. A. El-Sayed, J. Phys. Chem. B 110(39), 19220(2006).

[4] H. Zhang, Y. Sun, J. Wang, J. Zhang, H. Q. Zhang, H. Zhou, and D. Q. Song, Biosens. Bioelectron. 34(1), 137(2012).

[5] A. J. Morfa, K. L. Rowlen, T. H. Reilly, M. J. Romero, and J. van de Lagemaat, Appl. Phys. Lett. 92(1), 013504(2008).

[6] S. Kober, M. Salvador, and K. Meerholz, Adv. Mater. 23(41), 4725(2011).

[7] K. Ueno, S. Juodkazis, T. Shibuya, V. Mizeikis, Y. Yokota, and H. Misawa, J. Phys. Chem. C 113(27),

11720(2009).

[8] O. V. Sakhno, L. M. Goldenberg, T. N. Smimova, and J. Stumpe, Proc. SPIE 7487, 74870H(2009).
[9] A. T. Juhl, J. D. Busbee, J. J. Koval, L. V. Natarajan, V. P. Tondiglia, R. A. Vaia, T. J. Bunning, and P. V. Braun, ACS Nano 4(10), 5953(2010).
[10] R. A. Vaia, C. L. Dennis, L. V. Natarajan, V. P. Tondiglia, D. W. Tomlin, and T. J. Bunning, Adv. Mater. 13(20), 1570(2001).
[11] O. V. Sakhno, L. M. Goldenberg, J. Stumpe, and T. N. Smimova, J. Opt. A, Pure Appl. Opt. 11(2), 024013(2009).
[12] W. S. Kim, Y. C. Jeong, and 1. K. Park, Opt. Express 14(20), 8967(2006).
[13] C. Sanchez, M. J. Escuti, C. van Heesch, C. W. M. Bastiaansen, D. J. Broer, J. Loos, and R. Nussbaumer, Adv. Funct. Mater. 15(10), 1623(2005).
[14] C. Neipp, C. Pascual, and A. Belendez, J. Phys. D: Appl. Phys. 35(10), 957(2002).
[15] M. A. Ellabban, M. Fally, R. A. Rupp, and L. Kovacs, Opt. Express 14(2), 593(2006).
[16] A. S. Shcheulin, A. V. Veniaminov, Y. L. Korzinin, A. E. Angervaks, and A. I. Ryskin, Opt. Spectrosc. 103(4), 655(2007).
[17] T. Sato, H. Fujiwara, and K. Nakagawa, Opt. Rev. 11(1), 48(2004).
[18] N. Suzuki and Y. Tomita, Appl. Phys. Lett. 88(1), 011105(2006).
[19] H. Kogelnik, Bell Syst. Tech. J. 48(9), 2909(1969).
[20] S. Lal, S. Link, and N. J. Halas, Nat. Photonics 1(11), 641(2007).
[21] L. M. Goldenberg, O. V. Sakhno, T. N. Smimova, P. Helliwell, V. Chechik, and J. Stumpe, Chem. Mater. 20(14), 4619(2008).
[22] H. Y. Wei, L. C. Cao, C. Gu, Z. F. Xu, M. Z. He, Q. S. He, S. R. He, and G. F. Jin, Chin. Phys. Lett. 23(11), 2960(2006).
[23] S. H. Lin, P. L. Chen, and J. H. Lin, Opt. Eng. 48(3), 035802(2009).
[24] S. H. Lin, K. Y. Hsu, W. Z. Chen, and W. T. Whang, Opt. Lett. 25(7), 451(2000).
[25] S. Peng, Y. M. Lee, C. Wang, H. F. Yin, S. Dai, and S. H. Sun, Nano Res. 1(3), 229(2008).

光学仪器

ВП-4-ЭИ 型三向电感式车削测力仪的介绍[*]

研究金属切削过程和设计机床与刀具时，必须知道切削加工时的切削力。但截至目前，在理论上准确地计算切削力的公式还没有，即使有一些公式，也只能求出切削力的近似值；又因影响切削力的因素很多，应用范围狭小，故无论在生产上或试验室中都很少采用。目前要比较精确地测得切削力，都依靠测力仪去实际测量。测力仪的种类很多（见《机床与工具》1956 年第 23 期及 1957 年第 4 期），但其中所介绍的机械式和液压式测力仪，有很多缺点，在近年来很少使用。目前应用较广的是电工测力仪。但电工测力仪中，往往由于要用很多电子管、示波器等贵重的设备，使制造和使用复杂，且成本昂贵。下面介绍一种结构简单且不用电子管的 ВП-4-ЭИ 型三向电感式车削测力仪。此测力仪是由苏联古比雪夫航空学院工程师 В. Ф. 帕拉莫诺夫设计的。

1 对测力仪的要求

要设计出性能良好的测力仪，应满足下列几个基本的要求：
（1）应有小的弹性位移量，使该位移量保持在弹性元件的弹限以内，以提高测量的正确度。
（2）尺寸小，重量较轻并具有较高的刚性，以便增加系统的自然频率。这样可以作小惯性测量。
（3）结构要简单，这样保证了制造容易和使用可靠。
（4）工作要有足够的稳定性，即不受各种因素（如摩擦、磨损、温度和振动等）的影响。
（5）各别切削分力不影响其他方向切削分力的读数。
（6）安装和使用要简便。
过去广泛使用的液压式车削测力仪，就不能很好地满足上述的要求，因而就不可能被广泛采用。
在最近二三十年中出现了很多新型的测力仪，这种测力仪的基本特点就是利用电工仪表（如毫安培计、毫伏特计、示波器、应变仪等）来测量测力仪上的机械量发生变化，亦即把测力仪上的弹性元件的弹性位移量通过仪器上的辅助设备变成电量的变化，这个电量的变化也就是要测的机械量。但由于这些仪器还存在有下列缺点，因此还没有得到广泛的应用。
（1）由于这些仪器的辅助设备（电气）是要用振荡器、放大器、示波器等复杂的电器系统组成。因而制造困难而成本高，在使用和维护上也较困难，容易出错误而使测量不准确。故这种测力仪只能应用于动力载荷等测量。例如在测铣削力及研究振动

---

[*] 原发表于《机床与工具》，1957，7：46~48。

等情况中应用。

（2）这些仪器的结构中，机械系统的很多部分仍保持类似液压式测力仪的结构，当应用电工测量后，它有高的灵敏度，可以大大地减少弹性系统的位移量（比一般小10～20倍），因此机构中的惯性和接触变形量与被测量的弹性位移量相比也成了一个值得考虑的因素。这样就破坏了校正（Тарировка）的可靠性，也降低了测量的准确度。

В. Ф. 帕拉莫诺夫工程师考虑了这些仪器存在的缺点，以及对测力仪的基本要求，而设计出这个 ВП-4-ЭИ 型三向电感式车削测力仪。

2 测力仪结构和原理

测力仪的机械部分是一个弹性系统（图1），用件2、3、4、5各对称的小支板与本体联结成的"摇杆"组成。当被水平径向切削分力 P_y 作用时，小支板产生金属弹性的弯曲变形，使摇杆沿工件径向移动，此位移量可由电感传送器 D_y 测出。

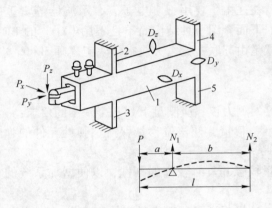

图 1　测力仪的机械部分

垂直切削分力 P_z 使杆1在垂直平面内产生 P_z 方向的位移，亦即摇杆产生金属弹性变形的弯曲量，此值可由电感传送器 D_z 测出。

轴向切削分力 P_x 使摇杆在工件轴向分力 P_x 方向产生弯曲，此弯曲量可由电感传送器 D_x 测出。

三个方向的位移量（摇杆的轴向移动和弯曲）都和三个方向上的切削力的大小成正比。

图2为测力仪的结构图，1为本体、2为摇杆、3为前支板、4为后支板、19为 P_z 的传送器 D_z、20为 P_y 的传送器 D_y、21为 P_x 的传送器 D_x。

仪器的电气部分是接在电压为220伏50周的交流电源上。电路系统是由电压稳定器和三个电桥组成（图3）。各电桥的两个桥臂是由两个缠绕在"T"形铁芯上的电感线圈组成，其中一个装在测力仪上，即为传送器 D_x、D_y、D_z 之一，其电感量为 L_1、L_3、L_5；另一线圈装在调节箱上，即为被传送器（Зад-атинк 见图4），电感量为 L_2、L_4、L_6。这些被传送器的电感是可以用手轮来调整的，用它来平衡电桥，并即调整调节箱上三个被传送器，使三个电桥分别平衡。

测量时利用这三个电桥，当测力仪的刀具上未加载荷以前（即尚未开始切削），先

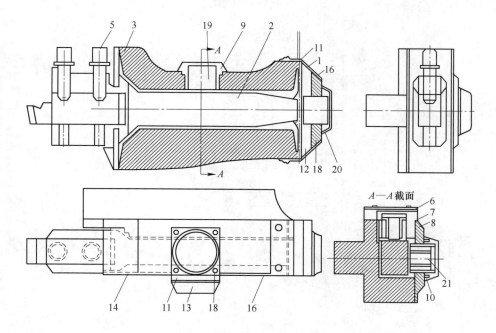

图 2 测力仪结构图

1—本体；2—摇杆；3—前支板；4—后支板；5—螺栓；6—内磁铁；7—外磁铁；8—线圈；9—套筒；10—螺帽；11—侧板；12—后盖板；13—盖子；14—前挡板；15—后挡板；16—上挡板；17—下挡板；18—螺钉；19—传送器 D_z；20—传送器 D_y；21—传送器 D_x

行调整调节箱上的三个手柄，即调节被传送器之 L_2、L_4、L_6 之电感，使它与测力仪上各传送器上的电感 L_1、L_3、L_5 相平衡（此时各微伏特计上的读数为零），在切削时刀具受到载荷之后，测力仪的原件产生弹性变形，使传送器的原始间隙改变。因而改变线圈的磁通量（亦即电感量 L_1、L_3、L_5），这就破坏了电桥的平衡，此时微伏特表上的指针指出的电感读数，也即与变形量相适应的切削力读数。

电路中是用氧化铜电桥整流器，按环形线路工作，整流原件受到线圈中的额定电流，使其能在伏安特性曲线上有利的一段（即直线性的一段）工作，即通入电压后有很小的电阻，此种整流线路有很高的输出效率，可达 0.8，大大的增加其灵敏度。

电阻 R_4-R_5；R_6-R_7；R_8-R_9 为桥臂的第二对参数，电阻 R_{10}-R_{11}-R_{12}；R_{13}-R_{14}-R_{15}；R_{16}-R_{17}-R_{18} 最后联结在微伏件表中的接头上，用来降低其灵敏度，即用来改变量程范围。电阻 R_1、R_2、R_3 用于确定该电桥合适的测量规范。

电路中应用了铁芯共振式电压稳定器，消除交流的电流中电压的波动对仪器读数的影响，电压稳定器的输出电压为 16V。

此外有三个微伏特计，可在电路中装一个换向开关，逐次联结在一个微伏特计上。

3 使用时应注意之点

首先把测力仪装在机床刀架上，而后把测力仪上带插头的电线接在调节箱的输入插头上，然后再接上电源，最后接通微伏特表。

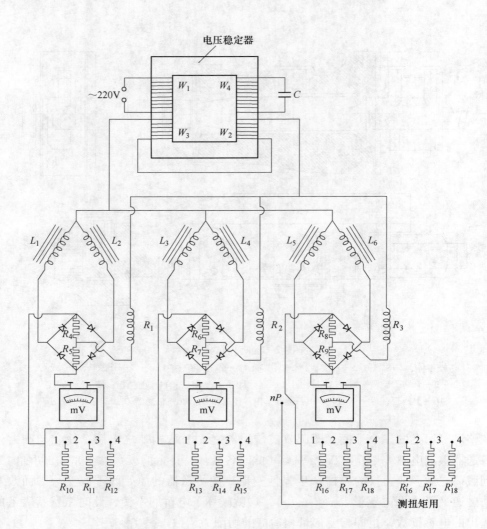

图 3 测力仪的电路图

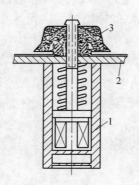

图 4 被传送器
1—被传送器线圈；2—调节器外壳；3—调节手轮

这样就可开始校准（Тарировка），校正可用普通方法，即用杠杆法加载荷或用校准仪来进行，校准的结果要作出各由 8~10 个点所组成的十二条校准曲线。

测量切削力时，应即在校准后原地不动来进行，测力仪进行测量时应在接通电源 5~10min 以后进行，要使仪器的电气部分的温度保持一定的条件下工作，否则当电流通过仪器的电路部分时，要消耗一部分电量变为热量，使电感示值产生误差。

应注意将电源及微伏特表接好后，不能使测力仪至调节箱的电线脱开，否则电桥将造成极大的不平衡，因流入微伏特表中电流过大，而把微伏特表烧毁。

改变量程范围可在工作时进行，要测量得更精确些，量程的选择应使微伏特表上的指针指示最大切削力时，它应在微伏特表最大读数（微伏特表的最大量程）附近。

4　ВП-4-ЭИ 型三向电感式车削测力仪的优缺点

这种车削测力仪根据古比雪夫市几个试验室在最近两年中使用的情况，以及我们清华大学机床试验室试制成功后使用的情况，可总结出下列几个主要的优点。

（1）ВП-4-ЭИ 型测力仪可以适用在各种车床上工作（车削、切端面、镗孔、切螺纹等），并在各种切削用量下能精确的测量各种车削加工中的三个方向的切削分力。

（2）测量的准确度高，因为影响仪器不准确度的因素较少。

（3）由于测力仪在 P_y 方向有很高的刚度（等于 750kg/mm），故用此测力仪时，即使切削深度很深，表示值也不受影响。

（4）仪器的尺寸小，可以迅速而方便的装夹在各种尺寸车床的刀架上。

（5）测力仪的结构简单，不用复杂的电气系统，因此制造容易和成本低。

（6）测力仪的维护方便，不需具有电学方面专业知识的人也可使用。

根据我校试验室使用中已发现有下列缺点：

（1）仪器的自然频率不够高并有焊接之处，因此不作小惯性测量（如测量切削时的振动等）。

（2）仪器上不带有放大器及示波器，故不能观测并录出切削力的变化规律。

参 考 资 料

[1] В. Ф. Парамонов-Трехкомпонентный Элекгроиндуктиный токарный Динамометр тил ВП-4-ЭИ 说明书.

[2] В. Ф. Парамонов-Трехкомпонентный Элекгроиндуктиный токарный Динамометр Станки и инст. 1953 年第 223 页.

[3] В. Ф. Парамонов-Свертнльный Элекгроиндулвный ДинамометрСтанки и инст. 1953 年第 8 期 28 页.

两种利用计算全息检测非球面的方法*

摘　要　计算全息检测非球面具有广泛的应用潜力。但在一般情况下，检测的非球面波差范围是有限的。本文提供和验证了两种扩大计算全息检测能力的方法，其一是计算全息和零位光学系统组合补偿的检测法，其二是像差平衡法制作计算全息图的检测法。此两法把检测非球面波差范围扩大几倍直至近百个 λ 的波差。所用的线型计算全息图能方便地再现位相型波面。实验证实了这两种方法是可行的。精度分析表明，检测非球面面形偏差时，精度可达 $\lambda/10$。

1　引言

实践证明，计算全息检测光学面形，特别是干涉检测非球面，具有广泛的应用前景[1]。计算全息检测非球面属于零位测试，因此易从干涉图上定量地弄清波差分布情况，从而直观地给出面形偏差信息。它较之光学补偿的零位测试法，有制作方便，调整简单的优点，测试精度也易提高。零位测试中的补偿系统不提供点像，甚至不要求提供同心光束（如非旋转对称波面的检测），故要制作高精度的光学补偿器，并要在测量上证实其补偿波面的准确，通常是相当困难的。计算全息检测法可以避免这些问题，只要给出补偿波面的数学描述，就能制成所需的全息图，这种全息图（大多数是二元的）抗干扰能力强，且能把测试误差的补偿事先计入全息图中，再现的波面就能降低甚至消除这些误差。计算全息检测法的这些特点，使它成为检测非球面的很有吸引力的新技术。

典型的计算检测非球面的光路见图 1。图 1 为修正型的泰曼-格林干涉仪。由被检镜返回的波面和参考平面波形成的干涉条纹，将在 CGH 平面上与计算全息干涉图叠加后产生莫尔条纹。带通滤波器 F 选出合适的衍射级，以改善条纹的对比，P 面上的干涉场分布就反映了被检镜的面形偏差信息。

但是，定量研究发现，应用计算全息检测非球面的波差范围是有限的。当被检镜的非球面度很大时，其波前相对于球面的波差可达成百个 λ。此时全息图平面上的波差梯度极度增长，计算全息图制作困难，精度不高，因此并不适用。所以必须设法扩大计算全息的检测能力使之更趋实用。本文讨论和验证了两种新的测试法。一种是计算全息和零位光学系统的组合补偿测试法，另一种是应用像差平衡法（又称离焦法）制作计算全息图的测试法。这两种方法可把计算全息检测能力提高三倍以上直至近百个 λ 的波差。误差分析表明，这两种方法都提高了测量精度，在我们的实验装置中，检测的面形精度可达 $\lambda/10$。

---

* 本文合作者：虞祖良，邹敏贤。原发表于《仪器仪表学报》，1981，2(4)：64~70。

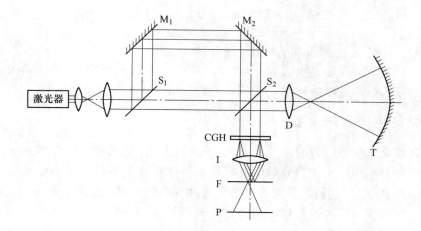

图 1 典型的计算全息检测光路图

M_1，M_2—反射镜；S_1，S_2—分束镜；D—发散镜；CGH—计算全息片；
F—空间滤波器；T—被检镜；I—成像透镜；P—干涉平面

2 线型计算全息图

应用计算全息检测非球面，关键是制备计算全息图。在面形测试中，希望有理想的非球面作参考波面，而这个波面一般是位相型的，故常选用线型计算全息图，它特别适宜再现位相型波面。线型计算全息图也叫计算全息干涉图，依据下述基本方程式制作[2]。

$$-q(x,y)\cdot\pi < 2\pi x/T - \phi(x,y) + 2\pi n \leq q(x,y)\cdot\pi \tag{1}$$

其中 $q(x,y) = \sin^{-1}A(x,y)/\pi$，$A(x,y)$，$\phi(x,y)$ 分别为物波 $u(x,y)$ 的振幅和位相函数，$n = 0$，± 1，± 2，…。

当物波为纯位相型时，可令 $q(x,y) = 0$。上式变为

$$2\pi x/T - \phi(x,y) = 2\pi n \tag{2}$$

式（2）是实际中常用的线型计算全息基本方程式。$1/T$ 是载频，其作用是在全息图再现时把一级衍射波（物波）和其他高级次的衍射波分开。当物波面在 x，y 方向的空间带宽分别为 B_x，B_y 时，必须有 $\frac{1}{T} > 1.5B_x$。在实际制作时，选 $\frac{1}{T} = 2B_x$ 作为载频。

计算机解基本方程式（1）时，有变量 x，y 及变参量 n，常用迭代法求解。依据抽样定理，在物波面上取一系列样点迭代。在 y 方向，常取 $B_y = B_x$，故 y 方向的取样间距为 $2T$；在 x 方向，由于要调制物波面的位相，为提高波面精度，应将取样间距进一步细分为 T/M，M 是整数。M 越大，波面位相的量化误差就越小，但要加长计算时间，故应选用适当。

由基本方程式解出的同一条纹的 x、y 值，输给绘图仪连接成线，即二元线型计算全息图。其透过率函数为：

$$h(x,y) = \sum_{m=-\infty}^{\infty} \frac{\sin[\pi m q(x,y)]}{\pi m} \exp\{jm[2\pi x/T - \phi(x,y)]\} \quad (3)$$

式（3）中 $m = -1$ 项再现了物波 $A(x,y)\exp[j\phi(x,y)]$。

在制作位相型波面时，也可取 $q = 1/2$，此时式（3）中全部偶数项消失，第一级衍射波有最高的衍射效率约 10%，经漂白处理后衍射效率可提高至 40%。

3 组合补偿法检测非球面

扩大计算全息检测能力的第一种方法，是应用计算全息和零位光学系统的组合补偿检测法[3]。首先用形式相当简单但具有足够制作精度的零位光学系统来完成初始的补偿（约 90% 的波差被补偿），剩余的波差可用简单的计算全息来完善补偿。经过这种组合补偿，能够检测很大的非球面波差，并有很好的测试效果。图 2 所示为二次曲面图形。

一个和 y-z 平面相切的二次回转曲面 Σ 的方程式为：

$$y^2 + z^2 = 2R_0 x - (1 - e^2)x^2 \quad (4)$$

其中 R_0 是顶点密切球面的曲率半径，e 是偏心率。二次曲面 Σ 的法线像差为

$$\Delta R_y = e^2 \cdot x \quad (5)$$

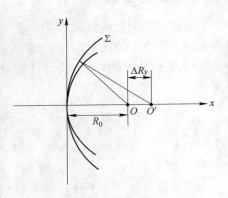

图 2　二次曲面图形

在组合补偿法中，只选用形式简单的透镜或反射镜产生的球差作初始补偿，不要求此球差值精确地和二次曲面的法线像差相匹配。用单透镜作零位透镜的测试光路可参见图 1。把发散镜 D 换成零位透镜，完成初始的补偿。然后对整个干涉仪系统（包括零位透镜和被测镜）进行光线追迹，求出在计算全息图放置的 CGH 平面上的波差 $W(x,y)$，波面函数可写为 $u(x,y) = \exp\left[j\frac{2\pi}{\lambda}W(x,y)\right]$。依据上节原理作线型计算全息图。全息图再现的波面是组合补偿后的波面，能完善补偿被检镜的波差。组合补偿法降低了各自的要求，特别是降低了对零位光学系统的要求。当检测非旋转对称波面时，零位光学系统仍可用回转对称系统，余留下的非对称波差可用计算全息图来补偿。否则，若要制作非对称波面的零位光学补偿器，在制作工艺上会有极大困难。实际计算可知，组合补偿法大大扩大了波差的检测范围，测量近百个 λ 波差的非球面，不会引起很大困难。而且制作的计算全息图，因为是对整个干涉仪系统进行光线追迹后确定物波面形成的，这就可把整个系统的误差计入全息图，再现的波面可全部补偿这些误差，故最后的测试结果，精度较高。

4 像差平衡法制作计算全息图

所谓像差平衡法制作全息图，就是在全息图平面的波差式中引入离焦项像差，进行像差平衡，降低整个物面的波差梯度，从而降低物面的空间带宽，相应地减少全息

图制作时的取样点数，简化计算全息图。这样，同样复杂（指取样点数相同）的一张计算全息图，应用离焦法时可检测的波差范围比非离焦时扩大好几倍，达到扩大计算全息检测能力的目的[4]。

测试光路排列仍如图 1 所示，发散镜 D 应对轴上点成理想像。在 CGH 处，由被测镜 T 返回的光波波差为：

$$W(x,y) = 2\left[A_4\left(\frac{x^2+y^2}{r_0^2}\right)^2 + A_6\left(\frac{x^2+y^2}{r_0^2}\right)^3 + \cdots\right] \quad (6)$$

式中，A_4、A_6 是为待检镜波差系数，R_0 是待检镜最大口径之半。移动被测镜 T 引进离焦量 d 后，式（6）变为

$$W(x,y) = 2\left[A_2\left(\frac{x^2+y^2}{r_0^2}\right) + A_4\left(\frac{x^2+y^2}{r_0^2}\right)^2 + A_6\left(\frac{x^2+y^2}{r_0^2}\right)^3 + \cdots\right] \quad (7)$$

式中，$A_2 = (d/2R_0^2\lambda)\cdot r_0^2$ 为离焦波差系数。引入离焦项像差可改变整个波面的波差大小。由式（7）可见，第二项以后是本征像差，量值不变，而离焦项像差却随 d 值的大小在较大范围内变化。可选 d 值使整个波差变小，或使波差的梯度变小。由于全息图制作时的取样点数与波差的梯度成比例，可用数值分析法（极大极小法）选择 d 值，使波差梯度的极值变小。对于回转对称波面，离焦系数为

$$A_2 = -1.5A_4 \quad (8)$$

离焦时全息图面的取样点数 N' 仅为非离焦时取样点数 N 的 25%。

离焦量 d 的值可由 A_2、A_4 求出：

$$d = -3e^2r_0^2/8R_0 \quad (9)$$

计算全息图是依据式（7）的波差方程制作的。由于离焦项像差的引入，同样复杂的一张全息图，比非离焦时扩大检测范围约三倍。

组合补偿检测法和离焦法测试各有特点。离焦法测试较简单，不用零位光学元件作补偿器。发散镜 D 的误差影响也可用光线追迹，即对整个干涉仪系统（包括发散镜和离焦的被检镜）用确定波差的办法来消除。离焦法的缺陷是检测的波差范围有一定的限制，故用于相对孔径较小的非球面测试中。组合补偿测试要附加零位光学系统，但由于零位光学元件相当简单，设计制作较容易，费用不高，精度也易于保证。并且，因已补偿了绝大部分波差，故计算全息图制作方便，精度可较高。组合补偿法检测非球面波差的范围可大大扩大，原则上可对应用中所提出的极复杂的非球面进行精确测试。

5 实验结果

分别应用组合补偿法和离焦法对抛物面反射镜作干涉检测。抛物镜焦距 $f' = 206$mm，外径 $D = 113$mm，相对孔径 $D/f' > 1/2$。

在组合补偿测试中，选用正单透镜作零位光学补偿件，其正球差大致和抛物镜各带法线像差互补，用光线追迹求出全息图平面孔径上七个带的波差值，再解联立方程找出波差的解析表达式，作成线型计算全息原图。

离焦法，选用优良像质的物镜作发散镜，相对孔径约 1/4.5，检测的抛物镜相对孔

径约 1/2.8，小于要求值。然后由式（7）的波差方程式，作成线型计算全息原图。

所有计算是在 NOVA-3/D 计算机上进行。用二分法解基本方程，物面在 x 方向的取样间距为 $T/66$，即 $M=66$，y 方向取样间距为 $6T$。计算结果记录在磁带上，输给 XY-NETICS1200 绘图仪，画成 $\phi=700mm$ 的原图。$q\simeq 1/2$，干涉条纹数各为 400 条和 500 条。全部计算时间约 20min，绘图时间（描二遍）约 1h。经光学缩版成计算全息片，其缩小倍率必须依据测试光路中物像关系求出。最后的全息底片尺寸，用组合法时为 $\phi=24.13mm$，用离焦法时为 $\phi=37.03mm$。

实验用的测试装置在有良好大气屏蔽的防震平台上搭成，其光路见图 3。

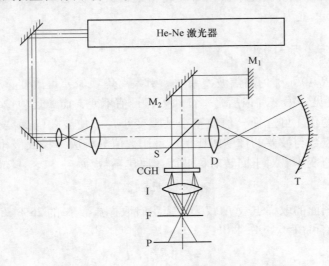

图 3　实验用的测试光路图

实验时，首先再现了计算全息图的非球面波前。这是在测试光路中选出全息图的一级衍射波和平面参考波相干涉，干涉图见图 4 中的照片 1、2。

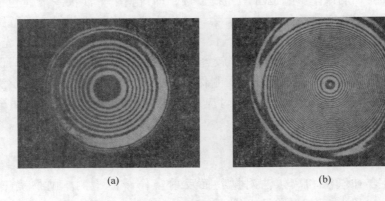

图 4　干涉图
（a）照片 1 组合补偿法再现波前干涉图；（b）照片 2 离焦法再现波前干涉图

由照片可见，再现的非球面波前的质量优良。

然后，在测试仪中放入被检镜 T，经仔细调整，在 P 面上可得被检镜返回波面的干涉图，分别见图 5 中的照片 3 和 4。

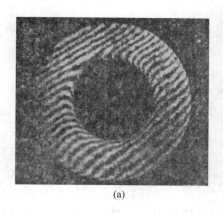
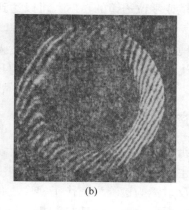

(a)　　　　　　　　　　　　　(b)

图 5　测试结果

（a）照片 3 组合补偿法测试结果（$D/f' = 1/1.8$）；（b）照片 4 离焦法测试结果（$D/f' = 1/2.8$）

由照片可见，两种方法的测试结果相近，说明这两种方法用于检测非球面是可行的。由于实验用的测试仪综合精度为 $\lambda/10$（见精度分析部分），故由测试结果的干涉条纹偏离直线的情况可分析出被检镜的面形偏差（双程），分析方法和泰曼-格林干涉仪完全一致。

6　精度分析

计算全息检测非球面，主要的误差源有以下 5 种。前 3 种是计算全息检测法特有的，而后两种是一般的球面干涉仪也会引进的。

（1）全息图再现波面的误差；
（2）全息图尺寸的不匹配；
（3）全息图在测试仪中横向位置失调；
（4）被检镜的不正确离焦；
（5）零位透镜或发散镜的误差。

全息图再现波面的误差，主要由解基本方程式的离散迭代，绘图仪的有限寻址单元等引入的位相量化误差，及描绘误差造成，抽样误差可忽略不计（此误差在实验中小于 $\lambda/10$）。

全息图尺寸的不匹配主要是在光学缩版时，微缩透镜有畸变，以及全息片没有正确放置在被检镜的成像位置而造成的。对组合补偿法结果分析表明，此项误差约为 $\lambda/15$。

当被测镜是旋转对称时，第 3 种误差主要由全息片和被测镜在测试中放置的横向匹配失调引起。其中尤以全息片的横向偏置引入的误差影响最大，而当被检镜有偏心时，其对波面的误差影响由于发散镜的作用会变小。借助于测试中的仔细装调，可把发散镜的装备误差影响减少。以组合补偿法为例，第 3 种误差为 $\lambda/15$。

第 4 种误差相当于引入新的离焦波差。但在测试中，我们可应用准直方法，仔细调整被检镜位置，观察 P 面形成的干涉条纹变化来降低不正确离焦的影响，实际中易

保证此项误差小于 $\lambda/10$。

发散镜或零位透镜的精度，可以达到 $\lambda/10$。

以上 5 种误差都能作随机误差处理。综合误差可取其均方根值，结果优于 $\lambda/5$。反映到被测镜面形的测量误差，精度可达 $\lambda/10$。

应指出，前 3 种测试误差都和全息图上波面的波差梯度成比例，本文提供的二种测试方法，因降低了波面的波差梯度，故相应地提高了测试精度。

7 结束语

应用组合补偿法和离焦法能扩大计算全息检测非球面的波差范围，使计算全息检测技术在生产中具有实际应用价值。精度分析表明，面形检测的精度达 $\lambda/10$。本文仅以抛物面反射镜的检测为例作了实验。但完全可以把计算全息检测法推广应用于其他类型的光学面形检测，包括检测折、反类型的旋转对称非球面，非旋转对称的光学面形，以及球面折、反射光学面形等。

参 考 文 献

[1] A. J MacGovern and J. C. Wyant, Appl. Opt., 10, (1971), 624~629.
[2] Wai-Hon Lee, Progress in Optics, XVI, Ed. E. Wolf, North-Holland Amsterdam, (1978).
[3] J. C. Wyant, Optical Shop Testing, John Wiley & Sons, Inc. (1978).
[4] Toyohiko Yatagai and Hiroyoshi Saito, Appl. Opt., 17, (1978), 558~565.

Two Methods for Optical Testing Aspherical Surface by Using a Computer-generated Hologram

Abstract Applying a computer-generated hologram (CGH) in measuring an optical wavefront is the new development in interferometry, particularly there is a situation with great potentialities in the field of measuring optical aspherical surfaces. As a general rule, the range of wave aberration to be measured by a CGH is limited. It affects the method to be widely used. In this paper, two methods for extending the capability of a CGH are discussed and investigated. One is the defocusing method, while the other is a combination of CGH with a null optical system. CGH is calculated and designed from rays tracing of the whole system. So all the aberrations on optical components of the system have been taken into consideration. By using these two methods, the measured range of wave aberration has been extended nearly three times up to 100λ of wave aberration. Interferometric computer-generated holograms are used in both cases. Experimental results proved these kinds of technique are available in practical work. From the error analysis, a precision of $\lambda/10$ can be obtained in measuring aspheric surfaces.

二维光学传递函数测量*

摘　要　本文提出了一个应用 Radon 反变换测量二维光学传递函数的方法。文中讨论了测量原理，给出了实验结果。实验表明，这种方法在原理上和技术上都是可行的。
关键词　光学传递函数；测量

1　引言

光学传递函数（OTF）作为对光学系统质量的综合描述，全面地反映了系统孔径、光谱成分、衍射情况及像差大小所引起的综合效果。随着现代光学理论和技术的不断发展，它在像质评价、光学设计、光学信息处理等方面都得到了一定的应用。光学传递函数是二维的。但是由于其计算和测量都比较复杂，测量所得的数据量很大，不易处理，一般只能对不同方向的 OTF 做一维分析。在做像质评价时，常以几个方向上的 OTF 平均值来描述光学系统的全面特性。这一平均值显然不能表征 OTF 的全貌，也不能全面地反映光学系统的实际情况。

今天，由于大容量高速度计算机的发展，二维 OTF 的计算和测量结果的处理已不再成为问题。但至今仍没有切实可行的二维 OTF 测量方法出现。这在一定程度上限制了光学传递函数的发展、应用和普及。为了解决这一问题，作者利用雷顿（Radon）分析法对线扩散函数做二维综合得到点扩散函数（PSF），对其做二维傅氏变换得到二维 OTF。

雷顿分析是医用断层扫描仪（CT）的数学基础。在 20 世纪 70 年代后期，Barrett 等人曾致力于光学模拟 CT 处理机的研究[1,2]，把雷顿分析引入到光学信息处理的研究领域。后来 Barrett 等人又借助雷顿分析原理，用一维空间光调制器实现了二维甚至三维卷积、相关、傅氏变换的光学模拟运算[3,4]。而运用雷顿分析方法测量光学系统的传递函数是作者首先尝试去做的。

2　数学原理

奥地利数学家 J. Radon 于 1917 年证明[5]：n 维物体可以由它的 $(n-1)$ 维投影得到。其再现公式即是著名的雷顿反变换公式。取 $n=2$，设有一个二维分布 $g(x,y)$。其一维投影称为雷顿变换：

$$h(s,\theta) = R[g(x,y)] = \int_{-\infty}^{+\infty} g(s\cos\theta - \mu\sin\theta, s\sin\theta + \mu\cos\theta) \mathrm{d}\mu \tag{1}$$

显然式（1）是沿投影线，见 Fig. 1：

---

* 本文合作者：杨向阳，邹敏贤。原发表于《光学学报》，1987，7(3)：242~246。

$$\begin{cases} x = s\cos\theta - \mu\sin\theta \\ y = s\sin\theta + \mu\cos\theta \end{cases} \tag{2}$$

的线积分。

反投影变换由下式定义：

$$B[h(s,\theta)] = \int_0^\pi h(x\cos\theta + y\sin\theta, \theta)\mathrm{d}\theta \tag{3}$$

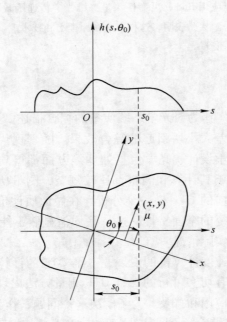

Fig. 1 Demonstration of Radon Transform

$BR[g(x,y)] \neq g(x,y)$，所以反投影变换与雷顿变换不能构成变换对。为了推导雷顿反变换，令：

$$g(x,y) = R^{-1}R[g(x,y)] \tag{4}$$

$$g(x,y) = F_2^{-1}F[g(x,y)] = \int_0^\pi \int_{-\infty}^{+\infty} [F_2 g(x,y)] \cdot |f_s| e^{\mathrm{i}2\pi f_s(x\cos\theta + y\sin\theta)} \mathrm{d}f_s \mathrm{d}\theta$$

其中 f_s 为变量 s 所对应的傅氏谱变量。

由中心切片定理[6]：

$$F_2[g(x,y)] = F_Y R[g(x,y)] \tag{5}$$

则：

$$g(x,y) = \int_0^\pi \int_{-\infty}^{+\infty} \{|f_s| \cdot F_Y R[g(x,y)]\} e^{\mathrm{i}2\pi f_s(x\cos\theta + y\sin\theta)} \mathrm{d}f_s \mathrm{d}\theta$$

$$= B F_Y^{-1}\{|f_s| \cdot F_Y R[g(x,y)]\} \tag{6}$$

$$g(x,y) = \frac{1}{2\pi} B H_Y D_Y R[g(x,y)] \tag{7}$$

其中 F_Y、H_Y、D_Y 分别是雷顿空间的傅氏变换算符、希尔伯特变换算符和微分算符。它们都是一维算符，只作用于变量 s。

式（7）就是 J. Radon 在 1917 年用算符所表示的雷顿反变换公式。尽管其数学形式很完美，却不适于计算机处理。在用计算机处理时多将其改写为：

$$g(x,y) = B\{F^{-1}[|f_s|] * R[g(x,y)]\} \tag{8}$$

这就是目前所有医用 CT 扫描仪采用的"滤波反投影公式"[6]。它表明：对投影数据做卷积滤波后再做反投影，即可不失真地再现出原分布。

在二维 OTF 测量中，用刀口扫描 PSF，如 Fig. 2 所示，接收到的透射信号为：

$$P(s) = \int_{-\infty}^{+\infty} \int_{-\infty}^{s} g(x,y) \, ds \, d\mu \tag{9}$$

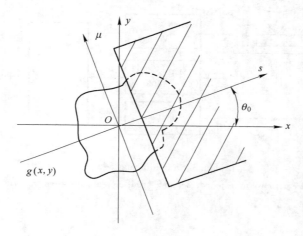

Fig. 2 Scanning

微分后得到线扩散函数：

$$\begin{aligned} I(s) &= \frac{dP(s)}{ds} \int_{-\infty}^{+\infty} g(x,y) \, d\mu \\ &= \int_{-\infty}^{+\infty} g(s\cos\theta_0 - \mu\sin\theta_0, s\sin\theta_0 + \mu\cos\theta_0) \, d\mu \\ &= h(s,\theta_0) = h(s,\theta)|_{\theta = \theta_0} \end{aligned} \tag{10}$$

因此，线扩散函数就是点扩散函数的一维投影。转动刀口重复测量，可以得到整个投影族。用公式（8）对投影数据做雷顿反变换，综合出二维 PSE。经计算机做二维傅氏变换就得到了二维 OTF（严格地讲，这里得到的是二维调制传递函数 MTF）。

3 实验研究

笔者实测了一个 40 倍显微物镜作为光盘读写微光斑聚焦头的光学传递函数。Fig. 3 为测量系统图。激光器经针孔滤波后由被测镜聚集为一微光斑。针孔看作点源则该光斑即为聚焦物镜的 PSF。刀口位于像平面上。计算机经高压驱动器带动电致伸缩微位移器件往复运动实现刀口的扫描。扫描运动的位移重复性优于 $0.03\mu m$。光电检测器件将透射光信号转化为电信号并送到计算机中。对这一信号微分就得到一个一维投影。步进电机带动刀口工作台转过某一角度，重复以上步骤，得到另一个方向上的投影。如此直到转过180°后完成全部投影测量工作。计算机对投影族做雷顿反变换和二维FFT，得出聚焦物镜的二维点扩散函数和二维调制传递函数。测量结果如 Fig. 4 和 Fig. 5 所示。

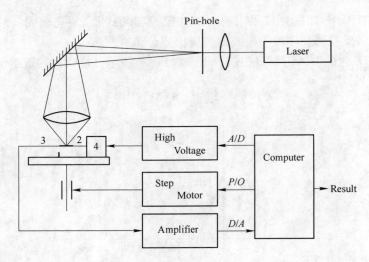

Fig. 3　Experimental Setup
1—objective to be tested；2—knife-edge；3—detector；4—piezocrystal

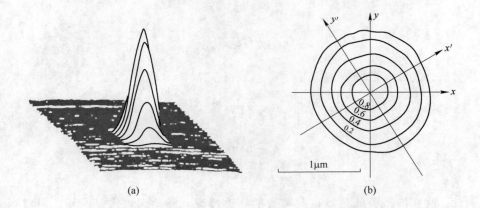

Fig. 4　Point Spread Function

Fig. 3 中针孔上的光场应为高斯分布。根据傅里叶光学理论[7]，被测镜聚焦后的点扩散函数为准高斯分布，次极大很小。所以 Fig. 4 中看不出有次极大存在。Fig. 4（b）表明 PSE 不是严格回转对称的，说明透镜存在像差。星点检查表明该透镜确有彗差存在，证明测量结果可信，PSF 的长短轴在 x'-y' 方向上，因此 MTF 在 x'-y' 方向差异最大，而在实验选定的 x-y 方向，即所谓子午-弧矢方向上则相差不多（如 Fig. 6 所示）。这也说明仅做两个正交方向上的 MTF 测量是远远不够的。

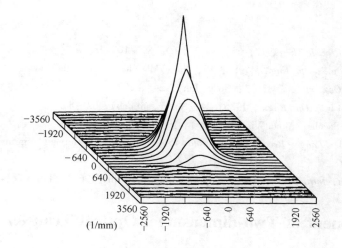

Fig. 5　Modulation Transfer Function

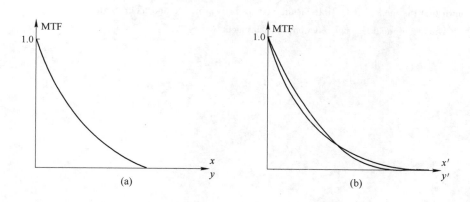

Fig. 6　MTF in x-y and x'-y' direction
（a）x-y direction；（b）x'-y' direction

　　用此法做多次重复测量，结果之间最大偏差小于5%，精度不是很高。对测量系统改进以后，比如选用高精度轴系，刀口和光电检测元件后，精度可以提高。

4　结束语

　　一个完善的二维光学传递函数测量仪应能测量不同波长、不同孔径的 OTF。作为原理实验，我们只对 Fig. 3 的光盘聚焦系统进行了测量。实验结果表明这一方法无论在原理上和技术上都是切实可行的。这一工作为雷顿分析法开辟了新的应用领域，解决了二维 OTF 测量问题，有助于光学传递函数的发展和应用普及。

　　国内许多单位现有的一维传函检查仪由于精度低，使用不便，利用率很低。若利用此原理，用一转动刀口代替原来的光栅扫描机构再配以微计算机就可以改装成高精度、高效率、操作简便的二维 OTF 检测仪。这一工作不但在技术上有所进步而且具有很大的经济效益。

参 考 文 献

[1] H. H. Barrett; *Proc. IEEE*, 1977, 65, No. 1(Jan), 89~107.
[2] A. G. Gmitro et al.; *Opt. Engi.*, 1980, 19, No. 3(May/Jun), 260~272.
[3] H. H. Barrett; *Opt. Lett.*, 1982, 7, No. 6(Jun), 248~252.
[4] A. J. Ticknor et al.; *Opt. Engi.*, 1985, 24, No. 1(Jan/Feb), 82~85.
[5] J. Radon; Ber. Saeohs, Akad. Wiss(Leizpzig), 1917, 69, 262~275.
[6] S. W. Rowland; *Image Reconstruction from Projection*, (G. T. Herman, Springer-Verlag, Berlin, 1979), 9~69.
[7] G. W. Goodman, *Introduction to Fourier Optics*, (McGraw-Hill, 1968), 115~117.

Measurement of Two-dimensional Optical Transfer Function

Abstract A method is discussed of measuring two-dimensional optical transfer function applying the principle of Radon transform. The measuring principle and experimental result are described. It is shown that the method is feasible on the view of both principle and technology.

Key words optical transfer function; measurement

中医舌诊自动识别方法的研究*

摘 要 本文分析了中医舌诊的主要内容,提出了实现自动识别的对策。在此基础上,以Munsell颜色系统为色标,运用色度学、近代光学技术、数字图像处理技术和计算机硬件技术,建立了中医舌诊自动识别系统。在该系统上,以中医辨症论治学说为指导,将计算机软件技术与临床辨舌经验结合,利用样本训练系统,根据模糊数学理论,确定有关舌像的定义域,进行纹理分析。临床研究了366例淡红舌、暗红舌、紫红舌、暗紫舌四种舌像,符合率86.43%。

关键词 中医舌诊;舌苔;舌质;数字图像处理;模糊聚类;纹理分析

1 引言

中医舌诊是中医临床辨症的主要客观指标之一,为历代医家所重视。名医杨云峰积多年行医经验,明确指出:"经络脏腑之病,不独伤寒发热,有胎可验,凡内外杂症,亦无不呈其形,著其色于舌,是以验舌一法,临症者不可不讲也"。但是,千百年来医生只能凭肉眼验舌,靠经验辨症,因人因时因地而异,影响了中医舌诊的继承和发展,迫切希望实现客观化、定量化、标准化。

近一二十年来,国内不少单位尝试用现代科学技术来解决舌诊客观化、标准化问题。有的运用微观观察法,观察舌黏膜上皮的细胞超微结构,分析各种舌苔的形成机理;有的采用微循环研究方法,观察舌尖的形态,探讨舌尖微循环变化与不同舌质形成间的关系;有的采用生化研究方法,观察唾液钠、钾、尿素氮的含量,探讨阴虚、阳虚病人舌像变化与唾液的生化关系;有的采用细胞学研究方法,观察舌上皮细胞角化锐度和白细胞对脂类染色反应,探讨阴虚、阳虚病的舌像变化与舌细胞的关系;有的采用物理方法,利用舌质光谱分析原理,分辨舌质的不同颜色,摸索舌质识别方法;有的采用生化及物理方法,测定舌组织的蛋白含量、血氧饱和度及舌苔厚度,进行舌像识别;还有的采用图像处理法,对舌像照片进行图像分析,探索舌质舌苔的计算机定量描述和分类[1]。无疑,这些研究都对中医舌诊客观标准化起了一定的促进作用。然而,这些方法基本上是采用西医的研究方法或者借鉴西医的研究成果,不能完全反映中医的宏观、整体、辨证的指导思想和达到中医的要求。如何使现代科学技术服务于中医,又使古老的中医向现代科学技术靠拢,是实现舌像自动识别的关键。

几年来,我们在国家自然科学基金委员会的资助下,以中医辨症论治学说为指导,运用色度学原理、近代光学技术、数字图像处理技术及计算机技术,结合丰富的临床经验,建成中医舌诊自动识别系统。在该系统上,既可从宏观出发辨识整个舌面,又

---

* 本文合作者:余兴龙,谭耀麟,竺子民,索忠莹,翁维良,许秀森,葛文津。原发表于《中国生物医学工程学报》,1994,13(4):336~344。

可兼顾局部区域辨识,甚至可以辨识一点,既可辨识舌苔,又可辨识舌质,既可辨识色彩,又可辨识形态,进行纹理、齿痕等分析,较全面地体现了中医诊断的特点。

2 中医舌诊的内容与实现自动识别的构思

舌像种类很多,有几百种[2]。舌诊的内容也很丰富(见图1),既有色彩又有纹理,既有形状又有状态,既有舌质又有舌苔。医生辨舌是通过肉眼观察,获得直观印象,再结合多年积累或从他人继承的经验进行判断,确定舌苔的类型,进行辨症施治。显而易见,实现自动识别首先必须由传感器获取舌像,提供原始信息,同时还应有反映医生临床辨舌经验的标准。在该基础上,才能由计算机进行识别,构思如图2所示。

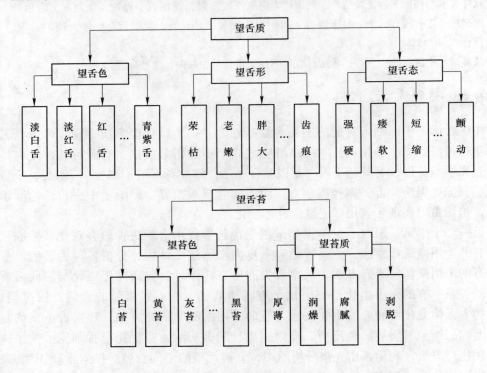

图1 中医舌诊的主要内容

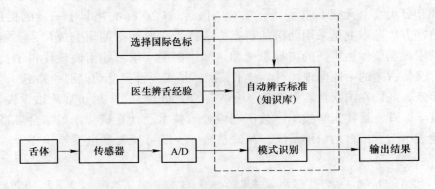

图2 实现自动识别的构思

综合分析舌诊内容，我们认为最基本、最活跃的因素是色彩，其他因素与色彩紧密相关，解决了舌色分类就得到了舌面状态的大致轮廓，为进一步细致分析打下了基础。从舌色识别入手，结合其他特征进行综合识别，这是我们的对策。识别色彩就面临色度标准，针对中医特点，该标准必须合乎人的视觉习惯，以利比较；同时，应该是国际通用的，便于统一和推广。考虑这两点，我们选择了 Munsell 颜色系统为色标[3]，根据是：

（1）该系统是按照视觉特点制定的颜色分类和标准；

（2）色样的编排在视觉上接近等间距；

（3）每一色标都给出色调（Value）、色彩（Hue）和彩度（Charoma）三属性，便于比较。

色调（V）、色彩（H）和彩度（C）完全描述了自然界中物体的绚丽多彩。彩色摄像机是靠 R（红）、G（绿）、B（蓝）三基色信号描述被摄物体的颜色，两者之间没有直接联系。要以 Munsell 系统为参照物，就必须完成 R、G、B 与 H、V、C 的转换，这已有另文发表，不再赘述。

3 自动识别系统的组成及工作过程

如图 3 所示，系统由照明系统、彩色摄像机、真彩色图像处理单元、IBM286 计算机（包括光电鼠标器和彩色监视器）组成。照明系统按柯勒照明原理设计，视场大于 150mm，照明不均匀性小于 1%，采用卤钨灯，保证摄像机对色温的要求。该系统已通

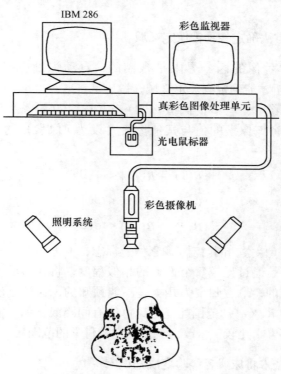

图3　中医舌诊自动识别系统结构示意图

过技术鉴定。

真彩色图像处理单元自行研制，有以下功能：

（1）图像获取，完成对彩色全电视信号的解码，对 R、G、B 三基色视频信号进行采样，量化为 8 比特/像素，40 毫秒/帧，图像获取格式为 512×512 或 256×256。

（2）图像存贮，可同时存贮一帧 512×512 或四帧 256×256 的真彩色图像，256 级灰度；还有 1K 比特的输出查找表，可一次写入四种不同类型的查找表供处理时选择。

（3）数据传输，采用 I/O 方式和优先排队，不受帧存显示刷新的时间限制。图像获取和显示时，100 微秒/像素；微机对帧存或查找表操作时采用 DMA 方式时，不大于 1.05 微秒/像素。

（4）显示，具有 256 级灰度，可显示 512×512 或 256×256 放大的真彩色图像，利用输出查找表可对任意指定的基色图像进行实时处理和伪彩色显示。在真、伪彩色显示时，均可实现图像、图形和字符的叠加显示。

工作过程是：被测者端坐，自然伸出舌头，照明系统将光均匀地投射在舌面上，调整摄像机的焦面和舌像画面的位置，摄入舌像。经图像单元处理，变成数字量信息。为了节约内存，该系统可用鼠标器从整幅像面中勾出舌体像，删除冗余信息。计算机从图像单元读取舌体信息进行识别，显示或打印结果，整个过程 5min 左右。

4 自动识别的方法

实现自动识别面临的第一个任务是建立知识库，即确定各种舌像的定义域，我们的方法如下。

4.1 运用模糊聚类方法确定各种舌像定义域

中医舌诊对各种舌像的区分界限是含糊的，自动识别也只能运用模糊数学原理[4,5]，进行聚类分析才能确定有关定义域，做法是：

预先设定某种舌像 R、G、B 的阈值 $td(0)$、$td(1)$、$td(2)$，令 i 像素的 R、G、B 为 $R(i)$、$G(i)$、$B(i)$，将该像素的值与其他像素（设为 j 像素）的对应值 $R(j)$、$G(j)$、$B(j)$ 相比较，若满足

$$|R(i)-R(j)|<td(0)$$
$$|G(i)-G(j)|<td(1)$$ (1)
$$|B(i)-B(j)|<td(2)$$

那么 j 像素就被归入 i 像素，同时对 j 像素做上标志，不再参加另一类 i 像素聚类。

模糊聚类实际上是通过比较每个像素的相似程度，将相似的像素归聚为一类。相似关系是预知或预定的，最后归聚为几类是不能预知的，这种方法较好地适应了中医的特点。各种舌像的 R、G、B 范围建立后，可能有间隔或交错，需要按总体规划作综合调整。经过反复摸索、比较、修改，就能确定较科学的范围。

4.2 运用样本训练技术将床辨舌经验变成知识库

建立舌像标准的过程是把医生临床辨舌经验变成一个知识库的过程，是人机相互

学习的过程，实现的途径是运用样本训练技术，如图4所示。

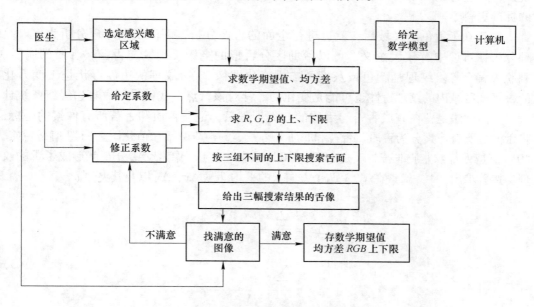

图4 样本训练系统示意图

采集的样本达到一定数量后，进行样本训练，过程如下：从硬盘或软盘中调出所存舌像，显示在监视器上，由医生操纵鼠标，将一个4×4个像素的小方框移到要研究的舌质或舌苔比较典型的区域内，获取其中的16个像素的R、G、B值，存入数据文件。一种类型的R、G、B范围变化较大，上述过程要进行多次，直到认为该类型有代表性的区域收集到为止。

接着，读数据文件，设共取了N个像素R、G、B的数学期望值$\overline{R}$、$\overline{G}$、$\overline{B}$，均方差ΔR、ΔG、ΔB和R、G、B的范围：

$$\overline{R}(\overline{G},\overline{B}) = \frac{\sum_{i=1}^{N} R(G,B)}{N}$$

$$\Delta R(\Delta G, \Delta B) = \frac{\sqrt{\sum_{i=1}^{N}[R_i(G_i,B_i)-\overline{R}(\overline{G},\overline{B})]^2}}{N} \quad (2)$$

$$R(G,B)_l = \overline{R}(\overline{G},\overline{B}) - \alpha \cdot \Delta R(\Delta G, \Delta B)$$

$$R(G,B)_u = \overline{R}(\overline{G},\overline{B}) - \alpha \cdot \Delta R(\Delta G, \Delta B)$$

式中，下标l表示下限；u表示上限；α是一个可调节的参数。令α为0.5，1.0或1.5，就可得到不同上、下限的范围，按照这三个不同的范围搜索舌像，以伪彩色显示出来。医生比较这三幅图像，挑选其中最能反映医生意图的一幅，所对应的期望值、均方差和范围就是该类型舌像的参数。若不满意，则需重新选区域或修改参数α，直到满意为止。

一个舌像还不能作为判断依据，只有对大量同种类型的舌像按上述过程进行综合统计，才能得到一个较合理的标准。同时，必须综合各名家的辨舌经验，才能使建立的标准更科学。

在建立了舌色定义域后，就可进行全面的舌像分析，下面仅以纹理分析为例。纹理分析在舌诊中有较大意义，可以帮助区分舌形的老嫩、裂纹与舌菌，舌质的腐腻、剥脱与偏全等。纹理是颜色或灰度的二维变化图像[6,7]。纹理的边界形状是千变万化的，不同区域内的纹理特性也可能是变化的。针对该特点，我们采用纹理边缘线条统计法，具体做法是：在完成舌苔与舌质的归类处理后，记下舌质和舌苔所有像素的坐标，将舌面上所有像素分为舌质、舌苔和非质非苔三类，构造两幅二值化图。"1"值是舌质，"0"值是舌苔或非质非苔，反之亦可。"0"像素与"1"像素交错分布就形成了舌的纹理。接着求出"1"像素在 x、y 两个方向上的平均间隔 Δx、Δy 以及其夹角 θ。

$$\Delta x = \frac{\sum_{i=1}^{N}(x_{i+1}-x_i)}{N}$$

$$\Delta y = \frac{\sum_{i=1}^{N}(y_{i+1}-y_i)}{N} \qquad (3)$$

$$\theta = \mathrm{atan}(\Delta y/\Delta x)$$

这三个参数反映了 x、y 方向上的粗糙度和纹理的大致走向。

如图 5 所示，小黑块代表舌质（舌苔）像素的位置，纹理分析限于实线所围区域，超出该范围是其形状问题而不是纹理问题。引入阀值 t_l 和 t_u（下标 l、u 的含义同前），$t_l < t_u$，x 或 y 方向的像素间隔满足

$$t_l < d < t_u \qquad (4)$$

这之间的像素才被计入纹理区域。在图 5 中，若令 $t_l=0$，$t_u=5$，则在 $y=17$ 的水平线上，从 $x=8$ 到 $x=11$ 的一段，从 $x=22$ 到 $x=24$ 的一段均被计入，而 $x=11$ 到 $x=22$ 的一段因 $d(=11)>t_u$ 而删去。纹理结构的形态是随机的，纹理分析只能建立在统计分析基础上。

5 自动识别程序设计

针对中医诊断特点，系统软件是在中文菜单的集成环境中使用，也可以英文形式显示结果。软件的底层采用汇编语言[8]，上层采用 C 语言[9]，运用结构化设计方法，构成模块。该软件能准确采集舌像，并从背景中拾取有效舌体像，划分区域分析。能以 Munsell 颜色系统为色标，进行模糊聚类分析，以伪彩色显示结果。能进行纹理分析与其他分析，具有学习和完善功能，操作方便，诊断流程如图 6 所示。

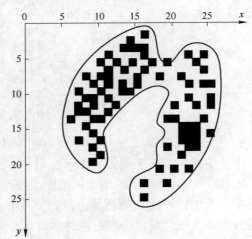

图 5　纹理区域与纹理分析示意图

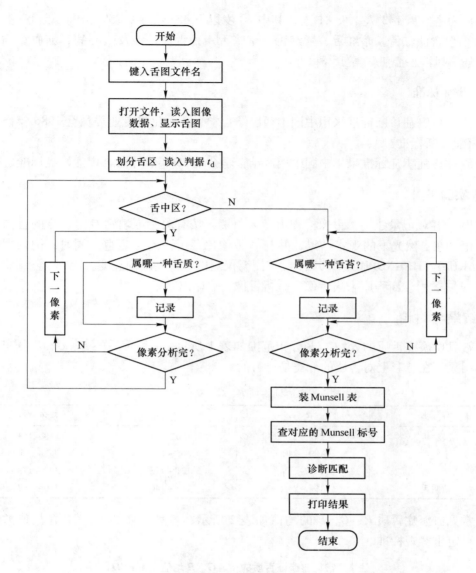

图 6　诊断流程图

6　实验研究

6.1　观察对象

（1）对照组：142 名清华大学学生及教工，男 95 人，女 57 人，年龄 16～61 岁，舌像为淡红舌，观察前经体检及必要理化检查，符合健康人条件[10]：

1）检查当日或数日内无不适症状；
2）以往虽有各种慢性病史，但一年内未出现症状者，肝脾肿大也不作计数；
3）近一个月内无发热外感等急性病史。

（2）观察组：224 名 1991 年 1 月至 1992 年 3 月在西苑医院住院及门诊的患者，以血瘀症为主。男 143 人，女 81 人，其中 40 岁以下 45 人，41~59 岁 96 人，60 岁以上 83 人，包含淡红舌、暗红舌、紫红舌、紫暗舌 4 种舌像，涉及冠心病、脑血管病、高血压病、风湿性心脏病等 16 种。

6.2 辨症标准

（1）血瘀症诊断标准采用中国中西医结合学会活血化瘀专业委员会 1986 年修订的《血瘀症诊断标准》。

（2）虚症辨症标准采用中国中西医结合学会 1986 年修订的《虚症辨症标准》。

6.3 实验步骤

（1）客观记录建档：舌像检查由专人负责，按有关标准填写专门设计的舌像调查表，记录在自然光下的观察结果，并与"舌诊比色版"对号定色。同时，记录用西德 UTRAKUST THERNOPHIL M202 型红外测温仪测得的舌表即刻温度。

（2）运用诊断程序采集舌像、分析舌像、输印出结果。

7 结果与讨论

常见血瘀症舌质的 R、G、B 分布范围如表 1 所示，4 种舌质的 R、G、B 分布范围区别明显，在色彩上对舌质的分类是可行的。

表 1 血辨证舌质 R、G、B 分布范围表

分类	红色分量	绿色分量	蓝色分量
淡红舌	233~255	120~153	118~153
暗红舌	189~240	105~124	80~115
紫红舌	231~255	131~173	132~178
紫暗舌	194~234	113~140	110~142

常见血瘀症舌质 R、G、B 的均值如表 2 所示，各种舌质的 R、G、B 均值差异清楚，可定出客观标准。

表 2 常见血瘀证舌质的 R、G、B 均值（$X \pm SD$）

分类	例数	红色分量（R）	绿色分量（G）	蓝色分量（B）
淡红舌	43	250.4±4.59	136.86±8.36	135.9±9.11
暗红舌	62	206.1±13.9*	112.87±6.01*	102±7.99*
紫红舌	66	248.2±7.43*△	51.8±10.65*△	154.8±11.73*△
紫暗舌	53	213.4±9.68*△▲	126.8±6.98*△▲	125.9±7.03*△▲
淡红舌对照组	142	253.4±1.07	138.7±6.48	136.9±9.74

注：*为与淡红舌比，$P<0.01$；△为与暗红舌比，$P<0.01$；▲为与紫红舌比，$P<0.01$。

肉眼观察与计算机自动识别结果比较如表 3 所示，符合率达 86.34%，结果较满意。

表3　血瘀证舌像的肉眼观察与计算机自动识别结果比较表

分 类	肉眼观察/例	系统识别		符合率/%
		相符/例	不相符/例	
淡红舌对照组	142	127	15	89.44
淡红舌	43	37	6	86.05
暗红舌	62	53	9	85.48
紫红舌	66	54	12	84.38
紫暗舌	53	45	8	84.91
合　计	366	316	50	86.34

上述结果说明，"中医舌诊自动识别系统"为舌像的客观标准化研究创造了条件，自动识别的方法既能较准确地反映医生的判断，又能定出客观的标准，是舌像定量描述、界限划分及标准确定的较理想方法。我们深入分析了错误识别的舌质，发现造成的因素有：

（1）肉眼分辨各类舌质时并没有严格的界限，因而有时舌质 R、G、B 的分布范围相互重叠，难以准确划分；

（2）样本量还不够大，舌色分类还不够细，容易造成相互交错；

（3）受条件限制，摄像机分辨率不高，影响识别的准确性。

这些因素有待我们在今后的研究中逐步改进，加以解决。

参 考 文 献

[1] 赵荣莱，等. 舌质舌苔的计算机定量描述和分类. 中医杂志. 1989；(2)：47.
[2] 宋天彬. 中医舌苔图谱. 北京：人民卫生出版社. 1984.
[3] Wyszecki Günter, Stiles W S. Color Science. USA：A Wiley Interscience Publication. 1982.
[4] Dubois D, et al. Fuzzy set and System-Theory and Application. USA. Academic Press. 1980.
[5] 区奕勤，张先迪. 模糊数学原理及应用. 四川．成都电讯工程学院出版社. 1988.
[6] Norman Katz, et al. An Image Processing System for Automatic Retina Diagnosis. SPIE Vol. 902 Three-Dimensional Imaging and Remote Sensing Image. 1988；131.
[7] 普拉特. 数字图像处理学. 北京：科学出版社. 1984.
[8] 沈美明，等. IBM PC [0520] 汇编语言程序设计. 北京：清华大学出版社. 1987.
[9] 宗丽苹，等. Microsoft C5.0 技术丛书. 北京：北京联想计算机集团公司. 1988.
[10] 陈泽霖，等. 5403 例正常人舌像检查分析. 中医杂志. 1981；(2)：18.

Study on Method of Automatic Diagnosis of Tongue Feature in Traditional Chinese Medicine

Abstract　The paper analyzed data of diagnosis by tongue feature in Traditional Chinese Medicine (TCM) and presented a measure of realizing automatic diagnosis. Munsell color system was used as standard of color. A system of automatic diagnosis by tongue feature in TCM was established using color science, modern optical, digital image processing technique and computer hardware technique. Besides the paper described structure and operation of the system, then on theory of dialectical therapy of TCM, experiences of clinical diagnosis by tongue feature were processed with computer software technique, and software of automatic diagnosis developed. Based on theory of Fuzzy mathematics, definition domain of image of tongues was provided by sample training technique. 366 clinical cases were observed, including four types of tongue features (nature, coating) of pale red, dark red, violet red and dull purple. The clinical experimental results were essentially conformed the diagnosis of TCM doctors with correspondence rate of 86.34%.

Key words diagnosis by tongue feature in Tradition Chinese Medicine; tongue nature; tongue coating; digital image processing; fuzzy clustering; texture analysis

Resolution Enhancement by Combination of Subpixel and Deconvolution in Miniature Spectrometers*

Abstract The resolution of a miniature spectrometer with a multichannel detector is limited by its throughput. A subpixel deconvolution method is proposed to enhance resolution without physically reducing the throughput. The method introduces subpixel reconstruction to overcome undersampling during deconvolution processing. The experimental result has shown a 36.6% reduction in FWHM of spectral lines, indicating the effectiveness of the method.

1 Introduction

It has long been realized that the width of the entrance slit has brought a trade-off between spectral resolution and throughput to a grating spectrometer[1-3]. The structure of a typical reflective grating spectrometer is shown in Fig. 1. Light from the entrance slit S is collimated by the mirror M_1 and dispersed by the plane grating G. The mirror M_2 focuses a series of spectral lines on the detector plane P. The spectral lines are, in fact, monochromatic images of S. Narrowing the slit indicates a reduction in FWHM of the spectral lines and consequently an enhancement in resolution. However, it reduces the throughput and degrades the signal-to-noise ratio in the mean time.

In a miniature spectrometer, the spectral resolution is relatively low due to the limitations in the linear dispersion and the aperture size. For a fixed

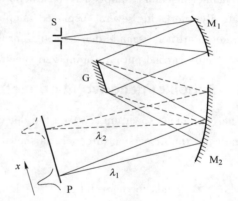

Fig. 1 Structure of a typical reflective grating spectrometer: S, entrance slit; M_1, collimating mirror; G, plane grating; M_2, camera mirror; P, detector plane

slit width w, practical methods to improve either resolution or throughput are desirable. Common methods to improve throughput, such as the application of a long curved slit[4] and further reduction of the F numbers of the system[5], are limited by the aperture size. Comparatively, applying deconvolution processing to spectral intensity distribution seems to be a potential way to improve resolution[6], for it has little requirement on system size. By mathematically eliminating

---

* Copartner: Li Xu, Huaidong Yang, Kexin Chen, Qiaofeng Tan, Qingsheng He. Reprinted from *Applied Optics*, 2007, 46 (16):3210-3214.

the impact of the entrance slit expanding line width, deconvolution algorithms could recover spectral details without physically reducing throughput.

Multichannel detectors are widely adopted in miniature spectrometers for their convenience and cheapness. However, they lead to sampling insufficiency in the practical application of deconvolution algorithms[7]. The spatial sampling rate of a multichannel detector is determined by its pixel size d. Low cost detectors for broadband used in spectrum detection generally have large pixel sizes, and therefore the sampling rate $1/d$ is limited. According to the Nyquist criterion, the spatial sampling rate of $1/d$ ought to be higher than twice that of the spatial cutoff frequency of the spectrum to avoid overlap in the frequency domain. When a deconvolution algorithm is applied, subtle high-frequency components of the spectrum are amplified and the cutoff frequency is increased. If the original sampling rate fails the requirement of the Nyquist criterion for the deconvolved spectrum, overlap will still occur, and the discrete data sequence will not recover the details correctly.

In this paper, the combination of the subpixel reconstruction technique[8] and deconvolution processing is presented as a method to enhance resolution in miniature spectrometers with multichannel detectors. Subpixel reconstruction is introduced here to increase the spatial sampling rate. By interlacing several displaced sampling frames of the spectrum, we would expect to generate an artificial sample dense enough for the deconvolution algorithm. An experiment is consequentially carried out on a minisized spectral device to prove the effectiveness of the method.

2 Subpixel deconvolution method

The continuous spectrum $t(x)$ on the detector plane could be expressed as (coordinate axis x indicates the dispersion direction, as shown in Fig. 1)

$$t(x) = s(x) \otimes r(x) \tag{1}$$

where $\otimes$ stands for convolution, $s(x)$ indicates a blurfree slit image, and $r(x)$ represents a spectrum generated by an infinitely narrow slit. Under uniform illumination, a blurfree slit image is approximately a rectangular impulse function, whose width is equal to the product of the slit width w and the system magnification factor κ. The height of the rectangular impulse function is $1/(w \times \kappa)$, if the area is set to be 1. In a real spectrometer, κ varies gradually with wavelengths. The whole spectrum can therefore be separated into several sections in each of which different wavelengths share a common value of κ.

After sampling, the spectrum is expressed by a series of data, whose values are the same as values of $t(x)$ at discrete sampling points. We use

$$T_0(n) = t(n \times d), \quad n = 1,2,3,\cdots,N \tag{2}$$

to represent the original sampling results (a frame), where N is the total pixel number.

Subpixel reconstruction provides extra information that allows deconvolution to recover details generally in three steps (as shown in Fig. 2): (i) acquire m frames of $t(x)$, each of which has fractional shifts of a pixel relative to adjacent frames, (ii) estimate the shift values, and (iii) reconstruct a densely sampled series of $t(x)$.

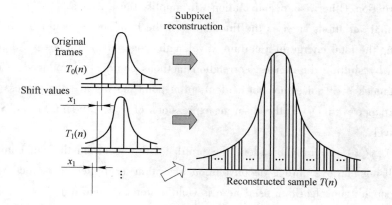

Fig. 2 Subpixel reconstruction process

The m frames, $T_0(n)$ to $T_{m-1}(n)$, could be acquired by mechanically shifting the detector along the dispersion direction. Similar to Eq. (2), they are

$$T_0(n) = t(n \times d)$$
$$T_1(n) = t(n \times d + x_1)$$
$$\vdots \qquad (3)$$
$$T_{m-1}(n) = t(n \times d + x_1 + x_2 + \cdots + x_{m-1})$$
$$n = 1,2,3,\cdots,N$$

where x_1 to x_{m-1} are shift values between frames. To simplify the procedure, the second step is omitted by setting all the shift values to be d/m. Then we have

$$T_0(n) = t(n \times d)$$
$$T_1(n) = t(n \times d + d/m)$$
$$\vdots \qquad (4)$$
$$T_{m-1}(n) = t[n \times d + (m-1) \times d/m]$$
$$n = 1,2,3,\cdots,N$$

As the distance between the sampling point of $T_{m-1}(n)$ and that of $T_0(n+1)$ is also d/m, by directly arranging data in the following pattern:

$T_0(1), T_1(1), \cdots; T_{m-1}(1), T_0(2), T_1(2), \cdots, T_{m-1}(2), \cdots, T_0(n), T_1(n), \cdots, T_{m-1}(n)$

theoretically forms an artificial sample sequence:

$$T(n) = t(n \times d/m), \quad n = 1,2,3,\cdots,m \times N \qquad (5)$$

whose spatial sampling rate is m/d, m times increased from $T_0(n)$'s.

Practically, the simplified subpixel method described above has two preconditions: (1) stable environment while acquiring different frames and (2) equal shift values. Noises would occur if either of these conditions is not satisfied. Filtering is a necessary part at the end of subpixel reconstruction. Otherwise, deconvolution will amplify these noises.

Environmental variations such as the fluctuation of the light source could be roughly corrected by unifying the total energy in a certain wavelength range among frames. While dealing with noises caused by shifting differences, correlation methods could be introduced to estimate the real shift distances[9]. If noises are not evident, an m-point moving average process will work as well. In accordance with $T(n)$, the discrete expressions of $s(x)$ and $r(x)$ are $S(n)$ and $R(n)$, respectively.

In Eq. (1), $s(x)$ generally broadens the profile of the spectrum through convolution. The resolution will not improve until a deconvolution algorithm is applied to deduct $S(n)$ from $T(n)$. The deconvolution algorithm used here is Gold's iterative algorithm[6]:

$$R^{(k+1)}(n) = R^{(k)}(n) \frac{T(n)}{s(n) \otimes R^{(k)}(n)} \tag{6}$$

where $R^{(k)}(n)$ and $R^{(k+1)}(n)$ indicate the kth and $(k+1)$th estimate of $R(n)$. $R^{(0)}(n) = T(n)$.

Compared with "inverse filter" deconvolution algorithms, such as the Wiener method [10] and the linear Bayesian deconvolution filter[11-13], which have no iteration, "constrained methods" such as Gold's algorithm can avoid the Gibb's phenomenon and will not give out negative values, which are impossible in spectrum detection.

3 Experiment

To validate the subpixel deconvolution method, we carried it out on a commercially designed prototype of a miniature Czerny-Turner spectrometer. The spectrometer has a 600 grooves/mm grating. Its optical structure is within 50mm × 70mm. The detector adopted is a complementary metal-oxide semiconductor (CMOS) detector with 1024 pixels, each with a 25μm pitch, whose perceptible wavelength ranges from 200 to 1000 nm. To fulfill subpixel shifting, a precision linear stage is used to carry the detector. The resolution of the stage is 0.1μm. The spectrum of a standard Hg-Ar source has been tested, and slit width w is 40μm. The throughput of the system is directly indicated by the detected voltage of the CMOS detector.

As shown in Fig. 3(a), the original sample of the 546.074nm peak has only three points above its FWHM. Having the detector shifted 5μm each for four times, we get five equally distantly displaced frames of the spectrum. After subpixel reconstruction, the artificial sample sequence $T(n)$ is acquired, whose sampling rate is improved five times. As can be seen in Fig. 3(b), noise introduced by a tiny disagreement among frames is successfully filtered by a five-point moving average process. The FWHM of the 546.07nm peak is ~71μm, having no evident change from the value before subpixel reconstruction.

From 550 to 600nm, $\kappa = 1.60$. The slit image width is then 64μm. The deconvoluted result $R(n)$ is shown in Fig. 4. The FWHM is reduced to ~45μm, 63.4% of that of $T(n)$.

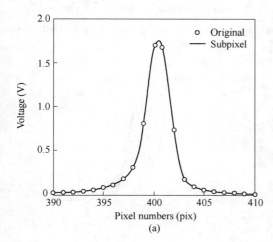
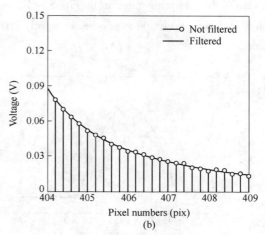

Fig. 3 Subpixel reconstruction at 546.074nm
(a) data points of one original frame and the subpixel result;
(b) a five-point moving average process to filter the noise

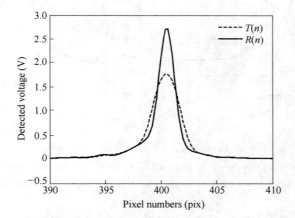

Fig. 4 Subpixel deconvoluted result at 546.074nm

4 Comparison and discussion

Fig. 5 illustrates the effectiveness of the subpixel deconvolution method. The two peaks being explored are the spectral lines at 576.960 and 579.066nm in the Hg spectrum. Figs. 5(a), 5(b), and 5(c) are the results with a 40μm slit. In Fig. 5(a), Gold's deconvolution algorithm is applied directly to one of the frames. Due to an insufficient sampling rate, the low-frequency components of the spectrum are folded into its high-frequency parts, and the algorithm cannot give reasonable results. In Fig. 5(b), subpixel reconstruction increases the sampling rate by five times. The two peaks are successfully separated, and the ratio between them after deconvolution remains similar to that before. Aliasing errors are generally excluded in Fig. 5(b). Further increasing the sampling rate, as in Fig. 5(c), where 25 equally distantly displaced frames (the shift between one frame to another is merely 1μm) are used, does not apparently offer better results.

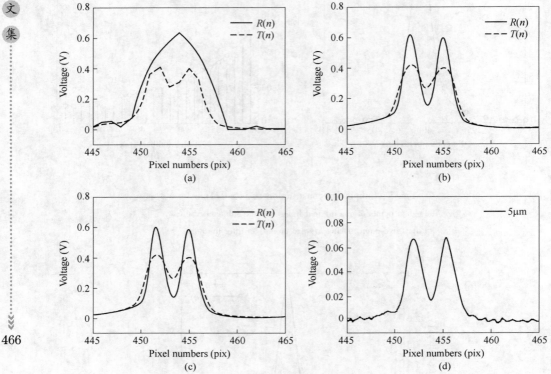

Fig. 5   Comparison at 576.960 and 579.066nm
(a) 40μm slit, deconvoluted only (one frame); (b) 40μm slit, subpixel deconvoluted, five frames;
(c) 40μm slit, subpixel deconvoluted, 25 frames; (d) 5μm slit, subpixel reconstructed only, five frames

Fig. 5(d) shows the subpixel reconstructed (not deconvoluted) result with a 5μm slit. Compared with Fig. 5(b), where the subpixel deconvoluted result with a 40μm slit is shown, the resolution is similar. The FWHM of the 546.074nm peak with the 5μm slit is ~ 50μm[not shown in Fig. 5(d)], which is close to the deconvoluted result. However, because of low throughput, applying a narrower entrance slit leads to much lower detected voltage.

For subpixel reconstruction and deconvolution, there are many successful concrete algorithms applied in multiple fields. Equal-distant shifting and Gold's algorithm are chosen here for simplicity, and they do prove the effectiveness of the method. So long as other concrete algorithms would satisfy the basic functional requirement, they are able to substitute for the present ones. There is no mention if they offer extra benefits.

5 Conclusion

We have presented a subpixel deconvolution method to achieve resolution enhancement in miniature spectrometers with multichannel detectors. First subpixel reconstruction is applied to increase the sampling rate, then deconvolution follows to improve resolution. By increasing the sampling rate five times, significant reduction in the FWHM is experimentally perceived while the throughput is unchanged.

Acknowledgements

This work was supported by the National Natural Science Foundation of China (grants 60578002 and 60378016) and the Key Project of the Chinese Ministry of Education (grant 106014).

References

[1] P. Jacquinot, "The luminosity of spectrometers with prisms, gratings, or Fabry-Perot etalons", J. Opt. Soc. Am. A 44, 761-765 (1954).
[2] J. F. James and R. S. Sternberg, *The Design of Optical Spectrometers* (Chapman and Hall, 1969).
[3] A. P. Thorne, *Spectrophysics*, 2nd ed. (Chapman and Hall, 1988).
[4] R. W. Esplin, "Use of curved slits to increase throughput of a Hadamard spectrometer", Opt. Eng. 19, 623-627 (1980).
[5] M. Futamata, T. Takenouchi, and K. -I. Katakura, "Highly efficient and aberration-corrected spectrometer for advanced Raman spectroscopy", Appl. Opt. 41, 4655-4665 (2002).
[6] P. A. Jansson, *Deconvolution with Applications in Spectroscopy* (Academic, 1984).
[7] H. Yang, L. Xu, K. Chen, Q. He, S. He, Q. Tan, and G. Jin, "The effect of the photoelectric detector on the accuracy of the spectrometer", spectrosc. Spectral Anal. (Beijing) 25, 1520-1523 (2005) (in Chinese).
[8] C. Pernechele, L. Poletto, P. Nicolosi, and G. Naletto, "Spectral resolution improvement technique for a spectrograph mounting a discrete array detector", Opt. Eng. 35, 1503-1510 (1996).
[9] S. S. Young and R. G. Driggers, "Superresolution image reconstruction from a sequence of aliased imagery", Appl. Opt. 45, 5073-5085 (2006).
[10] Z. Mou-yan, *Deconvolution and Signal Recovery* (National Defence Industry Press, 2001) (in Chinese).
[11] R. A. L. Tolboom, N. J. Dam, H. ter Meulen, J. Mooij, and H. Maassen, "Quantitative imaging through a spectrograph. 1. Principles and theory", Appl. Opt. 43, 5669-5681 (2004).
[12] R. A. L. Tolboom, N. J. Dam, and H. ter Meulen, "Quantitative imaging through a spectrograph. 2. Stoichiometry mapping by Raman scattering", Appl. Opt. 43, 5682-5690 (2004).
[13] R. A. L. Tolboom, N. J. Dam, N. M. Sijtsema, and J. J. ter Meulen, "Quantitative spectrally resolved imaging through a spectrograph", Opt. Lett. 28, 2046-2048 (2003).

Design of Freeform Mirrors in Czerny-Turner Spectrometers to Suppress Astigmatism*

Abstract Astigmatism is left uncorrected in traditional Czerny-Turner spectrometers with spherical mirrors, which leads to low throughput of modern instruments applying line-array detectors. By gradually varying the sagittal curvature, freeform mirrors are introduced to suppress astigmatism for a wide wavelength range simultaneously. So as not to reduce spectrum resolution, further calculations of the light path are performed for extra coma compensation. A design example is presented with optimized parameters. The ray-tracing result has revealed a reduction in sagittal spot size from several hundred micrometers to around $10 \mu m$ in the wavelength range from 200nm to 800nm.

1 Introduction

The performance of a Czerny-Turner spectrometer is significantly influenced by its geometrical aberrations, especially coma and astigmatism. Coma expands the point spread function (PSF) of the optical system along its dispersive direction (defined as the tangential/X direction in this paper), thus it limits the spectrometer's resolution. Astigmatism, on the other hand, expands the PSF in the direction perpendicular to dispersion (defined as the sagittal/Y direction, which is always vertical in Fig. 1). To make the resolution as high as possible, typical optical systems are

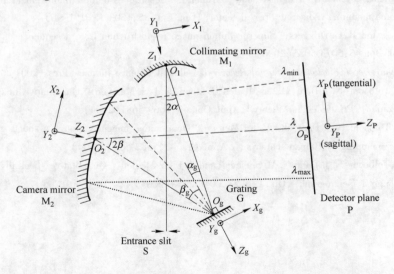

Fig. 1 Optical structure of a typical Czerny-Turner spectrometer with coordinates for each element

---

* Copartner: Li Xu, Kexin Chen, Qingsheng He. Reprinted from *Applied Optics*, 2009, 48(15):2871-2879.

arranged in a way that minimizes their PSF in the tangential direction, leaving astigmatism uncorrected and consequently a large scale in the sagittal direction as a kind of sacrifice[1,2].

In order to compensate for the diffusion of light energy due to large sagittal size, traditional monochromators require long curved slits at both the entrance and the exit of the optical system[1,3]. However, this is not feasible when common line-array detectors such as CCDs and complementary metal oxide semiconductors are used. In such cases, each single pixel is treated as an exit slit of the system. But when compared with a real exit slit of a traditional monochromator, it is neither curved nor long enough. Therefore, spectral energy that does not fall on the photosensitive surface is lost.

As energy loss leads to low throughput, which is yet another important index for an instrument, many efforts have been made to reduce astigmatism. In 1966, Dalton suggested a way in a low-resolution case, by distorting the plane grating into a concave cylinder[4]. Gil and Simon[5] have reported a monochromator design employing off-axis parabolic mirrors to eliminate astigmatism, and the design was recently realized by Schieffer et al.[6]. Moreover, toroidal camera mirrors are widely accepted[7,8] for their capability of evidently reducing the astigmatism of a single wavelength. Nevertheless, all these changes are mainly for monochromators, which examine spectral lines one after another, or spectrometers with scanning structures, which examine only a limited spectral range simultaneously. When a wide spectrum range is to be examined at one time, differences in light path among various wavelengths reduce the effectiveness of these methods. A similar phenomenon happened in a recent spectrometer design by Gulis and Kupreev, who have suggested the insertion of a tilted plane-parallel plate in the light path to suppress astigmatism[9].

In this paper, a design applying freeform mirrors is proposed to fulfill astigmatism suppression in Czerny-Turner spectrometers. This design evidently compresses the PSF in the sagittal direction for a wide wavelength range simultaneously. It can be directly applied to structure parameters chosen for spherical mirrors, i.e., no variations in distances or angles are required. Furthermore, the application of toroidal camera mirrors is proved to be a simplified special case of this design. Without decreasing the resolution, the ray-tracing example of the design has shown a significant reduction in RMS spot size in the sagittal direction, which is advantageous for line-array detectors.

2 Surface shape analysis for astigmatism suppression

The structure of a typical Czerny-Turner reflective grating spectrometer is shown in Fig. 1. Light from the entrance slit S is collimated by the collimating mirror M_1, then dispersed by the plane grating G, and finally refocused by the camera mirror M_2.

On the detector plane P, lights of different wavelengths form a serial of spectral lines along the dispersive direction.

For a chosen wavelength λ (usually around the middle of the spectral range $\lambda_{min} - \lambda_{max}$), in order to diminish the coma, the wavelength's light path should obey the well-known $\cos^3$ relation[2,10]

$$\frac{\sin\beta}{\sin\alpha} = \frac{R_{2T}^2 \cos^3\beta \cos^3\alpha_g}{R_{1T}^2 \cos^3\alpha \cos^3\beta_g} \tag{1}$$

where α and β stand for the off-axis incident angles of M_1 and M_2, respectively, and R_{1T} and R_{2T} are their curvature radii along their tangential directions (indicated by X_1 and X_2), respectively. α_g and β_g are the incident and diffraction angle of the plane grating G. For the same light path, astigmatism is due to different focal lengths of the mirrors in the tangential and sagittal directions. The difference in focal length Δf could be written as

$$\Delta f = (f_{1S} - f_{1T}) + (f_{2S} - f_{2T}) \tag{2}$$

where

$$f_{1T} = (R_{1T}/2)\cos\alpha, \quad f_{2T} = (R_{2T}/2)\cos\beta$$
$$f_{1S} = R_{1S}/(2\cos\alpha), \quad f_{2S} = R_{2S}/(2\cos\beta) \tag{3}$$

In Eqs. (2) and (3), f stands for focal length and R stands for curvature radius. Subscripts 1 and 2 refer to M_1 and M_2, while T and S stand for the tangential and sagittal direction, respectively.

If Δf is set to be zero by adjusting the parameters in Eq. (3), then for the chosen wavelength, astigmatism will be suppressed[7]. Nevertheless, according to the grating equation, β_g (and consequently β) varies from wavelength to wavelength. Hence there are usually residual aberrations for wavelengths other than λ.

In order to reduce astigmatism for a wide wavelength range at the same time, Δf should be kept close to zero for the whole spectrum. As light beams of different wavelengths cast on M_2 are separated along its tangential direction X_2, gradually varying the sagittal curvature radius R_{2S} of the camera mirror along axis X_2 would be a possible solution to set $\Delta f \approx 0$ all over.

A detailed description of this proposed freeform surface shape goes as follows (Fig. 2): The origin of $O_2 - X_2 Y_2 Z_2$ coordinate system is taken in the surface center of M_2. Along the tangen-

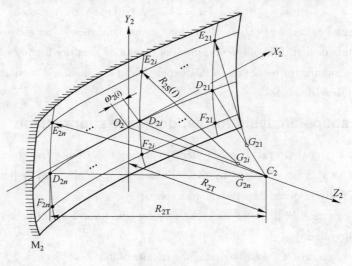

Fig. 2　Freeform surface shape of M_2

tial direction, in order to follow the cos3 relation as well as possible, the intersection of M_2 and plane $X_2O_2Z_2$, i. e., the curve $D_{21}\hat{O}_2D_{2n}$, should be exactly a circular arc of radius R_{2T}. The curve center of this arc is C_2. In the sagittal direction, the surface of M_2 is constructed by a set of circular arcs $E_{2i}\hat{D}_{2i}F_{2i}$, whose radii $R_{2S}(i)$ depend on the X positions $\omega_2(i)$ of D_{2i}, and centered on points G_{2i} along C_2D_{2i}. Therefore, the shape of M_2 can be characterized by $R_{2S} = R_{2S}(\omega_2)$, which can be given out in a polynomial form:

$$R_{2S}(\omega_2) = b_0 + b_1 \cdot \omega_2 + b_2 \cdot \omega_2^2 + b_3 \cdot \omega_2^3 + \cdots \quad (4)$$

where $b_0, b_1, \ldots$ are coefficients. In order to achieve sagittal curvature variation, at least a first-order polynomial expression is required of $R_{2S}(\omega_2)$:

$$R_{2S}(\omega_2) = b_0 + b_1 \cdot \omega_2 \quad (5)$$

However, such gradual variation of sagittal curvature also brings a gradual change to the tangential curvature along Y_2: $E_{21}\hat{E}_{2n}$ and $F_{21}\hat{F}_{2n}$ are no longer circular arcs as in the spherical or toroidal cases. In fact, they are now two-dimensional curves with varying curvature values along axis X_2. This consequent divergence from the $\cos^3$ relation introduces extra coma. Evident expansion of PSF due to extra coma is expected at every wavelength.

Further compensation of extra coma should be made based on calculation of the light path. Considering the symmetry of M_1 and M_2 in the light path, a reasonable method to fulfill this compensation would be varying R_{1S} gradually as well and hence setting M_1 to be a freeform mirror of the same kind as M_2. In this way, extra length in the light path introduced by M_2 would hopefully be canceled out by the extra length introduced by M_1, and thus the system will be brought back to the $\cos^3$ relation. Similar to $R_{2S}(\omega_2)$, $R_{1S}(\omega_1)$ is expressed in a polynomial form with coefficients $a_0, a_1, a_2, \ldots$,

$$R_{1S}(\omega_1) = a_0 + a_1 \cdot \omega_1 + a_2 \cdot \omega_1^2 + a_3 \cdot \omega_1^3 + \cdots \quad (6)$$

The minimum requirement of order is also one, i. e.,

$$R_{1S}(\omega_1) = a_0 + a_1 \cdot \omega_1 \quad (7)$$

where ω_1 stands for the coordinate along axis X_1 over the surface of M_1.

For further analysis, discrete expressions of freeform surfaces are given out in accordance with the definition of "Grid Sag Surface" in ZEMAX optical system design software[11]. Surface shapes are determined by a bicubic spline interpolation of sag values[12]. Based on a spherical plane of radius R_{1T}/R_{2T}, the sag values of freeform M_1/M_2 are given by adding additional sag terms Δ_1/Δ_2 defined on a rectangular array $(\omega_1, l_1)/(\omega_2, l_2)$ in coordinate system $O_1 - X_1Y_1/O_2 - X_2Y_2$ (see Appendix A for a detailed deduction):

$$\Delta_1 = \Delta_1(\omega_1, l_1) \approx \sqrt{R_{1T}^2 - \omega_1^2 - l_1^2} - \sqrt{\left(R_{1T} - R_{1S}(\omega_1)\left(1 - \sqrt{1 - \frac{l_1^2}{R_{1S}(\omega_1)^2}}\right)\right)^2 - \omega_1^2}$$

$$(8)$$

$$\Delta_2 = \Delta_2(\omega_2, l_2) \approx \sqrt{R_{2T}^2 - \omega_2^2 - l_2^2} - \sqrt{R_{2T} - R_{2S}(\omega_2)\left(1 - \sqrt{1 - \frac{l_2^2}{R_{2S}(\omega_2)^2}}\right)^2 - \omega_2^2} \tag{9}$$

In the following section, a detailed expression of extra coma, as well as formulas to compensate it, are given out through comparison with the traditional case applying spherical mirrors (where $R_T = R_S = R$) at both M_1 and M_2.

3 Formulas for Extra Coma Compensation

Each of the two parts of the light path in the spectrometer contains both a beam traveling from the mirror and a beam to the mirror: the collimating part consists of an input beam from the slit and an output beam toward the grating; the camera part (of each wavelength) consists of an input beam from the grating and an output beam toward the detector. Please see Fig. 3, in which the collimating part and the camera part are marked by different types of lines. Light of different wavelengths shares the same collimating part but has different camera parts of its own. For each wavelength, the total coma K can be expressed as a weighted sum:

$$K = -\frac{\cos\alpha_g}{\cos\beta_g}K_1 + K_2 \tag{10}$$

where K_1 and K_2 are the respective coma aberrations of the collimating part and the camera part. The weight factor $-\cos\alpha_g/\cos\beta_g$ stands for the influence of the plane grating, which can be derived by differentiating the grating equation[10].

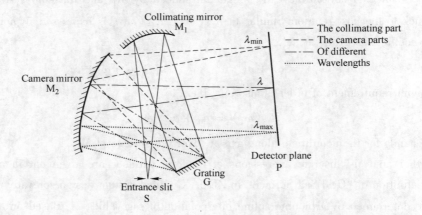

Fig. 3 Collimating part and the camera part of the light path

In the case of spherical mirrors, the $\cos^3$ relation makes comas generated by the collimating part and the camera part (denoted as K_1' and K_2', respectively, in the following equations) canceled out in Eq. (10), and thus the total coma K' is set to be almost zero. When freeform mirrors are applied, extra comas δK_1 and δK_2 are introduced in both parts, i.e.,

$$K_1 = K_1' + \delta K_1, \quad K_2 = K_2' + \delta K_2 \tag{11}$$

Hence, in order to compensate for the total coma K, total extra coma δK should also be zero:

$$\delta K = -\frac{\cos\alpha_g}{\cos\beta_g}\delta K_1 + \delta K_2 = 0 \quad (12)$$

As coma can be expressed as the partial derivative of Beutler's light-path function F [13], following calculations are focused on extra light-path generated by freeform mirrors. Fig. 4 shows the light traveling condition in the collimating part. Light from the object point $A(x_A, y_A, z_A)$ falls on a point $P_1(\omega_1, l_1, \xi_1)$, which is on the freeform surface of M_1 and is then reflected to the image point $B(x_B, y_B, z_B)$. $P'_1(\omega_1, l_1, \xi'_1)$ is a point on the "spherical base plane," a virtual sphere whose radius $R_1 = R_{1T}$ and whose center of curvature is also at $C_1(0, 0, R_{1T})$.

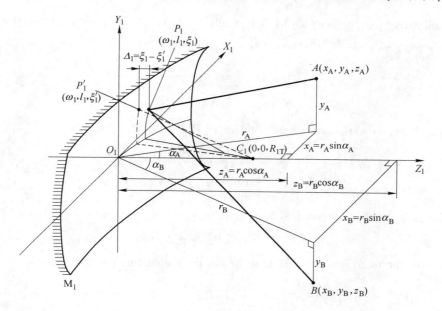

Fig. 4 Light traveling from the freeform collimating mirror

The coordinates of P'_1 must satisfy the equation

$$(R_{1T} - \xi'_1)^2 + \omega_1^2 + l_1^2 = 2R_{1T2} \quad (13)$$

The relation between P'_1 and P_1 can be expressed by $\xi_1 = \xi'_1 + \Delta_1$; ξ'_1, ξ_1, and Δ_1 are relatively tiny compared with z_A and z_B. Δ_1 is just the "additional sag term" expressed in Eq. (8).

The light path from A to P_1 (denoted as $[AP_1]$) can be written as

$$[AP_1] = \{(z_A - \xi_1)^2 + (x_A + \omega_1)^2 + (y_A - l_1)^2\}^{1/2} \quad (14)$$

Similarly, $[AP'_1]$, the light path in the virtual spherical case, is

$$[AP'_1] = \{(z_A - \xi'_1)^2 + (x_A - \omega_1)^2 + (y_A - l_1)^2\}^{1/2} \quad (15)$$

When the spherical surface is "substituted" with a freeform one, the variation in light path is calculated by the partial derivative of Eq. (15):

$$[AP_1] - [AP'_1] \approx \frac{\partial[AP'_1]}{\partial \xi'_1}\Delta_1$$

$$\approx \frac{-\Delta_1 z_A}{[(z_A - \xi'_1)^2 + (x_A - \omega_1)^2 + (y_A - l_1)^2]^{1/2}} (\text{as } z_A - \xi'_1 \approx z_A) \tag{16}$$

For convenience, denote the position of point A in cylindrical coordinates (r_A, α_A, y_A), as shown in Fig. 4. Then,

$$z_A = r_A \cos\alpha_A, \quad x_A = r_A \sin\alpha_A, \quad y_A = y_A \tag{17}$$

Substitute Eq. (17) and $\zeta'^2_1 + \omega_1^2 + l_1^2 = 2R_{1T}\xi'_1$ [derived from Eq. (13)] into Eq. (16):

$$[AP_1] - [AP'_1] \approx \frac{-\Delta_1 \cos\alpha_A}{\left[1 + \frac{y_A^2}{r_A^2} - \frac{2}{r_A}(\xi'_1 \cos\alpha_A + \omega_1 \sin\alpha_A) - 2\frac{y_A l_1}{r_A^2} + 2\frac{R_{1T}\xi'_1}{r_A^2}\right]^{1/2}} \tag{18}$$

In the collimating part, object point A is typically set to be at the meridional focus of off axis points, where $y_A = 0$, $r_A = 1/2 R_{1T}\cos\alpha$, and $\alpha_A = \alpha$. Hence from Eq. (19), as $\xi'_1 \ll r_A$,

$$[AP_1] - [AP'_1] \approx \frac{-\Delta_1 \cos\alpha}{\left(1 - 4\tan\alpha \frac{\omega_1}{R_{1T}}\right)^{1/2}} \tag{19}$$

In the same way, the variation in light path from $[P'_1 B]$ to $[P_1 B]$ is

$$[P_1 B] - [P'_1 B] \approx \frac{-\Delta_1 \cos\alpha_B}{\left[1 + \frac{y_B^2}{r_B^2} - \frac{2}{r_B}(\xi'_1 \cos\alpha_B + \omega_1 \sin\alpha_B) - 2\frac{y_B l_1}{r_B^2} + 2\frac{R_{1T}\xi'_1}{r_B^2}\right]^{1/2}} \tag{20}$$

As the image point B is supposed to be at infinity, $y_B = 0$, $r_B = \infty$, $\alpha_B = -\alpha$. Hence,

$$[P_1 B] - [P'_1 B] \approx -\Delta_1 \cos\alpha \tag{21}$$

Consequently, the total light path variation δF_1 in the collimating part is

$$\delta F_1 = ([AP_1] - [AP'_1]) + ([P_1 B] - [P'_1 B]) \approx -\Delta_1 \cos\alpha \left[1 + \frac{1}{\left(1 - 4\tan\alpha \frac{\omega_1}{R_{1T}}\right)^{1/2}}\right] \tag{22}$$

The angular aberration form of extra coma δK_1 is just the partial derivative of δF_1 with respect to ω'_1[10], i. e.

$$\partial K_1 = \frac{\partial \delta F_1}{\partial \omega'_1} \tag{23}$$

where ω'_1 stands for the perpendicular width of the parallel light beam. The relation among ω'_1, ω_1 (the width measured over the surface of mirror M_1), and ω_g (the width measured over grating G) is

$$\omega'_1 = \omega_1 \cos\alpha = \omega_g \cos\alpha_g, \quad \omega'_1 \in [-\omega'_1, \omega'_1]$$
$$\omega_1 \in [-\omega_1, \omega_1], \quad \omega_g \in [-\omega_g, \omega_g] \tag{24}$$

where ω'_1, ω_1, and ω_g are used to denote the maximum width of the beam, as shown in Fig. 5.

The deduction of δK_2 is parallel to the above derivation of δK_1, and thus it is not repeat-

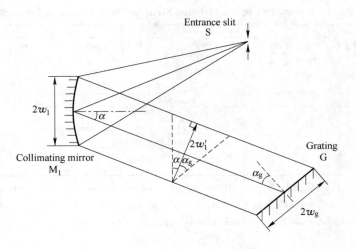

Fig. 5　Relationship among among $\omega'_1, \omega_1,$ and ω_g

ed. The useful equations related to δK_2 are given as Eqs. (25)-(28). Their major difference from Eqs. (8) and (23)-(25) of δK_1 is that, after dispersion, the light paths of different wavelengths are separated, thus parameters β_g and β vary from wavelength to wavelength. Also, as each wavelength only covers a part of the total width over the camera mirror M_2, the width of ω_2 is expressed as $[\omega_{2\lambda-}, \omega_{2\lambda+}]$, rather than the whole width of M_2:

$$\Delta_2 = \Delta_2(\omega_2, l_2)$$
$$\approx \sqrt{R_{2T}^2 - \omega_2^2 - l_2^2} - \sqrt{\left[R_{2T} - R_{2S}(\omega_2)\left(1 - \sqrt{1 - \frac{l_2^2}{R_{2S}(\omega_2)^2}}\right)\right]^2 - \omega_2^2} \quad (25)$$

$$\delta F_2 = -\Delta_2 \cos\beta \left[1 + \frac{1}{\left(1 - 4\tan\beta \dfrac{\omega_2}{R_{2T}}\right)^{1/2}}\right] \quad (26)$$

$$\partial K_2 = \frac{\partial \delta F_2}{\partial \omega'_2} \quad (27)$$

$$\omega'_2 = \omega_2 \cos\beta = \omega_g \cos\beta_g, \quad \omega'_2 \in [-\omega'_2, \omega'_2]$$
$$\omega_2 \in [-\omega_{2\lambda-}, \omega_{2\lambda+}], \quad \omega_g \in [-\omega_g, \omega_g] \quad (28)$$

4　Sample calculation and parameter estimation

The following example is based on a structure designed for spherical mirrors, whose parameters are shown in Table 1. To replace the spherical mirrors M_1 and M_2 with freeform ones, tangential and sagittal surface curvature radii $R_{1T}, R_{2T}, R_{1S}(\omega_1),$ and $R_{2S}(\omega_2)$ need to be decided. When the other parameters in the $\cos^3$ equation are not changed, R_{1T} and R_{2T} are set equal to R_1 and R_2, respectively. To get the expressions for $R_{1S}(\omega_1)$ and $R_{2S}(\omega_2)$ in Eqs. (5) and (7), four parameters are required, $a_0, a_1, b_0,$ and b_1. As shown in Fig. 6, discrete values of R_{2S} required for astigmatism suppression at different wave lengths (i.e., different ω_2 values) can be derived form Eqs. (2) and (3), and hence

$$b_0 = 121.13, \quad b_1 = -1.1821 \quad (29)$$

Table 1 Parameters of the Original Structure with Spherical Mirrors*

Parameter	Value	Parameter	Value
Wavelength range $(\lambda_{min} - \lambda_{max})$/nm	200 – 800	Design Wavelength (λ)/nm	500
M_1		M_2	
R_1/mm	100.00	R_2/mm	150.00
$\alpha/(°)$	10.00	$\beta_{-500}/(°)$	23.47
Grating			
$\alpha_g/(°)$	5.00	Groove/mm	600
$\beta_{g-500}/(°)$	22.78	Grating size/mm × mm	10 × 10
Distance between Elements/mm			
$S - O_1$	49.24	$O_1 - O_g$	55.00
$O_g - O_2$	49.09	$O_2 - O_p$	68.70

* As shown in Fig. 1, $S-O_1, O_1-O_g, O_g-O_2,$ and O_2-O_p stand for the distances from the entrance slit S to the center of M_1, from M_1 to the center of grating G, from G to the center of M_2, and from M_2 to the center of the detector plane, respectively. The detector plane is 2.50° inclined from perpendicular towards $O_2 - O_p$.

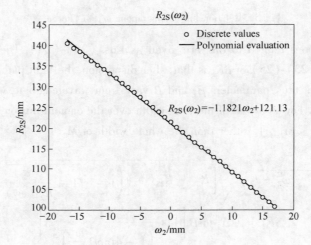

Fig. 6 Polynomial evaluation of $R_{2S}(\omega_2)$

are given out via polynomial evaluation. In Eq. (7), a_0 and a_1 are free parameters, which need to be optimized in order to satisfactorily compensate for extra coma. a_1 is introduced to compensate for the variation caused by $R_{2S}(\omega_2)$, and thus should be closely related to b_1; In an extreme case, if b_1 vanishes, then M_1 ought to be back to a spherical one, where $R_{1S} = R_{1T} = R_1$. Therefore it is reasonable to assume

$$a_0 = R_{1T} = R_1, \quad a_1 = q \cdot b_1 \tag{30}$$

where q is a real factor, indicating the proportion of sagittal curvature variation of M_1 to that of M_2. Following the discussion above, q is the key factor deter mining how well can M_1 compensate for the extra coma introduced by M_2.

It is worth noticing that, if b_1 in Eq. (5) is selected to be zero, there will be no variation in sagittal curvature radius for both M_1 and M_2. In this special case, $R_{1S} = a_0 = R_{1T}$ and $R_{2S} = b_0 \neq R_{2T}$, which is just the toroidal camera mirror setup for single wavelength astigmatism suppression.

Now q is the only parameter to be decided. Its optimized value can be based on the estimation of the general effect of extra coma compensation all over the surface of M_2. Since beams of different wavelengths dispersed by the grating have the same light-path model as that of a single beam sweeping over the whole surface of M_2, S, a weighted sum of δK^2, is developed in Eq. (31) to roughly estimate the q value:

$$S = \sum_i \sum_j \sum_k \left\{ V_{ijk} \times \left[-\frac{\cos\alpha_g}{\cos\beta_g(k)} \delta K_1(i,j) + \delta K_2(k,j) \right]^2 \right\} \quad (31)$$

As shown in Fig. 7, i, j, k are indices of uniform grid points in the X direction of $M_1(X_1)$, the Y direction of M_1 and $M_2(Y_1 \text{ and } Y_2)$, and the X direction of $M_2(X_2)$, respectively. As long as spatial sampling intervals are fixed in these directions, their ranges are limited by the size of the mirrors. The relationship between the sampling distances in directions X_1 and X_2 derives from Eqs. (24) and (28),

$$\omega_1 = \frac{\cos\alpha_g \cos\beta}{\cos\beta_g \cos\alpha} \omega_2 \approx 1.0064 \omega_2 \, (\text{at } \lambda = 500\text{nm}) \quad (32)$$

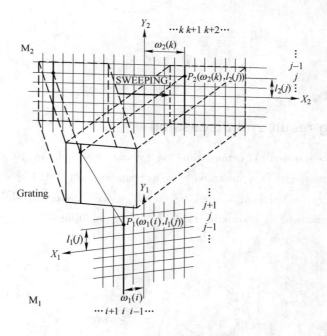

Fig. 7 Discrete grid points on M_1 and M_2 for extra coma estimation

Therefore they are set equal to 1.0064mm and 1mm, respectively. As there is no dispersion in the Y direction, the Y coordinates of every pair of points on M_1 and M_2 in the same light path, for example, $P_1(\omega_1(i), l_1(j))$ and $P_2(\omega_2(k); l_2(j))$ in Fig. 7, must be equal. Therefore, the sampling distances along both Y_1 and Y_2 are set equal to 1mm, which guarantees that for the same j, $l_1(j) = l_2(j)$.

For each pair of points, $\left[-\frac{\cos\alpha_g}{\cos\beta_g(k)} \delta K_1(i,j) + \delta K_2(i,j) \right]^2$ represents the squared amount of residual coma of the light passing them. V_{ijk} are the weights for different light paths and are here set to always be 1 for simplicity.

Fig. 8 reveals that when $q = 0.67$, the sum S has reached its minimum point, indicating an optimized condition of extra coma compensation. Nevertheless, considering all the factors neglected and approximations made before, the design needs to be verified by ray tracing.

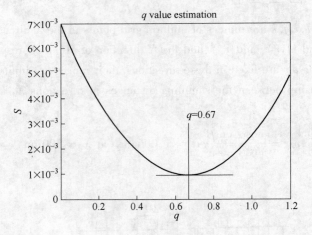

Fig. 8 q value estimation

5 Ray-tracing result and discussion

With the parameters selected in Section 4, the ray-tracing result obtained by ZEMAX is given in Fig. 9. The spectrum quality is represented by root-mean-square (RMS) spot size in X/Y direction. As previously explained, spot X sizes are mainly affected by coma, while spot Y sizes are dominated by astigmatism. As shown in Table 2, Y sizes of different wavelengths are around 10 μm on average.

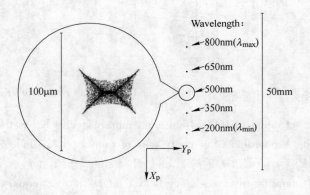

Fig. 9   Ray-tracing result of the design with free form M_1 and M_2

In Fig. 10, in order to test the effectiveness of factor q on coma compensation, the variation of spot X sizes with q values is summarized. When $q = 0.67$, almost all wavelengths reach their minimum X sizes, indicating the optimum case for extra coma compensation. The designed parameters are hence verified to be effective.

Table 2 RMS Spot Sizes for Different Wavelengths

Wavelength/nm	RMS Spot X Size/μm	RMS Spot Y Size/μm
800	27.1	10.6
650	14.7	13.2
500	8.4	12.4
350	16.5	8.7
200	27.5	11.9

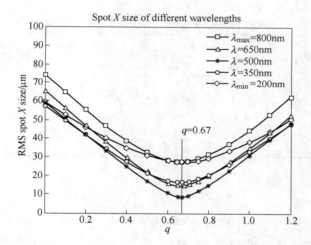

Fig. 10 Variation of spot X sizes with q

The design result of this paper successfully limits both the X and Y spot sizes for the whole wavelength range to be observed simultaneously. For further illustrations of this point, Fig. 11 and Table 3 give the spectrum generated by (a) spherical M_1 and M_2, (b) spherical M_1 and toroidal M_2, and (c) spherical M_1 and freeform M_2 (i. e., $q = 0$). Structure parameters other than surface shapes are kept unchanged in all of them. In Fig. 11(a), the application of traditional spherical mirrors results in evident astigmatism, as the RMS Y sizes are hundreds of micrometers in general. In Fig. 11(b), when a toroidal M_2 is introduced to replace the spherical one, astigmatism of the selected wavelength ($\lambda = 500$nm) is perfectly suppressed, as the Y size reduces to only 7.79μm. However, due to residual astigmatism, spot size changes sharply with wavelength: at λ_{min} and λ_{max}, Y sizes return to hundreds of micrometers. In Fig. 11(c), although applying freeform M_2 alone can suppress astigmatism, extra coma expands the X sizes as expected, which reduces resolution directly.

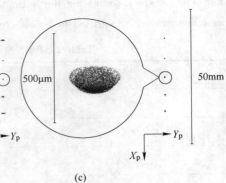

Fig. 11 Ray-tracing results of (a) spherical M_1 and M_2, (b) spherical M_1 and toroidal M_2, and (c) spherical M_1 and freeform M_2 ($q=0$)

Table 3 Spot Size Data for the Results in Fig. 11

Wavelength/nm	(a) Spherical		(b) Toroidal		(c) $q=0$	
	$X/\mu m$	$Y/\mu m$	$X/\mu m$	$Y/\mu m$	$X/\mu m$	$Y/\mu m$
800	32.5	325.0	30.1	331.4	74.0	53.9
650	18.1	438.4	16.8	164.6	65.3	57.4
500	7.1	559.1	5.0	7.8	59.5	60.2
350	14.4	687.4	14.0	142.3	57.4	62.7
200	27.6	823.9	28.6	283.6	59.3	66.4

6 Summary

A design using freeform mirrors with a gradually varying sagittal curvature is suggested in Czerny-Turner spectrometers to suppress astigmatism. In order to compensate for extra coma generated by the freeform camera mirror, the collimating mirror should be a freeform one as well. An example is given out based on coma compensation formulas. The ray-tracing result has shown that such a design has the ability to reduce RMS spot Y sizes from several hundred micrometers to around $10\mu m$, without increasing RMS spot X sizes, for a large wavelength range simultaneously. Energy concentration can be guaranteed via this design, which benefits the application of line-array detectors in modern instruments.

Appendix A Deduction of the expression of Δ_2

As the derivations of Δ_1/Δ_2 are mathematically the same, only that of Δ_2 is given here. For simplicity, move the origin of the Cartesian coordinate in Fig. 2 from O_2 to C_2 and focus on the area where $X_2 > 0$ and $Y_2 > 0$. P_2 is a point on the freeform surface of M_2; P'_2 is a point on the "spherical base plane", a virtual spherical surface whose radius $R_2 = R_{2T}$ and is also centered at C_2. As shown in Fig. 12, the distance from C_2 to any point on the intersection curve ($O_2 \frown D_2$) of the freeform surface and the $X_2O_2Z_2$ plane equals R_{2T}. The positions of P_2 and P'_2 in the new coordinate system are $(\omega_P, l_2; \xi_2 - R_{2T})$ and $(\omega_P, l_2, \xi'_2 - R_{2T})$, respectively.

Δ_2, the additional sag term, is defined as the distance from P'_2 to P_2 along axis Z_2:

$$\Delta_2 = \xi_2 - \xi'_2 = (\xi_2 - R_{2T}) - (\xi'_2 - R_{2T}) \tag{A1}$$

As P'_2 is on the sphere whose radius is R_{2T}

$$\xi'_2 - R_{2T} = -\sqrt{R_{2T}^2 - \omega_p^2 - l_2^2} \tag{A2}$$

On the other hand, P_2 is on the sagittal circular arc whose radius is $R_{2S}(\omega_2)$. ω_2 denotes the X_2 position of D_2, which is the intersection point of the circular arc and the $X_2 O_2 Z_2$ plane. Express the position of P_2 in terms of $R_{1S}(\omega_2)$, τ and θ, as denoted in Fig. 12:

$$\begin{aligned}
\xi_2 - R_{2T} &= -[R_{2T} - R_{2S}(\omega_2)(1 - \cos\tau)]\cos\theta \\
\omega_p &= [R_{2T} - R_{2S}(\omega_2)(1 - \cos\tau)]\sin\theta \\
l_2 &= R_{2S}(\omega_2)\sin\tau
\end{aligned} \tag{A3}$$

Eliminate τ and θ by substitution,

$$\xi_2 - R_{2T} = -\sqrt{\left[R_{2T} - R_{2S}(\omega_2)\left(1 - \sqrt{1 - \frac{l_2^2}{R_{2S}(\omega_2)^2}}\right)\right]^2 - \omega p^2} \tag{A4}$$

Neglecting the tiny difference between ω_P and ω_2,

$$\begin{aligned}
\Delta_2 &= (\xi_2 - R_{2T}) - (\xi'_2 - R_{2T}) \\
&\approx \sqrt{R_{2T}^2 - \omega_2^2 - l_2^2} - \sqrt{\left[R_{2T} - R_{2S}(\omega_2)\left(1 - \sqrt{1 - \frac{l_2^2}{R_{2S}(\omega_2)^2}}\right)\right]^2 - \omega_2^2}
\end{aligned} \tag{A5}$$

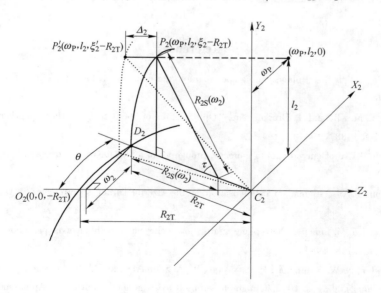

Fig. 12　Geometrical relationship of Δ_2

Although Δ_2 is also fairly small, considering that it is slowly and monotonically increasing with ω_2, the above approximation makes good sense. Fig. 13 shows Δ_2 values over M_2 of the given parameters in this paper. For the extreme case, where $\Delta_{2max} = 0.0407$mm, the maximum error due to this approximation is merely 4.38nm, or 0.0011%.

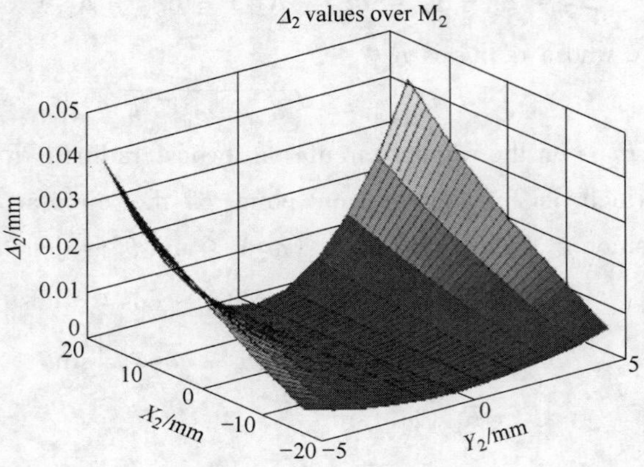

Fig. 13 Δ_2 values over M_2

Finally, expand (A5) to all over the $X_2 - Y_2$ plane. As in Eq. (A5), $R_{2S}(\omega_2)$ is always positive, and all other parameters related to ω_2 and l_2 are squared, this equation definitely suits the case whenever $(\omega^2, l^2) \in R^2$.

Acknowledgements

The authors are greatly indebted to Professor Minqiang Wang for inspiring discussions and for his foresight.

References

[1] W. G. Fastie, "High speed plane grating spectrograph and monochromator", U. S. patent 3,011,391(5 December 1961).

[2] A. Shafer, L. Megill, and L. Droppleman, "Optimization of the Czerny Turner spectrometer", J. Opt. Soc. Am. 54,879-887(1964).

[3] R. W. Esplin, "Use of curved slits to increase throughput of a Hadamard spectrometer", Opt. Eng. 19,623-627(1980).

[4] M. L. Dalton, Jr., "Astigmatism compensation in the Czerny-Turner spectrometer", Appl. Opt. 5, 1121-1123(1966).

[5] M. A. Gil and J. M. Simon, "New plane grating monochromator with off-axis parabolical mirrors", Appl. Opt. 22,152-158(1983).

[6] S. L. Schieffer, N. W. Rimington, V. P. Nayyar, W. A. Schroeder, and J. W. Longworth, "High-resolution, flat-field, plane grating, $f/10$ spectrograph with off-axis parabolic mirrors", Appl. Opt. 46, 3095-3101 (2007).

[7] M. Futamata, T. Takenouchi, and K. -I. Katakura, "Highly efficient and aberration-corrected spectrometer for advanced Raman spectroscopy", Appl. Opt. 41,4655-4665(2002).

[8] P. Villoresi, P. Nicolosi, and M-G. Pelizzo, "Design and experimental characterization of a high-resolution

instrument for measuring the extreme-UV absorption of laser plasmas", Appl. Opt. 39,85-93(2000).
[9] I. M. Gulis and A. G. Kupreev, "Astigmatism correction for a large-aperture dispersive spectrometer", J. Appl. Spectrosc. 75,150-155(2008).
[10] J. Reader, "Optimizing Czerny-Turner spectrographs: a comparison between analytic theory and ray tracing", J. Opt. Soc. Am. 59,1189-1196(1969).
[11] ZEMAX is a trademark of Zemax Development Corporation, Bellevue, Washington 98004, USA.
[12] ZEMAX User's Guide,225-226,Version:24 July 2002.
[13] H. G. Beutler, "The theory of the concave grating", J. Opt. Soc. Am. 35,311-350(1945).

Generalized Method for Calculating Astigmatism of the Unit-magnification Multipass System*

Abstract A generalized method to accurately calculate astigmatism of the unit-magnification multipass system(UMS) is proposed. A practical coaxial optical transmission model is developed for the UMS. Astigmatism analysis is then made convenient by a 4 by 4 general transfer matrix. Astigmatism correction is significantly promoted, and hence further improvement in imaging quality can be expected. Good agreement between numerical simulations and Zemax ray tracing results verifies the effectiveness of this method. The resulted RMS spot size of this method is only 25% to 64% of other previous methods based on the golden section search for minimum astigmatism in real design cases. This method is helpful for the optical design of the UMS.

1 Introduction

Multipass systems with multiple reflections have been widely used for decades in various scientific research and industrial fields. They can be generally classified into three main categories[1]: Herriott-type[2], White-type[3-10], and Chernin-type systems, i. e. , a multipass matrix system(MMS)[11-13]. The White-type system and MMS are well known because of their large numerical aperture and the ease with which they can be constructed. However, they do, in general, have large astigmatism that dominates the system's aberration[13,14], which is something that should be corrected for a high throughput in such multipass systems.

The astigmatism of several multipass systems has been studied by previous researchers[5,13-17]. Assuming that all the reflections are in the same plane, astigmatism of the Bernstein-Herzberg White cell(BHWC)[18] was approximated by summation[5,15]. A more accurate astigmatism estimation of the symmetrical BHWC was derived by assuming that the entire multipass trace could be split into symmetrical pairs whose astigmatism was additive, as reported by Kohn[16]. This analytic expression was not suitable for White-type systems, which, like the Pickett-Bradley White cell(PBWC)[10], violated the symmetrical condition. Astigmatism correction for the PBWC was refined by extra ray tracing but not accurately enough as reported by Tobin et al.[17]. Another method based on coordinate equations was implemented to correct imaging aberration for a symmetrical BHWC and MMS[13,14]. Nevertheless, astigmatism cannot be calculated effectively, since the chief ray was not in consideration. In general, methods for astigmatism calculation mentioned above depend more or less on the assumption that astigmatism was additive, which is actually not the truth. During our recent work to examine astigmatism of the

---

* Copartner: Kexin Chen, Huaidong Yang, Liqun Sun. Reprinted from *Applied Optics*, 2010, 49(10) : 1964-1971.

BHWC[19], a more accurate method is desired for both White-type systems and MMS.

We developed an improved model of the unit-magnification multipass system (UMS) representing White-type systems and MMS. Then astigmatism is accurately calculated based on the LDU factorization for the row transposition of a general 4×4 transfer matrix. Finally, numerical simulations and Zemax ray tracing results for both the BHWC and the PBWC are presented. The effectiveness of our method is proved by comparison with the previous methods mentioned above.

2 Principle and model

UMS is mainly composed of spherical mirrors whose radii of curvature are all R. As shown in Fig. 1.1, mirrors in the UMS could be distinguished as two opposite blocks, objective mirrors and field mirrors, with an interval close to R. The objective mirrors are all spherical mirrors. The entrance and exit apertures are placed by the side of the field mirrors. All the images of the entrance aperture by the objective mirrors are formed close to the surface of the field mirrors at unit magnification. Several UMSs are presented as shown in Figs. 1.2 and 1.3. The BHWC is constituted by two objectives [Fig. 1.2(a)] and one spherical field mirror with a T-shape aperture [Fig. 1.3(a)]. The PBWC is constituted by two objectives [Fig. 1.2(a)] and one spherical field mirror with a rectangular aperture [Fig. 1.3(b)]. UMS with matrixed images of the entrance aperture are constituted by more objectives and field mirrors. The modified White cell is constituted by two objectives [Fig. 1.2(a)] and one spherical field mirror combined with auxiliary prisms or corner mirrors [Fig. 1.3(c)]. A MMS is constituted by three or more objectives [Fig. 1.2(b)] and two spherical field mirrors with a rectangular aperture [Fig. 1.3(d)].

All the UMS are noncoaxial, noncoplanar, and partly asymmetric. It is difficult to analyze their astigmatisms because the meridional planes of reflections are not in the same plane. A coaxial optical transmission model is developed to make astigmatism calculation easier based on two equivalences with the same astigmatic imaging property as follows. First, the center of the entrance aperture is equal to an anastigmatic point source. Rays from the entrance aperture could be transformed to parallel rays that pass through an anastigmatic lens with a focal length of $R/2$ and then propagate at a distance of $R/2$. Second, individual reflection on a spherical mirror with an off-axis incident angle of θ is converted to transmission through an astigmatic thin lens with its optical axis in the direction of a chief ray. The two focal lengths of the astigmatic thin lens in the meridional and sagittal planes are $f_m = R\cos\theta/2$ and $f_s = R/(2\cos\theta)$, respectively.

The tilt direction of the meridional plane changes at each reflection in the UMS. Consecutive reflections as shown in Fig. 2(a) are equivalent to transmissions through multiple coaxial astigmatic thin lenses rotating at different angles about the optical axis as shown in Fig. 2(b). The rotated astigmatic thin lens is denoted as a general astigmatic thin lens (GATL) as in Ref. [20]. The parameters of the kth GATL are $(\theta_k; \alpha_k)$, where θ_k is the off-axis incident angle for the corresponding kth reflection and α_k is the rotation angle of this GATL about the optical axis. ε_k denotes the dihedral angle between the two incident planes of the kth and $(k+1)$th reflections. $z_k = O_{k-1}O_k$, which is equal to the distance between the $(k-1)$th and kth GATLs. In the corresponding partial model for astigmatism of the UMS as shown in Fig. 2(b), X—Y—Z is the global coordinate system whose Z axis is the optical axis. X_k—Y_k—Z is the local coordinate system for the kth GATL whose optical axis is the Z axis and the meridional plane is the Y_k—Z plane. α_k denotes the rotation angle about the Z axis from X—Y—Z to X_k—Y_k—Z for the kth GATL. It is clear that $\alpha_k - \alpha_{k-1} = \varepsilon_{k-1}$ ($k=2,\cdots,m$, where m is the total number of reflections in a UMS). Thus

$$\alpha_k = \alpha_1 + \sum_{j=1}^{k-1} \varepsilon_j \tag{1}$$

where the initial angle value α_1 depends on the direction of X—Y—Z and does not impact the calculation of total astigmatism.

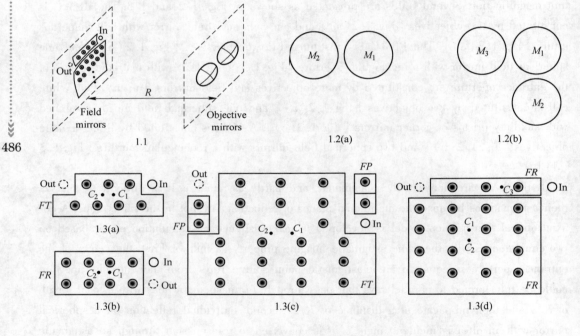

Fig. 1 Diagram of UMS

1.1—system layout of the UMS; 1.2—frontal views of the objective mirrors in the UMS; 1.2(a)—two objectives; 1.2(b)—three objectives; 1.3—frontal views of the field mirrors in the UMS; 1.3(a)—the BHWC; 1.3(b)—the PBWC; 1.3(c)—modified White cell; 1.3(d)—MMS. C_x, the curvature center of the objective mirror M_x where subscripts $x = 1, 2, 3$ represent the index of objective mirrors; In—the entrance aperture; Out—the exit aperture; filled circles—images of the entrance aperture; FT—spherical field mirror with T-shape aperture; FR—spherical field mirror with rectangular aperture; FP—prisms or corner mirrors

The narrow beam along the chief ray of the UMS is transformed to the rays on axis of the model composed of one anastigmatic thin lens and a series of GATLs coaxially separated by cer-

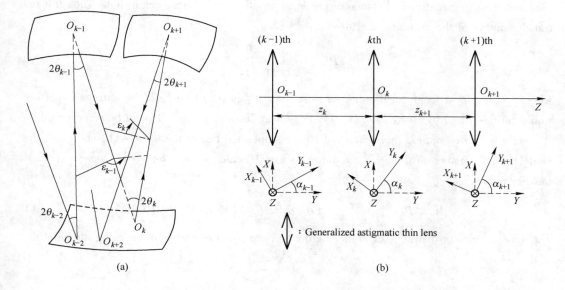

Fig. 2 Rotation angle of the kth generalized astigmatic thin lens
(a) reflections in the UMS; (b) corresponding partial model for astigmatism of the UMS

tain distances as shown in Fig. 3. The property of this narrow beam can be investigated by the 4×4 ray transfer matrices as follows.

It is well known that the 4×4 ray transfer matrix for a free optical propagation in a homogeneous medium is denoted as

$$P(z) = \begin{bmatrix} \mathbf{I} & z\mathbf{I} \\ \mathbf{0} & \mathbf{I} \end{bmatrix} \tag{2}$$

where z is the propagation distance, $\mathbf{I}$ is the 2×2 unit matrix, and $\mathbf{0}$ is the 2×2 zero matrix. The 4×4 ray transfer matrix for a GATL with parameters $(\theta; \alpha)$ is denoted as in Ref. [20]:

$$\begin{aligned}
\Lambda_g(\theta,\alpha) &= R(-\alpha);\Lambda(\theta)R(\alpha) \\
&= \begin{bmatrix} S(-\alpha) & \mathbf{0} \\ \mathbf{0} & S(-\alpha) \end{bmatrix}\begin{bmatrix} \mathbf{I} & \mathbf{0} \\ L(\theta) & \mathbf{I} \end{bmatrix}\begin{bmatrix} S(\alpha) & \mathbf{0} \\ \mathbf{0} & S(\alpha) \end{bmatrix} \\
S(\alpha) &= \begin{bmatrix} \cos\alpha & \sin\alpha \\ -\sin\alpha & \cos\alpha \end{bmatrix} \\
L(\theta) &= \begin{bmatrix} -f_s^{-1} & 0 \\ 0 & -f_m^{-1} \end{bmatrix} \\
&= \begin{bmatrix} -\dfrac{2\cos\theta}{R} & 0 \\ 0 & -\dfrac{2}{R\cos\theta} \end{bmatrix}
\end{aligned} \tag{3}$$

The entrance aperture is the object field of the UMS. The chief ray emitted from the center of the entrance aperture can be regarded as a parallel ray first passing through an anastigmatic lens and then propagating a distance of the focal length of this anastigmatic lens. Thus the ray transfer matrix of the entrance aperture is

$$P(f)\begin{bmatrix} \mathbf{I} & \mathbf{0} \\ -f^{-1}\mathbf{I} & \mathbf{I} \end{bmatrix}$$

where f is the focal length of the anastigmatic lens. If f is set to be $R/2$, the entrance aperture would be described by $P(R/2)\Lambda_g(0;0)$. The kth reflection in the UMS can be represented by $\Lambda_g(\theta_k;\alpha_k)$. A free optical propagation of $P(z_k)$ is operated before the kth GATL. Therefore, the model for astigmatism of the UMS as shown in Fig. 3 could be represented by a general 4×4 transfer matrix as

$$M = P(z_{m+1})\left\{\prod_{k=1}^{m}\Lambda_g(\theta_k,\alpha_k)P(z_k)\right\}P(R/2)\Lambda_g(0,0) \tag{4}$$

The general transfer matrix M exhibits imaging properties of the narrow beam along the chief ray of the UMS.

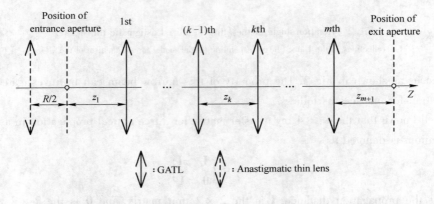

Fig. 3 Overall model for astigmatism of the UMS. The kth reflection is equivalent to the kth GATL; z_k, the distance between $(k-1)$th and kth GATLs; z_1, the distance between the position of the entrance aperture and the first GATL; z_{m+1}, the distance between the mth GATL and the position of the exit aperture

We implement the LDU factorization of the row transposition of M proposed by Macukow and Arsenault[21-23]:

$$M = \begin{bmatrix} A & B \\ C & D \end{bmatrix} = \begin{bmatrix} \mathbf{I} & E \\ 0 & \mathbf{I} \end{bmatrix}\begin{bmatrix} 0 & F \\ F^* & 0 \end{bmatrix}\begin{bmatrix} \mathbf{I} & G \\ 0 & \mathbf{I} \end{bmatrix} \tag{5}$$

where A, B, C, D, E, F, and G are all 2×2 sub-block matrices. The sub-block matrix E (E-matrix) of the factorized form of Eq. (5) is $E = AC^{-1}$. E is symmetrical and can be orthogonalized because M is symplectic[20]. Macukow and Arsenault reveal that the two eigenvalues of E are exactly the positions of foci with respect to the output plane[21-23].

For the UMS, two eigenvalues of the E-matrix as λ_s and λ_m are exactly equal to the distances between the exit aperture and the locations of the sagittal and meridional images. λ_s (or λ_m) is positive if the sagittal (or meridional) image is before the exit aperture along the chief ray and

negative if behind. Therefore, astigmatism of the UMS could be accurately computed as
$$l_{sm} = (-\lambda_s) - (-\lambda_m) = \lambda_m - \lambda_s \tag{6}$$
Supposing the E-matrix could be orthogonalized as
$$E = \begin{bmatrix} e_{11} & e_{12} \\ e_{12} & e_{22} \end{bmatrix} = S(-\beta) \begin{bmatrix} \lambda_s & 0 \\ 0 & \lambda_m \end{bmatrix} S(\beta)$$
$$= \begin{bmatrix} \cos^2\beta\lambda_s + \sin^2\beta\lambda_m & (\lambda_s - \lambda_m)\sin\beta\cos\beta \\ (\lambda_s - \lambda_m)\sin\beta\cos\beta & \sin^2\beta\lambda_s\cos^2\beta\lambda_m \end{bmatrix} \tag{7}$$

where β is the tilting angle of a synthetic meridional plane as to the $Y-Z$ plane in the model for astigmatism of the UMS. Obviously, from Eq. (7) we can obtain
$$(\lambda_s - \lambda_m)\cos 2\beta = e_{11} - e_{12}, \quad (\lambda_s - \lambda_m)\sin 2\beta = 2e_{12} \tag{8}$$
And Eq. (6) would now become
$$l_{sm} = \lambda_m - \lambda_s = -(e_{11} - e_{12})\cos 2\beta - 2e_{12}\sin 2\beta$$
$$2\beta = \arctan\left(\frac{2e_{12}}{e_{11} - e_{22}}\right) \tag{9}$$

Considering the sign of $e_{11} - e_{22}$, astigmatism of Eq. (9) can be simplified as
$$l_{sm} = \sqrt{(e_{11} - e_{22})^2 + 4e_{12}^2}$$
$$\beta = \begin{cases} \dfrac{1}{2}\arctan\left(\dfrac{2e_{12}}{e_{11} - e_{22}}\right) + \dfrac{\pi}{2}, & \text{for } e_{11} - e_{22} \geq 0 \\ \dfrac{1}{2}\arctan\left(\dfrac{2e_{12}}{e_{11} - e_{22}}\right), & \text{for } e_{11} - e_{22} < 0 \end{cases} \tag{10}$$

In summary, astigmatism analysis for any UMS can be implemented by the following steps: First, compute the chief ray for the parameters of $(\theta_k; \alpha_k)$ and z_k and build the model for astigmatism of this UMS. Second, obtain the E-matrix based on the LDU factorization of the row transposition of the general transfer matrix for the model. Then astigmatism can be accurately calculated by Eq. (10).

3 Simulation results

Both the BHWC and the PBWC are basic cells of the UMS. Other UMS with matrixed images, e.g., MMS and the modified White cells are recognized as recirculations of a beam based on the BHWC or the PBWC. Here the BHWC and the PBWC are taken as two examples to validate our method.

3.1 Astigmatism of the Bernstein-Herzberg White cell

The BHWC is a typical symmetric UMS. The system layout of the BHWC is presented as shown in Fig. 4. A three-dimensional Cartesian coordinate system $X_r—Y_r—Z_r$ with origin O is built to further describe the parameters. The subscript r is used to distinguish the coordinate from that in the model for astigmatism. C_F and the center of F are on the Z_r axis. C_1 and C_2 are on the Y_r

axis. For the n-pass BHWC, the coordinate of the center of the entrance aperture is $(-h/2, p, 0)$, C_F is $(0,0,R)$, C_1 is $(0, c/2, 0)$, and C_2 is $(0, -c/2, 0)$, where $c = 4p/n$. The distance between the two rows of images of the entrance aperture is h. The center of the entrance pupil is P $(0, \Delta + c/2, (R^2 - \Delta^2)^{1/2})$ and the stop is at M_1. The plane of $X_r O Z_r$ is the symmetrical plane for the BHWC and bisects the images of the entrance aperture.

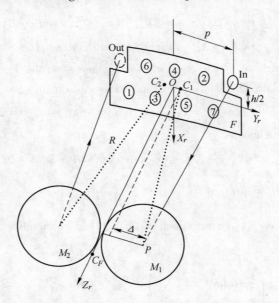

Fig. 4 System layout of the 16-pass BHWC

F—the field mirror with the center of curvature C_F; M_1 and M_2—two spherical objective mirrors with centers of curvature C_1 and C_2, respectively; In—the entrance aperture; Out—the exit aperture; Tags 1 to 7—images of the entrance aperture in sequence

To validate the accuracy of our method, a ray tracing model in the mixed sequential and non-sequential mode by Zemax is built and the narrow beam is traced. Orthogonal focusing meridional and sagittal image lines can be observed near the exit aperture. Then astigmatism can be obtained by computing the distance between the locations of the two focusing images. Simulations for the BHWC by Zemax show that astigmatism by ray tracing decreases slowly by the increase of the half-angle divergence. When it is no larger than 0.1°, astigmatism by ray tracing would become smaller than 3×10^{-5} mm. To ensure the precision of astigmatism by ray tracing, the half-angle of divergence for the traced narrow beam is set to $\leq 0.1°$ in all our simulations below.

Astigmatism of the BHWC is known to be corrected by optimizing h, therefore here we focus on the relation between astigmatism and h. Astigmatism of a 40-pass BHWC with fixed parameters of $R = 625$ mm, $p = 40$ mm, and $\Delta = 50$ mm is calculated varying with h as shown in Fig. 5. Meanwhile the results by Kohn's method and ray tracing results are also given. The maximum absolute deviation between the discrete ray tracing results and calculations by our method is 1.4×10^{-4} mm, which is much less than 4.6 mm between the discrete ray tracing results and results by the Kohn method. It can be seen from Fig. 5 that our results show excellent agreement with the ray tracing results by Zemax and is more accurate than that by the Kohn method.

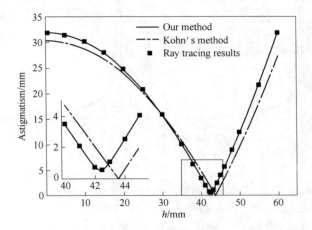

Fig. 5　Astigmatism of a 40-pass BHWC varying with h
solid curve—simulations by our method; dot-dash curve—simulations by
the Kohn method; discrete cubic—ray tracing results by Zemax;
subplot in figure—enlarged part in rectangle

Astigmatism first decreases then increases when h increases. Astigmatism can be minimized by optimizing h to improve the imaging quality. Since astigmatism versus h is simply unimodal as shown in Fig. 5, the golden section search method[24] is used for astigmatism optimization. The initial bracketing of h is $[0, 1.5p]$, and the minimum bracketing of h for the optimum point is set to 0.01 mm. The optimum h can be obtained by both methods, $h = 42.38$ mm by our method and $h = 43.55$ mm by the Kohn method. Besides, the method based on coordinate equations (MCEs) could also optimize h to 46.13 mm for good imaging quality, although astigmatism could not be calculated by the MCEs. To compare the imaging quality optimized by the three methods for the BHWC in a real case, this 40-pass BHWC with a large relative aperture, 0.05 NA, is examined. The MTF from the objective point in the center of the entrance aperture to the output focusing image plane is presented as shown in Fig. 6. The MTF of the BHWC optimized by our method has the best response to spatial frequency, especially at low and high frequency bands.

To further illustrate the optimization effects by the three methods, the spot diagrams and sizes of the output focusing image for the objective point in the center of the entrance aperture are compared as shown in Fig. 7 and Table 1. The intensity of the output focusing image optimized by our method is uniform and better concentrated. The RMS spot radius optimized by our method is only 64% of that by the Kohn method and 25% of that by the MCEs. The BHWC optimized by our method has better imaging quality.

3.2　Astigmatism of the Pickett-Bradley White cell

In contrast with the BHWC, the PBWC is asymmetrical and cannot be calculated effectively by

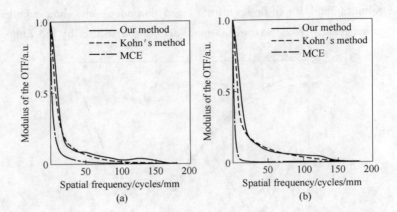

Fig. 6 MTF of a 40-pass BHWC optimized by our method (solid curves), the Kohn method (dashed curves), and MCE (dash-dot curves)

(a) MTF in the meridional plane; (b) MTF in the sagittal plane

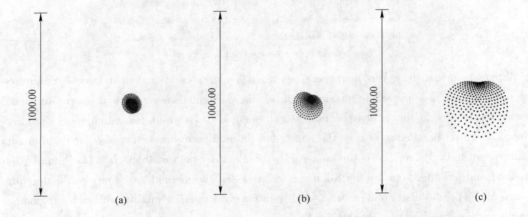

Fig. 7 Spot diagram in Zemax for output focusing image of the BHWC optimized by the three methods
(a) h = 42.38mm optimized by our method; (b) h = 43.55mm optimized by the Kohn method;
(c) h = 46.3mm optimized by MCE. Scale bars are 1000μm

Table 1 Spot Size of Output Focusing Image for the BHWC by Zemax Ray Tracing Optimized by Three Methods

Method	RMS Spot Radius/μm	RMS Spot X Size/μm	RMS Spot Y Size/μm
Ours	27.8	18.5	20.8
Kohn	43.7	30.1	31.7
MCE	113.1	79.0	80.9

the Kohn method and MCEs. Astigmatism of the PBWC has not yet been analyzed comprehensively. The system layout of the PBWC is presented with the coordinate X_r—Y_r—Z_r whose origin is O as shown in Fig. 8. C_1 and C_2 are on the Y_r axis, O coincides with C_1, and C_F is on the

Z_r axis. For the n-pass PBWC, the coordinate of the center of the entrance aperture is $(-h/2, p, 0)$, C_F is $(0, 0, R)$, C_1 is $(0, 0, 0)$, and C_2 is $(0, -c, 0)$, where $c = 4p/(n-2)$. The distance between two rows of images of the entrance aperture is h. The center of the entrance pupil is $P(0, \Delta, (R^2 - \Delta^2)^{1/2})$. The plane of $X_r OZ_r$ is no longer the symmetrical plane for the PBWC.

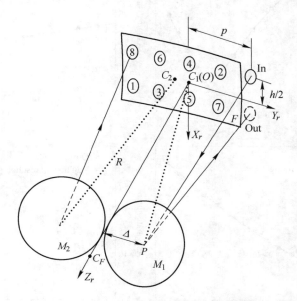

Fig. 8 System layout of the 18-pass PBWC

F—field mirror with the curvature center C_F; M_1 and M_2—two spherical objective mirrors with curvature centers C_1 and C_2, respectively; In—the entrance aperture; Out—the exit aperture; Tags 1 to 8—images of the entrance aperture in sequence

Astigmatism varying with h, Δ, p, and n for the PBWC is presented in four cases as shown in Fig. 9. The maximum of the absolute deviations for the results by our method as to Zemax ray tracing results are 1×10^{-4} mm, 1×10^{-4} mm, 1×10^{-4} mm, and 5×10^{-4} mm in the four cases, respectively. Astigmatism increases by the increase of p or n, but decreases by the increase of Δ with approximate linearity as shown in Figs. 9(b)-(d), respectively. Furthermore, astigmatism of the PBWC also first decreases and then increases by the increase of h such as the BHWC as shown in Fig. 9(a).

We minimize astigmatism of the 30-pass PBWC with fixed parameters of $R = 625$ mm, $p = 40$ mm, and $\Delta = 50$ mm by optimizing h to 46.27 mm as shown in Fig. 9(a). Nevertheless optimum h for astigmatism correction obtained by the expression of Tobin et al.[17] is 49.37 mm; the optimization is similar to that discussed in Subsection 3.1. Similar to the BHWC in Subsection 3.1, the real design case of a 30-pass BHWC with a large relative aperture such as 0.05 NA has been examined for optimizations by Tobin et al. and us. The MTF from the objective point in the center of the entrance aperture to the output focusing image plane is presented as shown in Fig. 10. The MTF of the PBWC optimized by our method has a better response to spatial frequency than that by Tobin et al.

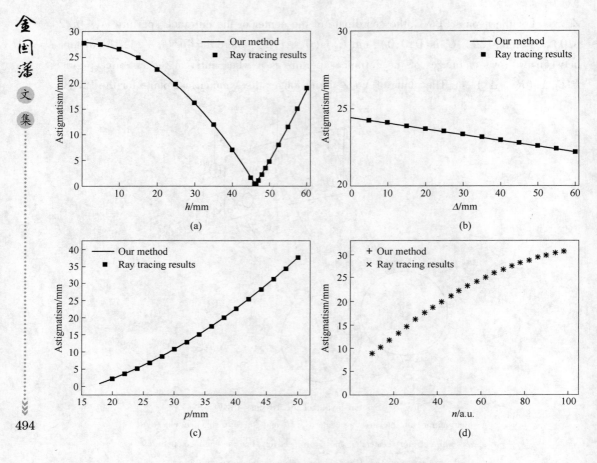

Fig. 9　Astigmatism of the PBWC for four cases

(a) astigmatism varying with h in a 30-pass PBWC with $R = 625$mm, $p = 40$mm, $\Delta = 50$mm; (b) astigmatism varying with Δ in a 30-pass PBWC with $R = 625$mm, $p = 40$mm, $h = 20$mm; (c) astigmatism varying with p in a 30-pass PBWC with $R = 625$mm, $h = 20$mm, $\Delta = 50$mm; (d) astigmatism in the varying n-pass PBWC with $R = 625$mm, $p = 40$mm, $h = 30$mm and $\Delta = 50$mm

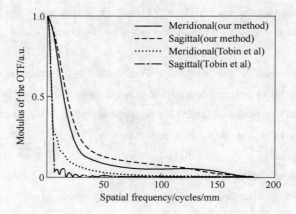

Fig. 10　MTF of a 30-pass PBWC optimized by Tobin et al. and us

The spot diagrams and sizes of the output focusing image for the objective point in the center of the entrance aperture are also shown in Fig. 11 and Table 2. The intensity of the output focusing image optimized by our method is uniform and better concentrated. The RMS spot radius optimized by our method is only 26% of that by Tobin et al. [17]. The PBWC optimized by our method has a better imaging quality. Astigmatism of other asymmetric UMSs can also be calculated and well corrected by our method following the two steps outlined in Section 2.

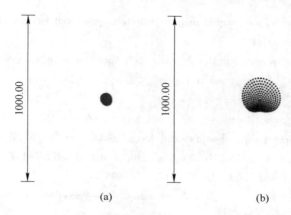

Fig. 11 Spot diagram in Zemax to focus the image of the PBWC optimized by Tobin et al. and us (a) $h = 46.27$ mm optimized by our method; (b) $h = 49.37$ mm optimized by Tobin et al. Scale bars are 1000 μm

Table 2 Spot Size of Output Focusing Image for the PBWC by Zemax Ray Tracing Optimized by Tobin et al. and Our Method

Method	RMS Spot Radius /μm	RMS Spot X Size/μm	RMS Spot Y Size/μm
Ours	19.2	12.6	14.5
Tobin et al.	74.7	52.4	52.3

4 Conclusion

We proposed a generalized method to accurately calculate astigmatism for any type of UMS. Based on the model for astigmatism of the UMS, the total astigmatism can be obtained based on the E matrix of LDU factorization for the row transposition of the 4×4 general transfer matrix. Results validate the effectiveness of our method for astigmatism calculation and correction. The intensity of the output focusing image optimized by our method is uniform and better concentrated. In our examples designed for a realistic system, the RMS spot radius optimized by our method is only 64% of that by the Kohn method and 25% of that by the MCEs for a symmetrical BHWC; the RMS spot radius optimized by our method is only 26% of that by Tobin et al. [17] for an asymmetrical PBWC. This method improves the imaging quality to increase the system throughput, which further benefits the optical design of the UMS. The analytic solution of our method with reasonable approximations under consideration will be further studied to improve its efficiency.

Acknowledgements

This research is supported by the Ph. D. Programs Foundation of the Ministry of Education of China, 20090002120010.

References

[1] C. Robert, "Simple, stable, and compact multiple-reflection optical cell for very long optical paths", Appl. Opt. 46, 5408-5418(2007).

[2] J. Silver, "Simple dense-pattern optical multipass cells", Appl. Opt. 44, 6545-6556(2005).

[3] R. P. Blickensderfer, G. E. Ewing, and R. Leonard, "A long path, low temperature cell", Appl. Opt. 7, 2214-2217(1968).

[4] J. U. White, "Very long optical paths in air", J. Opt. Soc. Am. 66, 411-416(1976).

[5] T. H. Edwards, "Multiple-traverse absorption cell design", J. Opt. Soc. Am. 51, 98-102(1961).

[6] A. L. Vitushkin and L. F. Vitushkin, "Design of a multipass optical cell based on the use of shifted corner cubes and right-angle prisms", Appl. Opt. 37, 162-165(1998).

[7] D. Horn and G. C. Pimentel, "2. 5km Low-temperature multiple-reflection cell", Appl. Opt. 10, 1892-1898 (1971).

[8] J. -F. Doussin, R. Dominique, and C. Patrick, "Multiple-pass cell for very-long-path infrared spectrometry", Appl. Opt. 38, 4145-4150(1999).

[9] J. T. K. McCubbin and R. P. Grosso, "White-type multiple-pass absorption cell of simple construction", Appl. Opt. 2, 764-765(1963).

[10] H. M. Pickett, G. M. Bradley, and H. L. Strauss, "White type multiple pass absorption cell", Appl. Opt. 9, 2397-2398(1970).

[11] S. M. Chernin, "Promising version of the three-objective multipass matrix system", Opt. Express 10, 104-107(2002).

[12] S. M. Chernin and E. G. Barskaya, "Optical multipass matrix systems", Appl. Opt. 30, 51-58(1991).

[13] S. M. Chernin, S. B. Mikhailov, and E. G. Barskaya, "Aberrations of a multipass matrix system", Appl. Opt. 31, 765-769(1992).

[14] Y. G. Barskaya, "Aberrations of a multipass cell", Opt. Technol. 38, 278-280(1971).

[15] T. R. Reesor, "The astigmatism of a multiple path absorption cell", J. Opt. Soc. Am. 41, 1059-1060 (1951).

[16] W. H. Kohn, "Astigmatism and White cells: theoretical considerations on the construction of an anastigmatic White cell", Appl. Opt. 31, 6757-6764(1992).

[17] D. C. Tobin, L. L. Strow, W. J. Lafferty, and W. B. Olson, "Experimental investigation of the self- and N_2-broadened continuum within the N_2 band of water vapor", Appl. Opt. 35, 4724-4734(1996).

[18] H. J. Bernstein and G. Herzberg, "Rotation-vibration spectra of diatomic and simple polyatomic molecules with long absorbing paths", J. Chem. Phys. 16, 30-39(1948).

[19] C. Kexin, Y. Huaidong, S. Liqun, and J. Guofan, "Astigmatism analysis by matrix methods in White cells", Proc. SPIE 7156, 71560G(2008),

[20] X. Liu and K. -H. Brenner, "Minimal optical decomposition of ray transfer matrices", Appl. Opt. 47, E88-

E98(2008).

[21] H. H. Arsenault and B. Macukow, "Factorization of the transfer matrix for symmetrical optical systems", J. Opt. Soc. Am. 73,1350-1359(1983).

[22] B. Macukow and H. H. Arsenault, "Matrix decompositions for nonsymmetrical optical systems", J. Opt. Soc. Am. 73,1360-1366(1983).

[23] B. Macukow and H. H. Arsenault, "Extension of the matrix theory for nonsymmetrical optical system", J. Opt. (Paris)15,145-151(1984).

[24] G. V. Reklaitis, A. Ravindran, and K. M. Ragsdell, *Engineering Optimization: Methods and Applications* (Wiley,1983), pp. 43-47.

Approximate Analytic Astigmatism of Unit-magnification Multipass System*

Abstract   We develop a way to estimate the approximate analytic astigmatism with a high accuracy for any unit-magnification multipass system (UMS). The coaxial optical transmission model for UMS is simplified based on the system's features. Furthermore, astigmatism is derived as a distinct form of vector addition and, thus, feasible analytic astigmatism can be obtained. The effectiveness of our method is verified by simulations for a Bernstein-Herzberg White cell. In our cases, the relative error of optimization for astigmatism correction by our method is smaller than 5‰, which is only one-tenth of that by Kohn's method. Our method significantly improves the efficiency for astigmatism correction, and further benefits the optical design of a UMS.

1 Introduction

Unit-magnification multipass systems (UMSs)[1], e.g., White-type systems[2-5] and multipass matrix systems (MMSs)[6,7] are widely used in spectroscopy analysis requiring long optical absorption paths, for their high imaging quality and easy construction. A UMS is mainly composed of multiple spherical mirrors with the same curvature radii R. Mirrors in a UMS can be divided into objective and field mirrors face to face with an interval of approximate R. The entrance aperture is placed at the side of the field mirrors. Therefore, images of the entrance aperture are formed near the field mirrors, in turn, by successive reflections on the objective mirrors at a unit magnification. The imaging quality of the last image output at the exit aperture is dominated by astigmatism[8,9].

Astigmatism of UMS has been analyzed numerically and analytically[1, 8-13]. Recently, we proposed a generalized numerical method to calculate accurate astigmatism for any UMS[1]: the UMS is equivalent to a coaxial optical transmission model by transforming reflection on the spherical mirror to transmission through a generalized astigmatic thin lens (GATL), and then astigmatism can be calculated accurately based on the system transfer matrix. This numerical method is accurate and generally valid. However, in some situations, such as for the purpose of quickly analyzing the influence of system parameters, an approximate analytic astigmatism analysis is preferred. Many researchers have studied the approximate analytic astigmatism for some UMS[8, 9, 11-13]. Assuming that all the reflections lay in the same plane, astigmatism of a Bernstein-Herzberg White cell (BHWC)[14] was approximated by scalar summation with a large deviation, as reported by Reesor[11, 12]. In addition, the analytic astigmatism of symmetrical

---

* Copartner: Kexin Chen, Huaidong Yang, Liqun Sun. Reprinted from *Applied Optics*, 2010, 49(12):2277-2287.

BHWC derived by Kohn's method[13] is not accurate enough; the method based on analytic coordinate equations[8,9] can only compensate astigmatism partly for symmetrical BHWC and MMS, but it cannot calculate astigmatism. Furthermore, these two methods are limited to symmetric systems, and the accuracy must be improved for astigmatism correction.

In this paper, we propose a new way to obtain an approximate analytic astigmatism with better accuracy for any UMS. First, the model for astigmatism in our previous generalized method[1] is simplified based on features of UMS. Second, a recursive proof is implemented to show that astigmatism of UMS is vector additive. Then, a feasible way to obtain analytic astigmatism with high accuracy is developed. Finally, we apply this method to BHWC as a typical application. The effectiveness of our method is demonstrated by simulations.

2 Principles

2.1 Features of unit-magnification multipass systems

Modeling astigmatism in our previous generalized method[1] is accurate but complicated. In fact, there are several characteristics of UMS that can be used to simplify the modeling and facilitate analytical astigmatism calculation: the entrance aperture is placed close to the curvature centers of the objective mirrors in order to gain a high imaging quality. The size of the mirrors in UMS is usually smaller than one-fifth of the curvature radii R due to limitations of volume and cost. Therefore, four conditions can be developed from these features, as follows.

Ⅰ. The image of entrance aperture by reflection on the objective mirror is the object for the next reflection on the field mirror. The object for the field mirror is very close to the surface. The aberration theory of optical systems shows[15] that it is aberration free when the object is at the surface (the object and image distance for the field mirror is about zero, while that for the objective mirror is about R). Astigmatism contributed from the field mirrors is much smaller than that from the objective mirrors. Therefore, astigmatism of UMS mainly results from the reflections on the objective mirrors;

Ⅱ. The distance of the free propagations between the field mirrors and the objective mirrors is approximately R.

Ⅲ. Both the meridional and the sagittal images of the entrance aperture are near the surface of field mirrors, and the astigmatism is much smaller than R.

Ⅳ. All the images of the entrance aperture are near the center of curvature of the objective mirrors. The off-axis incident angles for reflections on the objective mirrors are small, usually smaller than 5°.

The four conditions are reasonable for most real UMS. Conditions Ⅰ and Ⅱ contribute to the simplification of modeling for astigmatism of UMS, as shown in Subsection 2.2, and conditions Ⅲ and Ⅳ contribute to analytic astigmatism derivation, as shown in Subsection 2.3.

2.2 Simplification of modeling for astigmatism of unit-magnification multipass systems

Based on condition Ⅰ in Subsection 2.1, reflection on the objective mirrors can be transformed to a transmission through GATL, while reflection(s) on the field mirrors between two reflections on the objective mirrors can be transformed to transmission through an anastigmatic lens with a focal length of $R/2$. As shown in Fig. 1, the model for an n-pass UMS comprises coaxial $n/2$ GATLs, which are separated by anastigmatic thin lenses with an equal interval R based on condition II. The parameters of the kth ($k = 1, \cdots, n/2$) GATL are $(\theta_k; \alpha_k)$, as shown in Fig. 2. θ_k is the off-axis incident angle for the corresponding kth reflection on the objective mirrors. The meridional and sagittal focal lengths of the kth GATL are $(f_m)_k = R\cos\theta_k/2$ and $(f_s)_k = R/(2\cos\theta_k)$, respectively. α_k is the rotation angle of this GATL about the optical axis z as

$$\alpha_k = \alpha_1 + \sum_{j=1}^{l-1} \varepsilon_j \tag{1}$$

where ε_j is the dihedral angle between the incident planes of the jth and $(j+1)$th reflections. $j = 1, \cdots, l-1$, where l is the total reflection from the entrance aperture to the kth reflection on the objective mirrors, and $l \geqslant 2k - 1$. For UMS made from all spherical mirrors, e.g., BHWC and MMS, $l = 2k - 1$.

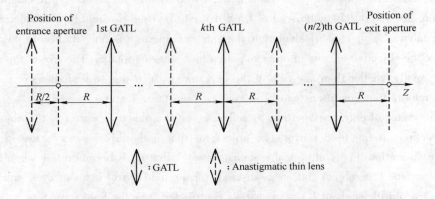

Fig. 1 Simplified model for astigmatism of UMS. The kth reflection on the objective mirrors is equivalent to the kth GATL

Denote the 4×4 ray transfer matrix $P(z)$ for free propagation with a distance z as

$$P(z) = \begin{bmatrix} \mathbf{I} & z\mathbf{I} \\ \mathbf{0} & \mathbf{I} \end{bmatrix} \tag{2}$$

and $\Lambda_g(\theta, \alpha)$ for GATL with the parameters (θ, α) as

$$\Lambda_g(\theta,\alpha) = \begin{bmatrix} \mathbf{I} & \mathbf{0} \\ L_g(\theta,\alpha) & \mathbf{I} \end{bmatrix} = \begin{bmatrix} S(-\alpha) & \mathbf{0} \\ \mathbf{0} & S(-\alpha) \end{bmatrix}\begin{bmatrix} \mathbf{I} & \mathbf{0} \\ L_g\theta & \mathbf{I} \end{bmatrix}\begin{bmatrix} S(\alpha) & \mathbf{0} \\ \mathbf{0} & S(\alpha) \end{bmatrix}$$

$$S(\alpha) = \begin{bmatrix} \cos\alpha & \sin\alpha \\ -\sin\alpha & \cos\alpha \end{bmatrix} \tag{3}$$

$$L_g(\theta,\alpha) = S(-\alpha)L(\theta)S(\alpha)$$

$$L(\theta) = \begin{bmatrix} -f_s^{-1} & 0 \\ 0 & -f_m^{-1} \end{bmatrix} = \begin{bmatrix} -2\cos\theta/R & 0 \\ 0 & -2/R/\cos\theta \end{bmatrix}$$

where $\mathbf{I}$ is the 2×2 unit matrix and $\mathbf{0}$ is the 2×2 zero matrix.

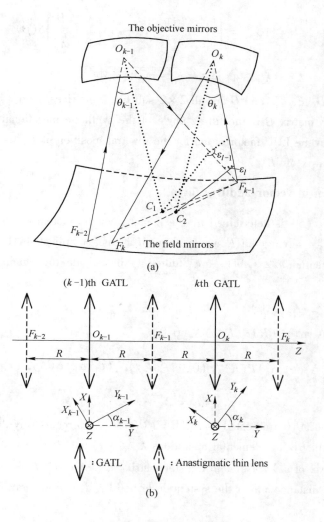

Fig. 2　Parameters for GATL
(a) reflections in UMS; (b) corresponding part of model for astigmatism of UMS

$X\text{—}Y\text{—}Z$, global coordinate in the model whose optical axis is Z;

$X_k\text{—}Y_k\text{—}Z$, local coordinate of kth GATL whose meridional plane is $Y_k\text{—}Z$

Obviously, the ray transfer matrix of the anastigmatic thin lens is $\Lambda_g(0, \alpha)$, where α can be arbitrary. For simplicity, it is denoted as $\Lambda_g(0, 0)$. Then, the general transfer matrix for the model would be

$$M = P(R) \left\{ \prod_{k=2}^{n/2} \Lambda_g(\theta_k, \alpha_k) P(R) \Lambda_g(0,0) P(R) \right\} \Lambda_g(\theta_1, \alpha_1) P(R) P(R/2) \Lambda_g(0,0) \quad (4)$$

Astigmatism can be obtained by determining the differences in the eigenvalues of matrix E in the LDU factorization for the row transposition of M, as in Ref. [1]:

$$M = \begin{bmatrix} A & B \\ C & D \end{bmatrix} = \begin{bmatrix} \mathbf{I} & E \\ \mathbf{0} & \mathbf{I} \end{bmatrix} = \begin{bmatrix} \mathbf{0} & F \\ F^* & \mathbf{0} \end{bmatrix} = \begin{bmatrix} \mathbf{I} & G \\ \mathbf{0} & \mathbf{I} \end{bmatrix}$$

$$E = AC^{-1} = \begin{bmatrix} e_{11} & e_{12} \\ e_{12} & e_{22} \end{bmatrix} = S(-\beta) \begin{bmatrix} \lambda_s & 0 \\ 0 & \lambda_m \end{bmatrix} S(\beta)$$

$$l_{sm} = \lambda_m - \lambda_s = \sqrt{(e_{11} - e_{22})^2 + 4e_{12}^2} \tag{5}$$

where A, B, C, D, E, F, and G are all 2×2 sub block matrices, and λ_s and λ_m are two eigenvalues of the E matrix. β is the tilting angle of the synthetic meridional plane in the model. The matrix E in the LDU factorization for the row transposition of M, as shown in Eq. (5), is called the E matrix of M.

2.3 Astigmatism in vector-additive form

Astigmatism in Eq. (5) is determined by the E matrix of the system transfer matrix M; nevertheless, analytically solving the E matrix from M is too complicated. Therefore, E matrices changing by sequential $n/2$ GATLs are studied to investigate the characteristic of astigmatism. Indicate

$$\begin{aligned} M_0 &= P(R/2)\Lambda_g(0,0) \\ M_1 &= \Lambda_g(\theta_1,\alpha_1)P(R)P(R/2)\Lambda_g(0,0) \\ M_k &= \left\{ \prod_{k=2}^{n/2} \Lambda_g(\theta_k,\alpha_k)P(R)\Lambda_g(0,0)P(R) \right\} \Lambda_g(\theta_1,\alpha_1)P(R)P(R/2)\Lambda_g(0,0) \end{aligned} \tag{6}$$

$$(K = 2,\cdots,n/2)$$

and the E matrix of M_k is indicated as E_k. $M = P(R)M_{n/2}$. To recursively derive E_k from E_{k-1}, changes of the E matrix by elemental operations $\lambda_g(\theta, \alpha)$ and $P(z)$ are studied as follows.

(1) The E matrix of a 4×4 system transfer matrix $M = [A\ B;C\ D]$ is $E = AC^{-1}$. If the ray propagates with a distance z after the system described by M, the general transfer matrix would be

$$M' = P(z)M = \begin{bmatrix} A + zC & B + zD \\ C & D \end{bmatrix} \tag{7}$$

The E matrix of M' would be

$$E' = (A + zC)C^{-1} = E + z\mathbf{I} \tag{8}$$

(2) If the ray passes through the GATL described by $\Lambda_g(\theta, \alpha)$ after the system of M, the general transfer matrix would be

$$M' = \Lambda_g M = \begin{bmatrix} \mathbf{I} & \mathbf{0} \\ L_g & \mathbf{I} \end{bmatrix} \begin{bmatrix} A & B \\ C & D \end{bmatrix} = \begin{bmatrix} A & B \\ L_g A + C & L_g B + D \end{bmatrix} \tag{9}$$

Thus, the E matrix of M' could be denoted as

$$E' = A(L_g A + C)^{-1} = (E^{-1} + L_g)^{-1} \tag{10}$$

The matrices Λ_g and L_g follow the definition in Eq. (3).

It is clear that the E matrix of M_0 is $E_0 = \mathbf{0}$. Suppose E_{k-1} ($k = 1, \cdots, n/2$) is known and can be orthogonally decomposed as follows:

$$E_{k-1} = S(-\beta_{k-1}) \begin{bmatrix} (\lambda_s)_{k-1} & 0 \\ 0 & (\lambda_m)_{k-1} \end{bmatrix} S(\beta_{k-1}) \qquad (11)$$

where $(\lambda_s)_{k-1}$ and $(\lambda_m)_{k-1}$ are two eigenvalues of E_{k-1}. E_k can be obtained from Eq. (6) based on Eqs. (8) and (10):

$$E_k = [(\{[E_{k-1} + R \cdot \mathbf{I}]^{-1} + L_g(0,0)\}^{-1} + R \cdot \mathbf{I})^{-1} + L_g(\theta_k, \alpha_k)]^{-1} \qquad (12)$$

From Eq. (12), the inverse matrix of E_k is denoted as

$$E_k^{-1} = S(-\beta_{k-1}) \begin{bmatrix} (e_{11})'_k & (e_{12})'_k \\ (e_{12})'_k & (e_{22})'_k \end{bmatrix} S(\beta_{k-1})$$

$$(e_{11})'_k = s_k \cos^2\gamma_k + m_k \sin^2\gamma_k + \{[-2R^{-1} + (R + (\lambda_s)_{k-1})^{-1}]^{-1} + R\}^{-1} \qquad (13)$$
$$(e_{12})'_k = (s_k - m_k)\sin\gamma_k \cos\gamma_k$$
$$(e_{22})'_k = s_k \sin^2\gamma_k m_k \cos^2\gamma_k + \{[-2R^{-1} + (R + (\lambda_m)_{k-1})^{-1}]^{-1} + R\}^{-1}$$

where $\gamma_k = \alpha_k - \beta_{k-1}$, $s_k = -(R/2/\cos\theta_k)^{-1}$, and $m_k = -(R\cos\theta_k/2)^{-1}$. Based on condition III as mentioned in Subsection 2.1,

$$(\lambda_{s(m)})_k \approx -R \qquad (14)$$

where $(\lambda_s)_k$ and $(\lambda_m)_k$ are two eigenvalues of E_k. From Eq. (14), reasonable approximations can be obtained as

$$|R + (\lambda_{s(m)})_{k-1}| \ll R \qquad (15)$$
$$(\lambda_s)_k (\lambda_m)_k \approx R^2 \qquad (16)$$

We can derive from Eq. (15) by Taylor expansion that

$$[2R + (\lambda_{s(m)})_{k-1}]^{-1} = [R + (R + (\lambda_{s(m)})_{k-1})]^{-1}$$
$$\approx R^{-1}[1 - (R + (\lambda_{s(m)})_{k-1})R^{-1}] \qquad (17)$$
$$\approx -(\lambda_{s(m)})_{k-1}/R^2$$

From Eq. (15), $2R^{-1} \ll |R + (\lambda_{s(m)})_{k-1}|^{-1}$, then we have

$$[-2R^{-1} + (R + (\lambda_{s(m)})_{k-1})^{-1}]^{-1} \approx R + (\lambda_{s(m)})_{k-1} \qquad (18)$$

Inserting Eq. (18) into Eq. (13), and then inserting Eq. (17) into Eq. (13), Eq. (13) would be

$$(e_{11})'_k = s_k \cos^2\gamma_k + m_k \sin^2\gamma_k - (\lambda_s)_{k-1}/R^2$$
$$(e_{12})'_k = (s_k - m_k)\sin\gamma_k \cos\gamma_k \qquad (19)$$
$$(e_{22})'_k = s_k \sin^2\gamma_k + m_k \cos^2\gamma_k - (\lambda_m)_{k-1}/R^2$$

The eigenvalues of $[(e_{11})'_k \ (e_{12})'_k; (e_{12})'_k \ (e_{22})'_k]$ are $(\lambda_{s(m)})_k^{-1}$, thus

$$\begin{bmatrix} (e_{11})'_k & (e_{12})'_k \\ (e_{12})'_k & (e_{22})'_k \end{bmatrix}^{-1} = (\lambda_s)_k (\lambda_m)_k \begin{bmatrix} (e_{22})'_k & -(e_{12})'_k \\ -(e_{12})'_k & (e_{11})'_k \end{bmatrix} \qquad (20)$$

Inserting Eqs. (16) and (19) into Eq. (20), E_k can be obtained from Eq. (13) as

$$E_k = S(-\beta_{k-1})[(e_{11})'_k \quad (e_{12})'_k; (e_{12})'_k \quad (e_{22})'_k]S(\beta_{k-1})$$
$$= S(-\beta_{k-1})[(e_{11})_k \quad (e_{12})_k; (e_{12})_k \quad (e_{22})_k]S(\beta_{k-1})$$

with

$$(e_{11})_k = (\lambda_s)_k (\lambda_m)_k (e_{22})'_k = R^2(s_k \sin^2\gamma_k + m_k \cos^2\gamma_k) - (\lambda_m)_{k-1}$$
$$(e_{22})_k = (\lambda_s)_k (\lambda_m)_k (e_{11})'_k = R^2(s_k \cos^2\gamma_k + m_k \sin^2\gamma_k) - (\lambda_s)_{k-1} \quad (21)$$
$$(e_{12})_k = -(\lambda_s)_k (\lambda_m)_k (e_{12})'_k = -R^2[(s_k - m_k)\sin\gamma_k + \cos\gamma_k]$$

Substitute Eq. (21) into Eq. (5), and the difference between $(\lambda_s)_k$ and $(\lambda_m)_k$ is

$$\sigma_k = (\rho_k^2 + \sigma_{k-1}^2 + 2\rho_k \sigma_{k-1} \cos 2\gamma_k)^{\frac{1}{2}} \quad (22)$$

where $\sigma_k = (\lambda_m)_k - (\lambda_s)_k$ and $\rho_k = R^2(s_k - m_k)$. β_k would be determined by

$$2(\beta_k - \beta_{k-1}) = \begin{cases} \arctan\left(\dfrac{\rho_k \sin(2\gamma_k)}{\rho_k \cos(2\gamma_k) + \sigma_{k-1}}\right) + \pi, & \text{for } \rho_k \cos(2\gamma_k) + \sigma_{k-1} \leq 0 \\ \arctan\left(\dfrac{\rho_k \sin(2\gamma_k)}{\rho_k \cos(2\gamma_k) + \sigma_{k-1}}\right), & \text{for } \rho_k \cos(2\gamma_k) + \sigma_{k-1} > 0 \end{cases} \quad (23)$$

Astigmatism is equal to $\sigma_n/2$, because the last premultiplied $P(R)$ in M does not impact the astigmatism. Based on condition IV in Subsection 2.1, with the approximations $\cos\theta_k \approx 1$, we obtain

$$\rho_k = 2R\sin^2\theta_k \quad (24)$$

Eqs. (22) and (23) represent a form of vector addition as shown in Fig. 3:

$$\boldsymbol{\sigma}_k = \boldsymbol{\sigma}_{k-1} + \boldsymbol{\rho}_k \quad (25)$$

We specify the vectors $\boldsymbol{\rho}_k = (\rho_k, 2\alpha_k)$ and $\boldsymbol{\sigma}_k = (\sigma_k; 2\beta_k)$ in a polar coordinate system, where ρ_k and σ_k are the vector lengths and the angles $2\alpha_k$ and $2\beta_k$ represent the direction of the vector as to the pole axis of $2\alpha = 0$. The direction of the pole axis is determined by the global coordinate $X—Y—Z$ in the model for astigmatism. Therefore, astigmatism of an n-pass UMS can be expressed in the form of vector addition:

$$l_{sm} = \sigma_{n/2} = \left|\sum_{k=1}^{n/2} \boldsymbol{\rho}_k\right| \quad (26)$$

Eq. (26) reveals that the contributions of each reflection on the objective mirrors to the total astigmatism of UMS are vector additive. The vector addition of system astigmatism can be explained from a different point of view by the vector-addition form of wavefront aberration in asymmetric systems[16-21]. Astigmatic contribution from each reflection on the objective mirrors can be

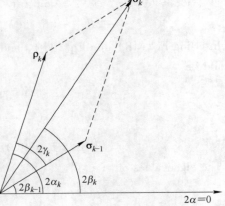

Fig. 3 Vector addition of astigmatism

equivalent to a vector in planar polar coordinates, whose length is $2R \times \sin^2\theta_k$ and whose direction is the rotation angle of the corresponding GATL. For the limiting case that all the reflections in UMS are in the same plane, α_k are equal to zero. Then Eq. (26) would be simplified as $\sum 2R^2 \sin\theta_k$, which is the same as Reesor's theory[11] with $\sin\theta_k \approx \theta_k$. In most UMS designs, all α_k are not zero. Eq. (26) is more general than Reesor's theory. Further comparisons in accuracy are shown in Section 3.

Obviously, astigmatism can be changed by θ_k and α_k. α_k can be changed by adjusting the distance between the two rows of images of the entrance aperture in BHWC (or other simple White-type systems with two-row images), or the size of the lattice of the matrixed images in MMS. Eq. (26) develops a feasible way to obtain analytic astigmatism and cancel it for any UMS by following the steps shown in Fig. 4 (a). First, solve the parameters (θ_k, α_k) for GATLs in our model. Then obtain the analytic astigmatism by Eq. (26). Compared to our previous generalized method for accurate numerical astigmatism, as shown in Fig. 4 (b), our method for approximate analytic astigmatism is more convenient. The accuracy of our method for approximate analytic astigmatism would be verified in the next section.

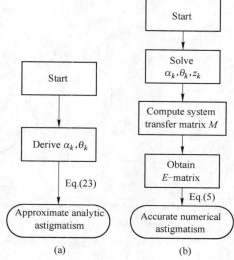

Fig. 4　Overview of astigmatism calculation procedure
(a) our method for approximate analytic astigmatism;
(b) our previous generalized method for accurate numerical astigmatism

3　Application: analytic astigmatism of BHWC

A BHWC is the basic cell of UMS. Other UMS with matrixed images, e.g., MMS and the modified White cells, are recognized as recirculations of the beam into BHWC. In this section, a typical application to BHWC is fulfilled to validate our method.

A BHWC is constituted by three concave spherical mirrors with equal curvature radii R, as shown in Fig. 5. A three-dimensional Cartesian coordinate system X_r—Y_r—Z_r with origin O is built to further describe the BHWC. The subscript r is used to distinguish the coordinate from that in the model for astigmatism. C_F and the center of F are on the Z_r axis. C_1 and C_2 are on the Y_r axis. For the n-pass BHWC, the coordinate of the center of the entrance aperture is $(-h/2, p, 0)$, C_F is $(0,0,R)$, C_1 is $(0,c/2,0)$, and C_2 is $(0,-c/2,0)$, where $c = 4p/n$. The distance between the two rows of images of the entrance aperture is h. The center of the entrance pupil is $O_1(0, \Delta+c/2, (R^2-\Delta^2)^{1/2})$, while the stop is at M_1. The plane of X_rOZ_r is the symmetrical plane for the BHWC and it bisects the images of the entrance aperture.

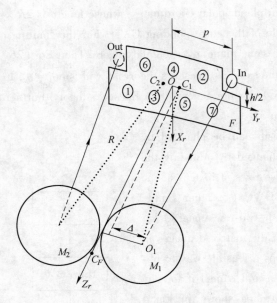

Fig. 5 System layout of the 16-pass BHWC

F—the field mirror with the center of curvature C_F; M_1 and M_2—two spherical objective mirrors with centers of curvature C_1 and C_2, respectively; In—the entrance aperture; Out—the exit aperture; tags 1 to 7—the images of the entrance aperture in sequence

The parameters (θ_k, α_k) are solved as follows. The incident point on the objective mirrors O_k slightly changes for each reflection. As shown in Fig. 6(a), the coordinate of O_k is (δ_k, $-\Delta - c/2$, H) for M_2 and ($-\delta_k$, $\Delta + c/2$, H) for M_1, where $H = (R^2 - \Delta^2 - \delta_k^2)^{1/2}$. As shown in Fig. 6(b), followed by the derivation in Eq. (A1), we have

$$\delta_k = (k-1)\frac{h\Delta^2}{2R^2} \tag{27}$$

For the kth reflection on the objective mirrors (odd k for M_1 and even k for M_2), l_k is the offset of F_k from the center of curvature of the objective mirrors (C_1 for odd k and C_2 for even k) along Y_r, and h_k is that along X_r. Considering the quadratic term for the reflection on the spherical mirror, as shown in Eq. (A2), there is

$$l_k = (l_{k-1} - c)(1 + r_k), \quad h_k = h_{k-1}(1 + r_k) \tag{28}$$

where $r_k = \dfrac{2l_{k0}\Delta}{R^2}$ with $l_{k0} = p - (k - 1/2)c$. Since r_k is very small, with the approximation the $(1 + r_i)(1 + r_j) \approx 1 + r_i + r_j$, Eq. (28) could be rewritten as

$$l_k = l_{k0} + (p - c/2)\sum_{i=1}^{k} r_i - c\sum_{i=1}^{k}(i-1)r_i$$

$$h_k = \frac{h}{2} + \frac{h}{2}\sum_{i=1}^{k} r_i \tag{29}$$

Then l_k and h_k can be expressed in p, c, h, Δ, R, and k only as

$$l_k = p - \left(k - \frac{1}{2}\right)c + \frac{\Delta k}{R^2}\left[2p(p - kc) - \frac{c^2}{6}(k^2 - 4)\right]$$
$$h_k = \frac{h}{2} + \frac{h\Delta k}{2R^2}(2p - kc) \tag{30}$$

As derived in Eq. (A2), the off-axis incident angle θ_k would approximately be

$$\sin\theta_k \approx \tan\theta_k \approx \frac{\sqrt{l_k^2 + h_k^2}}{R^2}\left(1 + \frac{l_k\Delta}{R^2}\right) \tag{31}$$

Therefore, we have

$$\sin^2\theta_k \approx \frac{l_k^2 + h_k^2}{R^2}\left(1 + \frac{2l_k\Delta}{R^2}\right) = \frac{[p - (k - 1/2)c]^2 + h^2/4}{R^2} + \tau R^{-4} + O(R^{-6}) \tag{32}$$

with τ as the coefficient of the R^{-4} term, which is not displayed here because it is too long.

As shown in Fig. 6(a), from Eq. (1), the rotation angle of equivalent kth GATL of BHWC is

$$\alpha_k = \alpha_1 + \sum_{i=1}^{k-1}(\varepsilon_{k-1} + \varepsilon'_{k-1}) \tag{33}$$

where ε_{k-1} is the dihedral angle between the incident planes of the $(k-1)$ reflection on the objective mirrors and the $(k-1)$ reflections on the field mirrors, and ε'_{k-1} is that of $(k-1)$ reflection on the field mirrors and kth reflections on the objective mirrors. As derived in Eq.

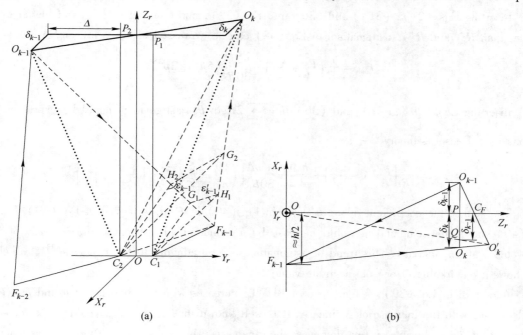

Fig. 6 Reflections in BHWC

(a) $(k-1)$ and kth reflections on the objective mirrors; (b) side view against the Y_r axis

O_k—incident point of the kth reflection on the objective mirrors;

F_k—incident point of the kth reflection on the field mirror

(A3), we can obtain

$$\cos\varepsilon'_{k-1} \approx \frac{l_k}{\sqrt{l_k^2 + h_k^2}}\left(1 + \frac{h_k \delta_k}{l_k \Delta}\right) \tag{34}$$

Similar to Eq. (34), ε_{k-1} would be

$$\cos\varepsilon_{k-1} \approx \frac{l_{k-1}(2\Delta + c) + 2h_{k-1}\delta_{k-1}}{\sqrt{l_{k-1}^2 + h_{k-1}^2}(2\Delta + c)} \approx -\frac{l_{k-1}}{\sqrt{l_{k-1}^2 + h_{k-1}^2}}\left(1 + \frac{h_{k-1}\delta_{k-1}}{l_{k-1}\Delta}\right) \tag{35}$$

Since $\delta_k \approx \delta_{k-1}$, by Eqs. (34) and (35), $\varepsilon'_{k-1} + \varepsilon_k \approx \pi$. It is clear that α_k is in a period of π. If we make the initial rotation angle $\alpha_1 = \pi - \varepsilon_1$, then $\alpha_k = \varepsilon'_{k-1}$. As $\delta_k \ll \Delta$, there could be

$$\cos\alpha_k \approx \frac{l_k}{\sqrt{l_k^2 + h_k^2}}\left(1 + \frac{h_k \delta_k}{l_k \Delta}\right) \tag{36}$$

From Eq. (26), the astigmatism of BHWC should be

$$l_{sm} = \left|\sum_{i=1}^{n/2} \boldsymbol{\rho}_k(2R\sin^2\theta_k, 2\alpha_k)\right| \tag{37}$$

As the component of l_{sm} in the direction of the pole axis of $2\alpha = 0$ is much larger than that of the direction perpendicular to the pole axis, Eq. (37) could be approximated as follows:

$$l_{sm} = \sum_{i=1}^{n/2} 2R\sin^2\theta_k(2\cos^2\alpha_k - 1) \tag{38}$$

Inserting Eqs. (29), (31), and (36) into Eq. (38), and omitting the terms of higher order than R^{-1} and R^{-3}, astigmatism in Eq. (38) can be expressed with p, h, n, and Δ only:

$$l_{sm} = \frac{n}{R}\left(\frac{p^2}{3} - \frac{h^2}{4}\right) - \frac{n^2 p \Delta}{30R^3}(5h^2 + 2p^2) \tag{39}$$

Inserting Eqs. (29), (31), and (36) into $\sum_{i=1}^{n/2} 2R\sin^2\theta_k$, astigmatism could be expressed, based on Reesor's theory[11], as

$$l_{sm} = \frac{n}{R}\left[\frac{p^2}{3}\left(1 - \frac{4}{n^2}\right) + \frac{h^2}{4}\right] - \frac{n^2 p \Delta}{30R^3}\left[\left(\frac{10}{n^2} - \frac{5}{2}\right)h^2 + \left(2 - \frac{32}{n^4}\right)p^2\right] \tag{40}$$

The sign of the coefficient of the h^2 term in Eq. (40) is contrary to that of Eq. (39). Eq. (40) shows that astigmatism would increase by h monotonically, because the increase of h would make θ_k increase. This trend of astigmatism is contradictory to the real case[13], which shows a remarkable error from accurate values.

As shown in Eq. (39), astigmatism of BHWC increases with the increase of p and n, yet decreases with the increase of Δ linearly. It is well known that astigmatism could be corrected by optimizing h, because it first decreases then increases when h increases. Therefore, astigmatism versus h is first focused. As shown in Fig. 7, astigmatism of a 40-pass BHWC with $p = 40$mm, $R = 625$mm, $\Delta = 50$mm, and varying h is presented by our analytical results from Eq. (39), the analytical results from Eq. (40) by Reesor's theory, analytical results by Kohn's method, and accurate numerical results by our previous generalized method proposed in Ref. [1]. Here, the results by Reesor's theory are partly drawn because the calculated values are much larger than the accurate results at $p > 20$mm. And we do not take Reesor's theory into account for all the simulations hereafter because of its poor accuracy.

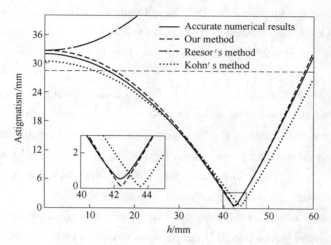

Fig. 7 Astigmatism of a 40-pass BHWC

The astigmatism by Eq. (39) and Kohn's method has negative values, where the synthetic meridional and sagittal planes are reversed as to our previous generalized method. We thus take absolute value for the results by our method and Kohn's method to compare with accurate numerical values. The deviations of our analytical results and Kohn's analytical results are presented in Fig. 8. The accuracy of analytic astigmatism by our method is better than that by Kohn's method.

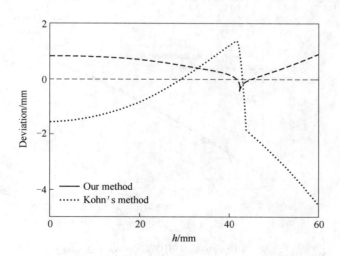

Fig. 8 Deviation of analytical results from accurate astigmatism by two methods

Astigmatism correction is the most important purpose for astigmatism calculation. Although analytic astigmatism by our method does not have the best accuracy at h near about 42.5mm, nearby astigmatism correction is better controlled (local maximum is smaller than 0.5mm) than by Kohn's method. Furthermore, more accurate h for astigmatism correction can be obtained by our method, because the trend of analytic astigmatism versus h by our method more closely coincides with the accurate numerical results. The optimum h for free of astigmatism is 42.43mm by Eq. (39) and 43.55 mm by Kohn's method. As to 42.38mm by accurate numerical results, the relative error of optimum h for astigmatism correction is improved from 3% to 1‰ by our method.

The accuracy of our method is further verified by astigmatism changing versus p, n, and Δ, as shown in Figs. 9 and 10. The errors of our method slowly increase when p increases, because BHWC with a large p would violate the four conditions mentioned in Subsection 2.1. As shown in Fig. 10(b), the relative error of optimum h for astigmatism correction by our method increases to 4.8‰ when $2p$ increases to $0.32R$, smaller than the 5.4% by Kohn's method. Our method maintains enough accuracy for astigmatism correction.

Analytic astigmatism based on the vector-additive form of astigmatism in UMS depends only on the four conditions in Subsection 2.1 and does not depend on whether the system is symmet-

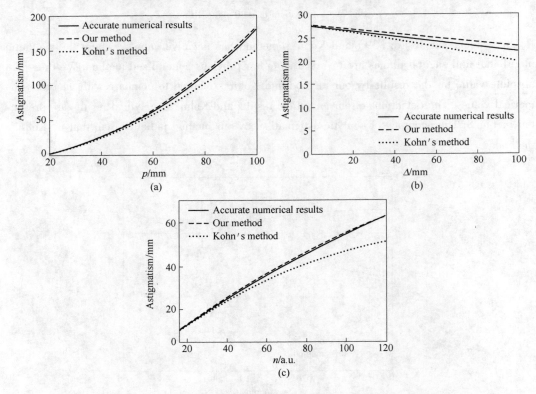

Fig. 9   Astigmatism of BHWC versus p, Δ, and n

(a) astigmatism varying with p in a 40-pass PBWC with $R = 625$mm, $h = 20$mm, and $\Delta = 50$mm;

(b) astigmatism varying with Δ in a 40-pass PBWC with $R = 625$mm, $p = 40$mm, and $h = 20$mm;

(c) astigmatism in the varying n-pass PBWC with $R = 625$mm, $p = 40$mm, $h = 20$mm, and $\Delta = 50$mm

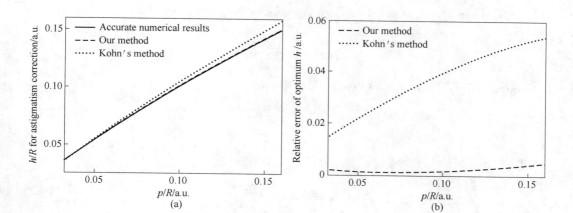

Fig. 10 Optimizing h for astigmatism correction versus p in a 40-pass BHWC with $\Delta = 0.08R$
(a) optimum h for astigmatism correction by two methods; (b) relative error of optimum h for astigmatism correction by two methods

ric (e.g., plane symmetry in BHWC applied by Kohn[13]). Analytic astigmatism can be estimated for any UMS by our method.

4 Conclusion

We proposed an effective way to derive approximate analytic astigmatism for any type of UMS. Based on four conditions developed from the characteristics of UMS, modeling for astigmatism is practically simplified. Then, based on a recursive proof, astigmatism is derived as a distinct form of vector addition. Astigmatism mainly results from the astigmatic reflections on the objective mirrors, each of which could be equivalent to a vector in planar polar coordinates. The vector length is the square of the off-axis incident angle times $2R$, and the vector direction is the rotation angle of the corresponding GATL. Analytic astigmatism for BHWC is obtained as an example for typical applications. The effectiveness of our method for analytic astigmatism is verified by simulation results. When the distance between the entrance and the exit apertures is smaller than $0.32R$, the relative error of optimizing h for astigmatism correction by our method is smaller than 5‰, which is only one tenth of that by Kohn's method. Our method improves the efficiency of astigmatism correction and further benefits the optical design of UMS.

Appendix A: Derivation of several equations

1. Equation (27)

As shown in Fig. 6(b), all lines are the projection of lines in Fig. 6(a) on plane $X_r O Y_r$. Let the ray $F_{k-1} O_k$ meet $O_{k-1} C_F$ at O'_k, and line OO'_k meet $O_{k-1} O_k$ at Q. For the reflection on the sphere centered at C_F, $PO_{k-1} \approx PQ = \delta_{k-1}$, $QO_k \approx \delta_k - \delta_{k-1}$, and $PC_F = R - (R^2 - \Delta^2)^{1/2} \approx \Delta^2/2/R$. The length of the perpendicular from O'_k to $O_{k-1} O_k$ is about $2PC_F$, and that from O'_k to OF_{k-1}

is $R + PC_F$. Line $O_{k-1}O_k$ is parallel to axis X_r, thus we have the proportion

$$\frac{QO_k}{OF_{k-1}} \approx \frac{2PC_F}{R + PC_F} \tag{A1}$$

Ignoring the higher-order term than R^{-2}, QQ_k would be

$$\delta_k - \delta_{k-1} \approx \frac{h\Delta^2}{2R^2} \tag{A2}$$

Obviously, $\delta_1 = 0$, therefore

$$\delta_k = (k-1)\frac{h\Delta^2}{2R^2} \tag{A3}$$

2. Equations (28) and (31)

If we do not consider the quadratic term for the reflection on the spherical mirror, there is $C_2F_{k-2} = C_2F_{k-1}$ and $C_1F_{k-1} = C_1F_k$:

$$l_k = l_{k-1} - c, \quad h_k = h_{k-1} = h/2 \tag{A4}$$

with $l_1 = p - c/2$.

As shown in Fig. 11, the perpendicular from F_k meets axis Y_r at A_k, and that from F_{k+1} at A_{k+1}. $F_kA_k = h_k$, $A_kC_1 = l_k$, $A_kC_2 = l_k - c$, $A_{k+1}C_2 = l_{k+1}$, and $F_{k+1}A_{k+1} = h_{k+1}$. The perpendicular from F_k meets C_2O_{k+1} at B_2, and F_kB_2 meets ray $O_{k+1}F_{k+1}$ at B. BB_3 is parallel to C_2O_{k+1}.

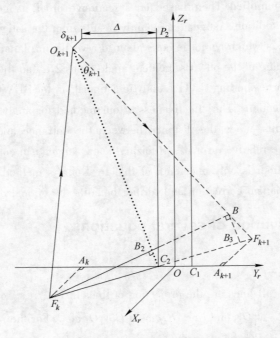

Fig. 11 Reflection on the spherical mirror M_2

(The symbols follow the convention in Fig. 6)

In triangle $C_2F_kO_{k+1}$, by the cosine theorem, we have

$$\cos\theta_{k+1} = [R^2 + (F_kO_{k+1})^2 - (C_2F_k)^2]/(2R \cdot F_kO_{k+1}) \quad (A5)$$

Because

$$C_2F_k = [(l_k - c)^2 + h_k^2]^{1/2}$$
$$F_kO_{k+1} = [(h_k - \delta_{k+1})^2 + (l_k - c - \Delta)^2 + (R^2 - \Delta^2 - \delta_{k+1}^2)]^{1/2} \quad (A6)$$

we have

$$B_2C_2 = R - F_kO_{k+1}\cos\theta_{k+1} = [(l_k - c)\Delta + h_k\delta_{k+1}]/R \approx (l_k - c)\Delta/R \quad (A7)$$

Since $B_2F_k = B_2B, C_2F_k = C_2B_3, BB_3 = 2C_2B_2$, and the proportion

$$\frac{BB_3}{C_2O_{k+1} - BB_3} = \frac{B_3F_{k+1}}{C_2B_3} \quad (A8)$$

there is

$$B_3F_{k+1} \approx \frac{2B_2C_2 \cdot F_kC_2}{C_2O_{k+1}} = \frac{2(l_k - c)\Delta}{R^2}F_kC_2 \quad (A9)$$

Thus we have

$$h_{k+1} = \left[1 + \frac{2(l_k - c)\Delta}{R^2}\right]h_k$$
$$l_{k+1} = \left[1 + \frac{2(l_k - c)\Delta}{R^2}\right](l_k - c) \quad (A10)$$

Based on condition Ⅳ in Subsection 2.1, $l_k \ll R$ and $h_k \ll R$ are satisfied. We denote that $l_{k0} = p - (k - 1/2)c$, then the difference between $2(l_k - c)\Delta/R^2$ and $2(l_{k0} - c)\Delta/R^2$ is the high-order term of $O(R^{-4})$. Thus Eq. (A10) can be approximated as

$$h_{k+1} = h_k(1 + r_{k+1}), \quad l_{k+1} = (l_k - c)(1 + r_{k+1}) \quad (A11)$$

where

$$r_{k+1} = \frac{2(l_{k0} - c)\Delta}{R^2} = \frac{2(l_{(k+1)0} - c)\Delta}{R^2}$$

with $l_{k0} = p - (k - 1/2)c$. As shown in Fig. 11,

$$\tan\theta_{k+1} = \frac{B_2F_k}{R - B_2C_2} \approx \frac{C_2F_{k+1} - B_3F_{k+1}}{R(1 - B_2C_2/R)} \quad (A12)$$

From Eq. (A9), B_3F_{k+1} is much smaller than C_2F_{k+1}, thus we have

$$\tan\theta_{k+1} \approx \frac{C_2F_{k+1}}{R}\left(1 + \frac{B_2C_2}{R}\right) = \frac{\sqrt{l_{k+1}^2 + h_{k+1}^2}}{R}\left[1 + \frac{\Delta l_{k+1}}{R^2}(1 + r_{k+1})^{-1}\right] \quad (A13)$$

The term including r_{k+1} is a high-order term of $O(R^{-5})$, thus ignore r_{k+1} and Eq. (A13) would be

$$\tan\theta_{k+1} \approx \frac{\sqrt{l_{k+1}^2 + h_{k+1}^2}}{R}\left(1 + \frac{l_{k+1}\Delta}{R^2}\right) \quad (A14)$$

3. Equation (34)

As shown in Fig. 6(a), in triangle $C_1F_{k-1}O_k$, by the cosine theorem we have

$$\cos(\angle C_1F_{k-1}H_1) = \frac{(C_1F_{k-1})^2 + (F_{k-1}O_k)^2 - (C_1O_k)^2}{2C_1F_{k-1} \cdot F_{k-1}O_k} \tag{A15}$$

Then C_1H_1 and $F_{k-1}H_1$ can be obtained as

$$\begin{aligned} C_1H_1 &= C_1F_{k-1}\sin(\angle C_1F_{k-1}H_1) \\ F_{k-1}H_1 &= C_1F_{k-1}\cos(\angle C_1F_{k-1}H_1) \end{aligned} \tag{A16}$$

Similarly, in triangle $O_{k-1}F_{k-1}O_k$ and $C_1F_{k-1}O_{k-1}$, we have

$$\begin{aligned} \cos(\angle G_1F_{k-1}H_1) &= \frac{(O_{k-1}F_{k-1})^2 + (F_{k-1}O_k)^2 - (C_1O_k)^2}{2C_1F_{k-1} \cdot F_{k-1}O_k} \\ \cos(\angle C_1F_{k-1}G_1) &= \frac{(C_1F_{k-1})^2 + (F_{k-1}O_{k-1})^2 - (C_1O_{k-1})^2}{2C_1F_{k-1} \cdot F_{k-1}O_{k-1}} \end{aligned} \tag{A17}$$

Thus,

$$\begin{aligned} F_{k-1}G_1 &= F_{k-1}H_1/\cos(\angle G_1F_{k-1}H_1) \\ C_1H_1 &= F_{k-1}H_1/\tan(\angle G_1F_{k-1}H_1) \end{aligned} \tag{A18}$$

In triangle $C_1G_1F_{k-1}$, we have

$$(C_1G_1)^2 = (C_1F_{k-1})^2 + (F_{k-1}G_1)^2 - 2C_1F_{k-1} \cdot F_{k-1}G_1 \cdot \cos(\angle C_1F_{k-1}G_1) \tag{A19}$$

and in triangle $C_1G_1H_1$,

$$\cos\varepsilon'_{k-1} = \frac{(C_1H_1)^2 + (G_1H_1)^2 - (C_1G_1)^2}{2C_1H_1 \cdot G_1H_1} \tag{A20}$$

There is a series of known quantities

$$\begin{aligned} C_1P_1 &= (R^2 - \Delta^2 - \delta_k^2)^{1/2}, \quad C_1O_k = R \\ C_1F_{k-1} &= (l_k^2 + h_k^2)^{1/2}/(1 + r_k) \\ F_{k-1}O_k &= [(C_1P_1)^2 + (h_k/(1 + r_k) - \delta_k)^2 + (l_k/(1 + r_k) - \Delta)^2]^{1/2} \\ O_{k-1}O_k &= [(2\Delta + c)^2 + (\delta_k + \delta_{k-1})^2]^{1/2} \\ F_{k-1}O_{k-1} &= [(\Delta + c + l_k/(1 + r_k))^2 + (C_1P_1)^2 + (h_k/(1 + r_k) - \delta_{k-1})^2]^{1/2} \\ C_1O_{k-1} &= [(C_1P_1)^2 + (\Delta + c)^2 + \delta_{k-1}^2]^{1/2} \end{aligned} \tag{A21}$$

Inserting Eq. (A21) into Eqs. (A15)-(A20), and ignoring the influence of r_k and the difference between δ_k and δ_{k-1}, we simplify Eq. (A20) as

$$\cos\varepsilon'_{k-1} = \frac{[(2\Delta + c)l_k + 2h_k\delta_k]R^2 + \xi_1 R^0}{\{[(2\Delta + c)^2 + \delta_k^2]R^2 + \xi_2 R^0\}^{1/2} \cdot [(l_k^2 + h_k^2)R^2 + \xi_3 R^0]^{1/2}} \tag{A22}$$

The terms $\xi_n R^0$ ($n = 1, 2, 3$) are much smaller than that of R^2 in Eq. (A22) and thus can be omitted. Equation (A22) would be further simplified by dropping the term δ_k^2 as

$$\cos\varepsilon'_{k-1} = \frac{l_k}{\sqrt{l_k^2 + h_k^2}}\left[1 + \frac{2h_k\delta_k}{(2\Delta + c)l_k}\right]$$

$$\approx \frac{l_k}{\sqrt{l_k^2 + h_k^2}}\left(1 + \frac{h_k\delta_k}{l_k\Delta}\right) \quad (A23)$$

The derivation of Eq. (35) is similar to Eq. (34); thus, it is not detailed here.

Acknowledgements

This work is supported by the Ph. D. Programs Foundation of the Ministry of Education of China, 20090002120010.

References

[1] K. Chen, H. Yang, L. Sun, and G. Jin, "Generalized method for calculating astigmatism of unit-magnification multipass system", Appl. Opt. 49, 1964-1971 (2010).

[2] R. P. Blickensderfer, G. E. Ewing, and R. Leonard, "A long path, low temperature cell", Appl. Opt. 7, 2214-2217 (1968).

[3] H. M. Pickett, G. M. Bradley, and H. L. Strauss, "A new White type multiple pass absorption cell", Appl. Opt. 9, 2397-2398 (1970).

[4] D. Horn and G. C. Pimentel, "2.5 Km low-temperature multiple-reflection cell", Appl. Opt. 10, 1892-1898 (1971).

[5] J. F. O. Doussin, R. Dominique, and C. Patrick, "Multiple-pass cell for very-long-path infrared spectrometry", Appl. Opt. 38, 4145-4150 (1999).

[6] S. M. Chernin, "Promising version of the three-objective multipass matrix system", Opt. Express 10, 104-107 (2002).

[7] D. R. Glowacki, A. Goddard, and P. W. Seakins, "Design and performance of a throughput-matched, zero-geometric-loss, modified three objective multipass matrix system for FTIR spectrometry", Appl. Opt. 46, 7872-7883 (2007).

[8] S. M. Chernin, S. B. Mikhailov, and E. G. Barskaya, "Aberrations of a multipass matrix system", Appl. Opt. 31, 765-769 (1992).

[9] Y. G. Barskaya, "Aberrations of a multipass cell", Opt. Technol. 38, 278-280 (1971).

[10] C. Kexin, Y. Huaidong, S. Liqun, and J. Guofan, "Astigmatism analysis by matrix methods in White cells", Proc. SPIE 7156, 71560G (2008).

[11] T. R. Reesor, "The astigmatism of a multiple path absorption cell", J. Opt. Soc. Am. 41, 1059-1060 (1951).

[12] T. H. Edwards, "Multiple-traverse absorption cell design", J. Opt. Soc. Am. 51, 98-102 (1961).

[13] W. H. Kohn, "Astigmatism and White cells: theoretical considerations on the construction of an anastigmatic White cell", Appl. Opt. 31, 6757-6764 (1992).

[14] H. J. Bernstein and G. Herzberg, "Rotation-vibration spectra of diatomic and simple polyatomic molecules with long absorbing paths", J. Chem. Phys. 16, 30-39 (1948).

[15] W. T. Welford, *Aberrations of Optical Systems* (Adam Hilger, 1986), pp. 158-161.

[16] R. V. Shack and K. Thompson, "Influence of alignment errors of a telescope system on its aberration field", Proc. SPIE 251, 146-153 (1980).

[17] K. P. Thompson, "Practical methods for the optical design of systems without symmetry", Proc. SPIE 2774, 2-12 (1996).

[18] J. R. Rogers, "Design techniques for systems containing tilted components", Proc. SPIE 3737, 286-300 (1999).

[19] J. R. Rogers, "Techniques and tools for obtaining symmetrical performance from tilted-component systems", Opt. Eng. 39, 1776-1787 (2000).

[20] K. Thompson, "Description of the third-order optical aberrations of near-circular pupil optical systems without symmetry", J. Opt. Soc. Am. A 22, 1389-1401 (2005).

[21] L. B. Moore, A. M. Hvisc, and J. Sasian, "Aberration fields of a combination of plane symmetric systems", Opt. Express 16, 15655-15670 (2008).

CGH Null Test of a Freeform Optical Surface with Rectangular Aperture*

Abstract In null computed generated hologram (CGH) test of optical elements, fitting method is needed in null CGH design to generate continuous phase function from the ray-traced discrete phase data. The null CGH for freeform testing usually has a deformed aperture and a high order phase function, because of the aberrations introduced by freeform wavefront propagation. With traditional Zernike polynomial fitting method, selection of an orthogonal basis set and choosing number of terms are needed before fitting. Zernike polynomial fitting method is not suitable in null CGH design for freeform testing. In this paper, a novel CGH design method with cubic B-spline interpolation is developed. For a freeform surface with 18mm × 18mm rectangular aperture and 630μm peak-to-valley undulation, the null CGH with a curved rectangular aperture is designed by using the method proposed. Simulation and experimental results proved the feasibility of the novel CGH design method.

Key words null test; freeform; computed generated hologram; B-spline

1 Introduction

The superiorities of optical freeform elements[1-6] in illuminating and imaging systems are attracting increasingly interest. With the development of design tools and fabrication methods, for optical freeform elements, the wide application is restricted by ultra-precision surface test technology. Null test with computer generated hologram (CGH) has been used in test of aspheric surfaces since 1970s[7], and is predicted to test freeform surfaces. Generally speaking, the design procedure of null CGH includes, ray tracing to obtain the discrete phase data[7], fitting the continuous phase function[8], and finally, calculation of fringe position[9]. However, the general study of the null test with CGH of optical freeform surface elements is still not sufficient. Freeform wavefront introduces aberrations while propagation[10]. Such aberrations cause aperture deformation and introduce even higher order terms into the wavefront. The null CGH is used to compensate the aberrations; therefore the null CGH commonly has deformed aperture and much higher order phase function. Design of null CGH with deformed aperture and high order phase function becomes an important issue in the general study of null test of freeform surfaces.

The traditional null CGH design method uses Zernike polynomials to fit the discrete phase data of the CGH obtained by ray tracing with the optical test configuration, and then CGH fringes positions can be solved for the fabrication of null CGH. During freeform testing, the null CGH commonly has a deformed aperture and high order phase function. When using the Zerni-

---

* Copartner: Ping Su, Jianshe Ma, Qiaofeng Tan, Guoguo Kang, Yi Liu. Reprinted from *Optical Engineering*, 2012, 51(2):025801.

ke polynomial fitting method, before fitting implementation, the selection of an orthogonal basis set for a deformed aperture and the number of used terms is needed to determine proper basis functions[11-13]. For traditional apertures such as circle, rectangle, or hexagon, the orthogonal basis sets have been reported[11-12]. However, Zernike basis set for deformed aperture, such as a shape of rectangle with four curves edges (defined as curved rectangle) and other irregular shapes has not been investigated yet. On the other hand, for surfaces with high order type, Zernike polynomials with much higher order are needed for precise fitting, and the number of used terms needs investigation for a particular case. For example, during the null test of an 8 order off-axis aspheric mirror, Wang M. used 87 terms 27 order Zernike polynomial to fit the phase function of the null CGH[8]. When higher order terms are used to fit more complex function, successive algorithms are more complicated and unstable. Moreover, a freeform surface is always expressed by parametric functions such as B-spline function[2-5], which makes the analysis of the aberration order very difficult. In a word, Zernike polynomial fitting is no longer suitable for the design of null CGH of freeform surface because such null CGH usually has a deformed aperture and high order phase function.

We propose to use cubic B-spline function, a more universal fitting function with low order to design null CGH for freeform test. B-spline functions are piecewise polynomials with good smoothness and cubic B-spline function can meet the needs of majority occasions. Cubic B-spline function has been the most commonly used function in modern 3D modeling and numerical control machining. Recent work has shown that for a complex wavefront, cubic B-spline with a high number of breakpoints (NBP) value performs much better than Zernike polynomial fitting[14]. Our simulation results show that the method of bicubic B-spline interpolation has high precision, low order, and sequence convergence[15]. Ref. [16] proposed the matrix formula of non-uniform bicubic B-spline function, which has the advantages of simple computation and easy analysis of the geometric properties comparing to traditional de Boor-Cox recursion formula representation[17].

In this paper, a novel CGH design method is developed. The method uses cubic B-spline interpolation method to fit the continuous phase function of null CGH. For a non-axisymmetrical surface with $18mm \times 18mm$ rectangular aperture and $630\mu m$ peak-to-valley undulation, the null CGH with a curved rectangular aperture is designed by using the novel method proposed. We use the matrix formula of bicubic non-uniform B-spline function to express the phase function of the null CGH. According to the monotone property of the fabrication phase function of null CGH, we simplify the precision intersection algorithm of a parameter surface and a plane surface, to obtain the fringes position of the null CGH. The calculation of fringes position only need simple judgment and solution of univariate cubic equations. Simulation and experimental results show the validity of the novel method.

2 Methods

2.1 Bicubic B-spline interpolation

Bicubic B-spline curved surfaces are segmental parametric surfaces, and each segment is determined by 16 control vertices and 2 basis functions of bi-cubic parametric polynomials. To solve a bicubic B-spline interporlation surface means to inversely compute the control vertices of the B-spline surface with the discrete data points. Parameterized method and boundary conditions are needed for inverse calculation, and would affect fitting accuracy which needs further research. We used bi-direction average length method and free end boundary condition[17] in this paper.

The phase data of null CGH obtained by ray tracing program is a data lattice denoted by $\{Q_{i,j}\}_{i=1,j=1}^{M,N} \in \Re^3$, the aperture of which is a curved rectangle. The knot vectors obtained by bidirection average chord length method according to the phase data are assumed to be $\boldsymbol{u} = [u_1, u_2, u_3, \cdots, u_{M-2}, u_{M-1}, u_M]$ and $\boldsymbol{v} = [v_1, v_2, v_3, \cdots, v_{N-2}, v_{N-1}, v_N]$, and the extended knot vectors are $\boldsymbol{U} = [u_1, u_1, \boldsymbol{u}, u_M, u_M]$, $\boldsymbol{V} = [v_1, v_1, \boldsymbol{v}, v_N, v_N]$. There are $M \times N$ B-spline segmental curve surfaces to express the continuous phase function of the null CGH. We denote the segmental surface with the number i, j to be $c^{i,j}(s,t)$, the matrix form of which is shown as Eq. (1).

Adding boundary conditions and according to the interpolation function $c^{i,j}(0,0) = Q_{ij}$ (the expression of $c_{i,j}$ is shown as Eqs. (1) and (2)), the control vertex matrix of the segmental surface with the number i, j can be obtained, and the analytic form of every bicubic B-spline interpolation surface can be obtained by traversing all the values of i and j and solving all the element values of control vertex matrix $\{P_{i,j}\}_{i=1,j=1}^{M+2,N+2}$ [16,17].

$$c^{i,j}(s,t) = \begin{bmatrix} 1 & s & s^2 & s^3 \end{bmatrix} L_U(i) \begin{bmatrix} P_{i,j} & P_{i,j+1} & P_{i,j+2} & P_{i,j+3} \\ P_{i+1,j} & P_{i+1,j+1} & P_{i+1,j+2} & P_{i+1,j+3} \\ P_{i+2,j} & P_{i+2,j+1} & P_{i+2,j+2} & P_{i+2,j+3} \\ P_{i+3,j} & P_{i+3,j+1} & P_{i+3,j+2} & P_{i+3,j+3} \end{bmatrix} L_V(j) \begin{bmatrix} 1 \\ t \\ t^2 \\ t^3 \end{bmatrix} \quad (1)$$

where $s = (U - U_{i+2})/(U_{i+3} - U_{i+2})$, and $t = (V - V_{j+2})/(V_{j+3} - V_{j+2})$, $j = 1, 2, \cdots, N-1$.

The expression of $L_U(i)$ is

$$L_U(i) = \begin{bmatrix} \frac{(U_{i+3} - U_{i+2})^2}{(U_{i+3} - U_{i+1})(U_{i+3} - U_i)} & 1 - L_U(i)_{1,1} - L_U(i)_{1,3} & \frac{(U_{i+2} - U_{i+1})^2}{(U_{i+4} - U_{i+1})(U_{i+3} - U_{i+1})} & 0 \\ -3L_U(i)_{1,1} & 3L_U(i)_{1,1} - L_U(i)_{2,3} & \frac{3(U_{i+3} - U_{i+2})(U_{i+2} - U_{i+1})}{(U_{i+4} - U_{i+1})(U_{i+3} - U_{i+1})} & 0 \\ 3L_U(i)_{1,1} & -3L_U(i)_{1,1} - L_U(i)_{3,3} & \frac{3(U_{i+3} - U_{i+2})^2}{(U_{i+4} - U_{i+1})(U_{i+3} - U_{i+1})} & 0 \\ -L_U(i)_{1,1} & L_U(i)_{1,1} - L_U(i)_{4,3} - L_U(i)_{4,4} & L_U(i)_{4,3} & \frac{(U_{i+3} - U_{i+2})^2}{(U_{i+5} - U_{i+2})(U_{i+4} - U_{i+2})} \end{bmatrix} \quad (2)$$

where $L_U(i)_{4,3} = -L_U(i)_{3,3}/3 - L_U(i)_{4,4} - (U_{i+3} - U_{i+2})^2/[(U_{i+4} - U_{i+2})(U_{i+4} - U_{i+1})]$. The expression of $L_V(j)$ is to substitute the U and i to V and j in Eq. (2).

Recent work has shown that for a complex wavefront, cubic B-spline interpolation performs much better than Zernike polynomial fitting[14]. Our simulation results also show that the method of bicubic B-spline interpolation in null CGH design has high precision, low order, and sequence convergence[15]. To further show the advantages of B-spline interpolation, here we give an example. For a complex phase function with the expression as Eq. (3), 36 terms 10 order Zernike polynomials fitting and bicubic B-spline interpolation are used for reconstruction of the phase function. The reconstruction errors are compared, as shown in Table 1 and Fig. 1. The results show that, compared to Zernike polynomials fitting, B-spline interpolation has the advantage of higher precision, lower order and sequence convergence. Therefore we choose to use bicubic B-spline interpolation instead of Zernike polynomials fitting during null CGH design for freeform testing.

$$\Phi(x,y) = 8y - 0.5 \times \exp(-x - 0.005y^2) - 150 \times \sqrt{1 - \left(\frac{x-5}{15}\right)^2} + 130$$

$$-5\mathrm{mm} \leqslant x, y \leqslant 5\mathrm{mm} \tag{3}$$

Table 1 Errors comparison of using bicubic B-spline interpolation
and 10 order Zernike polynomials fitting to fit Eq. (3)

NBP	Error of bicubic B-spline interpolation		Error of 10 order Zernike polynomials fitting	
	PV/rad	RMS/rad	PV/rad	RMS/rad
213 × 10	1.1×10^{-1}	3.7×10^{-5}	2.7	1.1×10^{-1}
226 × 18	9.3×10^{-2}	4.2×10^{-6}	2.3	1.2×10^{-1}

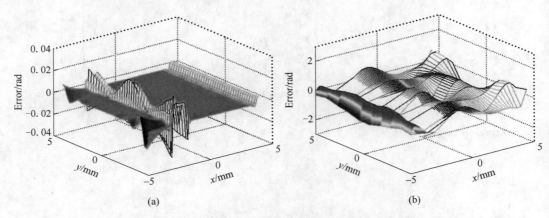

Fig. 1 Comparison of fitting errors using bicubic B-spline interpolation (a)
and 10 order Zernike polynomials fitting (b) to fit Eq. (3)

2.2 CGH fringes generation method

The surface under test is chosen to be an ellipsoid surface with a linear term and a quadratic term as determined by Eq. (4), which is non-axisymmetric. The unit of x, y, and z axis is mm. The peak-to-valley undulation of the surface is 630μm.

$$z = 135 \times \sqrt{1 - \left(\frac{x}{100}\right)^2 - \left(\frac{y}{500}\right)^2} + 0.002x^2 + 0.0001y, \quad -9 \leqslant x, y \leqslant 9 \tag{4}$$

The phase data of the null CGH is calculated by the ray tracing proposed in Ref. [18], which is discrete, and distributed in a curved rectangular region. Because of the advantages of higher precision, lower order and sequence convergence to Zernike polynomials fitting proved above, bicubic non-uniform B-spline interpolation is chosen to obtain the continuous phase function of null CGH. The interpolation method is presented in the above paragraphs.

After the interpolation, the phase function of the CGH is expressed by bi-cubic B-spline functions. As CGH is a diffractive element, separation of the diffraction orders must be concerned. A linear carrier frequency should be added into the phase function to separate and eliminate disturbing diffraction orders[7]. If the phase function expressed by bicubic B-spline function is $\varphi(x,y)$, assuming the carrier frequency is along x-axis and the value is $2\pi/T$, then the phase function with carrier frequency (defined as fabrication phase function of the null CGH) is $2\pi x/T - \varphi(x,y)$. For easy fabrication, the null CGH is designed to be a binary phase-only element, which has the highest diffraction efficiency 41% when the duty circle is 0.5 [9]. Under such condition, the coordinates of the CGH fringes position can be obtained by solving the equation

$$\frac{2\pi x}{T} - \varphi(x,y) = 2\pi\left(m \pm \frac{1}{4}\right), \quad m \text{ is an integer} \tag{5}$$

where the value range of m is determined by the value ranges of x and $\varphi(x,y)$. When the "±" in the equation is set as plus sign, the position coordinates obtained (x_1, y_1) are the coordinates of failing edges of CGH fringes, and when as minus sign, the position coordinates obtained (x_2, y_2) are the coordinates of rising edges of CGH fringes. To solve Eq. (5), is to compute the intersection curves between B-spline surface $(z_1(x,y) = \frac{2\pi x}{T} - \varphi(x,y))$ and plane surfaces $(z_2(x,y) = 2\pi(m \pm \frac{1}{4}))$. B-spline surface is a kind of parametric surface, and the intersection algorithm[17] is different from the traditional raster pattern generator.

Furthermore, the fabrication phase function $2\pi x/T - \varphi(x,y)$ is monotonous along x-axis because of the carrier frequency[9], so that the intersection algorithm can be simplified. For a certain fabrication phase value between the knots $Q_{k,j}$ and $Q_{k+1,j}$, the u parameter value must be between u_i and u_{i+1}. Therefore for a certain fringe order of CGH, the coordinates of the CGH fringe position can be calculated as follows (as shown in Fig. 2).

(1) For all $j = 1 : N$, find all k values satisfy $Q_{k,j} \leq 2\pi\left(m + \frac{1}{4}\right) \leq Q_{k+1,j}$;

(2) Find the control lattice of the curved surface piece with the parameter knots of (u_k, v_j), (u_k, v_{j+1}), (u_{k+1}, v_j) and (u_{k+1}, v_{j+1});

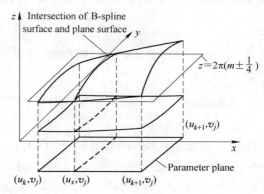

Fig. 2 Graph of calculation of CGH's fringes position

(3) Solve the cubic equation of $c_z^{k,j}(s_x, t) = 2\pi\left(m \pm \dfrac{1}{4}\right)$ with the variable of u_x ($u_k \leqslant u_x \leqslant u_{k+1}$), where $s_x = (u_x - u_i)/(u_{i+1} - u_i)$, t can be obtained from Eq. (1);

(4) Substitute the (u_x, v_j) values to the matrix expressions of the piece of B-spline curve surface $c_x^{k,j}$ and $c_x^{k,j}$, then the coordinate value (x, y) are obtained.

As described above, according to the properties of the bicubic B-spline interpolation curve surface and the monotone property of the fabrication phase function of null CGH, by means of simple judgment and solution of univariate cubic equations, accurate intersection points of plane surfaces and parametric surfaces can be calculated. Compared to traditional Zernike polynomial fitting method, which needs to solve high order equations to calculate the positions of fringes, the new design method is more stable and easier.

The expression of the phase function and the shape of the null CGH aperture cannot be estimated according to the expression of the surface under test, therefore the fitting accuracy cannot be estimated during null CGH design. However, according to the research results of Refs. [14] and [15], and the sequence convergence property of B-spline function fitting[15], if we choose the number of the sampling rays in ray tracing (corresponding to a high NBP in Ref. [14]) more than the number needed to satisfy sampling theorem, the fitting accuracy will be high enough. The subsequent simulation and experimental results verify the validity of the new design method of null CGH.

3 Simulation result

The null CGH is designed using the novel design method proposed above. To verify the validity of the null CGH design method, the test configuration with the standard surface is established in Zemax. The standard surface means the surface under test without fabrication error. After grid sampling and importing, the phase function of the CGH and the standard surface are modeled in Zemax with the surface types of Grid Phase and Grid Sag, respectively. After setting the optical layout according to the test system shown in Fig. 3, it can be observed that the residual wavefront error is 0.11λ, which verifies that the designed null CGH has successfully compensated the wavefront aberration brought by the standard surface under test.

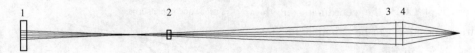

Fig. 3 The optical configuration to verify the design of null CGH

1—the standard surface under test; 2—null CGH; 3—standard lens; 4—collimating lens

The diffraction properties of the null CGH are also analyzed. The fabrication phase function of the null CGH is taken sample by grids, and the sampling results are imported to surface type of Grid Phase in Zemax. We use footprint diagram to analyze the diffraction properties of the null CGH. Fig. 4(a) shows the computed diffraction orders distribution of the null CGH at the back focal plane of the convergence lens, where the −1 working order spot is separated from the other disturbing orders. Fig. 4(b) shows the −1 order spot in the designed position of the surface under test, which is in accordance with the aperture shape of the surface under test.

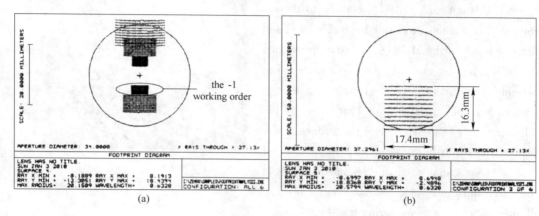

Fig. 4 Analysis of the diffraction properties of the null CGH
(a) diffraction spots at the back focal plane of the convergence lens; (b) −1 order diffraction spot in the designed position of the surface under test

When calculating an arbitrary point on the B-spline function expression of the phase function, and importing the phase data into Zemax (the shape of the Grid Phase surface type is rectangular while the aperture of the null CGH is curved rectangular), there are errors introduced to the model of the null CGH, because of the aperture difference and fit method difference. The errors caused the residual wavefront error of 0.11λ and the slight difference between the shape of the spot in Fig. 4(b) and the standard surface under test expressed by Eq. (4).

The simulated result shows the validity of B-spline interpolation for acquiring continuous phase function of the null CGH. The validity of fringes generation method needs to be confirmed by experiment.

4 Experimental result

A multifunctional diffractive optical element (M-DOE) with the null CGH, a reflective Fresnel zone plate (FZP) for CGH position calibration and two transmittive FZPs for surface under test alignment are designed and fabricated. The details of the design method of FZP can be found in Ref. [18]. Fig. 5 shows the image of the M-DOE.

We performed the null test with a phase-shifting Fizeau interferometer (Veeco RTI6100). The optical layout is the same as shown in Fig. 3, except for an extra spatial filter, which is used to block the disturbing diffraction orders of the null CGH. The null test result is shown in Fig. 6(a), where wv means the wavelength of the light source (633nm). The peak-to-valley and rms fabrication error of the surface under test are $1.7\mu m$ (2.78wv) and $0.27\mu m$ (0.433wv), respectively. The test result is the fabrication error wavefront propagating to the plane of the null CGH, therefore the aperture shape of the test result is the same as the aperture shape of the null CGH.

The substrate of the M-DOE is tested by the interferometer. The groove depth error and fringe distortion of the M-DOE are measured by a WYKO NT1100 optical profiler. According to the error analysis method proposed by Ref. [19], and the test result of the M-DOE, the error introduced by the fabrication error of CGH is analyzed. The optical layout model of the test built in Zemax is used to simulate the test error introduced by the adjusting error of the CGH and the surface under test[8]. The result of the error analysis of the null test in Table 2 shows that the alignment errors of the CGH and the freeform surface under test are the biggest error sources during the test.

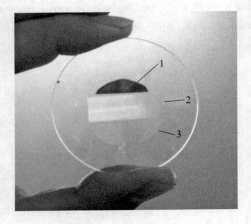

Fig. 5 Picture of the M-DOE
1—the null CGH; 2—the reflective FZP;
3—transmittive FZPs

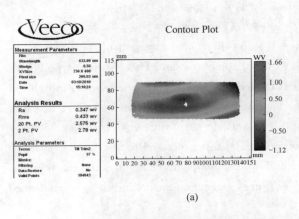

(a)

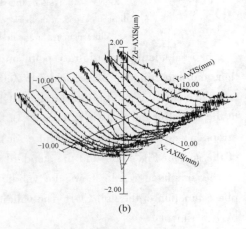

(b)

Fig. 6 Surface fabrication error test result of the surface under test
(a) by the null test; (b) by the 3D profiler (original data)

Table 2 Error analysis of the null test

Error source		Error value	Error introduced into test result
Substrate error	Thickness error	0.010mm	~0.08λ
	Angularity error	<30″	
	Flatness error	$\lambda/30$	
Fabrication error of CGH	Groove depth error	5%	0.004λ
	Fringe distortion	7.5μm	0.002λ
Alignment error of CGH	Defocus, decenter of x, y, tilt	—	~0.4λ
Alignment error of surface under test	Defocus, decenter of x, y, tilt, rotation around z	—	~0.5λ
System error of interferometer	—	—	$1/15\lambda$
RSS error of test		~0.7λ/0.44μm	

To verify the new design method of the null CGH, we compare the null test result to the test result of a Panasonic ultra-precision 3D profiler (UA3P-5). The test result of the UA3P-5 is shown in Fig. 6(b). The peak-to-valley value of the fabrication error is 1.7μm, and the valley value is 0.5μm, which is in good accordance with the result of the null test. There is an obvious pit both in the result of the null test and the UA3P-5 test. Comparing to the result by null test, the position of the pit (in the middle of the test area) is a little different, but the area divided by the whole area is in proportion with it by the null test. The reason of the pit deviation in test result of null test is that there is tilt error when aligning, and then the interference fringes deviate accordingly in the exit aperture of the interferometer.

According to the error analysis and the comparison result presented above, we can conclude that the null test result is reasonable in the range of allowable error. The null test error is quite large because of the alignment difficulties introduced by the non-axisymmetrical surface under test, but the experimental result confirmed the feasibility of the new design method of the null CGH. The surface under test in this paper was with a rectangular aperture and bicubic non-uniform B-spline interpolation was used during null CGH design. For surfaces with more complex apertures and expressions, non-rectangular topology and triangular B-spline surfaces[20] can be used for null CGH design.

5 Conclusion

In this paper, for null test of a non-axisymmetrical freeform surface with 18mm × 18mm rectangular aperture and 630μm peak-to-valley undulation, a novel null CGH design method using bicubic non-uniform B-spline interpolation was illustrated in detail. According to the properties of the bicubic B-spline interpolation curve surface and the monotone property of the fabrication phase function of null CGH, the procedure to calculate the fabrication data of null CGH, i.e., accurate intersection points of plane surfaces and parametric surfaces was simplified.

A null CGH with a curved rectangular aperture for the proposed surface test was designed and fabricated using the novel method proposed. The optical layout of the test was established in Zemax, and the simulation results showed the validity of the novel fitting method during the null CGH design. The null test was carried out, and the test result was in accordance with the test result of the ultra-precision 3D profiler. The comparison result again proved the feasibility of the new null CGH design method.

The new design method of null CGH is an efficient and reliable method with simple procedures. For null CGH with nontraditional aperture and high order phase function, there is no need to orthogonalize and to choose the number of the used terms to determine a proper basis function set. The calculation of fringes position in the novel design method only needs simple judgment and solution of univariate cubic equations. With the new CGH design method, the research work in this paper shows good insight in CGH null test to be an effective and ultra-precise non-contact test method for optical freeform surface, especially the surfaces with high order and nontraditional apertures.

Acknowledgements

The authors thank Prof. Fengzhou Fang and Dr. Xiaodong Zhang for fabrication of the freeform surface, and Xiang Ding of National Institute of Metrology, China for surface test using UA3P 3D profiler. The research was supported by the National Basic Research Program of China under grant No. 2011CB706701.

References

[1] D. Cheng, Y. Wang, H. Hua, and M. Talha, "Design of an optical see-through head-mounted display with a low f-number and large field of view using a freeform prism", Appl. Opt. 48(14), 2655-2668 (2009).

[2] K. Wang, F. Chen, Z. Liu, X. Luo and S. Liu, "Design of compact freeform lens for application specific light-emitting diode packaging", Opt. Exp. 18(2), 413-425 (2010).

[3] Y. Luo, Z. Feng, Y. Han, and H. Li, "Design of compact and smooth free-form optical system with uniform illuminance for LED source", Opt. Exp. 18(9), 9055-9063 (2010).

[4] M. Moiseev, L. Doskolovich, N. Kazanskiy, "Design of high-efficient freeform LED lens for illumination of elongated rectangular regions", Opt. Exp. 19(S3), A225-A233 (2011).

[5] Z. Zheng, X. Hao, X. Liu, "Freeform surface lens for LED uniform illumination", Appl. Opt. 48(35), 6627-6634 (2009).

[6] T. Ma, J. Yu, P. Liang, and C. Wang, "Design of a freeform varifocal panoramic optical system with specified annular center of field of view", Opt. Exp. 19(5), 3843-3853 (2011).

[7] J. Wyant, V. Bennett, "Using computer generated holograms to test aspheric wavefronts", Appl. Opt. 11(12), 2833-2839 (1972).

[8] M. Wang, D. Asselin, P. Topart, J. Gauvin, P. Berlioz, and B. Harnisch, "CGH null test design and fabrication for off-axis aspherical mirror tests", in International Optical Design Technical Digest(CD), paper ThA3 (2006).

[9] W. Lee, "Binary synthetic holograms", Appl. Opt. 13(7), 1677-1682 (1974).

[10] Y. Xu, X. Zhang, Y. Zhang, "Using computer-generated holograms to test cubic surfaces", Opt. Commun. 282(12), 2327-2331 (2009).

[11] G. Dai, V. Mahajan, "Orthonormal polynomials in wavefront analysis: error analysis", Appl. Opt. 47(19), 3433-3445(2008).

[12] W. Swantner and W. Chow, "Gram-Schmidt orthonormalization of Zernike polynomials for general aperture shapes", Appl. Opt. 33(10), 1832-1837 (1994).

[13] X. Hou, F. Wu, L. Yan, and Q. Chen, "Comparison of annular wavefront interpretation with Zernike circle polynomials and annular polynomials", Appl. Opt. 45(36), 8893-8901 (2006).

[14] M. Ares and S. Royo, "Comparison of cubic B-spline and Zernike-fitting techniques in complex wavefront reconstruction", Appl. Opt. 45(27), 6954-6964 (2006).

[15] P. Su, Q. Tan, G. Kang, and G. Jin, "B-spline interpolation of scattered phase data of computer generated hologram for null test of freeform surface (in Chinese)", Acta Opt. Sin. 30(6), 1767-1771 (2010).

[16] K. Qin, "General matrix representations for B-splines", Vis. Comput. 16(3-4), 177-186 (2000).

[17] L. Piegl L and W. Tiller, The NURBS Book. Springer, Berlin & New York, (1997).

[18] T. Kim, J. Burge, Y. Lee, and S. Kim, "Null test for a highly paraboloidal mirror", Appl. Opt. 43(18), 3614-3618(2004).

[19] Y. Chang, J. Burge, "Error analysis for CGH optical testing", Proc. SPIE 3782, 358-366 (1999).

[20] E. Catmull and J. Clark, "Recursively generated B-spline surfaces on arbitrary topological meshes", Comput. Aided Des. 10(6), 350-355 (1978).

Fast Statistical Measurement of Aspect Ratio Distribution of Gold Nanorod Ensembles by Optical Extinction Spectroscopy[*]

Abstract Fast and accurate geometric characterization and metrology of noble metal nanoparticles such as gold nanorod (NR) ensembles is highly demanded in practical production, trade, and application of nanoparticles. Traditional imaging methods such as transmission electron microscopy (TEM) need to measure a sufficiently large number of nanoparticles individually in order to characterize a nanoparticle ensemble statistically, which are time-consuming and costly, though accurate enough. In this work, we present the use of optical extinction spectroscopy (OES) to fast measure the aspect ratio distribution (which is a critical geometric parameter) of gold NR ensembles statistically. By comparing with the TEM results experimentally, it is shown that the mean aspect ratio obtained by the OES method coincides with that of the TEM method well if the other NR structural parameters are reasonably pre-determined, while the OES method is much faster and of more statistical significance. Furthermore, the influences of these NR structural parameters on the measurement results are thoroughly analyzed and the possible measures to improve the accuracy of solving the ill-posed inverse scattering problem are discussed. By using the OES method, it is also possible to determine the mass-volume concentration of NRs, which is helpful for improving the solution of the inverse scattering problem while is unable to be obtained by the TEM method.

1 Introduction

Metal nanoparticles (NPs), especially noble metal NPs, have important applications nowadays in various fields such as catalysis, medical diagnosis and therapy, biosensing, and drug delivery and release[1]. Reliable, fast, and accurate measurement methods and the related metrology standards are highly demanded for the production, characterization, and commercial use of metal NPs. Since the properties of metal NPs highly depend on their geometric characteristics due to the strong shape- and size-dependent localized surface plasmon resonance (LSPR) of the NPs[2], the dimensional metrology of metal NPs is of high importance.

So far, the most commonly used dimensional metrological methods for metal NPs are microscopic imaging methods such as transmission electron microscopy (TEM), scanning electron microscopy, and scanning probe microscopy. These methods, though with high precision, can only measure individual NPs or a small number of NPs locally. Hence, these methods are slow and costly when they are used to measure large amount of NPs (or the so-called NP ensembles). In addition, when the NPs are prepared for microscopic measurement, the NPs may aggregate

---

[*] Copartner: Ninghan Xu, Benfeng Bai, Qiaofeng Tan. Reprinted from *Optics Express*, 2013, 21(3):2987-3000.

strongly after the NP colloid is coated on a substrate and the solvent is evaporated, which is disadvantageous for accurate characterization of the NP geometry. Furthermore, for NPs with non-uniform geometries (i.e., the so-called polydisperse NP ensembles), it is often needed to characterize the distribution function of some geometric parameters of the NPs statistically, which is obviously hard to do with the microscopic methods due to the required large amount of sampling NPs. To achieve this goal, some methods based on scatterometry (i.e., the technique of retrieving the geometrical parameters of NPs from their scattering spectra) have been proposed, such as optical extinction spectroscopy (OES), small-angle x-ray scattering method, and dynamic light scattering (DLS) method. Among these methods, the DLS method is probably the most widely used one because of its versatility of measuring various materials of NPs. However, since the DLS method measures the hydrodynamic size of NPs in a liquid environment by detecting and analyzing the Brownian motion of the NPs, it can only give the equivalent spherical diameter of the measured NPs no matter what practical shape the NPs may have. Therefore, it cannot measure the shape of the NPs. For the characterization of non-spherical NPs, some other scatterometric methods such as the OES method have to be developed.

Gold nanospheres and nanorods (NRs) are typical noble metal NPs that are widely used nowadays in various applications such as biomedical diagnosis and therapy[3]. An ensemble of NPs are referred to as monodisperse if the NPs have the same size and shape, or otherwise as polydisperse. For polydisperse NP ensembles, a probability density function (PDF, which refers to the percentage of the NPs with certain size and shape specifications in the whole NP ensemble) is used to describe the size and shape distribution of the NPs. The standard deviation σ of the PDF therefore characterizes the polydispersity of the sample. In practice, when the standard deviation of the PDF is small enough, the NP ensemble can be regarded as monodisperse. According to previous studies[4,5], for monodisperse gold nanospheres with $\sigma < 0.1$, the mean diameter can be determined accurately by the OES method in a broad range of diameters (3-100nm). However, when the NP geometry deviates from an ideal sphere, the shape deviation should be taken into account[6]. On the other hand, for a polydisperse nanosphere ensemble with $\sigma > 0.1$, the polydispersity should also be taken into account[6]. Recently, Peña et al.[7] proposed a multivariate optimization algorithm to retrieve the average diameter and the diameter PDF of polydisperse metal nanoshpere ensembles. They showed that the OES method can measure the PDF of a large-amount NP ensemble accurately enough, while the TEM method is more useful for the characterization of a small amount of gold nanospheres directly.

For gold NRs, the light scattering behavior and LSPR property are more complicated than those of nanospheres. In the past few years, several electromagnetic methods have been used for simulating the LSPRs of metallic NRs, such as the Rayleigh-Gans approximation (RGA) method[8,9], the T-matrix method[10,11], the discrete dipole approximation method[12-14], the finite-difference time-domain method[15], the integral boundary element method[16,17], and the finite element method[18]. The advantages and disadvantages of these methods in simulating NPs are described detailedly in Refs. [19-21]. Previous research works[11,14] have shown that the geometric parameters of gold NRs such as the width, the aspect ratio (the length of the NR divided by its width), and the end-cap shape can affect the LSPR of NRs. Specifically, when the aspect ratio of gold NR is larger than about 1.5, there would appear two distinct LSPR peaks in the extinction spectra of the NRs [1]: a transverse-mode LSPR peak (named T-LSPR) and a longitudinal-mode LSPR peak (L-LSPR, which is usually stronger) due to the resonant oscillations of electrons along the short and long axes of the NR, respectively. Since the L-LSPR peak of gold NRs is usually more sensitive to the geometry change than the T-LSPR peak, it can be measured to estimate the mean width and aspect ratio of a monodisperse NR ensemble.

For polydisperse gold NR ensembles, Susie Eustis et al.[22] have used a theoretical fit of the measured L-LSPR to quickly determine the aspect ratio distribution (ARD). When retrieving the structural parameters, they used the RGA method that is only applicable to small spheroidal NPs[23] to simulate the LSPRs of NRs, so that the influences by the other geometric parameters of NRs such as the width and the end-cap shape were ignored. Instead of the RGA method, Boris Khlebtsove et al.[11] applied the T-matrix method (which is a rigorous semi-analytical method) to simulate the LSPR of the gold NPs ensemble based on thousands of TEM images. It shows that the width, the end-cap shape, and the surface electron scattering constant of the NRs may also affect the LSPRs, though weaker than the aspect ratio. Owing to the high calculation efficiency and high accuracy of the T-matrix method, they suggested to apply it to determine the ARD of NRs from extinction experiments by using an optimization procedure. Furthermore, in a previous report of them[24], they applied a modified approach of Susie Eustis et al.[22] to obtain the ARD by fitting the depolarization spectra in addition to the extinction spectra. When retrieving the ARD by optical data, they used the TEM-based data as *a priori* information and obtained an excellent agreement between the simulated and measured spectra.

These works show that the OES method is more convenient than the TEM method to characterize the size and shape distribution of polydisperse gold NR ensembles statistically. Here, we apply the same strategy by using the T-matrix method to rigorously calculate the extinction spectra of the NR ensembles when solving the inverse scattering problem. However, the main difference between our approach and the previous works[11,24] is that we use a fast and reliable optimization procedure so that no *a priori* information about the geometric values (such as the TEM-based width of the NRs) is needed beforehand. We study the use of the OES method to perform fast measurement of the ARD of polydisperse gold NR ensembles statistically, where the constrained nonnegative regularized least-square procedure was applied. The influences by the width, the end-cap shape, and the surface electron scattering constant of the NRs on the ARD measurement are thoroughly analyzed. The measurement results are compared with those obtained by the TEM method, showing the reliability of the OES method. The measures for further improving the solution accuracy of the ill-posed inverse scattering problem are discussed.

2 Theoretical method

The objective of the OES method is to retrieve the geometric parameters of the gold NRs from the measured extinction spectra. It is in essence an inverse scattering problem, which is solved by using the methods below.

2.1 Calculation of the extinction cross section of a single NR

To solve the inverse scattering problem, the premise is that the related directly problem is well solved, i. e., the extinction response of a NR can be precisely modeled and calculated. The NR geometry is defined in Fig. 1, which is a cylinder with its end caps in oblate spheroidal, prolate spheroidal, or spherical shapes, depending on the different generation methods of the NRs. Three parameters are defined to describe the NR geometry, i. e., the width D, the aspect ratio $AR = H/D$ (where H is the length of the NR), and the eccentricity of the end cap $e = 2L/D$. Clearly, when $e = 0$, the NR is a perfect cylinder; and when $e = 1$, the NR is a cylinder with two semi-spherical end caps.

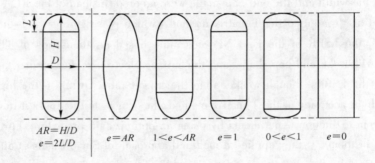

Fig. 1 Geometric model of the NR

Several NRs with the same width D and aspect ratio AR but different end-cap factor e are demonstrated

The extinction cross section C_{ext} is defined as the ratio of the radiant power being extinct by a particle to the radiant power incident on the particle in the process of scattering[8]. To rigorously calculate C_{ext} of a single NR or an ensemble of randomly oriented discrete gold NRs in a monodisperse system, some numerical methods such as the T-matrix method[10] can be used. The T-matrix method, a rigorous semi-analytical method, is used in our simulation because it is much faster for modeling randomly oriented NR ensembles than the other methods[10,11].

We developed our own T-matrix numerical codes based on the algorithms reported in Ref. [10], which were calibrated with the benchmark results reported in Ref. [25-27]. In our calculation, the real and imaginary parts of the complex dielectric function of bulk gold were taken, which were calculated from the experimental values given by Johnson and Christy[28]. According to previous works[1,3-6,11,24], this dielectric function should be corrected of gold NRs. Due to the mean free path limitation for electrons, the damping constant γ of gold NR is increased and can be modified by[4]

$$\gamma = \gamma_{bulk} + A_s \frac{v_F}{L_{eff}} \tag{1}$$

where γ_{bulk} is the damping constant of bulk gold, v_F is the electron velocity at the Fermi surface,

A_s is the surface electron scattering constant, and L_{eff} is the effective mean free path for collisions with the boundary. We take the values $\gamma_{bulk} = 1.64 \times 10^{14}$ s^{-1} and $v_F = 1.41 \times 10^{15}$ nm/s from Ref. [4]. The expression $L_{eff} = 4V/S$ was given by Ref. [29], where V and S are the volume and the surface area of the NR, respectively. The surface electron scattering constant A_s can be considered as a free parameter with the value varying around 1. The value of $A_s = 0.3$[30] was used in our calculations and its influence on the measurement results are discussed in subsection 4.4. The refractive index of the solvent, i.e., the water at room temperature, was calculated by[31]

$$n_s = 1.32334 + \frac{3479}{\lambda^2} - \frac{5.111 \times 10^7}{\lambda^4} \quad (2)$$

where λ is the wavelength of light in nanometers.

2.2 Calculation of the absorbance of NR ensembles

In the OES measurement, the measurand is the absorbance A of the sample, by which to retrieve the structural parameters of the NRs. Therefore, we should get the relation between A and the extinction cross section C_{ext} of a single NR. For a monodisperse ensemble of NRs, the transmittance (defined as the ratio of the transmitted intensity over the incident light intensity I/I_{inc}), the absorbance A, and the total extinction A_{ext} of the sample medium with length l are related by[8]

$$\frac{I}{I_{inc}} = 10^{-A} = e^{-A_{ext}l} \quad (3)$$

where A_{ext} is related to the extinction cross section of a single NR by $A_{ext} = N_v C_{ext}$ and N_v is the number of NRs per unit volume. Therefore, the extinction cross section C_{ext}, in turn, can be obtained from the experimentally measured absorbance A by

$$A = \frac{l}{\ln 10} A_{ext} = \frac{lN_v}{\ln 10} C_{ext} \quad (4)$$

Obviously, Eq. (4) is valid only for a monodisperse system. If the NR ensemble is polydisperse, the measured absorbance spectrum is the superposition of the absorbance spectra of the composing NRs with different sizes and shapes[22]. Therefore, for a polydisperse NR ensemble, the total absorbance A is calculated by integrating the contribution from the composing NRs according to the PDF $p(D, AR, e)$ as[10]

$$A(\lambda, D, AR, e) = \frac{lN_v}{\ln 10} \int_{D_{min}}^{D_{max}} \int_{AR_{min}}^{AR_{max}} \int_{e_{min}}^{e_{max}} p(D, AR, e) C_{ext}(\lambda, D, AR, e) \, dD \, dAR \, de \quad (5)$$

2.3 Solution of the inverse scattering problem

In the inverse scattering problem of determining the NR geometry by scatterometry, the three structural parameters of NR as well as the PDF are unknowns. Therefore, the inverse problem is generally ill-posed so that the solution is not unique or a small perturbation of the measurand may result in a large variation of the retrieved parameters. Furthermore, Eq. (5) usually cannot be solved analytically and should be discretized for numerical solution. If we discretize it with three variables AR, D, and e and use, for example, an optimization process to search for the solution without any *a priori* information of the PDF, the size of the matrix in calculation would be very large so that the condition number of the linear system is too large to give an accurate and stable solution. To avoid this problem, a model of the PDF $p(D, AR, e)$ of the NRs may be adopted, either according to preliminary experimental statistics (by, for example, TEM or dark-field microscopy) or by reasonable assumption based on the production method of the NPs. Then an optimization process can be launched to search for the solution. Therefore, the accuracy and stability of the solution are dependent on the adopted PDF model.

On the other hand, according to our study and many previous works[1,14,22], the aspect ratio AR is the primary parameter affecting the extinction of the NR ensemble. Hence, to a first approximation, we may fix the width D and the end-cap factor e as their mean values $\overline{D}$ and $\overline{e}$, so that these two variables can be separated from the integral equation:

$$A(\lambda, \overline{D}, AR, \overline{e}) = \frac{lN_v}{\ln 10} \int_{AR_{\min}}^{AR_{\max}} p(\overline{D}, AR, \overline{e}) C_{\text{ext}}(\lambda, \overline{D}, AR, \overline{e}, \lambda) \, \mathrm{d}AR \tag{6}$$

By this treatment, we just need to discretize Eq. (6) with respect to AR and λ. Then the condition number of the linear system would be much smaller and the optimization process would be faster and more stable.

The discretization of Eq. (6) results in the following system of linear algebraic equations:

$$\mathbf{A} = \mathbf{CP} \tag{7}$$

where $\mathbf{A}$ and $\mathbf{P}$ are $M \times 1$ and $N \times 1$ vectors, respectively, and $\mathbf{C}$ is a $M \times N$ matrix. The vector $\mathbf{A}$ contains the measured extinction values at different λ_m and the matrix $\mathbf{C}$ consists of the calculated extinction cross sections C_{ext} for NRs with each pair of λ_m and AR_n. The vector $\mathbf{P}$ is the PDF to be solved. Their specific expressions are shown below

$$\mathbf{A} = [A(\lambda_1) \quad A(\lambda_2) \quad \cdots \quad A(\lambda_m) \quad \cdots \quad A(\lambda_M)]^{\mathrm{T}}, \quad m = 1, 2, \cdots, M \tag{8}$$

$$\mathbf{P} = \Delta AR \cdot [p(AR_1) \quad p(AR_2) \quad \cdots \quad p(AR_n) \quad \cdots \quad p(AR_N)]^{\mathrm{T}}, \quad n = 1, 2, \cdots, N \tag{9}$$

$$\mathbf{C}_{mn} = \frac{lN_v}{\ln 10}[C_{\text{ext}}(\lambda_m, \overline{D}, AR_n, \overline{e})] \tag{10}$$

$$\Delta AR = \frac{AR_{\max} - AR_{\min}}{N} \tag{11}$$

m and n are integers, and the superscript T means the transpose of the vectors. Here we consider $M > N$ so that Eq. (7) is an overdetermined system with N unknowns. $p(AR_n)$ has two physical constraints: the non-negativity constraint $p(AR_n) \geq 0$ and the standard normalization condition $\sum_n p(AR_n) = 1$.

Since the discretized matrix $\mathbf{C}_{mn}$ is usually ill-conditioned, the inverse problem formulated in terms of Eq. (6) is ill-posed[32]. In order to find a unique and accurate solution of the inverse problem, one of the commonly used numerical techniques is Tikhonov regularization[33]. Here we just briefly summarize the process of Tikhonov regularization in our problem and more details can be found in Refs. [32-35].

The regularized least-squared solution $\mathbf{P}_{RLS}$ of Eq. (7) is given as[32]

$$\mathbf{P}_{RLS} = \min\{\|\mathbf{A} - \mathbf{CP}\|_2^2 + \gamma^2 \|\mathbf{L}(\mathbf{P} - \mathbf{P}^*)\|_2^2\} \quad (12)$$

where $\|\cdot\|_2$ is the Euclidean norm, γ is the regularization factor, $\mathbf{P}^*$ is an assumed *a priori* assumed solution (taken as $\mathbf{P}^* = 0$ here), $\mathbf{L}$ is typically either an identity matrix (as we take here) or a discrete approximation of the derivative operator[32]. Eq. (12) can be written in another equivalent form as follows[33]

$$\mathbf{P}_{RLS} = \min\left\{\frac{1}{2}\mathbf{P}^T\mathbf{Q}\mathbf{P} + \mathbf{q}^T\mathbf{P}\right\} \quad (13)$$

where $\mathbf{Q} = 2(\mathbf{C}^T\mathbf{C} + \gamma^2\mathbf{L}^T\mathbf{L})$ is a symmetrical matrix of size $N \times N$, and $\mathbf{q} = -2\mathbf{C}^T\mathbf{A}$ is a N-dimensional column vector. We use the active set method[34] to find the solution of Eq. (13). A mean square error (*MSE*) defined below is used to evaluate the quality of the solution

$$MSE = \frac{1}{M}\sum_m \left[\frac{A(\lambda_m) - A_{cal}(\lambda_m)}{A(\lambda_m)}\right]^2, \quad m = 1, 2, \cdots, M \quad (14)$$

where $A_{cal} = \mathbf{CP}_{RLS}$ is the fitted optimal solution of the absorbance. We set the criteria that if *MSE* is smaller than 1×10^{-3}, the solution $\mathbf{P}_{RLS}$ is acceptable.

3 Experiment and results

We applied the OES method to measure the ARD $p(AR)$ of gold NR ensemble samples and compared the results with those directly obtained by the TEM method. 30 samples of gold NR ensembles were measured and analyzed, in which each sample contains approximately 10^{10} NRs per millilitre. Here, without loss of generality, the results of three samples with different *D*, *AR* and *e* are demonstrated. The three samples designated as NR-40-700, NR-20-700, and NR-10-750 were obtained from NanoSeedz Ltd., which have the nominal width *D* of 40nm, 20nm, and 10 nm and the expected L-LSPR wavelengths of 700nm, 700nm, and 750nm, respectively.

In the OES measurement, a UV-VIS spectrophotometer (PekinElmer LAMBDA 950) was used to measure the extinction spectra of the samples. For each sample, the measurement was repeated six times in one hour and the average value is used. The measurement range of wavelength λ was 400-1000nm and the step was taken as 1nm. The most time-consuming process of the OES method is to prepare the extinction spectra database [corresponding to the matrix $\mathbf{C}$ in Eq. (7)] of the gold NRs with different values of the width, the aspect ratio and the end-cap factor *D*, *AR*, and *e*. Fortunately, the database just need to be calculated once. By using a dual-core 2.13GHz Intel Xeon CPU with 80Gb RAM, it takes about 12 seconds to calculate a single extinction spectrum of NRs in the wavelength range of the 400-1000nm, with a 1 nm spectral resolution, and a relative calculation accuracy of better than 1%. Based on the measurement data, the optimization process described above was implemented to retrieve the ARD $p(AR)$ of the samples, where *AR* was discretized in the range of 1 to 5, with a step of 0.1. The inverse algorithm was run on a 3.00GHz Intel Core2 Duo CPU with 4Gb RAM and the average time consumption is about 0.25 seconds for a single measured spectra.

In the TEM experiment, a transmission electron microscope (Hitachi H-7650B) was used to get the images of the NRs. For each of the three samples, ten TEM images were taken. Therefore, we analyzed altogether 788, 896, and 804 NRs in samples NR-400-700, NR-20-700, and NR-10-750, respectively to get the mean width $\overline{D}$, the mean end-cap factor $\overline{e}$ and the ARD $p(AR)$.

The TEM images of the three samples as well as their extinction spectra measured by the OES method are shown in Fig. 2. In Figs. 2(a)-(c), it is clearly seen that the gold NR ensembles are polydisperse. In Fig. 2(d), the measured L-LSPR extinction peaks of the three samples are very close to their nominal values. NR-40-700 (red line) and NR-20-700 (black line) have the same resonance wavelength but different linewidths, i.e., the full width at half maximum (FWHM) of the resonance peak, because of their different $\overline{D}$ and dispersancy.

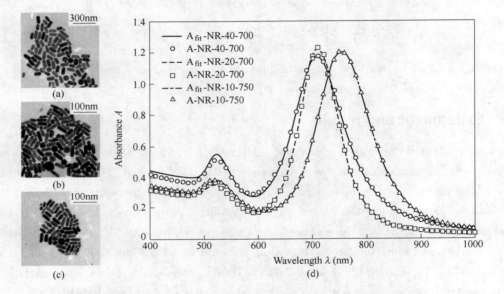

Fig. 2 TEM images of the three gold NR ensemble samples
(a) NR-40-700; (b) NR-20-700; (c) NR-10-750; (d) experimentally measured extinction spectra (dots) of the samples as well as the corresponding numerically reproduced extinction spectra (lines) according to the retrieved ARD functions $p(AR)$ based on the OES results

The AR distributions $p(AR)$ of the samples were retrieved from the OES results, with the procedure presented in Section 2. In Fig. 3, the red columns and curves are the retrieved results obtained by the OES method while the black ones are the results obtained by TEM. We use a sum of Gaussian functions to fit the discrete results:

$$p(AR) = \sum_i \frac{w_i}{\sigma_{i,AR}\sqrt{2\pi}} \exp\left[-\frac{(AR - \overline{AR_i})^2}{2\sigma_{i,AR}^2}\right] \tag{15}$$

where $\overline{AR_i}$ and $\sigma_{i,AR}$ are the mean value and standard deviation of the ith Gaussian function $p(AR)$, respectively. The constant w_i was chosen such that $p(AR)$ satisfies the standard normalization condition $\sum_n p(AR_n) = 1$.

Furthermore, with these retrieved AR distribution functions $p(AR)$, the extinction spectra of the three samples were numerically reproduced by Eq. (6), as shown in Fig. 2(d), which coincide with the measured extinction spectra quite well. Therefore, both Fig. 3 and Fig. 2(d) show that the retrieved results by the OES method are reliable in our characterization.

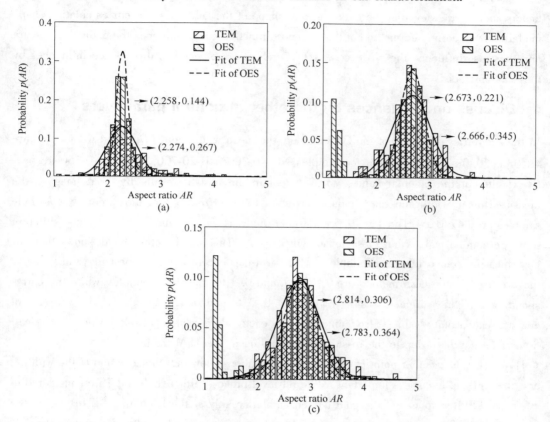

Fig. 3 Comparison of the measured AR distribution functions of three gold NR ensemble samples obtained by the OES method (red) and those obtained by the TEM method (black). In each subfigure, both the discrete AR distribution and a Gaussian fit of it are given. The values in parentheses ($\overline{AR}, \sigma_{AR}$) give the mean AR and the standard deviation of the PDF obtained by the two methods
(a) NR-40-700; (b) NR-20-700; (c) NR-10-750

By comparing the OES results with those obtained by the TEM method in detail, it is seen that the mean AR values derived by the two methods coincide with each other well, with their relative difference as 0.70%, 0.26% and 1.10% for samples NR-40-700, NR-20-700, and NR-10-750, respectively. The relative difference here is calculated by $(\overline{AR}_{OES} - \overline{AR}_{TEM})/\overline{AR}_{OES}$. It is worth noting that in Figs. 3(b) and (c), the OES results show significant ARD between 1 and 1.5, while the TEM results have few NPs in this range. The main reason is that in the counting process of the TEM method, we ignored most byproducts (such as spheres, cubes etc.) in the samples, as shown in the TEM images of Figs. 2(b) and (c). However, these byproducts also contribute to the extinction spectra and thus can be detected by the OES method. The standard deviation of he OES results ($\sigma_{OES,AR}$ = 0.144, 0.221, and 0.306) are smaller than those of the TEM results ($\sigma_{TEM,AR}$ = 0.267, 0.345, and 0.364), with their relative difference as 46.1%, 35.9%, and 15.9% for the three samples. The possible reason is that many factors such as the deviation between the real shape of the gold NRs and our calculation shape model and the correction method of the dielectric function could influence the extinction spectra and thus also influence the retrieved ARD. Therefore, we proceed to discuss the influences by these parameters.

4 Discussion: influences by the other structural parameters

In the AR retrieval process of the OES method, the mean width values $\overline{D}_{OES}$ that we adopted are 46.0nm, 20.0nm, and 22.0nm for samples NR-40-700, NR-20-700, and NR-10-750, respectively, but not the nominal values, so as to obtain the best-fit results. By TEM imaging, the measured mean width and the standard deviation ($\overline{D}_{TEM} \pm \sigma_{TEM,D}$) of the three NR ensemble samples are (47.0 ±4.1nm), (19.5 ±3.0nm), and (18.7 ±1.9nm), which are also different from the nominal values but close to our OES results. This, on the other hand, shows that our OES measurement results are reliable. For the end-cap eccentricity, the best-fit values of $\overline{e}_{OES}$ that we used in the OES method are 0.9, 0.6, and 0.3 for the three samples, while the corresponding TEM measurement results $\overline{e}_{TEM}$ are 0.8, 0.6, and 0.4, respectively. The two sets of end-cap eccentricity values also coincide with each other relatively well (where the small difference may be owing to the in-sufficient sampling in the TEM method).

However, from another point of view, these calculations show that the selection of the width $\overline{D}$ and the end-cap factor $\overline{e}$ is important in the retrieval process, although $\overline{D}$ and $\overline{e}$ are considered to affect the LSPR response of the gold NR ensembles weakly at the beginning[14]. In the following, we detailedly analyze the influences of $\overline{D}$ and $\overline{e}$ on the retrieval results. We analyze the value of mean width $\overline{D}$ ranged from 5nm to 50nm with a step 5nm and the mean end-cap factor $\overline{e}$ ranged from 0 to 1 with a step of 0.1. Without loss of generality, we choose sample NR-20-700 in the following analysis.

4.1 Influence by the selection of mean width $\overline{D}$

Here, the NRs have a fixed end-cap eccentricity $\overline{e}$ = 0.6 and the mean width $\overline{D}$ is varied. In the retrieval calculation, ten different values of $\overline{D}$ were selected to solve the inverse problem. The obtained mean square error MSE values are summarized in Fig. 4(b). It shows that four of them (for $10 \leqslant \overline{D} \leqslant 25$nm) are acceptable with $MSE \leqslant 1 \times 10^{-3}$ while the others (for $\overline{D} \geqslant 30$nm or $\overline{D} \leqslant 5$nm) are unacceptable.

Fig. 4(a) compares the retrieved aspect ratio distribution $p(AR)$ obtained by the OES method using five different assumed mean width $\overline{D}$ with the $p(AR)$ directly measured by the TEM method. It is seen that when $\overline{D}$ increases from 10nm to 30nm, the retrieved $p(AR)$ is left shifted and the FWHM decreases. By linear fitting, we find that the shift bears a linear relation with re-

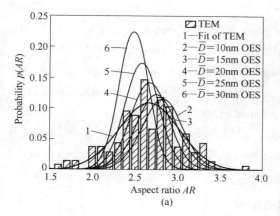
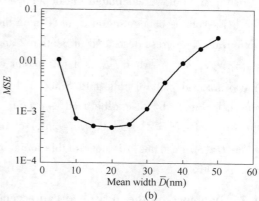

Fig. 4 (a) Comparison of the retrieved ARD $p(AR)$ obtained by the OES method using different assumed mean width $\overline{D}$ and the $p(AR)$ directly measured by the TEM method; (b) Dependence of the mean square error MSE on the assumed mean width $\overline{D}$

spect to $\overline{D}$, as shown in Fig. 5(a),

$$\overline{AR} = 3.01 - 0.0176\overline{D}\,(R^2 = 0.9971) \tag{16}$$

where R^2 means the coefficient of determination of the linear fit. According to Eq. (16), we know that a change of 10 nm in $\overline{D}$ leads to around 7% change in the retrieved $\overline{AR}$.

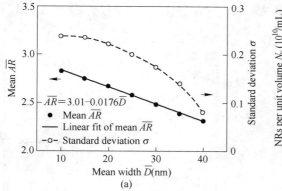
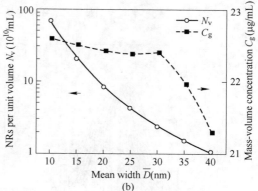

Fig. 5 (a) Dependence of the retrieved mean aspect ratio $\overline{AR}$ and the standard deviation σ on the assumed mean width $\overline{D}$ for sample NR-20-700 with assumed $\bar{e} = 0.6$; (b) Dependence of the number of NRs per unit volume N_v and the mass-volume concentration C_g of NRs on the assumed mean width $\overline{D}$

In addition, the standard deviation σ_{AR} of the retrieved $p(AR)$ also decreases with the increase of $\overline{D}$, as shown in Fig. 5(a). These are owing to the ill-posedness of the inverse problem. Therefore, if we want to get accurate solution of $p(AR)$ without knowing the value of $\overline{D}$, some other *a priori* information about the gold NR ensembles should be determined beforehand.

In Fig. 5 (b), we calculated the dependence of the number of NRs per unit volume N_v and the mass-volume concentration C_g of NRs on the mean width $\overline{D}$. Here, N_v can be obtained by the optimization progress described in Section 2 and the mass-volume concentration C_g is derived as $C_g = \rho N_v \mathbf{V}_n \cdot \mathbf{P}_{RLS}$, where $\mathbf{V}_n$ is a row vector consisting of the volume of each nanorod of AR_n and ρ is the density of bulk gold. Since N_v has an evident dependence on the mean width $\overline{D}$, as shown in Fig. 5(b), it can be used as *a priori* information for the retrieval process. In contrast, the value of C_g (= 22.45 ± 0.07 μg/mL) changes only a little with respect to the change of $\overline{D}$ when $\overline{D} \leqslant 30$ nm. Thus it is not suitable to act as *a priori* information for the determination of the mean width.

4.2 Influence by the mean end-cap eccentricity $\overline{e}$

To study the influence by the mean end-cap factor $\overline{e}$, we fix the mean width value $\overline{D} = 20$ nm and vary $\overline{e}$. In the calculation, 11 different values of $\overline{e}$ were adopted to solve the inverse problem and the obtained *MSE* values are summarized in Fig. 6(b). It is seen that eight of them (for $0.3 \leqslant \overline{e} \leqslant 1$) are acceptable and the others are unacceptable. Fig. 6(a) shows the comparison of the retrieved ARD $p(AR)$ obtained by the OES method using eight assumed mean end-cap eccentricities $\overline{e}$ and the measured $p(AR)$ obtained by the TEM method. With the increase of $\overline{e}$ from 0.3 to 1, the retrieved $p(AR)$ has a right shift while the FWHM only changes a little. The shift also bears a linear relation with respect to $\overline{e}$, following the fitted equation:

$$\overline{AR} = 2.404 + 0.43\overline{e}\ (R^2 = 0.9941) \tag{17}$$

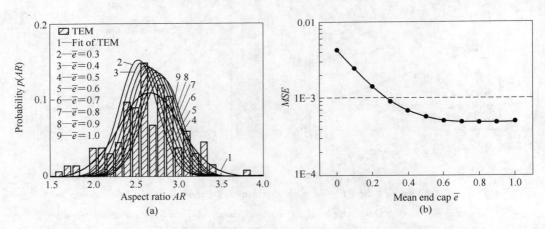

Fig. 6 (a) Comparison of the retrieved ARD $p(AR)$ obtained by the OES method using different assumed mean end-cap eccentricity $\overline{e}$ and the measured $p(AR)$ obtained by the TEM method; (b) Dependence of the *MSE* on the assumed $\overline{e}$

Similarly as the influence by $\overline{D}$, the fitted $\overline{AR}$ is affected by the change of $\overline{e}$ evidently. According to Eq. (17), a change of $\overline{e}$ by 0.1 may lead to the change of $\overline{AR}$ by 1.6%. It means that the retrieved PDF is dependent not only on the value of mean width $\overline{D}$, but also on the value of $\overline{e}$. However, different from the influence by $\overline{D}$, in this case the standard deviation σ

and the number of NRs per unit volume N_v only depend on $\bar{e}$ slightly, as shown in Fig. 7. Although N_v is also dependent on the end-cap factor e, the relativity is much smaller, compared with the influence of D. Thus it is difficult to use N_v as *a priori* information to determine the mean end-cap factor $\bar{e}$. Meanwhile the value of the mass-volume concentration can also be obtained (as $C_g = 22.39 \pm 0.26\,\mu g/mL$), which changes only a little with respect to $\bar{e}$ and the value coincides well with $C_g = 22.45 \pm 0.07\,\mu g/mL$ obtained in subsection 4.1. Therefore, we can conclude that the mass-volume concentration C_g of the gold NR ensembles can be determined accurately by the OES method, without knowing the other structrual parameters ($\bar{e}$ and $\bar{D}$) beforehand.

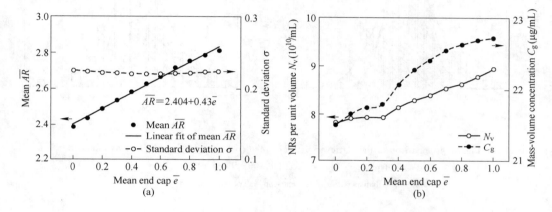

Fig. 7 (a) Dependence of the retrieved mean aspect ratio $\overline{AR}$ and the standard deviation σ on the assumed mean end-cap eccentricity $\bar{e}$ for sample NR-20-700 with assumed $\bar{D} = 20$ nm;
(b) Dependence of the number of NRs per unit volume N_v and the mass-volume concentration C_g of NRs on the assumed mean end-cap eccentricity $\bar{e}$

On the other hand, if we want to get accurate solution of $\overline{AR}$ without knowing the value of $\bar{e}$, we need to find another sensitive measurand related to the mean end-cap factor $\bar{e}$. For this aim, we may carry out some auxiliary measurements (such as scattering cross section measurement, and the polarization-dependent or incident-angle-dependent scattering measurements of individual NRs or well-aligned NR array) to facilitate the determination of $\bar{e}$. Or, alternatively, a reliable value of the mean end cap should be obtained beforehand, for example, by TEM imaging of a few sampling NRs.

4.3 Influence by the polydispersity of the width D and end-cap eccentricity e

So far, we have been always considering the fixed values of $\bar{D}$ and $\bar{e}$. In this subsection, we analyze the influence by the polydispersity of the width D and end-cap eccentricity e, i.e, σ_D and σ_e that are defined as the standard deviations of the PDFs of D and e, respectively. The retrieved $p(AR)$ was obtained by integrating the retrieval results with respect to each discrete pair of D and e.

Fig. 8 shows the retrieved ARD $p(AR)$ obtained by the OES method, by taking different σ_D

and σ_e. The *MSE* is always smaller than 1×10^{-3} so that the retrieved results are acceptable. It is seen that the changes of the mean aspect ratio $\overline{AR}$ and standard deviations σ_{AR} of $p(AR)$ are around 1% when σ_D and σ_e are significantly increased from 0 to 5.4 and to 0.143m, respectively. This shows that the influences by the polydispersity σ_D and σ_e are pretty small and can be ignored, compared with the influences by the mean width $\overline{D}$ and the mean end cap $\overline{e}$.

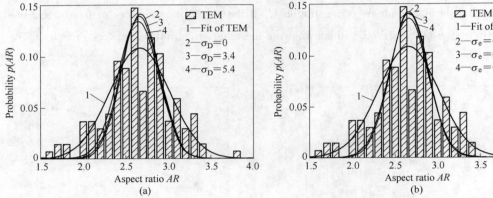

Fig. 8　Comparison of the retrieved $p(AR)$ obtained by the OES method by assuming
(a) different polydispersities of the width D and a fixed $e = 0.6$;
(b) different polydispersities of the end-cap eccentricity e and a fixed $D = 20$nm

4.4　Influence by the surface electron scattering constant A_s

To study the influence by the surface electron scattering constant A_s, we fix $\overline{D} = 20$nm and $\overline{e} = 0.6$. In the calculation, six different values of A_s were used to solve the inverse problem and the obtained *MSE* values are summarized in Fig. 9(b). It is seen that when A_s increases from 0.3 to 1.3, the values for *MSE* also increases. The range of the acceptable values ($0.3 \leq A_s \leq 0.6$) are consistent well with the measurement values determine by Ref. [11,30].

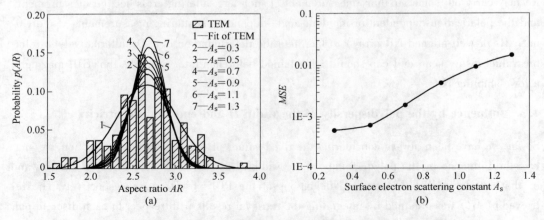

Fig. 9　(a) Comparison of the retrieved ARD $p(AR)$ by the OES method using different surface electron scattering constant A_s; (b) Dependence of the *MSE* on the A_s

Fig. 9(a) shows the retrieved ARD $p(AR)$ by the OES method, by taking different A_s. It is seen that when A_s is significantly increased from 0.3 to 1.3, the mean aspect ratio $\overline{AR}$ only changes negligibly (smaller than 1%) while the standard deviation σ_{AR} decreases by 20%. The decrease of σ_{AR} can be explained by Eq. (1): the increase of A_s increases the damping constant γ, which leads to the broadening of the L-LSPR[30]. Thus the collection of these spectra would be broadened and the σ_{AR} of the retrieved ARD would decrease. It is worth noting that only the gold NRs narrower than ~20nm may have broader resonance due to the surface scattering[30,36]. However, most of our calculations were performed for NRs with width and lengths exceeding 20 nm, so that the influences by A_s are pretty small (compared with the influences by $\overline{D}$ and $\overline{e}$) and can also be ignored.

5 Conclusions

We have studied the use of the OES method to measure the ARD of polydisperse gold NR ensembles statistically. To solve the inverse scattering problem, the extinction of the polydisperse NR ensemble is modeled rigorously by the T-matrix method and the AR parameter retrieval is performed by an optimization process with data fitting to the measured extinction spectra. We have shown that, for different NR samples that we have prepared, the retrieved PDF results coincide well with those obtained by the TEM method. The comparison results indicate that the OES method is fast, cost effective, and accurate enough if the mean width $\overline{D}$ and end-cap shape $\overline{e}$ of the NR ensembles are reasonably assumed or pre-determined. Furthermore, the C_g of NRs can also be measured by the OES method, which is useful for improving the solution of the inverse problem while cannot be obtained by the imaging methods.

Detailed analyses of the influences of NR parameters on the retrieval results have shown that the measured mean aspect ratio $\overline{AR}$ depends on the assumed mean width $\overline{D}$ and mean end-cap factor $\overline{e}$ linearly. A change of 10nm in $\overline{D}$ may lead to around 7% change in the retrieved $\overline{AR}$ and a change of $\overline{e}$ by 0.1 may lead to the change of $\overline{AR}$ by 1.6%. The influences by the polydispersity σ_D and σ_e, however, are pretty small and can be ignored. For gold NRs with the width larger than ~20nm, the influence by the surface electron scattering constant A_s is also very small and can be ignored. Based on the analyses, we suggest that the measurement accuracy can be further improved if some *a priori* information of the NRs can be obtained beforehand. A good guess of the mean width $\overline{D}$ can be obtained by measuring the number of NRs per unit volume N_v, which can be achieved by the OES method itself. To get a good guess of the end-cap shape e, some auxiliary measurements (such as scattering cross section measurements, and polarization-or incident-angle-dependent scattering measurements of NRs) could be taken, which are the tasks of our further work.

Acknowledgements

We acknowledge the support by the Ministry of Science and Technology of China (Project No. 2011BAK15B03) and the Natural Science Foundation of China (Project No. 61161130005).

References

[1] N. G. Khlebtsov and L. A. Dykman, "Optical properties and biomedical applications of plasmonic nanoparticles", J. Quant. Spectrosc. Radiat. Transfer 111, 1-35(2010).

[2] S. A. Maier, *Plasmonics*: Fundamentals and Applications(Springer, 2007).

[3] X. Huang, S. Neretina, and M. A. El-Sayed, "Gold nanorods: From synthesis and properties to biological and biomedical applications", Adv. Mater. 21, 4880-4910(2009).

[4] L. B. Scaffardi, N. Pellegri, O. de Sanctis, and J. O. Tocho, "Sizing gold nanoparticles by optical extinction spectroscopy", Nanotech. 16, 158-163(2005).

[5] W. Haiss, N. T. K. Thanh, J. Aveyard, and D. G. Fernig, "Determination of size and concentration of gold nanoparticles from uv-vis spectra", Anal. Chem. 79, 4215-4221(2007).

[6] N. G. Khlebtsov, "Determination of size and concentration of gold nanoparticles from extinction spectra", Anal. Chem. 80, 6620-6625(2008).

[7] O. Peña, L. Rodríguez-Fernández, V. Rodríguez-Iglesias, G. Kellermann, A. Crespo-Sosa, J. C. Cheang-Wong, H. G. Silva-Pereyra, J. Arenas-Alatorre, and A. Oliver, "Determination of the size distribution of metallic nanoparticles by optical extinction spectroscopy", Appl. Opt. 48, 566-572(2009).

[8] C. F. Bohren and D. R. Huffman, Absorption and Scattering of Light by Small Particles(Wiley, 1983).

[9] S. Link and M. A. El-Sayed, "Simulation of the optical absorption spectra of gold nanorods as a function of their aspect ratio and the effect of the medium dielectric constant", J. Phys. Chem. B 109, 10531C10532 (2005).

[10] M. I. Mishchenko, L. D. Travis, and A. A. Lacis, Scattering, Absorption, and Emission of Light by Small Particles(Cambridge University Press, 2002).

[11] B. Khlebtsov, V. Khanadeev, T. Pylaev, and N. Khlebtsov, "A new t-matrix solvable model for nanorods: Tembased ensemble simulations supported by experiments", J. Phys. Chem. C115, 6317-6323(2011).

[12] B. T. Draine and P. J. Flatau, "Discrete-dipole approximation for scattering calculations", J. Opt. Soc. Am. A11, 1491-1499(1994).

[13] V. L. Y. Loke and M. P. Mengüc, "Surface waves and atomic force microscope probe-particle near-field coupling: discrete dipole approximation with surface interaction", J. Opt. Soc. Am. A27, 2293-2303 (2010).

[14] S. W. Prescott and P. Mulvaney, "Gold nanorod extinction spectra", J. Appl. Phys. 99, 123504(2006).

[15] W. Yanpeng and N. Peter, "Finite-difference time-domain modeling of the optical properties of nanoparticles near dielectric substrates", J. Phys. Chem. C114, 7302-7307(2010).

[16] U. Hohenester and J. Krenn, "Surface Plasmon resonances of single and coupled metallic nanoparticles: A boundary integral method approach", Phys. Rev. B72, 195429(2005).

[17] V. Myroshnychenko, J. Rodriguez-Fernandez, I. Pastoriza-Santos, A. M. Funston, C. Novo, P. Mulvaney, L. M. Liz-Marzan, and F. J. Garcia de Abajo, "Modelling the optical response of gold nanoparticles", Chem. Soc. Rev. 37, 1792-1805(2008).

[18] M. Karamehmedović, R. Schuh, V. Schmidt, T. Wriedt, C. Matyssek, W. Hergert, A. Stalmashonak, G. Seifert, and O. Stranik, "Comparison of numerical methods in near-field computation for metallic nanoparticles", Opt. Express 19, 8939-8953(2011).

[19] T. Wriedt and U. Comberg, "Comparison of computational scattering methods", J. Quant. Spectrosc. Radiat. Transfer 60, 411-423(1998).

[20] T. Wriedt, "Light scattering theories and computer codes", J. Quant. Spectrosc. Radiat. Transfer 110, 833-

843(2009). Light Scattering:Mie and More Commemorating 100 years of Mie's 1908 publication.
[21] M. Karamehmedović, R. Schuh, V. Schmidt, T. Wriedt, C. Matyssek, W. Hergert, A. Stalmashonak, G. Seifert, and O. Stranik, "Comparison of numerical methods in near-field computation for metallic nanoparticles", Opt. Express 19, 8939-8953(2011).
[22] S. Eustis and M. A. El-Sayed, "Determination of the aspect ratio statistical distribution of gold nanorods in solution from a theoretical fit of the observed inhomogeneously broadened longitudinal plasmon resonance absorptionspectrum", J. Appl. Phys. 100, 044324(2006).
[23] R. Gans, "Über die form ultramikroskopischer goldteilchen", Annalen der Physik 342, 881-900(1912).
[24] B. N. Khlebtsov, V. A. Khanadeev, and N. G. Khlebtsov, "Observation of extra-high depolarized light scattering spectra from gold nanorods", The J. Phys. Chem. C112, 12760-12768(2008).
[25] M. I. Mishchenko, "Light scattering by randomly oriented axially symmetric particles", J. Opt. Soc. Am. A 8, 871-882(1991).
[26] F. Kuik, J. F. Dehaan, and J. W. Hovenier, "Benchmark results for single scattering by spheroids", J. Quant. Spectrosc. Radiat. Transfer 47, 477-489(1992).
[27] I. R. Ciric and F. R. Cooray, "Benchmark solutions for electromagnetic scattering by systems of randomly oriented spheroids", J. Quant. Spectrosc. Radiat. Transfer 63, 131-148(1999).
[28] P. B. Johnson and R. W. Christy, "Optical-constants of noble-metals", Phys. Rev. B6, 4370-4379(1972).
[29] E. A. Coronado and G. C. Schatz, "Surface plasmon broadening for arbitrary shape nanoparticles: A geometrical probability approach", J. Chem. Phys. 119, 3926-3934(2003).
[30] C. Novo, D. Gomez, J. Perez-Juste, Z. Zhang, H. Petrova, M. Reismann, P. Mulvaney, and G. V. Hartland, "Contributions from radiation damping and surface scattering to the linewidth of the longitudinal plasmon band of gold nanorods: a single particle study", Phys. Chem. Chem. Phys. 8, 3540-3546(2006).
[31] N. G. Khlebtsov, V. A. Bogatyrev, L. A. Dykman, and A. G. Melnikov, "Spectral extinction of colloidal gold and its biospecific conjugates", J. Colloid Interface Sci. 180, 436-445(1996).
[32] P. C. Hansen, "Regularization tools: A matlab package for analysis and solution of discrete ill-posed problems", NUMER ALGORITHMS 6, 1-35(1994).
[33] J. Mroczka and D. Szczuczynski, "Improved regularized solution of the inverse problem in turbidimetric measurements", Appl. Opt. 49, 4591-4603(2010).
[34] P. Gill, W. Murray, and M. Wright, Numerical Linear Algebra and Optimization(Addison Wesley, 1991).
[35] J. Mroczka and D. Szczuczynski, "Simulation research on improved regularized solution of the inverse problem in spectral extinction measurements", Appl. Opt. 51, 1715-1723(2012).
[36] B. N. Khlebtsov and N. G. Khlebtsov, "Multipole plasmons in metal nanorods: Scaling properties and dependence on particle size, shape, orientation, and dielectric environment", J. Phys. Chem. C111, 11516-11527(2007).

Phase Extraction from Interferograms with Unknown Tilt Phase Shifts Based on a Regularized Optical Flow Method[*]

Abstract A novel method is presented to extract phase distribution from phase-shifted interferograms with unknown tilt phase shifts. The proposed method can estimate the tilt phase shift between two temporal phase-shifted interferograms with high accuracy, by extending the regularized optical flow method with the spatial image processing and frequency estimation technology. With all the estimated tilt phase shifts, the phase component encoded in the interferograms can be extracted by the least-squares method. Both simulation and experimental results have fully proved the feasibility of the proposed method. Particularly, a flat-based diffractive optical element with quasi-continuous surface is tested by the proposed method with introduction of considerably large tilt phase shift amounts (i. e., the highest estimated tilt phase shift amount between two consecutive frame reaches 6.18λ). The phase extraction result is in good agreement with that of Zygo's MetroPro software under steady-state testing conditions, and the residual difference between them is discussed. In comparison with the previous methods, the proposed method not only has relatively little restrictions on the amounts or orientations of the tilt phase shifts, but also works well with interferograms including open and closed fringes in any combination.

1 Introduction

The temporal phase-shifting interferometry (PSI) has been generally accepted as the most accurate technique for wave-front reconstruction based on automated interferogram analysis[1]. PSI electronically records a series of interferograms while the reference phase of the interferometer is changed, so as to extract the phase information encoded in the variations in the intensity pattern of the recorded interferograms[2]. However, the accuracy of PSI is limited by the phase-shifting uncertainties resulting from the imperfect conditions in the real testing environments, such as the miscalibration, non-linear responses, aging of the phase shifter[3], and the mechanical vibration[1].

As a result, much effort has been made to improve the PSI methods, to alleviate the influence of phase-shifting uncertainties on phase extraction accuracy[1,3-15]. The papers [3-10] assume the phase steps are constant for all pixels in one interferogram but typically different between interferograms. However, sometimes such an assumption may be invalid, for instance, due to the unbalanced piezoelectric effect in the phase shifter or instability of the optical platform, the phase shifter may probably introduces non-negligible orientation errors during the shift[1,11-15];

---

[*] Copartner: Fa Zeng, Qiaofeng Tan, Huarong Gu. Reprinted from *Optics Express*, 2013, 21(14):17234-17248.

or as the extreme situations discussed in the paper [1], the strong environmental vibration will induce quite large tilt phase shift due to the relative motions between the test surface and the interferometer. In all those cases, the phase steps related to a specific interferogram are not longer constant for all pixels but vary with a tilt function across the field.

Several methods have been proposed to compensate the tilt-shift errors[11-15]. The method given in the paper [11] is based on a first-order Taylor series expansion of the phase shift errors, which iteratively update the parameters related to the tilt phase shifts until numerical convergence and the required accuracy are achieved. However, as the ratio of the tilt-shift errors to the unknown phase steps increases the first-order approximation will gradually lose the accuracy, then the compensation performance of the tilt-shift errors will get worse. In the papers [12,14] the interferograms are divided into a set of small blocks, and the phase steps within an individual block are viewed as an unknown constant. In the paper [12] these constants are solved by using calculated contrast maps, while in the paper [14] they are optimized by the least-squares method[4]. The size of the block in the papers [12,14] should be reasonably defined beforehand. The size of the block would be very small if the tilt-shift amount is large. In that case, the estimation of tilt-shift parameters will be much coarse and even untruthful. The paper [13] proposes a method to extract the tilt-shift parameters from the set of zero-crossing points of the difference between the fringe intensities. However, the amounts and orientations of the tilt phase shifts are crucial to the performance of that method (for instance, if the amount of the tilt phase shift is less than 1λ, it is impossible to estimate the related tilt-shift parameters by that method), which may not be carefully controlled in the real testing environments. In the papers [1,15] each interferogram is processed with the Fourier transform method (spatial carriers are assumed in the interferograms) so as to solve the related tilt phase shift parameters, and then the modulation phase can be extracted with the least-squares method. The paper [1] even has implemented experiments using a Mach-Zehnder interferometer, where the phase-shifts are introduced by induced vibrations without the need of any phase-shifter device. As the paper [1] includes affine registration preprocessing of the interferograms regarding for the camera movement due to strong vibration, it is expected to be applicable in a very hostile testing environment. However, if the interferograms are composed of closed fringes or the carrier frequency is inadequate, the methods given in the papers [1,15] will be erroneous.

In this paper a novel method is proposed to cope with unknown tilt phase shifts. Firstly, a set of interferograms are captured, and the coarse wrapped phase shift distributions across the whole field between the selected interferograms are estimated in order, based on the regularized optical flow method[10]. Secondly, those phase shift distributions are subsequently filtered and unwrapped, the parameters related to the tilt phase-shifting planes are determined with the help of an efficient 2-D frequency estimation method[16], and the sign ambiguity of the tilt phase shifts among interferograms is eliminated by comparing the coarse phase demodulation results obtained with the regularized optical flow method. Finally, accurate demodulation phase can be extracted from all the interferograms by the least-squares method, with all the estimated tilt phase shifts. The proposed method has some more flexibility than the previous methods[1,11-15], i. e. , it not only has relatively little restrictions on the amounts or orientations of the tilt phase shifts, but also works well with interferograms including open and closed fringes in any combination.

It should be pointed out that the extracted phase by the proposed method would have indetermination in the global sign if no prior knowledge about the phase shifts is available. However, such indetermination is not a particular problem of our method, because it is common to any asynchronous approach, if no information is given about the phase-shifts[8].

The rest of the paper is organized as follows: the principle of the proposed method is provided in detail in section 2, simulation and experimental results are reported in section 3 and section 4, respectively. Finally, conclusions are drawn in Section 5.

2 Principle

As for the set of interferograms with unknown tilt phase shifts, the intensity located at pixel (x, y) can be represented as[11,13]

$$I_n(x,y) = A(x,y) + B(x,y)\cos[\varphi(x,y) + \delta_n(x,y)] \\ = A(x,y) + B(x,y)\cos[\varphi(x,y) + (k_{xn}x + k_{yn}y + d_n)] \quad (1)$$

where $A(x, y)$ is the background, $B(x, y)$ is the modulation of the fringe pattern, $\varphi(x, y)$ is the phase to be determined, and $\delta_n(x, y)$ represents the phase shift related to the nth interferogram. $\delta_n(x, y)$ is located on a so-called phase-shifting plane, which can be described with the tilt-shift parameters k_{xn}, k_{yn}, and d_n. Without loss of generality, we can define $\delta_1(x, y) \equiv 0$. In principle, once the phase shifts $\delta_n(x, y)$ are known, all the unknowns including $A(x, y)$, $B(x, y)$, $\varphi(x, y)$ can be uniquely determined with a minimum of three interferograms[13]. For instance, with the interferograms I_1, I_2, I_3, the wrapped phase φ can be solved as[13]

$$\varphi = \arctan\frac{I_3 - I_2 + (I_1 - I_3)\cos\delta_2 + (I_2 - I_1)\cos\delta_3}{(I_1 - I_3)\sin\delta_2 + (I_2 - I_1)\sin\delta_3} \quad (2)$$

Here the pixel coordinates have been omitted for simplicity.

In practice, if possible, it is always preferable to take more than three interferograms for the phase extraction. The reasons for it can be easily understood. On one hand, for pixels where one or both of the phase shifts are equal or close to integer numbers of 2π, Eq. (2) will be invalid and the corresponding phase extraction result will be untruthful, as the equivalent number of interferograms is reduced for those pixels; On the other hand, as a rule of thumb, the PSI methods' resistance ability to noise can be improved as more interferograms are involved.

Fig. 1 shows the flow chart of the proposed method in this paper.

Obviously, the accurate estimation of all the phase-shifting plane parameters $\{k_{xn}, k_{yn}, d_n; n = 2, 3, \cdots, N\}$ [see Eq. (1)] would be crucial to the performance of the phase extraction procedure. As we will see below, the proposed method would provide an accurate and fast solution for it.

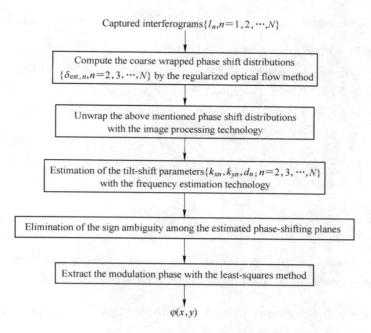

Fig. 1 The flow chart of the proposed method

Next, we will first briefly review the regularized optical flow method for the two-step interferometry, and then further introduce the procedures of the proposed method step by step. To be simple and intuitive, the descriptions in sections 2.1-2.4 are based on the case of two interferograms.

2.1 The regularized optical flow method for the two-step interferometry

Supposing there are two interferograms I_1, I_2, and their high-pass filtered background-suppressed versions are denoted as $\tilde{I}_1$ and $\tilde{I}_2$. Then we have: $\tilde{I}_1 \approx B(x,y) \cos[\varphi(x,y)]$, $\tilde{I}_2 = B(x,y) \cos[\varphi(x,y) + \delta(x,y)]$, where $\delta(x,y)$ represents the phase shift distribution between the I_1 and I_2, which is located at the phase-shifting plane: $\delta(x,y) = k_x x + k_y y + d$.

The paper [10] has put forward a regularized optical flow method to extract the wrapped demodulation phase from two interferograms with an arbitrary unknown but constant phase shift inside the range $(0, 2\pi)$, with a single exception of $\delta = \pi$, i.e. $k_x = k_y = 0$ and $d \neq \pi$ are assumed. First, the fringe direction is obtained with a regularized optical flow method in an iterative way

$$u^{k+1} = \bar{u}^k - I_x \frac{I_x \bar{u}^k + I_y \bar{v}^k + I_t}{\rho^2 + I_x^2 + I_y^2}$$

$$v^{k+1} = \bar{v}^k - I_y \frac{I_x \bar{u}^k + I_y \bar{v}^k + I_t}{\rho^2 + I_x^2 + I_y^2} \qquad (3)$$

$$\eta = \arctan\left(\frac{v}{u}\right) \qquad (4)$$

where I_x and I_y represent the derivatives of $\tilde{I}_1$ with respect to x, y, while I_t represents the difference between the two background-suppressed interferograms, i. e. $I_t = \tilde{I}_2 - \tilde{I}_1$; u^{k+1} and v^{k+1} are the velocity components obtained in the iteration $k+1$, while $\bar{u}^k$ and $\bar{v}^k$ correspond to the mean value of u and v in a defined neighborhood; ρ is the regularizing parameter that weighs the smoothness of u and v; η represents the fringe direction map. Then, the modulation phase can be extracted using the spiral phase transform

$$SPT\{\cdot\} = FT^{-1}\left\{\left(\frac{\omega_x + i\omega_y}{\sqrt{\omega_x^2 + \omega_y^2}}\right)FT\{\cdot\}\right\} \tag{5}$$

$$\varphi_{est} = \arctan\left(\frac{-i\exp(-i\eta)SPT\{\tilde{I}_1\}}{\tilde{I}_1}\right) \tag{6}$$

where $SPT\{\cdot\}$ denotes the spiral phase transform; ω_x and ω_y are the coordinates in the spectral domain; φ_{est} is the extracted phase by the regularized optical flow method.

2.2 Computation of the wrapped phase shift distribution between two interferograms

Actually, the phase extraction method given in the paper [10] can provide us with more information than the authors revealed.

On one hand, the phase shift between the two interferograms can be easily obtained with minor extension to that method, if its distribution is uniform across the field (i. e. $k_x = k_y = 0$). We find that the shape of the direction map η is almost invariant as the phase shift changes, but the global sign of it is dependent on the range value of the phase shift δ. Specifically, if we denote the resultant direction map as $\eta_1(\delta \in (0, \pi))$, and $\eta_2(\delta \in (\pi, 2\pi))$, respectively, then we will have $\eta_2 \approx -\eta_1$. As a result, we further have

$$\varphi_{est}(x,y) \approx \begin{cases} \mathrm{mod}(\varphi(x,y), 2\pi), & 0 < \delta < \pi \\ -\mathrm{mod}(\varphi(x,y), 2\pi), & \pi < \delta < 2\pi \end{cases} \tag{7}$$

Here, if we suppose $0 < \delta < \pi$, and exchange the order of the two interferograms, i. e. letting

$$\begin{aligned}\tilde{I}'_1(x,y) &= \tilde{I}_2(x,y) \approx B(x,y)\cos[\varphi'(x,y)] \\ \tilde{I}'_2(x,y) &= \tilde{I}_1(x,y) \approx \cos[\varphi'(x,y) - \delta] = \cos[\varphi'(x,y) + \delta']\end{aligned} \tag{8}$$

where $\varphi'(x,y) = \varphi(x,y) + \delta$, and $\delta' = 2\pi - \delta \in (\pi, 2\pi)$; $\mathrm{mod}()$ is a function related to the modulo operation. Then we can obtain another estimation result $\varphi'_{est}(x,y)$ of the demodulation phase with Eqs. (3)-(6). Meanwhile, from previous analysis, it is easy to deduce that: $\varphi'_{est}(x, y) \approx -\mathrm{mod}(\varphi(x,y) + \delta, 2\pi)$. Thus we can further get the following equation

$$\delta_{est}(x,y) = \mathrm{mod}(\varphi_{est}(x,y) + \varphi'_{est}(x,y), 2\pi) \approx 2\pi - \delta \tag{9}$$

Similarly, if we suppose $\pi < \delta < 2\pi$, we will have

$$\delta_{est}(x,y) = \mathrm{mod}(\varphi_{est}(x,y) + \varphi'_{est}(x,y), 2\pi) \approx \delta \tag{10}$$

From Eqs. (9) and (10), we can conclude that by adding the two extracted phase results obtained with the regularized optical flow method, an estimation of the phase shift can be obtained which is specially wrapped inside the approximate range $(\pi, 2\pi)$. The accuracy of the phase shift estimation can be improved by averaging it across the whole field, i.e., letting δ_{est} = mean $[\delta_{est}(x, y)]$, where the function mean ( ) is related to the average operation. Additionally, at some pixels the estimated phase shift $\varphi_{est}(x, y)$ will fall outside the range $(\pi, 2\pi)$, mainly due to the noise in the interferograms and the truncation effect of the modulo operation. To compensate for it, we will add 2π to these pixels where $\varphi_{est}(x, y) \in \left(0, \dfrac{\pi}{2}\right)$.

On the other hand, the phase shift distribution can also be analyzed as described above even if it is non-uniform across the whole field, since the regularized optical flow method is implemented in a local way. However, if the phase shift is equal to $k\pi (k \in Z)$ somewhere in the field, the computed wrapped phase shift distribution $\delta_{est}(x, y)$ would include some flip lines. In this case, it should be correctly unwrapped inside the approximate range $(0, 2\pi)$, which can be implemented as follows.

2.3 Unwrap the phase shift distribution

First, the map $\delta_{est}(x, y)$ is filtered with a mean filter (the size of filter adopted in this paper is 10×10 pixels) to generate a map denoted as $\bar{\delta}_{est}(x, y)$, then a binary map denoted as $BM1$ can be further produced as

$$BM1(x,y) = \begin{cases} 1, & |\bar{\delta}_{est}(x,y) - \pi| \leq \dfrac{\pi}{2} \\ 0, & |\bar{\delta}_{est}(x,y) - \pi| > \dfrac{\pi}{2} \end{cases} \quad (11)$$

Subsequently, the connected-component labeling and the morphological operations including erosion and dilation[17,18] are successively applied, to remove the spurious pixels and to fill small holes in the map $BM1$. Then the map $\bar{\delta}_{est}(x, y)$ can be divided into several sub-regions according to the value of $BM1(x, y)$. In each sub-region of the map $\bar{\delta}_{est}(x, y)$ the extreme-value points are searched by scanning along the row or column direction (the selection of scanning direction is dependent on the orientation of the sub-regions), and then the corresponding flip line can be fixed by linear fitting with the coordinates of the extreme-value points. The estimated flip line related to the border sub-region would be rechecked to ensure its validity, by referring to the fitting residual error and comparing the slope of it with the counterparts in other sub-regions. Supposing totally n estimated flip lines are found, the whole field will be divided into $n+1$ sub-regions, which will be labeled as $sub_i (i = 1, 2, \cdots, n+1)$ in order by their spatial locations. Then a new binary map $BM2$ can be produced by letting

$$BM2(x,y) = \begin{cases} 1, & p(x,y) \in sub_{2m}, \quad m = 1, 2, \cdots, \dfrac{n+1}{2} \\ 0, & p(x,y) \notin sub_{2m}, \quad m = 1, 2, \cdots, \dfrac{n+1}{2} \end{cases} \quad (12)$$

or

$$BM2(x,y) = \begin{cases} 1, & p(x,y) \notin sub_{2m}, \quad m = 1,2,\cdots,\dfrac{n+1}{2} \\ 0, & p(x,y) \in sub_{2m}, \quad m = 1,2,\cdots,\dfrac{n+1}{2} \end{cases} \quad (13)$$

where $p(x,y)$ denotes the pixel point (x,y). Then, the map $\bar{\delta}_{est}(x,y)$ can be unwrapped inside the approximate range $(0,2\pi)$ as

$$\tilde{\delta}_{est}(x,y) = \begin{cases} \bar{\delta}_{est}(x,y), & BM2(x,y) = 1 \\ 2\pi - \bar{\delta}_{est}(x,y), & BM2(x,y) = 0 \end{cases} \quad (14)$$

2.4 Estimation of the tilt phase shift parameters

From the expression $\delta(x,y) = k_x x + k_y y + d$ related to the phase-shifting plane, we can deduce that the data in any row (column) of the complex signal $\exp[-j\delta(x,y)]$ is single-tone, with frequency k_x rad/pixel (k_y rad/pixel). Thus, we can estimate the phase-shifting plane parameters k_x, k_y from $\exp[-j\tilde{\delta}_{est}(x,y)]$ using an efficient 2-D frequency estimation method given in the paper [16]. If the estimated results are denoted as $k_{x,est}$ and $k_{y,est}$, then the estimated parameter d can be further obtained as

$$d_{est} = \text{angle}\left\{ \sum_{x,y} \exp[j\tilde{\delta}_{est}(x,y)] \exp[-j(k_{x,est}x + k_{y,est}y)] \right\} \quad (15)$$

where angle$\{\cdot\}$ denotes the operation of extracting the phase angle from a complex number. Then the estimated phase-shifting plane between the interferograms I_1, I_2 can be obtained as

$$\tilde{\delta}_2(x,y) = k_{x,est}x + k_{y,est}y + d_{est} \quad (16)$$

2.5 Elimination of the sign ambiguity among the estimated phase-shifting planes

Supposing there are totally N phase-shifted interferograms, then all the phase-shifting planes $\delta_i(x,y)$, $2 \leq i \leq N$ can be estimated in order with the procedures introduced in section 2.1-2.4. However, it can be noted that as to the specific phase-shifting plane, the selection of Eq. (12) or Eq. (13), will give rise to two different estimation results with the same phase-shifting distribution but with an opposite global sign. As a result, the estimated phase-shifting planes will have at most 2^{N-1} possible arrangements $\{\pm\tilde{\delta}_2(x,y), \pm\tilde{\delta}_3(x,y), \cdots, \pm\tilde{\delta}_N(x,y)\}$, and the majority of them will give erroneous phase extraction results, i.e., only the two arrangement which are closest to the true phase-shifting arrangement $\{\delta_2(x,y), \delta_3(x,y), \cdots, \delta_N(x,y)\}$ or its opposite version $\{-\delta_2(x,y), -\delta_3(x,y), \cdots, -\delta_N(x,y)\}$, will achieve the accurate estimation of the modulation phase $\varphi(x,y)$ or its opposite version $-\varphi(x,y)$. Here we put forward a simple way to find the interested arrangement. Considering the case of three phase-shifted interferograms I_i, I_j, I_k ($1 \leq i,j,k \leq N$), where the estimated coarse phase distributions [see Eq. (6)], the intermediate binary maps [see Eqs. (12) or (13)], and the estimated phase-shifting planes from the two-step phase-shifted interferograms $\{I_i, I_j\}$, $\{I_i, I_k\}$ are denoted as $\{\varphi_{est,ij}$

$(x,y), \varphi_{est,ik}(x,y) \}$, $\{BM2_{ij}, BM2_{ik}\}$, and $\{\tilde{\delta}_{ij}(x,y), \tilde{\delta}_{ik}(x,y)\}$, respectively. Then we can define

$$\tilde{\varphi}_{est,ij}(x,y) = \begin{cases} \varphi_{est,ij}(x,y) & BM2_{ij}(x,y) = 1 \\ 2\pi - \varphi_{est,ij}(x,y), & BM2_{ij}(x,y) = 0 \end{cases} \quad (17)$$

The $\tilde{\varphi}_{est,ik}(x,y)$ is defined with $\varphi_{est,ik}(x,y)$ and the map $BM2_{ik}$ in the similar way. Subsequently, we further define

$$\begin{aligned} val1 &= \left| \sum_{x,y} \{\exp[-j\tilde{\varphi}_{est,ij}(x,y)]\exp[-j\tilde{\varphi}_{est,ik}(x,y)]\} \right| \\ val2 &= \left| \sum_{x,y} \{\exp[j\tilde{\varphi}_{est,ij}(x,y)]\exp[-j\tilde{\varphi}_{est,ik}(x,y)]\} \right| \end{aligned} \quad (18)$$

if $val1 > val2$, the arrangements $\{\tilde{\delta}_{ij}(x,y), -\tilde{\delta}_{ik}(x,y)\}$ or $\{-\tilde{\delta}_{ij}(x,y), \tilde{\delta}_{ik}(x,y)\}$ will be adopted, otherwise the arrangements $\{\tilde{\delta}_{ij}(x,y), \tilde{\delta}_{ik}(x,y)\}$ or $\{-\tilde{\delta}_{ij}(x,y), -\tilde{\delta}_{ik}(x,y)\}$ will be adopted. When there are more than three phase-shifted interferograms, the phase-shifting arrangement can be searched in the similar way.

2.6 Phase extraction with the least-squares method

Eq. (1) can be rewritten as

$$I_n(x,y) = a(x,y) + b(x,y)\cos[\delta_n(x,y)] + c(x,y)\sin[\delta_n(x,y)] \quad (19)$$

where $A(x,y) = a(x,y), b(x,y) = B(x,y)\cos[\varphi(x,y)], c(x,y) = -B(x,y)\sin[\varphi(x,y)], \delta_n(x,y) = k_{xn}x + k_{yn}y + d_n$. Supposing there are totally N phase-shifted interferograms ($N \geq 3$), after all the phase-shifting planes $\delta_{est,n}(x,y)$ have been determined, the phase distribution $\varphi(x,y)$ can be extracted with the least-squares method[1,4,15] as follows

$$\begin{bmatrix} a_{est}(x,y) \\ b_{est}(x,y) \\ c_{est}(x,y) \end{bmatrix} = \begin{bmatrix} N & \sum_{n=1}^{N}\cos[\delta_{est,n}(x,y)] & \sum_{n=1}^{N}\sin[\delta_{est,n}(x,y)] \\ \sum_{n=1}^{N}\cos[\delta_{est,n}(x,y)] & \sum_{n=1}^{N}\cos^2[\delta_{est,n}(x,y)] & \sum_{n=1}^{N}cs_n(x,y) \\ \sum_{n=1}^{N}\sin[\delta_{est,n}(x,y)] & \sum_{n=1}^{N}cs_n(x,y) & \sum_{n=1}^{N}\sin^2[\delta_{est,n}(x,y)] \end{bmatrix}^{-1} \times \begin{bmatrix} \sum_{n=1}^{N}I_n(x,y) \\ \sum_{n=1}^{N}I_n(x,y)\cos[\delta_{est,n}(x,y)] \\ \sum_{n=1}^{N}I_n(x,y)\sin[\delta_{est,n}(x,y)] \end{bmatrix} \quad (20)$$

$$\tilde{\varphi}(x,y) = \arctan\left[\frac{-c_{est}(x,y)}{b_{est}(x,y)}\right] \quad (21)$$

where in Eq. (20), $cs_n(x,y) = \cos[\delta_n(x,y)]\sin[\delta_n(x,y)]$. $\tilde{\varphi}(x,y)$ is the phase extraction result; while the estimated contrast parameters $\tilde{B}(x,y) = \text{sqrt}(c_{est}^2 + b_{est}^2)$ are dependent both on $B(x,y)$ and the phase shifts related to the specific pixel (x,y), which can be taken as the reliability measure of the phase extraction result at that pixel.

3 Simulation results

As the accurate estimation of the phase-shifting plane parameters is crucial to the performance of the phase extraction procedure, a series of computation simulations have been carried out to verify the effectiveness of phase-shifting plane estimation by the proposed method. Some results will be given in the following paragraphs.

In the first simulation, the expressions for the two background-suppressed inteferograms are $\tilde{I}_1 = B(x,y)\cos[\varphi(x,y)] + n_1(x,y)$ and $\tilde{I}_2 = B(x,y)\cos[\varphi(x,y) + \delta_2(x,y)] + n_2(x,y)$, where $\varphi(x,y) = 2 \times \text{peaks}(256) + 0.0586\pi \times (x+y)$, $\delta_2(x,y) = 0.0179x + 0.0164y + 1.583$, and $B(x,y) = 60\exp(-0.005^2 x^2 - 0.005^2 y^2)$; the units of $\varphi(x,y)$ and $\delta_2(x,y)$ are both in radians, and the image sizes of $\tilde{I}_1$ and $\tilde{I}_2$ are both 256×256, i.e., x and $y \in [-127, -126, \cdots, 128]$; $n_1(x,y)$ and $n_2(x,y)$ represent zero-mean additive Gaussian white noise, the standard deviation of which are both equal to 6; while peaks ( ) is a built-in function of MATLAB, which is obtained by translating and scaling Gaussian distributions. The simulated phase-shifting plane $\delta_2(x,y)$ is shown in Fig. 2(c), the tilt-shift amount across the whole field is 1.40λ (8.78rad); the estimated phase-shifting plane $\tilde{\delta}_2(x,y)$ is shown in Fig. 2(i), and the residual PV (peak to valley) error and RMS (root mean square) error are as small as 0.019λ (0.012rad) and 0.004λ (0.026rad), respectively [see Fig. 2(j)]. The total processing time of this simulation is 1.32s using a 2.5GHz laptop and MATLAB, and the main allocations of the runtime are as follows: about 0.50s is spent associated with the regularized optical flow method (that method is run two times to estimate the phase shift); about 0.54s is spent associated with the spatial imaging processing of the intermediate binary maps BM1 and BM2, including the connected-component labeling, the morphological operations, and division of sub-regions operations; and about 0.04s is spent on solving the parameters k_x, k_y by the frequency estimation method. We noted that the residual error distribution of the phase-shifting plane is linear proportional to the estimation errors of the parameters k_x, k_y and d, thus the residual error will accumulate to its highest values at the border region of the field. If we divide the field into 2×2 blocks and solve the parameter d [see Eq. (15)] locally for each block, then the resultant residual PV error and RMS error related to this simulation will be decreased by 33% and 38%, respectively. However, such improvement ratios are case-dependent, so we will not have further discussions on it.

We also have evaluated the performance of our method by testing some other phase-shifting planes with different tilt-shift amounts and different orientations. The simulation parameters are all the same as in Fig. 2, except for the simulated phase-shifting plane. Specifically, four different phase-shifting planes are considered, where $\delta_3(x, y) = \delta_2(x, y)/2, \delta_4(x, y) = 2\delta_2(x, y)$, $\delta_5(x, y) = 0.0011x + 0.0243y + 1.583$, and $\delta_6(x, y) = -0.0164x + 0.0179y + 1.583$, i.e., the corresponding tilt-shift amounts across the field are 0.70λ, 2.80λ, 1.40λ and 1.40λ, respectively. Actually, $\delta_5(x, y)$ and $\delta_6(x, y)$ are obtained by rotating $\delta_2(x, y)$ anticlockwise in the x-y coordinate system with 0.25π and 0.5π rad, respectively. Herein, the definition of $\delta_2(x, y)$ is the same as shown in Fig. 2(c). The residual errors of the estimation results in *PV* and *RMS* measures are shown in Table 1. As can be seen from table 1, all the residual *PV* errors

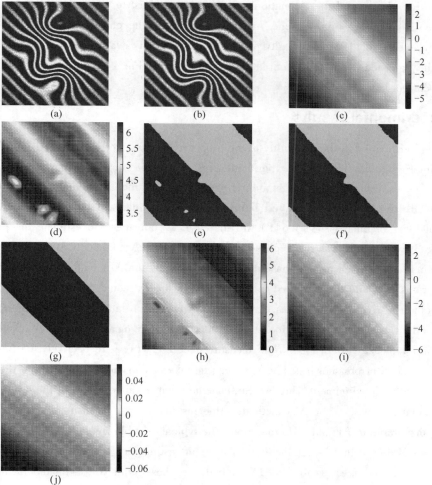

Fig. 2 (a), (b) The simulated background-suppressed inteferograms $\tilde{I}_1$ and $\tilde{I}_2$; (c) The simulated phase-shifting plane $\delta_2(x, y)$; (d) The map $\bar{\delta}_{est}(x, y)$; (e), (f): The binary map *BM*1 *via* Eq. (11) before and after being processed with the connected-component labeling as well as the morphological operations; (g) The binary map *BM*2; (h) The map $\tilde{\bar{\delta}}_{est}(x, y)$ computed *via* Eq. (14); (i) The estimated phase-shifting plane $\tilde{\delta}_2(x, y)$; (j) The residual error of the estimated phase-shifting plane, i.e. the wrapped difference between (c) and (i)

(The data shown in (c), (d), (h)-(j) are all in radians. In (e)-(g), the black and gray color represent the values of zero and one, respectively) and *RMS* errors are within 0.016λ and 0.004λ, respectively.

Table 1 The residual errors of the estimated phase-shifting planes

Simulated phase-shifting planes	PV(estimation error)/rad	RMS(estimation error)/rad	Simulated phase-shifting planes	PV(estimation error)/rad	RMS(estimation error)/rad
$\delta_3(x,y)$	0.040	0.015	$\delta_5(x,y)$	0.056	0.021
$\delta_4(x,y)$	0.055	0.022	$\delta_6(x,y)$	0.097	0.020

The above simulations demonstrate that the proposed method can well estimated the tilt phase-shifting planes, and the estimation performance is robust to noise, the orientation and the amplitude of the phase-shifting plane. As explained before, the phase-shifting plane estimation result will face a global sign ambiguity problem. To reasonably evaluate the estimation performance of the proposed method, the simulation results given in Fig. 1 and Table 1 are all assigned with the correct signs.

4 Experimental results

For further verification of the feasibility of the proposed method, we have applied it to the experimental interferograms. Two set of experimental results will be provided below.

In the first experiment, the data of interferograms are obtained from the paper [10], which include totally nine real phase-shifted interferograms with unknown but constant phase shifts (the image size of them is 481×641). In this case, we only need to estimate the parameters d_i. In Fig. 3 we show the reference phase obtained by the AIA method[4] using all the nine interferograms (b), as well as the phase extraction results by the AIA method but with only eight interferograms (c) and the proposed method (d). As the extracted phase by the AIA method would include a trivial constant bias, the residual errors of the latter two results after removing the bias would be 0.05λ in *PV* measure, 0.006λ in *RMS* measure [see Fig. 3(e)], and 0.024λ in *PV* measure, 0.007λ in *RMS* measure [see Fig. 3(f)], respectively. Particularly, we think the residual errors shown in Fig. 3(e) can be taken as an indirect measure of the non-ideality of the testing environment. Then we can conclude that in this experiment the proposed method has similar accuracy to the AIA method, as the residual errors shown in Fig. 3(e) and Fig. 3(f) are comparable in *PV* and *RMS* measures. The typical "double-frequency" fringe error shown in Fig. 3(f) is due to the discrepancies in phase shift estimations between the AIA method and the proposed method, i.e., as to the AIA method, the phase shifts are estimated in an iterative way and the information of different interferograms will be coupled in the least-squares sense during the iteration process, while for the proposed method, the estimation of phase shift between two interferograms is independent of the information encoded in other interferograms.

In another experiment, a circular flat-based diffractive optical element (DOE) with aperture of 100mm is tested, by measuring its transmitted phase. This DOE has quasi-continuous surface fabricated by the ion-beam etching technology, and it was made on the substrate of K9 glass. Firstly, the measurements are performed with a ZYGO Fizeau interferometer with the active vibration isolation workstation turned off. The recorded interferograms include moderate

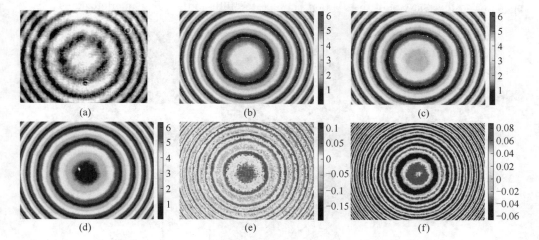

Fig. 3 (a) The first interferogram used in this experiment; (b) The wrapped reference phase map by the AIA method using all the interferograms; (c) The wrapped phase extraction result by the AIA method with eight interferograms (i. e., one of the interferograms is excluded); (d) The wrapped phase extraction result by the proposed method using all the interferograms; (e) The wrapped phase difference between (b) and (c), after removing the bias; (f) The wrapped phase difference between (b) and (d), after removing the bias

(The data shown in (b)-(f) are in radians)

amount of tilt phase shift induced by environmental vibration and considerably large tilt phase shift by purposely rotating the reflective mirror (located behind the DOE) in the test arm around the axis of the PZT. As the rotation operation is implemented manually, in our experiment the interferograms are captured with a time interval of about 3s, to eliminate the contrast reduction in them due to the sudden "rotation" actions. We randomly pick out six consecutively captured frames to extract the phase component, where two of the frames are shown in Figs. 4 (a) and (b), and the wrapped phase extraction result with the proposed method is given in Fig. 4(c). On the other hand, we have also tested the DOE by the Zygo interferometer under vibration-isolated conditions with a calibrated PZT. The phase of the test surface extracted by the Zygo's MetroPro software is shown in Fig. 4(d). Since noticeable lighting scattering would take place at the staircase locations of the DOE surface, the phase data are missing somewhere with the Zygo's software. The difference between the phase extraction results by the proposed method and the Zygo's software is presented in Fig. 4(e), which amounts to 0. 028λ in *RMS* measures with the piston and tilt components removed; it must be pointed out that from the sense of surface test the relative difference between them would decrease nearly by half, as all the phase extraction results given in the following figures relate to the "double-pass" transmitted wave-front. In addition, the estimated results of the phase-shifting planes with the proposed method are shown in Fig. 5. It can be seen that the tilt-shift amounts between consecutive interferograms reach 4. 15λ, 3. 80λ, 2. 33λ, 4. 07λ, 6. 18λ across the field, respectively; as to the accumulated phase shifts relative to the first interferogram, the highest tilt-shift amount across the field would be 14. 34λ, corresponding to the phase shift distribution between the fifth and the first interferograms.

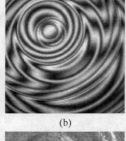
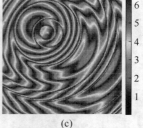
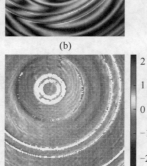
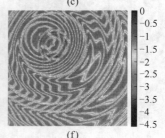

Fig. 4 (a), (b) The first two interferogram used in this experiment with a image size of 240 × 240, corresponding to a part of the DOE area; (c) The wrapped phase extraction result by the proposed method; (d) The phase extraction result by the Zygo's MetroPro software with the piston and tilt components removed, where the phase data related to the pixels in white color are missing; (e) The wrapped phase difference between (c) and (d), with the piston and tilt components removed; (f) The normalized estimated contrast parameters $\tilde{B}(x, y)$ shown in the log10 scale (the definition of $\tilde{B}(x, y)$ can be found in section 2.6)

(The data shown in (c)-(e) are in radians)

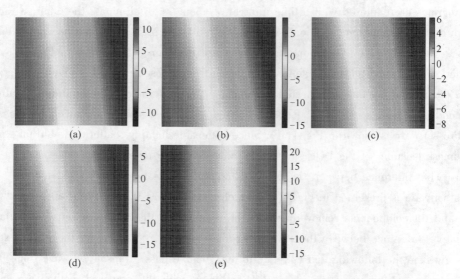

Fig. 5 The estimated phase-shifting planes
(All the data shown in these figures are in radians)
(a) between the first and the second interferograms; (b) between the second and the third interferograms;
(c) between the third and the fourth interferograms; (d) between the fourth and the fifth interferograms;
(e) between the fifth and the sixth interferograms

The residual error as shown in Fig. 4 is mainly due to the following factors: (1) the retrace errors; (2) the errors accompanied with the low estimated contrast parameters; (3) the fluctuations of the laser power; and (4) the dynamic variations of the testing environment, such as the air turbulence, which will be discussed in more detail. Firstly, as the wave-front of the DOE is far from the ideal plane wave, the introduction of tilt phase shifts will also give rise to non-negligible retrace errors. To demonstrate it, the DOE is measured at different status under vibration-isolated conditions. As shown in Fig. 6(c), the difference between the phase extraction results is mainly composed of tilt component, which equals to 7.25λ in the PV measure, and the resultant relative retrace error between them is shown in Fig. 6(d), which equals to 0.023λ in the RMS measure. As the tilt shift amount increases, the retrace error will be larger, and vice versa. As to our experiment, different retrace errors have been introduced in the interferograms we used, and the phase extraction result by the Zygo's software also include some retrace error itself. Therefore, the residual error as shown in Fig. 4(e) will inevitably contain the contribution of retrace errors. Secondly, as shown in Figs. 4(e) and (f) the areas with high residual error values are found to be related with extremely low estimated contrast parameters. It is easy to understand that in such areas the accuracy of phase extraction result is much more susceptible to the noise in the interferograms, as well as the residual estimation errors of the phase-shifting planes. This problem can be alleviated by using more interferograms for the phase extraction. Thirdly, as the interferograms used in this experiment were captured during a relative long time (about 16s), the parameters $A(x, y)$, $B(x, y)$ [see Eq. (1)] will be time dependent as a result of the fluctuations in laser power, which will introduce some error into the phase extraction result. Finally, the dynamic variations of the testing environment, such as the air turbulence will also make some differences between measurements.

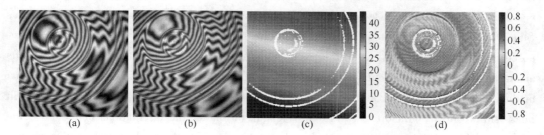

Fig. 6   Difference between phase extraction results by the Zygo's MetroPro software
at different testing status, to demonstrate the relative retrace error
(The data shown in (c), (d) are in radians)
(a) the typical interferogram related to the testing status 1; (b) the typical interferogram related to the testing status 2;
(c) the difference in phase extraction results by the Zygo's MetroPro software, between the testing status 1
and the testing status 2 (the phase data related to the pixels in white color are missing);
(d) the same as (c), but with the piston and tilt components removed

As to the second experiment, since the tilt phase shift amounts related to the phase-shifting planes are considerably large, the AIA method will be failed; and the block-wise methods given in the papers [12,14] also can hardly retrieve the correct phase, as in this case the interferograms should be divided into extremely many blocks with small sizes (probably more than one thousand blocks are required), so that the estimated phase shifts related to each small block are prone to be inaccurate. Besides, the methods given in the papers [1,15] also can hardly be applied in this experiment, which rely on introduction of adequate spatial carriers into the interferograms. Additionally, we also found that if the tilt-shift amount across the field get too large (for instance, larger than 10λ), the direct estimation of the phase-shifting plane by the proposed method is prone to be inaccurate, due to the probable failure in the correct sub-region division of the intermediate binary map $BM2$ (as to the definition of the map $BM2$, please referring to the section 2.3).

5 Conclusion

We have proposed a novel method for extracting phase distribution from interferograms with unknown tilt phase shifts. The proposed method can estimate the unknown tilt phase shift between two temporal phase-shifted interferograms, by extending the regularized optical flow method provided in the paper [10] with the spatial image processing and frequency estimation technology. With all the estimated tilt phase shifts, we can further obtain the phase extraction result with the least-squares method. Both simulation and experimental results have proved the feasibility of the proposed method. The proposed method is expected to be used in a testing environment with low frequency and high amplitude vibration, where costly and accurate phase-shifting devices are not longer required for steady-state measures.

Additionally, the proposed method has shown some more flexibility in comparison with the previous methods[1,11-15]. On one hand, the proposed method has relatively little restrictions on the amounts or orientations of the tilt phase shifts. The methods proposed in [11,12,14] are mainly adaptable to the phase-shifted interferograms with relatively low tilt-phase shift amounts (the typical amounts reported in those papers are much less than 1λ); and the method proposed in [13] is quite sensitive to the amounts and orientations of the unknown tilt phase shifts among interferograms, particularly, if the amount of the tilt phase shift is less than 1λ, that method will fail. While, the method proposed in this paper can handle with the unknown tilt phase shifts with amount up to several wavelength, no matter the orientations of them. On the other hand, unlike the methods proposed in the papers [1,15], the method proposed in this paper requires no spatial carriers in the interferograms, i.e., it can work well with interferograms including open and closed fringes in any combination.

Acknowledgements

We are grateful to Professor Javier Vargas at Centro Nacional de Biotecnología-CSIC, Spain, for his kind help about the phase-shifting methods and providing us with partial experimental data of interferograms used in this paper. The research was partially supported by the National Basic Research Program of China under grant No. 2011CB706701. Thanks also go to the anonymous reviewers for their valuable comments and suggestions.

References

[1] J. Vargas, J. A. Quiroga, A. Álvarez-Herrero, and T. Belenguer, "Phase-shifting interferometry based on induced vibrations", Opt. Express 19, 584-596(2011).

[2] D. Malacara, Optical Shop Testing, 3rd Edition(Wiley, 2007), Chap. 14, 547-666.

[3] R. Juarez-Salazar, C. Robledo-Sánchez, C. Meneses-Fabian, F. Guerrero-Sánchez, and L. M. Arévalo Aguilar, "Generalized phase-shifting interferometry by parameter estimation with the least squares method", Opt. Laser. Eng. 51, 626-632(2013).

[4] Z. Wang and B. Han, "Advanced iterative algorithm for phase extraction of randomly phase-shifted interferograms", Opt. Lett. 29, 1671-1673(2004).

[5] X. Xu, L. Cai, Y. Wang, X. Meng, W. Sun, H. Zhang, X. Cheng, G. Dong, and X. Shen, "Simple direct extraction of unknown phase shift and wavefront reconstruction in generalized phase-shifting interferometry: algorithm and experiments", Opt. Lett. 33, 776-778(2008).

[6] P. Gao, B. Yao, N. Lindlein, K. Mantel, I. Harder, and E. Geist, "Phase-shift extraction for generalized phase-shifting interferometry", Opt. Lett. 34, 3553-3555(2009).

[7] B. Li, L. Chen, W. Tuya, S. Ma, and R. Zhu, "Carrier squeezing interferometry: suppressing phase errors from the inaccurate phase shift", Opt. Lett. 36, 996-998(2011).

[8] J. Vargas and C. O. S. Sorzano, "Quadrature Component Analysis for interferometry", Opt. Laser. Eng. 51, 637-641(2013).

[9] H. Guo, "Blind self-calibrating algorithm for phase-shifting interferometry by use of cross-bispectrum", Opt. Express 19, 7807-7815(2011).

[10] J. Vargas, J. A. Quiroga, C. O. S. Sorzano, J. C. Estrada, and J. M. Carazo, "Two-step interferometry by a regularized optical flow algorithm", Opt. Lett. 36, 3485-3487(2011).

[11] M. Chen, H. Guo, and C. Wei, "Algorithm immune to tilt phase-shifting error for phase-shifting interferometers", Appl. Opt. 39, 3894-3898(2000).

[12] A. Dobroiu, D. Apostol, V. Nascov, and V. Damian, "Tilt-compensating algorithm for phase-shift interferometry", Appl. Opt. 41, 2435-2439(2002).

[13] O. Soloviev and G. Vdovin, "Phase extraction from three and more interferograms registered with different unknown wavefront tilts", Opt. Express 13, 3743-3753(2005).

[14] J. Xu, Q. Xu, and L. Chai, "Iterative algorithm for phase extraction from interferograms with random and spatially nonuniform phase shifts", Appl. Opt. 47, 480-485(2008).

[15] J. Xu, Q. Xu, and L. Chai, "Tilt-shift determination and compensation in phase-shifting interferometry", J. Opt. A: Pure Appl. Opt. 10, 075011(2008).

[16] S. Ye and E. Aboutanios, "Two dimensional frequency estimation by interpolation on Fourier coefficients", in Proc. of IEEE Int. Conf. on Acoustics, Speech and Signal Processing, 3353-3356(2012).

[17] R. C. Gonzalez, R. E. Woods, and S. L. Eddins, Digital Image Processing Using MATLAB(Prentice Hall, Upper Saddle River, NJ, 2004).

[18] L. G. Shapiro and G. C. Stockman, Computer Vision(Prentice Hall, Upper Saddle River, New Jersey, USA, 2001).

Accurate Geometric Characterization of Gold Nanorod Ensemble by an Inverse Extinction/Scattering Spectroscopic Method*

Abstract Aspect ratio, width, and end-cap factor are three critical parameters defined to characterize the geometry of metallic nanorod (NR). In our previous work [Opt. Express 21, 2987 (2013)], we reported an optical extinction spectroscopic (OES) method that can measure the aspect ratio distribution of gold NR ensembles effectively and statistically. However, the measurement accuracy was found to depend on the estimate of the width and end-cap factor of the nanorod, which unfortunately cannot be determined by the OES method itself. In this work, we propose to improve the accuracy of the OES method by applying an auxiliary scattering measurement of the NR ensemble which can help to estimate the mean width of the gold NRs effectively. This so-called optical extinction/scattering spectroscopic (OESS) method can fast characterize the aspect ratio distribution as well as the mean width of gold NR ensembles simultaneously. By comparing with the transmission electron microscopy experimentally, the OESS method shows the advantage of determining two of the three critical parameters of the NR ensembles (i.e., the aspect ratio and the mean width) more accurately and conveniently than the OES method.

1 Introduction

Fast and accurate geometric characterization of noble metal nanoparticles such as gold nanorod (NR) ensembles is highly demanded in practical production and trade of nanoparticles, for their important applications in various fields such as catalysis, medical diagnosis and therapy, biosensing, and drug delivery and release[1]. The NR geometry is defined in Fig. 1, which is a cylinder with its end caps in different shapes, depending on the different generation methods of the NRs. Usually, three critical parameters (i.e., aspect ratio, width, and end-cap factor) are defined to characterize the geometry of gold NRs.

In our previous work[2], the aspect ratio distribution (ARD) of gold NR ensembles was determined effectively and statistically by an optical extinction spectroscopic (OES) method. It is found that with the OES method, the measured ARD would depend on the nominal mean width and mean end-cap factor of the NRs, although both of them are considered to affect the magnitude and spectral position of the localized surface plasmon resonance (LSPR) of the NRs weakly. Thus we suggested that the measurement accuracy of the ARD can be further improved if the mean width and the mean end-cap factor of the NRs could be pre-determined as *a priori* information.

---

* Copartner: Ninghan Xu, Benfeng Bai, Qiaofeng Tan. Reprinted from *Optics Express*, 2013, 21(18):21639-21650.

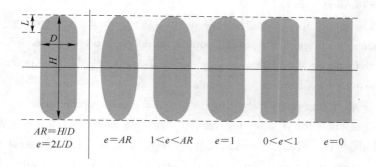

Fig. 1 Geometric model of the NR

(Several NRs with the same width D and aspect ratio AR but different end-cap factor e are demonstrated)

To achieve this goal, an intuitive idea is to use a microscopic imaging method such as transmission electron microscopy (TEM) to characterize the width and end-cap shape of individual NRs directly[3,4]. However, with such imaging methods it is hard to obtain the probability distribution function (PDF) of polydisperse NR ensembles statistically, because a large amount (to the order of 10^3) of sampling NRs should be characterized, which is time consuming and expensive. Therefore, in this work we aim to develop a method based on scatterometry to achieve this goal more conveniently.

It is known that the extinction (which is contributed by both absorption and scattering) property of metallic NRs is strongly dependent on the shape and size of the NRs. However, the geometric parameter dependence of the extinction property and that of the scattering property are different, which has been investigated in many previous works[5,6]. For the extinction spectra of NRs, the aspect ratio is the most sensitive parameter affecting the magnitude and position of the longitudinal mode LSPR (named in this work as L-LSPR) peak. However, for the scattering spectra, the NR width is the most sensitive parameter affecting the magnitude of the L-LSPR peak while the aspect ratio is still the most sensitive parameter affecting the position of the L-LSPR[5]. According to these properties, since the OES method has been developed[2,7,8] to retrieve the ARD of the polydisperse gold NR ensembles, a method based on optical scattering spectroscopy (OSS) has the potential to be used as an auxiliary method to determine the mean width of the gold NR ensembles.

Compared with the OES method that can be applied conveniently based on, for example, a UV-VIS spectrophotometer[2], the OSS setup is more complicated. To measure the scattering cross section of the gold nanoparticle ensembles, two alternative approaches have been reported. One is to measure the total scattering energy integrated over the 4π solid angle[9], while the other approach is to measure the angular scattering energy within a small solid angle[10,11]. It is obvious that the latter approach is easier to implement, where the integrating-sphere is not required. However, for the randomly oriented gold NRs, the simulation of the differential scattering cross section in the second approach is more complicated than the simulation of total scattering cross section in the first approach[12,13].

In this work, we apply the second scheme of the OSS method. By combining the OES and OSS measurements, we propose a so-called optical extinction/scattering spectroscopic (OESS) method to fast characterize the ARD as well as the mean width of gold NR ensembles simultaneously. The main difference between our approach and the previous works[10,11,14,15] is that we solve the inverse scattering problem by using a fast and reliable optimization procedure so that there is no need to know the *a priori* information about the geometric values of NRs (such as the nominal width of the NRs) beforehand. When constructing the database of the scattering and extinction spectra of the NRs, the T-matrix method[12] was used to simulate the extinction and differential scattering cross sections of gold NR ensembles rigorously. By using the OESS method to characterize different NR ensemble samples experimentally, it is shown that the measurement results coincide with those of the TEM method quite well, while the OESS method is much faster and more cost effective. The OESS method is also more accurate than the OES method to determine the two critical parameters (aspect ratio and mean width) of NRs.

2 Methods

It is in essence an inverse scattering problem by using the OESS method to retrieve the geometric parameters of the gold NRs from the measured extinction and differential scattering spectra. The objective is to retrieve the PDF $p(D, AR, e)$ with respect to the three structural quantities: the width D, the aspect ratio AR and the end-cap factor e of the NRs as well as determining the estimates of the parameters D, AR and e. The measurement method and the numerical retrieving algorithm applied in the OESS method are presented below.

2.1 Measurement method and setup

We apply a commercial UV-VIS spectrophotometer (PekinElmer LAMBDA950) to measure the extinction spectra of the sample, as shown in Fig. 2(a). The setup is the same as that used in the OES method presented in our previous work[2]. The measurand of this optical setup is the absorbance $A(\lambda)$ of the sample medium with a length l, which can be expressed as

$$A(\lambda) = -\log\left[\frac{I_e(\lambda)I_{r0}(\lambda)}{I_{r1}(\lambda)I_{m0}(\lambda)}\right] = \frac{lN_v}{\ln 10}\langle C_{ext}(\lambda)\rangle \tag{1}$$

where λ is the wavelength of light, N_v is the number of NRs per unit volume, $\langle C_{ext}(\lambda)\rangle$ is the average extinction cross section of the sample, and $I(\lambda)$ with different subscripts represent the intensities of different light beams shown in Fig. 2(a).

To perform the OSS measurement, we modify the spectrophotometer setup as shown in Fig. 2(b), which can measure the angular scattering spectra at different angles[10]. Here we measure the scattering at an angle of 90°. Two reflective mirrors are added in the measurement beam to change the incident angle of the beam onto the sample. The same sample is inserted in the reference beam with a neutral density (ND) filter to attenuate the beam. The measurand is

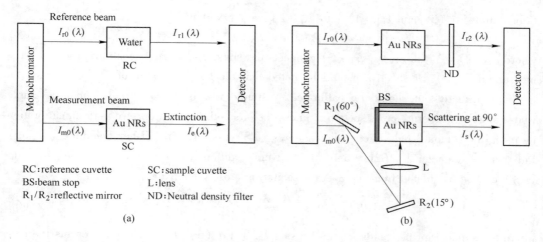

Fig. 2 Optical setups used in the OESS method for measuring (a) the optical extinction spectra and (b) the angular light scattering spectra at an angle of 90° of the gold NRs

the transmittance $T_0(\lambda)$ and can be expressed by

$$T_0(\lambda) = \frac{I_s(\lambda)I_{r0}(\lambda)}{I_{r2}(\lambda)I_{m0}(\lambda)} \tag{2}$$

where $I(\lambda)$ with different subscripts represent the intensities of different light beams shown in Fig. 2(b). $I_{r2}(\lambda)$ and $I_s(\lambda)$ in Eq. (2) can be expressed as

$$\begin{aligned} I_{r2}(\lambda) &= I_{r0}(\lambda)T_{nd}(\lambda)T_{ext}(\lambda) \\ I_s(\lambda) &= I_{m0}(\lambda)R_1(\lambda)R_2(\lambda)T_{lens}(\lambda)T_{ext}(\lambda)\alpha_{sca}^{90°}(\lambda) \end{aligned} \tag{3}$$

where $T_{nd}(\lambda)$, $T_{ext}(\lambda)$, and $T_{lens}(\lambda)$ denote the transmittances of the ND filter, the sample, and the lens, respectively, $R_1(\lambda)$ and $R_2(\lambda)$ are the reflectances of the mirrors, and $\alpha_{sca}^{90°}(\lambda)$ is the angular scattering efficiency.

When measuring $\alpha_{sca}^{90°}(\lambda)$, the same sample was inserted in the reference beam to correct the extinction of the gold NR sample in the measurement beam, as shown in Fig. 2(b). This is because that the scattering spectra measured without correction would differ radically from the real spectra if the extinction of the measurement volume is not taken into account[10]. Besides, here we use a lens to enhance the measurement beam and use a ND filter to attenuate the reference beam. The reason is that the value of $\alpha_{sca}^{90°}(\lambda)$ is quite small (due to the weak scattering of NRs at 90°) compared with the value of $I_{r2}(\lambda)$ so that the signal-to-noise ratio of the detector is poor when measuring the transmittance $T_0(\lambda)$ directly. By using the lens and the ND filter, the measurement values of $I_s(\lambda)$ and $I_{r2}(\lambda)$ can be adjusted to the same order of magnitude so as to get the best detection response.

In the OSS measurement, the angular scattering efficiency $\alpha_{sca}^{90°}(\lambda)$ is in proportion to the solid angle $\Omega_{90°}$ centered around the direction of scattering, the number of NRs per unit volume N_v, and the average differential scattering cross section $\langle dS(\lambda)\rangle$, i.e.[12],

$$\alpha_{sca}^{90°}(\lambda) \propto N_v \Omega_{90°} \langle dS(\lambda)\rangle \tag{4}$$

Substituting Eq. (3) and Eq. (4) into Eq. (2), we can obtain

$$T_0(\lambda) \propto S_{90}(\lambda) \tag{5}$$

where $S_{90}(\lambda) = N_v \langle dS(\lambda) \rangle$. Eq. (5) connects the angular scattering spectra $S_{90}(\lambda)$ with the measurand $T_0(\lambda)$. However, it should be noted that the absolute value of $S_{90}(\lambda)$ cannot be measured directly but should be calibrated. A simple way is to measure a standard sample whose angular scattering cross section is known beforehand. In this work, we perform calibration by using the standard polystyrene microspheres (PS) from Thermo Fisher Scientific (China) Ltd., whose diameter is 102nm and whose average differential scattering cross section $\langle dS(\lambda) \rangle$ can be calculated from the average scattering cross section $\langle C_{sca}(\lambda) \rangle$ by [12,13]:

$$\langle dS^{ps}(\lambda) \rangle = \frac{a_1^{ps}(\lambda, 90°) \langle C_{sca}^{ps}(\lambda) \rangle}{4\pi} \tag{6}$$

where the superscript ps stands for the PS sample. In Eq. (6), $a_1(\lambda, 90°)$ represents the element at the first row and the first column of the Mueller matrix (or phase matrix), which can be calculated conveniently by the T-matrix method[12]. The detailed calculation steps can be found in, for example, Ref. [13].

For the PS sample, the imaginary part of the refractive index is so small that the average absorption cross section can almost be ignored, which means $\langle C_{sca}^{ps}(\lambda) \rangle = 0$. Therefore, the extinction of the PS sample is contributed only by scattering. Taking into account Eq. (1), we can easily obtain

$$\langle C_{sca}^{ps}(\lambda) \rangle = \langle C_{ext}^{ps}(\lambda) \rangle - \langle C_{abs}^{ps}(\lambda) \rangle = \langle C_{ext}^{ps}(\lambda) \rangle = \frac{\ln 10}{l N_v^{ps}} A^{ps}(\lambda) \tag{7}$$

Combining Eqs. (5)-(7), we can obtain the absolute value of the angular scattering spectra of the gold NR ensemble as

$$S_{90}^g(\lambda) = N_v^g \langle dS^g(\lambda) \rangle = \frac{\ln 10}{l} \frac{T_0^g(\lambda)}{T_0^{ps}(\lambda)} \frac{A^{ps}(\lambda) a_1^{ps}(\lambda)}{4\pi} \tag{8}$$

where the superscripts g and ps stand for the gold NRs and the PS sample, respectively.

2.2 Inverse-problem solution algorithm

With the measured extinction spectra and angular scattering spectra, the estimates of the geometric parameters as well as the PDF $p(D, AR, e)$ of the gold NR ensemble can be retrieved. To solve this inverse scattering problem, we perform a similar retrieving procedure as that presented in our previous work[2]. The main steps are still based on the constrained nonnegative regularized least-square procedure. But the integral discrete method and the objective function of the optimization are modified as follows.

For a polydisperse NR ensemble, the total absorbance A is calculated by integrating the contribution of all the composing NRs according to the PDF $p(D, AR, e)$[2]:

$$\begin{aligned} A^g(\lambda) &= \frac{l N_v^g}{\ln 10} \langle C_{ext}^g(\lambda) \rangle \\ &= \frac{l N_v^g}{\ln 10} \int_{D_{min}}^{D_{max}} \int_{AR_{min}}^{AR_{max}} \int_{e_{min}}^{e_{max}} p(D, AR, e) C_{ext}^g(\lambda, D, AR, e) dD dAR de \end{aligned} \tag{9}$$

Similarly, the total angular scattering cross section $S_{90}(\lambda)$ of a polydisperse NR ensemble can be calculated by

$$S_{90}^g(\lambda) = N_v^g \int_{D_{min}}^{D_{max}} \int_{AR_{min}}^{AR_{max}} \int_{e_{min}}^{e_{max}} p(D,AR,e) \mathrm{d}S^g(\lambda,D,AR,e) \mathrm{d}D \mathrm{d}AR \mathrm{d}e \qquad (10)$$

In general, the integrals in Eq. (9) and Eq. (10) cannot be solved analytically and should be discretized for numerical solution. If they are discretized with respect to all the three variables AR, D, and e, the condition number of the reduced linear system is usually too large to produce an accurate and stable solution. Thus in many previous works[2,4,7], the width D and the end-cap factor e were usually fixed as their mean values (as we have done in our previous work[2]) or even were ignored, for the reason that the aspect ratio AR is the primary parameter affecting the extinction of the NR ensemble[6]. However, as we have pointed out, the estimate of D and e is significant to the retrieving result of the PDF of AR.

Here, by taking into account the differential scattering spectra of the NR ensemble, the width D is possible to be determined[5] so that we can fix only the end-cap factor e and discretize Eq. (9) and Eq. (10) with respect to AR, D, and λ. Similar with the OES method[2], the adoption of different end-cap factor e in the OESS method can also influence the retrieved results. Even by the OESS measurement, the mean end cap e cannot be determined accurately and should be assumed or be measured by some other methods beforehand.

The discretization results in the following system of linear algebraic equations:

$$\mathbf{A} = \mathbf{CP}, \quad \mathbf{S} = \mathbf{S}_d \mathbf{P} \qquad (11)$$

where $\mathbf{A}$ and $\mathbf{S}$ are $M \times 1$ vectors, $\mathbf{C}$ and $\mathbf{S}_d$ are $M \times N$ matrices, $\mathbf{P}$ is a $N \times 1$ vector, and M and N are integer numbers. The vectors $\mathbf{A}$ and $\mathbf{S}$ contain the values of the measured extinction $A(\lambda)$ and angular scattering cross section per unit volume $S_{90}(\lambda)$ at different λ, respectively. The values of extinction cross section of gold NRs $C_{ext}^g(\lambda, D, AR, e)$ are stored in the matrix $\mathbf{C}$ for various wavelengths λ (rows), various aspect ratios AR (columns) and various widths D (columns) for fixed end-cap shape e. By the same way, the values of differential scattering cross section $\mathrm{d}S^g(\lambda, D, AR, e)$ of gold NRs at the scattering angle $90°$ are stored in the matrix $\mathbf{S}_d$. The vector $\mathbf{P}$ is the PDF to be solved and it has two physical constraints: the non-negativity constraint $\mathbf{P} \geq 0$ and the standard normalization condition $\mathbf{UP} = 1$ where $\mathbf{U} = \underbrace{[1,1,\cdots,1]}_{1 \times N}$.

Usually the condition number of the linear system Eq. (11) have the order ~ 10, thus the problem is an ill-posed inverse problem. Here the constrained non-negative least-square procedure[16] is applied to solve it. The least-squared solution $\mathbf{P}_{RLS}$ can be expressed as[17]

$$\begin{aligned}\mathbf{P}_{RLS} &= \min_{\mathbf{P}}\{\|\mathbf{A}-\mathbf{CP}\|_2^2 + \omega_S\|\mathbf{S}-\mathbf{S}_d\mathbf{P}\|_2^2\}\\ &= \min_{\mathbf{P}}\{(\mathbf{A}-\mathbf{CP})^T(\mathbf{A}-\mathbf{CP}) + \omega_S(\mathbf{S}-\mathbf{S}_d\mathbf{P})^T(\mathbf{S}-\mathbf{S}_d\mathbf{P})\} \qquad (12)\\ &= \min_{\mathbf{P}}\{\mathbf{P}^T(\mathbf{C}^T\mathbf{C}+\omega_S\mathbf{S}_d^T\mathbf{S}_d)\mathbf{P} - \mathbf{A}^T\mathbf{CP} - (\mathbf{A}^T\mathbf{CP})^T - \omega_S\mathbf{S}^T\mathbf{S}_d\mathbf{P} - \omega_S(\mathbf{S}^T\mathbf{S}_d\mathbf{P})^T\}\end{aligned}$$

where $\|\ \|_2$ is the Euclidean norm, the superscript T means the transpose of the vectors, and the non-negative weight coefficient ω_S means the weight between the OES and OSS data. The value $\omega_S = \max(\mathbf{A})/\max(\mathbf{S})$ is used here to balance the weight of the OES data and the OSS data, or the amplitude of the OSS data would be much smaller than the amplitude of the OES data and we would not obtain a good optimization result. Note that in Eq. (12), $\mathbf{A}^T\mathbf{CP}$ and $\mathbf{S}^T\mathbf{S}_d\mathbf{P}$ result in scalars so that their transposes are themselves. Consequently, $\mathbf{A}^T\mathbf{CP} = (\mathbf{A}^T\mathbf{CP})^T$, $\mathbf{S}^T\mathbf{S}_d\mathbf{P} = (\mathbf{S}^T\mathbf{S}_d\mathbf{P})^T$, and Eq. (12) can be written in another equivalent form

$$\mathbf{P}_{RLS} = \min_{\mathbf{P}}\{\mathbf{P}^T(\mathbf{C}^T\mathbf{C} + \omega_S\mathbf{S}_d^T\mathbf{S}_d)\mathbf{P} - 2(\mathbf{A}^T\mathbf{C} + \omega_S\mathbf{S}^T\mathbf{S}_d)\}$$
$$= \min_{\mathbf{P}}\{\frac{1}{2}\mathbf{P}^T\mathbf{Q}\mathbf{P} + \mathbf{q}^T\mathbf{P}\}$$
(13)

where $\mathbf{Q} = 2(\mathbf{C}^T\mathbf{C} + \omega_S\mathbf{S}_d^T\mathbf{S}_d)$ is a symmetric matrix of size $N \times N$, and $\mathbf{q} = -2(\mathbf{C}^T\mathbf{A} + \omega_S\mathbf{S}_d^T\mathbf{S})$ is a N-dimensional column vector. We use the active set method[16] to find the solution of Eq. (13). Then the retrieved ARD and the mean width D_m can be calculated easily by reshaping the vector $\mathbf{P}_{RLS}$ to a two dimensional matrix which consist of the PDF $p(AR, D)$ and averaging the row and the column of the PDF matrix, respectively.

3 Experimental results and discussions

3.1 Comparison of the OESS and TEM measurements

We applied the OESS method presented above to measure the ARD function $p(AR)$ and the mean width D_m of gold NR ensemble samples. The results are compared with those directly obtained by the TEM method (which is considered as a benchmark). In our comparison experiment, altogether 20 NR samples were measured and analyzed, each of which contains approximately 10^{10} NRs per millilitre. Without loss of generality, the results of four samples with different D, AR and e are demonstrated here. The four samples, designated as NR-10, NR-20, NR-30, and NR-40, were obtained from the National Center for Nanoscience and Technology, Beijing, China, whose nominal mean width D_m are 10nm, 20nm, 30nm, and 40nm, respectively.

In the TEM experiment, a transmission electron microscope (Hitachi H-7650B) was used. In order to characterize the samples statistically, about 20 TEM images were taken for each sample and altogether 1522, 1222, 876, and 933 NRs in samples NR-10, NR-20, NR-30, and NR-40 were analyzed, respectively. In the OESS experiment, the measurement of each sample was repeated six times in one hour and the average value of the results was adopted. The measurement range of wavelength λ was 400-850nm, with a step of 1nm. The extinction spectra database and the angular scattering spectra database of the gold NRs [corresponding to the matrices $\mathbf{C}$ and $\mathbf{S}_d$ in Eq. (11)] were calculated with the T-matrix method. It takes about 30 seconds to calculate a single scattering spectrum of NRs in the wavelength range of 400-850nm, with a 1nm spectral resolution, by using a dual-core 2.13GHz Intel Xeon CPU with 80Gb RAM. After that, the optimization procedure described above was implemented to retrieve the ARD function p

(AR) and the mean width D_m of the samples, where AR was discretized in the range of 1 to 5, with a step of 0.1. The inverse-problem solution algorithm was run on a 3.00GHz Intel Core2 Duo CPU with 4Gb RAM and the time consumption is about 5 seconds for each sample.

Table 1 shows the comparison values of the geometric parameters of the four samples measured by the two methods, as well as their relative differences. For the mean end-cap factor e_m, six values with the range 0-1 and the step 0.2 were assumed beforehand. The values of e_m shown in the table 1 were adopted for the reason that the retrieved ARDs by the OESS method coincide with the measured ARDs by the TEM method best.

Table 1 Measurement results of the four gold NR ensemble samples

Sample No.	D_m/nm			AR_m			σ_{AR}		e_m	
	TEM	OESS	RD*	TEM	OESS	RD	TEM	OESS	TEM	OESS
NR-10	12.5	13.2	5.6%	3.112	3.105	0.2%	0.56	0.54	0.7	0.6
NR-20	17.1	16.8	1.6%	2.536	2.519	0.7%	0.32	0.24	0.8	0.6
NR-30	31.6	30.4	4.1%	2.659	2.610	1.9%	0.35	0.51	0.4	0.8
NR-40	40.9	38.5	5.8%	1.641	1.619	1.3%	0.13	0.16	0.7	0.6

* RD: Relative difference between the OESS and TEM results.

It can be seen that the mean width D_m and the mean aspect ratio AR_m derived by the two methods coincide with each other well, with their relative difference smaller than 6% and 2%, respectively. These show that our OESS measurement results are reliable. For the standard deviation σ_{AR} (which represents the polydispersity of the ARD), the two sets of values also coincide with each other relatively well, where the small difference may be owing to the deviation between the real shape of the gold NRs and the geometric model adopted in our calculation.

Besides the geometric parameters, the mass-volume concentration C_g of the NRs can also be determined by the OESS method. It can be derived as $C_g = \rho N \nu \mathbf{V} \cdot \mathbf{P}_{RLS}$, where $\mathbf{V}$ is a row vector consisting of the volume of each nanorod geometry in the sample and ρ is the density of bulk gold. The values of C_g measured by the OESS method for the four samples NR-10, NR-20, NR-30, NR-40 are $14.54 \pm 0.26 \mu g/mL$, $15.78 \pm 0.17 \mu g/mL$, $18.03 \pm 0.14 \mu g/mL$, and $15.34 \pm 0.24 \mu g/mL$, respectively. Since these concentration values cannot be obtained by the TEM method, we cannot make a comparison of the two methods here.

The measurement results of the four samples are shown in detail in Fig. 3-Fig. 6. From the four TEM images, it is clearly seen that the gold NR ensembles are polydisperse. In the four extinction spectra, the L-LSPR peaks are at different wavelengths, while the transverse mode LSPR (named by T-LSPR here) peaks are nearly at the same wavelength. However, in the four 90° scattering spectra, the L-LSPR peaks are slightly different from those in the extinction spectra and the T-LSPR peaks are much weaker than their counterparts in the extinction spectra. Furthermore, for different samples with different widths, the maximum intensities of the extinction spectra are in a small range (~0.3-0.5) and are close to each other, but the maximum intensities of the scattering spectra are in a relative bigger range (~0.006-0.05) and deviate from each other significantly. These different features of the extinction spectra and the scattering spectra show the potential for retrieving both the ARD function $p(AR)$ and the mean width D_m by the OESS method, as we mentioned before.

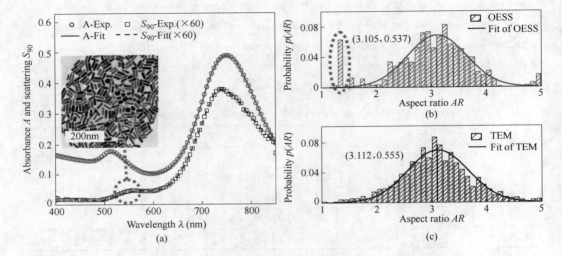

Fig. 3 Measurement results of the sample NR-10

(a) the absorbance A measured by the OES method and the 90° scattering intensity S_{90} (cm^{-1}) measured by the OSS method; (b) the ARD function retrieved by the OESS method; (c) the ARD function measured by the TEM method. In (a), the original measurement data of the extinction spectra A-Exp. (circle dots) and the scattering spectra S_{90}-Exp. (square dots, multiplied by 60), as well as the corresponding numerically reproduced extinction spectra A-Fit (solid line) and scattering spectra S_{90}-Fit (dashed line, multiplied by 60) according to the retrieved NR parameters are given. The inset in (a) shows the TEM image of the sample In (b) and (c), both the discrete ARD and a Gaussian fit of it are given; the values in parentheses give the mean AR and the standard deviation of the ARD

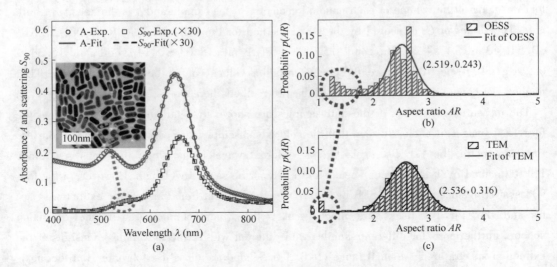

Fig. 4 The same as Fig. 3, but for sample NR-20

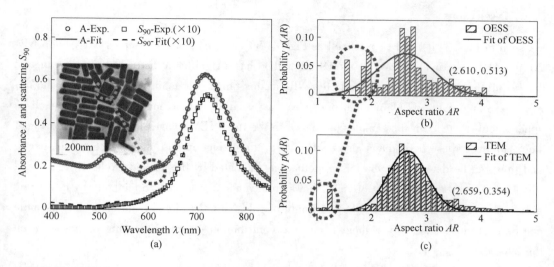

Fig. 5 The same as Fig. 3, but for sample NR-30

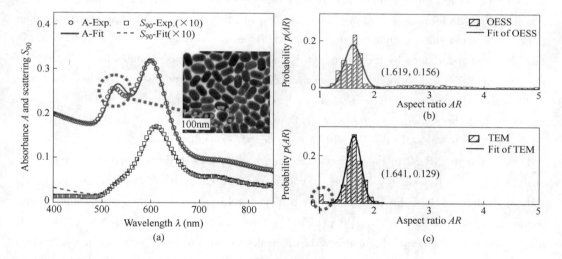

Fig. 6 The same as Fig. 3, but for sample NR-40

By comparing the OESS results with those obtained by the TEM method in Fig. 3-Fig. 6, it is seen that the ARD function $p(AR)$ derived by the two methods in general coincide with each other well. With the retrieved ARD function $p(AR)$ and the mean width D_m, the extinction spectra and 90° angular scattering spectra of the four samples were also numerically reproduced by Eq. (9) and Eq. (10), as shown in Fig. 3(a)-Fig. 6(a), which also coincide with the measured extinction spectra and the angular scattering spectra quite well. Thus Table 1 and Fig. 3-Fig. 6 have shown that the retrieved results by the OESS method are reliable in the characterization of the gold NRs ensembles. However, there are still some small differences between the measurement results derived by the OESS method and the TEM method. In the following, we discuss in detail these differences and their possible causes.

3.2 Discussions

As seen in Figs. 3(b) and (c) as well as Figs. 4(b) and (c), for the samples NR-10 and NR-20, the ARD function obtained by the OESS method has a significant distribution in the range of 1-2, while the TEM results have quite few NRs in this range. The main reason is that in the image processing of the TEM method, one may (as we did) ignore most byproducts (such as spheres, cubes etc.) in the samples, as shown in the inset TEM images in Fig. 3(a) and Fig. 4(a). However, these byproducts also contribute to the extinction and angular scattering spectra and thus can be detected by the OESS method, as indicated by the dashed circles in Fig. 3 and Fig. 4. This, from another point of view, shows an advantage of the OESS method: by measuring the scattering and extinction of a large amount of gold NRs, the global features of the ensemble can be captured statistically without losing the contribution of any composing particles (even the byproducts).

For the sample NR-40 in Fig. 6, the difference between the measured and reproduced extinction spectra of the sample is more significant (especially around the T-LSPR peak). The main reason is that the mean aspect ratio of this sample (which is around 1.6) is much smaller than those of the other three samples so that the L-LSPR peak is very close to the T-LSPR peak. In this case, the disturbance by the byproducts (whose LSPR mainly contributes to the T-LSPR of the extinction spectra) is more significant and is more difficult to be distinguished by the OESS method, as indicated in Figs. 6(b) and (c). Therefore, we can conclude that by using the OESS method, the detection of the byproducts would be more sensitive if the LSPR of the byproducts is not too close to the L-LSPR of the gold NR ensemble.

For the sample NR-30 in Fig. 5, there are three LSPR peaks in the extinction spectra. Obviously, the left LSPR peak (~520nm) is mainly contributed by the T-LSPR of the gold NRs (although the LSPR of the byproducts, such as the gold spheres of diameter around 30nm, would also contribute to this T-LSPR peak). The right LSPR peak (~720nm) is the L-LSPR of the gold NRs. These two characteristic LSPR peaks can also be seen in the extinction spectra of the other three samples, as shown in Fig. 3(a), Fig. 4(a), and Fig. 6(a). The LSPR peak in the middle (~610nm), however, is likely to be caused by the LSPR of the cubic byproducts with width 60nm and aspect ratio around 1.2, as indicated by the dashed circles in Fig. 5(a). In addition to these byproducts, some longer gold NRs with aspect ratio between 1.5 and 2 are also detected by the OESS method (which do not actually exist), as indicated in Fig. 5(b). The main reason is that in our extinction and scattering spectra database, only gold NRs but no rectangular cuboid nanoparticles were calculated. Consequently, the LSPR in the middle was actually fitted by the contribution of longer gold NRs. To confirm this inference, we have calculated the extinction spectra of two gold nanoparticle ensembles: an ensemble of rectangular cuboid gold nanoparticles with $D = 60$nm, $AR = 1.2$, and an ensemble of gold NRs with $D = 30$nm, $AR = 1.8$, $e = 0.8$, as shown in Fig. 7(a). Indeed, we can see that the LSPR peak around 610nm is reproduced in both spectra, where the peak of the cubic nanoparticles is even stronger.

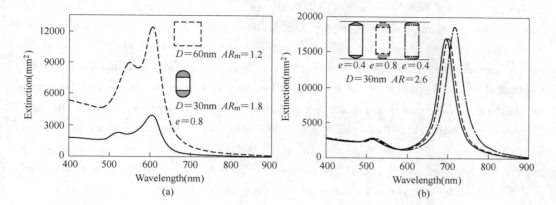

Fig. 7 Comparison of the extinction cross sections of different gold nanoparticles
(a) rectangular cuboid gold nanoparticles and gold NRs; (b) gold NRs with different end caps

For the sample NR-30, one may note that the mean end-cap factor e_m obtained by the OESS method ($e_m = 0.8$) deviates from the value obtained by the TEM method ($e_m = 0.4$) significantly, as shown in Table 1. The possible reason is that the end cap shape of the NRs in the sample NR-30 is sharper, as shown in the inset of Fig. 5(a), which deviates significantly from our cylindrical NR model with semi-spheroidal end cap. To corroborate this, we have calculated three NR ensembles with different end-cap shapes, as shown in Fig. 7(b). It is seen evidently that for the same value of end-cap factor $e_m = 0.4$, the L-LSPR peak of gold NRs with a cone-like end cap is blue shifted compared with that of gold NRs with a semi-spheroidal end cap. As a consequence, the extinction spectrum of the NRs with cone-like end cap of factor 0.4 matches the spectrum of the NRs with semi-spheroidal end cap of factor 0.8 better. For this reason, the fitted value of the end-cap factor that we obtained by the OESS measurement is 0.8, but not 0.4. These imply that in the practical characterization of gold NR ensembles, the database of extinction and scattering spectra should be expanded, by calculating not only NRs but also nanoparticles of other shapes (such as spheres, cubes, and NRs with special end-cap shapes). By taking these measures, the measurement accuracy of the OESS method can be further improved.

4 Conclusion

We have proposed an OESS method by combining the OES and OSS measurements to fast characterize the critical geometric parameters AR and D, as well as the probability density functions $p(AR)$ of gold NR ensembles statistically. To perform the OESS method, the angular scattering spectra at the angle of 90° are measured in addition to the extinction spectra measured by the UV-VIS spectrometer. To solve the inverse scattering problem, the extinction cross section and the differential scattering cross section of polydisperse NR ensembles are calculated rigorously by the T-matrix method. Then the critical parameters are retrieved by an optimization process with data fitting to the measured spectra.

By characterizing different samples of gold NR ensembles experimentally with the OESS method, it is shown that the measured ARD and mean width D_m coincide well with those obtained by the TEM method. By using the OESS method, it is also possible to determine the mass-volume concentration of NRs, which is unable to be measured by the TEM method. The comparison results indicate that the OESS method is fast, accurate, and cost effective to measure the critical geometric parameters of the gold NR ensemble. Further improvement is to enlarge the database of the extinction and scattering spectra of gold nanoparticles of different shapes (such as rectangular cuboid nanoparticles and NRs with special end-cap shapes) by considering the byproducts of practical samples, by which the retrieving accuracy can be further improved. The proposed OESS method has the potential to be developed for characterizing not only gold NR ensembles but also other non-spherical metal nanoparticles such as silver NRs.

Acknowledgements

We would like to thank Prof. Xiaochun Wu in the National Center for Nanoscience and Technology, China for preparing the gold nanorod samples. This work was supported by the Ministry of Science and Technology of China (Project No. 2011BAK15B03) and the Natural Science Foundation of China (Project No. 61161130005).

References

[1] N. G. Khlebtsov and L. A. Dykman, "Optical properties and biomedical applications of plasmonic nanoparticles", J. Quant. Spectrosc. Radiat. Transfer 111, 1-35(2010).

[2] N. Xu, B. Bai, Q. Tan, and G. Jin, "Fast statistical measurement of aspect ratio distribution of gold nanorod ensembles by optical extinction spectroscopy", Opt. Express 21, 2987-3000(2013).

[3] B. Khlebtsov, V. Khanadeev, T. Pylaev, and N. Khlebtsov, "A new t-matrix solvable model for nanorods: Tem-based ensemble simulations supported by experiments", J. Phys. Chem. C115, 6317-6323(2011).

[4] B. N. Khlebtsov, V. A. Khanadeev, and N. G. Khlebtsov, "Observation of extra-high depolarized light scattering spectra from gold nanorods", The Journal of Physical Chemistry C112, 12760-12768(2008).

[5] K. S. Lee and M. A. El-Sayed, "Dependence of the enhanced optical scattering efficiency relative to that of absorption for gold metal nanorods on aspect ratio, size, end-cap shape, and medium refractive index", The Journal of Physical Chemistry B109, 20331-20338(2005).

[6] S. W. Prescott and P. Mulvaney, "Gold nanorod extinction spectra", J. Appl. Phys. 99, 123504(2006).

[7] S. Eustis and M. A. El-Sayed, "Determination of the aspect ratio statistical distribution of gold nanorods in solution from a theoretical fit of the observed inhomogeneously broadened longitudinal plasmon resonance absorption spectrum", J. Appl. Phys. 100, 044324(2006).

[8] O. Peña, L. Rodríguez-Fernández, V. Rodríguez-Iglesias, G. Kellermann, A. Crespo-Sosa, J. C. Cheang-Wong, H. G. Silva-Pereyra, J. Arenas-Alatorre, and A. Oliver, "Determination of the size distribution of metallic nanoparticles by optical extinction spectroscopy", Appl. Opt. 48, 566-572(2009).

[9] D. D. Evanoff and G. Chumanov, "Size-controlled synthesis of nanoparticles. 2. measurement of extinction, scattering, and absorption cross sections", The Journal of Physical Chemistry B108, 13957-13962(2004).

[10] V. A. Bogatyrev, L. A. Dykman, K. B. N. , and N. G. Khlebtsov, "Measurement of mean size and evaluation of polydispersity of gold nanoparticles from spectra of optical absorption and scattering", Optics and Spectroscopy 96, 128-135(2004).

[11] G. S. He, J. Zhu, K. T. Yong, A. Baev, H. X. Cai, R. Hu, Y. Cui, X. H. Zhang, and P. N. Prasad, "Scattering and absorption cross-section spectral measurements of gold nanorods in water", The Journal of Physical Chemistry C114, 2853-2860(2010).

[12] M. I. Mishchenko, L. D. Travis, and A. A. Lacis, *Scattering, Absorption, and Emission of* Light by Small Particles(Cambridge University, 2002).

[13] M. I. Mishchenko, "Light scattering by randomly oriented axially symmetric particles", J. Opt. Soc. Am. A 8, 871-882(1991).

[14] A. V. Alekseeva, V. A. Bogatyrev, L. A. Dykman, B. N. Khlebtsov, L. A. Trachuk, A. G. Melnikov, and N. G. Khlebtsov, "Preparation and optical scattering characterization of gold nanorods and their application to a dotimmunogold assay", Appl. Opt. 44, 6285-6295(2005).

[15] B. Khlebtsov, V. Khanadeev, and B. N. Khlebtsov, "Tunable depolarized light scattering from gold and gold/silver nanorods", Physical Chemistry Chemical Physics 12, 3210(2010).

[16] P. Gill, W. Murray, and M. Wright, *Numerical Linear Algebra and Optimization*(Addison Wesley, 1991).

[17] J. Mroczka and D. Szczuczynski, "Simulation research on improved regularized solution of the inverse problem in spectral extinction measurements", Appl. Opt. 51, 1715-1723(2012).

论仪器仪表科技发展

国外光计算的进展*

摘　要　本文介绍了国外光计算科研工作方面的进展，其中包括美国、欧洲、苏联在这方面的研究工作及主要研究内容。介绍了光学与电子学相比的优势所在及有关双稳态器件、光学线路、光学算法、关联存贮、神经网络式光计算机、光学互连方面的进展。最后指出了目前和长远的研究目标和对光计算机前景的估计。

近年来，世界上科技先进的国家对光计算机的研究工作都非常重视。在新兴起的这场"光计算热潮"中，许多国家的政府及各大企业均不惜以巨额投资，加入到这场高技术的竞争行列中，各国的光学及电子学家更是以极大的热情迎接这场信息革命的挑战。有人断言，在新的一代（"第五代"，亦有人称是"第六代"）超高速并行计算机的研制中，光计算机将以其独特的优势，成为计算机中的佼佼者；光的时代即将到来！它有可能将导致一场新的计算机革命。这场使全球科学家们关注的"光计算热"的背景是什么？有什么优越性？国外光计算的研究到底达到何等水平？带着这些问题，笔者于1987年4月至10月有幸访问了欧洲的光计算中心——位于英国爱丁堡的Heriot-Watt大学，了解到一些光计算领域中的最新进展情况。下面就所了解到的一些情况作一个简单的介绍。

1 国外最近状况

随着现代科学技术的迅猛发展，现有电子计算机的能力已无法满足要求。如美国总统里根就职后提出的星球大战计划，其中的反洲际导弹系统要求信号处理的速度高达每秒几十亿次。因此，美国战略防御部的科技革新办公室要求各方面的研究工作者朝着把光学用于高速信号处理及计算的目标协同工作。从此以后光计算研究热潮席卷了美国几乎所有涉及光学信息处理的部门。与此同时，苏联为保持其在世界上的战略地位，也在这方面的基础研究中大量投资。欧洲、日本等亦不甘落后。现就所了解的各国研究情况概括如下。

1.1 美国

美国在光计算领域中一直居世界领先地位，主要的研究机构有：

（1）Arizona大学的光学中心（Optical Science Center）。此中心于1984年在National Science Foundation支持下成立一个研究小组，由M. Gibbs教授领导，并Bell实验室有合作关系。主要工作是"寻找新材料，在最有希望的材料下找出光学逻辑运算元件的合理设计，做出能用于并行阵列器件的雏形"。

---

* 原发表于《仪器仪表学报》，1988，9(3)：271~283。

（2）南加利福尼亚大学的 Center for Photonic Technology。该中心 1985 年成立。他们的重点放在跨学科的研究及材料、器件、系统的研究上。课题包括光计算、光连接和光通讯。并与 ITT、TRW、NBS 等有合作关系。

（3）Alabama 大学应用光学研究中心。此中心由 H. John Caulfield 教授领导，他们的研究重点放在计算机结构、图像处理和元器件上。

（4）Stanford 大学电机系光学图像处理研究室。该研究室由 J. Goodman 教授领导，一直接受美国陆军经费。早期曾搞过光计算矢量乘法等，现正进行一些光连接，光时钟方面的研究。

（5）AT&T Bell 实验室。该实验室可称是世界上研究光计算实力最强的单位，负责人 Alan Huang。据 Huang 说"Bell 在光计算的投资比世界上任何地方都要多"，"Bell 将成为一个光学公司"，"Bell 的饭碗将要靠光了"。但他们的研究项目与投资均严格保密。

1.2 欧洲

在欧洲共同体欧洲科技发展委员会（Committee for the Development of European Science and Technology）组织了欧洲联合光双稳态（European Joint Programme of Optical Bistability）项目，由英国 Heriot-Watt 大学的 D. Smith 教授和比利时 Libré 大学的 Mandel 教授领导，英、比、德、意、法参加。协定内容是建立光双稳态的光学逻辑元件与线路，发展光如何控制光的开关理论、逻辑门及存贮理论，进行非线性材料性质的研究、试制与新器件的测试。对此项计划欧洲共同体每年拨款 180 万欧洲共同体货币单位。此外 Heriot-Watt 大学还接受美国 SDI 计划项目，经费数目不详。

总之，在西方，目前光计算的中心主要集中在：

（1）AT&T Bell 实验室，Alan Huang 领导；

（2）Arizona 大学，Gibbs 领导；

（3）欧洲联合项目，Heriot-Watt 大学 Smith 领导。

1.3 苏联

据估计，从 1984 年起苏联在光计算方面投入的人力与经费大约比美国多四至十倍。有三个中心在进行有关的研究：

（1）列宁格勒的 A. F. Ioffe 物理技术研究所；

（2）莫斯科的 P. N. Lebedev 物理研究所；

（3）Novosibirisk 的无线电研究所。

他们的工作集中在材料、器件、结构、算法及应用方面。他们在全息方面的研究为世人瞩目。诺贝尔奖金获得者尼古拉·柏绍夫博士积极参与了超高速光处理器以及超高密度数据存贮系统的研制工作，可以说他们的工作决不比美国逊色。

2 光学优势

比较图 1 经典有限状态时序机和图 2 冯·诺依曼型计算机可以看到，后者由于使用了寻址器信号只能一维传输，即对每一个存贮单元在某一时刻只能有一个信号通过，其通迅是串行的，受到寻址方向的限制；而前者允许逻辑电路与存贮器分别连接，是并行处理系统。在冯·诺依曼系统中，虽然可以通过编码将内联次数从 N 减少至 $\log N$，但"瓶颈"作用已由结构决定无法消除，在一个芯片上做上百万只引脚是不可想象的。另外，即使这一结构中采用了时分多路方式来弥补其缺陷，但速度不断地提高又会带来新的问题——"时钟歪斜"，这会影响处理器的同步性能；再者计算速度有一个上限，也不可能一味地去提高。而使用并行处理的计算机，特别是选用光线来做大量的互联工作，将会使并行系统成为可能。不仅如此，使用了并行处理对元件开关速度的要求也降低了。

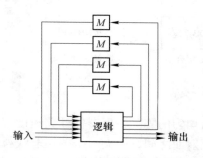 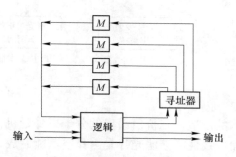

图1　经典有限状态时序机　　　　　图2　冯·诺依曼型计算机

总之，光学与电子相比，有以下几个明显优点：
（1）传播速度快，频带宽；
（2）不受电磁干扰；
（3）没有时钟歪斜问题；
（4）可以并行处理。

3　最新进展

目前有关光计算的研究主要集中在以下几个方面：
（1）材料的光学双稳态性质及光逻辑元件（Optical Bistability and Devices）；
（2）光学线路（Optical Circuits）；
（3）算法（Algorithms）；
（4）连接（Interconnections）。
下面就以上几个大方面将国外最近的一些进展作一介绍。

3.1　器件

主要是光学双稳态器件的研究，分为两类（见图3）：
（1）主动光学双稳态（OB）系统——有源器件，如 InGaAsP/InP、GaAs/GaAlAs 双稳态激光器；
（2）被动光学双稳态系统——无源器件，如 GaAs、InSb、ZnS、ZnSe。
由于有源器件的阵列化目前还存在较多的困难，研究工作多集中在无源器件上。其中美国贝尔实验室将工作重点放在 GaAs/GaAlAs 多量子阱双稳器件上，而英国 Heriot-Watt 大学和美国亚利桑那大学集中在 ZnSe 和 ZnS 等干涉滤光片器件上。多量子阱结

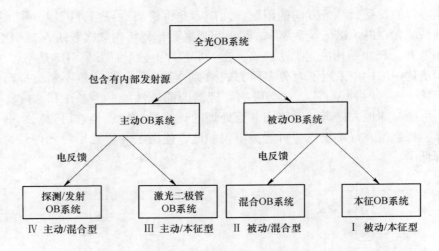

图3 光学双稳态器件分类表

构的制造需用价值昂贵的分子束外延机，不易做出均匀的陈列，但器件开关速度较高。而非线性干涉滤光片是由真空蒸镀制成，在两个反射面之间，被夹入了一层非线材料。反射面通常为10~20层，交替镀有高低折射率的介质膜。被用作非线性材料的主要有InSb、ZnS、ZnSe等。以ZnSe干涉滤光片为例，其结构为$HLHL\cdots\cdots(mH\cdot H)\cdots\cdots LHLH$，如图4所示。其特性曲线见图5。这种形式的非线性器件易于制成大面积均匀的器件，且易于切割成陈列。

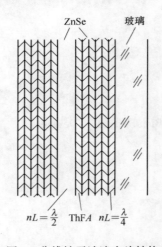

图4 非线性干涉滤光片结构

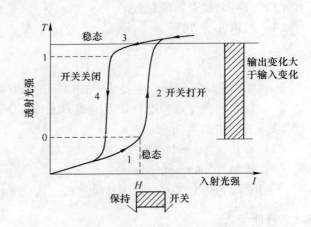

图5 滤光片特性及开关作用

各种双稳态器件的比较见图6。其中ATT为美国贝尔实验室，HW为英国Heriot-Watt大学，OSC为美国亚利桑那光学中心。图中SEED（Self Electro-optic Effect Device）为贝尔实验室最新研制的一种自电光效应器件，该器件利用对非线性层所加的平行或垂直的电场，可获得相当大的电吸收特性。它由几百层的GaAs与AlGaAs所组成。约10nm厚，通常工作在850~860nm下，开关速度400ns，功率比一般少5/6。大光斑直径下，开关功率与其半径成正比，开关时间与其面积成正比。

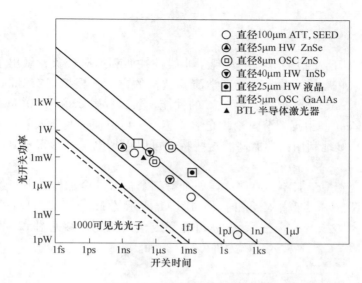

图 6 各种双稳态器件的比较

作为逻辑开关元件或存贮元件，使用时多在器件上加偏置光束（或称保持光），（见图 7）使器件处于开关点附近（见图 5），然后加上信号光束（或称控制光束）使其具有开关作用。不仅透射光，反射光也可用来获得一些特殊的逻辑功能，其反射特性和光路见图 8。以上只是单个门的研究，实际应用时，希望将它切成逻辑单元阵列，达到并行处理的目的（见图 9），这一过程称为阵列化（Pixellation）。总之发现新的功率小速度快，可在室温下工作的光学双稳态材料与制成阵列形器件仍为目前光计算研究的关键。

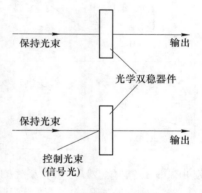

图 7 基本光路

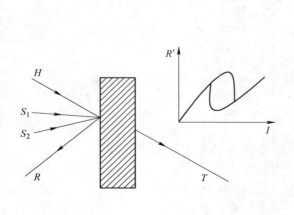

图 8 利用反射光

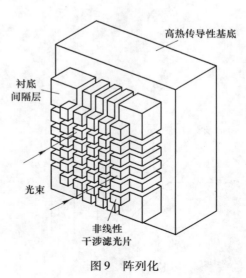

图 9 阵列化

3.2 光学线路（Optical Circuit）

目前科学家们已在很多材料中发现了非线性性质和迟滞现象，做出了一些全光型的数字处理器件，提出了不少有关光学演算结构的方案。根据现有 ZnSe 或 ZnS 器件（工作波长：$0.5\mu m$，工作温度：室温）的性能，若其存贮密度为 $10^4/cm^2$，传输速率可达 10^{10} Pit-Hz/cm^2。根据上节讨论，光学"芯片"可以通过改变入射角得到不同的特性，见图10。巧妙地利用光学器件的这些特性可以得到电子芯片得不到的效果。下面是一些例子。

（1）单门全光型全加器。通常这样一个有三个输入的全加器在电子线路中需要单路反相器以及双、三、四输入反相器。如图11所示的 Motorola MC996 全加器就为一例。而采用光学方法，一个光学元件就可能完成全加功能。参见图12，图13和表1，表2。图中 A、B、C 代表输入，C' 为进位，S 为和。

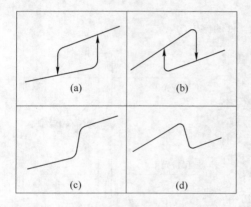

图10 （a）透射有迟滞环；（b）反射有迟滞环；
（c）透射无迟滞环；（d）反射无迟滞环

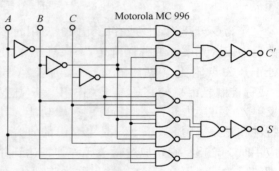

图11 电学方法实现全加器

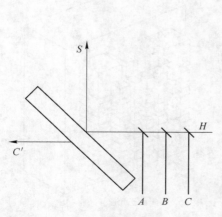

图12 光学全加器

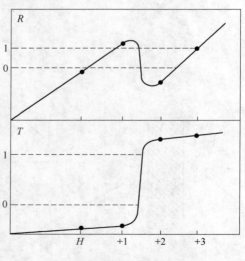

图13 器件性能曲线

表 1 全加器真值表

A	B	C	S	C'
0	0	0	0	0
0	0	1	1	0
0	1	1	0	1
1	1	1	1	1

表 2 多位数加法

项目		二进制值				
		2^4	2^3	2^2	2^1	2^0
初始数据项	B_1	0	1	1	1	0
初始数据项	C_1	0	1	1	0	0
进位项	C'_1	0	1	1	0	0
和 项	S_1	0	0	0	1	0
初始数据项	$B_2 = S_1$	0	0	0	1	0
进位循环	$C_2 = 2C'_1$	1	1	1	0	0
进位项	C_2	0	0	0	0	0
和 项	S_2	1	1	0	1	0

(2) 延迟循环存贮器。图 14 所示为：通过控制 H_1，H_2，H_3 使信号 S 到 S_1，S_1 到 S_2，S_2 到 S_3 得以控制，这样回路中可保存两个信号而不会相互干扰，这实际上即为延迟循环存贮器。

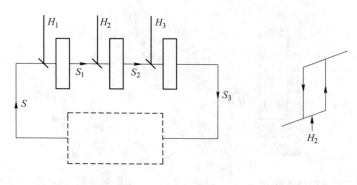

图 14 延迟循环存贮器

(3) 全光积分器。利用上图的延迟循环，可以构成一个全光的积分器，见图 15。图中有两个延迟循环光路：一个是和，称为累加寄存器；一个是进位延迟，这样积分器就可以将不同时间送来的信号累加起来输出。

(4) 三或门全光回路。图 16 所示是非线性干涉滤光片双稳态器件用于光学数字处理中一个成功的例子。图中使用的是离轴寻扯方式，每个器件的保持光为经声光调制器调制的氩离子激光束。所有的声光调制器均由计算机控制，这样的系统构成可以有不同的形态，用来解决不同的实际问题。尤其是让它们发挥并行处理的功能就可能成为实时处理系统。图 17 是三或门回路的输出特性。当输入为逻辑 1 时，三个或门按照

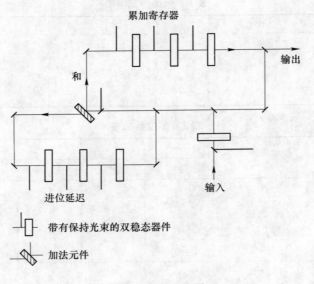

图 15 全光积分器

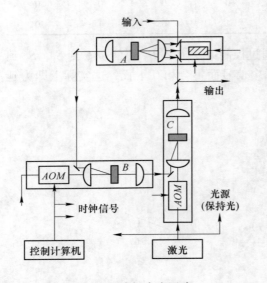

图 16 三或门全光回路

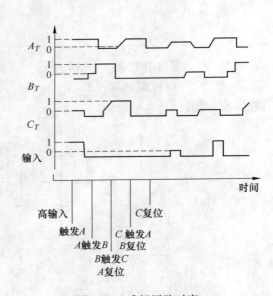

图 17 三或门回路时序

时序依次被打开,高输入打开开关 A,A 输出开启 B,B 又开启 C,C 最后又返回再打开 A,这样完成一个循环;而当输入为 0 时,三个门将保持原来的低透状态。在各个循环间控制器应使三个或门恢复零状态,然后再进入下一个循环。具体过程见时序图 17。以上回路在 Heriot-Watt 大学均已实现。

3.3 算法

算法在计算机中的重要地位是不言而喻的。在近 40 年计算机的发展中我们看到:逻辑门的延迟时间只减少了三个数量级,而浮点相乘运算的速率却提高了七个数量级。

见图18。由此可见使用合理的算法和系统结构（Architecture）可以有效地提高运算速度。特别是采用并行处理，可以大大降低对器件响应速度的要求。如图18中列在IBM-7090之后的计算机均不同程度地采用了并行结构。但要达到全并行使用电子恐怕就很困难。所以目前在算法和结构方面的研究也是一个热门。这里各选一个例子扼要介绍一下。一是算法方面的符号代换法；另一是关联存贮，又称神经网络计算机。

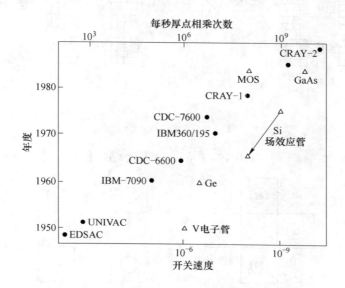

图18　开关速度与计算速度的发展

（1）符号代换（Symbolic Substitution）。冯·诺依曼曾设计过一个细胞自动机（Cellular automata），即将信息存贮在20000个细胞中，如每一个细胞单元有29种状态，设定一定的变化规则，则可以再现任意所需的信息。例如，运用这样一种变换规则（见图19）就可使状态20变为1。但只有得到匹配的左边图案，才可使用这一规则用右边的替代。

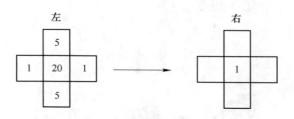

图19　一种变换规则

Alan Huang建议用光学方法来实现上述方案，他建议分为两步：1）复制和移位；2）重叠。实际上第一步是图形识别，第二步为代换。见图20。

可以做到只使用一种变换规则实现一整套拓扑逻辑连接。如原有四数据单元在一确定位置，现有四控制单元与之匹配（见图21），则可由左方图像变化为右方图像。图22所示为利用此原理实现信号向北传输。由此我们可以利用一些结构作逻辑单元。如AND见图23及图24。图23只有上面一路有输入，第二步无法匹配，AND运算无法继

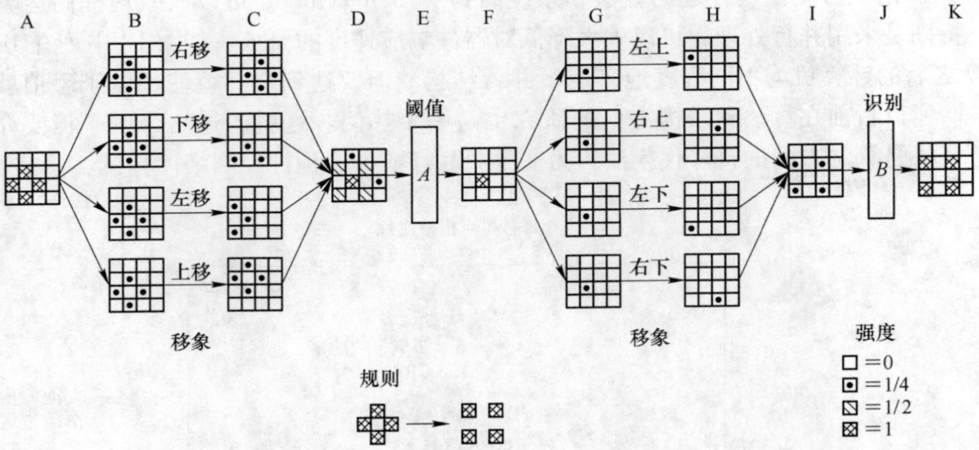

图 20 4×4 系统的符号代换过程

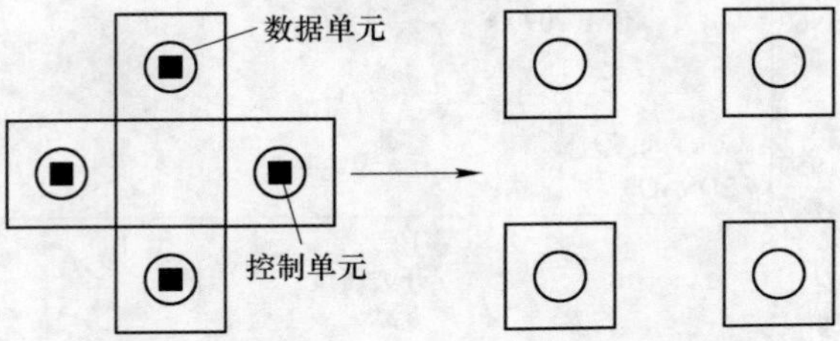

强度
□ =0
⊡ =1/4
▨ =1/2
▧ =1

图 21 变换规则

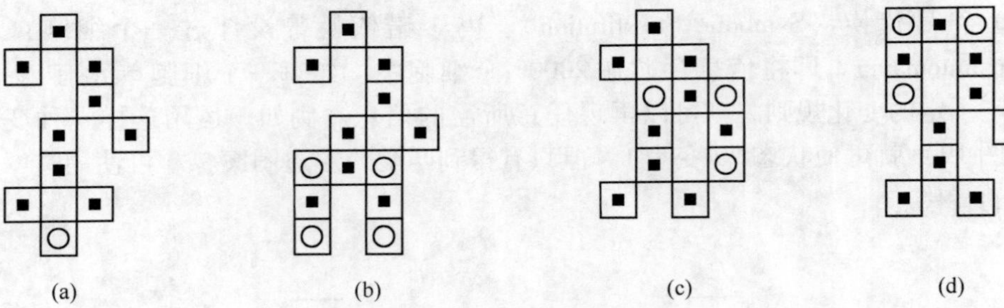

图 22 向北传播的时序
○—数据单元；■—控制单元

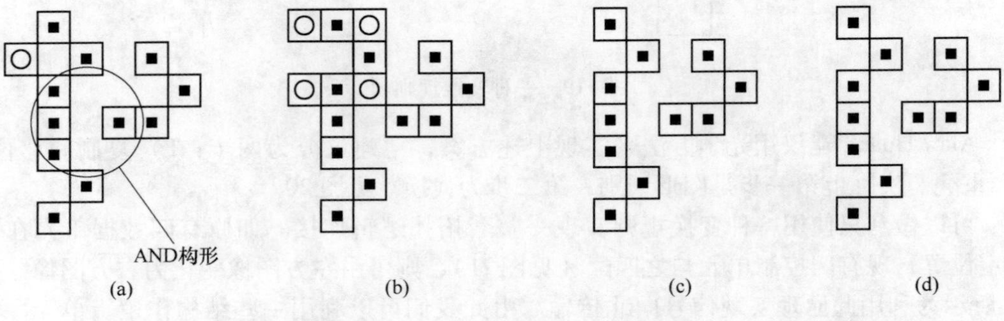

AND构形

图 23 AND（只有上一路有效）

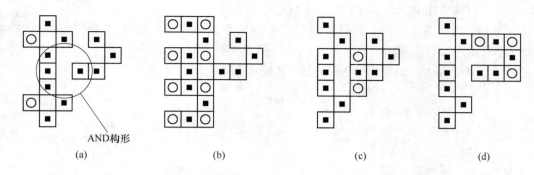

图 24 AND（上、下两路均有效）

续进行。图 24 中上、下两路均有输入，则可以得到运算结果。图 23 和图 24 圈起部分为 AND 的结构，可以看出要使之与图 21 的变换规则左图 LHS 匹配，运算才能实现；图中已给出四个数据中的两个，上下两通道必须提供另外两个数据。图 25 是双轨 AND 的结构示意图和真值表，原理同上所述。依靠这样一些简单的线路，可组合成复杂线路，实现各种复杂运算。至于如何用光学方法实现复制、移象及阈值控制和重叠这里不再赘述。由于光学具有高度的平行性，特别适用于这类运算，处理速度极快。再者由于这种代换是二维的，它亦比通常计算机所用的布尔逻辑具有更丰富的功能，不失为一个值得探讨的课题。

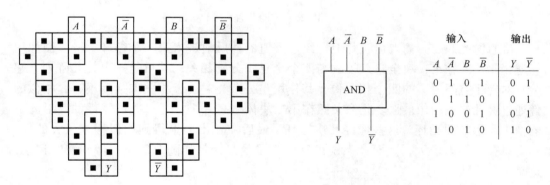

图 25 双轨 AND 逻辑

（2）关联存贮或神经网络计算机（Associative Memory or Optical Neural Computer）。随着电子计算机技术的飞速发展，器件向高速度、高密度发展，它们的功能也愈来愈强。就一个小小的计算器来说，用它来计算两个十位数的乘积，显然要比人脑快得多。但是即使是一个功能极强的电子计算机，让它去识别两种不同的槐树是困难的。相比之下，人脑的功能就要强得多。一个通用计算机只能机械地执行算法，但不具备人脑选取和综合信息的能力。

近年来，许多计算机科学家和数学家在医学工作者的配合下，研究人脑的结构，想解开思维之谜。计算机之所以不能像人脑一样加工处理信息是因为它缺乏解决随机问题的能力。如何解决呢？一个有益的尝试就是新兴的神经网络计算机的构想，它运用大量的处理器以光学方法实现内连接。图 26 是用光来实现这一方案的一个系统光路图。其中有两个主要元件：

（1）二维光开关元件。此器件用于模拟神经细胞的功能，它有10000个尺寸微小的光学双稳态器件组成的阵列，超过阈值光强，反射率将增加。

（2）体全息元件。在图26系统中，这种元件既作存贮又作连接。

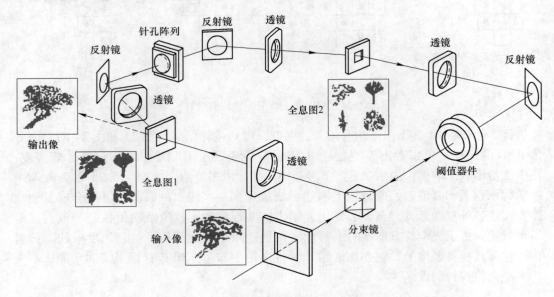

图26　实验装置

准直激光照在一输入透明片上，经过阈值器件反射进入第一个循环。输入图像用一透镜聚焦照在第一块全息片上，与四个全息图分别相关；其后是一个针孔阵列，它分别对应于空间分离的四个相关像，其作用是去杂光。由于相关像强度不一，相似的像通过针孔产生的像最强。此像又照在第二块体全息片上，通过阈值器件的反射滤去弱像进入第二次循环。这些像经几次循环，最后在输出面上得到相似程度最大的像。体全息存贮能力极大，因而此系统可存贮大量信息，且能随时取出，做到实时识别，这是一个飞跃。

3.4　连接（Interconnections）

一般我们对光学连接提出如下要求：

（1）高度并行性；

（2）无串扰；

（3）高光学效率——高反射或透射性能；

（4）可控制方向——可变或固定方向；

（5）高对比度或高SNR；

（6）光束间间隙小；

（7）具有扇入或扇出功能。

目前提出的光学连接方案有以下几种：其一是直接用光学元件，如透镜、反射镜、棱镜等，见图26；其二是使用光纤；其三是使用全息图，见图27；另外也可使用光学或计算全息图作为光互联元件或分束元件,其制作方法见图28。目前已可做到 25×25 阵列。最后,还有一种值得一提的连接方案,即用光折射晶体,如BSO,利用它的位相共轭波,实现实时互联。限于材料,需加极强电场才能产生高的非线性效应,但这不失为一个很有前途的方案。

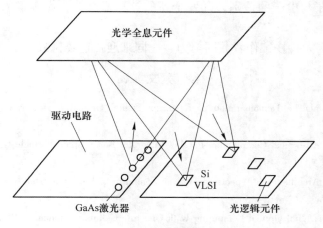

图 27　使用计算全息的连接方案

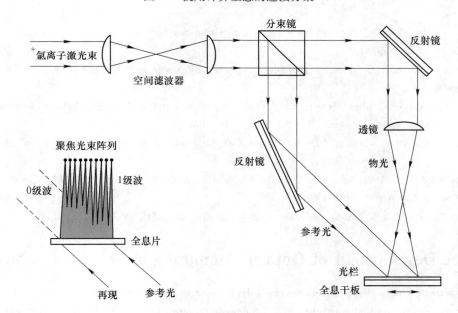

图 28　全息分光零件制作方法

4　光计算的目标

目前光计算机的研究正处于上升时期,各种新设想、新方案层出不穷,但这是一个遥远的目标,需要全世界的科学家齐心协力。正如 Caulfield 教授所说,"光计算只是一个方向,不是目标"。我们只有沿着这个方向,不断突破难关,改革创新,才能有所收获。就现在来看,要研究出全光型具有商业价值的计算机至少需要 10~15 年,道路还很长。

目前研究光计算有以下三个目标：

(1) 使电子计算机能工作得更快一些，做到不受电磁或核磁干扰；

(2) 做超高速的光学计算机；

(3) 寻找光子与电子的差别，开发全新的计算机结构，做出像人脑似的闪电计算机。

希望我国的广大科技工作者共同努力，一同来迎接这场挑战。

参 考 文 献

[1] S. D. Smith, et al. The Demonstration of Restoring Digital Optical Logic, Nature, Vol. 325, (Jan. 1987).

[2] B. S. Wherrent, All-optical Computation-Parallel Integrator Based upon A Single Gate Full Adder, Opt. Comm. Vol. 50, No. 2, (1985).

[3] B. S. Wherrent, Architectural Aspects of Optical Computing, Proc. Workshop on Optical Bistability in Optical Computing, (Nov. 1986).

[4] M. J. Murdocca, Digital Optical Computing With One-rule Cellular Automata, AT&T Bell Lab.

[5] Y. S. Abu-Mostafa & Demetri Psaltis, Optical Neural Computers Scientific American, March, 66, (1987).

[6] Karl-Heinz Brenner, New Implementation of Symbolic Substitution Logic, Applied Optics, Vol. 25, No. 18, (Sept. 1986).

[7] A. Sawchuk and T. C. Strand, Digital Optical Computing Proceedings of the IEEE, Vol. 72, No. 7, (July 1984).

[8] S. D. Smith, et al. Room Temperature, Visible Wavelength Optical Bistability InZnSe Interference Filters, Optics Communications, Vol. 5, No. 5, (Oct,1984).

[9] Karl-Heinz Brenner, et al, Digital Optical Computing with Symbolic Substitution, Applied Optics, Vol. 25, No. 18, (Sept. 1986).

[10] Hameed A. Al-Attar, et al. The Gain-Bandwidth of An InSb Transphasor, IEEE J. of Quantum Elect. Vol. QE-22, No. 5, (May 1986).

[11] H. M. Gibbs, et al. Non-linear Etalons and Optical Computing. SPIE Vol. 634, (1986).

The Development of Optical Computing in Foreign Countries

Abstract This paper introduces the newly development of optical computing in foreign countries. It includes the research work in U. S. A., European countries and U. S. S. R., the Priority of optics over electronics, the optical bistable devices, the optical circuits, the algorithm, the associative memory or optical neural computer, and the interconnections in optical computing. Finally, the author gives his comments to the aim of the research in the field of optical computing.

信息时代的光学器件——透镜仅针尖那么大*

集成电路的集成度愈来愈高，$1cm^2$ 上要集成几千或几万个器件，它们之间怎么连接呢？如果用导线连接那将像蜘蛛网似地绞在一起，对于多层结构的集成电路而言，难度就更大了。现在，科学家们终于找到了一个最有前途的方法，那就是用光束互连。这种互连不仅用在芯片的器件之间，也用在芯片与芯片之间，或线路板与线路板间的互连。因为光可在空气中自由传输，并不受电磁干扰。用光互连就需要许多微小的透镜和棱镜，这些小的透镜和棱镜就叫二元光学器件。

二元光学器件，是现代科学技术发展的产物。今天普通的光学仪器或传感器由于体积大、笨重，往往在系统中不易安装，特别是与密度日益提高的集成电路不能匹配。工业上已提出需要制作几毫米或小于 1mm 直径的透镜等元件。为了能和电子元件耦合并做到多通道，需要阵列化的光学元件。有的应用领域，需要将光学元件与电子元件集成在一个基体上，实现真正的光电结合。

那么怎样实现光学元件的微型化、阵列化和光电集成化呢？近年来兴起了一种用计算机辅助设计，使用大规模集成电路制作衍射光学元件的技术，统称二元光学元件。

衍射光学元件不同于透镜、棱镜等的工作原理，因一般的透镜、棱镜都是基于光学折射原理，而衍射光学则是利用浮雕的面形，基于光学中的衍射（光的绕射）而得到聚焦和分束等功能。

二元光学元件的优点是：衍射效率高，具有色散特性，提高了元件设计的自由度。过去在透镜系统的设计中往往需配置不同镜片的材料及改变镜片的曲率去消除像差和色差，因此可变的结构参数较少，而二元光学元件中可改变光栅的栅距、相位阶数、阶高、栅的宽度等，故设计的自由度大。此外还有材料的选择灵活和易于简化光学系统等优点。二元光学元件具有如此多的优点，所以它一出现就受到各方面的重视。美国 1984 年军事战略领先计划中就有二元衍射光学研究项目。

光学仪器与传感器的进一步发展不仅是个别器件做成光电集成器件，而是使整个系统成为紧凑的光电系统。如近年来国外开始制作仿生近距光电组合传感器，其中使用二元光学器件作光的权重分配和与接受器阵列相连接的器件，该系统做得非常紧凑，并有神经网络功能。由于尺寸小，并具有智能功能，故广泛用于航天器中测量各种变化参量。

总之，二元光学器件将和集成电路、光电子器件一样，在通讯、航天、医学、工业等各个领域发挥巨大作用。

---

* 原发表于《人民日报》，1995 年 8 月 14 日，第十一版。

仪器仪表的微小化、集成化和智能化*

1 前言

仪器仪表是对信息进行测量与控制的基础手段和设备，是现代社会不可缺少的部分。仪器仪表工业是一个国家科技发展水平的标志，也是现代化的综合因素之一。

现代仪器仪表的发展有许多新的特点，值得研究归纳。例如：

（1）电子和计算机技术已成为现代仪器仪表的重要组成部分，光、机、电一体化，数字化、自动化和智能化，计算机软件及仪器管理系统形成新的体系（如虚拟仪器、总线技术等）。

（2）新型功能材料、新型传感元件和新的测试机理促使仪器仪表发展。激光技术、超导技术、纳米技术、信号处理、图像处理和存储技术等高新技术在现代仪器仪表的开发和应用。

（3）现代的仪器仪表已不再局限于对被测量物进行简单的测量，它对信号的后续处理、分析显示及控制都有很高的要求。仪器仪表的功能更加完善和拓宽了。

（4）仪器用的元器件向微小化、集成化和智能化方向发展，如将测量流量、流速、流向、压力和气体种类的元件集成于一块硅片上的多功能传感器系统。

（5）仪器系统设计采用现代产品设计技术，如模块化设计、可靠性设计、并行工程技术等，以提高质量。在仪器仪表制造技术方面，现代制造技术得以应用。

（6）21世纪的仪器仪表在高技术和现代科学的基础上继续发展，将更多的科学技术融合到其自身中来，成为综合的学科和技术，如：物理、机械、光学、电子、计算机、材料、化学和生物等。

我国的仪器仪表研究开发单位有较强的实力，主要集中于大专院校和科研院所。

今年中国仪器仪表学会起草了"振兴我国仪器仪表工业的再建议"，指出面临的问题为：

1.1 缺少统一规划和宏观调控

我国仪器仪表工业起步于工业检测仪表，归属机械工业。机械工业部主要发展工业自动化仪表。随着科学技术和国民经济的不断发展，社会对仪器仪表的需求与日俱增。许多工业部门，甚至教育和科研部门相继建立了自己的数千家仪器仪表企业，生产经营各种通用和专用的仪器仪表。企业规模不大，产值不高，很难形成"团队"，得不到重视和支持。这些问题严重地制约了我国仪器仪表工业的顺利发展。

---

* 本文合作者：周兆英。原发表于《现代科学仪器》，2000，3：7~8。

1.2 不够重视基础研究

仪器仪表，尤其是大型科学仪器、医疗仪器等，涉及多种学科和先进技术，开发难度大，周期长，经济效益不高，基础研究得不到应有的重视。国家投入的科研经费不足，很少立项组织过重大的科技攻关。引进大型工程项目配套的仪器仪表大多由外商供应，国内立项的科技攻关项目主要经费也用来进口仪器仪表，两项费用每年高达近百亿美元。

大家希望加强统一归口管理，制定好国家规划，认真解决这些问题。这次香山会议会有更多的讨论和建议。本文则从一个角度谈谈仪器仪表发展的方向。

无论是从元器件或系统集成上，还是从材料与制造技术发展上，仪器仪表发展的一个重要方向可以用下面的"三化"来概括：微小化、集成化、智能化。未来的仪器仪表可以把微光学器件、微结构、微传感器、微制动器、信号处理器等集成在一起，能够对外界的各种物理、化学、生物等各种信号进行实时采样、处理、操作和控制的智能化信息系统。

2 关于微小化、集成化和智能化

2.1 微小化

提到微小化，不能不提到微系统的研究，也就是通常所说的 MEMS（Micro Electro-Mechanical System）或 MST（Micro System Technology）。由此可以产生一种将仪器仪表的传感器及其处理、控制和后续电路等都集成于芯片上的思想。微电子学在使信息技术飞速发展的同时，也促进着 MEMS 技术的飞速发展。例如以硅表面加工和体硅加工为主的硅微细加工，利用 X 射线、光刻、电铸的 LIGA 工艺，精密机械加工，以及微装配和封装（如使用粘接材料的粘接、硅玻璃静电封装、硅硅键合、玻璃玻璃键合和自对准组装技术等）、微系统控制和集成等关键技术等技术。在进行适当的改进后，以硅为基础材料的结构在制作工艺上能与 IC 工艺兼容，能大批量生产，能大幅降低成本，为仪器仪表实现集成化和智能化提供支持。

2.2 集成化

集成化的优点是众所周知的，目前国际上兴起研究热潮的生物芯片就是集成化的典范。

一种 Lab on chip 的结构。在芯片上实现混合、化学反应、分离等宏观上不连续的物理化学过程，使这些过程连续化，并提高系统的性能。

微全分析系统（μTAS）将样品的分析和信息的处理结合在一起，这要求将微流体单元、检测单元、控制电路集成在一起。以硅微细加工为主的 MEMS 加工技术和微电子工艺有着良好的兼容性，能够实现各种微生化功能单元和电路的集成。

2.3 智能化

以传感器为例，其智能化过程基本上可以分为三个阶段。有人认为敏感元件集成了信号处理电路就可称为智能传感器。按照这种定义，绝大部分的传感器都可称为智能型，但实际上这些传感器的智能化程度很低，通常被称为集成传感器，这是第一阶段。第二阶段的传感器集成了信号预处理部分。所谓智能传感器就是部分或全部集成了主要的处理单元，这是第三阶段。

传感器到现在至少已经经过了四代接口形式。第一代产品基本上没有电子器件，产生的响应结果实质上没有经过信号处理。第二代包含了放大器及一些温度补偿。到了第三代（现在大多数器件都属于这一情况），一些放大和信号缓冲利用离散和混成电子技术实现了模块化。传感器集成于一个包含模数转换器（ADC）和微处理器的远程信号处理包中，其通讯是单向高位模拟形式。第四代传感器达到了高度的集成，传感器芯片本身上就集成了部分或全部的单片传感器电路。现在正在发展作为系统成分的第五/六代的传感器。

3 结束语

关于仪器仪表工业和科研，必须加强统一归口管理，应当制定好国家规划，特别是"十五"计划。对于其中的"微小化、集成化和智能化"问题，建议：

3.1 认识"微小化、集成化和智能化"的重要性

仪器仪表在微小化、集成化和智能化方面的发展是迅速和必然的，是仪器仪表的共性问题，它的发展将会对21世纪的仪器仪表产生极其深远的影响。其发展的潜力和市场前景巨大。预计到2003年的微小系统的市场约为400亿美元，是商用航空业的一半。

3.2 "微小化、集成化和智能化"对仪器仪表交叉学科发展的要求

仪器仪表要实现微小化、集成化和智能化，必须要将各种最新的科学和技术应用到其研究和生产中来。仪器仪表的发展要综合多门学科研究成果，组织学科交叉的研究队伍。

3.3 加强企业的开发力量，提高我国仪器仪表行业的总体水平

制造技术是仪器仪表，特别是微小化、集成化和智能化发展的基础，企业研究开发力量薄弱应引起充分重视。

3.4 认真分析我国经济和科学发展对仪器仪表的需求

未来科技和经济的发展将促进各类用于分析、测量等检测与控制的仪器仪表开发研制，并提出更高的要求。微小化、集成化和智能化要紧密结合我国需求，才能发展。

21 世纪是信息与生命科学的时代

　　ICO（International Commission for Optics）即世界光学委员会，是一个国际性组织，目前包括有 48 个国家和地区的光学学会，美国、英国、德国、日本、俄国、中国、意大利等主要光学大国都是其成员。

　　ICO 第 18 次学术会议于 1999 年 8 月 2 日至 6 日在美国旧金山召开，会议的主题是"下一千年的光学"。与会者约 400 余人。会议内容涉及全光学领域，分 16 个分会，覆盖衍射光学、自适应光学、光通讯、光子晶体、气象光学、光学器件、光纤与集成光学、相干性与电磁场、光学网络与光孤子、图像与图像处理、干涉技术、瞬态光学、光学工程、激光与非线性光学、光计算、半导体激光器、分数光学、生物医学光学、层析成像、光存储、量子光学、全息与双光子存储、显微术与粒子分析、光学微机械、颜色与视觉等学科。共发表口头报告 51 篇，书面报告 287 篇。

　　从会议文章的数目上可看到有些领域是现代光学的研究热点，如衍射光学有 20 余篇，非线性光学有 10 余篇，自适应光学，光通讯，光存储光学工程，层析成像，光互连，光学微机械有多篇，因此这些领域将是 21 世纪的重要发展领域。

　　会上我宣读了两篇论文，题目为《用小波变换来改善体全息相关器的识别精度》和《人脸识别处理器的进展》，以及一篇张贴论文《使用光学本征滤波器解决畸变不变性的目标识别》。我的《人脸识别处理器的进展》一文报告后，立即有法国马赛物理研究院信号与图像实验室的教授表示愿与我们合作；我在讲解《使用光学本征滤波器解决畸变不变性的目标识别》一文时，得到前世界光学委员会的主席 Lohmann 和 Goodman 的称许，他们认为人脸识别应用光学图像处理与数字图像处理结合才是正确的出路。

　　会后我深深感到 21 世纪确实是信息与生命科学的时代，我们的专业必须适应时代的需要，向这两个方向倾斜。

---

　　* 原发表于《国际学术动态》，2000，5：38。

二元光学*

1 导言

现在希望仪器能做得尽量小,尤其是在空间上,这样拿到各地使用就很方便。因此,仪器的发展首先是小型化。但大家可以看到:任何一种光学仪器都比较大。现在除了我的眼镜以外,所有的光学仪器大都是光电结合,很少完全是目视仪器。光电结合,对光学器件有一定的要求,即光学器件应小型化;其次是希望能有多个通道,把它做成阵列化;第三,希望把它集成在一起,集成化,好在光、机、电都能集成在一起。小型化、阵列化、集成化是仪器,特别是光学仪器发展的方向。

什么是二元光学?它没有一个确定的定义。二元光学是一种衍射光学器件,一般的光学器件,如透镜、反射镜等,利用光的折射、反射原理,而二元光学利用的是光的衍射原理。更重要的一点是:二元光学是利用大规模集成电路(VLSI)的工艺或类似工艺制造出的。

二元光学是一种什么样的光学器件呢?稍微年长的同志都有这个经验,从前,看9寸的电视机,在前面放一块玻璃板似的器件,它叫菲涅耳透镜,能把图像放大。菲涅耳透镜的形成原理见图1,就是把一块普通透镜中间等相位的区域去掉了,把对相位有影响的一部分连接起来。用车、铣的方法加工出一个模子,然后用塑料板夹一下,就得到一块菲涅耳透镜。

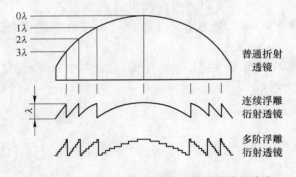

图 1 折射透镜到二元光学器件浮雕结构的演变

如果菲涅耳透镜的斜边不是加工成圆弧状,而是由一个一个的台阶来代替,当台阶数很多的话,就趋近于斜边,二元光学就是做成一个个台阶的,这样可利用 VLSI 工艺。利用一块模板照相一次,再利用另一块模板照一次,连续照几次,就能加工成一个二元光学器件,如图2所示。

---

* 原发表于《物理与工程》,2000,10(5):2~5,16。

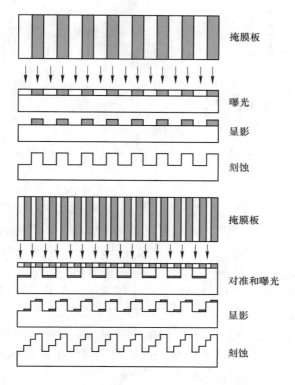

图 2　刻蚀法套刻制作四台阶二元光学器件工艺流程

在玻璃、石英或任意基底材料上涂上感光胶,然后利用一块掩膜曝光,这个工艺和 VLIS 工艺是一样的。曝光以后,被掩膜挡住的部分没有曝光,而曝光的部分可以进行刻蚀。刻蚀可以用化学湿刻或离子束干刻。刻蚀以后就得到两台阶器件。最初,就是刻蚀了两台阶,所以叫 Binary Optics——二元光学。实际上可以做更多的相位台阶,即 Multi-phase levels。在两台阶器件上,再利用掩膜曝光,就能得到四台阶器件,依次类推。这种器件很薄,而且尺寸可做得很小。所以说,二元光学是一种衍射光学器件,diffractive optical element designed by CAD,利用 CAD 进行设计,利用 VLSI 工艺制造。这种衍射光学器件就是二元光学器件。

二元光学器件有如下优点。

第一,有比较高的衍射效率。经过理论计算,两台阶,衍射效率达 41%;四台阶,可达 81%;而八台阶,高达 95%。利用更多的掩膜,四块、六块或八块时,能得到更高的衍射效率。但这样做是很不值的,八台阶以后,衍射效率已趋于饱和,如图 3 所示。况且,每套刻一次,就失掉一次精度,不可能每次套刻那么准,所以一般做成八台阶器件。

第二个优点是其独特的色散性能。对于一般的折射光学器件,由于光波不同,红光与蓝光聚焦的地方不同,即存在色散。而二元光学器件,控制相位台阶的位置与高度,使红光与蓝光聚焦在不同的地方,但红光、蓝光的相对位置正好同折射器件相反,若将折射器件与二元光学器件叠加在一起,红光、蓝光可聚焦在同一个点上,甚至白

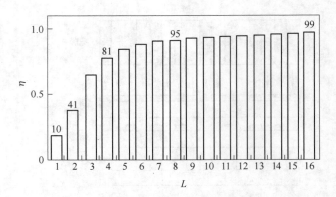

图 3　衍射效率 η 随相位阶数 L 的变化

光也可完全聚焦在一个点。因此，可利用二元光学器件去补偿折射器件的色散，形成折衍混合系统。需要指出的是，一般的光学器件，光焦度常常与波长无关，但对于二元光学器件，光焦度与波长是很有关系，对波长很敏感；另外，一般光学器件的阿贝值是大于零的，而二元光学器件的阿贝值是小于零的。因此可利用这些特性，构造混合系统，可大大简化光学系统，减轻重量，减小体积。

第三，有更多的参数选择性。在一般的光学系统设计中，为校正像差，需弯曲一些曲面；而校正色差，需用两种不同性质的玻璃。但校正参数的选择余地是很小的。例如，想消除像差时，只能弯曲曲面，而现在就不同了，利用二元光学器件，只需在平面或曲面上做一些台阶，而台阶的宽度、深度、位置等都是可变的，所以选择性就大多了。

第四，宽广的材料选择性。有一些材料，在普通光学设计时很难选用，如 ZnSe 等，还有一些半导体材料。现在材料的选择性就很大，无论选用什么材料，只要在上面加工出一些台阶，就可得到一定的光学性质。

最后一点，希望将器件小型化，而且形成阵列一个含有 2500（50×50）个微透镜阵列，局部如图 4 所示，过去任何传统工艺都是不可能的，而二元光学器件能实现这种功能，不仅如此，还能做在不同的材料上。

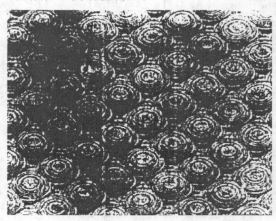

图 4　二元光学微透镜阵列

依据上述二元光学的一些优点，可以做各种各样的器件。例如分光器件，将不同波长的光聚焦在不同的地方。可形成任意波面，例如形成一个非球面波，或校正一个非球面波为平面波等。二元光学一出现，就受到美国军方的高度重视。美国军方有一个 DARPA 计划，其中计划里面就有一个二元光学的项目。这个项目中就提到利用制造电子线路的工艺发展光学。这样一来可节省劳动力，大大增加材料的选择性，另外可做出一些新光学功能器件；二来可促进 CAD 的设计，更重要的是将光电结合，整个系统进行优化；第三，在美国整个工业范围内掀起一个使用二元光学技术的高潮。

因此，可以得出一些结论，衍射光学器件或二元光学器件是光学领域革命的新方向，也可以说是光学和电子嫁接的桥梁。从设计上来说，是一个整体的优化；从工艺上来说，是利用制造半导体的工艺来制造光学器件。二元光学是一种非常有效的简化光学系统的办法，对整个光学系统可以优化，二元光学还是光机电系统中的一个重要元件。

2 二元光学的设计

当所设计的二元光学器件的特征尺度大于波长时，可利用"标量衍射理论"进行设计。但当其特征尺寸越来越小，接近或小于波长时，"标量衍射理论"就不能得到一个正确的结论，需采用"矢量衍射理论"。"矢量衍射理论"是利用麦克斯韦尔方程组加上一定的边界条件来求解，这方面还处于研究阶段，所设计、制造出的器件比较少。现在大部分的二元光学器件是基于"标量衍射理论"。

二元光学的设计，是一个优化过程。优化在工程上应用比较多，下降法、模拟退火法、误差递减法、输入输出法等，都是优化方法。以设计"达曼光栅"为例，"达曼光栅"可以将激光束变成多个等光强的光束（例如，33×33），这一点对通信非常重要。设计目标是寻找一个相位函数，使得在激光束入射时，能高衍射效率地将光束变成多个等光强的光束。设计中不断调整台阶的数目、位置、宽度及深度，就得到不同的相位函数对其进行傅立叶变换，变换后幅值与相位不一定满足"达曼光栅"的要求，这种差距用评价函数来定量描述。设计的目的是使评价函数最小，是一个优化过程。

3 二元光学的制造

在导言中已经讲过了，利用掩膜来加工二元光学器件。须说明的是，二元光学的制造与 VLSI 工艺有很大的差别。在制造大规模集成电路时，仅仅精确控制二维尺寸，刻蚀深度就是刻透和不刻透的区别。而在制造二元光学时，还需严格控制刻蚀深度，所以二元光学是一个三维制造过程。掩膜越多，对套刻精度影响越大，所以二元光学制造的发展方向是无掩膜的激光直写或电子束直写。在制作二元光学器件时，器件一直放在工作台上，通过控制器件每个区域内的激光曝光量来控制该位置处的刻蚀深度。直写最大的缺点是需要花费很长的时间，且设备非常昂贵。激光直写在我国只有一台，在成都光电所；电子束直写也只有一台，在无锡。因此，只利用直写实现单件或少数几件生产，一般情况还是利用掩膜套刻。另外一种制造工艺是全息。现在有许多器件，要做成亚波长结构，亚波长结构在许多方面有应用，尤其在军事上。例如防反膜，传统的方法是镀膜，现在可以用二元光学器件实现。如果二元光学器件的特征尺寸接近波长或达到亚波长，就可以直接把它设计成一个防反膜。这种器件的特征尺寸非常小，用直写、套刻的方法很难加工，因此直写、套刻能达到 1μm 精度就很不容易了。这种亚波长量级的器件加工可以用全息来实现，因为全息感光胶 1mm 内可以加工出几千条线，在感光胶上产生干涉条纹，再进行感光、刻蚀，能加工出特征尺寸非常小的二元光学器件。二元光学器件制造发展的一个方向是光机电结合，在基片上，可以做成上面是光学器件，下面是电子元件。二元光学的进一步发展普及，必须解决复制问题。因为二元光学器件的加工很昂贵，做一套掩膜，需上万元，很难将其推广。因此要解决复制问题，大批量生产，把成本降低。

4　二元光学的应用

二元光学是一个新生事物，现在实际应用场合还不是很多，在此介绍几种二元光学器件的应用。

4.1　波前改造

正如导言中所介绍的，利用二元光学器件能得到一个非球面波或任意波前，也可以将非球面波或任意不规则波面校正为平面波。这种波前改造是用传统光学元件无法实现的。例如在光学实验、光学测量中，入射激光难以满足使用要求，因为发出的激光，即使是气体激光，其光强分布是一种高斯分布，用聚焦透镜将其聚焦在一点，实际上是不可能的。因此需将波面改造成平顶分布，即 Top Hat，过去用计算全息图校正波前，现在用二元光学器件，二元光学器件的衍射效率比计算全息高多了。

波前改造在许多方面有实际应用，例如激光热处理，利用激光进行扫描，但由于只有一个焦点，非常耗时，利用（国内已经开始使用）二元光学器件，就是前面所说的"达曼光栅"，将一个光束分成几十个等光强点，大大提高加工效率，如图5所示。

气体激光发出的是高斯分布，但半导体的光束质量更差，发出的光是双高斯分布，传播不了多长距离就发散了。在土木建筑上，需要水平仪，水平仪其实很简单，就是旋转棱镜将气体激光器发出的激光束扫描一圈，便找到一个平面，但这种水平仪在工地上非常不好用，因为气体激光器很容易被毁坏。因此需要利用半导体激光器，但其光束质量不

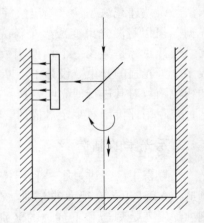

图5　激光热处理中应用二元光学器件

好，所以需对其波面进行整形，将其发出的光束改造为准直性非常好的光束，这可以利用二元光学器件实现。如图6所示，准直范围达几十米。

利用二元光学器件还可进行波面整形，例如我们现在的一个工作是利用二元光学实现惯性约束核聚变均匀照明。惯性约束核聚变（ICF）中，几十束激光穿过一个小孔，打到一个靶丸上，靶丸只有 1mm 左右。若光束不能精确打在位置上，将不能产生聚变。美国正在建立一个国家点火装置（NIF），就是专门从事 ICF 研究。这项工作对

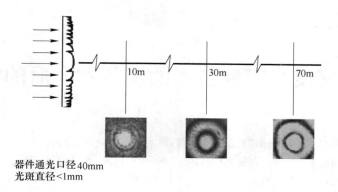

器件通光口径40mm
光斑直径<1mm

图6　二元光学准直器件

一个国家是非常有意义的。现在不允许核爆炸，不管是空爆还是地面爆炸。但各国对核武器的研究工作都没停止。ICF的研究成功还将提供下一代取之不尽、用之不竭的清洁能源。我国上海光机所等单位也在从事ICF研究工作。其中一个关键问题就是几十束激光如何以所需的光强分布穿过小孔，这就要利用二元光学器件去实现这种功能。

4.2　光束分束与合束

前面的"达曼光栅"是一种分束器件。能分，就能合。例如，利用二元光学器件将半导体激光阵列器件上多个激光器发出的光束合在一起，提高激光的能量。

4.3　二元光学视网膜

在图像处理中，常常需对图像进行分割，例如将地图中的河流、房屋、桥梁及树丛等分开。图像分割的一种办法是对图像扫描，再通过滤波器来判别。但计算机扫描是串行工作，非常慢。如果用光学的方法，可以并行处理，提高处理效率。一种系统是这样的，图像输出到一个液晶电视上，经达曼光栅分成多个像，再通过盖伯（Gabor）滤波器，进行图像分割。这种系统比较复杂，后来利用二元光学器件来简化系统，设计加工出一个二元光学器件，同时集成了棱镜、滤波器、达曼光栅等功能。可以实现图像分割，即做成了一个"二元光学视网膜"，大大简化系统、提高了效率。

二元光学还有其他一些用途，在此就不多说了。下面简单谈谈二元光学的前景。二元光学元件的发展已经经历了三代。第一代，是利用二元光学技术来改进传统的折射光学元件，以提高它们的常规性能，并实现普通光学元件无法实现的特殊功能。这类元件主要用于像差校正和消色差、波面整形等。第二代，主要应用于微光学元件和微光学阵列。在光通信、光学信息处理、光存储和激光束扫描等许多领域中有重要的应用。第三代，即目前正在发展的一代，二元光学瞄准了多层或三维集成微光学，在成像和复杂的光互连中进行光束变换和控制。多层微光学能够将光的变换、探测和处理集成在一体，构成一种多功能的集成化光电处理器，这一进展将使一种能按照不同光强进行适应性调整，探测出目标的运动并自动确定目标在背景中的位置的图像传感器成为可能，为传感器的微型化、集成化和智能化开辟了新的途径。

在可预见的将来，二元光学必将获得越来越广泛的应用。

超高密度光存储技术的现状和今后的发展*

摘　要　文章综述了光存储领域的研究进展，主要包括体全息存储、近场光学存储和双光子双稳态存储技术。在介绍各种存储技术发展现状的同时，分析了各自的优势和存在的问题。从整个光存储学科发展的角度给出了未来的趋势。

关键词　光存储；体全息；近场光学；双光子

1　信息时代的光学存储

21 世纪人类进入信息社会，知识经济成为推动社会进步、促进科技发展的强大动力，信息存储、传输与处理是提高社会整体发展水平最重要的保障条件之一。全球的信息量今后几年会以更快的速度增长。由于信息的多媒体化，人们需要处理的不仅是数据、文字、声音、图像，而且是活动图像和高清晰的图像等。一页 A4 文件为 2KB（千字节），而一张 A4 彩色照片就占 5MB（兆字节），放一分钟广播级的 FMV 就要占 40MB，可见信息量与日俱增。在信息技术的几个环节（获取、传输、存储、显示、处理）中，信息存储是关键。20 世纪 80 年代到 90 年代，人们最关心的是信息处理，即如何提高计算机芯片的处理速率和效率，全球掀起的计算机主处理器竞争已使本世纪可达 1GHz 的处理速度；随后通信网络的掀起及数据共享和通信使人们认识到了网络时代的到来；面对 21 世纪，人们又在考虑如何有效地存储和管理越来越多的数据和如何应用这些数据，信息存储空间日益拥挤，信息数据的采集和数据管理体系的复杂性越来越高，以及网络的普及，导致 21 世纪信息技术的浪潮将在存储领域兴起。

光信息存储（简称光存储）作为继磁存储之后新兴起的重要信息存储技术（目前以光盘为代表的光学数字数据存储技术）已成为现代信息社会中不可缺少的信息载体。与磁存储技术相比，现有的光盘存储技术具有许多特点：（1）数据存储密度高、容量大、携带方便。目前普通的 ϕ120mm 的光盘能存储 650MB，是硬磁盘的几十倍，软盘的几百倍。（2）寿命长、功能多。在常温环境下数据保存寿命在 100 年以上，且可根据用途采用不同介质制成只读型、一次写入型或可擦除型等不同功能的光盘。（3）非接触式读/写和擦。（4）信息的载噪比高，光盘的载噪比可达 50dB 以上。（5）生产成本低廉、数据复制工艺简单、效率高。

以 CD 系列为代表的第一代光盘技术产品的存储容量仍为十年前的 650MB；第二代 DVD 系列，由于激光波长的减小和物镜数值孔径的增大，DVD 盘面的坑点尺寸可以减小而提高了 DVD 的存储密度，单面双层存储容量为 8.5GB，盘容量为 17GB[1]；2000 年日本 Sony 公司采用蓝光激光器实现单面存储容量达 25GB 的高密度 DVR 已见报道[2]。尽管如此，作为计算机科学中的关键研究领域的高密度数据存储，为了满足预

---

* 本文合作者：张培琨。原发表于《中国计量学院学报》，2001，2：6～12，15。

计到2005年新型网络系统和第三代多媒体出现时计算机外部存储容量至少应为100GB，数据传输率至少为50MB/s的需求，则必须运用新原理，启用新材料才有可能研究出新一代超高密度、超快速存储系统。

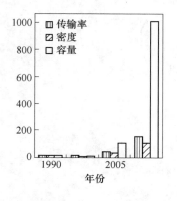

图1　时代对信息存储量的要求　　　　图2　光存储的未来发展趋势

实际上，各发达国家都已投入了大量人力财力开展超高密度、超快速数据存储方面的研究。尽管人们在开展各式各样的高密度存储研究，但一致看好短期有实用前景的存储方法主要集中在三维体全息、近场光学存储、双光子效应存储等方面。美国 Jet Propulsion 实验室、Rockwell 科学中心、Stanford 大学、亚利桑那大学光学中心、Carnegie Mellon 大学数据存储系统中心、IBM、AT&T、NIST、日本松下、NTT、SONY、SEIKO 等研究机构都在开展三维体全息、近场光学存储、双光子效应存储等方面的研究。我国也将这方面的研究列入了国家重点基础研究发展规划（973）项目中，以便跟上国际高新技术发展的步伐并获得自主的知识产权。

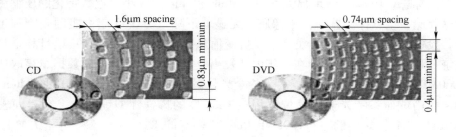

图3　DVD 与 CD 的对比

2　光学体全息存储的发展现状

光学体全息存储是超高密度存储最重要的研究领域之一，见图4。早在1963年美国科学家 Pieter J. van Heerden 就曾提出利用全息术进行数据存储的概念[3]，起初由于缺少合适的记录材料以及当时的光学及光电子元器件技术还不成熟，因而在随后很长一段时间内，体全息存储的研究工作进展很小。目前，随着计算机科学和现代信息处理技术的不断发展，一方面，对于具有大容量、高传输率、可快速存取的数据存储系统提出了日益迫切的要求，另一方面，随着新型优良全息记录材料（如光折变晶体和

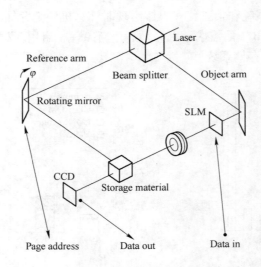

图 4 体全息存储原理图

光聚合物）的研制出现以及相关元器件，如高密度空间光调制器（SLM）和 CCD 光电探测阵列制造技术的不断进步，可满足各种实际应用要求的体全息数据存储系统正逐步成为可能，人们对体全息存储的兴趣又重新高涨起来。一般的光学体全息数据存储机理可简单描述为：待存储的数据（数字或模拟）经空间光调制器（SLM）被调制到信号光上，形成一二维信息页，然后与参考光在记录介质中干涉形成体全息图并被介质记录，利用体全息图的布拉格选择性，改变参考光的入射角度或波长以实现多重存储。

1991 年加州理工学院（CIT）的 F. H. Mok 等人[4]在 1 立方厘米掺铁铌酸锂晶体中存储并高保真地再现了 500 幅高分辨率军用车辆全息图成为再度掀起体全息存储技术研究热潮的标志。1994 年斯坦福大学 Hesselink 领导的研究小组把数字化的压缩图像和视频数据存储在全息存储器中，图像质量无显著下降[5]。1995 年由美国政府和工业部门主持，投资约 7000 万美元，实施了光折变信息存储材料（PRISM）项目和全息数据存储系统（HDSS）项目，预期在 5 年内开发出具有容量为 1 万亿位数据、存取速率为每秒 1000MB 的一次写入或可重复写入的全息数据存取系统。1997 年 CIT 的 Allen Pu 和 D. Psaltis 使用球面参考光通过移位复用在 1mm 厚的掺铁铌酸锂晶体上获得面密度为 100bits/μm^2 的体全息存储[6]；1998 年 Bell 实验室的 K. Curtis 等利用相关复用技术（Correlation Mutiplexing）在掺铁铌酸锂中的存储面密度超过了 350bits/$\mu m^{2[7]}$。1999 年 D. Psaltis 等人又在铌酸锂晶体中记录了 160000 幅全息图；H. J. Eichler（Berlin Technical University, Germany）提出采用微全息盘式存储方案，使用 1 微米光腰的激光采用 10 个波长复用以及 16 个角度复用可以在 CD 大小的盘上两层共存储 100Gbyte 的容量[8]。以上研究结果表明体全息存储面密度至少能达到几百 bits/μm^2，而现有的二层四面的 DVD 总的面密度也仅近似为 20bits/μm^2。显然，体全息存储可使存储容量较目前光盘呈数量级地提高。在克服光折变晶体固有的读写时已有存储信息被部分擦除的缺点方面，研究焦点集中在光致聚合物全息存储和双光子无挥发双掺光折变晶体的全息存储上。2000 年 A. Adibi 和 D. Psaltis 等用紫外和红光在双掺锰和铁的铌酸锂晶体上实现了无挥发全息存储[9]。

我国在光折变非线性光学材料与效应的基础研究中也已取得许多成果，非线性光学晶体生长技术在国际上取得较高的地位。1998 年以来清华大学相继完成单一公共体积中存储 1000 幅[10]，动态散斑全息存储[11] 及系统小型化的研究；北京工业大学实现了盘式单轨 2000 幅和盘式多轨 10000 幅全息图像存储的验证性实验[12]。

光学体全息数据存储具有如下几个显著特征：

（1）数据冗余度高：信息是以全息图的形式存储在一定的扩展体积内，因而具有高度的冗余性。在传统的磁盘或光盘存储中，每一数据比特占据一定的空间位置，当存储密度增大，存储介质的缺陷尺寸与数据单元大小相当时，必将引起对应数据丢失，而对全息存储来说，缺陷只会使得所有的信号强度降低，而不致引起数据丢失。

（2）数据并行传输：信息以页为单位，并行读写，因而可具有极高的数据传输率，其极限值将主要由 I/O 器件（SLM 及 CCD）来决定。目前多信道 CCD 探测阵列的运行速度已可达 128MHz，采用巨并行探测阵列的全息存储系统的数据传输率将有望达 1Gbyte/s[13]。

（3）存储密度高：利用体全息图的布拉格选择性或其他选择特性，可在同一存储体积内多重存储很多全息图，因而系统的有效存储密度很高。存储密度的理论极限值为 $1/\lambda^3$，其中 λ 为光波波长，在可见光谱区中，该值约为 10^{12}bits/cm^3。现已发展了多种复用存储技术，例如，1991 年 C. Denz 等采用相位编码复用技术[14]、1992 年 A. Yariv 等采用波长复用技术[15]、1993 年 F. H. Mok 采用角度复用技术[16]、D. Psaltis 等 1995 年采用移位复用技术[17]、1996 年 C. C. Sun 等采用随机相位编码复用技术[18] 以及随后其他不同的混合复用技术分别实现了多重全息存储。

（4）寻址速度快：参考光可采用声光、电光等非机械式寻址方式，因而系统的寻址速度很快，数据访问时间可降至亚毫秒范围或更低。例如，美国 Rockwell 于 2000 年提出的两种分别存储 100MB 和 1GB 的系统都是利用声光调制器来实现小于 50μs 的寻址功能[19]，而传统磁盘系统的机械寻址需要 10ms。

（5）具有关联寻址功能：对于块状角度复用体全息存储，如果在读出时不用参考光而改由物光中的某幅图像（或其部分）照射公共体积内由角度多重法存储的多重全息图，那么将会读出一系列不同方向的"参考光"，各光的强度大小代表对应存储图像与输入图像之间的相似程度，利用此关联特性，可以实现内容寻址操作，该功能对基于图像相关运算的快速目标识别（如原 Holoplex 公司利用体全息存储技术已做成一种高速"全息指纹识别系统"，其样机存储 1000 幅指纹图像，在一秒钟内便可完成输入指纹与所有存储内容的快速准确比较[5]；清华大学建立了基于体全息存储的快速人脸识别光电混合系统，在 1s 内实现 200 幅人脸快速识别[20~22]）、自动导航（如 CIT 的 D. Psaltis 利用体全息存储器特有的内容关联存储特性，构成的快速车辆导航系统[5]）、卫星星图匹配定位、大型数据库的检索与管理等应用十分重要。

目前制约体全息存储的关键仍在于获取合适的存储记录材料。

3 近场光学存储的发展现状

近场光学存储是超高密度光存储的另一重要研究领域。基于超衍射分辨的近场光学原理和方法的近场存储将可能使存储密度提高几个数量级。存储线宽可以达到 10 ~

50nm，相应的理论存储密度可以达到 1000GB/in² 以上。近场固体浸没透镜（SIL）虽然只能够得到 100nm 左右存储光斑，但存储密度也提高了十倍，达到 40GB/in² 以上。国外所建立已能够进行存取数据操作的实验系统可分为三种：

（1）探针型近场存储（见图5），它是将激光束通过直径非常小的孔对存储介质进行记录和读取，当记录介质距小孔相当近，则通过小孔的光便在光盘上形成尺寸与小孔相当的记录点。1996年 Hosaka 用这种方法以 785nm 的激光在相变介质上获得了 60nm 的记录点[23]，而经典光学显微镜的衍射受限分辨率约为 250nm 左右；1999年贝尔实验室的 A. Partovi 小组抛弃传统的光纤探针，采用 250nm 大小孔径的微小孔径激光（Very-small-aperture Laser，VSAL）（波长为 980nm）获得了 250nm 的记录点[24]。

（2）超分辨率近场结构存储（super-RENS）是在盘片中距记录层 20nm 处加掩模层（见图6），基于近场增强效应和近场表面等离子波效应，掩模层在激光照射下产生纳米尺寸隐失场，在近场区域内所产生的光斑直径要小于衍射极限分辨尺寸，从而实现超分辨率的记录点。目前日本的 Tominaga 利用这种方法已得到 81nm 的记录点[25]。

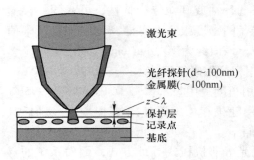

图5 探针型近场存储

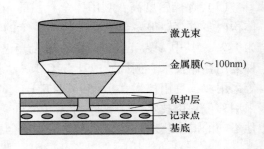

图6 超分辨率近场结构存储

（3）固体浸没透镜（Solid Immersion Lens，SIL）近场存储（见图7），是通过使用高数值孔径的固体浸没透镜来减小读写光斑的直径。SIL 底面和记录介质之间距离保持在近场范围内，聚焦在 SIL 底面的光斑通过近场耦合将隐失场光能量传到记录介质中实现高密度的记录，其理论上可获得直径为 125nm 的光斑。1999年丰田科技学院的 A. Chekanov 等人用 SIL 方案在磁光介质上获得了 150nm 的记录点[26]。

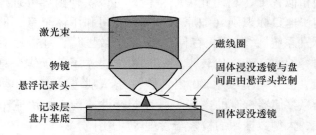

图7 固体浸没透镜近场存储

国内近场光学的研究大多集中于近场光学成像、近场光学荧光探测等。北京大学曾使用探针式近场光学显微镜系统进行了量子阱、量子线、激光器近场光谱和生物样品成像实验。清华大学建立了近场光学显微镜系统，自制了纳米光纤探针对多种样品进行成像,分辨率达到 50～100nm。设计了固体浸没透镜式近场光学超高密度存储系统,同时在进行 super-RENS 光盘的研究。中国科技大学对有源和无源光纤探针进行了研究。

近场光学存储的优势突出在读写光斑的减小大大提高存储的面密度和容量,同时可以充分利用已有的相关技术,与现有的光盘存储系统兼容,如硬盘驱动器的空气悬浮技术等而无需另行重新设计开发新的系统。但如何控制高速旋转的记录盘片表面与近场光学读写头间距保持在近场范围是一难题。

4 光学双光子双稳态三维数字记录的发展现状

光学双光子双稳态三维数字记录基本原理是根据两种光子同时作用于原子时,能使介质的原子中某一特定能级上的电子激发至另一稳态,并使其光学性能发生变化,所以若使两个光束从两个方向聚焦至材料的空间同一点时,便可实现三维空间的寻址与写入,读出。利用材料折射率、吸收度、荧光或电性质的改变来实现存储。信号由于是荧光读出,在未写入点无荧光,是零背景过程,所以读出灵敏度很高。由于此反应属于原子对光量子的吸收过程,反应速度为皮秒级,而最小记录单元,理论上可达到原子级。这种方法能实现 Tbits/cm³ 的体密度,可达到 40MB/s 的传输率。国际上最有代表性的是美国加州大学 San Diego 分校及 Call & Recall 公司,其 100 层的记录方法已见报道。1997 年,A. S. Dvornikov 等人(University of California, USA)采用双光束写入、单光束读出的方案,材料为罗丹明 B,制成立方体形状:10mm×10mm×10mm,存储 100 层共 1Mb,10000bit/层,信息单元的间距为 30μm。写入时,掩膜被 Nd: YAG 锁模脉冲(波长 1064nm,脉宽 35ps)照射,成像在立方体材料上形成 4mm×4mm 平面。532nm 倍频的激光束变形聚焦成 80μm×5mm 的片状,与立方体内的 IR 图像平面对齐,从而记录信息。读出采用 200μW、CW 型、543nm 的 He-Ne 激光器[27]。

1998 年,Y. Kawata, S. Kawata 等人(Osaka University, Japan)用双光子吸收技术,采用单光束写入、单光束读出方案,在光折变晶体 $LiNbO_3$(10×10mm×800μm)上进行了三维光学记录。层间距约 20μm,信息单元的间距是 5μm,记录了 7 层。写入采用 Ti: Sapphire 锁模脉冲(波长 762nm,脉宽 130fs,峰值光强 0.4kW/μm²)和 NA=0.85 的物镜。读出采用背面光照方式,利用 Zernike 相衬显微镜,物镜 NA=0.75。推测密度为 33Gbits/cm³(1.2μm×1.8μm×14.2μm)[28]。

1999 年,H. E. Pudavar 等人(State University of New York)同样采用单光束写入、单光束读出方案,材料为掺杂 AF240(2%)光色变分子的有机聚合物(来源于 US Airforce Research Laboratory),结构为多层盘片式。实验存储密度为 100Gbits/cm³(推算依据:层间距 10μm,位间距 1μm)。写入时,采用 Ar 离子泵浦的 Ti: Sapphire 锁模脉冲(波长 798nm,脉宽 70fs,重复频率 90MHz,平均功率 200mW),60 倍油浸物镜和 XYZ 扫描平台。读出时采用同样的激光器,但平均功率低,为 10～20mW[29]。此外,俄罗斯的 N. I. Koroteev 等人(Moscow State University),使用 NP 光色变分子材料搭建了单光束写入、单光束读出装置。存储材料每层厚 1μm,层间距 30μm,点间距 1.7μm[30]。

国内清华大学从 1995 年开始这一研究,初步建立了针对有机介质的记录物理模型，并完成了对双光子记录介质特性测试专用设备的研制，获国家发明专利。

对于已有的双光子存储方案，我们可以看出：（1）在双光束记录结构中，由于对各自光束的峰值功率要求不太高，可以采用皮秒级的 Nd: YAG 锁模脉冲激光器。（2）在单光束记录结构中，由于对光束的峰值功率要求很高，必须采用飞秒级的 Ti: Sapphire 锁模脉冲激光器。然而，大型和昂贵的飞秒 Ti: Sapphire 锁模脉冲激光器成为制约双光子存储实用化的一个主要因素。（3）存储体的形状多采用立方体（cube）、或多层盘片结构，且大都采用 XYZ 平台寻址。（4）记录信息的读取，普遍采用"共焦显微"系统以及 CCD 摄像头。（5）对于光色变材料的信息，可以采用双光子读出或者单光子读出方案。单光子方案易于采用"page By page"的读出系统。但是单光子读出方案的层间窜扰要大于双光子读出方案，因此必须采用"共焦显微"结构。（6）在光色变存储方案中，AF240 材料的存储密度可达到 100Gbits/cm^3 以上。而在光折变材料方案中，由于球差和擦除作用，使在 LiNbO$_3$（铌酸锂）晶体中仅能达到 33Gbits/cm^3。另外，由于是荧光读出也就对弱信号检测提出更高的要求。

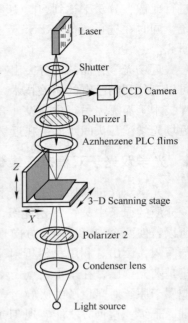

图 8　双光子数据存储实验装置

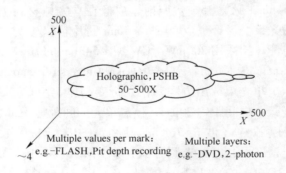

图 9　提高光学存储密度不同可行方案

5　超高密度光存储技术的发展趋势

除了上述各种光存储技术外，还有许多其他存储技术也在发展之中，如光谱烧孔技术（目前其工作温度要求低温是最主要的障碍）、激光微爆存储技术和电子俘获存储技术等等。所以这些技术都是以提高存储容量、密度、可靠性和数据传输率作是其高密度光存储技术的主要发展目标。只要是利用光来改变某种稳定物质的物理或化学状态记录信息的各种方法都属于光存储技术的研究范畴。但从整个学科发展的角度预测，今后高密度光存储技术的主要发展可能着重于：

（1）数字式记录仍是最基本和有效的方式，人们对信息的使用仍然主要基于计算机和网络，各类信息都要数字化。

（2）进一步缩小记录单元是发展高密度光存储的一有效途径。近场超分辨存储就是典型的尝试。随着精密技术及弱信号处理等相关技术的进步，相信信息的记录单元将从目前的分子团逐渐减小到单原子或分子量级。

（3）从目前的二维存储向多维存储发展。所谓的多维包括两方面的含义，一方面是指记录单元空间尺度的多维，即平面存储拓展到三维体存储，已有的努力如双光子多层存储及 1999 年 V. Markov 等人（Metrolaser Inc. , California USA）利用散斑全息实现的多层全息存储等[31]；另一方面是指复用维数的多维，例如 H. J. Eichler 提出的微全息存储技术就是将传统的光盘存储位用微全息光栅来表示，利用全息的波长或角度选择特性使实际存储复用维数得到增加[9]。

（4）并行读写逐步代替串行读写提高数据的读取传输率。并行读写功能是体全息页面存储的一个固有特性，也是体全息存储被普遍重视的原因之一。

（5）改善和发展存储系统的寻址方法，努力实现无机械寻址功能的实用化，从根本上解决目前难以提高随机寻址速度的问题。这方面体全息存储系统中的声光调制寻址和相位编码寻址就是很好的尝试。

（6）光学信息存储同光学信息处理精密结合以提高信息系统整体性能及功能，充分利用光学特性实现信息存储、传输、处理和计算的集成。

参 考 文 献

[1] 徐端颐. 光盘存储系统设计原理. 北京：国防工业出版社, 2000.
[2] NIKKEI ELECTRONICS, 2000 年 10 月 9 日号, No. 780, 33～34.
[3] P. van Heerden, Appl. Opt. , 1963, 2：393.
[4] F. H. Mok, et al. , Opt. Lett. , 1992, 16：605.
[5] D. Psaltis, et al. , Scientific American, November, 1995, 70.
[6] A. Pu and D. Psaltis. Topical Meeting on Optical Data Storage-Digest of Technical Papers Apr 7-9 1997, 1997 Sponsored by：IEEE Optical Soc of America, 48～49.
[7] K. Curtis, W. L. Wilson, U. S. Patent, PN：5, 719, 691, granted 1998.
[8] H. J. Eichler, P. Kuemmel, S. Orlic, et al. , IEEE J. of Selected Topics in Quantum Electron. , 1998, 4(5).
[9] A. Adibi, K. Buse, D. Psaltis, Opt. Lett. , 2000, 25(8)：539～541.
[10] 李晓春, 等. 光学学报, 1998, 18 (6)：722～725.
[11] P. Zhang, Q. He, G. Jin, et al. SPIE's Photonics Taiwan 26-28 July 2000, SPIE 4081.
[12] S. Tao, et al. The Joint International Symposium on Optical Memory and Optical Data Storage. 1999（11-15 July 1999, Hawaii）.
[13] G. T. Sincerbox. Holographic storage revisited. J. C. Dainty Eds. , London Academic Press 1994.
[14] C. Denz, G. Pauliat, G. Roosen, Opt. Commun. , 1991, 85：171～176.
[15] G. A. Rakuljic, V. Leyva, A. Yariv, Opt. Lett. , 1992, 17：1471～1473.
[16] F. H. Mok, Opt. Lett. , 1993, 18：915～917.
[17] D. Psaltis, M. Levene, A. Pu, et al. , Opt. Lett. , 1995, 20：782～784.
[18] C. C. Sun, R. H. Tsou, W. Chang, et al. , Opt. Quantum Electron. , 1996, 28：1509～1520.

[19] J. Ma, T. Chang, S. Choi and J. Hong, Opt. and Quantum Electron., 2000, 32: 383~392.
[20] D. Psaltis, et al. Scientific American, November, 1995, 70~76.
[21] Q. He, H. Liu, M. Wu, et al. SPIE, 1999, 3804: 241~248.
[22] 刘海松, 邹敏贤, 金国藩, 等. 中国激光, 1999, A26(11): 1031~1035.
[23] Sumio Hosaka, et al. Jpn. J. Appl. Phys., 1996, 35(1B): 443~447.
[24] Afshin Partovi, David Peale, Matthias Wuttig, et al., Appl. Phys. Lett., 1999, 75(11): 1515~1517.
[25] Junji Tominaga. SPIE 3864: 372~374.
[26] A. Chekanov, M. Birukawa, Y. Itoh, et al., J. of Appl. Phys., 1999, 85(8 Part 2B): 5324~5326.
[27] A. S. Dvornikov, et al. IEEE Transactions on Computer Spart A. 1997, 20(2): 203~210.
[28] Y. Kawata. Opt. Lett., 1998, 23(10): 756~758.
[29] H. E. Pudavar, et al. Appl. Phys. Lett., 1999, 74(9): 1338~1340.
[30] N. I. Koroteev, et al., Jpn. J. of Appl. Phys., 1998, 37(4B): 2279~2280.
[31] V. Markov, J. Millerd, J. Trolinger, et al. Opt. Lett., 1999, 24: 265~267.

Trends in Research on Super High Density Optical Storage

Abstract Novel optical storage is an indispensable important storage technology in modern information society. In this paper, the recent research progress in optical storage, including the volume holographic storage and near-field optical storage as well as two-photon optical storage, is reviewed. At the same time the advantage and disadvantage of each storage technology are analyzed. The developing tendency on super-high density storage is given from the point of view of whole optical storage.

Key words optical storage; volume hologram; near-field optics; two-photon

我国仪器仪表产业发展之路*

1 仪器仪表在当代社会的重要作用

当今世界已进入信息时代,信息技术成为推动科学技术和国民经济高速发展的关键技术;仪器仪表是对物质世界的信息进行测量与控制的基础手段和设备,是信息产业的源头和组成部分。现代仪器仪表在当今社会的重要作用怎样评估都不为过。

在工业生产中,仪器仪表是"倍增器"。美国仪器仪表产业占社会总产值的4%,而它拉动的相关经济产值却达到社会总产值的66%。在现代化大生产中,如果没有只占企业固定资产10%~15%的各种测量与控制仪器仪表的正常运行,发电厂、炼油厂、化工厂、钢铁厂等各种现代化企业都不能维持稳定的生产,更不会创造巨额的产值。仪器仪表拉动工业生产的作用犹如"四两拨千斤"。

在科学研究中,仪器仪表是"先行官"。近年来我国航天科技事业发展神速成就骄人,一个重要原因是抓住并解决了测量与控制这个必须先行的关键技术。在重大科技攻关项目中,几乎一半是研究和制作专用测量与控制仪器设备。诺贝尔奖设立至今,有众多科学家得奖都是借助于先进仪器的诞生才获得重大的科学发现;甚至许多物理学家、化学家、医学家,直接因发明仪器仪表而获奖。据统计,自1980年以来同仪器仪表有关的获奖者高达38人。

在军事上,仪器仪表是"战斗力"。聂荣臻元帅当年领导研制两弹一星时曾深刻地指出,必须抓好三件大事:一是新材料,二是仪器仪表,三是大型试验设备。离开现代仪器仪表,就没有两弹一星。现代武器装备,几乎无一不配备相关的测量控制仪器仪表。

此外,现代仪器仪表还发挥出"物化法官"的重要作用。在检查产品质量、监测环境污染、查服违禁药物、识别指纹假钞等各种判案过程中,现代仪器仪表都成为不可或缺的最具权威的"物化法官"。同时,仪器仪表在试验教学、气象预报、大地测绘、诊治疾病、指挥交通、探测灾情等社会生活许多领域都有广泛应用,已经遍及"农轻重、海陆空、吃穿用"无所不在。

因此,现代仪器仪表的发展水平是国家科技水平和综合国家力的重要体现,仪器仪表的制造水平反映出国家的文明程度。

2 我国仪器仪表工业的现状

根据国际发展的潮流和我国的现状,现代仪器仪表可以界定为工业自动化仪表与控制系统,科学仪器、医疗仪器、信息技术电测仪器,及其相关的传感器、元器件和

---

* 原发表于《自动化信息》,2002,1:4~7。

材料。

根据初步掌握的数据,我国现有各类仪器仪表企业6000多家,职工总数88万人,2000年总销售额1200多亿元。其中,工业自动化仪表与控制系统生产企业2000多家,销售额近500亿元;科学仪器生产企业约1500家,销售额近300亿元;医疗仪器生产企业约1200家,销售额近200亿元;其他各类仪器仪表及元器件材料等生产企业近1000家,销售额约200亿元。我国仪器仪表已经形成门类品种比较完全,具有一定技术基础和生产规模的工业体系,成为亚洲除日本以外的第二大仪器仪表生产国。

"九五"以来,我国仪器仪表工业总的发展形势是好的,主要表现在以下方面:

(1) 仪器仪表工业销售收入以年平均增长率8%不断递增,科学仪器年平均增长率超过25%。

(2) 涌现出一批技术先进的新型产品,如黑体空腔式钢水连续测温仪、微波等离子体炬光谱仪、柔性控制系统、高强度聚焦超声肿瘤治疗系统等多项产品,技术上处于领先国际水平。

(3) 仪器仪表产品出口创汇有明显增长。1999年我国仪器仪表出口创汇超过40亿美元,比1998年增长12%。

(4) 一批具有相当规模和发展前景的民营企业的崛起,是我国仪器仪表产业发展的新生力量。这些企业有浙大中控、北京和利时、深圳德维森、浙江华立、舜禹、北京普析、上海天美、深圳华谊等公司。

应当清醒地看到,虽然我国仪器仪表工业有了一定的发展,但远远不能满足国民经济各方面日益增长的迫切需要。主要存在问题如下:

(1) 我国仪器仪表产品绝大部分属于中低档技术水平,而且可靠性、稳定性等关键性指标尚未全部达到要求。高档、大型仪器设备几乎全部依赖进口。据海关统计,除随成套项目配套引进的仪器仪表外,2000年进口各类仪器仪表总额近70亿美元,接近我国仪器仪表工业总产值的50%。

(2) 低水平重复生产严重。比如,全国有近百家企业重复生产涡轮流量计,有几十家企业重复生产色相色谱仪,有300多家企业重复生产热电偶,等等。低水平重复生产,耗费了大量人力、物力、财力,难以摆脱长期分散落后的局面;低水平重复生产,破坏了市场竞争的有效秩序,导致出现不公平竞争和暗箱操作等丑陋现象;低水平重复生产的结果,必然出现大量质量低劣产品,给用户造成恶劣影响,使民族仪器仪表工业的发展受到沉重打击。

3 我国仪器仪表产业的发展面临着三重压力

我国仪器仪表产值1200亿元人民币,不足美国产值3200亿美元的5%,在世界市场的占有率不到1%。如此脆弱和幼稚的仪器仪表产业却面临着三重压力,形势非常严峻。

(1) 国际仪器仪表正向着微型化、集成化、智能化和总线化方向迅速发展,必须快速跟上国际仪器仪表迅猛发展的潮流,否则就会被远远地抛在后面。

(2) 欧美日等发达国家的仪器仪表厂商,多年来一直窥视和挤占中国迅速增长的仪器仪表市场。我国即将加入WTO,国内外产品为争夺仪器仪表的竞争将更为激烈。

（3）我国仪器仪表产业的发展面临"瓶颈"的制约。制约我国仪器仪表产业发展的"瓶颈"主要表现在以下四个方面：

1）科技创新及其产业化进展滞缓。我国仪器仪表产业在科技创新及产业化与国际相比差距很大，形势非常严峻。在跨入21世纪的今天，我国仪器仪表的水平还停留在20世纪80年代国际水平上。大型和高档仪器设备几乎全部依赖进口；许多急需的专用仪器还是空白；中低档产品在质量上还有许多难关需要攻克。

制约我国仪器仪表产业科技创新及产业化发展滞缓的主要因素有三个：第一，科技投入严重不足。国际著名仪器仪表企业用于科技创新的开发资金一般都超过年销售额的10%，而我国仪器仪表企业不仅销售金额不高，用于科技创新的开发资金一般不超过年销售额的3%，相比之下企业科技投入少得几乎无济于事。第二，人才匮乏。仪器仪表科技创新需要一批既有学识又有经验的边缘科学和应用技术的人才。仪器仪表行业这类人才本来就不多，近年来又大量流失。人才匮乏成为仪器仪表产业科技创新及产业化的严重障碍。第三，缺乏政、产、学、研、金的有效结合。仪器仪表产业科技创新及产业化必须有政府、企业、高校、科研院所和金融界及用户的介入。没有政府和金融界的介入，实现产业化几乎是不可能的。

2）产品的稳定性和可靠性没有根本解决。我国生产的仪器仪表产品在技术性能上与国外同类产品相比差距不是很大，但稳定性和可靠性问题却长期得不到根本解决。在对近1000名专家和用户的调查中，有80%以上的人认为由于产品稳定性和可靠性不高，极大地限制了国产仪表的应用范围和可信程度。其主要原因是：长期忽视基础技术的研究和开发；国产通用件和基础件质量不过关；企业对产品的质量控制和管理不力。产品质量不过关，加强仪器仪表产业发展只能是纸上谈兵。

3）旧有体制束缚了企业的发展。体制问题是制约我国经济，特别是国有企业发展的一个共性问题，仪器仪表行业也不例外。目前有相当数量的一批国家投资建设起来的仪器仪表骨干企业，由于在旧有体制的束缚下长期不能从沉重的历史包袱中挣脱出来，在市场竞争中丧失活力，生产和经营出现严重滑坡，一批骨干企业在生死线上苦苦挣扎，怎么能保证我国仪器仪表产业有快速发展呢！而一批民营、合资和股份制企业。由于体制合理，运行灵活，在市场竞争中迅速崛起，成为我国仪器仪表产业的新生力量。

4）仪器仪表产业的发展受到客观环境的制约。主要表现在：第一，税赋过重。仪器仪表企业一般规模不大，生产批量较小，产值和经济效益总量不高。但是现代仪器仪表对国民经济有巨大的拉动作用，可产生难以估量的"倍增"效应。对具有如此特殊属性的仪器仪表产业，如同其他产业一样征收17%增值税、33%所得税及相同比例的关税，则显得税赋过重，在相当程度上制约了企业的科技开发和扩大再生产。第二，多年来，各级政府包括仪器仪表产业主管部门，以及银行、税务、工商等部门对发展仪器仪表工业的重要性和紧迫性认识不足，支持不够。第三，缺少支持民族仪器仪表发展的采购政策。我国仪器仪表市场被外商挤占，除了国内产品存在稳定性和可靠性毛病外，跟国家保护不力有直接关系。由于国家没有制定出保护民族仪器仪表工业发展的明确的采购政策，加之对审批项目经费控制不严，除去正常和必需的引进之外，出现大量购进可用国产仪表替代的国外产品的现象，对我国仪器仪表产业造成很大的冲击。第四，我国基础产业能力差，包括产品质量、服务能力和信誉程度较差，直接影响到仪器仪表产业的发展。市场环境相关配套的改革跟不上，以产品的国家标准为例，至今有关稳定性和可靠性的标准尚无，很多标准从20世纪70年代制订以来30年一成不变，这样怎么能刺激市场的需求和促进企业的发展？

尽快使我国仪器仪表产业从三重压力下解放出来，振兴我国仪器仪表产业，是我国政府和仪器仪表战线在21世纪都紧迫的历史使命。

4 我国仪器仪表产业发展的战略目标和重点领域

4.1 战略目标

通过政策引导，鼓励资金和人才等资源投向仪器仪表产业，加强仪器仪表产业发展，力争在5~10年内实现如下战略目标：

（1）我国仪器仪表产业，包括工业自动化仪表与系统、科学仪器、医疗仪器、信息技术电测仪器、其他各类测量仪器仪表以及相关的传感器、元器件和材料，研究开发和生产能力达到或接近21世纪初期国际水平。

（2）2005年，我国仪器仪表产业销售总收入达到国民经济总值的1.5%，工业自动化仪表与系统，科学仪器能够占领国内市场45%以上的份额，医疗仪器和其他仪器仪表能够占领国内市场的30%以上的份额；2010年，我国仪器仪表产业销售总收入达到国民经济总值的2%，工业自动化仪表与系统，科学仪器能满足国内市场60%以上的需求，医疗仪器和其他仪器仪表能满足国内市场50%以上的需求。

（3）在仪器仪表各主要领域，新建18~20个工程技术中心和产业化基地，大力推进仪器仪表科技创新，加强实现创新成果产业化。

（4）在仪器仪表各主要领域，培养和发展不少于30个具有综合实力与相当规模的生产发展基地和重点企业。

（5）支持和发展一批仪器仪表系统集成公司，5~10年内迅速集聚力量，能在钢铁、石油、化工、电力、环保等多种大型重要的工程中承包自动化项目，促进我国仪器仪表产业的迅速发展。

4.2 重点领域

"十五"期间，我国仪器仪表产业发展的重点领域如下。

（1）工业自动化仪表与控制系统：

1）新一代主控系统及其综合自动化开发和产业化，主要包括分散控制系统、现场总线控制系统和以工业计算机为基础的开放式控制系统等。

2）先进控制、优化软件开发与产业化，主要包括先进控制技术、过程优化技术、实时监控软件平台、信息集成软件平台、系统集成技术等。

3）智能仪表、采用现场总线技术的检测仪表、执行器与变送器、成套专用控制装置和成套专用优化系统的开发与产业化。

（2）科学仪器：

1）重点解决色谱、光谱、电化学等各类通用仪器的稳定性和可靠性，开发高灵敏检测器和高精度传感器。

2）研制特定领域的专用仪器，如农产品品质、食品营养成分、成套环境监测、灾害监测、生命科学用分析仪器、医院用生化分析仪器、过程在线分析仪器、计量仪器等。

3）研制有自主知识产权和特色的新型仪器，重点发展微分析仪器、智能仪器、联用仪器、成像仪器等。

4）研制科学仪器软件和支撑系统，重点发展应用软件、标准化数据处理软件、科学仪器开发平台、测试数据系统及支撑系统等。

（3）医疗仪器：

1）开发研制医用光学仪器，包括内窥镜、眼科仪器、手术显微镜等。

2）研制以数字成像、高档黑白超、彩超换能器等关键技术的超声医用仪器。

3）研究开发 X 射线图像处理系统，开放式超导型核磁共振系统等大型仪器。

4）研究开发高能智能化肿瘤治疗大型仪器系统，包括数字化系统，高能管技术、放疗模拟定位机改造，以及多页光阑系统等关键技术。

（4）信息技术电测仪器：

1）集成电路自动测试技术与系统。

2）通信、计算机、网络测量技术、测量仪器及测试系统。

3）微波、毫米波测量仪器及测试系统。

4）数字电视、广播、音响、多媒体测试技术、测量仪器与系统。

5）电工自动测试系统及设备。

6）可大量出口的电工仪器仪表，如电度表、数字万用表等。

（5）传感器、元器件及仪表材料：

1）用于现场总线及智能化仪表的温度、压力、流量、物位传感器；用于环保等领域的多功能传感器；用于航天、航空领域的微传感器等。

2）特殊弹性元件：用于数控机床、纺织机械、自动化仪表及汽车等领域的各种计数器，半导体专用电路（ASIC）和厚膜电路。

3）重点研究开发薄膜化、小型化、纤维化、粉体化、复合化、多功能化、材料——元件一体化、智能化各种新材料。

4）加强仪器仪表基础技术，尤其是提高稳定性和可靠性等共性技术研究。

5 振兴我国仪器仪表产业的对策和建议

5.1 振兴仪器仪表产业需要国家鼓励发展政策

为了振兴我国仪器仪表产业，国务院决定由国家计委、国家经贸委、科技部、财政部等有关部门共同协商制定必要的扶植政策，促进仪器仪表产业发展。在扶植政策中建议包括以下内容：

1）建立仪器仪表产业风险投资机制，鼓励仪器仪表产业的风险投资。

2）降低仪器仪表企业的税赋，尤其是增值税，使企业能够积累较多的资金用于科技创新和扩大再生产。允许企业提取上年度销售总额的 10% 用于科技开发，计入当年成本，并能够滚动使用。

3）在产业技术政策方面，支持仪器仪表通用和基础技术的开发。

4）鼓励仪器仪表产品出口，尤其鼓励技术含量较高的产品出口，制定相应鼓励出口的有关政策。

5）明确制定有利于民族工业发展的采购政策。

6）大力培养人才，吸引和使用人才要制定相关政策给予保证。

7）享受鼓励政策的企业必须认定。

5.2 扶植和发展一批仪器仪表发展基地和重点企业

基地建设具有十分重要的意义，美国硅谷的建设和它带动信息产业发展所发挥的非常作用就是最好的例证。要加快仪器仪表产业的发展，扶植和发展一批生产基地具有战略性意义。基地建设要选择行业内高校、科研院所、尤其是生产企业相对集中，具有较好基础和发展前景的城市或地区。基地建设必须纳入国家规划，国家给予扶植。基地建设要充分调动地方的积极性，允许竞争。

现代仪器仪表综合了多种高新技术成果，发展现代仪器仪表企业必须集中优秀人才，投入巨额资金，需要一批有相当经济实力和现代化生产经营开发规模以及著名品牌的大型公司作为仪器仪表行业的"龙头"企业。在5~10年内扶植发展一批重点企业对振兴我国仪器仪表产业具有战略性意义。重点企业的建设和发展，必须纳入国家规划，国家给予重点扶植。

5.3 建立仪器仪表发展专家指导委员会

现代仪器仪表集成了光、机、电、计算机等各种新技术和应用了多种基础学科的研究成果。现代仪器仪表应用在国民经济各种产业、科学研究、国防建设以及社会生活的各个领域，因此仪器仪表不应当归属在某个具体的产业部门。国家今后对产业的管理主要通过制定政策和统一规划来实现。针对仪器仪表产业，建立一个专家指导委员会，在制定发展政策和统一规划协调方面作为政府的参谋和咨询组织是十分必要的，也符合国家改变管理职能，充分发挥中介组织在行业发展中起协调管理作用的方针。专家指导委员会对全行业的发展，包括发展方向和重大项目的建议，基地建设和重点企业的发展，组织政、产、学、研、金、用相结合等许多方面都可以发挥指导作用。

5.4 加快国有企业的体制改革

仪器仪表国有企业同其他行业国有企业一样，由于旧体制的制约，积累了大量问题，严重阻碍着企业的发展。但是，仪器仪表企业一般规模不大，历史不长，"包袱"较轻，产品结构容易调整，因此仪器仪表国有企业改制难度相对较小。目前，4000余家国有企业，特别是一些骨干企业，如果能够实现尽快改制，对仪器仪表工业的发展无疑注入了新的活力，将极大地产生推动和促进作用，建议国家对仪器仪表国有企业加快改制步伐，提出明确要求，这对促进仪器仪表国有企业改制将产生积极的推动作用。

5.5 支持一批建设项目作为发展仪器仪表产业的依托工程

五年内，可以在钢铁、石油、电力、化工、环保、轻工等领域重点选择一批建设项目，如1000MW核电站、600MW火电站、50万吨炼油、1350m^3高炉及连铸连轧工程，50万吨/年污水处理工程等作为发展仪器仪表产业的依托工程。建设一批现代仪器仪表科研和工程发展中心作为发展仪器仪表产业的技术依托，推动创新成果产业化。"十五"期间，仪器仪表产业必须建立10~12个科研和工程发展中心。

5.6 军工与民用相互渗透，促进仪器仪表发展

我国军工领域仪器仪表的研发和生产具有很强的实力，但长期同民用脱钩，对我国仪器仪表产业的发展是巨大的损失。建议国家要制定政策，全面部署，鼓励军转民；同时鼓励民用企业接受军工任务，相互渗透，促进发展。

"雄关漫道真如铁，而今迈步从头越"。我国仪器仪表与国外先进国家的差距约20年，只要我们认真总结经验，扩大改革开放，制定优惠政策，学习国外先进技术，利用加入WTO的有利时机，加快发展我国仪器仪表产业，定会在不久的将来赶上和超过国外先进水平。

测量技术是信息技术的源头
——谈王大珩院士的仪器科学思想[*]

王大珩院士从事科技事业，尤其是光学和仪器科学事业已经整整 70 年。王大珩院士对科技事业的贡献是多方面、全方位、战略性的。在这里，我想谈谈王大珩院士关于仪器科学创新发展的一些重要思想和杰出贡献。

王老非常强调仪器仪表在当今社会具有重要的作用和地位，对此他具有深刻的、科学的认识。王老指出："仪器不是机器，仪器是认识和改造物质世界的工具，而机器只能改造却不能认识物质世界。"王老又说"测量技术是信息技术的重要组成部分，是信息技术的源头。"并指出"仪器仪表是工业生产的'倍增器'，科学研究的'先行官'，军事上的'战斗力'和社会生活中的'物化法官'。""仪器仪表产业是国民经济和科学技术发展'卡脖子'的产业。""科学技术是第一生产力，而现代仪器设备则是第一生产力的三大要素之一。"又概括地指出"仪器仪表对促进精神文明建设和提高全民科学素质也具有重要的作用。"王老还有一个非常经典的比喻，那就是"中国科学技术要像蛟龙一样腾飞，这条蛟龙的头是信息技术，仪器仪表则是蛟龙的眼睛，要画龙点睛。"等等。王老是一位使命感和责任心极强的科学家，他把仪器仪表重要性的这些思想和认识无数次地发出呼吁和广泛宣传，是我国仪器仪表发展历程中值得珍惜的宝贵精神财富。

王老的一系列重要思想为仪器仪表的发展正了名，指了路，鼓了气。现在，再也听不到仪器仪表不重要的声音，国家发改委、科技部、教育部等政府部门对发展仪器仪表高度重视，在政策上、财力上给予了大力支持，我国仪器仪表产业 10 年来每年产值以超过 20% 的增长率迅速增长，仪器科学与产业面临大好形势，王大珩院士重要思想的影响功不可没。

王老为科学与产业的发展发挥了重要的作用和影响。我只列举一个例子说明。2000 年王大珩院士联合 11 位院士提出加快仪器仪表产业发展的建议提出后，受两委一部委托在全国开展了调查研究，王老又主持撰写了《振兴我国仪器仪表产业的对策和建议》的调查报告。这份报告赶在 2001 年 2 月送到国家计委。在 3 月召开的全国人大九届四次会议上，计委提出的国民经济和社会发展第十个五年计划纲要中明确提出"把发展仪器仪表放到重要位置"，这是新中国成立以来的第一次。这就把"发展仪器仪表放到重要位置"列入了国家的发展政策。接着，王老又约同杨嘉墀院士和我向当时计委主任曾培炎同志提出建议，希望计委立专项支持仪器仪表的发展。几年来，国家发改委对工业自动化仪表控制系统、科学仪器和医疗仪器立了若干专项，拨出了很大一笔资金支持产业化发展。科技部对科学仪器的发展非常重视，年年都立专项给予

---

[*] 原发表于《光明日报》，2007 年 7 月 9 日，第十版。

支持。

王老多次参与科学技术发展规划的制定,这对我国仪器仪表的发展贡献很大。王老一生中多次参与了国家和学科以及部门科技发展规划的制定,其中最重要的有两次。一次是参与制定我国1956~1967年科学技术发展远景规划,由他主持编写了仪器仪表和计量技术的专业项目规划,另一次是参与制定我国科学技术中长期发展规划的工作。

2003年5月,王老得知要准备制定我国科学技术中长期发展规划,当时还是非典在北京肆虐横行的时候,王老便告知中国仪器仪表学会要立即组织专家开展调查研究,提出关于仪器科学中长期发展规划的建议。在提出建议书的过程中,王老几次参与讨论提出意见。王老还建议向农业、环境、交通等10个专题提出相关领域发展仪器仪表的建议。为此,中国仪器仪表学会组织专家为我国制定中长期科学技术发展规划提出了10份建议书,这份热情,这份心意和所作的如此大量的工作,受到了科技部、工程院、中国科协等许多部门和组织的肯定和赞扬。不仅如此,王老担任中长期科技发展规划的顾问,还亲自撰写了一份发展科学仪器的建议。在我国制定出的科学技术中长期发展规划中,装备制造、信息技术、环境科学、农业、交通、安全等许多学科和产业领域中,都对发展相关的仪器仪表给予了充分的重视,我国仪器科学与产业未来的发展处在一个良好的外部条件下,前景非常美好。

王老一生为科技事业付出的心血,做出的贡献,受到了党和政府与人民的尊重和褒奖。他被授予"两弹一星元勋"称号,获得了各种奖励,他被选为全国人大代表,党和国家领导人多次到他家中看望他,他在人民心中是一位卓越的战略科学家。

我国当代仪器仪表的发展*

20世纪80年代,人类进入了信息时代。信息技术成为推动科学技术高速发展的关键技术。信息技术的快速发展,产生了新兴的庞大的信息产业,信息产业已经成为带动世界经济发展的龙头产业。

1 仪器仪表在当代社会中的地位和作用

仪器仪表与信息技术和信息产业是什么关系呢?

著名科学家钱学森明确指出,"信息技术包括测量技术、计算机技术和通信技术。测量技术是基础。"王大珩院士也一再强调,"测量技术是信息技术的源头。"美国商业部1999年年度报告关于新兴数字经济部分提出,信息产业包括计算机软硬件行业、通信设备制造及服务行业、仪器仪表行业。这就是说,测量技术是信息技术的基础和源头,仪器仪表行业是信息产业的重要组成部分。不言而喻,仪器仪表在当代社会信息时代对推动科学技术和国民经济的发展具有何等重要的地位。

那么,仪器仪表对推动科学技术和国民经济的发展究竟发挥着什么作用呢?

有四句话是大家非常熟悉并广为采用的,那就是"仪器仪表是工业生产的'倍增器',是科学研究的'先行官',是军事上的'战斗力',是现代生活的'物质法官'。"这四句话对仪器仪表的重要作用作了高度地概括,又是形象地比喻,是仪器仪表工作者多年实践的总结。我们无需再对四句话作详细的解释,只提出一个案例就足够了。2008年发生了奶粉中含过量三聚氰胺的事件,多少婴幼儿因此而中毒,甚至被夺走了生命,这是关系到人民健康生命安全的大事。由此政府决定对食品安全严格检查。判定食品是否安全的"法官",就是分析仪器。

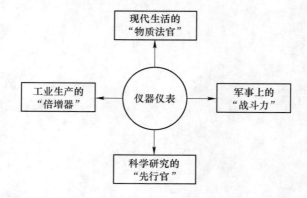

有位著名的光通信专家,就是美国贝尔实验室的"波分复用"的发明人,厉鼎毅

---

* 原发表于《电气时代》,2009,10:35~38。

先生，也是中国工程院的外籍院士，他说"中国发展到现在，必须重视 Metrology（计量）和 Inspection（检测）。我国目前的现实也已充分地说明仪器仪表的重要性。

由于仪器仪表的重要地位和作用受到了社会的高度重视，近些年来国家采取了一系列重大措施加快发展。

2001年3月在七届四次全国人大会议上批准"国家经济与社会发展第十个五年计划纲要"中明确提出"把发展仪器仪表放到重要位置"。随之，国家计委、经贸委和科技部等许多部委列出若干专项，动用大笔资金支持仪器仪表发展。

2005年，国家发改委下达了"加快振兴装备制造业的若干意见"，提出了在各行业中选出16项重点发展领域立专项支持发展，其中第11项就是重大工程自动化控制系统和精密测试仪器。

2006年制定《国家中长期科学与技术发展规划纲要》，多项仪器仪表发展项目被列其中。

2008年4月，科技部、国家发改委、教育部和中国科协联合发出"关于加强创新方法工作的若干意见"，正式启动创新方法工作在全国开展。这份文件中，明确提出创新方法包括创新思维、创新方法和创新工具三个要素，创新工具主要就是指推动科技创新的新型科学仪器。科学仪器作为科技创新"先行官"的作用得到进一步肯定和提升。

此外，在863计划，特别是航天计划等国家科技发展计划中，支持加快发展仪器仪表也都放到了重要位置。

就是在这样一个十分有利的形势下，我国仪器仪表科技与产业得到了迅速的发展。我们期待着丹东市仪器仪表产业基地尽早建成，它将是我国仪器仪表产业大军中一支重要的力量。

2　仪器仪表是一门独立的学科体系

为什么要谈这个问题？过去在许多人的认识中，只把仪器仪表看作一类工业产品，习惯地划入到机械产品大类，不把它认为是一门独立的学科，这个问题争论了多年。随着仪器仪表的发展，它的确切含义应当指仪器仪表与测量控制。如果说仪器仪表指工业产品，那么测量控制就是它的技术基础与内涵。今天我们可以肯定地说，仪器仪表与测量控制是一门独立的学科体系，因为有四条充足的理由。

第一，仪器仪表与测量控制学科具有自己特定的一整套基础理论和技术，其中主要包括传感器技术、检测计量技术和信号处理理论和技术，但它们都是建立在自动控制理论、信号处理理论和误差分析等理论上的。在这些基础理论和技术的研究开发中，培育和造就一批杰出的科学家和大量优秀的科技工作者，他们的研究开发，为学科的创立和发展作出了重大的贡献。

第二，综观科学技术发展史，一门新兴学科在形成和发展过程中，教育体系，尤其是高等教育一定会应运而生形成和发展学科教育，培养新的学科人才。多年来，我国教育部已经围绕着仪器仪表学科设立了不少相关的专业，2008年正式确定为一级学科，取名为"仪器科学与技术"，成立了学科专业指导委员会，指导学科教育的发展。目前全国有近250所大专院校设置了相关的专业，在校本科生约3万名，研究生约1万名。高等教育为我国仪器仪表学科已经培养了几十万人才。

第三，仪器仪表是一门工程应用的学科，与之相适应的产业的形成和发展是学科发展的物质基础和技术支撑。我国仪器仪表产业已经具备相当规模，仪器仪表学科也不是在象牙塔内研究，而且构成了学科产业密切联系的体系，有着强大的生命力和发展空间。

第四，学科发展必须伴随成立自己独有学术组织。中国仪器仪表学会成立30年来同仪器仪表学科和产业并肩发展，为促进仪器仪表学科和产业的发展做了大量工作，取得了光辉的业绩。

因此，仪器仪表今天已经发展成为一门独立的学科体系应当不再引发争议了。之所以谈这个问题，是为了今后在发展仪器仪表产业的同时，要十分注意发展仪器仪表学科，不然产业的发展将会成为无本之木。中国仪器仪表学会近几年来不断开展学科进展的研究，撰写出学科进展研究报告，对促进仪器仪表产业发展具有十分重要的意义。

3 我国仪器仪表产业的现状

经过多年来的发展，特别是近10年来的快速发展，我国仪器仪表产业已经形成门类品种基本齐全，布局比较合理，具有相当技术基础和生产规模的产业体系。在亚洲我国是除日本以外第二大仪器仪表生产国，是发展中国家综合实力最强的仪器仪表生产国。2007年，我国仪器仪表产业规模以上企业为3954家，实现总产值3018亿元，销售收入3005亿元，其中出口88亿美元。在3000亿元的总产值中，工业自动化仪表与控制系统约占28%，科学仪器占25%，医疗仪器占13%，电子与电磁测量仪器仪表占11%，其余23%为各类专用仪器和传感器及仪表元器件与材料。

应当看到，近些年来我国仪器仪表产业发展很快，每年年产值增长率超过12%，分析仪器更是超过18%。我国生产的中低档产品国内市场占有率很高，电工产品超过90%，深圳市生产的数字万用表每年700万台，销售到世界90多个国家。国产仪器仪表技术水平也得到很大提高。在工业控制系统方面，国产DCS系统已经实现大型超临界火电机组现场控制。2007年1月，上海自动化仪表股份有限公司的DCS系统在襄樊电厂首台600MW超临界机组上投运成功移交生产。浙江中控公司和和利时公司的DCS系统在交通和石化工程中得到了很好应用。在科学仪器方面，最值得提出的是，我国嫦娥1号卫星携带的8种探测仪器都是我国科技人员自主研制设计的，总体技术上达到了国际先进水平，同时有自己的特点和创新。创新包括实现首次对月球表面进行全月面三维立体照相，使用的γ射线谱仪探测的分辨率和灵敏度都高于国际上以往使用同类仪器。我国仪器仪表技术进步的实例很多，这里不能一一介绍。四川汶川大地震，全国各地很快送去了大量环境、水文测量便携式分析仪器，表明了我国科学仪器产业的进步和实力。

但是我们更要清醒地认识到，虽然我国仪器仪表产业有了很大的发展和进步，与发达国家相比，至少还有10~15年的差距。差距是全方位的，最主要有三点。

第一，我国仪器仪表产业规模小，产值低，无论就整个行业或是各个企业来看都是这样。目前，我国仪器仪表产业总产值不过3000亿元，只占工业总产值2.5%。10

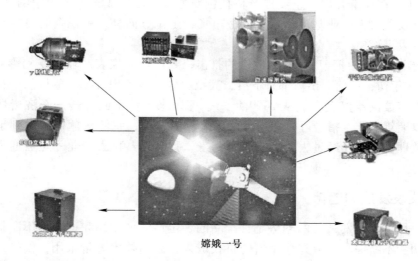

嫦娥一号

年前美国仪器仪表产业总产值占工业总产值4%，达到2000亿美元，是我们的5倍。美国仪器仪表生产企业年产值超过20亿美元不少于50家，我国年产值很高的仪器仪表企业，京仪集团80亿人民币，川仪集团60亿人民币，超过10亿人民币也不到30家，其他大多为中小企业。两相比较，差距太大。行业和企业产值低，规模小就会直接影响到产业的竞争力、活力与发展。要缩小和消除这个差距，需要我们努力奋斗10～20年。据了解，重庆、北京和杭州等一些城市近年来花大力气支持仪器仪表产业的发展，成效很大。如今丹东市要建设仪器仪表城，据说规划宏伟，决心很大。如果全国再多一些地方这么做，相信消除差距也许用不上10年了。

第二，我国仪器仪表产品质量上，品种上还存在不少问题。质量上最致命的问题是可靠性、稳定性差，长期以来没有得到根本解决。质量问题严重影响到市场竞争力，也束缚了我国仪器仪表的发展。品种上的问题是大型精密仪器制造不出来，国内需求几乎全部依赖进口。2007年我国仪器仪表产品出口达到了88亿美元，然而进口却达到了172亿美元，逆差84亿美元，成为装备制造业之最。这个问题不解决，仪器仪表产业将永远摆脱不了落后被动的局面。

第三，我国仪器仪表企业自主创新能力不强，无法承担起科技创新主体的使命。国际上仪器仪表产业发展极快，产品更新换代的周期只有2～4年，多数企业销售额一半以上几乎都来自于5年内上市的新产品。我国仪器仪表产品不少还沿自于20世纪80年代技术引进的产物。更有甚者，相当多企业产品是10年一贯制，几乎没有任何改进和创新。企业自主创新能力不强有多方面原因，影响最大的不外乎两条：一是科技创新资金投入太少。国外企业用于科技创新资金的投入一般占销售额8%～10%，我国企业本来销售额就少，科技创新资金投入多数不到5%，很少有达到10%的。资金不足就办不成大事，极大地限制了科技创新有效的开展；二是人的因素。有的企业领导缺乏创新意识，更多的是企业创新人才匮乏。党的十七大提出，提高自主创新能力，建设创新型国家是国家发展战略的核心，是提高综合国力的关键。为了提升企业自主创新能力，国家在全国启动了创新方法工作，支持建立以企业为核心的产学研战略联盟。中国仪器仪表学会接受中国科协的委托，在仪器仪表企业中开展了科技创新方法培训，计划 2010 年将在辽宁开展。企业提升科技创新能力，已经成为刻不容缓的历史使命。丹东市建立仪器仪表产业基地，除了把握市场，定位方向外，重视企业自主创新能力应当是一件大事。

2008 年以来，全球经历了金融危机的打击，我国仪器仪表产业也受到影响，2008 年底至 2009 年初，产值、出口都有下滑，甚至少数企业出现亏损。危机的阴影已经过去，我国大多数仪器仪表企业从 2009 年二季度以来形势不断好转，经济指标逐渐上升，10% 以上年增长率指日可待。经历一场危机的洗礼，我们从危机中学到了生存发展的许多道理。严冬已经过去，春天即将到来，我国仪器仪表产业一定会迎来一个快速发展的春天。

4 仪器仪表发展的趋势

当今，仪器仪表发展总的趋势是：产品的稳定性、可靠性和适应性不断提高，科技指标和功能不断提高，最先采用新的科技成果，高新技术大量采用，仪器及测控单元微小型化、智能化日趋明显，要求仪器及测控单元可独立使用、嵌入式使用和联网使用，仪器测控范围向立体化、全球化扩展，测控功能向系统化、网络化发展，便携式、手持式以至个性化仪器大量发展。

发展的技术特点是：综合各种新技术，在研究仪器仪表相关类型传感器、元器件和材料及技术的基础上，创新开发新型微弱信号敏感、传感、检测及融合技术，复杂组成样品的联用分析技术，生命科学的原位、在位、实时、在线、高灵敏度、高通量和高选择性检测技术，创建各类新型检测仪器仪表；综合系统论、控制论的发展，在开发工业自动化测控的在线分析和控制、原位分析及控制、高可靠性、高性能和高适应性等技术的基础上，创新发展工业自动化仪表与控制系统；结合生命科学、人体科学的发展，在研发医疗诊治的健康状况监测、早期诊治、无损诊断、无创和低创直视诊疗、精确定位治疗技术的基础上发展医疗仪器；同时跟踪新学科领域和各类应用领域的发展，开发各种专用、快捷、自动化检测、计量技术及专用仪器仪表。

丹东市在辽宁省是工业布局中重要的仪器仪表的生产地区，也是丹东经济的重要部分。丹东 2008 年全市仪器仪表产业实现工业产值 15.9 亿元，销售收入 14.6 亿元。因此，可以说丹东在仪器仪表产业上有很好的基础。特别是在冶金、矿山生产过程的测量控制仪表，燃气表（2008 年生产 240 万台，占国内市场 35%），X 射线探伤（丹东测控 2008 年产值 5.2 亿元，丹东奥龙 5127 万元），其他还有无损探伤，粒度仪等。

下一步如何建设成一个仪器仪表城？有下列几点值得注意。

（1）仪器仪表种类繁多，不能求全。要在有基础的领域中做出有特色，做出品牌，不断创新。不但要更大地占领国内市场还要远销国外。今后应再选择适合本地区发展一些其他仪器领域。

（2）要重视国内外仪器发展的趋势，如目前国内急需的科学仪器，特别是分析仪器、节能减排仪器等。

（3）重视官、产、学、研和用的结合，发挥科研机构与高等学校的作用，这是多快好省地创新的源泉。

（4）重视人才的引进和培养，高质量、高水平的管理和技术人员就能使企业发展和兴旺起来。

  总之，辽宁省委、省政府下决心要举全省之力支持丹东市仪器仪表产业基地建设。举全省之力，我想除了资金上的保证外，最重要的是充分利用好辽宁省仪器仪表产业的资源。辽宁省是工业大省，科技教育实力也很雄厚，对仪器仪表的需求迫切，市场潜力极大，这是发展仪器仪表极为有利的资源和条件。我们要认真深入地分析，找准基地建设的方向，定位准确才不会走弯路。辽宁省现有的仪器仪表产业门类比较齐全，具有一定的技术实力，这也是重要的资源。建议在建设丹东基地的同时，全省要合理布局，要协调发展。辽宁省人才资源充足，要集中人才搞几个大项目，重点突破，产业发展也就带动起来了。科技创新是发展仪器仪表的灵魂，建设产业基地要把提倡和加速自主创新放在重要位置，最重要的出路是建设以企业为核心的产学研产业联盟。这方面要下大力气培养和创建一批联盟，并作出实际的成果。吸引外来资金和人才很重要，省里已经有了政策和措施，要坚持不渝地做下去。我们做好了这些工作，就一定能获得成功。任重道远，前景无限。祝愿丹东仪表基地早日出现在我们面前。

我国平板显示产业面临的迫切科学技术问题*

1 平板显示产业面临重大转变

平板显示技术在当代人生活与工作中变得越来越重要,"大屏小屏人人有"这句话充分表述了这种状况。我国是世界上人口最多的大国,平板显示产业的发展在我国产业发展中的重要性是毋庸置疑的。平板显示技术目前的主流是液晶显示技术。我国已是世界上液晶产业规模最大的国家,20 年来我国在液晶显示技术上的进步是有目共睹的。但是,由于我国一直未能自主掌握相关产业的核心技术,所以在这个重大产业的国际竞争中,我国一直处于被动的"挨整"地位,尽管有很大的发展,但经济上的损失还是相当巨大的。这个经验教训是业内都承认的。

目前,平板显示技术又一次处在发生重大转变的前夜。我国又一次处在相对被动而且缺乏应有准备的状况。如果不及时采取强有力的措施,适当集中力量,努力争取避免重蹈过去 20 年的覆辙,而是继续在国际竞争中处在被动的"引进"状态,不能自主掌握核心技术,今后我国国民经济和科学技术的发展将会面临严重的损失。

平板显示产业技术将要面临两个重大的带有本质性的转变:

(1) 电视屏技术("大屏"技术)将要从以液晶(LC)显示屏为主转变到以有机发光二极管(OLED)显示屏为主。

(2) 手机、笔记本电脑等"小屏"技术将要转向以"杂化"的全息激光投影(holographic laser projection,HLP)技术为主。

从国际发展的局面来预测,大约在 3~5 年内前者将会大规模地出现在市场上,后者也许要慢一些,但在 3~5 年内也会有一定规模的出现。因此,摆在我们面前的任务是极端迫切的。

2 基于氧化物电子学的 TFT 技术事关全局

经过 20 多年的发展,OLED 作为平板显示的基础元件已经成熟了。从亮度、色彩、寿命、工艺等方面看,尽管还会不断有所改进,但是用 OLED 屏代替 LC 屏在技术上已经成熟了。OLED 屏较之 LC 屏,从耗能、色彩、视感方面看,优势太大了,这也已经为实践所证实。那么,为什么这个代替还没有被普遍接受呢? 问题出在驱动 OLED 显示屏工作的电路系统上。在显示系统中,LC 屏实质上是起光开关作用的,它的运行本质上是电压驱动的。目前使用用于液晶显示的非晶硅(a-硅)技术制备的 TFT 电路,是能够满足 LC-TFT 规模生产的需求的。对 OLED 屏系统来说,OLED 器件本身就是光源,它们的运作需要较大的电流。但由于目前用于液晶 TFT 的非晶硅(a-硅)薄膜的

---

* 本文合作者:甘子钊,欧阳钟灿,范守善,王恩哥. 原发表于《中国电子报》,2012 年 8 月 17 日,第十一版。

载流子迁移率很低，所以电路系统无法提供给OLED器件足够的电流。为了提高a-硅薄膜的迁移率，发展了基于晶粒再结晶效应的低温多晶硅（LTPS）技术。利用这种改进的LTPS技术，可以用来制备OLED显示屏。于是，数年前市场上开始出现了使用OLED屏的手机、笔记本电脑以及小型的彩色电视。但是估计由于成品率在面积较大时难以做到较高的水平，所以除手机屏外，其他都没有真正形成较有规模的市场。

事实上，在光电子学技术中广泛使用的氧化物导电材料，它们的载流子迁移率不难做到多晶硅薄膜的几倍到几十倍。因此，发展适用于OLED显示屏的基于氧化物材料的TFT电路技术也在国际上得到重视。在三年前我们写的一份关于OLED屏的咨询意见中，我们建议，做好平板显示产业从LC屏转移到OLED屏为主的科学技术工作的关键是发展相应的TFT电路技术。但是报告中强调的当前重点还是放在发展改进了的LTPS技术，争取进一步提高其均匀性和大面积的成品率上。同时也提出应重视氧化物TFT电路技术的发展，加强研究力量。可是近期的国际动态表明，我们原来的这个估计太保守了。毫无疑问，建立在基于氧化物的TFT电路技术上的OLED屏显示技术，现在已经处在成熟和规模化地实现产业化的前夜，已经开始有产品出售。毫无疑问，3~5年内，在电视产业中，OLED屏的显示技术将要代替LC屏的显示技术成为主流，而OLED屏显示技术用的将是基于氧化物电子学的TFT电路技术。我们认为这正是做好准备实现这个重大转变的科学技术关键。

近20多年来，围绕氧化物的物理、化学、材料科学以及它们在电子学和光电子学等方面的应用，国际上进行了大量的工作，我国也有相当数量的各方面研究工作，应该说已经有了一定的基础。氧化物电子学（oxide electronic）在国际国内也逐渐成为一个常用的词。如果我们以发展自主的基于氧化物的大面积TFT电路技术，为我国的OLED屏显示技术产业的自主发展提供基础为奋斗目标，结合国内原有的较好基础的OLED材料和器件技术的力量，结合国内企业界引进平板显示技术的积极性，精心组织物理、化学、材料科学、材料工艺学和集成电子学的科技力量，大力协同，力求在3~5年内掌握有关的科学技术关键，包括在工艺装备的研制上有自主的能力，从而在这场高新产业的竞争中，为我国争到一定意义上的主动，将是很有意义，也是值得去努力的。

这场竞争，不仅对OLED屏平板显示产业的发展有帮助，也为我国在氧化物电子学上争到较前沿的位置。估计氧化物电子学在信息存储、光电子学、磁电子学、超导电子学等方面还会有较大的应用前景，这也是值得去努力的。

3 应重视基于全息激光投影技术的显示屏技术

近几年，把网络、摄影、电视等功能都综合在移动通信（手机）上的发展趋势非常引人注目，但也暴露出电子学的巨大进步受限于最后需要一个显示屏作为人机界面的局限性。如果我们考察一下以iPhone和iPad为代表的这一类电子产品的发展，就很容易理解这个问题了。有没有可能基于激光投影来解决这个问题？如果发展出一种利用激光投影，用虚拟键盘和投影屏来代替平板显示屏作为人机界面，就能把笔记本电脑（文字处理）、通信、摄影、电视等全都综合在可以放在上衣口袋中的手机上！这是多么诱人的设想。由于半导体激光器和微机械技术的发展，这种设想已经完全可能，而且经济上也不是很昂贵。例如这样的手机投影仪已经在市场上出现了。问题出在利用传统的像素到像素的投影方式，光的利用效率较低，屏的亮度不够。再加上激光由于相干性产生的"闪斑"效应，对人眼有损害，为了避免这个损害的措施又进一步降低了光的利用效率，因此上述设想的激光投影显示技术一直难以成为规模化的产业。

最近国际上发展出一种全新的激光投影技术，这种技术是基于激光全息投影（holographic laser projection，HLP）的概念，把传播来的图像信息通过数值技术处理，将图像的长波部分转变成相位调制的全息图（全息光栅），利用衍射光学的方法来实现投影，这是投影光能的大部分。短波部分则还是通过像素到像素的方式扫描投影，来提供图像的"细节"，它只是投影光能的小部分。这样做的结果是可以大大提高光能利用的效率，因而提高了图像的亮度，同时也把"闪斑"问题解决了。不仅是理论计算，而且实验也证明了利用这种可以称作"杂化"的HLP技术，确实可把手机改造成一台笔记本电脑，使得上网、通信、照相、电视、文字处理等功能综合在可以放入上衣口袋的手机上，随时随地都可以使用。这将是一个平板显示的全新时代！更不用说，HLP技术本身就含着三维显示的可能，预示着3D图像技术的未来。

衍射光学，特别是全息光学是现代光学的重要部分，我国在自适应光学技术、图像数据处理和传输技术等学科方面也已经有一定基础。半导体激光器和微机械技术在我国也已经有一定基础。目前应该抓紧时间，组织国内有关力量，从发展杂化的HLP显示技术的角度，把微机械技术、半导体紫外和蓝绿红激光技术、衍射光学和计算全息学技术等学科协同发展起来，力争在3~5年内，我国在这个剧烈竞争的产业中，有一定意义上的自主发展位置。

这样做的结果也会推进我国在微机械技术、现代光学技术和激光技术的进步，而这些技术在推进高技术产业、国防技术方面的意义也是很大的。

上述两个方面的研发工作，和当代材料科学、现代光学、纳米科学、激光科学都有密切联系，会对这些学科部门发展有大的推动，可以认为这是学科发展的一个机会。由于产业部门对这两方面的发展都有强烈的兴趣，这也是中国科学院、研究型大学与产业部门协同发展的一个好机会。